Handbook of
Optical Engineering

Handbook of
Optical Engineering

edited by

Daniel Malacara
Centro de Investigaciones en Optica, A.C.
León, Mexico

Brian J. Thompson
University of Rochester
Rochester, New York

CRC Press
Taylor & Francis Group
Boca Raton London New York

CRC Press is an imprint of the
Taylor & Francis Group, an **informa** business

CRC Press
Taylor & Francis Group
6000 Broken Sound Parkway NW, Suite 300
Boca Raton, FL 33487-2742

© 1996 by Taylor & Francis Group, LLC
CRC Press is an imprint of Taylor & Francis Group, an Informa business

First issued in paperback 2019

No claim to original U.S. Government works

ISBN-13: 978-0-367-44725-0 (pbk)
ISBN-13: 978-0-824-79960-1 (hbk)

Visit the Taylor & Francis Web site at
http://www.taylorandfrancis.com

and the CRC Press Web site at
http://www.crcpress.com

Preface

This comprehensive and cohesive work includes all the relevant data to allow optical engineers worldwide to meet present and upcoming challenges in their day-to-day responsibilities. The thrust of the *Handbook of Optical Engineering* is toward engineering and technology rather than theoretical science.

The book has 26 chapters that cover most but not all topics in optics, beginning with a few chapters describing the principles of optics elements. These are followed by more technical and applied chapters.

All authors prepared their chapters with the following criteria in mind:

Descriptions are restricted to explaining principles, processes, methods, and procedures in a concise and practical way so that the reader can easily apply the topics discussed. Fundamental descriptions and a how-to-do-it approach are emphasized.

Useful formulas are provided wherever possible, along with step-by-step, worked-out examples, as needed, to illustrate applications and clarify calculation methods. Formulas are arranged in the best sequence for use on a computer or calculator.

The book is replete with tables, flow charts, graphs, schematics and line drawings in the tradition of useful reference books and major handbooks. National and ISO standards are included where appropriate, and permitted, in suitable abridgement for useful reference. Overlapping among different chapters has been avoided unless absolutely necessary.

Daniel Malacara
Brian J. Thompson

Contents

Contributors

Vicente Aboites   *Centro de Investigaciones en Optica, León, Mexico*

Sofia E. Acosta-Ortiz   *Centro de Investigaciones en Optica, Aguascalientes, Mexico*

David Anderson   *Rayleigh Optical Corporation, Tucson, Arizona*

Glenn D. Boreman   *University of Central Florida, Orlando, Florida*

Jim Burge   *The University of Arizona, Tucson, Arizona*

Sergio Calixto   *Centro de Investigaciones en Optica, León, Mexico*

Alberto Cordero-Davila   *Centro de Investigaciones en Optica, León, Mexico*

Alejandro Cornejo-Rodriguez   *Centro de Investigaciones en Optica, León, Mexico*

Dennis H. Goldstein   *Air Force Research Laboratory, Eglin AFB, Florida*

Mohammad A. Karim   *The City College of the City University of New York, New York, New York*

Malgorzata Kujawinska   *Institute of Precise and Optical Instruments Technical University, Warsaw, Poland*

Daniel J. Lougnot   *UMR CNRS, Mulhouse, France*

Daniel Malacara   *Centro de Investigaciones en Optica, León, Mexico*

Daniel Malacara, Jr.   *Centro de Investigaciones en Optica, León, Mexico.*

H. Zacarias Malacara   *Centro de Investigaciones en Optica, León, Mexico*

Angus Macleod   *Thin Film Center, Inc., Tucson, Arizona*

Duncan T. Moore   *University of Rochester, Rochester, New York*

Gonzalo Paez   *Centro de Investigaciones en Optica, León, Mexico*

Luis Efrain Regalado   *Centro de Investigaciones en Optica, León, Mexico*

Ramon Rodriguez-Vera   *Centro de Investigaciones en Optica, León, Mexico*

Manuel Servin   *Centro de Investigaciones en Optica, León, Mexico*

Cristina Solano   *Centro de Investigaciones en Optica, León, Mexico*

A. N. Starodumov   *Centro de Investigaciones en Optica, León, Mexico*

Orestes Stavroudis   *Centro de Investigaciones en Optica, León, Mexico*

Marija Strojnik   *Centro de Investigaciones en Optica, León, Mexico*

Chandra S. Vikram   *The University of Alabama in Huntsville, Huntsville, Alabama*

William Wolfe   *The University of Arizona, Tucson, Arizona*

Francis T. S. Yu   *The Pennsylvania State University, University Park, Pennsylvania*

Additional Volumes in Preparation

Handbook of
Optical Engineering

1

Basic Ray Optics

ORESTES STAVROUDIS

Centro de Investigaciones en Optica, León, Mexico

1.1 INTRODUCTION

Geometrical optics is a peculiar science. It consists of the physics of the 17th and 18th centuries thinly disguised by the mathematics of the 19th and 20th centuries. Its contemporary applications are almost entirely in optical design which, like all good engineering, remains more of an art even after the advent of the modern computer. This brief chapter is intended to convey the basic formulas as well as the flavor of geometrical optics and optical design in a concise and compact form. I have attempted to arrange the subject matter logically, although not necessarily in historical order.

The basic elements of geometrical optics are *rays* and *wavefronts*: neither exist, except as mathematical abstractions. A ray can be thought of as a beam of light with an finitesimal diameter. However, to make a ray experimentally by passing light through a very small aperture causes diffraction to rear its ugly head and the light spreads out over a large solid angle. The result is not a physical approximation to a ray but a distribution of light in which the small aperture is a point source. A wavefront is defined as a surface of constant phase to which can be attributed definite properties such as principal directions, principal curvatures, cusps, and other singularities. But, like the ray, the wavefront cannot be observed. Its existence can only be inferred circumstantially with interferometric methods.

However there is in geometrical optics an object that is observable and measurable: the *caustic surface*. [1] It can be defined in distinct but equivalent ways:

- As the envelope of an *orthotomic* system of rays; i.e., rays ultimately from a single object point.

- As the cusp locus of a wavefront train, or, equivalently, the locus of points where the element of area of the wavefront vanishes.

I think the most useful definition is that the caustic is the locus of principal centers of curvature of a wavefront. In general, every surface has two principal curvatures at each of its points. This definition then shows clearly that the caustic is a two-sheeted surface.

1.2 GAUSSIAN OPTICS À LA MAXWELL

Usually the formulas of Gaussian optics are derived from paraxial optics, a system based on approximations to the equations for ray tracing. These we will encounter in a subsequent section. Maxwell, on the other hand, took a global approach. He used a model of a perfect optical instrument and from that model, in a very elegant but straightforward way, deduced its properties, defined its parameters, and derived the various equations associated with Gaussian optics. Gauss actually found the equations for paraxial optics from the first-order terms of two power series expansions. While this is not a forum appropriate for a detailed discussion of the method Maxwell used, I will present an outline of his argument.

Maxwell [2] began by assuming that a perfect lens maps each point in object space into one and only one point in image space. Since a lens turned around is still a lens, the inverse of this mapping has to have exactly the same mathematical structure. Included in this mapping and its inverse are points at infinity whose images are the focal points of the instrument.

The mapping that Maxwell chose is the linear fractional transformation,

$$
\begin{aligned}
x' &= \frac{a_1 x + b_1 y + c_1 z + d_1}{ax + by + cz + d}, \\
y' &= \frac{a_2 x + b_2 y + c_2 z + d_2}{ax + by + cz + d}, \\
z' &= \frac{a_3 x + b_3 y + c_3 z + d_3}{ax + by + cz + d},
\end{aligned}
\tag{1.1}
$$

where (x, y, z) represents a point in object space and where (x', y', z') is its image. The inverse transform has an identical structure,

$$
\begin{aligned}
x &= \frac{A_1 x' + B_1 y' + C_1 z' + D_1}{Ax' + By' + Cz' + D}, \\
y &= \frac{A_2 x' + B_2 y' + C_2 z' + D_2}{Ax' + By' + Cz' + D}, \\
z &= \frac{A_3 x' + B_3 y' + C_3 z' + D_3}{Ax' + By' + Cz' + D},
\end{aligned}
\tag{1.2}
$$

another linear fractional transform. Here the coefficients, denoted by capital letters, are determinants whose elements are the coefficients that appear in Eq. (1.1).

The fractional-linear transformation maps planes into planes. This can be seen in the following way. Suppose a plane in object space is given by the equation,

$$
px + qy + rz + s = 0,
\tag{1.3}
$$

into which we substitute (x, y, z) from Eq. (1.2). The result is

$$(pA_1 + qA_2 + rA_3 + sA)x' + (pB_1 + qB_2 + rB_3 + sB)y'$$
$$+ (pC_1 + qC_2 + rC_3 + sC)z' + (pD_1 + qD_2 + rD_3 + sD) = 0, \tag{1.4}$$

clearly the equation of a plane in image space that is evidently the image of the plane in object space.

This transformation, therefore, maps planes into planes. Since a straight line can be represented as the intersection of two planes, it follows that this transform maps straight lines into straight lines.

From Eq. (1.1) we can see that the plane in object space, $ax + by + cz + d = 0$ is imaged at infinity in object space; from Eq. (1.2), infinity in object space is imaged into the plane $Ax' + By' + Cz' + D = 0$, in image space.

We have established coordinate systems in both object and image space. Now we impose conditions on the coefficients that bring the coordinate axes into correspondence. First we look at a plane through the coordinate origin of object space perpendicular to the z-axis, Eq. (1.3), with $r = s = 0$, as its equation. From this, and Eq. (1.4), we obtain the equation of its image,

$$(pA_1 + qA_2)x' + (pB_1 + qB_2)y' + (pC_1 + qC_2)z' + pD_1 + qD_2 = 0.$$

For this plane to pass through the image space coordinate origin and be perpendicular to the z'-axis, the coefficient of z' and the constant term must vanish identically, yielding

$$C_1 = C_2 = D_1 = D_2 = 0. \tag{1.5}$$

Again using Eq. (1.3), by setting $q = 0$, we get the equation of a plane perpendicular to the y-axis whose image, from Eq. (1.4), is

$$(pA_1 + rA_3 + sA)x' + (pB_1 + rB_3 + sB)y' + (rC_3 + sC)x' + rD_3 + sD) = 0.$$

For this to be perpendicular to the y'-axis the coefficient of y' must equal zero, yielding

$$B_1 = B_2 = B = 0, \tag{1.6}$$

The final step in this argument involves a plane perpendicular to the x-axis, obtained by setting $p = 0$ in Eq. (1.3). Its image, from Eq. (1.4), is

$$(qA_2 + rA_3 + sA)x' + (qB_2)y' + (rC_3 + sC)z' + rD_3 + sD = 0.$$

Now the coefficient of x' must vanish, yielding the last of these conditions,

$$A_2 = A_3 = A = 0. \tag{1.7}$$

These conditions assure that the coordinate axes in image space are the images of those in object space. Nothing has been done to change any of the optical properties of this ideal instrument.

Substituting these, from Eqs (1.5), (1.6), and (1.7), into Eq. (1.2) yields

$$x = \frac{A_1 x'}{Cz' + D}, \qquad y = \frac{B_2 y'}{Cz' + D}, \qquad z = \frac{C_3 z' + D_3}{Cz' + D} \tag{1.8}$$

It is a simple matter to invert this transformation to obtain

$$A_1 = \frac{cd_3 - c_3 d}{a_1}, \qquad B_2 = \frac{cd_3 - c_3 d}{b_2},$$

$$C_3 = -d, \qquad C = c, \qquad D_3 = d_3, \qquad D = -c_3,$$

(1.9)

so that Eqs (1.1) and (1.2) now read

$$x' = \frac{a_1 x}{cz + d}, \qquad y' = \frac{b_2 y}{cz + d}, \qquad z' = \frac{c_3 z - d_3}{cz + d},$$

(1.10)

and

$$x = \frac{x'(cd_3 - c_3 d)}{a_1(cz' - c_3)}$$

$$y = \frac{y'(cd_3 - c_3 d)}{b_2(cz' - c_3)}$$

$$z = \frac{dz' - d_3}{cz' - c_3}.$$

(1.11)

Now we impose a restriction on the instrument itself by assuming that it is rotationally symmetric with the z-axis, and its image, the z'-axis as its axis of symmetry. Then in Eqs (1.10) and (1.11), $b_2 = a_1$, so that we need only the y- and z-coordinates. They then degenerate into

$$y' = \frac{a_1 y}{cz + d}, \qquad z' = \frac{c_3 z + d_3}{cz + d},$$

(1.12)

and

$$y = \frac{y'(cd_3 - c_3 d)}{a_1(cz' - c_3)}$$

$$z = \frac{dz' - d_3}{cz' - c_3}.$$

(1.13)

To recapitulate, the image of a point (y, z) in object space is the point (y', z') as determined by Eq. (1.12). Conversely, the image of (y', z') in image space is the point (y, z) in object space obtained from Eq. (1.13).

The plane perpendicular to the z-axis, given by the equation $cz + d = 0$, has, as its image, the plane at infinity, as can be seen from Eq. (1.12). Therefore, $z_f = -d/c$ is the z-coordinate of the focal point of the instrument in object space. In exactly the same way, we can find the z'-coordinate of the focal point in image space is $z'_f = c_3/c$, from Eq. (1.13). To summarize, we have shown that

$$z_f = -d/c,$$

$$z'_f = c_3/c.$$

(1.14)

At this point we make a change of variables, shifting both coordinate origins to the two focal points; thus,

$$z = \bar{z} + z_f,$$

$$z' = \bar{z}' + z'_f,$$

so that, from the second equation of Eq. (1.12), we obtain

$$\bar{z}' = \frac{cd_3 - c_3 d}{c^2 \bar{z}}. \tag{1.15}$$

From the second equation of Eq. (1.13), using the same transformation, we obtain an identical result.

When the transformation is applied to the first equation of Eq. (1.12), we obtain

$$y' = \frac{a_1 y}{c z}, \tag{1.16}$$

while the first equation of Eq. (1.13) yields

$$y = \frac{(cd_3 - c_3 d)y'}{a_1 c \bar{z}'}. \tag{1.17}$$

Now define *lateral magnification* as $m = y'/y$. Then, from Eqs (1.16) and (1.17) it follows that

$$m = \frac{a_1}{c\bar{z}} = \frac{a_1 c \bar{z}'}{cd_3 - c_3 d}, \tag{1.18}$$

from which we can see that the conjugate planes of unit magnification are given by

$$\bar{z}_p = a_1/c,$$
$$\bar{z}'_p = \frac{cd_3 - c_3 d}{a_1 c}.$$

These are called the *principal planes* of the instrument. Now $\bar{z}_p$ and $\bar{z}'_p$ are the distances, along the axis of symmetry, between the foci and the principal points. These distances are called the *front* and *rear focal lengths* of the instrument and are denoted by f and f', respectively; thus,

$$f = a_1/c,$$
$$f' = (cd_3 - c_3 d)/a_1 c. \tag{1.19}$$

Next we substitute these relations into Eq. (1.15) and get Newton's formula,

$$\bar{z}\bar{z}' = ff', \tag{1.20}$$

while from Eqs (1.16) and (1.17) it follows that

$$y' = fy/\bar{z},$$
$$y = f'y'/\bar{z}'. \tag{1.21}$$

Suppose now that $\bar{y}$ and $\bar{z}$ define a right triangle with a corner at the focus in object space and let θ be the angle subtended by the z-axis and the hypotenuse. Then the first equation of Eq. (1.21) becomes the familiar

$$y' = f \tan \theta. \tag{1.22}$$

Finally, let e equal the distance of an axial point in object space to the first principal point and let e' be the distance between its conjugate and the second principal point. Then it follows that

$$e = \bar{z} - f,$$
$$e' = \bar{z}' - f'. \tag{1.23}$$

Substituting these relations into Newton's formula, Eq. (1.20), results in the familiar

$$\frac{f}{e} + \frac{f'}{e'} = 1. \tag{1.24}$$

We have seen that straight lines are mapped into straight lines. Now we complete the argument and assume that such a line and its image constitute a single ray that is traced through the instrument.

From these results we can find object–image relationships using a graphic method. In Fig. 1.1 the points z_f and z_f' are the instrument's foci and z_p and z_p' its principal planes. Let O be any object point. Let $\overline{OP}$ be a ray parallel to the axis, passing through P. Let its extention pass through P'. Since P and P' lie on the conjugate principal planes the ray in image space must pass through P'. Since this ray is parallel to the axis in object space its image must pass through z_f'. These two points determine completely this ray in image space. Now take a second ray, $\overline{Oz_fQ}$, through the object point O. Since it passes through z_f it must emerge in image space parallel to the axis. Since it passes through Q on the principal plane it must also pass through its image Q'. These two points determine this ray in image space. Where the two rays cross is I, the image of O.

With this concept we can find a most important third pair of conjugates for which the instrument's *angular magnification* is unity. Then a ray passing through one of these points will emerge from the instrument and pass undeviated through the other. These are the *nodal* points.

Refer now to Fig. 1.2. Suppose a ray passes through the axis at z_0, at an angle θ, and intersects the principal plane at y_p. After passing through this ideal instrument it intersects the axis in image space at z_0', at an angle θ', and passes through the

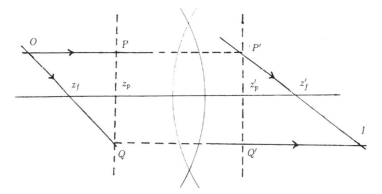

Figure 1.1 Graphical construction of an object–image relationship. The points z_f and z_f' are the instrument's foci and z_p and z_p' its principal planes. From object point O, ray $\overline{OP}$ is parallel to the axis. Since P is on the object principal plane, its image P' must be at the same height. Image ray must therefore pass through $\overline{P'z_f'}$. A second ray, $\overline{Oz_fQ}$, passes through the object focus and therefore must emerge in image space parallel to the axis. It must also pass through Q', the image of Q. The two rays cross at I, the image point.

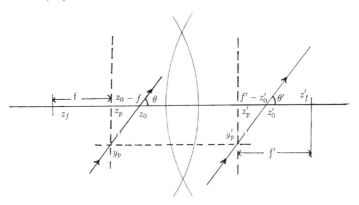

Figure 1.2 Graphical construction of the nodal points. The points z_f and z_f' are the two focal points and z_p and z_p' its two principal planes; z_0 and z_0' are the nodal points, where $\theta = \theta'$ and f and f' are the front and rear focal lengths.

principal plane at y_p'. Newton's formula, Eq. (1.20) provides a relationship between z_0 and z_0',

$$z_0 z_0' = ff'. \tag{1.25}$$

Moreover, y_1 and y_1' are equal, since they represent the heights of conjugate points on the principal planes. From Fig. 1.2 we can see that

$$
\begin{aligned}
y_p &= -(z_0 - f)\tan\theta, \\
y_p' &= -(f' - z_0')\tan\theta',
\end{aligned} \tag{1.26}
$$

so that the angular magnification is given by

$$M = \frac{\tan\theta'}{\tan\theta} = \frac{z_0 - f}{f' - z_0'}.$$

With the aid of Eq. (1.25) this becomes

$$M = \frac{z_0}{f'}.$$

For z_0 and z_0' to represent points where the angular magnification is unity it must be that

$$
\begin{aligned}
z_0 &= f', \\
z_0' &= f.
\end{aligned} \tag{1.27}
$$

These nodal points are important for two reasons. In optical testing there is an instrument called the nodal slide, which is based on the properties of the nodal points and is used to find, quickly and accurately, the focal length of a lens. A more subtle property is that images in image space bear the same perspective relationship to the second nodal point as do the corresponding objects in object space to the first nodal point.

With this in mind we make another change of variables: a translation of the z-axes to place the origins at the two nodal points. The new z-coordinates will be g and g'. The change is realized by

$$g = \bar{z} - \bar{z}_n = \bar{z} - f',$$
$$g' = \bar{z}' - \bar{z}'_n = \bar{z}' - f.$$

Again, using Newton's formula, Eq. (1.20), we obtain

$$gg' + gf + g'f' = 0,$$

from which comes

$$\frac{f}{g'} + \frac{f'}{g} + 1 = 0. \tag{1.28}$$

This concludes the study of the ideal optical instrument. We have found the six *cardinal points*, the foci, the principal points and the nodal points, solely from considering the Maxwell model of an ideal instrument. The model is a static one; there is no mention of wave fronts, velocities, or refractive indices. These characteristics will be introduced in subsequent sections of this chapter.

1.3 THE EIKONAL FUNCTION AND ITS ANTECEDENTS

This subject has a rather odd pedigree. It was first discovered by Hamilton, who called it the *characteristic function*. Then it was rediscovered by Bruns who dubbed it the *eikonal*. [3] Its origins lie much earlier. The law of refraction, discovered by Willebrord Snell using empirical methods, after his death, came into the hands of Descartes who derived for it what he claimed to be an analytic proof. Fermat disagreed. In his opinion, Snell's law was only an approximation and Descartes' proof was erroneous. He then set out to find the exact formula for refraction. But, to his surprise, Snell's law was indeed exact.

The approach to the derivation that he used has come down to us as Fermat's principle: light consists of a flow of particles, termed *corpuscles*, the trajectories of which are such that their time of transit from point to point is an extremum, either a maximum or a minimum. These trajectories are what we now call *rays*. Fermat's justification for this principle goes back to observations by Heron of Alexandria, but that is the subject of an entirely different story. What does concern us here is the interpretation of his principle in mathematical terms: its representation in terms of the variational calculus which deals specifically with the determination of extrema of functions.

To set the stage, let us consider an optical medium in which a point is represented by a vector $\mathbf{P} = (x, y, z)$ and in which the refractive index is given by a vector function of position: $n = n(\mathbf{P})$. We will represent a curve in this medium by the vector function $\mathbf{P}(s)$, where the parameter s is the geometric distance along the curve. It follows that $d\mathbf{P}/ds = \mathbf{P}'$ is a tangent vector to the curve. It can be shown that $\mathbf{P}'$ is a *unit* tangent vector.

Note that

$$d\mathbf{P} = \mathbf{P}'ds \tag{1.29}$$

so that

$$\mathbf{P}' \cdot d\mathbf{P} = \mathbf{P}'^2 ds = ds. \tag{1.30}$$

The velocity of light in this medium is c/n where c is its velocity *in vacuo*. The time of transit between any two points on a ray is therefore given by

$$\int (n/c)\,ds,$$

which is proportional to the *optical path length*,

$$I = \int n\,ds. \tag{1.31}$$

For $\mathbf{P}(s)$ to be a ray, according to Fermat's principle, I must be an *extremum*, a term from the variational calculus. A necessary condition for the existence of an extremum is that a set of differential equations, the *Euler equations*, must be satisifed. [4] The Euler equations that guarantee that I be an extremum, for geometric optics, takes the form, [5]

$$\frac{d}{ds}\left(n\frac{d\mathbf{P}}{ds}\right) = \nabla n, \tag{1.32}$$

which, when expanded becomes

$$n\mathbf{P}'' + (\nabla n \cdot \mathbf{P}')\mathbf{P}' = \nabla n. \tag{1.33}$$

What we have here is the general case, a ray in an inhomogeneous (but isotropic) medium which has come to be called a *gradient index medium* and is treated more broadly in a different chapter.

It is useful to look at this differential equation from the point of view of the differential geometry of space curves. We define the unit tangent vector to the ray $\mathbf{t}$, the unit normal vector, $\mathbf{n}$; and the unit binormal, $\mathbf{b}$, as follows: [6]

$$\begin{aligned} \mathbf{t} &= \mathbf{P}', \\ \mathbf{n} &= \rho\mathbf{P}'', \\ \mathbf{b} &= \mathbf{t} \times \mathbf{n}. \end{aligned} \tag{1.34}$$

With these, Eq. (1.31) can be rewritten as

$$\frac{n}{\rho}\mathbf{n} + (\nabla n \cdot \mathbf{t})\mathbf{t} = \nabla n, \tag{1.35}$$

which shows that the tangent vector, the normal vector, and the gradient of the refractive index must be collinear. It follows that the binormal vector must be perpendicular to the index gradient,

$$\nabla n \cdot \mathbf{b} = 0, \tag{1.36}$$

or, stated differently, that

$$\frac{n}{\rho}\mathbf{b} = \mathbf{t} \times \nabla n. \tag{1.37}$$

At any point on a space curve the vectors $\mathbf{t}$, $\mathbf{n}$, and $\mathbf{b}$ can be regarded as a set of orthonormal axes for a local coordinate system. As the parameter s varies, this point, along with the three associated vectors, slides along the curve (Fig. 1.3). Their motion is governed by the Frenet–Serret equations, [7]

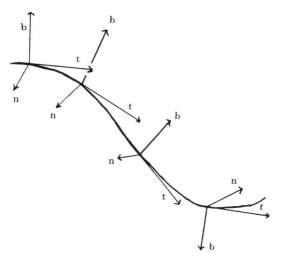

Figure 1.3   The sliding tetrahedron. As a point moves along the space curve the vectors **t**, **n**, and **b** slide along with it. The Frenet–Serret equations (Eq. (1.38)) describe their rates of change.

$$\mathbf{t}' = \frac{1}{\rho}\mathbf{n},$$

$$\mathbf{n}' = -\frac{1}{\rho}\mathbf{t} + \frac{1}{\tau}\mathbf{b}, \qquad\qquad\qquad\qquad (1.38)$$

$$\mathbf{b}' = -\frac{1}{\tau}\mathbf{n},$$

where $1/\rho$, as before, is the curve's curvature at the point in question and $1/\tau$ is its torsion. These formulas show that the curvature is the rate of change of **t** and that torsion is the rate of change of the **b** vector. Both these motions are in the direction of **n**. By squaring the expression in Eq. (1.37) we obtain a formula for the ray's curvature,

$$\frac{1}{\rho^2} = \frac{(\mathbf{t} \times \nabla n)^2}{n^2}. \qquad\qquad\qquad\qquad (1.39)$$

For the torsion refer to the third of the Frenet–Serret equations in Eq. (1.38) and obtain

$$\frac{1}{\tau} = \mathbf{n}' \cdot \mathbf{b} = \rho^2(\mathbf{P}' \times \mathbf{P}'') \cdot \mathbf{P}''', \qquad\qquad\qquad\qquad (1.40)$$

where we have used Eq. (1.34). By taking the derivative of Eq. (1.33) and multiplying the result by $\mathbf{P}' \times \mathbf{P}''$, we obtain

$$\frac{1}{\tau} = \frac{\rho^2}{n}(\mathbf{P}' \times \mathbf{P}'') \cdot (\nabla n)', \qquad\qquad\qquad\qquad (1.41)$$

where $(\nabla n)'$ represents the derivative of the gradient with respect to s.

To show how this works, consider Maxwell's fish eye [8] in which the refractive index function is given, in idealized form, by

$$n(\mathbf{P}) = \frac{1}{1 + \mathbf{P}^2},$$

so that its gradient is

$$\nabla n = \frac{-2\mathbf{P}}{(1 + \mathbf{P}^2)^2}.$$

Substituting these into the ray equation, Eq. (1.33), yields

$$(1 + \mathbf{P}^2)\mathbf{P}'' - 2(\mathbf{P} \cdot \mathbf{P}')\mathbf{P}' + 2\mathbf{P} = 0.$$

When this is differentiated, one obtains

$$(1 + \mathbf{P}^2)\mathbf{P}''' - 2(\mathbf{P} \cdot \mathbf{P}'')\mathbf{P}' = 0. \tag{1.42}$$

Now multiply by $\mathbf{P}''$ to get

$$\mathbf{P}''' \cdot \mathbf{P}'' = (1/2)(\mathbf{P}''^2)' = (1/2)(1/\rho^2)' = 0,$$

which shows that the curvature is constant. From Eqs (1.40) and (1.42) we get

$$\frac{1}{\tau} = 0. \tag{1.43}$$

The torsion is everywhere zero so that the ray is a plane curve. It follows that the ray path is the arc of a circle, exactly as it should be.

Let us return for the moment to Eq. (1.32), the ray equation. Its vector product with $\mathbf{P}$ results in

$$\mathbf{P} \times \frac{d}{ds}\left(n\frac{d\mathbf{P}}{ds}\right) = \frac{d}{ds}\left[\mathbf{P} \times \left(n\frac{d\mathbf{P}}{ds}\right)\right] = \mathbf{P} \times \nabla n. \tag{1.44}$$

Now assume that the refractive index function is symmetric with respect to the z-axis, so that $n(\mathbf{P}) = n(\rho, z)$, where $\rho^2 = x^2 + y^2$. Then its gradient is

$$\nabla n = \left(\frac{\partial n}{\partial \rho}\frac{x}{\rho}, \frac{\partial n}{\partial \rho}\frac{y}{\rho}, \frac{\partial n}{\partial z}\right), \tag{1.45}$$

and

$$\mathbf{Z} \cdot \left[\mathbf{P} \times \frac{d}{ds}\left(n\frac{d\mathbf{P}}{ds}\right)\right] = \frac{d}{ds}\left\{\mathbf{Z} \cdot \left[\mathbf{P} \times \left(n\frac{d\mathbf{P}}{ds}\right)\right]\right\}$$
$$= \mathbf{Z} \cdot (\mathbf{P} \times \nabla n) = \mathbf{P} \cdot (\nabla n \times \mathbf{Z}). \tag{1.46}$$

But

$$\nabla n \times \mathbf{Z} = \left(\frac{\partial n}{\partial \rho}\frac{x}{\rho}, \frac{\partial n}{\partial \rho}\frac{y}{\rho}, \frac{\partial n}{\partial z}\right) \times (0, 0, 1) = \frac{1}{\rho}\frac{\partial n}{\partial \rho}(y, -x, 0), \tag{1.47}$$

so that

$$\mathbf{P} \cdot (\nabla \times \mathbf{Z}) = 0, \tag{1.48}$$

showing that, from Eq. (1.38),

$$\frac{d}{ds}\left\{ \mathbf{Z} \cdot \left[\mathbf{P} \times \left(n\frac{d\mathbf{P}}{ds} \right) \right] \right\} = 0. \tag{1.49}$$

Thus

$$\mathbf{Z} \cdot \left[\mathbf{P} \times \left(n\frac{d\mathbf{P}}{ds} \right) \right] = \text{constant}. \tag{1.50}$$

This is therefore independent of s and is known as the *skewness invariant* or, more simply, the *skewness*. [9]

Now we return to Eq. (1.31), which, when we apply Eq. (1.30), becomes the line integral

$$I = \int n\mathbf{P}' \cdot d\mathbf{P}, \tag{1.51}$$

where $\mathbf{P}(s)$ is a solution of the ray equation, Eq. (1.32) or Eq. (1.33). Now let $\mathbf{P}_0$ be a starting point of a ray in this medium and let $\mathbf{P}_1$ be its end point. Then the line integral is

$$I(\mathbf{P}_0, \mathbf{P}_1) = \int_{\mathbf{P}_0}^{\mathbf{P}_1} n\mathbf{P}' \cdot d\mathbf{P}. \tag{1.52}$$

Define two nabla operators,

$$\nabla_0 = \left(\frac{\partial}{\partial x_0}, \frac{\partial}{\partial y_0}, \frac{\partial}{\partial z_0} \right); \qquad \nabla_1 = \left(\frac{\partial}{\partial x_1}, \frac{\partial}{\partial y_1}, \frac{\partial}{\partial z_1} \right).$$

Then

$$\nabla_0 I = -n_0 \mathbf{P}_0'; \qquad \nabla_1 I = n_1 \mathbf{P}_1' \tag{1.53}$$

where $n_0 = n(\mathbf{P}_0)$, $n_1 = n(\mathbf{P}_1)$, $\mathbf{P}_0' = \mathbf{P}'|_{\mathbf{P}_0}$, and $\mathbf{P}_1' = \mathbf{P}'|_{\mathbf{P}_1}$. The function I, given by Eq. (1.52), is known as *Hamilton's characteristic function* or, more simply, the *eikonal*, while the equations in Eq. (1.53) are *Hamilton's characteristic equations*. [10] By squaring either of the expressions in Eq. (1.53) we obtain the *eikonal equation*,

$$(\nabla I)^2 = n^2. \tag{1.54}$$

The eikonal equation can be derived from the Maxwell equations in several different ways. By assuming that a scalar wave equation represents light propagation, along with an application of Huygens' principle, Kirchhoff obtained a harmonic solution (discussed in Chapter 2) for light intensity at a point. He showed that the eikonal equation was obtained as a limit as wavelength approached zero. Kline and Kay [11] give a critique as well as a detailed account of this method.

Luneburg, [12] on the other hand, took a radically different approach. He started with an integral version of the Maxwell equations, regarded a wave front as a singularity in the solution of these equations and that radiation transfer consisted of the propagation of these singularities. Then he used Huygens' principle to obtain what we have called the eikonal equation.

This has led to speculation that geometric optics is a limiting case of physical optics as frequency becomes small. Perhaps. Suffice it to say that the eikonal equation, since it can be derived from sundry starting points remains a crucial point in optical theory.

Now suppose the optical medium is discontinuous: that there is a surface S where the index of refraction function has a jump discontinuity. To fix ideas, assume that light travels from left to right. Choose a point $\mathbf{P}_0$ to the left of S and a second, $\mathbf{P}_1$, to its right. Unless $\mathbf{P}_0$ and $\mathbf{P}_1$ are conjugates they will be connected by a unique ray path determined by Fermat's principle and, moreover, the segments of the ray path will be solutions of Eqs (1.32) or (1.33). Let $\bar{\mathbf{P}}$ be the point where this ray path crosses the discontinuity S and let n_- and n_+ be the left- and right-hand limits of the refractive index function along the ray path at $\bar{\mathbf{P}}$. For convenience let $\mathbf{S} = d\mathbf{P}/ds$ represent a ray vector and let $\mathbf{S}_-$ and $\mathbf{S}_+$ represent the left and right limits of $\mathbf{S}$ at the point $\bar{\mathbf{P}}$.

To best describe the consequences of a discontinuity with mathematical rigor and vigor one should apply the Hilbert integral [13] of the calculus of variations. For our purposes a schoolboy explanation is more appropriate.

Joos [14] uses the definition of the gradient in terms of limits of surface integrals to define a surface gradient as a gradient that straddles a surface of discontinuity so that

$$\nabla\phi = (\phi_+ - \phi_-)\mathbf{N}, \tag{1.55}$$

where $\mathbf{N}$ is a unit normal vector to the surface. Suppose we have Eq. (1.32) straddle S and replace the derivative by a difference quotient; then we obtain

$$\frac{d}{ds}(n\mathbf{S}) = \frac{1}{\Delta s}(n_+\mathbf{S}_+ - n_-\mathbf{S}_-) = (n_+ - n_-)\mathbf{N}. \tag{1.56}$$

When we take the vector product of this with $\mathbf{N}$ we obtain

$$n_+(\mathbf{S}_+ \times \mathbf{N}) = n_-(\mathbf{S}_- \times \mathbf{N}). \tag{1.57}$$

As we shall see in the next section this leads to the vector form of Snell's law for homogeneous media.

There is another use for the eikonal concept: the Cartesian oval, a refracting surface that images an object point perfectly onto an image point. [15] Let us assume that such a surface passes through the coordinate origin and that the two conjugate points be located at distances t and t' from that origin on the z-axis, so that their coordinates are $\mathbf{P} = (0, 0, -t)$ and $\mathbf{P}' = (0, 0, t')$. And let any point on the presumed refracting surface have the coordinates $(\bar{x}, \bar{y}, \bar{z})$. As usual, let n and n' be the refractive indices.

It turns out that for $\mathbf{P}$ and $\mathbf{P}'$ to be perfect conjugates the optical path length *along any ray* must be constant. [16] In other words,

$$n\sqrt{\bar{x}^2 + \bar{y}^2 + (\bar{z} + t)^2} + n'\sqrt{\bar{x}^2 + \bar{y}^2 + (\bar{z} - t')^2} = \text{constant}. \tag{1.58}$$

Now take a ray through the coordinate origin, so that $\bar{x} = \bar{y} = \bar{z} = 0$. From this we see that the constant is exactly equal to $nt = n't'$, so that the equation for the surface is

$$n\sqrt{\bar{x}^2 + \bar{y}^2 + (\bar{z} + t)^2} + n'\sqrt{\bar{x}^2 + \bar{y}^2 + (\bar{z} - t')^2} = nt + n't'. \tag{1.59}$$

By eliminating the square roots we get the rather formidable polynomial

$$[(n^2 - n'^2)(\bar{x}^2 + \bar{y}^2 + \bar{z}^2) + 2\bar{z}(n^2 t + n'^2 t')]^2$$
$$- 4nn'(nt + n't')[(n't + nt')(\bar{x}^2 + \bar{y}^2 + \bar{z}^2) + 2tt'(n - n')\bar{z}] = 0. \tag{1.60}$$

a quartic surface in the shape of an oval. This is shown in Fig. 1.4.

Kepler found this numerically early in the 17th century. Descartes, a generation later, found the mathematical formula. And it has been rediscovered over and over again, even by Maxwell, ever since.

An interesting (indeed, fascinating) consequence obtains when the object point approaches infinity. Divide this formula by $t^2 t'^2$ and then let t become large. The result is

$$n^2 \bar{z}^2 - n'^2(\bar{x}^2 + \bar{y}^2 + \bar{z}^2) - 2n't'(n - n')\bar{z} = 0, \tag{1.61}$$

which can be rearranged in the form

$$\left(\frac{n+n'}{n't'}\right)^2 \left(\bar{z} - \frac{n't'}{n+n'}\right)^2 + \frac{n+n'}{(n'-n)t'^2}(\bar{x}^2 + \bar{y}^2) = 1, \tag{1.62}$$

the equation of a conic section of revolution whose eccentricity is $\epsilon = n/n'$, whose center is at $z_c = n't'/(n+n')$, whose z intercepts are $z = 0$ and $z = 2n't'/(n+n')$ and whose vertex curvature is $c = 1/(1 - \epsilon)t'$. Clearly the surface is an ellipsoid when $n' > n$ and a hyperboloid when $n' < n$. This can be seen in Fig. 1.5.

Finally we come to the aplanatic surfaces of a sphere. [17] Let

$$t = -k(1 + n'/n), \qquad t' = k(1 + n/n'), \tag{1.63}$$

and substitute into Eq. (1.60). Since $nt + n't' = 0$, this degenerates into the equation of a sphere that passes through the origin and has a radius of k,

$$\bar{x}^2 + \bar{y}^2 + \bar{z}^2 + 2k\bar{z} = 0. \tag{1.64}$$

Since the refracting sphere has central symmetry, t and t' can be taken as the radii of two other spheres, that are perfect conjugates. These are the aplanatic surfaces and they are shown in Fig. 1.6. The aplanatic surfaces were the basis for a system for tracing meridional rays before the advent of computers.

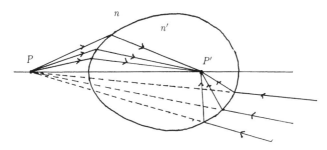

Figure 1.4 The Cartesian oval. Point P is imaged perfectly on point P'. Rays coming from the left are from the real object P. For rays coming from the right, P is a virtual object.

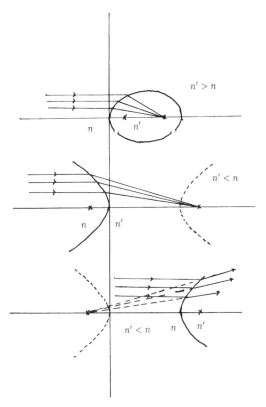

Figure 1.5 The Cartesian oval when the object is at infinity. In this case the quartic surface degenerates into a quadratic surface: an ellipse when $n' > n$; a hyperbola when $n' < n$. '×' indicates the location of a focus.

1.4   RAY TRACING AND ITS GENERALIZATION

Now we take up the very practical problem of tracing rays in a homogeneous, isotropic medium, a medium in which the refractive index, n, is constant. It follows that its gradient, ∇n, is zero so that Eq. (1.32) becomes

$$\frac{d^2 \mathbf{P}}{ds^2} = 0, \tag{1.65}$$

a second-order differential equation whose solution is a linear function of s; therefore, a ray in this medium must be a straight line. Note here that in media of constant refractive index,

$$\frac{d\mathbf{P}}{ds} = \mathbf{S} = (\xi, \eta, \zeta). \tag{1.66}$$

Snell's law, from Eq. (1.57), now takes the form

$$n'(\mathbf{S}' \times \mathbf{N}) = n(\mathbf{S} \times \mathbf{N}), \tag{1.67}$$

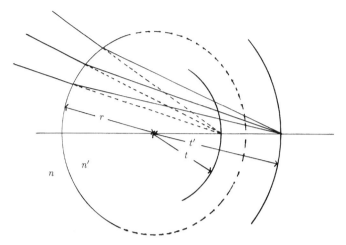

Figure 1.6 The aplanatic surfaces of a sphere. Here k is the radius of the refracting sphere; $t = k(1 + n'/n)$ is the radius of the object surface; $t = k(1 + n/n')$, that of the image surface.

where **S** and **S**$'$ are the direction cosine vectors of a ray before and after refraction, respectively; where **N** is the unit normal vector to the refracting surface at the point of incidence; and where n and n' are the refractive indices of the media before and after the refracting surface. Note that in the preceding section we used the *prime* symbol ($'$) to denote differentiation with respect to the parameter s; here we use it to signal refraction or (subsequently) transfer.

By taking the absolute value of Eq. (1.67) we get the more familiar form of Snell's law

$$n' \sin i' = n \sin i, \tag{1.68}$$

where i and i' are the angles of incidence and refraction, respectively. This statement, unlike its vector form, does not tell the whole story. Equation (1.67) provides the additional information that the vectors **S**$'$, **S**, and **N** are coplanar and determines the *plane of incidence*.

In what follows we will develop the equations for ray tracing. The form in general use today, developed by T. Smith over a period of several decades, [18] will be cast in vector form here. [19] An earlier scalar version designed particularly for computer use, is by Feder. [20] If we rearrange the terms of Eq. (1.67) to get

$$(n'\mathbf{S}' - n\mathbf{S}) \times \mathbf{N} = 0, \tag{1.69}$$

we can see that the vector $n'\mathbf{S}' - n\mathbf{S}$ is parallel to the unit normal vector **N** so that, for some γ,

$$n'\mathbf{S}' - n\mathbf{S} = \gamma\mathbf{N}, \tag{1.70}$$

from which we can get the *refraction equation*,

$$n'\mathbf{S}' = n\mathbf{S} + \gamma\mathbf{N}. \tag{1.71}$$

Note that $\cos i = \mathbf{S} \cdot \mathbf{N}$ and $\cos i' = \mathbf{S}' \cdot \mathbf{N}$, so that by taking the scalar product of Eq. (1.71) with $\mathbf{N}$ we find that γ is given by

$$\gamma = n'(\mathbf{S}' \cdot \mathbf{N}) - n(\mathbf{S} \cdot \mathbf{N}) = n' \cos i' - n \cos i. \qquad (1.72)$$

The convention for reflecting surfaces is only a convention, quite divorced from any physical reality. One sets $n' = -n$ in Eqs (1.71) and (1.72). Since $i' = i$, it follows that

$$\mathbf{S}' = \mathbf{S} + 2 \cos i \mathbf{N}. \qquad (1.73)$$

This takes care of the *refraction* or *reflection operation*. It involves only local properties of the refracting surface: the location of the point of incidence and the unit normal vector $\mathbf{N}$, at that point. On the other hand, the *transfer operation*, by means of which the point of incidence and the surface normal are found, involves the global properties of the surface.

Suppose the surface is given by a vector function of position,

$$\mathscr{F}(\mathbf{P}) = 0. \qquad (1.74)$$

A ray can be represented in parametric form by

$$\mathbf{P} = \mathbf{P}_0 + \lambda \mathbf{S}, \qquad (1.75)$$

where λ represents the distance along the ray from $\mathbf{P}_0$ to $\mathbf{P}$. If $\mathbf{P}$ is the *point of incidence*, then λ must be a solution for the equation

$$\mathscr{F}(\mathbf{P}_0 + \lambda \mathbf{S}) = 0. \qquad (1.76)$$

With the value of λ so obtained, Eq. (1.75) provides the point of incidence. The normal to a surface is best given by the gradient of its equation, $\nabla \mathscr{F}$, so that the unit normal vector is found from

$$\mathbf{N} = \frac{\nabla \mathscr{F}}{\sqrt{(\nabla \mathscr{F})^2}}, \qquad (1.77)$$

calculated, of course, at the point of incidence.

But there is a problem here. Equation (1.76) may have multiple roots. For the sphere, or for that matter for any other conic section, $\mathscr{F}$ is a quadratic and will have two solutions: either two real roots, indicating that the ray intersects the surface at two points; or two complex roots, in which case the ray misses the surface completely. More complicated surfaces produce more complicated solutions. A torus, a quartic surface, may have up to four real roots, corresponding to four points where ray and surface intersect. Deciding which is which is a daunting problem.

A particularly useful method is to identify that region of the surface that is of interest, then do a translation of coordinates to a point in that region, and then solve the equation for the reformulated $\mathscr{F}$ function and choose that solution that lies within that region or is nearest to the chosen point. Thus, the transfer operation becomes a two-step process.

To illustrate this, consider a rotationally symmetric optical system consisting of spherical refracting surfaces. Let each surface have a local coordinate system with the z-axis as the axis of symmetry and the x- and y-axes tangent to the sphere where it is intersected by the z-axis. This x, y-plane is called, for reasons that I do not understand, the *vertex plane*. Suppose $\mathbf{P}_0$ is the point of incidence of a ray with a refracting surface whose coordinates are relative to the local coordinates associated with that surface; suppose, further, that the distance along the z-axis, between this surface and the next succeeding surface, is t; then $\bar{\mathbf{P}}$, the point of intersection of the ray with the next vertex plane, is given by Eq. (1.75), which in scalar form is

$$\begin{aligned} \bar{x} &= x_0 + \bar{\lambda}\xi \\ \bar{y} &= y_0 + \bar{\lambda}\eta \\ \bar{z} &= z_0 + \bar{\lambda}\zeta = t, \end{aligned} \tag{1.78}$$

so that

$$\bar{\lambda} = -(x_0 - t)/\zeta. \tag{1.79}$$

This next sphere passes through the origin at its vertex and has the formula

$$\mathscr{F}(\mathbf{P}) = x^2 + y^2 + (x - r)^2 = (\mathbf{P} - r\mathbf{Z})^2 = r^2, \tag{1.80}$$

relative to its own coordinate system. Substituting from Eq. (1.75) yields the equation

$$[(\bar{\mathbf{P}} - r\mathbf{Z}) + \lambda\mathbf{S}]^2 = r^2, \tag{1.81}$$

a quadratic equation in λ,

$$\lambda^2 + 2\lambda(\bar{\mathbf{P}} - r\mathbf{Z}) \cdot \mathbf{S} + (\bar{\mathbf{P}} - r\mathbf{Z})^2 - r^2 = 0, \tag{1.82}$$

whose solution is

$$\lambda = -(\bar{\mathbf{P}} - r\mathbf{Z}) \cdot \mathbf{S} \pm \sigma, \tag{1.83}$$

where

$$\sigma^2 = r^2 - [(\bar{\mathbf{P}} - r\mathbf{Z}) \times \mathbf{S}]^2. \tag{1.84}$$

This constitutes the second part of this two-part transfer operation.

The ambiguity in sign in Eq. (1.80) has an easy explanation. In general, a ray will intercept a sphere at two points. We are almost always interested in the point of incidence nearest the vertex plane and therefore choose the appropriate branch of the solution.

The unit normal vector is easily obtained from the expression for the gradient in Eq. (1.77) and that for the sphere in Eq. (1.80), and is

$$\mathbf{N} = \frac{1}{r}(\mathbf{P} - r\mathbf{Z}) = \frac{1}{r}(x, y, z - r) = (cx, cy, cz - 1), \tag{1.85}$$

where $c = 1/r$ is the sphere's curvature.

For surfaces more complicated than the sphere, we need only substitute their formulas into Eq. (1.75) and proceed.

The *skewness invariant*, shown in Eq. (1.50), takes a slightly different form. In media of constant refractive index,

$$\frac{d\mathbf{P}}{ds} = \mathbf{S} = (\xi, \eta, \zeta),$$

so that

$$n\mathbf{Z} \cdot (\mathbf{P} \times \mathbf{S}) = \text{constant}. \tag{1.86}$$

Since $\mathbf{P} = (x, y, z)$, the skewness invariant becomes

$$n'(x'\eta' - y'\xi') = n(x\eta - y\xi), \tag{1.87}$$

valid for both the refraction and transfer operations.

Geometric wavefronts (we exclude wavefronts that arise from diffraction) are defined in several equivalent ways. [21] As a *surface of constant phase*, it is the locus of points that have the same optical path length from some object point. A system of rays originating from some common object is termed an *orthotomic system or a normal congruence*. In these terms a wavefront can be thought of as a *transversal surface* orthogonal to each of the rays in the system. A third definition is based on Huygens' principle, in which the wavefront is taken to be the envelope of a family of spherical wavelets centered on a preceding wavefront.

However they are defined, wavefronts have structures and properties that are best described using the language of the differential geometry of surfaces. [22] As we shall see, wavefronts are smooth surfaces that may possess cusps but which have continuous gradients.

In general, a smooth surface, at almost every point, possesses two unique directions, called the *principal directions*, that may be indicated by a pair of orthogonal vectors tagent to the surface. They have the property that curvatures of arcs embedded in the surface in these directions have curvatures that are extrema; the arc curvature in one principal direction is a maximum relative to that of all other arcs through the same point. The arc curvature in the other principal direction is a minimum. These two maximum and minimum curvatures are called the *principal curvatures*. Obvious exceptions are the plane, in which curvature is everywhere zero, and the sphere, where it is everywhere constant. In both cases principal directions cannot be defined. Another exception is the *umbilical point*, a point on a surface at which the two principal curvatures are equal. There the surface is best fit by a sphere.

What follows is a method for determining the changes in the principal directions and principal curvatures of a wavefront in the neighborhood of a traced ray. These calculations depend on and are appended to the usual methods of tracing rays. I have called them *generalized ray tracing*. [23]

Consider now a ray traced through an optical system. Through each of its points passes a wavefront that has two orthogonal principal directions and two principal curvatures. As before, let $\mathbf{S}$ be a unit vector in the direction of ray propagation and therefore normal to the wavefront. Suppose one of these principal directions is given by the unit vector $\mathbf{T}$ so that the other principal direction is $\mathbf{T} \times \mathbf{S}$. Let the two principal curvatures be $1/\rho_1$ and $1/\rho_2$. The quantities $1/\rho_1$, $1/\rho_2$, and $\mathbf{T}$ are found using general methods of differential geometry. These will not be treated here.

Suppose this ray is intercepted by a refracting surface that has, at the point of incidence, a unit normal vector $\mathbf{N}$, a principal direction $\bar{\mathbf{T}}$ and as principal curvatures, $1/\bar{\rho}_1$ and $1/\bar{\rho}_2$. Through this point passes one of the incident wavefronts, with parameters defined as above.

The equations for refraction, Eqs (1.71) and (1.72), define the plane of incidence. The unit normal vector $\mathbf{P}$ to this plane is defined by

$$\mathbf{P} = \frac{\mathbf{N} \times \mathbf{S}}{\sin i} = \frac{\mathbf{N} \times \mathbf{S}'}{\sin i'}, \tag{1.88}$$

where we have used Eqs (1.67) and (1.68). Note that **P** is invariant with respect to refraction. From this we may define three unit vectors lying in the plane of incidence:

$$\mathbf{Q} = \mathbf{P} \times \mathbf{S}; \qquad \bar{\mathbf{Q}} = \mathbf{P} \times \mathbf{N}; \qquad \mathbf{Q}' = \mathbf{P} \times \mathbf{S}'. \tag{1.89}$$

Taking the vector product of **P** and Eq. (1.71) gives us the refraction equations for the **Q** vectors,

$$n'\mathbf{Q}' = n\mathbf{Q} + \gamma\bar{\mathbf{Q}}, \tag{1.90}$$

where γ is given in Eq. (1.72)

The next step in these calculations is to find the curvatures of sections of the wavefront lying in and normal to the plane of incidence. Let θ be the angle between the wavefront principal direction **T** and the normal to the plane of incidence **P**, so that

$$\cos\theta = \mathbf{T} \cdot \mathbf{P}. \tag{1.91}$$

Then we find the desired curvatures, $1/\rho_p$ and $1/\rho_q$, as well as a third quantity, $1/\sigma$, related to the torsion of these curves. The equations are

$$
\begin{aligned}
\frac{1}{\rho_q} &= \frac{\cos^2\theta}{\rho_1} + \frac{\sin^2\theta}{\rho_2}, \\
\frac{1}{\rho_q} &= \frac{\sin^2\theta}{\rho_1} + \frac{\cos^2\theta}{\rho_2}, \\
\frac{1}{\sigma} &= \left[\frac{1}{\rho_1} - \frac{1}{\rho_2}\right]\frac{\sin 2\theta}{2}.
\end{aligned}
\tag{1.92}
$$

We do exactly the same thing for the refracting surface:

$$\cos\bar\theta = \bar{\mathbf{T}} \cdot \mathbf{P}. \tag{1.93}$$

$$
\begin{aligned}
\frac{1}{\bar\rho_p} &= \frac{\cos^2\bar\theta}{\bar\rho_1} + \frac{\sin^2\bar\theta}{\bar\rho_2}, \\
\frac{1}{\bar\rho_q} &= \frac{\sin^2\bar\theta}{\bar\rho_1} + \frac{\cos^2\bar\theta}{\bar\rho_2}, \\
\frac{1}{\bar\sigma} &= \left[\frac{1}{\bar\rho_1} - \frac{1}{\bar\rho_2}\right]\frac{\sin 2\bar\theta}{2}.
\end{aligned}
\tag{1.94}
$$

Note that Eqs (1.91) and (1.93) and Eqs (1.92) and (1.94) are identical. In a computer program both can be calculated with the same subroutine.

The refraction equations we use to relate all these are

$$
\begin{aligned}
\frac{n'}{\rho'_p} &= \frac{n}{\rho_p} + \frac{\gamma}{\bar\rho_p}, \\
\frac{n'\cos i'}{\sigma'} &= \frac{n\cos i}{\sigma} + \frac{\gamma}{\bar\sigma}, \\
\frac{n'\cos^2 i'}{\rho'_q} &= \frac{n\cos^2 i}{\rho_q} + \frac{\gamma}{\bar\rho_q},
\end{aligned}
\tag{1.95}
$$

where γ is given in Eq. (1.72).

The next step is to find θ', the angle between **P** and one of the principal directions of the wavefront after refraction,

$$\tan 2\theta' = \frac{2}{\sigma' \left[\dfrac{1}{\rho_q'} - \dfrac{1}{\rho_p'} \right]}. \tag{1.96}$$

The penultimate step is to find **T**$'$, the vector in a principal direction,

$$\mathbf{T}' = \mathbf{P}\cos\theta' + \mathbf{Q}'\sin\theta', \tag{1.97}$$

where $\mathbf{Q}'$ is found using Eq. (1.89), and the two principal curvatures,

$$\begin{aligned}
\frac{1}{\rho_1'} &= \frac{\cos^2\theta'}{\rho_p'} + \frac{\sin^2\theta'}{\rho_q'} + \frac{\sin 2\theta'}{\sigma'}, \\
\frac{1}{\rho_1'} &= \frac{\sin^2\theta'}{\rho_p'} + \frac{\cos^2\theta'}{\rho_q'} - \frac{\sin 2\theta'}{\sigma'}.
\end{aligned} \tag{1.98}$$

This takes care of refraction at a surface. The transfer operation is far simpler. The principal directions **T** and **T** × **S** are unchanged and the principal curvatures vary in an intuitively obvious way,

$$\rho_1' = \rho_1 - \lambda; \qquad \rho_2' = \rho_2 - \lambda, \tag{1.99}$$

where λ is obtained by adding Eqs (1.80) and (1.83).

Note that the first and third equations in Eq. (1.95) are exactly the Coddington equations. [24] If a principal direction of the wavefront lies in the plane of incidence, then θ and therefore $1/\sigma$ are zero. The same is true for the principal directions of the refracting surface at the point of incidence: $\bar{\theta}$ and $1/\bar{\sigma}$ both vanish. If both occur, if both the wavefront and the refracting surface have a principal direction lying in the plane of incidence, then Eqs (1.92), (1.94) and (1.98) become ephemeral, in which case Eq. (1.95) reduces to the two Coddington equations. However, generally speaking, the surface principal directions will not lie in the plane of incidence. Indeed this will happen only if the refracting surface is rotationally symmetric and the plane of incidence includes the axis of symmetry.

In a rotationally symmetric system a plane containing the axis of symmetry is called a *meridional plane*; a ray lying entirely in that plane is a *meridional ray*. A ray that is not a meridional ray is a *skew ray*. What we have shown here is that the Coddington equations are valid only for meridional rays in rotationally symmetric optical systems where a principal direction of the incident wavefront lies in the meridional plane. No such restriction applies to the equations of generalized ray tracing.

This concludes the discussion on generalized ray tracing except for a few observations. If the incident wavefront is a plane or a sphere, or if the traced ray is at an umbilical point of the wavefronts, then the principal curvatures are equal and the principal directions are undefined. In that case, as a *modus operandi*, the incident **T** vector may be chosen arbitrarily as long as it is perpendicular to **S**.

It is no secret that rays are not real and that the existence of wavefronts can be inferred only by interferometry. The only artifact in geometric optics that can be observed directly is the caustic surface. [25] Unlike waves and wavefronts the caustic can be seen, photographed, and measured. It also can be calculated using the methods of generalized ray tracing.

Like the wavefront the *caustic surface* can be defined in several different but equivalent ways, each revealing one of its properties. [26] As the envelope of an orthotomic system of rays it is clear that light is concentrated on its locus. The caustic also is where the differential element of area of a wavefront vanishes, showing that it is a cusp locus of the wavefront train. Where the wavefront and caustic touch the wavefront fold back on itself. Our final definition is that the caustic is the locus of the principal centers of curvature of a wavefront. Since there are for each ray two principal curvatures it is clear that the caustic is, in general, a complicated, two-sheeted surface.

It is the last of these definitions that is relevant to generalized ray tracing and that provides a means for calculating the caustic. Suppose that P represents a point on a ray where it intersects the final surface or the exit pupil of an optical system and that S is the ray's direction vector, both found using ordinary ray tracing. At that point ρ_1 and ρ_2 are found using generalized ray tracing. Then the point of contact of the ray with each of the two sheets of the caustic is given by

$$C_i = P + \rho_i S, (i = 1, 2). \tag{1.100}$$

The importance of the caustic cannot be underestimated. For perfect image formation, it degenerates into a single image point. Its departure from the ideal point then is a measure of the extent of the geometric aberrations associated with that image point; its location is an indicator of the optical system's distortion and curvature of field.

1.5   THE PARAXIAL APPROXIMATION AND ITS USES

In the conclusion of his two-volume opus in which he derived his equations for the local properties of a wavefront, Coddington [27] remarked that his formulas were OK (here I paraphrase) but were far too complicated to have any practical value and that he intended to continue to use his familiar, tried-and-true methods to design lenses. Such was the case until the advent of the modern computer. Tracing a single ray was a long tedious process. Tracing enough rays to evaluate an optical system was tedium raised to an impossible power. Indeed, most traced rays were meridian rays; skew rays were far too difficult for routine use. Folklore has it that only paraxial rays were used in the design process, which, when concluded, was followed by the tracing of fans of meridian rays to determine whether the design was good enough to make a prototype. If it was and if the prototype proved satisfactory the lens went into production; otherwise, it was destroyed or relegated to a museum and the designer tried again. It was cheaper to make and destroy unsatisfactory prototypes than to trace the skew rays required to make a decision as to the quality of a design.

The paraxial approximation is simplicity itself. [28] The quantities in the finite ray-tracing equations are expanded into a power series and then all but the linear turns are dropped. Thus the sine of an angle is replaced by the angle itself (in radians, of course) while the cosines become unity. Rays calculated using these approximations are termed *paraxial* rays because their domain of validity is an ϵ region that hugs the axis of symmetry of the optical system.

However, it needs to be mentioned that Gauss, whose name is often attached to these equations, experimented with a power series solution to the eikonal equation, shown in Eq. (1.54). [29] He got no further than terms of the first order but found that these enabled him to obtain the paraxial equations that we are about to derive. For that reason, these equations are frequently referred to as the *first-order equations*.

So, we make the following assumptions and approximations. First, we set $x = \xi = 0$, so that we are confined to meridional rays and we can write η and ζ as trigonometric functions of u, the angle between the ray and the axis. Here we need to set $\eta = -\sin u$ and $\zeta = \cos u$; the minus sign is needed to conform with the sign convention used in paraxial ray tracing. Next we assume the u is so small that $\eta^2 \sim 0$, so that $\eta \sim u$ and $\zeta \sim 1$. The ray vector now takes the form

$$\mathbf{S} \sim (0, -u, 1). \tag{1.101}$$

The point of incidence can be represented by the vector

$$\mathbf{P} = (0, y, z), \tag{1.102}$$

where we further assume that $y^2 \sim 0$ and $z^2 \sim 0$. From Eq. (1.85) the unit normal vector is

$$\mathbf{N} = \frac{1}{r}(\mathbf{P} - r\mathbf{Z}) = (0, cy, cz - 1). \tag{1.103}$$

where $\mathbf{Z}$ is the unit vector along the axis of rotation. It follows that

$$\mathbf{N}^2 = c^2 y^2 + (c^2 z^2 - 2cz + 1) \sim 1 - 2cz. \tag{1.104}$$

But $\mathbf{N}$ is a unit vector, so that $cz \sim 0$ and

$$\mathbf{N} = (0, cy, -1). \tag{1.105}$$

Finally, the sine of the angle of incidence is given by

$$\sin i = |\mathbf{N} \times \mathbf{S}| = cy - u, \tag{1.106}$$

with a similar expression valid for $\sin i'$. But these, too, are paraxial quantities and

$$\begin{aligned} i &\sim cy - u \\ i' &\sim cy - u'. \end{aligned} \tag{1.107}$$

When we apply the scalar form of Snell's law, Eq. (1.68), we get its paraxial equivalent,

$$n'i' = ni. \tag{1.108}$$

$$n'u' = nu + cy(n' - n), \tag{1.109}$$

the formula for paraxial refraction.

Next we take up the problem of paraxial transfer. Recall that transfer consists of two parts. The generic formula is given by Eq. (1.75)

$$\mathbf{P}' = \mathbf{P} + \lambda \mathbf{S}. \tag{1.110}$$

The first part, transfer from vertex plane to vertex plane, is represented by Eq. (1.79),

$$\bar{\lambda} = -(z_0 - t)/\zeta, \tag{1.111}$$

and the second part, transfer from vertex plane to sphere, by Eqs (1.83) and (1.84),

$$\lambda = -(\bar{\mathbf{P}} - r\mathbf{Z}) \cdot \mathbf{S} - \sigma,$$
$$\sigma^2 = r^2 - [(\bar{\mathbf{P}} - r\mathbf{Z}) \times \mathbf{S}]^2. \tag{1.112}$$

Consider the first transfer. To the paraxial approximation, $z_0 \sim 0$ and $\zeta \sim 1$, so that

$$\bar{\lambda} = t. \tag{1.113}$$

The second transfer is a little less straightforward. The second term in the expression for σ^2 in Eq. 1.112 is the square of a paraxial quantity and is therefore equal to zero so that $\sigma = r$. The distance λ then becomes

$$\begin{aligned}
\lambda &= -(\bar{\mathbf{P}} - r\mathbf{Z}) \cdot \mathbf{S} - r \\
&= -(0, \bar{y}, -r) \cdot (0, -u, 1) - r \\
&= \bar{y}u \sim 0,
\end{aligned} \tag{1.114}$$

which follows from the paraxial assumptions. Paraxial transfer is therefore given by

$$y' = y + tu. \tag{1.115}$$

Strictly speaking these formulas are valid only for rays exceedingly close to the optical axis. (Indeed, because of the underlying assumptions, they are not valid for systems lacking rotational symmetry.) Yet they are of immense value in defining properties of lenses from a theoretical point of view and at the same time are indispensable to the optical designer and optical engineer.

We can use these paraxial formulas for refraction to derive the Smith–Helmholtz invariant. [30] Figure 1.7 shows a single refracting surface $\bar{\mathbf{P}}$ and two planes, located at $\mathbf{P}$ and $\mathbf{P}'$, that are conjugates. Let t and t' be the distance of each of these planes from the surface. Because the two planes are conjugates, a ray from $\mathbf{P}$ to a point $\bar{\mathbf{P}}$ on the surface must then pass through $\mathbf{P}'$. Let the height $\mathbf{P}'$ be h and let the angles that these two rays make with the axis be u and u', respectively. Then from the figure, $u = h/t$ and $u' = h/t'$, so that

$$h = tu = t'u'. \tag{1.116}$$

Now take a point on the object plane at a height y and its image whose height is y' and consider a ray connecting these two points that intersects the refracting surface at the axis. The angle subtended by this ray and the axis is its angle of incidence i, so that $i = y/t$. A similar expression holds for the refracted ray, $i' = y'/t'$ with i' being the angle of refraction. This yields

$$y = it; \qquad y' = i't'. \tag{1.117}$$

Finally, we invoke the paraxial form of Snell's law, Eq. (1.108), $n'i' = ni$. We make the following cascade of calculations,

$$n'y'y' = n'i't'u' = nit'u' = nitu = nyu; \tag{1.118}$$

in other words,

$$n'y'u' = nyu. \tag{1.119}$$

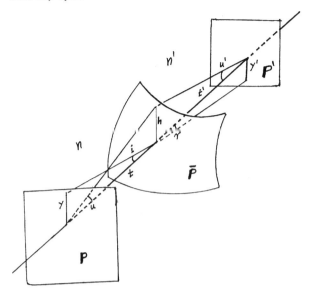

Figure 1.7 In the Smith-Helmholtz invariant, $\bar{P}$ is a refracting surface seperating two media of refractive index n and n', respectively. **P** and **P**$'$ are a pair of conjugate planes.

Suppose we are dealing with an optical system consisting of n surfaces. Then this relation can be iterated at each surface in the following way:

$$nyu = n_1 y_1 u_1 = \ldots = n_{k-1} y_{k-1} u_{k-1} = n_k y_k u_k = \ldots = n'y'u', \tag{1.120}$$

the Smith–Helmholtz invariant. This derivation is based on the paraxial ray tracing equations. An identical result can be obtained by using the general equations for a perfect optical system, as given in Section 1.1, applied to a single surface.

We say that paraxial rays are defined by Eqs (1.109) and (1.115). We also speak of the paraxial image of a point. Consider two paraxial rays emanating from some point in object space. Where they intersect in image space is the location of its image. All other paraxial rays from that point will also intersect at the same paraxial image point. We can conclude from this that any distinct pair of paraxial rays provides a basis for all other paraxial rays; any other paraxial ray can be written as a linear combination of these two. We can generalize this in an obvious way to paraxial images of surfaces.

Now consider an optical system that consists of a sequence of glass elements, all rotationally symmetric, all aligned on a common axis. Associated with each is a surface; each surface is limited by an edge that defines its boundary. By tracing paraxial rays through the surfaces preceding it, an image of that surface is formed in object space together with an image of its boundary. Any ray in object space that passes through this image and that lies inside the image of the boundary will pass through the surface itself; a ray that fails to clear the boundary of this image will not pass. All refracting surfaces in the lens will have images in object space; any ray that clears the boundary of each of these images will pass through the lens. Those that fail to do so will exceed the boundary of some surface within the lens and be blocked. [31]

In many cases, particularly for camera lenses, where the regulation of the light passing through the lens is important for the control of film exposure, an additional surface is introduced whose diameter is controlled externally. The paraxial image in object space of this *diaphragm* or *stop* is called the *entrance pupil*: in image space, the *exit pupil*. To be effective as a light regulator, the stop must be the dominant aperture. This is assured by locating the entrance pupil and setting its diameter so that it is smaller than the other aperture images in object space. (Its location is often dictated by a need to adjust values of the third-order aberrations.) Indeed, once its location and diameter have been established, the diameters of the other elements in the lens can be adjusted so that their images match that of the entrance pupil.

The entrance pupil defines the boundary of the aggregate of rays from any object point that pass through the lens; that ray that passes through its axial point is, ideally, the barycenter of the aggregate. Such a paraxial ray is termed a *chief ray* or a *principal ray*. The *marginal ray* is a ray from the object point that just clears the boundary of the entrance pupil. Almost always these two paraxial rays are taken to be the *basis* rays; every other paraxial ray can be represented as a linear combination of these two.

From Eqs (1.109) and (1.115), the two sets of formulas for marginal rays and chief rays, the latter indicated by a superior 'bar,' [32]

$$
\begin{aligned}
n'u' &= nu + cy(n' - n) \qquad n'\bar{u}' = n\bar{u} + c\bar{y}(n' - n) \\
y' &= y + tu \qquad \bar{y}' = \bar{y} + t\bar{u}.
\end{aligned}
\tag{1.121}
$$

It is easy to see that

$$
\mathscr{L} = n(y\bar{u} - \bar{y}u) - n'(y'\bar{u}' - \bar{y}'u')
\tag{1.122}
$$

is a paraxial invariant for the entire optical system and is called the Lagrange invariant. Note that if $\bar{y}$ is equal to zero the Lagrange invariant reduces to the Smith–Helmholtz invariant, which makes it clear that they are versions of each other.

The stop or diaphragm controls the amount of radiation that passes through the lens. The entrance pupil and the exit pupil can be likened to holes through which the light pours. Apart from this, there is another important property associated with these pupils. The larger the hole, for an ideal lens, the better its resolving power (see Chapters 2 and 16). But lenses are rarely ideal and certain aberrations grow with the stop aperture. (This is discussed in the following section.)

Several parameters are used to quantify the light transmission properties of a lens. [33] The most familiar of these is the $f/number$, defined as the ratio of the system's focal length to the diameter of the entrance pupil. It can be used directly in the formula for calculating the diameter of the Airy disk, for an equivalent ideal lens, which provides an estimate of the upper limit of the lens' resolving power. If the geometric aberrations of such a lens are of the same magnitude as the Airy disk then it is said to be *diffraction limited*. The details of this are treated in Chapter 6.

A second parameter, applied exclusively to microscope objectives is the *numerical aperature*, defined as the sine of one-half the angle subtended by the entrance pupil at the axial object point multiplied by the refractive index.

1.6 THE HUYGENS' PRINCIPLE AND DEVELOPMENTS

Christian Huygens (1629–1695) seemed to be a far different man than his renowned contemporaries Renè Descartes and Pierre de Fermat. They were scholastic types content to dream away in their ivory towers: the one constructed magnificent philosophies out of his reveries; the other, after his magisterial and political duties, retreated to his library where he entered notes in the margins of books. Huygens was not like that. He made things; among his creations were lenses and clocks. He was an observer. The legend is that he watched waves in a pond of water and noticed that when these waves encountered an obstruction they gave rise to circular ripples, centered where wave and obstruction met, that spread out and interfered with each other and the original waves.

The principle that he expounded was that each point on a water wave was the center of a circular wavelet and that the subsequent position and shape of that wave was the envelope of all those wavelets with equal radii. [34]

The extension of Huygens' observations on interfering wavelets to optical phenomena is the basis of what we have come to know as optical interference, interferometry, and diffraction, subjects that go far beyond the scope of this chapter. [35] However, his principle on the formation and propagation of waves, when applied to geometrical optics, provides an alternative route to the law of refraction.

Figure 1.8 shows the evolution of a train of wavefronts constructed as envelopes of wavelets developed from an arbitrary starting wavefront. The medium is assumed to be homogeneous and isotropic. There are two things to notice. As mentioned previously, in the neighborhood of a center of curvature of the initial wavefront, cusps are formed. The locus of these cusps constitutes the caustic surface associated with the wavefront train. In the same region, wavefronts in the train tend to intersect each other, giving rise to the interference patterns sometimes observed in the neighborhood of the caustic surface. [36]

A ray in this construction can be realized as the straight line connecting the center of a wavelet with the point where it touches the envelope. It is not possible to see but it turns out that the envelope of these rays is exactly and precisely the caustic and therefore coincides with the cusp locus of the wavefront.

Now for the law of refraction. Suppose a plane refracting surface separating media of refractive index n and n', lies on the x, y-plane of a coordinate system so that the surface unit normal vector $\mathbf{Z}$ is the z-axis. Take the incident ray vector to be $\mathbf{S} = (\xi, \eta, \zeta)$. If $\mathbf{W} = (p, q, r)$ represents a point on a wavefront normal to $\mathbf{S}$, then an equation for this wavefront is

$$\mathbf{S} \cdot \mathbf{W} = p\xi + q\eta + r\zeta = s, \tag{1.123}$$

where s is a parameter for the distance from the wavefront to the coordinate origin. The wavefront intersects the refracting surface on a line obtained from Eq. (1.123) by setting $p = \bar{x}$, $q = \bar{y}$, and $r = 0$, where $\bar{x}$ and $\bar{y}$ are coordinates on the refracting plane; thus

$$\bar{x}\xi = \bar{y}\eta = s. \tag{1.124}$$

The point $(\bar{x}, \bar{y}, 0)$ is to be the center of a spherical wavelet with radius s'. The total optical path length to the wavefront will be $ns + n's' = \text{const}$. Then $s' = \mu(s + k)$, where $\mu = n/n'$ and k is that constant.

Figure 1.8 Demonstrating Huygens' principle. The curve on the left side of the figure represents the initial state of a wavefront. To its right are four wavefronts generated by families of wavelets centered on the initial wavefront. Note that the second, third, and fourth of these have cusps and intersect themselves. The cusp locus is the caustic.

The equation of the wavelet is then

$$\mathscr{F} \equiv (x - \bar{x})^2 + (y - \bar{y})^2 + r^2 = \mu^2(s + k)^2 = \mu^2(\bar{x}\xi + \bar{y}\eta + k)^2, \tag{1.125}$$

an equation for a two parameter family of spherical wavelets all centered on the refracting surface and all with radii arising from an incident wavefront. To find the envelope of this family we take the partial derivatives of $\mathscr{F}$ with respect to $\bar{x}$ and $\bar{y}$, solve for those two parameters and then substitute their values back into the original equation in Eq. (1.125). These derivatives are

$$\begin{aligned}
\mathscr{F}_{\bar{x}} &\equiv x - \bar{x} + \mu^2(\bar{x}\xi + \bar{y}\eta - k)\xi = 0 \\
\mathscr{F}_{\bar{y}} &\equiv y - \bar{y} + \mu^2(\bar{x}\xi + \bar{y}\eta - k)\eta = 0
\end{aligned} \tag{1.126}$$

which rearranges itself into

$$\begin{aligned}
(1 - \mu^2\xi^2)\bar{x} - \mu^2\xi\eta\bar{y} &= x - \mu^2k\xi \\
-\mu^2\xi\eta\bar{x} + (1 - \mu^2\eta^2)\bar{y} &= y - \mu^2k\eta,
\end{aligned} \tag{1.127}$$

a simultaneous pair whose determinant of coefficients is

$$\Delta = 1 - \mu^2(\xi^2 + \eta^2) = 1 - \mu^2(1 - \zeta^2). \tag{1.128}$$

The solution is then

$$\Delta\bar{x} = (1 - \mu^2\eta^2)x + \mu^2\xi\eta y - \mu^2 k\xi$$
$$\Delta\bar{y} = \mu^2\xi\eta x + (1 - \mu^2\xi^2)y - \mu^2 k\eta. \tag{1.129}$$

From this we can show that

$$\bar{x}\xi + \bar{y}\eta - k = \frac{1}{\Delta}(x\xi + y\eta - k), \tag{1.130}$$

and with this and Eq. (1.126) we obtain

$$x - \bar{x} = -\frac{\mu^2}{\Delta}(x\xi + y\eta - k)\xi$$
$$y - \bar{y} = -\frac{\mu^2}{\Delta}(x\xi + y\eta - k)\eta. \tag{1.131}$$

When this is substituted into Eq. (1.125) we obtain

$$\Delta^2 z^2 - \mu^2(x\xi + y\eta - k)^2[1 - \mu^2(\xi^2 + \eta^2)] = 0, \tag{1.132}$$

which reduces to

$$\Delta z^2 - \mu^2(x\xi + y\eta - k)^2 = 0. \tag{1.133}$$

Here we have used Eq. (1.128).

So, the three components of the vector that goes from the center of the wavelet to its point of contact with the envelope are

$$x - \bar{x} = -\frac{\mu^2}{\Delta}(x\xi + y\eta - k)\xi$$
$$y - \bar{y} = -\frac{\mu^2}{\Delta}(x\xi + y\eta - k)\eta \tag{1.134}$$
$$z = \frac{\mu}{\sqrt{\Delta}}(x\xi + y\eta - k).$$

Normalizing this vector yields $\mathbf{S}'$, the direction cosine vector for the refracted ray,

$$\mathbf{S}' = (\mu\xi, \mu\eta, \sqrt{\Delta}) = \mu(\xi, \eta, \zeta) + (\sqrt{\Delta} - \mu\zeta)(0, 0, 1). \tag{1.135}$$

But $\mathbf{S} = (\xi, \eta, \zeta)$, the unit normal vector to the refracting surface is $\mathbf{N} = (0, 0, 1)$ and $\sqrt{\Delta} - \mu\zeta$ is exactly γ/n', this from Eq. (1.73). Thus, Eq. (1.135) is identical to the refraction equation, Eq. (1.71).

This concludes the section of Huygens' principle. We have shown how this sytem of elementary wavelets and their envelope can be used to derive the refraction equation. The same technique can be applied in media in which the wavelets are not spheres; for example, in birefringent media, where the wavelets are two-sheeted surfaces. [37]

1.7 THE ABERRATIONS, SEIDEL AND OTHERWISE

Aberration theory and optical design are so profoundly intertwined as to be almost indistinguishable. The most obvious difference is that mathematicians do one and optical engineers do the other with some physicists playing an intermediate roll. Perhaps another distinction is that what aberration theorists consider publishable, optical engineers guard jealously as trade secrets. In this day and age computers and their programmers yield wide control over both areas and to them accrue much of the glory and much of the funding.

The etiology of the concept of an aberration makes an interesting, and sometimes droll, history. Kepler, nearly 400 years ago, identified spherical aberration. Subsequently, it was realized that it was independent of object position, and that other aberrations, namely coma and astigmatism, were not. In this century we deal with a power series, that may not converge, whose terms are identified with the physical appearance of the aberrations. Much of this history is contained in a brief article by R. B. Johnson. [38]

The power series on which modern aberration theory is based is a solution of the eikonal equation, Eq. (1.54), in which Hamilton's equations, given in Eq. (1.53), play an important part. Under the assumptions of rotational symmetry the power series solution possesses only even terms; therefore, only odd terms appear in its derivatives. Gauss experimented with these odd series but his studies did not go beyond terms of the first order. These terms comprise the *first-order* equations which are identical to the equations for paraxial optics. For this reason they are often referred to as *Gaussian optics*. Seidel extended the work of Gauss to terms of the third order and correlated their coefficients with observable image errors. [38] The *third-order* or *Seidel* aberrations (in Europe frequently called the *primary* aberrations) are terms of the third order in this expansion and relate quantities obtained from the paraxial equations with other components of an optical system to observable phenomena associated with defects in optical image formation.

1.7.1 The Classical Third-Order Aberration Equations

As in Section 1.5, we distinguish two types of paraxial rays. The chief ray passes through the center of the entrance pupil and is indicated by a superior bar. The marginal ray is a ray parallel to the axis of the lens that just clears the edge of the aperature of the entrance pupil.

Further, we adapt the notation introduced by Feder. [39] A symbol with no subscript denotes a quantity at a particular surface or in the following medium. One with a subscript -1 lies on the next preceding surface or in the next preceding medium. Finally, a symbol with the subscript 1 is on the next following surface.

Finally, we modify the equations in Eqs (1.125), (1.127), and (1.129) to conform with the notation of T. Smith, [40]

$$u = cy(1 - \mu) + \mu u_{-1} \qquad \bar{u} = c\bar{y}(1 - \mu) + \mu \bar{u}_{-1}$$

$$y_1 = y - tu \qquad \bar{y}_1 = \bar{y} - t\bar{u} \tag{1.136}$$

$$i = cy - u_{-1} \qquad \bar{i} = c\bar{y} - \bar{u}_{-1},$$

where c is the curvature of the refracting surface; t is the separation between two adjacent surfaces; and $\mu = n/n_{-1}$, with n_{-1} being the refractive index of the medium next preceding the refracting surface and n that of the next following. The Lagrange invariant is of course, from Eq. (1.133):

$$\mathcal{L} = n(y\bar{u} - \bar{y}u) = n_{-1}(y_{-1}\bar{u}_{-1} - \bar{y}_{-1}u_{-1}) = n_1(y_1\bar{u}_1 - \bar{y}_1u_1), \tag{1.137}$$

In order to calculate the aberration contributions at each surface we need several auxiliary quantities. The first of these is the Petzval contribution

$$P = \frac{(\mu - 1)c}{N_{-1}}. \tag{1.138}$$

Two additional auxiliary quantities are

$$S = \frac{N_{-1}y(u - i)(1 - \mu)}{2\mathcal{L}}$$

$$\bar{S} = \frac{N_{-1}\bar{y}(\bar{u} - \bar{i})(1 - \mu)}{2\mathcal{L}}, \tag{1.139}$$

that are used, along with P, in calculating the aberration contributions from the individual surfaces. It is remarkable that P is the only quantity that does not depend on paraxial quantities but only on surface curvature and the surrounding refractive indices.

The surface contributions are

	Image contributions	Pupil contributions
Spherical aberration	$B = Si^2$	$\bar{B} = -\bar{S}\bar{i}^2$
Coma	$F = Si\bar{i}$	$\bar{F} = -\bar{S}i\bar{i}$
Astigmatism	$C = S\bar{i}^2$	$\bar{C} = -\bar{S}i^2$ (1.140)
Curvature	$D = C + P\mathcal{L}/2$	$\bar{D} = \bar{C} - P\mathcal{L}/2$
Distortion	$E = -\bar{F} + (\bar{u}^2 - \bar{u}_{-1}^2)/2$	$\bar{E} = -F + (u^2 - u_{-1}^2)/2.$

These are the third-order contributions from a single surface; the aberrations for the entire system are obtained by summing the individual terms; thus

Image errors	Field errors
$\mathcal{B} = \sum B$	$\mathcal{D} = \sum D$
$\mathcal{F} = \sum F$	$\mathcal{E} = \sum E$ (1.141)
$\mathcal{C} = \sum C$	

for the aberrations associated with image formation and similar 'bared' expressions for the aberrations associated with the pupils. It is remarkable that these third-order aberrations add algebraically. It is not the case for aberrations of degree greater than third.

Now take a ray in object space with coordinates on the object that are y and z and that passes through the point $\bar{y}$ and $\bar{z}$ on the entrance pupil plane. Let the nominal format radius on the object plane be h and let the radius of the entrance pupil aperture be $\bar{h}$. Finally, let

$$H = \mathcal{E}\frac{y^2+z^2}{h^2} + 2\mathcal{C}\frac{y\bar{y}+z\bar{z}}{h\bar{h}} + \mathcal{F}\frac{\bar{y}^2+\bar{z}^2}{\bar{h}}$$

$$K = \mathcal{D}\frac{y^2+z^2}{h^2} + 2\mathcal{F}\frac{y\bar{y}+z\bar{z}}{h\bar{h}} + \mathcal{B}\frac{\bar{y}^2+\bar{z}^2}{\bar{h}};$$

(1.142)

then

$$y' = h'[(1+H)(y/h) + K(\bar{y}/\bar{h})]$$

$$z' = h'[(1+H)(z/h) + K(\bar{z}/\bar{h})],$$

(1.143)

where y' and z' are the coordinates of this ray on the image plane. When the ray is a chief ray, then $\bar{y} = \bar{z} = 0$ and $y'/h' = (1+H)(y/h)$ and $z'/h' = (1+H)(z/h)$. In the absence of aberrations, $H = 0$, so that h' must be the radius of the image field.

As the object plane approaches infinity, the ratios y/h and z/h remain constant and less than unity and are replaced by appropriate trigonometric functions.

The third-order chromatic aberrations are treated in the same way. If ΔN is the partial dispersion of the medium after the refracting surface and if ΔN_{-1} is that for the next preceding medium, then the chromatic aberrations are given by

$$a = y(\Delta N_{-1} - \mu\Delta N)i$$

$$b = y(\Delta N_{-1} - \mu\Delta N)\bar{i}.$$

(1.144)

These represent the blurring of the image due to the dispersion of the component glasses and can be thought of as the displacement of an image point as the wavelength changes. A lateral displacement of the image of a point, called *lateral color* or *chromatic difference of magnification*, is given by $\sum b$; *longitudinal color* is the shift of the image plane along the axis and is given by $\sum a$.

1.7.2 Aberrations of the Fifth Order

The completion of Seidel's work represented an enormous step in the transformation of optical design from an ill-understood art to the beginnings of an engineering science. That it was only an improvement over paraxial optics was clear. Moreover, it suggested that it was only the second of an infinite number of steps in the power series expansion on which it was based.

The progression to the third step is an interesting story. In 1904, Schwarzschild [41] derived what he called the *Seidel eikonal*, a function of five variables whose first partial derivatives were, exactly, the five Seidel aberration coefficients. Four years later Kohlschütter [42] calculated the second partial derivatives. These he believed, by analogy, were the 15 fifth-order aberrations. Rightly or not, the formulas were well beyond the capability of routine computation at the time. Forty years later Wachendorf would describe them as *schrechlich*. Both Wachendorf [43] and Herzberger [44] attempted the derivation of these complicated expressions; both succeeded, but both failed to provide a significant improvement in accuracy that warranted their over-complexity. It was Buchdahl who realized that these fifth-order terms were not complete; that, at each surface, functions of third-order terms from it and all preceding surfaces needed to be added. We have come to identify the fifth-order contributions at a surface as the *intrinsic contributions* and the functions of the third-order contributions of the preceding surfaces as the *extrinsic contributions*.

And here arises a paradox. Within the context of geometric optics it is well known that rays are reversible; a ray traced from object space to image space is identical to the ray traced, with the same parameters, from image space to object space. It is also acknowledged that the aberration terms are approximations to real ray tracing. But in the case of the fifth order this is not so. The intrinsic contributions are the same in both cases but the extrinsic contributions are not.

The following formulas are from Buchdahl [45] by way of Rimmer [46] and Stavroudis. [47] Certain changes have been introduced to make them compatible with Eqs (1.136)–(1.140).

(a) Auxiliary Quantities

$$x_{73} = 3ii' + 2u^2 - 3u_{-1}^2 \qquad x_{42} = \bar{y}^2 ci - y\bar{i}(\bar{u} + \bar{u}_{-1})$$

$$x_{74} = 3i\bar{i}' + 2u\bar{u} - 3u_{-1}\bar{u}_{-1} \qquad x_{82} = -\bar{y}^2 cu_{-1} + y\bar{i}'(\bar{u} - \bar{u}_{-1})$$

$$x_{75} = 3\bar{i}\bar{i}' + 2\bar{u}^2 - 3\bar{u}_{-1}^2 \qquad \bar{x}_{42} = y^2 c\bar{i} + \bar{y}i(u + u_{-1})$$

$$x_{76} = -i(3u_1 - u) \qquad \bar{x}_{82} = -y^2 c\bar{u}_{-1} + \bar{y}i'(u + u_{-1})$$

$$x_{77} = -\bar{i}(2u_{-1} - u) - i\bar{u}_{-1}$$

$$x_{78} = -\bar{i}(3\bar{u}_{-1} - \bar{u})$$

$$\omega = (i^2 + i'^2 + u^2 - 3u_{-1}^2)/8$$

$$\hat{S}_{1p} = 3\omega Si/2$$

$$\hat{S}_{2p} = S(\bar{i}x_{73} + ix_{74} - \bar{u}x_{76} - ux_{77})/4$$

$$S_{3p} = n_{-1}(\mu - 1)[x_{42}x_{73} + x_{76}x_{82} + y(i + u)(ix_{75} - ux_{78})]/8$$

$$\hat{S}_{4p} = S(\bar{i}x_{74} - \bar{u}x_{77})/2$$

$$\hat{S}_{5p} = n_{-1}(\mu - 1)[x_{42}x_{72} + x_{77}x_{82} + y(i + u)(\bar{i}x_{75} - \bar{u}x_{78})]/4$$

$$\hat{S}_{6p} = n_{-1}(\mu - 1)(x_{42}x_{75} + x_{78}x_{82})/8$$

$$\hat{S}_{1q} = n_{-1}(\mu - 1)(\bar{x}_{42}x_{73} + x_{76}\bar{x}_{82})/8$$

(b) Intrinsic Fifth Order

$$\text{Spherical aberration} \quad \tilde{B}_5 = 2i\hat{S}_{2p}$$

$$\text{Coma} \quad \tilde{F}_1 = 2i\hat{S}_{1p} + i\hat{S}_{2p}$$

$$\tilde{F}_2 = i\hat{S}_{2p}$$

$$\text{Oblique spherical aberration} \quad \tilde{M}_1 = 2i\tilde{S}_{2p}$$

$$\tilde{M}_2 = 2i\tilde{S}_{3p}$$

$$\tilde{M}_3 = 2i\tilde{S}_{4p}$$

Elliptical coma $\tilde{N}_1 = 2i\tilde{S}_{3p}$

$\tilde{N}_2 = 2i\tilde{S}_{4p} + 2i\tilde{S}_{5p}$

$\tilde{N}_3 = 2i\tilde{S}_{5p}$

Astigmatism $\tilde{C}_5 = i\tilde{S}_{5p}/2$

"Petzval" $\tilde{\pi}_5 = 2i\tilde{S}_{6p} - i\tilde{S}_{5p}/2$

Image distortion $\tilde{E}_5 = 2i\tilde{S}_{6p}$

Pupil distortion $\tilde{\tilde{E}} = 2i\tilde{S}_{1q}$

In what follows $\sum'$ denotes the sum of the third-order aberrations calculated on all the refracting surfaces up to but not including the surface on which the fifth-order intrinsic aberrations are calculated.

(c) Extrinsic Fifth Order

Spherical aberration $B^o = 3\left[F\sum'B - B\sum'F\right]/2\mathscr{L}$

Coma $F_1^o = \left[(P + 4c)\sum'B + \left(5\sum'F - 4\sum'E\right)F\right.$

$\left. - \left(2\sum'P + 5\sum'\bar{C}\right)B\right]/2\mathscr{L}$

$F_2^o = \left[(P + 2c)\sum'B + 2\left(2\sum'F - \sum'\bar{E}\right)F\right.$

$\left. - \left(\sum'P + 4\sum'\bar{C}\right)B\right]/2\mathscr{L}$

Oblique spherical $M_1^o = \left[E\sum'B + \left(4\sum'F - \sum'\bar{E}\right)C\right.$

$\left. + \left(\sum'C - 4\sum'\bar{C} - 2\sum'P\right)F - B\sum'\bar{F}\right]/\mathscr{L}$

$M_2^o = \left[E\sum'B + (P + C)\left(2\sum'F - \sum'\bar{E}\right)\right.$

$\left. + \left(\sum'P + 3\sum'C - 2\sum'\bar{C}\right)F - 3B\sum'\bar{F}\right]/2\mathscr{L}$

$M_3^o = 2\left[(2C + P)\sum'F + \left(\sum'C - 2\sum'\right)\bar{F}\right.$

$\left. - B\sum'\bar{F}\right]/\mathscr{L}$

Astigmatism $C^o = \left[E\left(4\sum'C + \sum'P\right) - P\sum'\bar{F}\right.$

$\left. + 2C\left(\sum'E - 2\sum'\bar{F}\right) - 2F\sum'\bar{B}\right]/4\mathscr{L}$

"Petzval" $P^o = \left[E\left(\sum'P - 2\sum'C\right) + P\left(4\sum'E - \sum'\bar{F}\right)\right.$

$\left. + 2C\sum'E + \sum'\bar{F}\right) - 2F\sum'\bar{B}\right]/4\mathscr{L}$

Elliptical coma $N_1^o = \left[3E\sum'F - (P + C)\left(\sum'P + \sum'\bar{C}\right)\right.$

$\left. + 2C\left(\sum'P - \sum'\bar{C}\right) + F\left(\sum'E - 2\sum'\bar{F}\right)\right.$

$\left. - B\sum'B\right]/2\mathscr{L}$

$N_2^o = \left[3E\sum'F + (P + 3C)\left(3\sum'C - \sum'\bar{C} + \sum'P\right)\right.$

$\left. - C\left(\sum'P + \sum'C\right) + F\left(\sum'E - 8\sum'\bar{F}\right)\right.$

$\left. - B\sum'\bar{B}\right]/\mathscr{L}$

$$N_3^\circ = \left[E \sum {}' F + (P + C)\left(3 \sum {}' C - \sum {}' \tilde{C} + \sum {}' P\right) \right. $$
$$+ \left. C\left(\sum {}' C + \sum {}' P\right) + F\left(\sum {}' E - 4 \sum {}' \bar{F}\right) \right.$$
$$- \left. B \sum {}' \bar{B}\right]/2\mathscr{L}$$

Image distortion $\quad E^\circ = \left[3E \sum {}' E - (P + 3C) \sum {}' \bar{B}\right]/2\mathscr{L}$

Pupil distortion $\quad \tilde{E}^\circ = \left[-3E \sum {}' \bar{E} + (P + 3\tilde{C}) \sum {}' B\right]/2\mathscr{L}$

REFERENCES

Gaussian Optics à la Maxwell

1. Cagnet, M., M. Françon, and J. C. Thrier, *The Atlas of Optical Phenomina*. Prentice Hall, Englewood Cliffs, N. J., 1962.
2. Maxwell, J. C., *Scientific Papers I*, Cambridge University Press, 1890, p. 271. Cited in Born, M. and E. Wolf, *Principles of Optics*, 4th edn, Pergamon Press, London, 1970, pp. 150–157.

The Eikonal Equation and Its Antecedents

3. Herzberger, M., "On the Characteristic Function of Hamilton, the Eikonal of Bruns and Their Uses in Optics," *J. Opt. Soc. Am.*, **26**, 177 (1936); Herzberger, M., "Hamilton's Characteristic Function and Bruns' Eikonal," *J. Opt. Soc. Am.*, **27**, 133 (1937); Synge, J. L. , "Hamilton's Characteristic Function and Bruns' Eikonal," *J. Opt. Soc. Am.*, **27**, 138 (1937).
4. A concise account is in *Encyclopedic Dictionary of Mathematics*, 2nd edn, MIT Press, Cambridge, Mass, 1993, Section 46. A more detailed reference is, for example, Clegg, J. C., *Calculus of Variations*, John Wiley, New York, 1968.
5. Stavroudis, O. N., *The Optics of Rays, Wavefronts and Caustics*, Academic Press, New York, 1972, Chapter II. Klein, M., and I. W. Kaye, *Electromagnetic Theory and Geometrical Optics*, Interscience Publishers, New York, 1965, pp. 72–74. Herzberger, M., *Modern Geometrical Optics*, Interscience Publishers, New York, 1958, Chapter 35.
6. See *Encyclopedic Dictionary of Mathematics*, 1958, 2nd edn, MIT Press, Cambridge, Mass, 1993, Section 111F; App. A, Table 4.1. See also Struik, D. J., *Lectures on Classical Differential Geometry*, 2nd edn, Addison-Wesley, Reading, Mass., 1961, Chapter 1.
7. *Encyclopedic Dictionary of Mathematics*, 2nd edn, MIT Press, Cambridge, mass., 1993, Section 111D; 111H; App. A, Table 4.1. See also Struik, D. J., *Lectures on Classical Differential Geometry*, 2nd edn, Addison-Wesley, Reading, Mass., 1961, Chapter 2.
8. Luneburg, R. K., *Mathematical Theory of Optics*, Berkeley, University of California Press, 1966, pp. 172–182. Stavroudis, O. N., *The Optics of Rays, Wavefronts and Caustics*, Academic Press, New York, 1972, Chapters 4 and 11. Born, M., and E. Wolf, *Principles of Optics*, 4th edn, Pergamon Press, Oxford, 1970, pp. 147–149.
9. Welford, W. T., *Aberrations of the Symmetrical Optical System*, Academic Press, London, 1974, pp. 66–69. Stavroudis, O. N., *The Optics of Rays, Wavefronts and Caustics*, Academic Press, New York, 1972, Chapter 12.
10. Conway, A. W., and J. L. Synge (eds), *The Mathematical Papers of Sir William Roan Hamilton, Vol. 1, Geometrical Optics*, Cambridge University Press, London, 1931. Rund, H., *The Hamilton–Jacobi Theory in The Calculus of Variations*, Van Nostrand-Reinhold, Princeton, N. J., 1966.
11. Klein, M., and I. W. Kay, *Electromagnetic Theory and Geometrical Optics*, New York, Interscience Publishers, 1965, Chapter III.

12. Luneburg, R. K., *Mathematical Theory of Optics*, University of California Press, Berkeley, 1966, Chapters I and II.

13. Bliss, G. A., *Lectures on the Calculus of Variations*, University of Chicago Press, 1946, Chapter 1. Stavroudis, O. N., *The Optics of Rays, Wavefronts and Caustics*, Academic Press, New York, 1972, pp. 71–73.

14. Joos, G., *Theoretical Physics*, Tr. I. M. Freeman, Blackie and Son, London, 1934, pp. 40–42. Stavroudis, O. N., *The Optics of Rays, Wavefronts and Caustics*, Academic Press, New York, 1972, Chapter V.

15. Luneburg, R. K., *Mathematical Theory of Optics*, University of California Press, Berkeley, 1966, pp. 129–138. Herzberger, M., *Modern Geometrical Optics*, Interscience Publishers, New York, 1958, Chapter 17.

16. Born, M., and E. Wolf, *Principles of Optics*, 4th edn, Pergamon Press, Oxford, 1970, pp. 130–132.

17. Born, M., and E. Wolf, *Principles of Optics*, 4th edn, Pergamon Press, Oxford, 1970, pp. 149–150. Luneburg, R. K., *Mathematical Theory of Optics*, University of California Press, Berkeley, 1966, pp. 136–138. Herzberger, M., *Modern Geometrical Optics*, Interscience Publishers, New York, 1958, p. 190. Welford, W. T., *Aberrations of the Symmetric Optical System*, Academic Press, London, 1974, pp. 139–142.

Ray Tracing and Its Generalization

18. Smith, T., articles in Glazebrook, R., *Dictionary of Applied Physics*, MacMillan and Co, London, 1923.

19. Stavroudis, O. N., *The Optics of Rays, Wavefronts and Caustics*, Academic Press, New York, 1972, Chapter VI.

20. Feder, D. P., "Optical Calculations with Automatic Computing Machinery," *J. Opt. Soc. Am.*, **41**, 630–635 (1951).

21. Welford, W. T., *Aberrations of the Symmetric Optical System*, Academic Press, London, 1974, pp. 9–11. Herzberger, M., *Modern Geometrical Optics*, New York, Interscience Publishers, 1958, pp. 152–153, 269–271. Born, M., and E. Wolf, *Principles of Optics*, 4th edn, Pergamon Press, London, 1970, Chapter 3. Stavroudis, O. N., *The Optics of Rays, Wavefronts and Caustics*, Academic Press, New York, 1972, Chapter VIII. Klein, M., and I. W. Kay, *Electomagnetic Theory and Geometrical Optics*, Interscience Publishers, New York, 1965, Chapter V.

22. *Encyclopedic Dictionary of Mathematics*, 2nd edn, MIT Press, Cambridge, Mass., 1993, Section 111. Also Struik, D. J., *Lectures on Classical Differential Geometry*, 2nd edn, Addison-Wesley, Reading, Mass., 1961, Chapter 2.

23. Stavroudis, O. N., *The Optics of Rays, Wavefronts and Caustics*, Academic Press, New York, 1972, Chapter X. Stavroudis, O. N., "A Simpler Derivation of the Formulas for Generalized Ray Tracing," *J. Opt. Soc. Am.*, **66**, 1330–1333 (1976). Kneisley, J. A., "Local Curvatures of Wavefronts," *J. Opt. Soc. Am.*, **54**, 229–235 (1964).

24. Coddington, H., *A System of Optics*. In two parts. Simkin and Marshal, London, 1829, 1830–1839. Gullstrand, A., "Die Reelle Optische Abbildung." *Sven. Vetensk. Handl.*, **41**, 1–119 (1906). Altrichter, O., and G. Schäfer, "Herleitung der gullstrandschen Grundgleichungen für schiefe Strahlbuschel aus den Hauptkremmungen der Wellenfläsche," *Optik*, **13**, 241–253 (1956).

25. Cagnet, M., M. Françon, and J. C. Thrier, *The Atlas of Optical Phenomina*, Prentice Hall, Englewood Cliffs, N. J., 1962.

26. Born, M., and E. Wolf, *Principles of Optics*, 4th edn, Pergamon Press, London, 1970, pp. 127–130, 170. Herzberger, M., *Modern Geometrical Optics*, New York, Interscience Publishers, 1958, pp. 156, 196, 260–263. Stavroudis, O. N., *The Optics of Rays, Wavefronts and Caustics*, Academic Press, New York, 1972, pp. 78–80, 157–160, 173–

175. Klein, M., and I. W. Kay, in *Electromagnetic Theory and Geometrical Optics*, New York, Interscience Publishers, 1965, extend the idea of the caustic to include diffraction, pp. 498–499.

The Paraxial Approximation and Its Uses

27. Coddington, H., *A System of Optics*. In two parts. Simkin and Marshal, London, 1829, 1839.

28. Welford, W. T., *Useful Optics*, University of Chicago Press, 1991, Chapter 3. Welford, W. T., *Aberrations of the Symmetrical Optical System*, Academic Press, London, 1974, Chapter 3. Herzberger, M., *Modern Geometrical Optics*, Interscience Publishers, New York, 1958, Chapter 8. Luneburg, R. K., *Mathematical Theory of Optics*, University of California Press, Berkeley, 1966, Chapter IV.

29. Johnson, R. B., "An Historical Perspective on Understanding Optical Aberrations," in *Lens Design. Critical Reviews of Optical Science and Technology*, **CR41**, p. 24.

30. Born, M., and E. Wolf, *Principles of Optics*, 4th edn, Pergamon Press, London, 1970, pp. 164–166. Welford, W. T., *Aberrations of the Symmetrical Optical System*, Academic Press, London, 1974, p. 23.

31. Herzberger, M., *Modern Geometrical Optics*, Interscience Publishers, New York, 1958, pp. 101–105, 406. Born, M., and E. Wolf, *Principles of Optics*, 4th edn, Pergamon Press, London, 1970, pp. 186–188. Welford, W. T., *Aberrations of the Symmetrical Optical System*, Academic Press, London, 1974, pp. 27–29. Smith, W. J., *Modern Lens Design. A Resource Manual*. McGraw-Hill, Inc., New York, 1992, pp. 33–34.

32. Born, M., and E. Wolf, *Principles of Optics*, 4th edn, Pergamon Press, London, 1970, pp. 193–194. Feder, D. P., "Optical Calculations with Automatic Computing Machinery," *J. Opt. Soc. Am.*, **41**, 630–635 (1951).

33. Definitions of f/number and numerical aperature are to be found in Born, M., and E. Wolf, *Principles of Optics*, 4th edn, Pergamon Press, London, 1970, p. 187. The relationship between these quantities and resolution, as defined by Rayleigh, are in Ditchburn, R. W., *Light*, 3rd edn, *Vol. 1*, Academic Press, London, 1976, p. 277.

The Huygens' Principle and Developments

34. Herzberger, M., *Modern Geometrical Optics*, Interscience Publishers, New York, 1958, p. 482.

35. Born, M., and E. Wolf, *Principles of Optics*, 4th edn, Pergamon Press, London, 1970, pp. 370–375. Klein, M., and I. W. Kay, in *Electromagnetic Theory and Geometrical Optics*, Interscience Publishers, New York, 1965, pp. 328, 342–346.

36. Cagnet, M., M. Françon and J. C. Thrier, *The Atlas of Optical Phenomina*, Prentice Hall, Englewood Cliffs, N. J., 1962.

37. Stavroudis, O. N., "Ray Tracing Formulas for Uniaxial Crystals," *J. Opt. Soc. Am.*, **52**, 187–191 (1962).

The Aberrations, Seidel and Otherwise

38. Johnson, R. B., "Historical Perspective on Understanding Optical Aberrations," in *Lens Design*, W. J. Smith, ed., *SPIE Critical Reviews of Optical Science and Technology*, **CR24**, 18–29 (1992).

39. Feder, D. P., "Optical Calculations with Automatic Computing Machinery," *J. Opt. Soc. Am.*, **41**, 630–635 (1951).

40. Smith, T., sundry articles on optics in Glazebrook, R., *Dictionary of Applied Physics*, MacMillan and Co., London, 1923.

41. Born, M., and E. Wolf, *Principles of Optics*, 4th edn, Pergamon Press, London, 1970, pp. 207–211. Schwarzschild, K., "AstronomBeobachtungen mit elementaren Hilfmitteln," Teubner, Leipzig, 1904.

42. Kohlschütter, A., *Die Bildfehler fünfter Ordnung optischer Systeme*. Dissertation, Universität Göttingen, 1908.

43. Wachendorf, F., Bestimmung der Bildfehler 5, Ordnung in zentrierten optischer Systemce *Optik*, **5**, 80–122 (1949).

44. Herzberger, M., *Modern Geometrical Optics*, Interscience Publishers, New York, 1958, Part VII.

45. Buchdahl, H., *Optical Aberration Coefficients*, Dover Publications, New York, 1968. This is a reissue of the 1954 edition with a series of journal articles appended.

46. Rimmer, M., "Optical Aberration Coefficients," in *Ordeals II Program Manual*, M. B. Gold (ed.), Tropel, Fairport, N. Y., 1965, *Appendix IV.*

47. Stavroudis, O. N., *Modular Optical Design*, Springer-Verlag, Berlin, 1982, pp. 112–116.

2

Basic Wave Optics

GLENN D. BOREMAN

University of Central Florida, Orlando, Florida

2.1 DIFFRACTION

If one looks closely at the edges of shadows—the transition regions between bright and darkness—the transition is not abrupt as predicted by geometrical optics. There are variations in irradiance—interference fringes—seen at the boundary. As the size of the aperture is reduced, we expect that the size of the illuminated region in the observation plane will decrease also. This is true, but only to a certain point, where decreasing the dimensions of the aperture will produce spreading of the observed irradiance distribution.

Huygens' principle can be used to visualize this situation qualitatively. Each point on the wavefront transmitted by the aperture can be considered a source of secondary spherical waves. The diffracted wave is the summation of the secondary sources, and the envelope of these spherical waves is the new diffracted wavefront. For an aperture whose width is large compared with the wavelength of the radiation, the transmitted wavefront has very nearly the direction of the original wavefront, and the spreading caused by truncation of the spherical waves at the edges of the aperture is negligible. However, when the aperture is sufficiently small that it contains only a few Huygens' sources, the transmitted wavefront will exhibit a large divergence angle. A single Huygens' source would emit uniformly in all directions as a spherical wave.

We can put this phenomenon on a quantitative basis as follows. Let a scalar field V represent an optical disturbance (like the magnitude of the electric field, but a scalar rather than a vector), where V^2 is proportional to irradiance. This field V satisfies the scalar wave equation

$$\nabla^2 V(\bar{x}, t) = \frac{1}{c^2} \frac{\partial^2 V(\bar{x}, t)}{\partial t^2}. \tag{2.1}$$

If we assume that V is monochromatic, the scalar disturbance takes the form

$$V(\bar{x}, t) = \psi(\bar{x}) e^{-j2\pi vt}, \tag{2.2}$$

where $\psi(\bar{x})$ accounts for the spatial variation of the amplitude and phase. The spatial variation of the optical disturbance also satisfies a wave equation

$$\nabla^2 \psi + \left(\frac{2\pi v}{c}\right)^2 \psi = \nabla^2 \psi + k^2 \psi = 0. \tag{2.3}$$

An important point to note is that this wave equation is linear in ψ, which ensures the validity of the superposition principle.

We consider two coordinate systems, with subscript s for the source plane and subscript o for the observation plane. Any general point in the source plane $P(x_s, y_s)$ gives rise to a spherical wave that emits equally in all directions. For a general source distribution, the optical disturbance in the observation plane is just a weighted sum of spherical waves that originate at the various source points that comprise the source distribution. The proper expression for such a Huygens' wavelet is

$$\psi = \frac{1}{j\lambda} \frac{e^{jkr(\bar{x}_s, \bar{x}_o)}}{r(\bar{x}_s, \bar{x}_o)}, \tag{2.4}$$

where it is explicit that r is a function of position in both the source and observation planes. Thus the source distribution $\psi_s(x_s, y_s)$ produces a field distribution in the observation plane:

$$\psi(x_o, y_o) = \int_{\text{aperture}} \psi_s(x_s, y_s) \frac{e^{jkr(\bar{x}_s, \bar{x}_o)}}{j\lambda r(\bar{x}_s, \bar{x}_o)} \, dx_s dy_s. \tag{2.5}$$

In simplifying Eq. (2.5), we can approximate r in the denominator of the integrand by z, the axial separation between source and observation planes. In the exponential, we express r using the first two terms of the binomial expansion

$$\sqrt{1 + a} \approx 1 + \frac{a}{2} - \frac{a^2}{8} + \cdots \tag{2.6}$$

yielding

$$r(\bar{x}_s, \bar{x}_o) = \sqrt{(x_o - x_s)^2 + (y_o - y_s)^2 + z^2} = z\sqrt{1 + \left(\frac{x_o - x_s}{z}\right)^2 + \left(\frac{y_o - y_s}{z}\right)^2} \tag{2.7}$$

$$r(\bar{x}_s, \bar{x}_o) \approx z\left(1 + \frac{1}{2}\left(\frac{x_o - x_s}{z}\right)^2 + \frac{1}{2}\left(\frac{y_o - y_s}{z}\right)^2\right) \tag{2.8}$$

$$r(\bar{x}_s, \bar{x}_o) \approx z + \frac{x_o^2 + y_o^2}{2z} + \frac{x_s^2 + y_s^2}{2z} - 2\left(\frac{x_s x_o + y_s y_o}{2z}\right). \tag{2.9}$$

With these substitutions, Eq. (2.5) becomes

$$\psi(x_o, y_o) = \frac{1}{j\lambda z} \int\limits_{\text{aperture}} \psi_s(x_s, y_s) \exp[jkr(\bar{x}_s, \bar{x}_o)] dx_s dy_s \tag{2.10}$$

$$\psi(x_o, y_o) = \frac{1}{j\lambda z} \int\limits_{\text{aperture}} \psi_s(x_s, y_s)$$

$$\exp\left[jk\left(z + \frac{x_o^2 + y_o^2}{2z} + \frac{x_s^2 + y_s^2}{2z} - 2\left(\frac{x_s x_o + y_s y_o}{2z}\right) \right) \right] dx_s dy_s \tag{2.11}$$

$$\psi(x_o, y_o) = \frac{e^{jkz}}{j\lambda z} \exp\left[jk\frac{x_o^2 + y_o^2}{2z} \right] \int\limits_{\text{aperture}} \psi_s(x_s, y_s)$$

$$\exp\left[jk\left(\frac{x_s^2 + y_s^2}{2z} - 2\left(\frac{x_s x_o + y_s y_o}{2z}\right) \right) \right] dx_s dy_s. \tag{2.12}$$

Diffraction problems separate into two classes, based on Eq. (2.12). For Fresnel diffraction, the term $(k/2z)(x_s^2 + y_s^2)$ is sufficiently large that it cannot be neglected in the exponent. In Fresnel diffraction, the spherical waves arising from point sources in the aperture are approximated as quadratic-phase surfaces. Fresnel diffraction patterns change functional form continuously with distance z. The validity of the Fresnel approximation is ensured provided the third term in the binomial expansion above is negligible. This condition can be written as

$$z\frac{2\pi}{\lambda}\frac{1}{8}\left[\frac{(x_o - x_s)^2 + (y_o - y_s)^2}{z^2} \right]^2 \ll 1, \tag{2.13}$$

or equivalently

$$z^3 \gg \frac{\pi}{4\lambda}\left[(x_o - x_s)^2 + (y_o - y_s)^2 \right]^2. \tag{2.14}$$

Fraunhofer (far-field) diffraction conditions are obtained when the term $(k/2z)(x_s^2 + y_s^2)$ is small. Under this approximation, Eq. (2.12) for the diffracted amplitude becomes

$$\psi(x_o, y_o) = \frac{e^{jkz}}{j\lambda z} \exp\left[jk\frac{x_o^2 + y_o^2}{2z} \right] \int\limits_{\text{aperture}} \psi_s(x_s, y_s)$$

$$\exp\left[-j2\pi\left(x_s\left(\frac{x_o}{\lambda z}\right) + y_s\left(\frac{y_o}{\lambda z}\right) \right) \right] dx_s dy_s, \tag{2.15}$$

which aside from an amplitude factor $1/\lambda z$ and a multiplicative phase factor (which is of no importance when irradiance $|\psi(x_o, y_o)|^2$ is calculated) is just a Fourier transform of the amplitude distribution across the aperature, with identification of the Fourier-transform variables

$$\xi = \frac{x_o}{\lambda z} \text{ and } \eta = \frac{y_o}{\lambda z}. \tag{2.16}$$

Thus, Fraunhofer diffraction patterns will scale in size with increasing distance z, but keep the same functional form at all distances consistent with the far-field approximation. Because the distance from the aperture to the observation plane is so large, the Fraunhofer expression approximates the spherical waves arising from point sources in the aperture as plane waves.

2.1.1 Fresnel Diffraction

We must modify Eq. (2.9) for $r(\bar{x}_s, \bar{x}_o)$ in the situation of point-source illumination of the aperture from a point at a distance r_{10} away, as seen in Fig. 2.1. The aperture-to-observation-plane axial distance is r_{20}, and the distance from the source point to a general point $Q(x, y)$ in the aperture is r_1 and the distance from Q to the observation point P is r_2. We analyze the special case of a rectangular aperture that is separable in x- and y-coordinates. The diffraction integral, Eq. (2.10),

$$\psi(x_o, y_o) = \frac{1}{j\lambda z} \int_{\text{aperture}} \psi_s(x_s, y_s) \exp[jkr(\bar{x}_s, \bar{x}_o)] dx_s dy_s \tag{2.17}$$

becomes, in terms of on-axis amplitude in the observation plane at a point P,

$$\psi(P) = \frac{1}{j\lambda(r_{10} + r_{20})} \int_{y_1}^{y_2} \int_{x_1}^{x_2} \exp[jk(r_1 + r_2)] dx dy, \tag{2.18}$$

where x_1, x_2, y_1, and y_2 are the edge locations of the diffracting aperture. The typical procedure for computation of Fresnel diffraction patterns of separable aperatures involves calculation of irradiance at point P for a particular aperture position, then moving the aperture incrementally in x or y and calculating a new on-axis irradiance. For aperture sizes small compared with $r_{10} + r_{20}$, this procedure is equivalent to a calculation of irradiance as a function of position in the observation plane, but is simpler mathematically.

The distance terms in the exponent of Eq. (2.18) can be expressed as

$$r_1 = \sqrt{r_{10}^2 + x^2 + y^2} \approx r_{10} + \frac{x^2 + y^2}{2r_{10}} \tag{2.19}$$

$$r_2 = \sqrt{r_{20}^2 + x^2 + y^2} \approx r_{20} + \frac{x^2 + y^2}{2r_{20}}. \tag{2.20}$$

We can thus write, for Eq. (2.18), the amplitude in the observation plane as

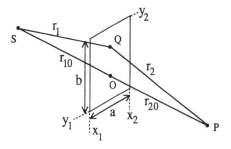

Figure 2.1 Geometry for Fresnel diffraction.

$$\psi(P) = 1\frac{1}{j\lambda(r_{10} + r_{20})}\int_{y_1}^{y_2}\int_{x_1}^{x_2}$$
$$\exp\left[jk\left\{(r_{10} + r_{20}) + (x^2 + y^2)\left(\frac{r_{10} + r_{20}}{2r_{10}r_{20}}\right)\right\}\right]dxdy. \tag{2.21}$$

With a change of variables

$$\xi = x\sqrt{\frac{2(r_{10} + r_{20})}{\lambda r_{10}r_{20}}} \tag{2.22}$$

and

$$\eta = y\sqrt{\frac{2(r_{10} + r_{20})}{\lambda r_{10}r_{20}}}, \tag{2.23}$$

we find, for Eq. (2.21), the diffracted wave amplitude

$$\psi(P) = \frac{1}{2j}\frac{e^{jk(r_{10}+r_{20})}}{(r_{10} + r_{20})}\int_{\xi_1}^{\xi_2}e^{y\pi\xi^2/2}\,d\xi\int_{\eta_1}^{\eta_2}e^{j\pi\eta^2/2}d\eta. \tag{2.24}$$

An important special case is that of plane-wave illumination, $r_{10} \to \infty$. Under those conditions, Eqs (2.22) and (2.23) become

$$\xi = x\sqrt{\frac{2}{\lambda r_{20}}} \tag{2.25}$$

$$\eta = y\sqrt{\frac{2}{\lambda r_{20}}}. \tag{2.26}$$

In general, we identify the term

$$\psi_0(P) = \frac{e^{jk(r_{10}+r_{20})}}{(r_{10} + r_{20})} \tag{2.27}$$

as the unobstructed wave amplitude at P in the absence of a diffracting screen. Hence, we can write for Eq. (2.24)

$$\psi(P) = \frac{\psi_0(P)}{2j}\int_{\xi_1}^{\xi_2}e^{j\pi\xi^2/2}d\xi\int_{\eta_1}^{\eta_2}e^{j\pi\eta^2/2}d\eta. \tag{2.28}$$

We identify the Fresnel cosine and sine integrals

$$C(w) = \int_0^w\cos(\pi\xi^2/2)\,d\xi \tag{2.29}$$

and

$$S(w) = \int_0^w\sin(\pi\xi^2/2)\,d\xi \tag{2.30}$$

seen in Fig. 2.2. Using Eqs (2.29) and (2.30), we can write for Eq. (2.28), the diffracted on-axis amplitude

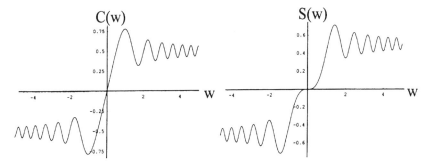

Figure 2.2 Fresnel cosine and sine integrals.

$$\psi(P) = \frac{\psi_o(P)}{2j}\big\{[C(\xi_2) - C(\xi_1)] + j[S(\xi_2) - S(\xi_1)]\big\}$$

$$\big\{[C(\eta_2) - C(\eta_1)] + j[S(\eta_2) - S(\eta_1)]\big\} \tag{2.31}$$

and for the diffracted on-axis irradiance

$$E(P) = \frac{E_o(P)}{4}\big\{[C(\xi_2) - C(\xi_1)]^2 + [S(\xi_2) - S(\xi_1)]^2\big\}$$

$$\big\{[C(\eta_2) - C(\eta_1)]^2 + [S(\eta_2) - S(\eta_1)]^2\big\}, \tag{2.32}$$

where $E_0(P)$ is the unobstructed irradiance. If the wavelength and geometry are specified, the above expression can be evaluated numerically for separable apertures.

As an example of the calculation method, let us compute the Fresnel diffraction pattern from an edge, illuminated from a source at infinity. Assuming that the edge is oriented vertically, $\eta_1 = -\infty$ and $\eta_2 = \infty$. If we let the left edge of the open aperture be at $-\infty$, we have $\xi_1 = -\infty$ and $\xi_2 = x_2(2/\lambda r_{20})^{1/2}$. Using the asymptotic forms $C(-\infty) = S(-\infty) = -0.5$ and $C(\infty) = S(\infty) = 0.5$, we can write for Eq. (2.32):

$$E(P) = \frac{E_o(P)}{4}\big\{[C(\xi_2) + 0.5]^2 + [S(\xi_2) + 0.5]^2\big\}\big\{[0.5 + 0.5]^2 + [0.5 + 0.5]^2\big\}. \tag{2.33}$$

The value of irradiance at the edge of the geometrical shadow region ($\xi_2 = 0$) is found to be one-quarter of the unobstructed irradiance. This result is intuitive, because the presence of the edge blocks half of the amplitude on-axis, resulting in one-quarter of the irradiance that would be present without the edge. The complete irradiance distribution is shown in Fig. 2.3.

Examples of other irradiance distribution calculations are shown in Fig. 2.4, for an aperture of dimensions 2 mm × 2 mm, with plane-wave illumination at 0.5-μm wavelength. The distance from the diffracting aperture to the observation screen, r_{20}, is the variable. In the first plot, cases are shown for $r_{20} = 0.4$ m and $r_{20} = 4$ m. The changes in the form of the irradiance pattern with changes in r_{20} are evident. There is not a simple scaling of size as one finds with Fraunhofer diffraction, but a change of functional form depending on the value of r_{20}. For the situation where $r_{20} = 0.4$ m, the geometrical shadow region is within $|\xi_2| < 3.16$; and for the case of $r_{20} = 4$ m, the

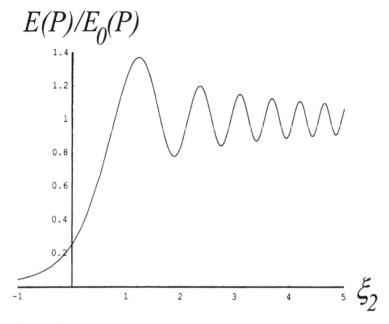

Figure 2.3   Fresnel diffraction pattern (irradiance) at an edge.

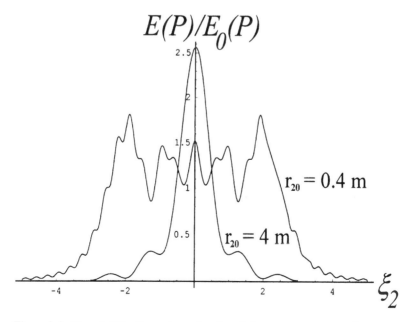

Figure 2.4   Fresnel diffraction patterns (irradiance) for an aperture at two distances.

shadow region is within $|\xi_2| < 1$. The shadow region is just the size of the original aperture, expressed in terms of ξ_2, for a given distance.

Figure 2.5 shows the irradiance distribution at a longer distance, for $r_{20} = 40\,\text{m}$. This is calculated by means of the Fresnel-integral technique above, but the dimensions involved are such that the observation plane is essentially in the Fraunhofer region of the aperture. Thus, the irradiance distribution approximates that of the Fourier transform squared of the aperture distribution. Note that the peak irradiance has decreased dramatically, consistent with the spreading of flux far beyond the boundaries of the geometrical shadow at $|\xi_2| = 0.32$.

Fresnel diffraction from a nonseparable aperture, even one as simple as a circular disk, is more difficult to describe analytically. The most effective calculation method is to compute the squared modulus of a two-dimensional Fourier transform, but with a multiplicative quadratic phase factor in the aperture. This phase factor accounts for the optical path difference between the axial distance $(r_{10} + r_{20})$ between source and observation points and the distance $(r_1 + r_2)$ that traverses the maximum extent of the aperture. The Fresnel pattern can then be calculated by the same means as a Fraunhofer pattern.

2.1.2 Fraunhofer Diffraction

For situations in which the distance from the aperture to the observation plane is sufficiently large, or for which the aperture is small, we can neglect the term $(k/2z)(x_s^2 + y_s^2)$. According to Eq. (2.15), the diffracted amplitude can be expressed in terms of the Fourier transform of the amplitude transmittance function of the aperture $\psi_s(x_s, y_s)$:

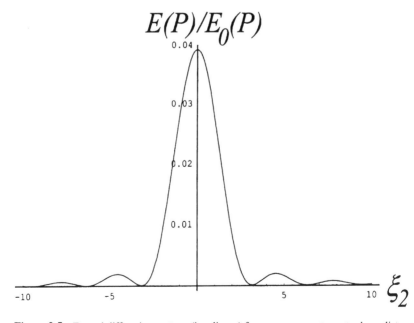

Figure 2.5 Fresnel diffraction pattern (irradiance) for a square aperture at a long distance.

$$\psi(x_o, y_o) = \frac{e^{jkz}}{j\lambda z} \exp\left[jk \frac{x_o^2 + y_o^2}{2z} \right] \int_{\text{aperture}} \psi_s(x_s, y_s)$$

$$\exp\left[-j2\pi \left(x_s \left(\frac{x_o}{\lambda z}\right) + y_s \left(\frac{y_o}{\lambda z}\right) \right) \right] dx_s dy_s \tag{2.34}$$

$$\psi(x_o, y_o) = \frac{1}{\lambda z} \exp\left[j \left\{ k \left(z + \frac{x_o^2 + y_o^2}{2z} \right) - \frac{\pi}{2} \right\} \right] \mathcal{F}\{\psi_s(x_s, y_s)\} \Big|_{\xi = \frac{x_o}{\lambda z}, \eta = \frac{y_o}{\lambda z}}. \tag{2.35}$$

The complex exponential does not affect the value of the diffracted irradiance, $E(x_o, y_o) = |\psi(x_o, y_o)|^2$, so that we find

$$E(x_o, y_o) = \frac{1}{\lambda^2 z^2} |\mathcal{F}\{\psi_s(x_s, y_s)\}|_{\xi = \frac{x_o}{\lambda z}, \eta = \frac{y_o}{\lambda z}}. \tag{2.36}$$

Note that the Fourier transform of a spatial-domain aperture function returns a function of spatial frequency, and thus a change of variables seen in Eq. (2.16), $\xi = x_o/\lambda z$ and $\eta = y_o/\lambda z$, is required for evaluation of the result, Eq. (2.36), in terms of position in the observation plane. The change of variables takes the slightly different form

$$\xi = \frac{x_o}{\lambda f} \quad \text{and} \quad \eta = \frac{y_o}{\lambda f}, \tag{2.37}$$

when the diffraction pattern is observed in the focal plane of a positive lens of focal length f, allowing convenient scaling of the pattern to any desired dimensions by the choice of lens focal length.

Our examples of calculation of Fraunhofer patterns begin with the one-dimensional rectangular aperture of full-width b. Using the standard notation of Gaskill, [1] we take the aperture distribution as $\psi_s(x_s) = \text{rect}(x_s/b)$, and for the usual expression of a diffraction pattern normalized to an on-axis value of unity,

$$\frac{E(x_o)}{E(x_o = 0)} = \text{sinc}^2 \left(\frac{\pi b x_o}{\lambda z} \right) = \left(\frac{\sin\left(\frac{\pi b x_o}{\lambda z}\right)}{\left(\frac{\pi b x_o}{\lambda z}\right)} \right)^2. \tag{2.38}$$

This function, seen in Fig. 2.6, has a maximum value of 1 and a first zero at $x_o = \lambda z/b$.

Another useful aperture distribution is that representing two point apertures separated by a distance l:

$$\psi_s(x_s) = \tfrac{1}{2}[\delta(x_s - l/2) + \delta(x_s + l/2)], \tag{2.39}$$

for which the corresponding irradiance distribution in the diffraction pattern is

$$\frac{E(x_o)}{E(x_o = 0)} = \cos^2 \left(\frac{\pi l x_o}{\lambda z} \right). \tag{2.40}$$

A circular aperture of full width D, $\psi_s(x_s) = \text{circ}(r_s/D)$, is another important case. The normalized diffracted-irradiance distribution is

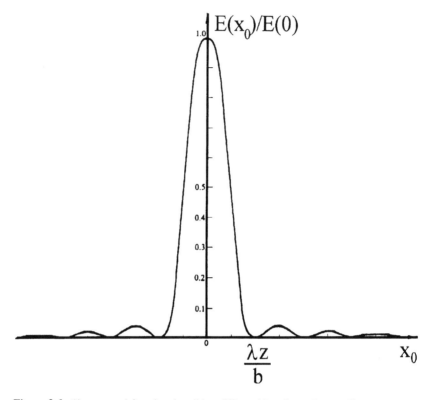

Figure 2.6 Sinc-squared function describing diffracted irradiance from a slit.

$$\frac{E(r_o)}{E(r_o = 0)} = \left[\frac{2J_1\left(\frac{\pi D r_o}{\lambda z}\right)}{\left(\frac{\pi D r_o}{\lambda z}\right)} \right]^2 \tag{2.41}$$

which, as seen in Fig. 2.7, has its first zero when

$$\frac{\pi D r_o}{\lambda z} = 1.22\pi = 3.83, \tag{2.42}$$

or in terms of radius at

$$r_o = \frac{1.22\lambda z}{D}. \tag{2.43}$$

Integrating the irradiance distribution, we find that 84% of the power contained in the pattern is concentrated into a diameter equal to $2.44\,\lambda z/D$. When the diffracting aperture is a lens aperture itself, the full width of the diffracted-irradiance distribution in the focal plane is then $2.44\,\lambda F/\#$ where the lens focal ratio is defined as $F/\# \equiv f/D$. This relationship determines the ability of an imaging sytem to resolve objects of a certain size. The angular resolution β, expressed as the full width of the diffraction spot divided by the focal length of the lens, is

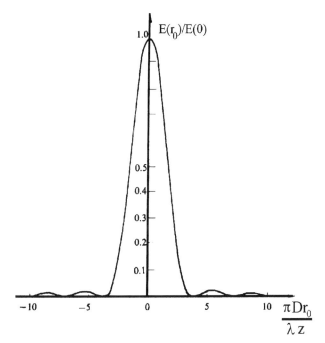

Figure 2.7 Bessel-squared function for diffracted irradiance from a circular aperture.

$$\beta = 2.4\frac{\lambda f/D}{f} = 2.4\frac{\lambda}{D}. \tag{2.44}$$

When the angle in Eq. (2.45) is multiplied by p, the distance from the lens to the object plane, the resulting dimension is the diffraction-limited resolution in the object, as seen in Fig. 2.8.

For example, a diffraction-limited system with an aperture size of 10 cm can resolve a 12-mm target at 1 km if the system operates at a wavelength of 0.5 µm, while a system of the same diameter operating at $\lambda = 10$ µm can resolve targets of 240-mm lateral dimension.

2.2 INTERFERENCE

The phenomenon of the interference depends on the principle of superposition of waves. Suppose that the following sinusoid corresponds to the electric-field magnitude of a light wave:

$$\mathscr{E}(x, t) = A\cos(kx - \omega t + \phi) \tag{2.45}$$

where $k = 2\pi/\lambda$ is the free-space wavenumber and $\omega = 2\pi/T$ is the radian frequency. Any number of waves can be added together coherently, and the magnitude of the resultant wave will depend on whether the interference is constructive or destructive. As an example, suppose that two waves are added together (with time dependence suppressed):

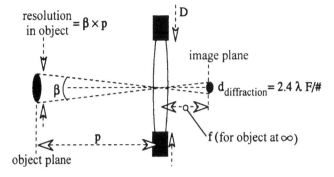

Figure 2.8 Diffraction-limited resolution in an optical system.

$$\mathcal{E}(x) = A_1 \cos(kx + \phi_1) + A_2 \cos(kx + \phi_2). \tag{2.46}$$

Constructive interference occurs when the amplitude of the resultant sum wave is larger than the magnitude of either component. This occurs when the relative phase difference between the two waves $\delta = \phi_2 - \phi_1$ is close to zero or an even multiple of 2π. Destructive interference occurs when the amplitude of the resultant is less than the magnitude of either component. This condition occurs when δ is close to an odd multiple of π. The relative phase difference between two waves derived from the same source depends on the optical path difference (OPD) between the two paths. In terms of the refractive index n and the physical path length d for the two paths, the phase difference is defined as

$$\delta = \frac{2\pi}{\lambda} \times \text{OPD} = \frac{2\pi}{\lambda} \left[\sum_{\text{path 2}} nd - \sum_{\text{path 1}} nd \right]. \tag{2.47}$$

Any detector of optical or infrared radiation responds to the long-term time average of the square of the resultant electric field, a quantity proportional to irradiance, $E(\text{W/cm}^2)$. Assuming that the $\mathcal{E}$ field vectors of the two waves are co-polarized, we can write an expression for the total irradiance:

$$E_{\text{total}} = E_1 + E_2 + 2\sqrt{E_1 E_2} \cos \delta. \tag{2.48}$$

If the interfering waves are of equal magnitude, the maximum value of $E_{\text{total}} = 4E$, and the minimum value is 0. We characterize the contrast of the fringes by the visibility V, a number between 0 and 1, defined as

$$V = \frac{E_{\text{max}} - E_{\text{min}}}{E_{\text{max}} + E_{\text{min}}}. \tag{2.49}$$

For equal-amplitude co-polarized light waves, the visibility is 1. Unequal wave amplitudes and any cross-polarized component (which does not interfere) will reduce the fringe visibility. We have assumed the phase difference between the two waves is stable with time. If δ changes rapidly over a measurement time, then the $\cos \delta$ term averages out, and an incoherent addition of irradiances is obtained.

Some special cases of interference are of particular interest for applications. If we interfere two plane waves of the same frequency and parallel polarization:

$$\bar{\mathcal{E}}_1 = \hat{z}\mathcal{E}_1 e^{i(\bar{k}_1 \cdot \bar{r} + \phi_1)} \tag{2.50}$$

$$\bar{\mathcal{E}}_2 = \hat{z}\mathcal{E}_2 e^{i(\bar{k}_1 \cdot \bar{r} + \phi_2)}, \tag{2.51}$$

where r is a position vector in the x–y plane, $\bar{r} = \hat{x}x + \hat{y}y$. Letting the initial phases both be equal to zero, we can write an expression for the constructive-interference condition

$$k[x(\cos\theta_2 - \cos\theta_1) + y(\sin\theta_2 - \sin\theta_1)] = 2\pi m \ (m \text{ integer}). \tag{2.52}$$

This fringe pattern consists of straight fringes of equal spacing, parallel to the line bisecting the ray directions. The fringe spacings along the x- and y-axes are

$$\Delta y = \frac{\lambda}{(\sin\theta_2 - \sin\theta_1)} \tag{2.53}$$

and

$$\Delta x = \frac{\lambda}{(\cos\theta_2 - \cos\theta_1)}. \tag{2.54}$$

As seen from Eqs (2.53) and (2.54), the fringe spacing increases as the angle between the interfering beams gets smaller.

Interference of two spherical waves is another important special case, with the two components given by

$$\bar{\mathcal{E}}_1 = \hat{z}\mathcal{E}_1 \frac{e^{i(k|\bar{r}-\bar{r}_1|+\phi_1)}}{|\bar{r}-\bar{r}_1|} \tag{2.55}$$

and

$$\bar{\mathcal{E}}_2 = \hat{z}\mathcal{E}_2 \frac{e^{i(k|\bar{r}-\bar{r}_2|+\phi_2)}}{|\bar{r}-\bar{r}_2|}. \tag{2.56}$$

Setting the phase offsets to zero in Eqs (2.55) and (2.56), we find, for a given fringe, that

$$k(|\bar{r}-\bar{r}_1| - |\bar{r}-\bar{r}_2|) = \text{constant}. \tag{2.57}$$

Because the distance between two points is a constant, this is the equation of a family of hyperboloids of revolution. On an observation screen perpendicular to the line of centers of the sources, one obtains circular fringes, and on an observation screen parallel to the line of centers, one obtains (at least near the center) equally spaced sinusoidal fringes. Taking the circular-fringe case (we will analyze the other case in the next section when dealing with Young's interference experiment), we begin with the condition for a bright fringe:

$$m\lambda = |\bar{r}-\bar{r}_1| - |\bar{r}-\bar{r}_2|. \tag{2.58}$$

Let source #1 be located on the x-axis at $x = -v$ and source #2 located at $x = -u$. Thus, we can write

$$|\bar{r}-\bar{r}_1| = \sqrt{y^2 + v^2} = v\sqrt{1 + \frac{y^2}{v^2}} \cong v\left[1 + \frac{y^2}{2v^2}\right] = v + \frac{y^2}{2v} \tag{2.59}$$

and

$$|\bar{r} - \bar{r}_2| = \sqrt{y^2 + u^2} = u\sqrt{1 + \frac{y^2}{u^2}} \cong u\left[1 + \frac{y^2}{2u^2}\right] = u + \frac{y^2}{2u}. \tag{2.60}$$

The bright-fringe condition, Eq. (2.58), becomes

$$m\lambda = |\bar{r} - \bar{r}_1| - |\bar{r} - \bar{r}_2| = v - u + \frac{y^2}{2}\left[\frac{1}{v} - \frac{1}{u}\right], \tag{2.61}$$

and the resulting fringe pattern corresponds to the case where the radii are proportional to the square root of the order number m.

The final special case to consider is interference of a plane wave and a spherical wave. The components are

$$\bar{\mathscr{E}}_1 = \hat{z}\mathscr{E}_1 e^{i(\bar{k}\cdot\bar{r} + \phi_1)} \tag{2.62}$$

and

$$\bar{\mathscr{E}}_2 = \hat{z}\mathscr{E}_2 \frac{e^{i(k|\bar{r} - \bar{r}_2| + \phi_2)}}{|\bar{r} - \bar{r}_2|}. \tag{2.63}$$

Assuming that the spherical wave arises from a point source located on the x-axis, at a point $x = -x_0$, the exponent in Eq. (2.63) can be written

$$k|\bar{r} - \bar{r}_2| = k\sqrt{(x - x_0)^2 + y^2 + z^2}. \tag{2.64}$$

The condition for a bright fringe is

$$k\left[x\cos\theta + y\sin\theta - \sqrt{(x - x_0)^2 + y^2 + z^2}\right] = 2\pi m. \tag{2.65}$$

Observing in the y–z ($x = 0$) plane, Eq. (2.65) reduces to

$$m\lambda = \left[y\sin\theta - \sqrt{x_0^2 + y^2 + z^2}\right]. \tag{2.66}$$

Assuming that we observe the fringe pattern in a small region of the y–z plane, $|x_0| \gg y$ and $|x_0| \gg z$, we can expand Eq. (2.66) as a binomial series:

$$m\lambda = y\sin\theta - x_0\sqrt{1 + \frac{y^2 + z^2}{x_0^2}} \cong y\sin\theta - x_0\left[\frac{1 + y^2 + z^2}{2x_0^2}\right]. \tag{2.67}$$

Considering the case where the plane wave propagates parallel to the x-axis, we find the condition for bright fringes, Eq. (2.67), reduces to

$$m\lambda = x_0 + \frac{r^2}{2x_0}. \tag{2.68}$$

Thus, the fringes are concentric circles where the radius of the circle is proportional to the square root of the order number m. The fringe spacing Δr thus decreases with radius, according to $\Delta r = \lambda x_0/r$.

2.2.1 Wavefront Division

Division-of-wavefront interferometers are configurations where the optical beam is divded by passage through apertures placed side by side. The most important example of this tyupe of interferometer is Young's double-slit configuration. Typically, a point source illuminates two point apertures in an opaque screen. We will consider later the effects of finite source size (spatial coherence) and finite slit width (envelope function). These points are sources of spherical waves (Huygens' wavelets) that will interfere in their region of overlap. The usual situation is that the fringes are observed on a screen that is oriented parallel to the line of centers of the pinholes. The basic setup is seen in Fig. 2.9. The bright-fringe condition is

$$m\lambda = |\bar{r} - \bar{r}_1| - |\bar{r} - \bar{r}_2|. \tag{2.69}$$

If the two pinholes are illuminated with the same phase (source equidistant from each), then there will be a bright fringe on the axis, along the line perpendicular to the line of centers. The expressions for the distances from each pinhole to a general point in the observation plane are

$$|\bar{r} - \bar{r}_1| = \sqrt{(y - l/2)^2 + x_0^2} \tag{2.70}$$

and

$$|\bar{r} - \bar{r}_2| = \sqrt{(y + l/2)^2 + x_0^2}. \tag{2.71}$$

Using Eqs (2.70) and (2.71), the bright-fringe condition, Eq. (2.69) becomes

$$m\lambda = \sqrt{(y - l/2)^2 + x_0^2} - \sqrt{(y + l/2)^2 + x_0^2}$$
$$= x_0\sqrt{1 + \frac{(y - l/2)^2}{x_0^2}} - x_0\sqrt{1 + \frac{(y + l/2)^2}{x_0^2}}. \tag{2.72}$$

Under the binomial approximation of Eq. (2.6) (equivalent to a small-angle condition), Eq. (2.72) becomes

$$m\lambda \cong x_0\left[1 + \frac{(y - l/2)^2}{2x_0^2}\right] - x_0\left[1 + \frac{(y + l/2)^2}{2x_0^2}\right] = \frac{ly}{x_0}. \tag{2.73}$$

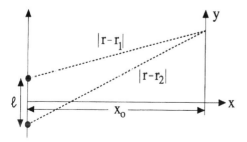

Figure 2.9 Young's double-slit configuration.

A simple memnonic construction that yields this condition is seen in Fig. 2.10. We assume that the rays drawn are essentially parallel and will thus interfere at infinity. The OPD between the two rays is the distance identified as $l\sin\theta$. When the OPD is equal to an even number of wavelengths, there will be constructive interference

$$m\lambda = l\sin\theta\tan\theta = \frac{ly}{x_0}. \qquad (2.74)$$

The expression for center-to-center fringe separation is

$$\Delta y = \frac{x_0\lambda}{l}, \qquad (2.75)$$

which means that as $\Delta y \propto x_0$, the fringe pattern scales with distance; as $\Delta y \propto \lambda$, the fringe pattern scales with wavelength; and as $\Delta y \propto 1/l$, the fringe pattern scales as the pinhole separation decreases.

It should also be noted from a practical point of view that it is not necessary to make x_0 a very long distance to achieve the overlap of these essentially parallel rays. As seen in Fig. 2.11, a positive lens will cause parallel rays incident at an angle of θ to overlap at a distance f behind the lens, at a y height of θf. This configuration is often used to shorten the distance involved in a variety of experimental setups for interference and diffraction.

It can be verified that the fringes obtained in a Young's double-slit configuration have a cosine-squared functional form. Taking a phasor model for the addition of light from the two slits, and remembering that the irradiance is the square of the field, we find

$$E \propto |e^{j0} + e^{j\delta}|^2 = 2 + 2\cos\delta = 4\cos^2(\delta/2), \qquad (2.76)$$

as seen in Fig. 2.12. The phase shift δ can be written as

$$\delta = \frac{2\pi}{\lambda} \times \mathrm{OPD} = \frac{2\pi}{\lambda}[l\sin\theta] \cong \frac{2\pi}{\lambda}l\frac{y}{x_0}. \qquad (2.77)$$

Thus, for small angles, Eq. (2.76) describing the irradiance becomes

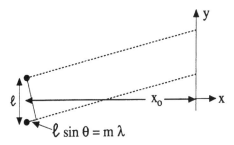

Figure 2.10 Simple construction for Young's fringes.

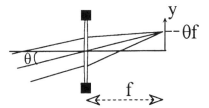

Figure 2.11 Use of a lens to produce the far-field condition for Fraunhofer diffraction.

$$E = 4\cos^2\left[\frac{\frac{2\pi}{\lambda}l\frac{y}{x_0}}{2}\right] = 4\cos^2\left[\frac{\pi l y}{\lambda x_0}\right]. \tag{2.78}$$

The first zero of $\cos^2(\beta)$ is at $\beta = \pi/2$, so the first zero of the double-slit irradiance pattern of Eq. (2.78) is at

$$y_{\text{first-zero}} = \frac{\lambda x_0}{2l}, \tag{2.79}$$

and the fringe spacing near the center of the pattern is

$$\Delta y = \frac{\lambda x_0}{l}. \tag{2.80}$$

It is useful to develop expressions relating the phase difference δ, the diffraction angle θ, and the position on the observation screen y. First, for small angles, where $\sin\theta \approx \tan\theta \approx \theta$, we have, in terms of diffraction angle θ,

$$l\theta \approx m\lambda. \tag{2.81}$$

Taking the derivative with respect to order number m:

$$\frac{\partial\theta}{\partial m} = \frac{\lambda}{l}. \tag{2.82}$$

Letting Eq. (2.82) be a finite difference

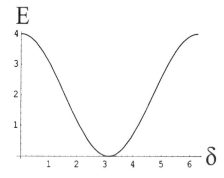

Figure 2.12 Irradiance as a function of phase shift δ for two-beam interference.

$$\Delta\theta = \Delta m \frac{\lambda}{l}, \tag{2.83}$$

we have, for the angular spacing between fringes, $\Delta\theta$ for $\Delta m = 1$:

$$\Delta\theta|_{\Delta m=1} = \frac{\lambda}{l}. \tag{2.84}$$

A similar development can be obtained in terms of position in the observation plane. Assuming a small angle,

$$y \approx x_0\theta. \tag{2.85}$$

Taking the derivative with respect to angle, we have

$$\frac{\partial y}{\partial\theta} = x_0. \tag{2.86}$$

Combining Eqs (2.85) and (2.86),

$$\frac{\partial y}{\partial m} = \frac{\partial y}{\partial\theta}\frac{\partial\theta}{\partial m} = x_0[\lambda/l]. \tag{2.87}$$

Finding the y-position spacing between fringes as the increment in y for which the increment in order number $m = 1$

$$\Delta y = \Delta m \times x_0[\lambda/l] \tag{2.88}$$

$$\Delta y|_{\Delta m=1} = x_0[\lambda/l]. \tag{2.89}$$

Now, leaving the domain of small angles ($l \ll x_0$ but y not $\ll x_0$), we begin the development with

$$l\sin\theta = m\lambda. \tag{2.90}$$

Taking the derivative

$$\frac{\partial\theta}{\partial m} = \frac{\lambda}{l\cos\theta}. \tag{2.91}$$

The angular fringe spacing is then

$$\Delta\theta|_{\Delta m=1} = \frac{\lambda}{l\cos\theta}. \tag{2.92}$$

We note that the fringe spacing increases as the angle increases.

Now, in terms of y position, we begin with

$$y = x_0\tan\theta. \tag{2.93}$$

Taking the derivative

$$\frac{\partial y}{\partial\theta} = x_0\frac{1}{\cos^2\theta} \tag{2.94}$$

$$\frac{\partial y}{\partial m} = \frac{\partial y}{\partial\theta}\frac{\partial\theta}{\partial m} = \frac{x_0}{\cos^2\theta}\frac{\lambda}{l\cos\theta} = \frac{x_0\lambda}{l\cos^3\theta}. \tag{2.95}$$

For the fringe spacing in the observation plane (setting $\Delta m = 1$)

$$\Delta y = \frac{x_0\lambda}{l\cos^3\theta}. \tag{2.96}$$

Again, the fringes have a larger spacing away from the center of the pattern, but with a different functional dependence than seen in Eq. (2.92) for the spacing in terms of angle.

Young's interference experiment can be performed with any number of slits, and the result can be easily obtained using phasor addition:

$$E \propto |e^{j0} + e^{j\delta} + e^{j2\delta}|^2, \tag{2.97}$$

where δ is the phase difference between successive slits. The three phasors are all in phase for $\delta = m \times 2\pi$, where m is an integer (including zero). The value of the irradiance for this condition is nine times the irradiance of a single slit. There are two values of δ, for a three-slit pattern, for which the irradiance equals zero. Using phasor diagrams, we can see that this condition is satisfied for $\delta = 2\pi/3$ and $\delta = 4\pi/3$. There is a subsidiary maximum (with irradiance equal to that from one slit alone) for the condition where $\delta = \pi$. The irradiance as a function of δ is shown in Fig. 2.13, for the three-slit and four-slit cases.

2.2.2 Amplitude Division

Division of amplitude interferometers is typically implemented when one surface of the system is a partial mirror—in other words, a beamsplitter—that allows for an optical wave to be split into two components which will travel different paths and recombine afterwards. These interferometers are useful for determining the flatness of optical surfaces or the thickness of transparent films and are widely used for measurement purposes.

To properly account for the phase shift on reflection, it is necessary to note that a "rare-to-dense" reflection will differ in phase by π radians from the ray undergoing a "dense-to-rare" reflection. That this is true for either polarization can be shown easily by application of the Stokes' reversibility principle, which states that all light rays are reversible if absorption is negligible. Using Fig. 2.14, we let the reflection and transmission coefficients for a ray incident from medium #1 be r and t, respectively. For a ray incident from the second medium, the reflection and transmission coefficients are r' and t'.

From Fig. 2.14, we have

$$rt + r't = 0 \tag{2.98}$$

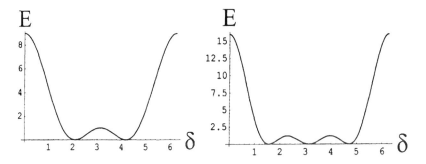

Figure 2.13 Irradiance as a function of phase shift δ for three- and four-beam interference.

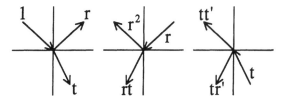

Figure 2.14 Stokes' reversibility principle.

and

$$tt' + r^2 = 1. \tag{2.99}$$

From Eq. (2.98), we can see that $r = -r'$, which verifies the required phase shift.

The first interferometer configuration we consider is that of fringes of equal thickness. There is a thin transparent film for which the thickness is a function of position. For example, the film can be an air film, contained between two pieces of glass, an oil film on the surface of water, or a freestanding soap film. We begin our analysis by considering interference in a plane-parallel plate of index n, immersed in air. We assume illumination by a single point source, and follow the ray trajectory as seen in Fig. 2.15.

We consider interference between the rays reflected from the top and bottom surfaces of the plate. If the plate is parallel, these two rays will be parallel, and will interfere at infinity, or at the focus of a lens. For this analysis, we ignore other rays that undergo multiple reflections in the plate. Taking the plate as having a refractive index of glass (around 1.5), the two rays shown each carry about 4% of the original power in the incident ray, and all other rays carry a much smaller power. We assume that the thickness d of the plate is small enough to be within the coherence length of the light source used. Typically for films of a few micrometers or less thickness, interference effects are seen even with white light, such as sunlight. As the thickness of the plate increases to several millimeters or larger, interference effects are seen only with coherent sources, such as lasers.

Taking into account the difference in phase shifts on reflection seen in Fig. 2.14, the condition for a bright fringe can be written as

$$\frac{2\pi}{\lambda}\left[n(\overline{AB} + \overline{BC}) - \overline{AD}\right] = (m + 1/2)2\pi \tag{2.100}$$

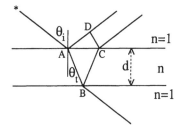

Figure 2.15 Interference in a plane-parallel plate.

$$\frac{2\pi}{\lambda}\left[n\frac{2d}{\cos\theta_t} - 2nd\tan\theta_t\sin\theta_t\right] = (m+1/2)2\pi \tag{2.101}$$

$$\frac{2\pi}{\lambda}\frac{2nd}{\cos\theta_t}[1 - \sin^2\theta_t] = (m+1/2)2\pi \tag{2.102}$$

$$2nd\cos\theta_t = (m+1/2)\lambda. \tag{2.103}$$

It should be noted that as θ increases, $\cos\theta_t$ decreases, and hence the resonance wavelength decreases. This effect can be important for the "off-angle" response for optical interference filters. In general, d, $\cos\theta$, and λ can all be variables in a given situation. If, by the configuration chosen, we arrange to keep θ_t constant, then we get interference maxima at certain wavelengths (if white light is used), depending on the local thickness of the film. If monochromatic light is used, we will see bright and dark Fizeau fringes across the film that correspond to thickness variations. The fringes are the loci of all points for which the film has a constant optical thickness.

Practically speaking, θ_t varies somewhat except when a collimated point source is used to provide the input flux. However, the variation in $\cos\theta_t$ is typically small enough to neglect this effect. This is especially true if the aperture of the collection lens is small, as is usually the case when viewing fringes directly by eye. Angular variations can also be neglected for the cases wherer the collection lens is far from the film being measured, or if the flux enters and leaves the film at nearly normal incidence, where the change of $\cos\theta_t$ with angle is minimum.

These conditions are satisfied in the Fizeau interferometer, often called a Newton interferometer, when the surface to be tested is very nearly in contact with a reference flat. As seen in Fig. 2.16, a slight tilt (α) between two planar glass surfaces forms a wedge-shaped air film ($n = 1$) between. For light incident at approximately $\theta = 0$, we can write the condition for a bright fringe (taking into account phase shifts on reflection as before) as

$$\frac{2\pi}{\lambda}[2h] = (m+1/2)2\pi, \tag{2.104}$$

where $2h$ is the round-trip path length inside the air film of local height h. Thus, bright fringes represent half-wave differences in film air-gap thickness, and are formed whenever

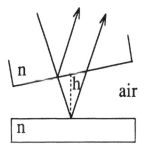

Figure 2.16 Interference in a wedge-shaped air gap between two flat surfaces.

$$h = (m + 1/2)\frac{\lambda}{2}.\tag{2.105}$$

Referring to Fig. 2.17, if we look down from the top, we will see straight fringes, assuming that both surfaces are flat. From the relationship

$$\tan\alpha \approx \alpha = \frac{\lambda/2}{\Delta x},\tag{2.106}$$

we find that the fringe separation is

$$\Delta x = \lambda/2\alpha.\tag{2.107}$$

These straight-line fringes do not indicate which way the wedge is oriented. However, by physically pushing down on the low end of the wedge, more fringes are added to the interferogram because the tilt angle is increased. Pushing on the high end of the wedge decreases the tilt angle, and hence fewer fringes are seen.

Fringes that are not straight indicate a departure from flatness in the surface being tested. Referring to Fig. 2.18, for a given peak departure from flatness Δ, the peak height error in the piece is

$$\text{height error} = \frac{\Delta}{\Delta x}\frac{\lambda}{2}.\tag{2.108}$$

The interferogram of Fig. 2.18 does not show whether the peak error is a high spot or a low spot. Remembering that a fringe is a locus of points of constant gap dimension, a knowledge of which end of the air wedge is open will allow determination of the sign of the height error. Compared with a situation where equally spaced fringes exist between two tilted flat surfaces, a high spot will displace the fringes toward the open end of the wedge, and a low spot will displace the fringes toward the closed end of the wedge. As seen in Fig. 2.19, this fringe displacement will keep the local gap height constant along a fringe.

Newton's rings are fringes of equal thickness seen between a flat and a curved surface. If the sphere and the flat are in contact at the center of the curved surrface, there will be a dark spot at the center when viewed in reflected light, consistent with

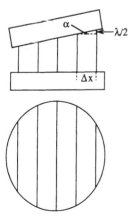

Figure 2.17   Relationship of wedge-shaped air gap to interferogram.

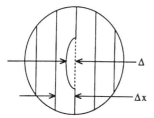

Figure 2.18 Interferogram of a surface with a departure from flatness.

the π phase difference between dense-to-rare and rare-to-dense reflections. Referring to Fig. 2.20, we can develop the expression for the fringe radii. By the Pythagorean theorem, we can write a relationship between the air gap height h, the fringe radius r, and the radius of curvature of the surface R:

$$(R - h)^2 + r^2 = R^2. \tag{2.109}$$

Under the approximation $h^2 \ll r^2$, we find that

$$h \approx \frac{r^2}{2R}. \tag{2.110}$$

We can thus write the condition for a bright fringe (in reflection) as

$$2h = (m + 1/2)\lambda. \tag{2.111}$$

Substituting, we find the fringe radii are proportional to the square root of integers:

$$r_{\text{bright fringe}} = \sqrt{(m + 1/2)R\lambda}. \tag{2.112}$$

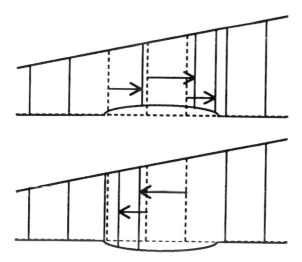

Figure 2.19 Fringe displacements for high and low areas.

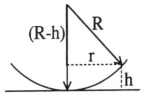

Figure 2.20 Construction for Newton's rings.

For surfaces that are not spherical, the shape of the resulting Fizeau fringes can provide a topographic characterization of the surface profile, because the fringes are loci of points of constant thickness of the air gap. Addition of a small tilt between the reference surface and the surface under test can allow quantitative measurement of departure from flatness on the order of $\lambda/20$ or less, which would otherwise be impossible to measure.

Using filtered arc sources, the Fizeau fringes must be observed in an air gap that is within the coherence length of the source, a few micrometers at most. This requires contact between the reference surface and the test surface. For situations where this is undesirable, a laser-based Fizeau interferometer allows substantial separation between the two surfaces, as seen in Fig. 2.21.

A wide variety of configurations for the laser-based Fizeau are possible, allowing characterization of surface figure of both concave and convex surfaces, as well as homogeneity of an optical part in transmission. Schematic diagrams of this instrumentation can be found in Malacara. [2] A Twyman–Green Interferometer, seen in Fig. 2.22, allows for a distinct separation of the optical paths of the reference arm and the test arm, and typically use a collimated point source such as a focused laser beam for illumination.

The interpretation of Twyman–Green interferograms is essentially the same as for Fizeau instruments, because in both cases light travels through the test piece twice. This double-pass configuration yields a fringe-to-fringe spacing that represents $\lambda/2$ of OPD. In a Mach–Zehnder interferometer, seen in Fig. 2.23, the test piece is only traversed once, so that fringes represent one wavelength of OPD.

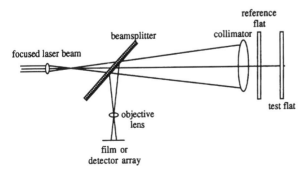

Figure 2.21 Laser-based Fizeau interferometer.

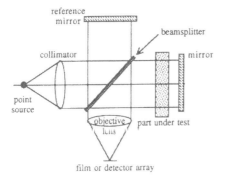

Figure 2.22 Twyman–Green interferometer.

2.2.3 Multiple-Beam Interference

In a typical fringes-of-equal-thickness configuration, where the refractive index difference is between air ($n = 1.0$) and uncoated glass ($n = 1.5$), only the first two reflected rays are significant, having powers referenced to the incoming beam of 4% and 3.69%. The next reflected beam has a power of 6×10^{-3}%. Only the two primary beams interfere, producing cosine-squared fringes. The multiple reflected beams do not effect the interference in any appreciable way. The situation changes when the reflectivity of the interfaces is high. The example shown in Fig. 2.24 is for surfaces having a power reflectivity of 90%. The falloff of power in the multiple reflected beams is slower, both in reflection and in transmission. More rays will add to produce the interference fringes. The fringe shape will be seen to be sharper than sinusoidal, with narrower maxima in reflection and narrower minima in transmission. In Fig. 2.25 we show the multiple-beam case for a general surface reflectivity.

By our previous analysis of interference in a plane-parallel plate, the difference in OPD between adjacent rays is

$$OPD = 2nd \cos\theta_t, \tag{2.113}$$

and the resulting round-trip phase difference between adjacent rays is

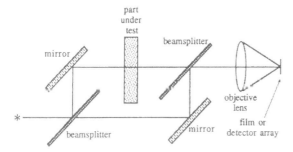

Figure 2.23 Mach–Zehnder interferometer.

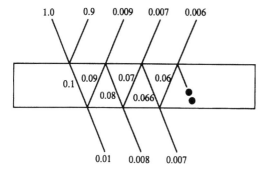

Figure 2.24 Multiple reflected beams for a surface power reflectivity of 90%.

$$\delta = \frac{2\pi}{\lambda} 2nd \cos \theta_t. \tag{2.114}$$

These rays are parallel if the plate is parallel and can be brought to interfere at infinity, or at the focus of a lens. The symbols t and r in Fig. 2.25 are the transmittance and reflectance for the case of incidence from a low-index medium. The symbols r' and t' apply for the case of incidence from a high-index medium. We can write expressions for the complex amplitude of the total reflected $\mathscr{E}_r$ and transmitted $\mathscr{E}_t$ beams, in terms of the amplitude of the incident beam $\mathscr{E}_0$ and the incremental phase shift δ:

$$\mathscr{E}_r = \mathscr{E}_0 \left(r + tt'r'e^{i\delta} + tt'(r')^3 e^{i2\delta} + tt'(r')^5 e^{i3\delta} + \cdots + tt'(r')^{2N-3} e^{i(N-1)\delta} \right), \tag{2.115}$$

$$\mathscr{E}_r = \mathscr{E}_0 \left(r + tt'r'e^{i\delta} \left(1 + (r')^2 e^{i\delta} + ((r')^2 e^{i\delta})^2 + \cdots + ((r')^2 e^{i\delta})^{N-2} \right) \right), \tag{2.116}$$

$$\mathscr{E}_t = \mathscr{E}_0 tt' \left(1 + (r')^2 e^{i\delta} + (r')^4 e^{i2\delta} + \cdots + (r')^{2(N-1)} e^{i(N-1)\delta} \right). \tag{2.117}$$

Letting $N \to \infty$, we can sum Eq. (2.117) for the reflected amplitude using a geometric series

$$\lim_{n \to \infty} (1 + a + a^2 + \cdots + a^n) \to \frac{1}{1-a} \tag{2.118}$$

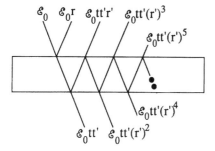

Figure 2.25 Multiple reflected beams for a general surface reflectivity.

to yield

$$\mathscr{E}_{\mathrm{r}} = \mathscr{E}_0 \left(r + tt'r'e^{i\delta} \left[\frac{1}{1 - (r')^2 e^{i\delta}} \right] \right). \tag{2.119}$$

Using the Stokes' relations from Fig. 2.14, and Eqs (2.98) and (2.99)

$$r = -r' \tag{2.120}$$

and

$$tt' = 1 - r^2 \tag{2.121}$$

we find, for the reflected amplitude

$$\mathscr{E}_{\mathrm{r}} = \mathscr{E}_0 \left(\frac{r(1 - e^{i\delta})}{1 - r^2 e^{i\delta}} \right). \tag{2.122}$$

We now express the ratio of reflected irradiance to incident irradiance:

$$\frac{E_{\mathrm{r}}}{E_0} = \frac{\mathscr{E}_{\mathrm{r}}\mathscr{E}_{\mathrm{r}}^*}{\mathscr{E}_0\mathscr{E}_0^*} = \frac{r(1 - e^{i\delta})}{1 - r^2 e^{i\delta}} \frac{r(1 - e^{-i\delta})}{1 - r^2 e^{-i\delta}} = \frac{2r^2(1 - \cos\delta)}{(1 + r^4) - 2r^2\cos\delta}$$

$$= \frac{\left(\dfrac{2r}{1 - r^2}\right)^2 \sin^2(\delta/2)}{1 + \left(\dfrac{2r}{1 - r^2}\right)^2 \sin^2(\delta/2)}. \tag{2.123}$$

Using the power reflectance $R = r^2$, and with identification of the coefficient of finesse F,

$$F = \left(\frac{2r}{1 - r^2} \right)^2 = \frac{4R}{(1 - R)^2}, \tag{2.124}$$

we can write

$$\frac{E_{\mathrm{r}}}{E_0} = \frac{F\sin^2(\delta/2)}{1 + F\sin^2(\delta/2)}. \tag{2.125}$$

If there is no absorption in the plate, we can write an expression for the transmittance

$$\frac{E_{\mathrm{t}}}{E_0} = 1 - \frac{E_{\mathrm{r}}}{E_0} \tag{2.126}$$

$$\frac{E_{\mathrm{t}}}{E_0} = \frac{1}{1 + F\sin^2(\delta/2)}. \tag{2.127}$$

Equation (2.127) for transmittance is called the Airy function. For small surface reflectivity, the reflection and transmission functions are nearly sinusoidal in nature. For higher values of reflectivity (leading to larger values of F) the functions become sharply peaked, as shown in the Fig. 2.26. It can be shown that the full width of the Airy function at the 50% level of transmittance is equal to $4/\sqrt{F}$.

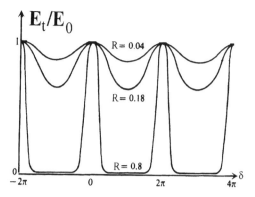

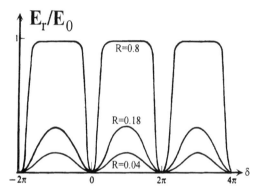

Figure 2.26 Multiple-beam transmission and reflection curves as a function of the power reflectance R.

2.3 COHERENCE

Issues of coherence are important in practical optical systems, because ideal sources of light are never actually obtained. No source is truly monochromatic; there is always a finite bandwidth, so we consider temporal coherence effects. Also, any source of interest has a finite spatial extent; there are no true point sources. Thus, we need to consider spatial coherence.

2.3.1 Temporal Coherence

The coherence time t_c is the average time interval over which we can predict the phase of a light wave. If the light wave were truly monochromatic (an infinite sinusoid), one could predict the phase at all times, given the phase at one time. Over time intervals that are short compared with t_c, the light waves behave as a coherent wavetrain, and interference fringes of good visibility can be formed. For time intervals on the same order as t_c, the fringe visibility begins to decrease. For longer time intervals, interference is not possible because the wavetrain does not have a stable phase relationship.

Beginning with the basic relationship between wavelength and frequency and speed of light c

$$\lambda \nu = c, \tag{2.128}$$

we take the derivative with respect to λ and find the following relationship between the bandwidths in terms of frequency and wavelength:

$$\Delta \nu = \Delta \lambda c / \lambda^2. \tag{2.129}$$

The coherence time can be expressed as the inverse of the frequency bandwidth:

$$t_c = 1/\Delta \nu. \tag{2.130}$$

Some typical magnitudes of these quantities are given in the following examples. An He–Ne laser with mean wavelength $\lambda = 0.63\,\mu m$ and spectral bandwidth $\Delta \lambda = 5 \times 10^{-15}\,m$ will have a frequency bandwidth $\Delta \nu \approx 4\,MHZ$ and a corresponding coherence time $t_c \approx 0.25\,\mu s$. For a filtered Hg-arc lamp with $\lambda = 0.546\,\mu m$ and $\Delta \lambda = 10^{-8}\,m$, the frequency bandwidth will be $\approx 10^{13}\,Hz$, corresponding to a coherence time of $\approx 10^{-13}\,s$.

Coherence length l_c is the distance that light travels during the coherence time:

$$l_c = ct_c = \lambda^2 / \Delta \lambda. \tag{2.131}$$

The coherence length is the maximum OPD allowed in an interference configuration if fringes of high visibility are to be obtained. For the examples above, the He–Ne laser had a coherence length of 75 m, while the filtered Hg-arc lamp had a coherence length of 30 µm. Another interesting example is visible light such as sunlight, with a mean wavelength of 0.55 µm and a wavelength range from 0.4 to 0.7 µm. We find a coherence length for visible sunlight on the order of 1 µm, which is just sufficient for the operation of thin-film interference filters, for which the round-trip path in the film is on the order of a micrometer.

2.3.2 Spatial Coherence

Spatial coherence is defined by the ability of two points on a source to form interference fringes. Using a Young's interference configuration seen in Fig. 2.27, we ask what is the visibility of the fringes that are formed in the observation plane, as a function of the size of the source, the separation of the pinholes, and the distance from the source to the pinhole plane.

Even for an extended incoherent source, where all points have an independent random-phase relationship, we find interference fringes of good visibility are formed in the observation plane for certain geometries. The key to understanding this phenomenon is that, while independent point sources are incoherent and are incapable of exhibiting mutual interference, the position of the fringes formed in a Young's interference setup will depend on the location of the point source that illuminates the pinholes. For an extended source, a number of separate source locations correspond to various point sources that comprise the extended source. If the radiation from each of two such points gives rise to interference fringes that have maxima and minima that are registered in the same location, the overall fringe pattern, while just an incoherent superposition of the two individual fringe patterns, will still have good fringe visibility. Because the mutual coherence of these two sources is measured

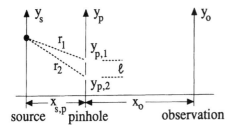

Figure 2.27 Configuration for definition of complex degree of coherence.

in terms of the visibility of the fringes that are formed, these two independent sources are said to exhibit correlation. It is not that the sources themselves are made more correlated by the process of propagation but that the corresponding fringe patterns become co-registered. If the fringes formed by one source had maxima that were located at the position of the minima of the other source, the resulting fringe pattern would be of low visibility and the sources would be said to have small mutual coherence.

We can put these ideas on an analytical footing using the VanCittert–Zernike theorem, which states that the complex degree of coherence γ is just the Fourier transform of the spatial irradiance distribution of the source. We begin the development by considering a particular off-axis source point at a location y_s in the source plane, as shown in Fig. 2.27. The distances from the source to each of the two pinholes are r_1 and r_2, respectively. The fact that these two distances are different gives rise to a phase shift in the fringes, so that the central maximum will not be located on axis. We can write an expression for the irradiance as a function of y_o in the observation plane:

$$E(y_o) = E_o + \mathrm{Re}\{E_o e^{j\Delta\phi} e^{j2\pi l y_o/\lambda x_o}\}, \tag{2.132}$$

where the phase shift is

$$\Delta\phi = \frac{2\pi}{\lambda}(r_1 - r_2). \tag{2.133}$$

Using the Pythagorean theorem, we have expressions for r_1 and r_2 in terms of the axial distance x_{sp} from the source plane to the pinhole plane, and the y position of the two pinholes y_{p1} and y_{p2}:

$$r_1 = \sqrt{x_{sp}^2 + (y_s - y_{p1})^2} \tag{2.134}$$

and

$$r_2 = \sqrt{x_{sp}^2 + (y_s - y_{p2})^2}. \tag{2.135}$$

We form an expression for $r_1 - r_2$:

$$r_1 - r_2 = x_{sp}^2 \sqrt{1 + \frac{(y_s - y_{p1})^2}{x_{sp}^2}} - x_{sp}^2 \sqrt{1 + \frac{(y_s - y_{p2})^2}{x_{sp}^2}}. \tag{2.136}$$

Using the binomial expansion of Eq. (2.6), we find

$$r_1 - r_2 \approx x_{sp}\left(1 + \frac{(y_s - y_{p1})^2}{2x_{sp}^2}\right) - x_{sp}\left(1 + \frac{(y_s - y_{p2})^2}{2x_{sp}^2}\right) \tag{2.137}$$

$$r_1 - r_2 \approx \frac{(y_s - y_{p1})^2 - (y_s - y_{p2})^2}{2x_{sp}} \approx \frac{y_{p1}^2 - y_{p2}^2 - 2y_s(y_{p1} - y_{p2})}{2x_{sp}}. \tag{2.138}$$

So for the phase shift, Eq. (2.133), we find

$$\Delta\phi = \frac{2\pi}{\lambda x_{xp}}(r_1 - r_2) \approx \frac{2\pi}{\lambda x_{xp}}\left[\frac{1}{2}(y_{p1}^2 - y_{p2}^2) - y_s(y_{p1} - y_{p2})\right]. \tag{2.139}$$

If we now allow for an extended source, many such point sources will exist in the source plane. Each source will independently form a Young's fringe pattern in the observation plane. Because these fringe patterns are each derived from a point source that is incoherent with all the others, the resulting irradiance distribution in the observation plane is simply the incoherent addition of the fringe patterns, each with some value of spatial shift corresponding to the phase shift $\Delta\phi$:

$$E(y_o) = \int_{source} E_o(y_s)\,dy_s + \mathrm{Re}\left\{\int_{source} E_o(y_s)e^{j\Delta\phi(y_s)}e^{j2\pi l y_o/\lambda x_o}\,dy_s\right\}. \tag{2.140}$$

Because the periodic phase term is independent of y_s,

$$E(y_o) = \int_{source} E_o(y_s)\,dy_s + \mathrm{Re}\left\{e^{j2\pi l y_o/\lambda x_o}\int_{source} E_o(y_s)e^{j\Delta\phi(y_s)}\,dy_s\right\}. \tag{2.141}$$

Let us introduce two new parameters, the total source exitance

$$E_{total} \equiv \int_{source} E_o(y_s)\,dy_s \tag{2.142}$$

and the normalized source irradiance distribution

$$\hat{E}_o(y_s) \equiv \frac{E_o(y_s)}{\displaystyle\int_{source} E_o(y_s)\,dy_s}. \tag{2.143}$$

Then, noting that $\Delta\phi$ is a function of the pinhole coordinates, the complex degree of coherence is defined as

$$\gamma(y_{p1}, y_{p2}) \equiv \int_{source} \hat{E}_o(y_s)e^{j\Delta\phi(y_s)}\,dy_s. \tag{2.144}$$

Substituting in the expression for $\Delta\phi$ from Eq. (2.139),

$$\gamma(y_{p1}, y_{p2}) \equiv \int_{source} \hat{E}_o(y_s)\exp\left[j\frac{2\pi}{\lambda x_{xp}}\left[\frac{1}{2}(y_{p1}^2 - y_{p2}^2) - y_s(y_{p1} - y_{p2})\right]\right]dy_s \tag{2.145}$$

and

$$\gamma(y_{p1}, y_{p2}) \equiv \exp\left[\frac{j\pi}{\lambda x_{xp}}(y_{p1}^2 - y_{p2}^2)\right] \int_{\text{source}} \hat{E}_o(y_s) \exp\left[\frac{-j2\pi}{\lambda x_{xp}} y_s(y_{p1} - y_{p2})\right] dy_s.$$

(2.146)

The first term in Eq. (2.146) is a purely imaginary phase factor; the integral following is the Fourier transform of the normalized source brightness distribution function. The change of variables

$$\xi = \frac{y_{p1} - y_{p2}}{\lambda x_{sp}} = \frac{l/x_{sp}}{\lambda} = \frac{\theta_{p,s}}{\lambda}$$

(2.147)

is used, where $\theta_{p,s}$ is the angular separation of the pinholes (with an in-plane separation l), as viewed from the source plane. With these substitutions, Eq. (2.141) becomes

$$E(y_o) = E_{\text{total}}\left[1 + \text{Re}\left\{\gamma(y_{p1}, y_{p2})e^{j2\pi l y_o/\lambda x_o}\right\}\right].$$

(2.148)

Comparing this to the expression for double-slit interference, from Eqs (2.76) and (2.78)

$$E(y_o) = 2E_o\left[1 + \text{Re}\left\{e^{j2\pi l y_o/\lambda x_o}\right\}\right],$$

(2.149)

we find the interpretation of the complex degree of coherence, $\gamma(y_{p1}, y_{p2})$. The magnitude of γ determines the contrast of the fringes produced in the observation plane, and the phase portion of γ determines the amount of phase shift relative to an on-axis point source. Given the source size in the y_s plane, γ will be in pinhole (y_p) coordinates, and will depend on the specific values for λ and x_{sp}. However, γ will be independent of x_o, because that distance scales the sizes of the fringe pattern and changing that distance does not affect the fringe visibility.

As the first example, we consider a one-dimensional slit source of total length L, which can be expressed in source-plane coordinates as

$$\hat{E}_o(y_s) = \text{rect}(y_s/L).$$

(2.150)

Taking the Fourier transform and making the required change of variables from Eq. (2.147), we find that

$$\gamma = \frac{\sin\left(\pi L \frac{l}{\lambda x_{sp}}\right)}{\left(\pi L \frac{l}{\lambda x_{sp}}\right)},$$

(2.151)

which exhibits a first zero of visibility at a pinhole separation $l = \lambda x_{sp}/L$.

The normalized source function for a two-dimensional uniform source can be written as

$$\hat{E}_o(r_s) = \frac{4}{\pi D^2} \text{cyl}(r_s/D),$$

(2.152)

which is Fourier transformed to yield

$$\gamma = \frac{2J_1\left(\dfrac{\pi D l}{\lambda x_{sp}}\right)}{\left(\dfrac{\pi D l}{\lambda x_{sp}}\right)}, \tag{2.153}$$

which has its first zero of visibility at

$$\frac{\pi D l}{\lambda x_{sp}} = \pi \times 1.22, \tag{2.154}$$

corresponding to a pinhole spacing $l = 1.22 x_{sp}\lambda/D$.

Two point sources separated by a distance b can be described in normalized form as

$$\hat{E}_o(y_s) = \tfrac{1}{2}[\delta(y_s + b/2) + \delta(y_s - b/2)], \tag{2.155}$$

which, upon performing the transformation and the change of variables from Eq. (2.147), can then be written as

$$\gamma = \cos\left(2\pi\frac{b/2}{\lambda x_{sp}}l\right). \tag{2.156}$$

In general, for a given size of an extended source, the farther away the source is, we find that γ will approach 1, for a given pinhole separation. This means that the light from the extended source, if allowed to interfere after passing through pinholes with that separation, would tend to form fringes with good visibility. This does not mean that the extended source itself acquires a coherence by propagation, but that the fringe patterns formed from each portion of the source would tend to become more nearly registered as the source-to-pinhole distance x_{sp} increases. Because the fringe patterns add on an irradiance basis, an increasing registration of the fringe maxima and minima will produce an overall distribution with higher visibility fringes. Similar arguments apply when the pinhole separation decreases, for a given source size and distance x_{sp}.

2.4 POLARIZATION

Polarization effects on optical systems arise from the vectorial nature of the electric field $\mathscr{E}$, which is orthogonal to the direction of propagation of the light. Because of the polarization dependence of the Fresnel equations, differences in the power distribution between reflected and transmitted rays occur when the electric field makes a nonzero angle with the boundary between two media. Other effects arise in anisotropic crystals from the fact that the refractive index of a material may depend upon the polarization of the incident light. In this case, the trajectory of rays may depend upon the orientation of the electric field to one particular preferred direction.

2.4.1 Polarizing Elements

Linear polarizers produce light polarized in a single direction from natural light, which has light polarized in all directions. The amount of power transmitted through a linear polarizer is at most 50% when the input light is composed of all polarizations. Light that is already linearly polarized will be transmitted through a linear polarizer with a power transmittance that depends on the angle θ between the electric field of the input light, and the pass direction of the polarizer, according to Malus' law:

$$\phi_t = \phi_i \cos^2 \theta. \tag{2.157}$$

The functions of a linear polarizer may be accomplished by long molecular chains or by metallic wires with spacing less than the wavelength. These act by allowing electric currents to flow in the long direction of the structure. Light with an incident electric field oriented in this direction will be preferentially absorbed because of joule heating arising in the currents, or may be reflected if the absorption losses are small. Light having a polarization orthogonal to the long direction will pass through the structure relatively unattenuated.

Another common means of producing linearly polarized light uses a stack of dielectric plates, all oriented at Brewster's angle to the original propagation direction. Each plate will transmit all of the p polarization component and some of the s component. By arranging to have the incident light traverse many such plates of the same orientation, the transmitted beam can be made to be strictly p polarized.

2.4.2 Anisotropic Media

Usually, optical materials are isotropic in that they have the same properties for any orientation of electric field. However, anisotropic materials have a different refractive index for light polarized in different directions. The most useful case is a so-called birefringent material that has two different refractive indices for light polarized in two orthogonal directions. These refractive indices are denoted n_o, the ordinary index, and n_e, the extraordinary index. By convention, the ordinary index is less than the extraordinary index. By properly orienting the polished faces of a slab of birefringent material, it is possible to produce a retarder, which introduces a phase delay φ between the two orthogonal polarizations of a beam

$$\varphi = \frac{2\pi}{\lambda_0} d(n_e - n_o). \tag{2.158}$$

By proper choice of d, a phase delay of $90°$ can be produced. If linearly polarized light is incident with its electric field at angle of $45°$ to the fast axis of such a slab, a phase delay is introduced between the component of light polarized along the fast axis and that polarized along the slow axis, producing circularly polarized light.

2.4.3 Optical Activity

Certain materials produce a continuous rotation of the electric field of polarized light as the light traverses the bulk of the material. The most common substances having this property are quartz and sugar dissolved in water, but the most important technologically are liquid crystals, which have the advantage that their optical activity can be controlled by an external voltage. By placing such a material between orthogonal linear polarizers, a variable amount of light can be made to pass through the structure, depending upon the amount of rotation of the polarization.

REFERENCES

1. Gaskill, J., *Linear Systems, Fourier Transforms, and Optics*, Wiley, New York, 1978.
2. Malacara, D., *Optical Shop Testing*, Wiley, New York, 1978.

BIBLIOGRAPHY

1. Born, M. and E. Wolf, *Principles of Optics*, Pergamon, New York, 1975.
2. Stone, J., *Radiation and Optics*, McGraw-Hill, New York, 1963.
3. Reynolds, G., J. DeVelis, G. Parrent, and B. Thompson, *Physical Optics Notebook*, SPIE Press, Bellingham, WA, 1989.

3

Basic Photon Optics

SOFIA E. ACOSTA-ORTIZ

Centro de Investigaciones en Optica, Aguascalientes, Mexico

3.1 NATURE OF QUANTUM OPTICS

The first indication of the quantum nature of light came in 1900 when M. Planck discovered he could account for the spectral distribution of thermal light by postulating that the energy of a harmonic oscillator is quantized. In 1905, Albert Einstein showed that the photoelectric effect could be explained by the hypothesis that the energy of a light beam was distributed in discrete bundles, later known as *photons*.

With the development of his phenomenological theory in 1917, Einstein also contributed to the understanding of the absorption and emission of light from atoms. Later it was shown that this theory is a natural consequence of the quantum theory of electromagnetic radiation.

A remarkable feature of the theory of light is the large measure of agreement between the predictions of classical and quantum theory, despite the fundamental differences between the two approaches. For example, classical and quantum theories predict identical interference effects and associated degrees of coherence for experiments that use coherent or chaotic light, or light with coherence properties intermediate between the two.

The vast majority of physical-optics experiments can be explained adequately using the classical theory of electromagnetic radiation based on Maxwell's equations. Interference experiments of Young's type do not distinguish between the predictions of classical theory and quantum theory. It is only in higher-order interference experiments involving the interference of intensities that differences between the predictions of both theories appear. Whereas classical theory treats the interference of intensities, in quantum theory the interference is still at the level of probability amplitudes. This is one of the most important differences between quantum theory and classical theory.

The main tool in the quantum description of a beam of light is its photon probability distribution, or more generally, its density operator. Photon-counting experiments provide a fairly direct measurement of the photon probability distribution for all kinds of light embraced by the quantum theory. Such experiments form the observational basis of quantum optics and play a leading role in the study of quantum phenomena in light beams.

The field of quantum optics occupies a central position involving the interaction of atoms with the electromagnetic field. It covers a wide range of topics, ranging from fundamental tests of quantum theory to the development of new laser light sources.

3.1.1 Young's Experiment

Young's experiment is one example of a phenomenon that can be explained by both classical and quantum theory. In this experiment (Fig. 3.1), monochromatic light is passed through a pinhole S so as to illuminate a screen containing two further identical pinholes or narrow slits placed close together. The presence of the single pinhole S provides the necessary mutual coherence between the light beams emerging from the slits S_1 and S_2. The wavefronts from S intersect S_1 and S_2 simultaneously in such a way that the light contributions emerging from S_1 and S_2 are derived from the same original wavefront and are therefore coherent. These contributions spread out from S_1 and S_2 as "cylindrical" wavefronts and interfere in the region beyond the screen. If a second screen is placed as shown in the figure, then an interference pattern consisting of straight line fringes parallel to the slits is observed on it.

The phase difference between the two sets of waves arriving at P from S_1 and S_2 depends on the path difference $(D_2 - D_1)$ as, in general, phase difference $= (2\pi/\lambda)$ (optical phase difference); then

$$\phi = \phi_2 - \phi_1 = (2\pi/\lambda)(D_2 - D_1), \tag{3.1}$$

where D_1 and D_2 are the distances from S_1 and S_2 to P, respectively.

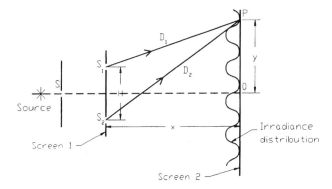

Figure 3.1 Schematic diagram showing Young's double-slit interference experiment.

Bright fringes occur when the phase difference is zero or $\pm 2n\pi$, where n is an integer; that is, when

$$(2\pi/\lambda)(D_2 - D_1) = \pm 2n\pi,$$

which is equivalent to $D_2 - D_1 = n\lambda$. Therefore, bright fringes will occur whenever the path difference is an integral number of wavelengths. Similarly, dark fringes will occur when $\phi = \pm(2n + 1)\pi$; that is, when the path difference is an odd integral number of half-wavelengths.

Bright fringes occur at points P a distance y from O such that

$$y = \pm(n\lambda x)/H, \tag{3.2}$$

provided both y and H are small compared with x, where H is the slit separation and x is the distance from the screen containing the slits to the observing screen (Wilson and Hawkes [1]).

The quantum theory of the experiment parallels the classical theory. The Heisenberg electric-field operator at position $\mathbf{r}$ on the screen is given by

$$\hat{E}^+(\mathbf{r}t) = u_1 \hat{E}^+(\mathbf{r}_1 t_1) + u_2 \hat{E}^+(\mathbf{r}_2 t_2), \tag{3.3}$$

where we have assumed for simplicity a single polarization direction for the light.

The superposition theory proceeds in a similar way to the classical treatment, except that the classical intensity, proportional to an ensemble average of E^*E, must be replaced by the quantum-mechanical detector response, proportional to an average of the intensity operator $\hat{E}^- \hat{E}^+$.

The intensity distribution is given by Loudon [2]:

$$\langle \hat{I}(\mathbf{r}t) \rangle = 2\varepsilon_0 c \Big\{ u_1^2 \langle \hat{E}^-(\mathbf{r}_1 t_1) \hat{E}^+(\mathbf{r}_1 t_1) \rangle + u_2^2 \langle \hat{E}^-(\mathbf{r}_2 t_2) \hat{E}^+(\mathbf{r}_2 t_2) \rangle$$
$$+ u_1^* u_2 \langle \hat{E}^-(\mathbf{r}_1 t_1) \hat{E}^+(\mathbf{r}_2 t_2) \rangle + u_1 u_2^* \langle \hat{E}^-(\mathbf{r}_2 t_2) \hat{E}^+(\mathbf{r}_1 t_1) \rangle \Big\}, \tag{3.4}$$

where the angle-bracket expressions are evaluated as

$$\langle \hat{E}^-(\mathbf{r}_1 t_1) \hat{E}^+(\mathbf{r}_2 t_2) \rangle = \mathrm{Tr}(\rho \hat{E}^-(\mathbf{r}_1 t_1) \hat{E}^+(\mathbf{r}_2 t_2)). \tag{3.5}$$

The two final expectation values in Eq. (3.4) are complex conjugates, and bearing in mind the purely imaginary character of u_1 and u_2, the intensity reduces to

$$\langle \hat{I}(\mathbf{r}t) \rangle = 2\varepsilon_0 c \Big\{ u_1^2 \langle \hat{E}^-(\mathbf{r}_1 t_1) \hat{E}^+(\mathbf{r}_1 t_1) \rangle + u_2^2 \langle \hat{E}^-(\mathbf{r}_2 t_2) \hat{E}^+(\mathbf{r}_2 t_2) \rangle$$
$$+ 2u_1^* u_2 \mathrm{Re} \langle \hat{E}^-(\mathbf{r}_1 t_1) \hat{E}^+(\mathbf{r}_2 t_2) \rangle \Big\}. \tag{3.6}$$

The electric-field operators in Eq. (3.6) are related to photon creation and destruction operators by

$$\hat{E}^+(\mathbf{R}t) = i \sum_k (\hbar\omega_k/2\varepsilon_0 V)^{1/2} \varepsilon_k \hat{a}_k \exp(-i\omega_k t + i\mathbf{k} \cdot \mathbf{R}) \tag{3.7}$$

and

$$\hat{E}^-(\mathbf{R}t) = -i \sum_k (\hbar\omega_k/2\varepsilon_0 V)^{1/2} \varepsilon_k \hat{a}_k^\dagger \exp(i\omega_k t - i\mathbf{k} \cdot \mathbf{R}), \tag{3.8}$$

where $\mathbf{R}$ is the position of the nucleus.

The first two terms in Eq. (3.6) give the intensities on the second screen that result from each pinhole in the absence of the other. The interference fringes result from the third term, whose normalized form determines the quantum degree of first-order coherence of the light.

Finer details of the quantum picture of the interference experiment are found in Walls [3] and Loudon [2].

In Young's experiment, each photon must be capable of interfering with itself in such a way that its probability of striking the second screen at a particular point is proportional to the calculated intensity at that point. This is possible only if each photon passes partly through both pinholes, so that it can have a knowledge of the entire pinhole geometry as it strikes the screen. Indeed, there is no way in which a photon can simultaneously be assigned to a particular pinhole and contribute to the interference effects. If a phototube is placed behind one of the pinholes to detect photons passing through, then it is not possible to avoid obscuring that pinhole, with consequent destruction of the interference pattern (Loudon [2]). These remarks are in agreement with the principles of quantum mechanics.

In the 1960s improvements in photon counting techniques proceeded along with the development of new laser light sources. Then, light from incoherent (thermal) and coherent (laser) sources could be distinguished by their photon counting properties.

It was not until 1975, when H. J. Carmichael and D. F. Walls predicted that light generated in resonance fluorescence from a two-level atom would exhibit photon antibunching, that a physically accessible system exhibiting nonclassical behavior was identified.

3.1.2 Squeezed States

Nine years after the observation of photon antibunching, another prediction of the quantum theory of light was observed: the squeezing of quantum fluctuations. To understand what squeezing of quantum fluctuations means remember that the electric field for a nearly monochromatic plane wave may be decomposed into two quadrature components with a time dependence $\cos \omega t$ and $\sin \omega t$, respectively. In a coherent state, the closest quantum counterpart to a classical field, the fluctuations in the two quadratures are equal and minimize the uncertainty product given by Heisenberg's uncertainty relation. The quantum fluctuations in a coherent state are equal to the zero-point vacuum fluctuation and are randomly distributed in phase. In a squeezed state the quantum fluctuations are no longer independent of phase. One quadrature phase may have reduced quantum fluctutions at the expense of increased quantum fluctuations in the other quadrature phase such that the product of the fluctuations still obeys Heisenberg's uncertainty relation.

Then, squeezed states have been considered as a general class of minimum-uncertainty states. A squeezed state may, in general, have less noise in one quadrature than a coherent state. To satisfy the requirements of a minimum-uncertainty state, the noise in the other quadrature must be greater than that of a coherent state. Coherent states are a particular member of this more general class of minimum uncertainty states with equal noise in both quadratures.

Some applications of squeezed light are interferometric detection of gravitational radiation and sub-shot-noise phase measurements (Walls and Milburn [4]).

3.1.3 The Hanbury-Brown and Twiss Experiment

The experiment carried on by Hanbury-Brown and Twiss [5] in 1957 let us see more clearly the differences between classical and quantum theories. The schematic diagram of the experiment is shown in Fig. 3.2. In the experiment, each photon that strikes the semitransparent mirror is either reflected or transmitted, and can only be registered in one of the phototubes. Therefore, the two beams that arrive at the detectors are not identical. The incident photons have equal probabilities of transmission or reflection and their creation and destruction operators can be written as

$$\hat{a}^\dagger = (\hat{a}_1^\dagger + \hat{a}_2^\dagger)/2^{1/2}$$
$$\hat{a} = (\hat{a}_1 + \hat{a}_2)/2^{1/2}, \tag{3.9}$$

where the $2^{1/2}$ factors ensure that all like pairs of creation and destruction operators have the same commutation rule:

$$[\hat{a}, \hat{a}^\dagger] = \hat{a}\hat{a}^\dagger - \hat{a}^\dagger\hat{a} = 1.$$

A state with n incident photons has the form

$$n >= (n!)^{-1/2}(\hat{a}^\dagger)^n 0 >,$$

and the probability $P_{n1,n2}$ that n_1 photons are transmitted through the mirror to detector 1 while n_2 are reflected to detector 2. $P_{n1,n2}$ is given by

$$P_{n1,n2} \equiv \langle n_1, n_2, n \rangle^2 = n!/n_1!n_2!2^n.$$

We can use the probability distribution to obtain average properties of the photon counts in the two arms of the experimental system. The mean numbers of counts n_1 and n_2 are both equal to $(1/2)$ (Loudon [2]).

$$\bar{n}_1 = (1/2)n \tag{3.10}$$

$$\bar{n}_2 = \langle n \, \hat{a}_2^\dagger \hat{a}_2 \, n \rangle = (1/2)n, \tag{3.11}$$

and the correlation is

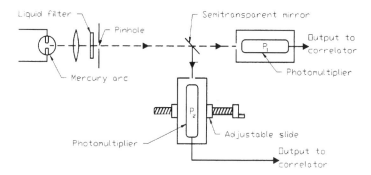

Figure 3.2 Experimental arrangement for an intensity interference experiment (Hanbury-Brown and Twiss [5]).

Table 3.1 Photon Distribution in a Hanbury-Brown and Twiss Experiment

n	n_1	n_2	$\bar{n}_1 = \bar{n}_2$	$\langle n_1 n_2 \rangle$	$g^{(2)}(\tau)$
1	$\begin{cases} 1 \\ 0 \end{cases}$	$\begin{matrix} 0 \\ 1 \end{matrix}$	1/2	0	0
2	$\begin{cases} 2 \\ 1 \\ 1 \\ 0 \end{cases}$	$\begin{matrix} 0 \\ 1 \\ 1 \\ 2 \end{matrix}$	1	1/2	1/2
3	—	—	3/2	3/2	2/3
4	—	—	2	3	3/4

$$\langle \hat{a}_1^\dagger \hat{a}_2^\dagger \hat{a}_2 \hat{a}_1 \rangle \equiv \langle n_1 n_2 \rangle = (1/4)n(n-1); \tag{3.12}$$

$\langle n_1 n_2 \rangle$ indicates the average of the product of the photon counts at the two photon photomultipliers.

Table 3.1 shows the photon-count distributions and the degrees of second-order coherence ($g^{(2)}(\tau)$), for small numbers n of incident photons.

The Hanbury-Brown and Twiss experiment was designed to produce the average

$$\frac{\langle (n_1 - \bar{n}_1)(n_2 - \bar{n}_2) \rangle}{\bar{n}_1 \bar{n}_2} = g^{(2)}(\tau) - 1, \tag{3.13}$$

where $g^{(2)}(\tau)$ is given by

$$g^{(2)}(\tau) = \langle n_1 n_2 \rangle / \bar{n}_1 \bar{n}_2. \tag{3.14}$$

The normalized correlation gives (Loudon [2]):

$$\frac{\langle (n_1 - \bar{n}_1)(n_2 - \bar{n}_2) \rangle}{\bar{n}_1 \bar{n}_2} = -1/n. \tag{3.15}$$

From the above analysis it is seen that nonclassical light with a degree of second order smaller than unity occurs in the Hanbury-Brown and Twiss experiment because the photons must make a choice between reflection and transmission at the mirror. This is an essential quantum requirement. The quantum analysis of the Hanbury-Brown and Twiss experiment is readily extended to incident beams that have some superposition of statistical mixture of the number states of a single mode.

Later in this chapter some of the most actual quantum effects and their applications will be discussed.

3.2 NONLINEAR OPTICS

Nonlinear optics is the study of the phenomena that occur as a consequence of the modification of the optical properties of matter due to the presence of light. In principle, only laser light has the intensity sufficient to modify the optical properties of matter. In fact, the start of the nonlinear optics field is frequently considered as the start of second-harmonic generation by Franken *et al.* [6] in 1961, soon after the demonstration of the first laser by Maiman in 1960. Nonlinearity is intended in the sense that nonlinear phenomena occur when the response of the material to an applied electric field depends in a nonlinear manner on the magnitude of the optical field.

As an example, second-harmonic generation occurs as a result of the atomic response that depends quadratically on the magnitude of the applied optical field. As a consequence, the intensity of the generated light at the frequency of the second harmonic tends to increase as the square of the intensity of the applied laser light.

To better understand the significance of nonlinear optics, let us consider the dependence of the dipolar moment by unitary volume: that is, the polarization $P(t)$ on the magnitude $E(t)$ of the applied optical field. In the linear case, the relation is given by

$$P(t) = \chi^{(1)} E(t), \tag{3.16}$$

where χ is the constant of proportionality known as "linear susceptibility." In non-linear optics, the expression (3.16) can be generalized by expressing $P(t)$ as a power series on the magnitude of the field $E(t)$:

$$P(t) = \chi^{(1)} E(t) + \chi^{(2)} E^2(t) + \chi^{(3)} E^3(t) + \cdots$$

$$P(t) \equiv P^1(t) + P^2(t) + P^3(t) + \cdots, \tag{3.17}$$

where $\chi^{(2)}$ and $\chi^{(3)}$ are known as the second- and third-order nonlinear optical susceptibilities, respectively.

In general, $P(t)$ and $E(t)$ are nonscalar quantities but vectorial ones. In such case, $\chi^{(1)}$ becomes a second-rank tensor, $\chi^{(2)}$ a third-rank tensor, $\chi^{(3)}$ a fourth-rank tensor, etc.

We refer then to $P^2(t) = \chi^{(2)} E^2(t)$ as second-rank nonlinear polarization and to $P^3(t) = \chi^{(3)} E^3(t)$ as third-rank nonlinear polarization, each of which gives rise to different physical phenomena.

Furthermore, second-order nonlinear optical interactions can only occur in noncentrosymmetric crystals—that is, in crystals that no possess inversion symmetry—while third-order nonlinear phenomena can occur in any medium regardless of whether it possesses inversion symmetry or not.

Optical nonlinearity manifests by changes in the optical properties of a medium as the intensity of the incident light wave increases or when two or more light waves are introduced into a medium.

Optical nonlinearity can be classified into two general categories: extrinsic and intrinsic (Baldwin [7]).

- **Extrinsic Optical Nonlinearity**. Extrinsic optical nonlinearity is the change in the optical properties of the medium that is directly related to a change in the composition of the medium as a result of the absorption or emission of light. This change can be a change in the relative population of the base and excited states or in the number of optically effective electrons. The laser medium itself, certain dyes used for laser Q-switching, and laser mirrors with semiconductor covers have this property.

- **Intrinsic Optical Nonlinearity**. The optical phenomena with intrinsic nonlinearity are violations to the superposition principle that arises from the nonlinear response of the individual molecules or unitary cells to the fields of two or more light waves. This category includes a nonlinear response to a single light beam, as it is possible to consider any light beam as the sum of two or more similar light waves, identical in polarization, frequency, and direction.

In any type of nonlinearity, the optical properties of the medium depend on the intensity of the light and therefore it is useful to classify them according to the intensity of the light involved. For example, the intensity of the second-harmonic light at $\lambda = 0.53\,\mu m$ generated in lithium niobatium by the laser radiation of neodymium is observed to be proportional to the square of the intensity of the fundamental $1.06\,\mu m$ and therefore is classified as a second-order nonlinear process.

Every advance in the development of optics has involved the development of advanced technology. The quantum theory, for example, rests to a high degree in highly sensitive and precise spectroscopic techniques. The key that opened the doors to nonlinear optics was the development of the *maser*. This device uses stimulated emission to generate radiation of narrow bandwidth in the range of microwaves from an adequate prepared medium. Later, the range was extended to optical frequencies by the development of the *laser*, which allows the generation of highly monochromatic light beams that can be concentrated at very high intensities.

A peculiar property of the laser, essential to nonlinear optics, is its high degree of coherence; i.e., the several monochromatic light waves are emitted in a synchronized way, being in phase both in time and space. This property allows us to concentrate the radiation from a laser into a small area, whose size is only limited by diffraction and by the optical quality of the laser and the focusing system. In this way, it is possible to obtain fields of local radiation that are extremely intense but in small volumes.

Coherence also allows us to combine small contributions of nonlinear interactions due to very separated parts of an extended medium to produce an appreciable result. For this reason, it is necessary to use laser light as the light power, to be able to observe optical nonlinear phenomena.

As examples of second- and third-order nonlinear processes, we have (the number in parenthesis indicates the order of the optical nonlinearity):

- Second-harmonic generation   (2)
- Rectification   (2)
- Frequency generation of sum and difference   (2)
- Third-harmonic generation   (3)
- Parametric oscillation   (2)
- Raman scattering   (2)
- Inverse Raman effect   (2)
- Brillouin scattering   (2)
- Rayleigh scattering   (2)
- Inverse Faraday effect   (3)
- Two photons absorption   (2)
- Parametric amplification   (2)
- Induced reflectivity   (2)

- Intensity-dependent refraction index (3)
- Induced opacity (2)
- Optical Kerr effect (3)
- Four-wave mixing (3)
- Electro-optic effect (2)
- Birefringence (2)

We now describe some of these nonlinear phenomena.

3.2.1 Second-Harmonic Generation

The essential feature of second-harmonic generation (SHG) is the existence of a perceptible dependence of the electronic polarization on the square of the electric field intensity at optical frequencies, in addition to the usual direct dependence.

Practically all SHG materials are birefringent crystals. When its SHG is under consideration, the ith component of the total charge polarization **P** in a birefringent medium contains two contributions:

$$P_i = P_i^{(\omega)} + P_i^{(2\omega)} \qquad i = 1, 2, 3 \tag{3.18}$$

where

$$P_i^{(\omega)} = \varepsilon_0 \chi_{ij}^{(\omega)} E_j^{(\omega)} \qquad i, j = 1, 2, 3 \tag{3.19}$$

and

$$P_i^{(2\omega)} = \varepsilon_0 \chi_{ijk}^{(2\omega)} E_j^{(\omega)} E_k^{(\omega)} \qquad i, j, k = 1, 2, 3. \tag{3.20}$$

The first term, $P_i^{(\omega)}$, accounts for the linear part of the response of the medium. The second one, $P_i^{(2\omega)}$, is quadratic in the electric field, and introduces the third-rank tensor $\chi_{ijk}^{(2\omega)}$. The superscripts (ω) and (2ω) are now necessary to distinguish the frequencies at which the respective quantities must be evaluated. Thus, two sinusoidal electric field components at frequency ω acting in combination exert a resultant containing the double frequency 2ω (and a dc term). The susceptibility factor χ_{ijk} must be evaluated at the combination frequency.

The components of the nonlinear optical coefficients for three SHG crystals are given in Table 3.2. The nonlinear optical coefficient is defined as $d_{ijk}^{(2\omega)} = (\varepsilon_0/2)\chi_{ijk}^{(2\omega)}$.

Only noncentrosymmetric crystals can possess a nonvanishing d_{ijk} tensor. This follows from the requirement that in a centrosymmetric crystal, a reversal of the signs of $E_j^{(\omega)}$ and $E_k^{(\omega)}$ must cause a reversal in the sign of $P_i^{(2\omega)}$ and not affect the amplitude (Yariv [8]).

It follows that since no physical significance is attached to the order of the electric field components, all the d_{ijk} coefficients that are related by a rearrangement of the order of the subscripts are equal. This statement is known as the Kleinman's conjecture (Kleinman [9]). The Kleinman conjecture applies only to lossless media, but since most nonlinear experiments are carried out in the lossless regime, it is a powerful practical relationship.

Table 3.3 gives a list of the nonzero components for the nonlinear optical coefficients of a number of crystals.

The first experiment carried to demonstrate optical second-harmonic generation was performed by Franken *et al.* [6] in 1961. A sketch of the original experiment

Table 3.2 Components of the Nonlinear Optical Coefficient for Several Crystals

Component i	Components j, k					
	x, x	y, y	z, z	y, z	x, z	x, y
Barium titanate						
x	0	0	0	0	d_{15}	0
y	0	0	0	d_{15}	0	0
z	d_{31}	d_{31}	d_{33}	0	0	0
Potassium dihydrogen phosphate (KDP)						
x	0	0	0	d_{14}	0	0
y	0	0	0	0	d_{14}	0
z	0	0	0	0	0	d_{36}
Quartz						
x	d_{11}	$-d_{11}$	0	d_{14}	0	0
y	0	0	0	0	$-d_{14}$	$-2d_{11}$
z	0	0	0	0	0	0

is shown in Fig. 3.1. In this experiment a ruby laser beam at 694.3 nm was focused on the front surface of a crystalline quartz plate. The emergent radiation was examined with a spectrometer and was found to contain radiation at twice the input frequency, that is at $\lambda = 347.15$ nm. The conversion efficiency in this experiment was very poor, only about 10^{-8}. In the last few years, the availability of more efficient materials, higher intensity lasers, and index-matching techniques have resulted in conversion efficiencies approaching unity.

Mathematical details to obtain the conversion efficiency as well as examples of second-harmonic generation can be found in Yariv [8]; see also Zyss *et al.* [10].

3.2.2 Raman Scattering

The Raman effect is one of the first discovered and best-known nonlinear optical processes. It is used as a tool in spectroscopic studies, and also in tunable laser development, high-energy pulse compression, etc. Several review articles exist that summarize the earlier work on the Raman effect (Bloembergen [11], Penzkofer *et al.* [12]).

Raman scattering involves the inelastic scattering of light from a crystal. The Raman effect belongs to a class of nonlinear optical processes that can be called quasi-resonant (Mostowsky and Raymer [13]). Although none of the fields is in resonance with the atomic or molecular transitions, the sum or difference between two optical frequencies equals a transition frequency.

Raman scattering is one of the physical processes that can lead to spontaneous light scattering. By spontaneous light scattering we mean light scattering under conditions such that the optical properties of the material system are unmodified by the presence of the incident light beam. Figure 3.4 shows a diagram for an incident beam on a scattering medium (a) and the typical observed spectrum (b) in which Raman, Brillouin, Rayleigh, and Rayleigh-wing features are present. By

Table 3.3 The Nonlinear Optical Coefficients of a Number of Crystals (Yariv [8])

Crystal	$d_{ijk}^{(2\omega)}$ in units of $(1/9) \times 10^{-22}$ MKS
$LiIO_3$	$d_{15} = 4.4$
$NH_4H_2PO_4$	$d_{36} = 0.45$
(ADP)	$d_{14} = 0.50 \pm 0.02$
KH_2PO_4	$d_{36} = 0.45 \pm 0.03$
(KDP)	$d_{14} = 0.35$
KD_2PO_4	$d_{36} = 0.42 \pm 0.02$
	$d_{14} = 0.42 \pm 0.02$
KH_2ASO_4	$d_{36} = 0.48 \pm 0.03$
	$d_{14} = 0.51 \pm 0.03$
Quartz	$d_{11} = 0.37 \pm 0.02$
$AlPO_4$	$d_{11} = 0.38 \pm 0.03$
ZnO	$d_{33} = 6.5 \pm 0.2$
	$d_{31} = 1.95 \pm 0.2$
	$d_{15} = 2.1 \pm 0.2$
CdS	$d_{33} = 28.6 \pm 2$
	$d_{31} = 30 \pm 10$
	$d_{36} = 33$
GaP	$d_{14} = 80 \pm 14$
GaAs	$d_{14} = 72$
$BaTiO_4$	$d_{33} = 6.4 \pm 0.5$
	$d_{31} = 18 \pm 2$
	$d_{15} = 17 \pm 2$
$LiNbO_3$	$d_{15} = 4.4$
	$d_{22} = 2.3 \pm 1$
Te	$d_{11} = 517$
Se	$d_{11} = 130 \pm 30$
$Ba_2NaNb_5O_{15}$	$d_{33} = 10.4 \pm 0.7$
	$d_{32} = 7.4 \pm 0.7$
Ag_3AsS_3	$d_{22} = 22.5$
(proustite)	$d_{36} = 13.5$
CdSe	$d_{31} = 22.5 \pm 3$
$CdGeAs_2$	$d_{36} = 363 \pm 70$
$AgGaSe_2$	$d_{36} = 27 \pm 3$
$AgSbS_3$	$d_{36} = 9.5$
ZnS	$d_{36} = 13$

definition, those components that are shifted to lower frequencies are known as Stokes' components, and those that are shifted to higher frequencies are known as anti-Stokes' components. Typically the Stokes' lines are orders of magnitude more intense than the anti-Stokes' lines. In Table 3.4 a list of the typical values of the parameter describing these light-scattering processes is given.

Raman scattering results from the interaction of light with the vibrational modes of the molecules constituting the scattering medium and can be equivalently described as the scattering of light from optical phonons.

Light scattering occurs as a consequence of fluctuations in the optical properties of a material medium. A completely homogeneous material can scatter light only in the forward direction (Fabelinskii [15]).

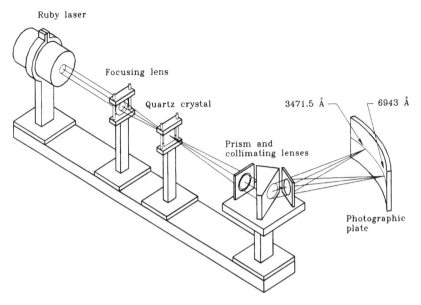

Figure 3.3 Arrangement used for the first experimental demonstration of optical second-harmonic generation (Franken *et al.* [6]).

Spontaneous Raman scattering was discovered in 1928 by Raman. To observe the effect, a beam of light illuminates a material sample and the scattered light is observed spectroscopically. In general, the scattered light contains frequencies dif-

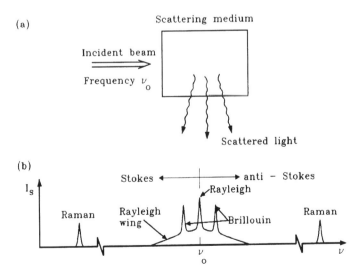

Figure 3.4 (a) Diagram for an incident beam on a scattering medium; (b) typical observed Raman, Brillouin, Rayleigh, and Rayleigh-wing spectra.

Table 3.4 Typical Values of the Parameters Describing Several Light-Scattering Processes (Boyd [14])

Process	Shift (cm^{-1})	Linewidth (cm^{-1})	Relaxation time (sec)	Gain* (cm/MW)
Raman	1000	5	10^{-12}	5×10^{-3}
Brillouin	0.1	5×10^{-3}	10^{-9}	10^{-2}
Rayleigh	0	5×10^{-4}	10^{-8}	10^{-4}
Rayleigh wing	0	3	10^{-12}	10^{-3}

*Gain of the stimulated version of the process.

ferent from those of the excitation source; i.e., contains Stokes' and anti-Stokes' lines. Raman–Stokes' scattering consists of a transition from the ground state i to a virtual level associated with the excited state f' followed by a transition from the virtual level to the final state f. Raman–anti-Stokes' scattering entails a transition from level f to level i with f' serving as the intermediate level (Fig. 3.5). The anti-Stokes' lines are typically much weaker than the Stokes' lines because, in thermal equilibrium, the population of level f is smaller than the population in level i by the Boltzmann factor: $\exp(-\hbar\omega_{ng}/kT)$.

The Raman effect has important spectroscopic applications because transitions that are one-photon forbidden can often be studied using Raman scattering. Figure 3.6 shows a spectral plot of the spontaneous Raman emission from liquid N_2 obtained by Clements and Stoicheff [16] in 1968. In Table 3.5 the Raman scattering cross sections per molecule of some liquids are given.

Typically, the spontaneous Raman scattering process is a rather weak process. The scattering cross section per unit volume for Raman–Stokes' scattering in condensed matter is only about $10^{-6} cm^{-1}$. Hence, in propagating through 1 cm of the scattering medium, only approximately 1 part in 10^6 of the incident radiation will be scattered into the Stokes' frequency. However, under excitation by an intense laser beam, highly efficient scattering can occur as a result of the stimulated version of the Raman scattering process.

The ability of lasers to produce light of extremely high intensity makes them especially attractive sources for Raman spectroscopy of molecules by increasing the intensity of the anti-Stokes' components in the Raman effect. Each resulting pair of lines, equally displaced with respect to the laser line, reveals a characteristic

Figure 3.5 Energy level diagrams for (a) Raman–Stokes' and (b) Raman–anti-Stokes' scattering.

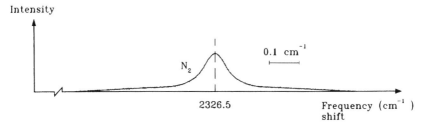

Figure 3.6 Spontaneous Raman emission from liquid N_2 (Clements and Stoicheff [16]).

vibrational frequency of the molecule. Compared to the spontaneous Raman scattering, stimulated Raman scattering is a very strong scattering process: 10% or more of the energy of the incident laser beam is often converted into the Stokes' frequency.

Stimulated Raman scattering was discovered by Woodbury and Ng [18] in 1962 and described by Eckhardt *et al.* [19] in the same year. Later some authors made a review of the properties of stimulated Raman scattering; see, for example, Bloembergen [11], Kaiser and Maier [20], Penzkofer *et al.* [12], and Raymer and Walmsley [21].

Table 3.5 Raman Scattering Cross Sections per Molecule of Some Liquids (Kato and Takuma [17])

Raman lines	Wavelength of the exciting light (nm)	Raman scattering cross section $(d\sigma/d\Omega)$ $(10^{-29}\,cm^2\,molecule^{-1} \cdot sr^{-1})$
C_6H_6	632.8	0.800 ± 0.029
992 cm^{-1}	514.5	2.57 ± 0.08
Benzene	488.0	3.25 ± 0.10
$C_6H_5CH_3$	632.8	0.353 ± 0.013
1002 cm^{-1}	514.5	1.39 ± 0.05
Chlorobenzene	488.0	1.83 ± 0.06
$C_6H_5NO_2$	632.8	1.57 ± 0.06
1345 cm^{-1}	514.5	9.00 ± 0.29
Nitrobenzene	488.0	10.3 ± 0.4
	694.3	0.755
CS_2	632.8	0.950 ± 0.034
656 cm^{-1}	514.5	3.27 ± 0.10
	488.0	4.35 ± 0.13
CCl_4	632.8	0.628 ± 0.023
459 cm^{-1}	514.5	1.78 ± 0.06
	488.0	2.25 ± 0.07

One of the earliest methods of accomplishing Raman laser action employed a repetitively pulsed Kerr cell as an electro-optical shutter, or "Q-switch," enclosed within the laser cavity, together with a polarizing prism, as shown in Fig. 3.7 (Baldwin [7]). This allows laser action to occur only during the brief time intervals when the Kerr cell is transmitting; the laser avalanche then discharges energy stored over the much longer time interval since the preceding pulse on the Kerr cell. It was observed that for sufficiently high laser-pulse intensity, the 694.3 nm ruby laser line is accompanied by a satellite line at 767 nm which originates in the nitrobenzene of the Kerr cell. The satellite line increases markedly in intensity as the laser output is increased above a threshold level of the order of 1 MW/cm^2, persists only while the laser output is above this threshold, shares the direction of the laser radiation, and becomes spectrally narrower at higher intensities. Its wavelength agrees with that of the known Raman–Stokes' shift in nitrobenzene. The conclusion was therefore reached that the phenomenon is a stimulated Raman scattering process, pumped by the laser beam and resonated by the end-reflectors of the laser.

Raman media are now widely used in conjunction with pulsed lasers to generate coherent radiation at frequencies other than those currently accessible to direct laser action. This is one of the powerful advantages of the Raman scattering process as a spectroscopic technique. Table 3.6 gives some properties of stimulated Raman scattering for several materials.

For detailed theoretical discussions of the Raman process, see Baldwin [7], Boyd [14], Yariv [8], Mostowsky and Raymer [13]. For experimental techniques, see Demtröder [22]; see also Ederer and McGuire [23].

3.2.3 Rayleigh Scattering

The spectrum of scattered light generally contains an *elastic* contribution, where the scattered frequency ω_s equals the incident frequency ω, together with several *inelastic* components, where ω_s differs from ω. The *elastic* component is known as Rayleigh scattering.

Rayleigh scattering is the scattering of light from nonpropagating density fluctuations. It is known as elastic or quasielastic scattering because it induces no fre-

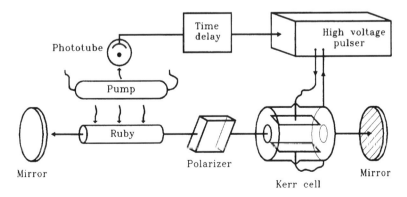

Figure 3.7 Experimental setup for pulsed operation of a laser employing a Kerr cell as an electro-optical shutter (Baldwin [7]).

Table 3.6 Frequency Shift v_v, Scattering Cross Section $N(d\sigma/d\Omega)$ of Spontaneous Raman Scattering (N is the number of molecules per cm^3) and Steady-State Gain Factor g_s/I_l of Stimulated Raman Scattering in Different Substances ($\lambda_1 = 694.3$ nm). Linewidths and Temperatures are also Indicated (Kaiser and Maier [20])

Substance	Frequency shift v_v (cm^{-1})	Linewidth Δv (cm^{-1})	Cross section $N(d\sigma/d\Omega)$ $\times 10^8$ (cm^{-1} ster^{-1})	Gain factor[a] g_s/I_l in units of 10^{-3} (cm/MW)	Temperature (K)
Liquid O_2	1552	0.117	0.48 ± 0.14	14.5 ± 4	
				16 ± 5	
Liquid N_2	2326.5	0.067	0.29 ± 0.09	17 ± 5	
				16 ± 5	
Benzene	992	2.15	3.06	2.8	300
		2.3	3.3	3.0	
			4.1	3.8	
CS_2	655.6	0.50	7.55	24.0	300
Nitrobenzene	1345	6.6	6.4	2.1	300
			7.9	2.6	
Bromobenzene	1000	1.9	1.5	1.5	300
Chlorobenzene	1002	1.6	1.5	1.9	300
Toluene	1003	1.94	1.1	1.2	300
$LiNbO_3$	256	23.0	381.0	8.9	300
	258	7.0	262.0	28.7	80
	637	20.0	231.0	9.4	300
	643	16.0	231.0	12.6	80
Li_6NbO_3	256			17.8	300
	266			35.6	80
	637			9.4	300
	643			12.6	80
$Ba_2NaNb_5O_{15}$	650			6.7	300
	655			18.9	80
$LiTaO_3$	201	22.0	238.0	4.4	300
	215	12.0	167.0	10.0	80
Li_6TaO_3	600			4.3	300
	608			7.9	80
SiO_2	467			0.8	300
				0.6	300
H_2 gas	4155			1.5 ($P > 10$ atm)	300

[a] To obtain the gain constant g_s(cm^{-1}) at v_s, multiply by v_s/v_l and by the intensity in MW/cm^2.

quency shift and can be described as scattering from entropy fluctuations (Boyd [14]).

Rayleigh-wing scattering (i.e., scattering in the wing of the Rayleigh line) is scattering from fluctuations in the orientation of anisotropic molecules. Since the molecular reorientation process is very rapid, this component is spectrally very broad. This type of scattering does not occur for molecules with an isotropic polarizability tensor.

Imagine a gaseous medium subdivided into volume elements in the form of small cubes of $\lambda/2$ on a side, as shown in Fig. 3.8 (Baldwin [7]). The radiation

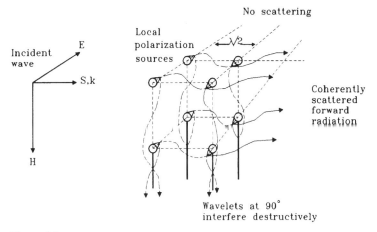

Figure 3.8 Coherent scattering of light by elementary dipoles induced in a medium by an electromagnetic wave incident from the left. The wavelets reradiated by the dipoles interfere constructively in the forward direction while those at 90° interfere destructively (Baldwin [7]).

scattered by two adjacent elements will be in phase in the forward direction but oppositely phased at 90°; if they contain exactly equal numbers of molecules, there will be no 90° scattering. However, for green light, $\lambda = 500$ nm, the volume of each element will be $\lambda^3/8 = 1.6 \times 10^{-20}$ m^3, and at atmospheric pressure, it will contain approximately

$$N\lambda^3/8 = [2.7 \times 10^{25}\,\text{m}^{-3}] \times [1.6 \times 10^{-20}\,\text{m}^3] = 4.3 \times 10^5\,\text{molecules},$$

where N is the density of molecules. The root-mean-square fluctuation, approximately 600 molecules, is 0.15% of the total number. At shorter wavelengths, these numbers are less, but the fluctuations become more pronounced. These produce the net scattering. This phenomenon is responsible for the blue color of the sky. The scattered power is proportional to the total number of scattering centers and to the fourth power of the frequency. The scattered light is incoherent.

The Rayleigh scattering reduces the intensity of a light beam without actually absorbing the radiation. Despite the regular arrangement of atoms in a crystalline media, which reduces the fluctuations, light can also be scattered by crystals, as thermal motion causes slight random fluctuations on density. Then we can observe Rayleigh scattering not only in gases and liquids but also in crystalline media.

Details of how to obtain the total average cross sections can be found in Loudon [2] and Chu [24]; other recommended texts are Bloembergen [25], Khoo *et al.* [26], and Keller [27].

3.3 MULTIPHOTON PROCESSES

Multiphoton processes refer to the processes where transitions between discrete levels of atoms are produced by the simultaneous absorption of n photons.

Multiphoton transitions are found whenever electrons bound in atoms or molecules interact with sufficiently intense electromagnetic radiation. Multiphoton ionization is the ultimate outcome of multiphoton transitions provided the radiation is reasonably intense, which is usually taken to mean about $10^6\,\mathrm{W/cm^2}$, or more, at optical frequencies. With intense sources and interaction times of the order of a nanosecond or more, ionization may be expected for atomic gases, while dissociation, with varying degrees of ionization, for molecular gases.

Multiphoton processes have features that are different from the traditional interaction of radiation with atoms and molecules. First, the interactions are nonlinear; secondly, a multiphoton transition probes atomic and molecular structure in a more involved and detailed way than can be imagined in single-photon spectroscopy; thirdly, owing to the high intensities available, extremely high levels of excitation can be reached, thus introducing qualitatively new physics; and fourthly, new applications such as isotope separation, generation of coherent radiation at wavelengths shorter than ultraviolet, and plasma diagnostics can be realized.

Multiphoton processes are one of the main sources of nonlinearity in the interaction of intense laser fields with atoms and molecules. Resonant multiphoton processes are of special interest because the multiphoton transition probabilities are enhanced under resonance conditions, and can be observed in fields of moderate intensity. The resonance is intended in the sense that the sum of the energies of the n photons is equal to the energy difference between the two concerned levels. It excludes the case where an intermediate level is resonantly excited during the process. However, some authors have to deal with the problem of processes involving both types of resonance (Feld and Javan 1969, Hänsch and Toschek [28]).

The first detailed theoretical treatment of two-photon processes was reported more than 60 years ago, by Göppert-Mayer [29] in 1931. However, experiments could not be realized until 30 years later, when lasers were available (Kaiser and Garret [30], Hopfield *et al.* [31]).

3.3.1   Two-Photon Absorption

Two-photon absorption can be described by a two-step process from initial level $|i>$ via a "virtual level" $|v>$ to the final level $|f>$ as shown in Fig. 3.9. The virtual level is represented by a linear combination of the wave functions of all real molecular

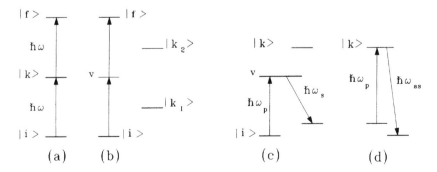

Figure 3.9   Energy level diagrams for several two-photon transitions: (a) resonant two-photon absorption with a real intermediate level $|k>$; (b) nonresonant two-photon absorption; (c) Raman transition; and (d) resonant and anti-Stokes' Raman scattering.

levels $|k_n\rangle$ which combine with $|i\rangle$ by allowed one-photon transitions. The excitation of $|v\rangle$ is equivalent to the off-resonance excitation of all these real levels $|k_n\rangle$. The probability amplitude for a transition $|i\rangle \rightarrow |v\rangle$ is represented by the sum of the amplitudes of all allowed transitions $|i\rangle \rightarrow |k\rangle$ with off-resonance detuning $(\omega - \omega_{ik})$. The same arguments apply to the second step $|v\rangle \rightarrow |f\rangle$.

The probability of two-photon absorption at low intensity is proportional to the square of the intensity of the light field I^2 (in the case of a single laser beam), the probability of absorption of each photon being proportional to I. In the case of n photon transitions, the probability of excitation is proportional to I^n. For this reason, pulsed lasers are generally used, in order to have sufficiently large peak powers. The spectral linewidth of these lasers is often comparable to or even larger than the Doppler width.

For a molecule moving with a velocity $\mathbf{v}$, the probability A_{if} for a two-photon transition between the ground state E_i and an excited state E_f, induced by the photons $\hbar\omega_1$ and $\hbar\omega_2$ from two light waves with wave vectors $\mathbf{k}_1$ and $\mathbf{k}_2$, polarization unit vectors $\hat{\mathbf{e}}_1$, $\hat{\mathbf{e}}_2$, and intensities I_1, I_2, respectively, can be written as (Demtröder [22]):

$$A_{if} \propto \frac{\gamma_{if} I_1 I_2}{[\omega_{if} - \omega_1 - \omega_2 - \mathbf{v} \cdot (\mathbf{k}_1 + \mathbf{k}_2)]^2 + (\gamma_{if}/2)^2} \tag{3.21}$$

$$\times \left| \sum_k \frac{\mathbf{R}_{ik} \cdot \hat{\mathbf{e}}_1 \; \mathbf{R}_{kf} \cdot \hat{\mathbf{e}}_2}{\omega_{if} - \omega_1 - \omega_2 - \mathbf{v} \cdot \mathbf{k}_1} + \frac{\mathbf{R}_{ik} \cdot \hat{\mathbf{e}}_2 \; \mathbf{R}_{kf} \cdot \hat{\mathbf{e}}_1}{\omega_{if} - \omega_1 - \omega_2 - \mathbf{v} \cdot \mathbf{k}_2} \right|^2 .$$

The first factor gives the spectral line profile of the two-photon transition of a single molecule. It corresponds exactly to that of a single-photon transition of a moving molecule at the center frequency $\omega_{if} = \omega_1 + \omega_2 + \mathbf{v} \cdot (\mathbf{k}_1 + \mathbf{k}_2)$ with a homogeneous linewidth γ_{if}.

The second factor describes the transition probability for the two-photon transition and can be derived quantum mechanically by second-order perturbation theory (Bräunlich [32], Worlock [33]). It contains a sum of products of matrix elements $\mathbf{R}_{ik}\mathbf{R}_{kf}$ for transitions between the initial state i and intermediate molecular levels k or between these levels k and the final state f.

The summation extends over all molecular levels k that are accessible by allowed one-photon transitions from the initial state $|i\rangle$. However, only those levels k which are not too far off resonance with one of the Doppler-shifted laser frequencies $\omega_1' = \omega_1 - \mathbf{v} \cdot \mathbf{k}_1$, $\omega_2' = \omega_2 - \mathbf{v} \cdot \mathbf{k}_2$ will mainly contribute, as can be seen from the denominator.

The frequencies ω_1 and ω_2 can be selected in such a way that the virtual level is close to a real molecular eigenstate. This greatly enhances the transition probability and then is generally advantageous to excite the final level E_f by two different photons with $\omega_1 + \omega_2 = (E_f - E_i)/h$ rather than by two photons out of the same laser with $2\omega = (E_f - E_i)/h$.

Multiphoton processes have several applications in laser spectroscopy. For example, two-quantum resonant transitions in a standing wavefield is an important method for eliminating Doppler broadening. One of the techniques of high-resolution laser spectroscopy is based on this approach (Shimoda [34], Letokhov and Chebotayev [35]).

3.3.2 Doppler-free Multiphoton Transitions

Doppler broadening is due to the thermal velocities of the atoms in the vapor. If $\mathbf{v}$ is the velocity of the atom and $\mathbf{k}$ is the wave vector of the light beam, the first-order Doppler shift is $\mathbf{k} \cdot \mathbf{v}$. If the sense of the propagation of light is reversed ($\mathbf{k} \rightarrow -\mathbf{k}$), the first-order Doppler shift is reversed in sign.

Suppose that a two-photon transition can occur between the levels E_i and E_f of an atom in a standing electromagnetic wave of angular frequency ω (produced for example by reflecting a laser beam onto itself using a mirror). In its rest frame, the atom interacts with two oppositely traveling waves of angular frequencies $\omega + \mathbf{k} \cdot \mathbf{v}$ and $\omega - \mathbf{k} \cdot \mathbf{v}$. If the atom absorbs one photon from each traveling wave, the energy conservation implies that

$$E_f - E_i = E_{fi} = h(\omega + \mathbf{k} \cdot \mathbf{v}) + h(\omega - \mathbf{k} \cdot \mathbf{v}) = 2h\omega. \tag{3.22}$$

The term depending on the velocity of the atom disappears, indicating that, at resonance, all the atoms, irrespective of their velocities, can absorb two photons (Vasilenko *et al.* [36], Cagnac *et al.* [37]).

In theory, the Doppler-free two-photon absorption resonance must have a Lorentzian shape, the width of the resonance being the natural one (Grynberg *et al.* [38]). However, the wings of the resonance generally differ from the Lorentzian curve because, if the frequency ω of the laser does not fulfill the resonant condition, Eq. (3.22), but is still close to it, the atoms cannot absorb two photons propagating in opposite directions, although some atoms of definite velocity can absorb two photons propagating in the same direction, provided that the energy defect ($E_{fi} - 2h\omega$) is equal to the Doppler shift $\pm 2\mathbf{k} \cdot \mathbf{v}$. For each value of ω, there is only one group of velocities which contribute to this signal, whereas at resonance (due to the absorption photons propagating in opposite directions) all the atoms contribute.

The two-photon line shape appears as the superposition of two curves, as shown in Fig. 3.10. A Lorentzian curve of large intensity and narrow width (natural width) corresponds to the absorption of photons from the oppositely traveling waves, while a Gaussian curve of small intensity and broad width (Doppler width) corresponds to the absorption of photons from the same traveling wave. Typically, the Doppler width of the Gaussian curve is 100 or 1000 times larger than the natural width of the Lorentzian curve and the Gaussian curve appears as a very small background. In some cases, the choice of different polarizations permits one to suppress completely the Doppler background, using the different selection rules corresponding to different polarization (Biraben *et al.* [39]).

The first experimental demonstrations of Doppler-free two-photon transitions were performed with pulsed dye lasers on the 3S–5S transition in sodium (Biraben *et al.* [39], Levenson and Bloembergen [40]). The precision of the measurements has been increased by the use of cw dye lasers in single-mode operation. A typical setup using a cw laser is shown in Fig. 3.11.

The cw laser is pumped by an argon ion laser. In order to obtain good control of the laser frequency, two servo loops are used. The purpose of the first one is to maintain the single-frequency oscillation of the dye laser. The Fabry–Perot etalon inside the laser cavity selects one particular longitudinal mode of the cavity. The

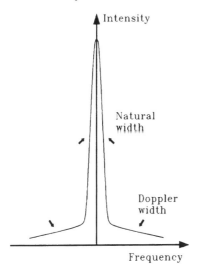

Figure 3.10 Theoretical Doppler-free two-photon absorption resonance. The two-photon line shape appears as the superposition of two curves (Grynberg *et al.* 1980).

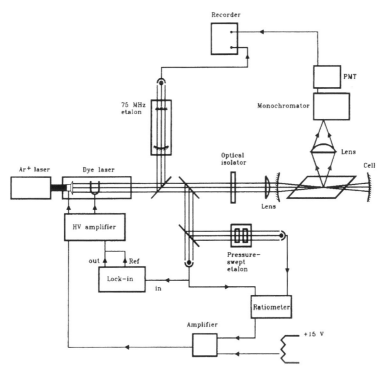

Figure 3.11 Typical setup for Doppler-free two-photon experiments using cw dye laser (Feld and Letokhov [80]).

second servo loop is used to control the frequency of the laser cavity and does not include any modulation. For details see Grynberg *et al.* [38].

The light coming from the laser is focused into the experimental cell in order to increase the energy density. The transmitted light is refocused from the other side into the cell using a concave mirror whose center coincides with the focus of the lens. In some experiments the energy is increased by placing the experimental cell in a spherical concentric Fabry–Perot cavity (Giacobino *et al.* [41]). The lens is chosen to match the radius of curvature of the wavefront to the radius of the first mirror. In order to reduce the losses in the cavity, the windows of the experimental cell are tilted to the Brewster angle. The length of the cavity is locked to the laser frequency to obtain the maximum transmitted signal. The two-photon resonance is detected by collecting photons emitted from the excited state at a wavelength λ_{vf} (see Fig. 3.9) different from the exciting wavelength λ. Sometimes it is still more convenient to detect the resonance on another wavelength λ_{ab} emitted by the atom in a cascade. The characteristic λ_{vf} is selected with an interference filter or a monochromator. The difference between λ_{vf} and λ allows the complete elimination of the stray light of the laser, despite its high intensity, and the observation of very small signals on a black background.

A simpler experimental arrangement for Doppler-free two-photon spectroscopy is shown in Fig. 3.12 (Demtröder [22]). The two oppositely traveling waves are formed by reflection of the output beam from a single-mode tunable dye laser. The Faraday rotator prevents feedback into the laser. The two-photon absorption is monitored by the fluorescence emitted from the final state E_f into other states E_m. The two beams are focused into the sample cell by the lens L and the spherical mirror M.

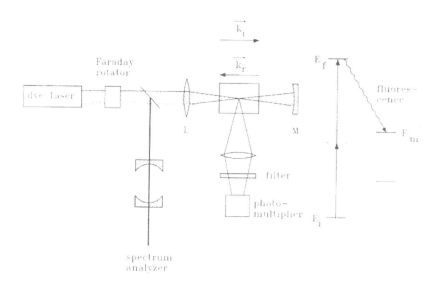

Figure 3.12 A simpler experimental arrangement for Doppler-free two-photon experiments (Demtröder [22]).

More examples of Doppler-free two-photon spectroscopy can be found in Demtröder [22], page 475.

3.3.3 Multiphoton Spectroscopy

If the incident intensity is sufficiently large, a molecule may absorb several photons simultaneously. Equation (3.21) can be generalized in order to obtain the probability for absorption of a photon $\hbar\omega_k$ on the transition $|i> \to |f>$ with $E_f - E_i = \sum \hbar\omega_k$. In this case, the first factor in Eq. (3.21) contains the product $\prod_k I_k$ of the intensities I_k, of the different beams. In the case of n-photon absorption of a single laser beam, this product becomes I^n. The second factor in Eq. (3.21) includes the sum over products of n one-photon matrix elements.

In the case of Doppler-free multiphoton absorption besides the energy conservation $\sum \hbar\omega_k = E_f - E_i$, the momentum conservation

$$\sum_k \mathbf{p}_k = h \sum_k \mathbf{k}_k = 0 \tag{3.23}$$

has also to be fulfilled. Each of the absorbed photons transfers the momentum $h\mathbf{k}_k$ to the molecule.

If Eq. (3.23) holds, the total transfer of momentum is zero, which implies that the velocity of the absorbing molecule has not changed. This means that the photon energy is completely converted into excitation energy of the molecule without changing its kinetic energy. As this is independent of the initial velocity of the molecule, the transition is Doppler-free.

Figure 3.13 shows a possible experimental arrangement for Doppler-free three-photon absorption spectroscopy while Fig. 3.14 shows the three-photon excited resonance fluorescence in Xe at $\lambda = 147\,\text{nm}$, excited with a pulsed dye laser at $\lambda = 441\,\text{nm}$ with $80\,\text{kW}$ peak power (Faisal *et al.* 1977).

3.3.4 Multiphoton Ionization Using an Intermediate Resonant Step

Resonant multiphoton excitation often occurs as an intermediate step in other processes. In the case of multiphoton ionization, the number of ions increases by a huge

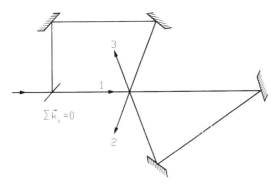

Figure 3.13 Possible experimental arrangement for Doppler-free three-photon absorption spectroscopy.

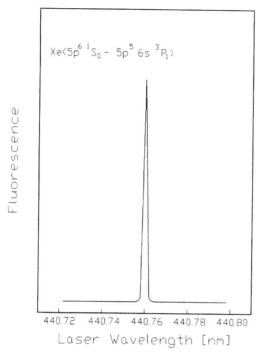

Figure 3.14   Three-photon excited resonance fluorescence in Xe at $\lambda = 147$ nm, excited with a pulsed dye laser at $\lambda = 441$ nm with 80 kW peak power (Faisal *et al.* 1977).

factor when the wavelength of the exciting laser is adjusted to obtain a resonant multiphoton with an intermediate level, as shown in Fig. 3.15.

The figure shows the four-photon ionization of cesium with an intermediate resonant level. The variation of the number of atomic cesium ions as a function of the laser frequency (Nd glass laser) is also shown. The curves have a big enhancement in the neighborhood of the three-photon transition $6S \rightarrow 6F$. The center of the resonance is shifted by an amount which is proportional to the intensity. For an intensity $I = 1$ GW/cm^2, the wavelength of excitation is reduced by an amount larger than 0.1 nm.

This light shift explains the strange behavior of the order of nonlinearity of a multiphoton ionization process near an *n*-photon resonance (Morellec *et al.* [42]).

The order of nonlinearity K is defined by (Grynberg *et al.* [38]):

$$K = \frac{\partial \log N_i}{\partial \log I}. \tag{3.24}$$

where N_i is the number of ions obtained in the multiphoton ionization and I is the light intensity. Far away from any intermediate resonance, K is equal to the number of photons K_0 which is needed for photoionization, but close to a resonance, this is not true.

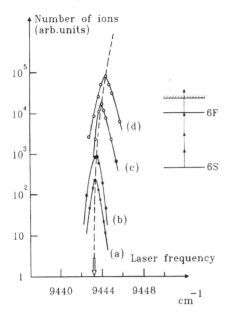

Figure 3.15 Four-photon ionization of cesium with an intermediate resonant level. The variation of the number of atomic cesium ions as a function of the laser frequency in the neighborhood of the three-photon transition 6S → 6F is also shown (Feld and Letokhov [80]).

Because of the light shift, the effective detuning from the resonance increases or decreases, depending on the sign of the laser detuning ($E_{fi} - n\hbar\omega$). This explains why, on one side of the resonance, K is much larger than K_0 while, on the other side, K is smaller.

The resonant multistep processes permit the selective photoionization of atoms. This approach is fundamental for laser methods of single-atom detection (Letokhov [43]).

3.4 PHASE CONJUGATE OPTICS

Phase conjugate optics refers in its most elemental basics to the conversion in *real time* of a monochromatic optical field $\mathbf{E}_1$ to a new field $\mathbf{E}_2$ such that

$$E_1(\mathbf{r}, t) = \text{Re}[\psi(\mathbf{r})\,e^{i(\omega t - kz)}]$$

$$E_2(\mathbf{r}, t) = \text{Re}[\psi^*(\mathbf{r})\,e^{i(\omega t + kz)}].$$

We refer to E_2 as the phase conjugate replica of E_1.

Suppose that a monochromatic optical beam E_1 propagates to the right and then incides into a lossless medium that distorts it. If the distorted beam is reflected by a mirror, then we will have a reflected beam traveling to the left that when it incides into the lossless medium is distorted in exactly the opposite way such that the emerging beam E_2 is the replica of the initial beam E_1 everywhere.

To show that it is possible that both beams E_1 and E_2 coincide everywhere, we refer and demonstrate the Distortion Correction Theorem (Yariv [44]), which may be stated as follows: "If a (scalar) wave $E_1(r)$ propagates from left to right through an arbitrary dielectric (but lossless) medium, then if we generate in some region of space (say, near $z = 0$) its phase conjugate replica $E_2(r)$, then E_2 will propagate backward from right to left through the dielectric medium remaining *everywhere* the phase conjugate of E_1."

Consider the (scalar) right-going wave E_1:

$$E_1 = \psi_1(r)\,e^{i(\omega t - kz)},$$

where k is a real constant. E_1 obeys in the paraxial limit the wave equation:

$$\nabla^2 E_1 + \omega^2 \mu \varepsilon(\mathbf{r}) E_1 = 0. \tag{3.25}$$

If we substitute E_1 into the wave equation (3.25), we obtain

$$\nabla^2 \psi_1 + [\omega^2 \mu \varepsilon(\mathbf{r}) - k]\,\psi_1(r) - 2ik\,\partial\psi_1/\partial z = 0. \tag{3.26}$$

The complex conjugate of Eq. (3.26) leads to:

$$\nabla^2 \psi_1^* + [\omega^2 \mu \varepsilon^*(\mathbf{r}) - k]\,\psi_1^*(r) + 2ik\,\partial\psi_1^*/\partial z = 0. \tag{3.27}$$

Now take the wave E_2 propagating to the left, into the wave Eq. (3.23). We obtain

$$\nabla^2 \psi_2 + [\omega^2 \mu \varepsilon(\mathbf{r}) - k]\,\psi_2(r) + 2ik\,\partial\psi_2/\partial z = 0. \tag{3.28}$$

We see from Eqs (3.27) and (3.28) that ψ_1^* and ψ_2 obey the same differential equation if $\varepsilon(\mathbf{r}) = \varepsilon^*(\mathbf{r})$; i.e., if we have a lossless and gainless medium.

If $\psi_2 = a\psi_1^*$, where a is an arbitrary constant over any plane, say $z = 0$, Eq. (3.26) is still valid. Then due to the uniqueness property of the solutions to second-order linear differential equations:

$$\psi_2(x, y, z) = a\psi_1^*(x, y, z) \text{ for all } x, y, z < 0. \tag{3.27}$$

This completes the proof of the distortion correction theorem.

Phase conjugate waves can be generated by means of nonlinear optical techniques. As second-order nonlinear optics gives rise to phenomena such as second-harmonic generation and parametric amplification, third-order nonlinear optics, which involves the third power of the electric field $P \propto E$, gives rise to phenomena such as third-harmonic generation and to the phenomenon of *four-wave mixing* (FWM) where, if three waves of different frequencies ω_1, ω_2, ω_3, incide into a medium material, this radiates with a frequency ω_4, such that $\omega_4 = \omega_1 + \omega_2 - \omega_3$.

We will show that if three waves A_1, A_2, A_3, are mixed into a medium as shown in Fig. 3.16, this medium generates and radiates a new wave A_4, which is the phase conjugate of A_1.

The nonlinear medium is crossed simultaneously by four waves at the same frequency:

$$E_1(\mathbf{r}, t) = (1/2)A_1'(z)\,e^{i(\omega t - kz)} + \text{c.c.}$$

$$E_2(\mathbf{r}, t) = (1/2)A_2'(z)\,e^{i(\omega t - k_2 \cdot z)} + \text{c.c.}$$

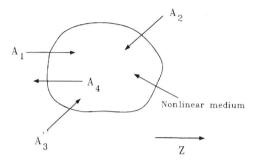

Figure 3.16 Four-wave mixing into a nonlinear medium.

$$E_3(\mathbf{r}, t) = (1/2)A_3'(z)\,e^{i(\omega t - \mathbf{k_3} \cdot \mathbf{z})} + \text{c.c.}$$

$$E_4(\mathbf{r}, t) = (1/2)A_4'(z)\,e^{i(\omega t + \mathbf{k_2})} + \text{c.c.}$$

where $k^2 \equiv \omega^2 \mu \varepsilon$.

The output power of the beams 1 and 4 increases as they cross the nonlinear medium to the expense of beams 2 and 3. The quantum mechanics description (Fisher [45]) of the processes shows that in atomic scale, two photons—one from the beam 2 and another one from beam 3—are simultaneously annihilated while two photons are created. One of these photons is added to the beam 1 and the other one to the beam 4.

The mathematical analysis gives us the expression (Yariv [8], Hellwarth [46]):

$$A_4(x, y, z < 0) = -i[\kappa^*/|\kappa|]\tan|\kappa|L]A_1^*(x, y, z < 0)$$

This is the phase conjugation basic result for the four-wave mixing. This expression shows that the reflected beam $A_4(\mathbf{r})$ to the left of the nonlinear medium ($z < 0$) is the phase conjugate of the input beam $A_1(\mathbf{r})$. Here $\kappa^* = (\omega/2)\sqrt{[(\mu/c)\chi A_2 A_3]}$ and χ is a fourth-rank tensor.

3.4.1 Optical Resonators with Phase Conjugate Reflectors

In the optical resonators with phase conjugate optics, one of the mirrors of the resonator is substituted by a phase conjugate mirror (PCM), as shown in Fig. 3.17.

Let us consider that the two mirrors are separated a distance l and that a Gaussian beam with quantum numbers m, n, is reflected by both mirrors. Call $\Phi_1(m, n)$ the phase shift suffered by the beam due to its propagation between the two mirrors; let Φ_R be the phase shift after reflection from the conventional mirror and α the phase shift after reflection from the PCM.

If ϕ_1 is the initial phase of the beam in, for example, the plane P, then the several phases of the beam after each reflection will be

$$\phi_2 = -\phi_1 + \alpha$$

$$\phi_3 = \phi_2 + \Phi_1(m, n) = -\phi_1 + \alpha + \Phi_1(m, n)$$

$$\phi_4 = \phi_3 + \Phi_R = -\phi_1 + \alpha + \Phi_1(m, n) + \Phi_R$$

$$\phi_5 = \phi_4 + \Phi_1(m, n) = -\phi_1 + \alpha + 2\Phi_1(m, n) + \Phi_R$$

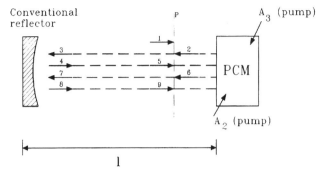

Figure 3.17 Optical resonator formed by a conventional mirror and a phase conjugate mirror (PCM).

$$\phi_6 = -\phi_5 + \alpha = \phi_1 - 2\Phi_1(m, n) - \Phi_R$$

$$\phi_7 = \phi_6 + \Phi_1(m, n) = \phi_1 - \Phi_1(m, n) - \Phi_R$$

$$\phi_8 = \phi_7 + \Phi_R = \phi_1 - \Phi_1(m, n)$$

$$\phi_9 = \phi_8 + \Phi_1(m, n) = \phi_1$$

We have shown that the phase of the beam reflected into the resonator reproduces itself after two round trips. The phase conjugate resonator has a resonance at the frequency of the pump beams. The resonance condition is satisfied independently of the length l of the resonator and the transversal order (m, n) of the Gaussian beam. The phase conjugate resonator is stable independently of the radius of curvature R of the mirror and the distance l between both mirrors.

3.4.2 Applications

We will consider only some of the most common applications of phase conjugation.

(a) Dynamic Correction of Distortion into a Laser Resonator

One important application of phase conjugate optics is the real-time dynamic correction of distortion in optical resonators. Let us consider the optical resonator shown in Fig. 3.17, but now with an amplifier and a distortion (it could be the gain medium itself or even a "bad" optics), as shown in Fig. 3.18. If a Gaussian beam is distorted when it passes through the distortion, it recovers its form when it is reflected by the PCM and passes the distortion again now in the opposite direction, according to the Distortion Correction Theorem. Then, at the output of the resonator, we will have a Gaussian beam again.

The experimental arrangement for a laser oscillator with phase conjugate dynamic distortion correction is shown in Fig. 3.19.

(b) Aberration Compensation

During the generation, transmission, processing, or imaging of coherent light, we can have aberrations due to turbulence, vibrations, thermal heating, and/or imperfec-

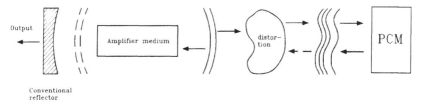

Figure 3.18   Laser oscillator with distortion.

tions of optical elements. Most of these aberrations can be compensated by the use of a PCM if we let the wavefront follow the path shown in Fig. 3.20.

As stated before, the complex amplitudes of forward- and backward-going waves are complex conjugates of each other and the function of the PCM is to generate a conjugate replica of a general incident field, having an arbitrary spatial phase and amplitude distribution.

One of the many experiments carried out to show aberration compensation is that of Jain and Lind [72]. In this experiment, the input light was that of a focused, pulsed ruby laser beam operating in a TEM_{00} mode. The phase aberrator was an etched sheet of glass, and the PCM was realized via degenerate four-wave mixing (DFWM), by using a semiconductor-doped glass as the nonlinear optical medium, pumped by a ruby laser.

This experiment shows the near-perfect recovery of a diffraction-limited focal spot using nonlinear phase conjugate techniques.

Another example of aberration compensation was carried out by Bloom and Bjorklund [47]. In this experiment, a resolution chart is illuminated by a planar probe wave and placed at plane (a) of Fig. 3.20. The PCM is realized using a DFWM process, with carbon disulfide as the nonlinear medium and a doubled Nd:YAG laser as the light source.

This example shows both the compensation ability of the system and its lensless imaging capabilities. Also it points out the importance of generating a true phase conjugate replica, both in terms of its wavefront contours and in terms of the propagation direction.

(c)   Lensless Imaging: Applications to High-Resolution Photolithography using Four-Wave Mixing

Nonlinear phase conjugation (NLPC) allows us not only to compensate for defects in optical elements, such as lenses, but permits the elimination of the lenses

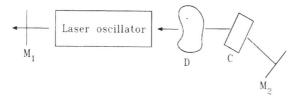

Figure 3.19   Experimental system for a laser oscillator with a distortion D, a phase conjugate crystal C, and two mirrors M_1 and M_2.

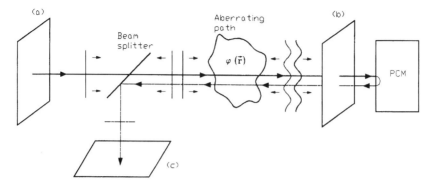

Figure 3.20 Experimental arrangement for aberration compensation. An input field at plane (a) becomes aberrated at plane (b) after propagation through a distortion path $\varphi(\mathbf{r})$. As a consequence of the conjugation process the conjugate of the initial field is recovered at plane (a). The beamsplitter allows one to view the same compensator field at the plane (c). PCM, phase conjugate mirror.

altogether. Further, this lensless approach permits a major reduction in the effective F number, thereby improving the spatial resolution of the system.

An example of an NLPC application on photolithography was carried out by Levenson [40] in 1980 and by Levenson *et al.* [48] in 1981. A typical photolithography requirement is diffraction-limited resolution over a wafer as large as 3–4 inches in diameter. Levenson *et al.* obtained a resolution of 800 lines/mm over a $6.8\,\text{mm}^2$ field and a features size of $0.75\,\mu\text{m}$. This is equivalent to a numerical aperture (NA) of 0.48, which far exceeds the NA of conventional imaging systems. A crystal of LiNbO_3 was used as the FWM phase conjugate mirror, pumped by a 413 nm Kr ion laser. As in all phase conjugators, the accuracy of the conjugated wave produced by the PCM is critically dependent upon the quality of the nonlinear medium and of the pump waves.

The speckle typical problem in laser imaging systems is also eliminated, and according to Levenson [40], the elimination is a consequence of using the PCM in conjunction with plane-wave illumination.

(d) Interferometry

Phase conjugate mirrors can also be used in several conentional and unconventional interferometers. For example, in the Mach–Zender type (Fig. 3.21) the object wavefront interferes with its phase conjugate instead of a plane-wave reference (Hopf [49]). In Fig. 3.21, the input signal $I(x, y)$ evaluated at plane x is given by

$$I(x, y) = A(x, y)\exp[i\phi(x, y)].$$

This field is imaged to the observation plane with an arbitrary transmission T, yielding an amplitude $I_T = TI(x, y)$. The conjugate of the input is generated by a four-wave mixing process in reflection from a phase conjugate mirror and arrives at the observation plane with an amplitude

$$I_R = RA(x, y)\exp[-i\phi(x, y)],$$

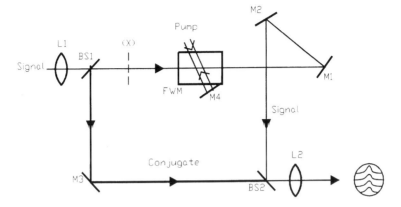

Figure 3.21 Mach–Zender interferometer using phase conjugate mirror (Hopf [49]).

where R and T depend on the reflectivity and transmittance of the PCM and other components within the interferometer. The time-averaged intensity at the observation plane will be proportional to

$$I = |I_T + I_R|^2 = [A(x, y)]^2 \{T^2 + R^2 + 2RT \cos[2\phi(x, y)]\}$$

An advantage of this system is that we can adjust the reflectivity of the components, including the PCM, so that $T = R$, yielding a fringe visibility of unity, independent of intensity variations in or across the sample beam.

(e) Nonlinear Laser Spectroscopy

Some spectroscopic studies have been carried out to obtain information about linewidths, atomic motion, excitation diffusion, etc., by measuring the four-wave mirrorfrequency response, angular sensitivity, or polarization dependence. Table 3.7 gives some examples of these spectroscopic studies.

In addition, the spatial structure of the nonlinear susceptibilities can be studied using nonlinear microscopy by three-wave mixing (TWM), as reported by Hellwarth and Christensen [50] or four-wave mixing, as has been done by Pepper *et al.* [51].

Some other interesting applications include optical computing, communications, laser fusion, image processing, temporal signal processing, and low-noise detection schemes. Also, the extension of these applications to other regions of the electromagnetic spectrum can provide new classes of quantum electronic processors.

3.5 ULTRASHORT OPTICAL PULSES

It has been more than three decades since the era of ultrashort optical pulse generation was ushered in with the report of passive model-locking of the ruby laser by Mocker and Collins [95] in 1965. One year later, the first optical pulses in the picosecond range were generated with an Nd:glass laser by Demaria [84]. Since then, the width of optical pulses has been reduced at an exponential rate, as shown in Fig. 3.22, where the logarithm of the shortest reported optical pulse width versus year is graphed.

Table 3.7 Laser Spectroscopy Using Four-Wave Mixing

Phase conjugate mirror reflectivity versus	Physical mechanism	Reference
Angle	Atomic-motional effects	a
Buffer gas pressure	Collisional excitation	b
	Pressure-broadening mechanisms	c
Pump probe polarization	Coherent-state phenomena	d
	Multiphoton transitions	e
	Quadrupole optical transitions	f
	Electronic and nuclear contributions to nonlinear optical susceptibility	g
Magnetic fields	Optical pumping	h
	Zeeman state coupling	i
	Liquid-crystal phase transitions	j
Electric fields	Stark effects	f
	Liquid-crystal phase transitions	k
RF and microwave fields	Hyperfine state coupling	f
Pump-probe detuning	Atomic populations relaxation rates	l
	Atomic linewidth effects	l
	Optical pumping effects	l
	Doppler-free one- and two-photon spectroscopy	l
Frequency ($\omega_{pump} = \omega_{probe}$)	Natural linewidth measurements	m
	Atomic-motionally induced nonlinear coherences	m
Pump frequency scanning	Laser-induced cooling of vapors	n
	Sub-Doppler spectroscopy	f
Transient regime (temporal effects)	Atomic coherence times	o
	Population relaxation rates	p
	Inter- and intramolecular relaxation	q
	Carrier diffusion coefficients	r
Pump-wave intensity	Saturation effects	s
	Inter- and intramolecular population coupling	t
	Optically induced level shifts and splittings	u

[a] Wandzura [52]; Steel *et al.* [53]; Nilsen *et al.* [54]; Nilsen and Yariv [55, 56]; Humphrey *et al.* [57]; Fu and Sargent [58]; Saikan and Wakata [59]; Bloch *et al.* [60].
[b] Liao *et al.* [61]; Fujita *et al.* [62]; Bodgan *et al.* [63].
[c] Raj *et al.* [64, 65]; Woerdman and Schuurmans [66]; Bloch *et al.* [60].
[d] Lam *et al.* [67].
[e] Steel and Lam [53]; Bloch *et al.* [60].
[f] No evidence regarding this mechanism has been reported to date.
[g] Hellwarth [46].
[h] Economou and Liao [68]; Steel *et al.* [69].
[i] Economou and Liao [68]; Yamada *et al.* [70]; Steel *et al.* [69].
[j] Khoo [71].
[k] Jain and Lind [72].

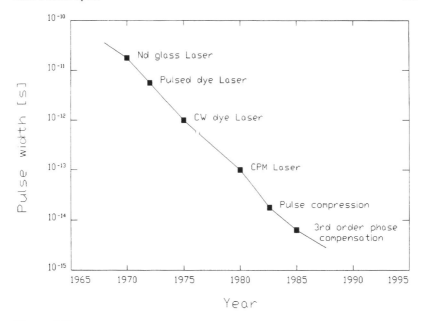

Figure 3.22 Historical development of the progress in generating ultrashort pulses (Demtröder [22]).

Each reduction in pulse width has been accompanied by an advance in technology. As an example, a pulse width of about 10^{-14} s was possible thanks to the development of pulse compression. Optical pulse widths as short as 6 fs have been generated, approaching the fundamental limits of what is possible in the visible region of the spectrum.

Progress in generating intense ultrashort laser pulses has made possible systematic studies of coherent interactions between picosecond laser pulses and molecular vibrations (Shapiro [85]). The first method used for this purpose (Von der

---

Table 3.7 Laser Spectroscopy Using Four-Wave Mixing (*contd.*)

[l] Bloch [73]; Bloch et al. [60]; Steel et al. [69].
[m] Bloom et al. [74]; Lam et al. [67].
[n] Palmer [75].
[o] Liao et al. [61].
[p] Fujita et al. [62]; Steel et al. [69].
[q] Steel and Lam [53]; Steel et al. [76].
[r] Eichler [77]; Hamilton et al. [78]; Moss et al. [79].
[s] Fu and Sargent [80]; Harter and Boyd [81]; Raj et al. [64, 65]; Woerdman and Schuurmans [66]; Steel and Lind [82].
[t] Dunning and Lam [83].
[u] Bloom et al. [74]; Nilsen et al. [54]; Nilsen and Yariv [55, 56]; Fu and Sargent [80]; Saikan and Wakata [59]; Steel and Lind [82].

Linde *et al.* [86], Alfano *et al.* [87]) has proved to be the most efficient. In this method, the sample is simultaneously irradiated by two coherent collimated ultra-short light pulses, whose frequency difference is equal to the molecular vibrational frequency. This induces an excitation, of the Raman-type, of the N molecules contained in a coherent interaction volume. An ultrashort pulse of variable delay then probes the state of the system as it decays. Both the intensity and direction of the scattered probe pulse can then be studied.

Because of the coherent nature of the interaction between the excitation and probe pulses, the interaction efficiency for short delay is proportional to N^2, and depends on the relative orientation of the wave vectors of the exciting and probe fields. However, as the molecular vibrations dephase, the interaction becomes incoherent, leading to isotropic efficiency proportional only to N. These features make it possible to separate coherent and incoherent processes occurring on a picosecond time scale (Feld and Letokhov [88]). The picosecond pulse techniques can also be used to study inhomogeneous broadening of the vibrational transitions and its internal structure.

Ultrashort optical pulses are related to ultrafast nonlinear optics: *ultrafast* is defined as referring to events which occur in a time of less than about 10 ps. The motivation for using ultrafast pulses can be either that they afford the time resolution necessary to resolve the process of interest, or they are required to obtain a high peak power at a relatively low pulse energy. Ultrafast pulses are not available at every wavelength. In fact, most of them have been obtained in the orange region of the spectrum where the most reliable sources exist. Some others have been obtained in the gallium arsenide semiconductor research region at 800 nm, and in the optical communications region at 1500 nm.

A major advance in the generation of ultrashort optical pulses has been the process of mode-locking. Another approach to generating short pulses is the process of pulse compression. Some other novel pulse-generation schemes have been recently developed. We will try to describe them briefly along with applications.

3.5.1 Mode-Locking

Pulses in the picosecond regime can be generated by mode-locking. The simplest way to visualize mode-locked pulses is as a group of photons clumped together and aligned in phase as they oscillate through the laser cavity. Each time they hit the partially transparent output mirror, part of the light escapes as an ultrashort pulse. The clump of photons then makes another round trip through the laser cavity before another pulse is emitted. Thus the pulses are separated by the cavity tround-trip time $2L/c$, where L is the cavity length and c is the speed of light.

Mode-locking requires a laser that oscillates in many longitudinal modes. Thus, it does not work for many gas lasers with narrow emission lines but can be used with argon or krypton ion, solid-state crystalline, semiconductor and dye lasers (which have exceptionally wide-gain bandwidth). The pulse length is inversely proportional to the laser oscillating bandwidth, so dye lasers can generate the shortest pulses because of their exceptionally broad-gain bandwidths.

The number of modes oscillating is limited by the bandwidth Δv over which the laser gain exceeds the loss of the resonator, as shown in Fig. 3.23. Unless some mode-selecting element is placed in the laser resonator, the output consists of a sum

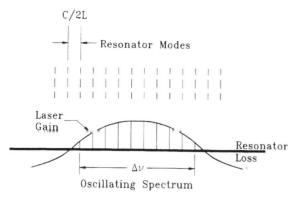

Figure 3.23 Resonator modes. The number of modes oscillating is limited by the bandwidth Δv and determined by the gain profile and the resonator loss.

of frequency components corresponding to the oscillating modes. The electric field may be written as

$$E(t) = \sum_n \alpha_n \exp i[(\omega_0 + n\delta\omega)t + \phi_n]$$

where α_n is the amplitude of the nth mode and $\delta\omega$ is the mode spacing. In general, the laser output varies in time, although the average power remains relatively constant. This is because the relative phases between the modes are randomly fluctuating. However, if the modes are forced to maintain a fixed phase and amplitude relationship, the output of the laser will be a well-defined periodic function of time. In thisway, we say that the laser is "mode-locked."

Both continuous and pulsed lasers can be mode-locked. In either case, mode-locking produces a train of pulses separated by the cavity round-trip time, as shown in Fig. 3.24.

In this case, the picture corresponds to a single pulse traveling back and forth between the laser resonator mirrors. It is also possible to produce mode-locking with

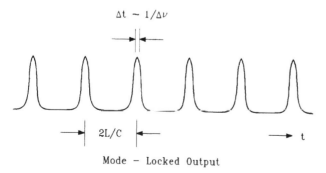

Figure 3.24 Train of pulses obtained at the output of the laser with all modes locked in the proper phase.

N pulses in the cavity, spaced by a multiple of $c/2L$. The pulses have a width $\Delta\tau$, which is approximately equal to the reciprocal of the total mode-locked bandwidth $\Delta\nu$, and the temporal periodicity is given by $T_p = 2L/c$. The ratio of the pulse width to the period is approximately equal to the number of locked modes.

For pulsed lasers, the result is a series of pulses that fall within the envelope defined by the normal pulse length. Single mode-locked pulses can be selected by passing the pulse train through grating modulators that allow only a single pulse to pass.

The modulation can be active, by changing the transmission of a modulator, or passive, by saturation effects. In either case, interference among the modes produces a series of ultrashort pulses.

Active Mode-Locking

As to date, a shutter than can be inserted into the optical cavity, which opens and closes at the approximate frequency to obtain the desired pulse duration has not yet been built; some approximations have been developed. For example, an intracavity phase or loss modulator has been inserted into the optical cavity driven at the frequency corresponding to the mode spacing (Hargrove *et al.* [89]). The principle of active mode-locking by loss modulation is as follows: an optical pulse is likely to form in such a way as to minimize the loss from the modulator. The peak of the pulse adjusts in phase to be at the point of minimum loss from the modulator. However, the slow variation of the sinusoidal modulation provides only a weak mode-locking effect, making this technique unsuitable for generating ultrashort optical pulses. Similarly, phase modulation can also produce mode-locking effects (Harris and Targ [90]).

Active mode-locking is particularly useful for mode-locking Nd:YAG and gas lasers such as the argon laser (Smith [91] and Harris [92]). Recently, a 10-fs pulse has been generated from a unidirectional Kerr-lens mode-locked Ti:sapphire ring laser (Kasper and Wittex [93]).

Passive Mode-Locking

Passive mode-locking works by the insertion of a saturable absorbing element inside the optical cavity of a laser. The saturable absorber can be an organic dye, a gas, or a solid, but the first one is the more common. The first optical pulses in the picosecond time domain (DeMaria *et al.* [84], as well as the first optical pulses in the femtosecond time domain (Fork *et al.* [94]), were obtained by this method.

Passively mode-locked lasers can be divided in two groups: (a) giant pulse lasers and (b) continuous or quasi-continuous lasers. Passive mode-locking was first observed for the first group in ruby lasers (Mocker and Collins [95]) and in Nd:glass lasers (DeMaria *et al.* [84]). Continuous passive mode-locking is observed primarily in dye lasers and was first theoretically described by New [96] in 1972.

(a) Giant Pulse Lasers

The optical configuration for a mode-locked giant pulse laser is shown in Fig. 3.25. The dye cell is optically contacted with the laser mirror in one end of the cavity in order to reduce the problem of satellite pulses (Weber [97], Bradley *et al.* [98], Von der Linde [99]). In designing the cavity, it is important to eliminate subcavity reso-

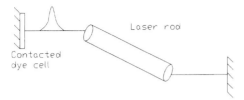

Figure 3.25   Optical configuration for a mode-locked giant pulse laser (Kaiser [101]),

nances and spurious reflections which may cause the formation of subsidiary pulse trains.

For a giant pulse to occur, the upper laser level lifetime must be long (as in ruby or Nd:glass lasers), typically hundreds of microseconds. Pulse generation occurs in a highly transient manner in a time much shorter than the upper level population response. The "fluctuation model" proposed by Letokhov [100] in 1969, describes the operation of these lasers. Briefly, the operation is as follows (Kaiser [101]): At the start of the flashlamp pumping pulse, spontaneous emission excites a broad spectrum of laser modes within the optical cavity. Since the modes are randomly phased, a fluctuation pattern is established in the cavity with a periodic structure corresponding to a cavity round-trip time $T = 2L/c$. When the gain is sufficient to overcome the linear and nonlinear losses in the cavity, the laser threshold is reached and the fields in the cavity initially undergo linear amplification. At some point, the field becomes intense enough to enter a phase where the random pulse structure is transformed by the nonlinear saturation of the absorber and by the laser gain saturation. As a result, one of the fluctuation spikes grows in intensity until it dominates and begins to shorten in time. As the short pulse gains intensity it reaches a point where it begins to nonlinearly interact with the glass host and the pulse begins to deteriorate. At the beginning of the pulse train, as recorded on an oscilloscope, the pulses are a few picoseconds in duration and nearly bandwith limited (Von der Linde *et al.* 1970, Zinth *et al.* [103]). Later, pulses in the train undergo self-modulation of phase and self-focusing, which leads to temporal fragmentation of the optical pulse.

The role of the saturable absorber in the fluctuation model is to select a noise burst that is amplified and ultimately becomes the mode-locked laser pulse. As a consequence, the relaxation time and the absorber, T_a, sets an approximate limit to the duration of mode-locked pulses (Garmire and Yariv [104], Bradley *et al.* [98]). Kodak dyes A9740 and A9860, with lifetimes of 7 and 11 picoseconds, respectively, are typically used. However other dyes with shorter lifetimes have been investigated (Kopinsky *et al.* [105], Alfano *et al.* [106]).

(b)   Continuous Lasers

In 1981, Fork *et al.* [94] described the generation of sub-100 femtosecond pulses for the first time, and coined the term "colliding-pulse-mode-locked laser" or CPM laser. The initial CPM laser consisted of a seven-mirror ring cavity with Rhodamine 6G in the gain jet and the dye DODCI in the absorber jet. These same two dyes have been used since 1972 in passively mode-locked cw dye lasers, but only generated picosecond pulses (Ippen *et al.* [107]).

Passive mode-locked continuous lasers involve a very different physics of pulse formation from that of giant pulse lasers. Random noise fluctuations due to longitudinal mode beating occur in the laser until one of the noise spikes is large enough to saturate the absorber. This pulse sees an increased transmission through the absorber jet due to this saturation and then encounters the gain jet. Here, it saturates the gain slightly and reduces the gain for the noise spikes that follow. This selection process continues until only one pulse survives. The pulse then shortens further due to the same saturable effects. Saturable absorption selectively removes energy from the leading edge of the pulse while saturable gain steepens the trailing edge. The pulse continues to shorten until a pulse-broadening effect, such as dispersion, can balance it (Walmsley and Kafka [108]).

The mode-locked pulse duration is typically much shorter than either the lifetime of the amplifying or gain medium, or the saturable absorber recovery time. In 1972, New [96] described for the first time, the conditions of pulse formation in continuous passively mode-locked lasers. Later, analytical and numerical techniques were applied to describe the transient formation of an ultrashort optical pulse (Garside and Lim [109, 110]; New and Rea [111]), while Haus obtained a closed formed solution by assuming a cavity bandwidth and a hyperbolic secant pulse shape (Haus [112]).

The shortest light pulses reported up to now are only 6 fs long (Fork *et al.* [113]). This corresponds to about three oscillation periods of visible light at $\lambda = 600$ nm!

Table 3.8 shows a short summary of different mode-locking techniques and the typical pulse duration obtained.

Now, we will see some examples of mode-locked pulses generated by different kinds of lasers.

Neodymium lasers. Neodymium lasers can generate mode-locked pulses (generally with continuous excitation), usually of tens to hundreds of picoseconds. The shorter the pulse, the higher the peak power for a given laser rod and pump source;

Table 3.8 Short Summary of Different Mode-Locking Techniques

Technique	Mode locker	Laser	Typical pulse duration	Typical pulse energy
Active mode-locking	Acousto-optic modulator	Argon cw	300 ps	10 nJ
		He–Ne cw	500 ps	0.1 nJ
	Pockels cell	Nd:YAG pulsed	100 ps	10 μJ
Passive mode-locking	Saturable absorber	Dye cw	1 ps	1 nJ
		Nd:YAG	1–10 ps	1 nJ
Synchronous pumping	Mode-locked pump laser and matching or resonator length	Dye cw	1 ps	10 nJ
		Color center	1 ps	10 nJ
CPM	Passive mode-locking and eventual synchronous pumping	Ring dye laser	< 100 fs	≈ 1 nJ

however, shrinking pulse length beyond a certain point can reduce the pulse energy. Pulse lengths are listed in Table 3.9 based on data from industry directories (Hecht [114]).

Because glass can be made in larger sizes and has lower gain than Nd-YAG, glass lasers can store more energy and produce more energetic pulses. However, its thermal problems limit repetition rate and thus lead to low average power. Nd-YLF also can store more energy than Nd-YAG because of its lower cross section, but it cannot be made in pieces as large as glass. Mode-locked lasers emit series of short pulses, with spacing equal to the round-trip time of the laser cavity. Many of these pulses may be produced during a single flashlamp pulse, and Q-switches can select single mode-locked pulses. Glass lasers have lower repetition rates than Nd-YAG or Nd-YLF, and commercial models typically produce no more than a few pulses per second. Repetition rates may be as low as one pulse every few minutes, or single-shot for large experimental lasers.

Both Nd-glass and Nd-YLF can generate shorter mode-locked pulses, because they have broader spectral bandwidths than Nd-YAG. The shortest pulses are produced by a technique called *chirped pulse amplification*, which is similar to that used to generate femtosecond pulses from a dye laser. A mode-locked pulse, typically from an Nd-YLF laser, is passed through an optical fiber, where nonlinear effects spread its spectrum over a broader range than the laser generates. Then that pulse is compressed in duration by special optics, and amplified in a broad-bandwidth Nd-glass laser. The resulting pulses are in the 1–3 picosecond range, and when amplified can reach the terawatt (10^{12} W) level in commercial lasers (although their brief duration means that pulse energy is on the order of only a joule). Pulses of 3×10^{12} W have been reported and work has commenced on a 10×10^{12} W laser (Perry [116]).

Table 3.9 Duration and Repetition Rates Available from Mode-Locked Pulsed Neodymium Lasers Operating at 1.06 μm

Type length	Modulation	Excitation source	Typical repetition rate	Typical pulse
Nd-glass	Mode-locked	Flashlamp	Pulse trains[a]	5–20 ps
Nd-YAG	Mode-locked	Flashlamp	Pulse trains[b]	20–200 ps
Nd-YAG	Mode-locked	Arc lamp	50–500 MHz	20–200 ps
Nd-YAG	Mode-locked and Q-switched	Arc lamp	Varies[c]	20–200 ps[d]
Nd-YLF	Mode-locked	Arc lamp	Same as YAG	about half YAG
Nd-YLF[e]	Mode-locked	Diode	160 MHz	7 ps
Nd-YLF/glass	Mode-locked and chirped pulse amplification	Lamp	0–2 Hz	1–5 ps

[a,b] Series of 30–200 ps pulses, separated by 2–10 ns, lasting duration of flashlamp pulse, on the order of a millisecond.

[c] Depends on Q-switch rate; mode-locked pulses.

[d] In pulse trains lasting duration of Q-switched pulses (100–700 ns).

[e] Laboratory results (Juhasz *et al.* [115]).

Dye lasers. Synchronous mode-locking of a dye laser to a mode-locked pump laser (either rare gas ion or frequency-doubled neodymium) can generate pulses as short as a few hundred femtoseconds. Addition of a saturable absorber can further reduce pulse length. Alternatively, a saturable absorber can by itself passively mode-lock a dye laser pumped by a cw laser. In each case, the saturable absorber allows gain only briefly while it is switching between off and on states.

The commercial dye laser which generates the shortest pulses is a colliding-pulse mode-locked ring laser. The schematic diagram is shown in Fig. 3.26. This type of ring laser does not include components to restrict laser oscillation to one direction, and produces pulses going in opposite directions around the ring. One-quarter of the way around the ring from the dye jet, the cavity includes a saturable absorber which has lowest loss when the two opposite-direction pulses pass through it at the same time (causing deeper saturation and hence lower loss). Pulse lengths can be under 100 fs in commercial versions and tens of femtoseconds in laboratory systems.

Pulses from a dye laser can be further compressed by a two-stage optical system which first spreads the pulse out in time (by passing it through a segment of optical fiber) and then compresses it spatially (by passing it between a pair of prisms or diffraction gratings). This requires extremely broad bandwidth pulses but can generate pulses as short as 6 fs (Fork *et al.* [113]).

The wavelength range availabe from passively mode-locked dye lasers has been extended with the use of different gain and absorber dyes, and subpicosecond pulses can be generated from below 500 nm to nearly 800 nm (French and Taylor [117], Smith *et al.* [118], French and Taylor [119]).

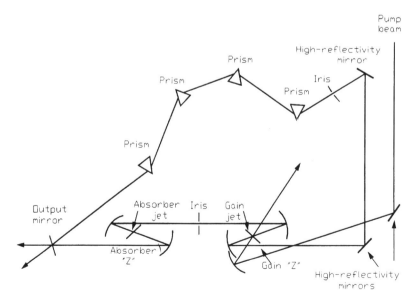

Figure 3.26 Schematic of a colliding-pulse mode-locked ring laser (Clark Instrumentation Inc.).

Ion lasers. Mode-locked ion lasers can produce trains of picosecond pulses from lasers emitting in multiple longitudinal modes. With the typical 5 GHz linewidth of ion laser lines, the resulting pulses are about 90–200 picoseconds long, produced at repetition rates in the 75–150 MHz range.

Diode lasers. Diode lasers can be mode-locked in external cavities, and experimental diode lasers have been made with internal structures that generate mode-locked pulses (Hecht [114]). For example, a passively mode-locked two-section GaAlAs multiple-quantum-well laser has generated 2.4 ps pulses at 108 GHz. However, mode-locked diode lasers are not available commercially.

Colliding-pulse mode-locking techniques have been applied to form a monolithic quantum-well semiconductor laser (Wu *et al.* [120]).

Passive mode-locking with the help of bulk semiconductors and quantum wells was applied to generate femtosecond pulses in color-center lasers (Islam *et al.* [121, 122]).

Most of the experiments on femtosecond pulses performed up to now have used dye lasers, Ti:sapphire-lasers, or color-center lasers. The spectral ranges were restricted to the regions of optimum gain of the active medium. New spectral ranges can be covered by optical mixing techniques. One example is the generation of 400 fs pulses in the mid-infrared around $\lambda = 5 \,\mu m$ by mixing the output pulses from a colliding-pulse mode dye laser at 620 nm with pulses of 700 nm in a $LiIO_3$ crystal (Elsässer Th. 1991).

During recent years, new techniques have been developed in order to generate still shorter pulses. One of these is *pulse compression*.

3.5.2 Pulse Compression

More than 30 years ago, Gires and Tournois [123] proposed that optical pulses could be shortened by adapting microwave pulse compression techniques to the visible spectrum. The shortest pulse duration that can be produced by a laser oscillator is generally limited by the wavelength of the gain medium and the group velocity dispersion into the cavity. However, if we give enough initial peak power, the technique of pulse compression can produce pulses one order of magnitude shorter. The physics involved in pulse compression, called *self-phase modulation*, also plays an important role in the majority of the ultrashort laser oscillators.

The general principles of pulse compression were first applied to radar signals and later (in the 1960s) to optical signals (Giordmaine *et al.* [124]) (see also Johnson and shank [125], for a review of pulse compression). Pulse compression technique consists of two steps: in the first step, a frequency sweep or "chirp" is impressed on the pulse; in the second step, the pulse is compressed by using a dispersive delay line.

Single-mode optical fibers were first used to create a frequency chirp (Nakatsuka and Grischkowosky [126]). The chirp can be impressed on an intense optical pulse by passing the pulse through an optical Kerr medium (Fisher *et al.* [127]). When an intense optical pulse is passed through a nonlinear medium, the refractive index, n, is modified by the electric field, E, as follows (Shank [128]):

$$n = n_o + n_2 \langle E^2 \rangle + \cdots$$

A phase change, $\delta\phi$, is impressed on the pulse:

$$\delta\phi \approx n_2 \langle E^2 \rangle (\omega z/c),$$

where ω is the frequency, z is the distance traveled in the Kerr medium, and c is the velocity of light.

As the intensity of the leading edge of the optical pulse rises rapidly, a time-varying phase or frequency sweep is impressed on the pulse carrier. Similarly, a frequency sweep in the opposite direction occurs as the intensity of the pulse falls on the trailing edge. This frequency sweep is given by

$$\delta\omega \approx (\omega z n_2/c) \, \mathrm{d}/\mathrm{d}t \langle E^2(t) \rangle$$

For a more rigorous approach to this problem, see Shank [128], pp. 26–29.

As optical fiber waveguides are a nearly ideal optical Kerr medium for pulse compression (Mollenauer *et al.* [129], Nakatsuka *et al.* [130]), one of the most common configurations of a pulse compressor uses self-phase modulation in an optical fiber to generate the chirped pulse and a pair of gratings aligned parallel to each other as the dispersive delay line.

Consider a Gaussian pulse

$$E(t)_{\mathrm{input}} = A \exp[-(t/b)^2] \exp(i\omega_0 t).$$

The effect of the pulse compressor on the Gaussian pulse has been calculated in the limited of no group dispersion velocity by Kakfa and Baer [131] in 1988. The self-phase modulation in the fiber transforms this pulse to

$$E(t)_{\mathrm{fiber}} = A \exp[-(t/b)^2] \exp i[\omega_0 t + \Theta(I)],$$

where $\Theta(I)$ is an intensity-dependent shift in the phase of the carrier. We can Fourier transform the pulse $E(t)$ to the frequency domain $E(\omega)$ in order to apply the grating operator. This operator causes a time delay, which depends on the instantaneous frequency of the pulse:

$$E(\omega)_{\mathrm{compressed}} = E(\omega)_{\mathrm{fiber}} \exp[-i\phi(\omega)].$$

The final pulse $E(t)_{\mathrm{compressed}}$ is obtained by taking the Fourier transform of $E(\omega)_{\mathrm{compressed}}$. The intensity can be calculated at any point by taking the square of this field.

The experimental set-up used for pulse compression in the femtosecond time domain is shown in Fig. 3.27.

Mollenauer *et al.* [129] were the first who experimentlaly investigated pulse compression using optical fibers as a Kerr medium. They worked on the soliton compression of optical pulses from a color center laser. As the wavelength of the optical pulses at $\lambda = 1.3\,\mu\mathrm{m}$ was in the anomalous or negative dispersion region, a separate compressor was not needed, because the dispersive properties of the fiber material self-compressed the pulse. Using this compression technique, a 7 ps optical pulse was compressed to 0.26 ps with a 100 m length of single mode fiber (Mollenauer *et al.* [132]).

Later in 1984, Mollenauer and Stolen [133] extended the ideas of fiber soliton pulse compression to form a new type of mode-locked color center laser: the soliton laser (Mitschke and Mollenauer [134]). In the soliton laser, an optical fiber of the appropriate length is added to the feedback path of a synchronously pumped color center laser. The soliton proeprties of the fiber feedback force the laser itself to

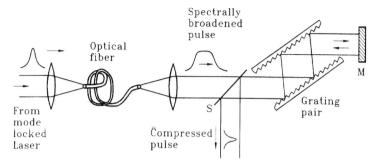

Figure 3.27 Diagram of the experimental arrangement for pulse compression in the femto-second time domain.

produce pulses of a definite shape and width. With the soliton laser pulses of less than 50 fs can be obtained.

The soliton mode-locking technique has been extended to the wavelength region where solitons cannot form. This new technique is called *additive pulse mode-locking* and has been described both theoretically (Ippen *et al.* [135]) and experimentally (Goodeberlet *et al.* [136], French *et al.* [137], Zhu *et al.* [138], Zhu and Sibbett [139]).

Nakatsuka and Grischkowosky [126] worked in the positive group velocity dispersion regime ($\lambda \leq 1.3\,\mu m$) using an optical fiber for the chirping process and anomalous dispersion from an atomic vapor as the compressor. Shank *et al.* [140] replaced the atomic vapor compressor with a grating pair compressor (Treacy [141]) and achieved pulse compression to 30 fs optical pulse width.

3.5.3 Applications of Ultrashort Optical Pulses

Measurement of Subpicosecond Dynamics

One application of optical pulses is to measure the relaxation of elementary excitations in matter, via the nonlinear optical response of the system under study. The time scale and form of the relaxation gives information concerning the microscopic physics of the dissipative forces acting on the optically active atom or molecule. The development of coherent light sources that produce pulses of a few tens of femtoseconds duration has greatly extended the possibility of realizing such measurements in liquid and condenses phases of matter, where relaxation time scales are of the order of pico- or femtoseconds (Walmsley and Karfa [108]).

There are several methods for the measurement of lifetimes [22] of excited atomic or molecular levels, including phase-shift method (Demtröder [22], Lakowvicz and Malivatt [142, 143]) and the delayed coincidence technique (O'Connor and Phillips [144]).

For measurements of very fast relaxation processes with a demanded time resolution below 10^{-10} s, the pump and probe technique is the best choice (Lauberau and Kaiser [145], Demtröder [22]). In this technique the molecules under study are excited by a fast laser pulse on the transition from the base state to the first excited state, as shown in Fig. 3.28. A probe pulse with a variable time

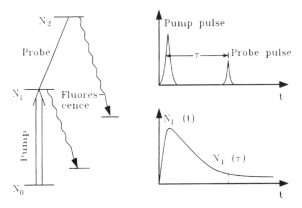

Figure 3.28   Pump and probe technique.

delay τ against the pump pulse probes the time evolution of the population density $N_1(t)$. The time resolution is only limited by the pulse width ΔT of the two pulses but not by the time constants of the detectors.

Strong-field Ultrafast Nonlinear Optics

The high peak powers obtained using short pulses have impacted significantly in the field of multiphoton atomic ionization and molecular dissociation. Three important applications are molecular photodissociation (Scherer *et al.* [145], Dantus *et al.* [146]), above-threshold ionization (Agostini *et al.* [147], Gontier and Trahin [148]) and the generation of high-order harmonics (Rhodes [149], Li *et al.* [150], Kulander and Shore [151]).

Study of Biological Processes

Ultrafast laser techniques have been applied to the study of biological processes, such as heme protein dynamics, photosynthesis, and the operation of rhodopsin and bacteriorhodopsin (Hochstrasser and Johnson [152]).

For other examples of applications see Khoo *et al.* [26]. Recent applications include the determination of the magnitude and time response of the nonlinear refractive index of transparent materials using spectral analysis after nonlinear propagation (Nibbering *et al.* [153]); the charcterization of ultrafast interactions with materials through the direct measurement of the optical phase (Clement *et al.* [154]) and a proposal for the generation of subfemtosecond VUV pulses from high-order harmonics (Schafer *et al.* [155]).

REFERENCES

1. Wilson, J. and J. F. B. Hawkes, *Optoelectronics. An Introduction*, 2nd edn, Prentice Hall International Series in Optoelectronics, Prentice Hall International (UK), 1989.
2. Loudon, R., *The Quantum Theory of Light*, 2nd edn, Oxford Science Publications, Oxford University Press, 1994.
3. Walls, D. F., *Am. J. Phys.*, **45**, 952 (1977).
4. Walls, D. F. and G. J. Milburn, *Quantum Optics*, Springer-Verlag, Berlin, 1994.

5. Hanbury-Brown, R. and R. Q. Twiss, *Nature*, **177**, 27 (1957). See also (1957) *Proc. R. Soc.*, **A242**, 300 (1957); *ibid.*, **A243**, 291 (1958).

6. Franken, P. A., A. E. Hill, C. W. Peters, and G. Weinreich, *Phys. Rev. Lett.*, **7**, 118 (1961).

7. Baldwin, G. C., *An Introduction to Nonlinear Optics*, Plenum Press, New York, 1969.

8. Yariv, A., *Quantum Electronics*, 3rd edn, J. Wiley & Sons, Inc., Chichester, 1989.

9. Kleinman, D. A., *Phys. Rev.*, **126**, 1977 (1962).

10. Kopinsky, B., W. Kaiser, and K. Drexhage, *Opt. Commun.*, **32**, 451 (1980).

10. Zyss, J., I. Ledoux, and J. F. Nicoud, "Advances in Molecular Engineering for Quadratic Nonlinear Optics," in *Molecular Nonlinear Optics. Materials, Physics and Devices*, Zyss, J. (ed.), Quantum Electronics, Principles and Applications, Academic Press, Inc., New York, 1994.

11. Bloembergen, N., *Am. J. Phys.*, **35**, 989 (1967).

12. Penzkofer, A. A., A. Laubereau, and W. Kaiser, *Quantum Electron.*, **6**, 55 (1979).

13. Mostowsky, J. and M. G. Raymer, "Quantum Statistics in Nonlinear Optics," in *Contemporary Nonlinear Optics*, Agrawal, G. P. and R. W. Boyd (eds), Quantum Electronics, Principles and Applications, Academic Press, Inc., New York, 1992.

14. Boyd, R. W., *Nonlinear Optics*, Academic Press, Inc., New York, Chapter 7, 1992.

15. Fabelinskii, L., *Molecular Scattering of Light*, Plenum Press, New York, 1968.

16. Clements, W. R. L. and B. P. Stoicheff, *Appl. Phys. Lett.*, **12**, 246 (1968).

17. Kato, Y. and H. Takuma, *J. Chem. Phys.*, **54**, 5398 (1971).

18. Woodbury, E. J. and W. K. Ng, *Proc. I.R.E.*, **50**, 2367 (1962).

19. Eckhardt, G., R. W. Hellwarth, F. J. McClung, S. E. Schwarz, D. Weiner, and E. J. Woodbury, *Phys. Rev. Lett.*, **9**, 455 (1962).

20. Kaiser, W. and M. Maier, *Stimulated Rayleigh, Brillouin and Raman Spectroscopy Laser Handbook*, Arecchi, F. T. and E. O. Schulz-Dubois (eds), North Holland, Amsterdam, 1972.

21. Raymer, M. G. and I. A. Walmsley, *Progress in Optics*, Vol. **28**, Wolf, E. (ed.), North Holland, Amsterdam, 1990.

22. Demtröder, W., *Laser Spectroscopy. Basic Concepts and Instrumentation*, 2nd (enlarged) edn, Springer-Verlag, Berlin, 1996.

23. Ederer, D. L. and J. H. McGuire (ed.), *Raman Emission by X-ray Scattering*, World Scientific Publishing Co., Inc., 1996.

24. Chu, B., *Laser Light Scattering. Basic Principles and Practice*, 2nd edn, Academic Press, Inc., New York, 1991.

25. Bloembergen, N., *Nonlinear Optics*, 4th edn, World Scientific Publishing Co. Inc., 1996.

26. Khoo, I. C., J. F. Lam, and F. Simoni (eds), *Nonlinear Optics and Optical Physics*, Series in Nonlinear Optics, Vol. 2, World Scientific Publishing Co. Inc., 1994.

27. Keller, O., *Notions and Perspectives of Nonlinear Optics*, Series in Nonlinear Optics, Vol. 3, World Scientific Publishing Co. Inc., 1996.

29. Göppert-Mayer, M., *Ann. Physik*, **9**, 273 (1931).

30. Kaiser, W. and C. G. Garret, *Phys. Rev. Lett.*, **7**, 229 (1961).

31. Hopfield, J. J., J. M. Worlock, and K. Park, *Phys. Rev. Lett.*, **11**, 414 (1963).

32. Bräunlich, P., "Multiphoton Spectroscopy," in *Progress in Atomic Spectroscopy*, Hanle, W. and H. Kleinpoppen (eds), Plenum, New York, 1978.

33. Worlock, J. M., "Two-Photon Spectroscopy," in *Laser Handbook*, Arecchi, F. T. and E. O. Schulz-Dubois (eds), North-Holland, Amsterdam, 1972.

34. Shimoda, K. (ed.), *High Resolution Laser Spectroscopy*, Springer-Verlag, Berlin, 1976.

35. Letokhov, V. S. and V. P. Chebotayev, *Nonlinear Laser Spectroscopy*, Springer-Verlag, Berlin, 1977.

36. Vasilenko, L. S., V. P. Chebotayev, and A. V. Shishaev, *JETP Lett.*, **12**, 161 (1973).

37. Cagnac, B., G. Grynberg, and F. Biraben, *J. Phys.*, **34**, 845 (1973).

38. Grynberg, G., B. Cagnac, and F. Biraben, "Multiphoton Resonant Processes in Atoms," in *Coherent Nonlinear Optics. Recent Advances*, Feld, M. S. and V. S. Letokhov (eds), *Serie Topics in Current Physcis*, Springer-Verlag, Berlin, 1980.

39. Biraben, F., B. Cagnac, and G. Grynberg, *Phys. Rev. Lett.*, **32**, 643 (1974).

40. Levenson, M. D., "High-Resolution Imaging by Wave-Front Conjugation," *Opt. Lett.*, **5**, 182 (1980).

41. Giacobino, E., F. Biraben, G. Grynberg, and B. Cagnac, *J. Physique*, **83**, 623 (1977).

42. Morellec, J., D. Normand, and G. Petite, *Phys. Rev. A*, **14**, 300 (1976).

43. Letokhov, V. S. (ed.), *Tunable Lasers and Applications*, Springer-Verlag, Berlin, 1976.

44. Yariv, A., *Optical Electronics*, 4th edn, international edition.

45. Fisher, R. A., *Optical Phase Conjugation*, Academic Press, New York, 1983.

46. Hellwarth, R. W., "Third Order Susceptibilities of Liquids and Gases," *Progr. Quant. Elec.*, **5**, 1 (1977).

47. Bloom, D. M. and G. C. Bjorklund, "Conjugate Wavefront Generation and Image Reconstruction by Four-Wave Mixing," *Appl. Phys. Lett.*, **31**, 592 (1977).

48. Levenson, M. D., K. M. Johnson, V. C. Hanchett, and K. Chaing, "Projection Photolithography by Wave-Front Conjugation," *J. Opt. Soc. Am.*, **71**, 737 (1981).

49. Hopf, F. A., "Interferometry Using Conjugate-Wave Generation," *J. Opt. Soc. Am.*, **70**, 1320 (1980).

50. Hellwarth, R. W. and P. Christensen, "Nonlinear Optical Microscope Using SHG," *Appl. Opt.*, **14**, 247 (1974).

51. Pepper, D. M., J. AuYeung, D. Fekete, and A. Yariv, "Spatial Convolution and Correlation of Optical Fields via Degenerate Four-Wave Mixing," *Opt. Lett.*, **3**, 7 (1978).

52. Wandzura, S. M., "Effects of Atomic Motion on Wavefront Conjugation by Resonantly Enhanced Degenerate Four-Wave Mixing," *Opt. Lett.*, **4**, 208 (1979).

53. Steel, D. G. and J. F. Lam, "Two-Photon Coherent-Transient Measurement of the Nonradiative Collisionless Dephasing Rate in SF6 via Doppler-Free Degenerate Four-Wave Mixing," *Phys. Rev. Lett.*, **43**, 1588 (1979).

54. Nilsen, J., N. S. Gluck, and A. Yariv, "Narrow-Band Optical Filter through Phase Conjugation by Nondegenerate Four-Wave Mixing in Sodium Vapor," *Opt. Lett.*, **6**, 380 (1981).

55. Nilsen, J. and A. Yariv, "Nearly Degenerate Four-Wave Mixing Applied to Optical Filters," *Appl. Opt.*, **18**, 143 (1979).

56. Nilsen, J. and A. Yariv, "Nondegenerate Four-Wave Mixing in a Doppler-Broadened Resonant Medium," *J. Opt. Soc. Am.*, **71**, 180 (1981).

57. Humphrey, L. M., J. P. Gordon, and P. F. Liao, "Angular Dependence of Line Shape and Strength of Degenerate Four-Wave Mixing in a Doppler-Broadened System with Optical Pumping," *Opt. Lett.*, **5**, 56 (1980).

58. Fu, T. Y. and M. Sargent III, "Theory of Two-Photon Phase Conjugation," *Opt. Lett.*, **5**, 433 (1980).

59. Saikan, S. and H. Wataka, "Configuration Dependence of Optical Filtering Characteristics in Backward Nearly Degenerate Four-Wave Mixing," *Opt. Lett.*, **6**, 281 (1981).

60. Bloch, D., E. Giacobino, and M. Ducloy, *5th Int. Conf. Laser Spectrosc.*, Jasper, Alberta, 1981.

61. Liao, P. F., N. P. Economou, and R. R. Freeman, "Two-Photon Coherent Transient Measurements of Doppler-Free Linewidths with Broadband Excitation," *Phys. Rev. Lett.*, **39**, 1473 (1977).

62. Fujita, M., H. Nakatsuka, H. Nakanishi, and M. Matsuoka, "Backward Echo in Two-Level Systems," *Phys. Rev. Lett.*, **42**, 974 (1979).

63. Bodgan, A. R., Y. Prior, and N. Bloembergen, "Pressure-Induced Degenerate Frequency Resonance in Four-Wave Light Mixing," *Opt. Lett.*, **6**, 82 (1981).

64. Raj, R. K., D. Bloch, J. J. Snyder, G. Carney, and M. Ducloy, "High-Sensitvity Nonlinear Spectroscopy Using a Frequency-Offset Pump," *Opt. Lett.*, **5**, 163, 326 (1980a).

65. Raj, R. K., D. Bloch, J. J. Snyder, G. Carney, and M. Ducloy, "High-Frequency Optically Heterodyned Saturation Spectroscopy via Resonant Degenerate Four-Wave Mixing," *Phys. Rev Lett.*, **44**, 1251 (1980b).

66. Woerdman, J. P. and M. F. H. Schuurmans, "Effect of Saturation on the Spectrum of Degenerate Four-Wave Mixing in Atomic Sodium Vapor," *Opt. Lett.*, **6**, 239 (1981).

67. Lam, J. F., D. G. Steel, R. A. McFarlane, and R. C. Lind, "Atomic Coherence Effects in Resonant Degenerate Four-Wave Mixing," *Appl. Phys. Lett.*, **38**, 977 (1981).

68. Economou, N. P. and P. F. Liao, "Magnetic-Field Quantum Beats in Two-Photon Free-Induction Decay," *Opt. Lett.*, **3**, 172 (1978).

69. Steel, D. G., J. F. Lam, and R. A. McFarlane, *5th Int. Conf. Laser Spectrosc.*, Jasper, Alberta, 1981a.

70. Yamada, K., Y. Fukuda, and T. Haski, "Time Development of the Population Grating in Zeeman Sublevels in Sodium Vapor-Detection of Zeeman Quantum Beats," *J. Phys. Soc. Jpn.*, **50**, 592 (1981).

71. Khoo, I. C., "Degenerate Four-Wave Mixing in the Nematic Phase of a Liquid Crystal," *Appl. Phys. Lett.*, **38**, 123 (1981).

72. Jain, R. K. and R. C. Lind, "Degenerate Four-Wave Mixing in Semiconductor-Doped Glass," *J. Opt. Soc. Am.*, special issue on phase conjugation (1983).

73. Bloch, D., PhD thesis, 1980.

74. Bloom, D. M., P. F. Liao, and N. P. Economou, "Observation of Amplified Reflection by Degenerate Four-Wave Mixing in Atomic Sodium Vapor," *Opt. Lett.*, **2**, 58 (1978).

75. Palmer, A. J., "Nonlinear Optics in Radiatively Cooled Vapors," *Opt. Commun.*, **30**, 104 (1979).

76. Steel, D. G., R. C. Lind, and J. F. Lam, "Degenerate Four-Wave Mixing in a Resonant Homogeneously Broadened System," *Phys. Rev. A*, **23**, 2513 (1981b).

77. Eichler, H. J., "Laser-Induced Grating Phenomena," *Opt. Acta.*, **24**, 631 (1977).

78. Hamilton, D. S., D. Heiman, J. Feinberg, and R. W. Hellwarth, "Spatial Diffusion Measurements in Impurity-Doped Solids by Degenerate Four-Wave Mixing," *Opt. Lett.*, **4**, 124 (1979).

79. Moss, S. C., J. R. Lindle, H. J. Mackey, and A. L. Smirl, "Measurement of the Diffusion Coefficient and Recombination Effects in Germanium by Diffraction from Optically-Induced Picosecond Transient Gratings," *Appl. Phys. Lett.*, **39**, 227 (1981).

80. Fu, T. Y. and M. Sargent III, "Effects of Signal Detuning on Phase Conjugation," *Opt. Lett.*, **4**, 366 (1979).

81. Harter, D. J. and R. W. Boyd, "Nearly Degenerate Four-Wave Mixing Enhanced by the ac Stark Effect," *IEEE J. Quant. Electron.*, **QE-16**, 1126 (1980).

82. Steel, D. G. and R. C. Lind, "Multiresonant Behavior in Nearly Degenerate Four-Wave Mixing: the ac Stark Effect," *Opt. Lett.*, **6**, 587 (1981).

83. Dunning, G. J. and R. C. Lind, "Demonstration of Image Transmission through Fibers by Optical Phase Conjugation," *Opt. Lett.*, 558, Elsässer, Th. and M. C. Nuss, *Opt. Lett.*, **16**, 411 (1991).

84. DeMaria, A. J., D. A. Stetser, and H. Heyman, *Appl. Phys. Lett.*, **8**, 22 (1966).

85. Shapiro, S. L. (ed.), *Ultrashort Laser Pulses*, Springer-Verlag, Berlin, 1977.

86. Von der Linde, D., A. Laubereau, and W. Kaiser, *Phys. Rev. Lett.*, **26**, 954 (1971).

87. Alfano, R. R. and S. L. Shapiro, *Phys. Rev. Lett.*, **26**, 1247 (1977).

88. Feld, M. S. and V. S. Letokhov (eds), *Coherent Nonlinear Optics, Recent Advances*, Springer-Verlag, Berlin, 1980.

89. Hargrove, L. E., R. L. Fork, and M. A. Pollack, *Appl. Phys. Lett.*, **5**, 4 (1964).

90. Harris, S. E. and R. Targ, *Appl. Phys. Lett.*, **5**, 205 (1964).

91. Smith, P. W., *Proc. IEEE*, **58**, 1342 (1970).

92. Harris, S. E., *Proc. IEEE*, **54**, 1401 (1966).

93. Kasper, A. and K. J. Witte, in *Generation, Amplification and Measurement of Ultrashort Laser Pulses III*, White, W. E. and D. H. Reitze (eds), *Proc. SPIE.*, **2701**, 2 (1996).

94. Fork, R. L., B. I. Greene, and C. V. Shank, *Appl. Phys. Lett.*, **38**, 671 (1981).

95. Mocker, H. and R. Collins, *Appl. Phys. Lett.*, **7**, 270 (1965).

96. New, G. H. C. *Opt. Commun.*, **6**, 188 (1972).

97. Weber, H., *J. Appl. Phys.*, **39**, 6041 (1968).

98. Bradley, D., G. H. C. New, and S. Caughey, *Phys. Lett.*, **30A**, 78 (1970).

99. Von der Linde, D., *IEEE J. Quant. Electron.*, **QE-8**, 328 (1972).

100. Letokhov, V., *Soviet Phys. JETP*, **28**, 562 and 1026 (1969).

101. Kaiser, W. (ed.), *Ultrashort Laser Pulses, Generation and Applications*, Topics in Applied Physics, Volume **60**, 2nd edn, Springer Verlag, Berlin, 1993.

102. Von der Linde, D., O. Bernecker, and W. Kaiser, *Opt. Commun.*, **2**, 149 (1970).

103. Zinth, W., A. Lauberau, and W. Kaiser, *Opt. Commun.*, **22**, 161 (1977).

104. Garmire, E. and A. Yariv, *IEEE J. Quant. Electron.*, **QE-3**, 222 (1967).

106. Alfano, R. R., N. Schiller, and G. Reynolds, *IEEE J. Quant. Electron.*, **QE-17**, 290 (1981).

107. Ippen, E. P., C. V. Shank, and A. Dienes, *Appl. Phys. Lett.*, **21**, 348 (1972).

108. Walmsley, I. A. and J. D. Kafka, "Ultrafast Nonlinear Optics," in *Contemporary Nonlinear Optics*, Agrawal, G. P. and R. W. Boyd (eds), Academic Press Inc., New York, 1992.

109. Garside, B. and T. Lim, *J. Appl. Phys.*, **44**, 2335 (1973).

110. Garside, B. and T. Lim, *Opt. Commun.*, **12**, 8 (1974).

111. New, G. H. C. and D. H. Rea, *J. Appl. Phys.*, **47**, 3107 (1976).

112. Haus, H., *IEEE J. Quant. Electron.*, **QE-11**, 736 (1975).

113. Fork, R. L., C. H. BritoCruz, P. C. Becker, and C. V. Shank, *Opt. Lett.*, **12**, 483 (1987).

114. Hecht, J., *The Laser Guidebook*, 2nd edn, McGraw-Hill, Inc., New York, 1992.

115. Juhasz, T., S. T. Lai, and M. A. Pessot, *Opt. Lett.*, **15**(24), 1458 (1990).

116. Perry, M. D., *Energy & Technology Review*, November (Published by Lawrence Livermore National Laboratory, Livermore, Calif.) (review), 1988, p. 9.

117. French, P. M. W. and J. R. Taylor, *Ultrafast Phenomena V*, Springer-Verlag, Berlin, 1986, p. 11.

118. Smith, K., N. Langford, W. Sibbett, and J. R. Taylor, *Opt. Lett.*, **10**, 559 (1985).

119. French, P. M. W. and J. R. Taylor, *Opt. Lett.*, **13**, 470 (1988).

120. Wu, M. C., Y. K. Chen, T. Tanbun-Ek, R. A. Logan, M. A. Chin, and G. Raybon, *Appl. Phys. Lett.*, **57**, 759 (1990).

121. Islam, M. N., E. R. Sunderman, I. Bar-Joseph, N. Sauer, and T. Y. Chang, *Appl. Phys. Lett.*, **54**, 1203 (1989a)

122. Islam, M. N., E. R. Sunderman, C. E. Soccoligh, *et al.*, *IEEE J. Quant. Electron.*, **QE-25**, 2454 (1989b).

123. Gires, F. and P. Tournois, *C. R. Acad. Sci. Paris*, **258**, 6112 (1964).

124. Giordmine, J. A., M. A. Duguay, and J. W. Hansen, *IEEE J. Quant. Electron.*, **QE-4**, 252 (1968).

125. Johnson, A. M. and C. V. Shank, in *The Supercontinuum Laser Source*, Springer-Verlag, New York, 1989, Chapter 10.

126. Nakatsuka, H. and D. Grischkowsky, *Opt. Lett.*, **6**, 13 (1981).

127. Fisher, R. A., P. L. Kelly, and T. K. Gustafson, *Appl. Phys. Lett.*, **37**, 267 (1969).

128. Shank, C. V., in *Ultrashort Laser Pulses. Generation and Applications*, Kaiser, W. (ed.), 2nd edn, Topics in Applied Physics, Volume **60**, Springer-Verlag, Berlin, 1993.

129. Mollenauer, L. F., R. H. Stolen, and J. P. Gordon, *Phys. Rev. Lett.*, **45**, 1095 (1980).

130. Natatsuka, H., D. Grischkowsky, and A. C. Balant, *Phys. Rev. Lett.*, **47**, 1910 (1981).

131. Kakfa, J. D. and T. Baer, *IEEE J. Quant. Electron.*, **QE-24**, 341 (1988).
132. Mollenauer, L. F., R. H. Stolen, J. P. Gordon, and W. J. Tomlinson, *Opt. Lett.*, **8**, 289 (1983).
133. Mollenauer, L. F. and R. H. Stolen, *Opt. Lett.*, **9**, 13 (1984).
134. Mitschke, F. M. and L. F. Mollenauer, *IEEE J. Quant. Electron.*, **QE-22**, 2242 (1986).
135. Ippen, E. P., H. A. Haus, and L. Y. Liu, *J. Opt. Soc. Am.*, **6**, 1736 (1989).
136. Goodberlet, J., J. Wang, J. G. Fujimoto, and P. A. Schultz, *Opt. Lett.*, **14**, 1125 (1989).
137. French, P. M. W., J. A. R. Williams, and J. R. Taylor, *Opt. Lett.*, **14**, 686 (1989).
138. Zhu, X., P. N. Kean, and W. Sibbett, *Opt. Lett.*, **14**, 1102 (1989).
139. Zhu, X. and W. Sibbett, *J. Opt. Soc. Am.*, **B7**, 2187 (1990).
140. Shank, C. V., R. L. Fork, R. Yen, R. H. Stolen, and W. J. Tomlinson, *Appl. Phys. Lett.*, **40**, 761 (1982).
141. Treacy, E. B., *IEEE J. Quant. Electron.*, **QE-5**, 454 (1969).
142. Lakowvicz, J. R. and B. P. Malivatt, *Biophys. Chem.*, **19**, 13 (1984).
143. Lakowvicz, J. R. and B. P. Malivatt, *Biophys. J.*, **46**, 397 (1984).
144. O'Connor, D. V. and D. Phillips, *Time Correlated Single Photon Counting*, Academic Press, New York, 1984.
145. Lauberau, A. and W. Kaiser, "Picosecond Investigations of Dynamic Processes in Polyatomic Molecules and Liquids," in *Chemical and Biochemical Applications of Lasers II*, Moore, C. B. (ed.), Academic Press, New York, 1977.
145. Scherer, N. F., J. Knee, D. Smith, and A. H. Zewail, *J. Phys. Chem.*, **89**, 5141 (1985).
146. Dantus, M., M. J. Rosker, and A. H. Zewail, *J. Chem. Phys.*, **37**, 2395 (1987).
147. Agostini, P., F. Fabre, G. Mainfray, G. Petite, and N. K. Rahman, *Phys. Rev. Lett.*, 42, 1127 (1979).
148. Gontier, Y. and M. Trahin, *J. At. Mol. Phys. B*, **13**, 4381 (1980).
149. Rhodes, C. K., *Physica Scripta*, **717**, 193 (1987).
150. Li, X., A. L'Huillier, M. Ferray, L. Lompre, G. Mainfray, and C. Manus, *J. Phys. B*, **21**, L31 (1989).
151. Kulander, K. and B. Shore, *Phys. Rev. Lett.*, **62**, 524 (1989).
152. Hochstrasser, R. M. and C. K. Johnson, "Biological Processes Studied by Ultrafast Laser Techniques," in *Ultrashort Laser Pulses, Generation and Applications*, Topics in Applied Physics, Volume **60**, 2nd edn, Springer Verlag, Berlin, 1993.
152. Hochstrasser, R. M. and C. K. Johnson, "Biological Processes Studied by Ultrasfast Laser Techniques," in *Ultrashort Laser Pulses, Generation and Applications*, Topics in Applied Physics, Volume **60**, 2nd edn, Springer-Verlag, Berlin, 1993.
153. Nibbering, E. T. J., H. R. Lange, J. F. Ripoche, C. Le Blanc, J. P. Chambaret, and M. A. Franco, in *Generation, Amplification and Measurement of Ultrashort Laser Pulses III*, White, W. E. and D. H. Reitze (eds), *Proc. SPIE*, **2701**, 222 (1996).
154. Clement, T. S., G. Rodriguez, W. M. Wood, and A. J. Taylor, in *Generation Amplification and Measurement in Ultrashort Laser Pulses III*, White, W. E. and D. H. Reitze (eds), *Proc. SPIE*, **2701**, 229 (1996).
155. Schafer, K. J., J. A. Squier, and C. P. J. Barty, in *Generation, Amplification and Measurement of Ultrashort Laser Pulses III*, White, W. E. and D. H. Reitze (eds), *Proc. SPIE*, **2701**, 248 (1996).
Cronin-Golomb, M., B. Fischer, J. Nilsen, J. O. White, and A. Yariv, "Laser with Dynamic Holographic Intracavity Distortion Correction Capability," *Appl. Phys. Lett.*, **41**(3), 219 (1982).
Cronin-Golomb, M., B. Fischer, J. O. White, and A. Yariv, "Theory and Applications of Four-Wave Mixing in Photorefractive Media," *IEEE J. Quant. Elect.*, **QE-20**(1), 12 (1984).
Grynberg, G. and B. Cagnac, *Rpt. Progr. Phys.*, **40**, 79 (1977).
Hänsch, T. W. and P. Toscheck, *Z. Phys.*, **236**, 213 (1970).

Levenson, M. D. and N. Bloembergen, *Phys. Rev. Lett.*, **32**, 645 (1974).

White, J. O. and A. Yariv, "Real-Time Image Processing via Four-Wave Mixing in a Photorefractive Medium," *Appl. Phys. Lett.*, **37**(1), 5 (1980).

Yariv, A. and D. M. Pepper, "Amplified Reflection, Phase Conjugation and Oscillation in Degenerate Four-Wave Mixing," *Opt. Lett.*, **1**, 16 (1977).

4

Prisms and Refractive Optical Components

DANIEL MALACARA
Centro de Investigaciones en Optica, León, Mexico

DUNCAN T. MOORE
University of Rochester, Rochester, New York

4.1 PRISMS

In this chapter we will describe only prisms made out of isotropic materials such as glass or plastic (Hopkins [1, 2]). There are two kinds of prisms: prisms that produce chromatic dispersion of the light beam and prisms that only deflect the light beam, changing its traveling direction and image orientation. Deflecting prisms usually make use of total internal reflection, which occurs only if the internal angle of incidence is larger than the critical angle (about $41°$) for a material whose index of refraction is 1.5. If an internal reflection is required with internal angles of incidence smaller than the critical angle, the surface has to be coated with silver or aluminum.

A deflecting prism or system of plane mirrors does not only deflect the light beam but also changes the image orientation (Berkowitz [3], Pegis and Rao [4], Walles and Hopkins [5], Walther [6]). We have four basic image transformations: a reflection about any axis at an angle θ; a rotation by an angle θ; an inversion, which is a reflection about a horizontal axis; and a reversion, which is a reflection about a vertical axis. Any mirror or prism reflection produces a reflection transformation. The axis for this operation is perpendicular to the incident and the reflected beams. Two consecutive reflections may be easily shown to be equivalent to a rotation, with the following rule:

reflection at α_1 + *reflection* at α_2 = *rotation* by $2(\alpha_2 - \alpha_1)$.

These transformations are illustrated in Fig. 4.1.

An image is said to be readable if it can be returned to its original orientation with just a rotation. It can be proved that an even number of inversions, reversions,

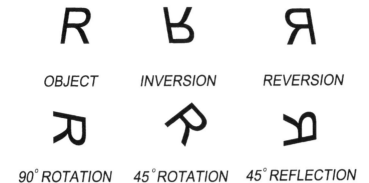

Figure 4.1   Image transformations.

and reflections is equivalent to a rotation, thus producing a readable image. On the other hand, an odd number of reflections always give a nonreadable image. Any single reflection of the observed light beam produces a reflection, reversion, or inversion of the image. An important conclusion is that an image observed through a deflecting prism is readable if it is reflected an even number of times. Two consecutive transformations may be combined to produce another transformation, as in the following examples:

Inversion + Reversion = Rotation by 180°

Inversion + Rotation by 180° = *Reversion*

Reversion + Rotation by 90° = *Reflection* at 45°

We may also show that if the axis of a reflection transformation rotates, the resulting image also rotates, in the same direction and with twice the angular speed. Thus, a practical consequence is that all inverting systems may be converted to reversing systems by a rotation by an angle of 90°.

Prisms and mirror systems with arbitrary orientations have many effects that must be taken into account when designing an optical system (Hopkins [1,2]). Among these effects we can mention the following:

(a)   A change in the direction of propagation of the light beam.
(b)   A transformation on the image orientation.
(c)   The image is displaced along the optical axis.
(d)   The finite size of their faces act as stops.
(e)   Some aberrations, mainly spherical and axial chromatic aberrations, are introduced.

The image displacement ΔL of the position on the image along the optical axis is given by

$$\Delta L = \left(n - \frac{\cos U}{\cos U'}\right)\frac{t}{n} \cong \frac{(n-1)}{n}t, \tag{4.1}$$

where U is the external angle of the meridional light ray with respect to the optical axis, U' is the internal angle of the meridional light ray with respect to the optical axis, t is the effective prism thickness, and n is its refractive index. The approximation is valid for paraxial rays.

Using this expression it is easy to conclude that the longitudinal spherical aberration (SphL) introduced to the system by the presence of the prism is given by

$$\text{SphL} = \frac{t}{n}\left(1 - \frac{\cos U}{\cos U'}\right) \cong \frac{tu^2}{2n}, \tag{4.2}$$

where u is the paraxial external angle for the meridional ray with the optical axis.

From Eq. (4.1) we can see that the axial longitudinal chromatic aberration introduced by the prism is

$$l_C - l_F = \frac{\Delta n}{n^2} t. \tag{4.3}$$

Thus, any optical design containing prisms must take into account the glass added by the presence of the prism while designing and evaluating its aberration by ray tracing. All studied effects may easily be taken into account by unfolding the prism at every reflection to find the equivalent plane-parallel glass block. By doing this we obtain what is called a tunnel diagram for the prism.

The general problem of the light beam deflection by a system of reflecting surfaces has been treated by many authors, including Pegis and Rao [4] and Walles and Hopkins [5]. The mirror system is described using an orthogonal system of coordinates x_0, y_0, z_0 in the object space, with z_0 being along the optical axis and pointing in the traveling direction of the light. For a single mirror we have the following linear transformation with a symmetrical matrix R

$$\begin{bmatrix} l' \\ m' \\ n' \end{bmatrix} = \begin{bmatrix} (1-2L^2) & (-2LM) & (-2LN) \\ (-2LM) & (1-2M^2) & (-2MN) \\ (-2LN) & (-2MN) & (1-2N^2) \end{bmatrix} \cdot \begin{bmatrix} l \\ m \\ n \end{bmatrix}, \tag{4.4}$$

where (l, m, n) and (l, m, n) are the direction cosines of the reflected and incident rays, respectively. The quantities (L, M, N) are the direction cosines of the normals to the mirror. This expression may also be written as $\bar{l}' = A\bar{l}$, where $\bar{l}'$ is the reflected unit vector and $\bar{l}$ is the incident unit vector. To find the final direction of the beam, the reflection matrices for each mirror are multiplied in the order opposite to that in which the light rays are reflected on the mirrors, as follows:

$$\bar{l}'_n = [R_N R_{N-1}, \ldots, R_2 R_1]\bar{l} \tag{4.5}$$

On the other hand, as shown by Walles and Hopkins [5], to find the image orientation, the matrices are multiplied in the same order that the light strikes the mirrors.

Now, let us consider the general case of the deflection of a light beam in two dimensions by a system of two reflecting faces with one of these faces rotated at an angle θ relative to the other, as shown in Fig. 4.2. The direction of propagation of the light beam is changed by an angle 2θ, independently of the direction of incidence with respect to the system, as long as the incident ray is in a common plane with the normals to the two reflecting surfaces.

In the triangle **ABC** we have

$$\phi = 2\alpha + 2\beta \tag{4.6}$$

and, since in the triangle **ABD**,

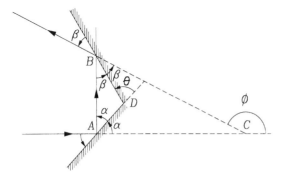

Figure 4.2 Reflection of a ray in a system of two reflecting surfaces.

$$\theta = \alpha + \beta, \tag{4.7}$$

then we can find

$$\phi = 2\theta. \tag{4.8}$$

Thus, if the angle between the two mirrors is θ, the light ray will deviate its trajectory by an angle ϕ, independently of the direction of incidence of the light ray.

Generalizing this result, we can prove that by means of three reflections in three mutually perpendicular surfaces a beam of light may also be deflected by an angle of 180°, reflecting it back along a trajectory in a parallel direction to the incident light beam. This is called a retroreflecting system.

The prisms with three mutually perpendicular reflecting surfaces is called a cube corner prism, which has been studied by Yoder [7], Chandler [8], and Eckhardt [9].

The general problem of prisms of systems of mirrors with a constant deviation independent of the prism orientation, also called stable systems, has been studied by Friedman and Schweltzer [10] and by Schweltzer *et al.* [11]. They found that in three-dimensional space this is possible only if the deflection angle is either 180° or 0°.

4.1.1 Deflecting Prisms

Besides transforming the image orientation, these prisms bend the optical axis, changing the direction of propagation of the light. There are many prisms of this kind as will now be described.

The *right angle* prism is the simplest of all prisms and in most cases it can be replaced by a flat mirror. The image produced by this prism is not readable, since there is only one reflection, as shown by Fig. 4.3(a). This prism can be modified to produce a readable image. This is accomplished by substituting the hypotenuse side by a couple of mutually perpendicular faces, forming a roof, to obtain an *Amici prism* (Fig. 4.3(b)). Rectangular as well as Amici prisms can be modified to deflect a beam of light 45° instead of 90°, as in the prisms shown in Fig. 4.4.

In the prisms previously described, the deflecting angle depends on the angle of incidence. It is possible to design a prism in which the deflecting angle is independent of the incidence angle. This is accomplished with two reflecting surfaces instead of

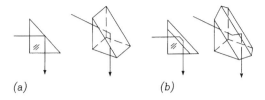

Figure 4.3 (a) Right-angle and (b) Amici prisms.

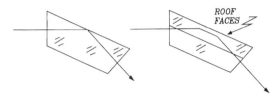

Figure 4.4 45° deflecting prism.

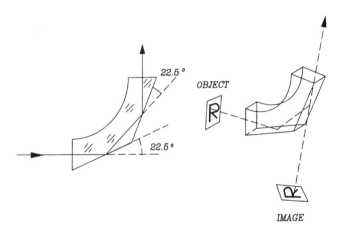

Figure 4.5 Wollaston prism.

just one, by using the property described above. The deflection angle is twice the angle between the two mirrors or reflecting surfaces.

This property is used in the *Wollaston* prism (Fig. 4.5) and in the *pentaprism* (Fig. 4.6). In the Wollaston prism both reflecting surfaces form a 45° angle and the deflecting angle is 90°. In the pentaprism both surfaces form an angle of 135° and thus the deflection angle is 270° in these two prisms; the image is readable, since there are two reflections. The pentaprism is more compact and simpler to build. Although both prisms can be modified to obtain a 45° deflection, it results in an impractical and complicated shape. To obtain a 45° deflection independent of the incidence angle, the prism in Fig. 4.7 is preferred. These prisms are used in microscopes, to produce a comfortable observing position.

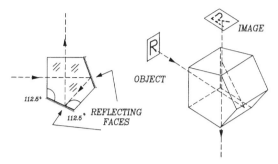

Figure 4.6 Pentaprism.

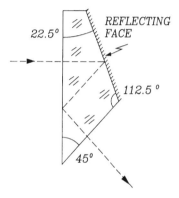

Figure 4.7 Constant 45° deflecting prism.

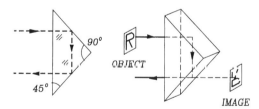

Figure 4.8 Rectangular prism in a two-dimensional retroreflecting configuration.

4.1.2   Retroreflecting Prisms

A *retroreflecting prism* is a particular case of a constant deviation prism, in which the deflecting angle is 180°. A right-angle prism can be used as a retroreflecting prism with the orientation shown in Fig. 4.8. In such a case, it is called a *Porro prism*. The Porro prism is a perfect retreflector; the incident ray is coplanar with the normals to the surfaces.

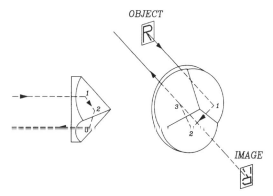

Figure 4.9 Cube corner prism.

Another perfect retroreflecting prism, made with three mutually perpendicular reflecting surfaces, is called a *cube corner prism*. This prism is shown in Fig. 4.9.

Cube corner prisms are very useful in optical experiments where retroreflection is needed. Uses for the cube corner retroreflector are found in applications where the prism can wobble or jitter or is difficult to align because it is far from the light source. Applications for this prism range from the common ones like reflectors in a car's red back light to the highly specialized ones like the reflectors placed on the surface of the moon in the year 1969.

4.1.3   Inverting and Reverting Prisms

These prisms preserve the traveling direction of the light beam, changing only the image orientation. In order to produce an image inversion or reversion, these prisms must have an odd number of reflections. We will consider only prisms that do not deflect the light beam. The simplest of these prisms has a single reflection, as shown in Fig. 4.10. This is a single rectangular prism, used in a configuration called a *dove prism*.

Although we have two refractions, there is no chromatic aberration since entrance and exiting faces act as in a plane-parallel plate. These prisms cannot be

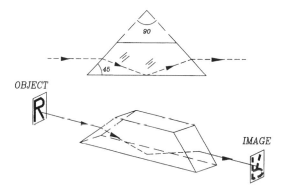

Figure 4.10   Dove prism.

used in strongly convergent or divergent beams of light because of the spherical aberration introduced, unless this aberration is compensated elsewhere in the system.

An *equilateral triangle prism* can be used as an inverting or reverting prism if used as in Fig. 4.11. On this configuration, we have two refractions and three reflections. Like the dove prism, this prism cannot be used in strongly convergent or divergent beams of light.

Figures 4.12, 4.13, and 4.14 show three reverting prisms with three internal reflections. The first one does not shift the optical axis laterally, while in the last two the optical axis is displaced. These prisms can be used in converging or diverging beams of light. The first two prisms can be made either with two glass pieces or a single piece.

The *Pechan prism*, shown in Fig. 4.15, can be used in converging or diverging pencils of light, besides being a more compact prism than the previous ones.

4.1.4 Rotating Prisms

A *half-turn rotating prism* is a prism that produces a readable image, rotated 180° as the real image produced by a convergent lens. A rotating prism can bring back the

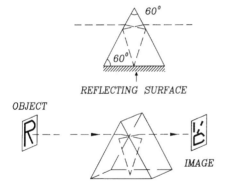

Figure 4.11   Inverting-reversing equilateral triangle prism.

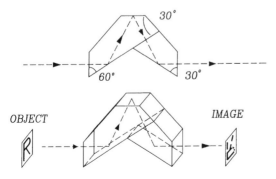

Figure 4.12   A reverting-inversing prism.

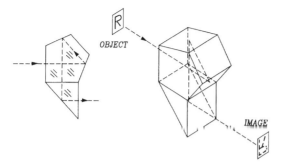

Figure 4.13   A reverting-inversing prism.

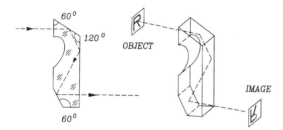

Figure 4.14   A reverting-inversing prism.

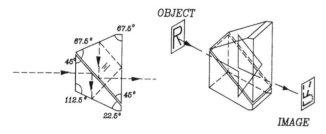

Figure 4.15   Pechan prism.

rotated image of a lens system to the original orientation of the object. These prisms are useful for monocular terrestrial telescopes and binoculars.

All reversing prisms can be converted to rotating prisms by substituting one of the reflecting surfaces by a couple of surfaces with the shape of a roof. With this substitution the *Abbe prism*, the *Leman prism*, and the *Schmidt–Pechan prism*, shown in Fig. 4.16, are obtained. This last prism is used in small hand telescopes. An advantage for this prism is that the optical axis is not laterally displaced.

A double prism commonly used in binoculars is the *Porro prism*, shown in Fig. 4.17.

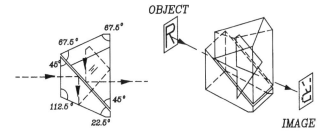

Figure 4.16 Schmidt–Pechan prism.

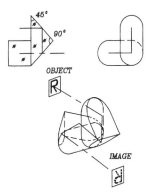

Figure 4.17 Porro prism.

4.1.5 Beamsplitting Prisms

These prisms divide the beam of light into two beams, with the same diameter as the original one, but the intensity is reduced for both beams that now travel in different directions. Beamsplitting prisms are used in amplitude division interferometers, binocular microscopes, and telescopes, where a single image must be observed simultaneously with both eyes. Basically, this prism is formed by a pair of rectangular prisms glued together to form a cube. One of the prisms has its hypotenuse face deposited with a thin reflecting film, chosen in such a way that, after cementing both prisms together, both the reflected and transmitted beam have the same intensity. Both prisms are cemented in order to avoid a total internal reflection. This prism and a variant are shown in Fig. 4.18.

4.1.6 Chromatic Dispersing Prisms

The refractive index is a function of the light wavelength and, hence, of the light color. This is the reason why chromatic dispersing prisms decompose the light into its elementary chromatic components, obtaining a rainbow, or spectrum.

Equilateral Prism. The simplest chromatic dispersing prism is the equilateral triangle prism illustrated in Fig. 4.19. This prism is usually made with flint glass, because of its large refractive index variation with the wavelength of the light.

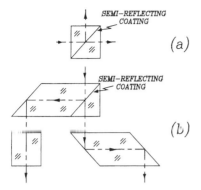

Figure 4.18   Binocular beamsplitting system.

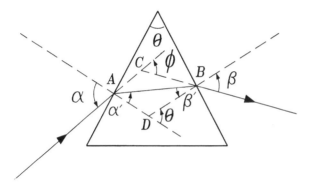

Figure 4.19   Triangular dispersing prism.

As shown in Fig. 4.19, ϕ is the deviation angle for a light ray and θ is the prism angle. We can see from this figure that

$$\phi = (\alpha - \alpha') + (\beta - \beta'); \tag{4.9}$$

also

$$\theta = \alpha' + \beta', \tag{4.10}$$

From this we obtain

$$\phi = \alpha + \beta - \theta. \tag{4.11}$$

From Snell's law, we also know that

$$\frac{\sin \alpha}{\sin \alpha'} = n \tag{4.12}$$

and

$$\frac{\sin \beta}{\sin \beta'} = n. \tag{4.13}$$

From this we conclude that the deviation angle is a function of the incidence angle α, the apex angle θ, and the refractive index n. The angle ϕ as a function of the angle α for a prism with an angle $\theta = 60°$ and $n = 1.615$ is shown in Fig. 4.20.

The deviation angle ϕ has a minimum magnitude for some value of α equal to α_m. Assuming, as we easily can, that there exists a single minimum value for ϕ, we can use the reversibility principle to see that this minimum occurs when $\alpha = \beta = \alpha_m$. It may be shown that

$$\sin \alpha_m = n \sin \theta/2. \tag{4.14}$$

Assuming that for yellow light $\alpha = \alpha_m$ in a prism with $\theta = 60°$ made from flint glass, the angle ϕ changes with the wavelength λ, as shown in Fig. 4.21.

Let us now suppose that the angle θ is small. It can be shown that the angle ϕ is independent from α and is given by

$$\phi = (n - 1)\theta. \tag{4.15}$$

Pellin-Broca or constant deviation prism. Taking as an example the prism shown in Fig. 4.22, we can see that the beam width for every color will be different and with an elliptical transverse section. The minor semi-axis for the ellipse for the refracted beam will be equal to the incident beam only when the angle α is equal to the angle β.

For precise photometric spectra measurements, it is necessary that the refracted beam width be equal to the incident beam for every wavelength. This condition is only met when the prism is rotated so that $\alpha = \beta$ (minimum deviation). Usually, these measurements are uncomfortable, since both the prism and the observer have to be rotated.

A dispersing prism that meets the previous condition with a single rotation of the prism for every measurement and does not require the observer to move is the *Pellin-Broca* [12] or *constant deviation* prism, shown in Fig. 4.23. This prism is built in a single piece of glass, but we can imagine it as the superposition of three rectangular prisms, glued together as shown in the figure. The deflecting angle ϕ is con-

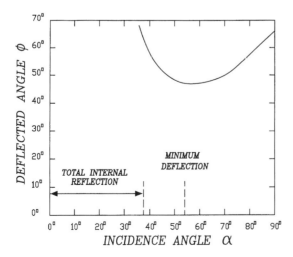

Figure 4.20 Angle of deflection vs. the angle of incidence in a dispersing prism.

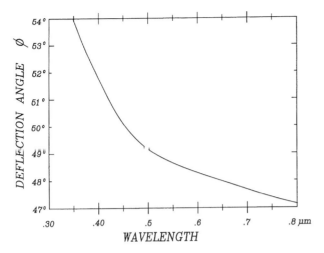

Figure 4.21 Deflection angle vs. the wavelength in a dispersing prism.

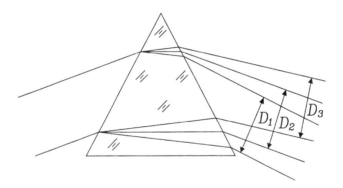

Figure 4.22 Variation in the beam deflection for different wavelengths in a triangular prism.

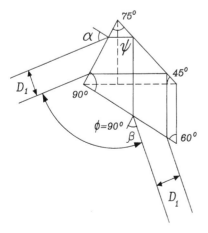

Figure 4.23 Pellin-Broca or constant deviation prism.

stant, equal to 90°. The prism is rotated to detect each wavelength. The reflecting angle must be 45° and, hence, angles α and β must be equal.

The Pellin-Broca prism has many interesting applications besides its common use as a chromatic dispersive element (Moosmüller [13]).

4.2 LENSES

In this section we will study isotropic lenses of all kinds—thin and thick, as well as Fresnel and gradient index lenses.

4.2.1 Single Thin Lenses

A lens has two spherical concave or convex surfaces. The optical axis is defined as the line that passes through the two centers of curvature. If the lens thickness is small compared with the diameter it is considered a thin lens. The focal length f is the distance from the lens to the image when the object is a point located at an infinite distance from the lens, as shown in Fig. 4.24. The object and the image have positions that are said to be conjugate to each other and related by

$$\frac{1}{f} = \frac{1}{l'} - \frac{1}{l} \tag{4.16}$$

where l is the distance from the lens to the object and l' is the distance from the lens to the image, as illustrated in Fig. 4.25. Some definitions and properties of the object and image are given in Table 4.1.

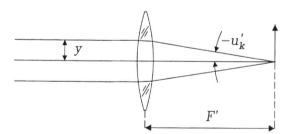

Figure 4.24 Focal length in a thin lens.

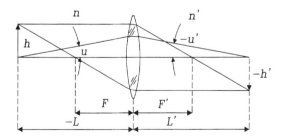

Figure 4.25 Image formation in a thin lens.

Table 4.1 Some Properties of the Object and Image formed by a Lens

		Real		Virtual
Object	$l < 0$	Left of lens	$l > 0$	Right of lens
Image	$l' > 0$	Right of lens	$l' < 0$	Left of lens

The focal length is related to the radii of the curvature r_1 and r_2 and the refractive index n by

$$\frac{1}{f} = (n - 1)\left(\frac{1}{r_1} - \frac{1}{r_2}\right), \tag{4.17}$$

where r_1 or r_2 are positive if its center of curvature is to the right of the surface and negative otherwise.

4.2.2 Thick Lenses and Systems of Lenses

A thick lens has a thickness that cannot be neglected in relation to its diameter. In a thick lens we have two focal lengths; i.e., the effective focal length F and the back focal length F_B, as shown in Fig. 4.26. The effective focal length is measured from the principal plane, which is defined as the plane where the rays would be refracted in a thin lens whose focal length is equal to the effective focal length of the thick lens. There are two principal planes: one when the image is at the right-hand side of the lens and another when the image is at the left-hand side of the lens (Fig. 4.27).

In an optical system where the object and image media are air, the intersections of the principal planes with the optical axis define the principal or nodal prints. A lens can rotate about an axis, perpendicular to the optical axis and passing through the second modal point N_2 and the image from a distant object remains stationary. This is illustrated in Fig. 4.28.

In a thick lens or system of lenses the object and image positions are given by

$$\frac{1}{f} = \frac{1}{L'} - \frac{1}{L}. \tag{4.18}$$

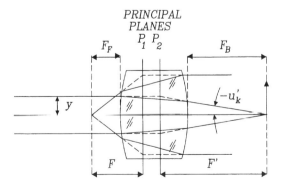

Figure 4.26 Effective and back focal lengths in a thick lens or system of lenses.

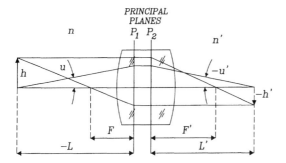

Figure 4.27 Image formation in a thick lens or system.

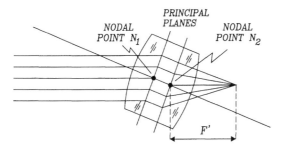

Figure 4.28 Rotating a thick lens system about the nodal point.

Thick Lenses

The effective focal length f of a thick lens is the distance from the principal plane $\mathbf{P}_2$ to the focus. It is a function of the lens thickness, and it is given by

$$\frac{1}{f} = (n - 1)\left(\frac{l'}{r_1} - \frac{l}{r_2}\right) + \frac{(n - 1)t}{nr_1r_2}. \tag{4.19}$$

The effective focal length is the same for both possible orientations of the lens. The back focal length is the distance from the vertex of the last optical surface of the system to the focus, given by

$$\frac{1}{F_B} = (n - 1)\left|\frac{1}{r_1 - t\dfrac{(n - 1)}{n}} - \frac{1}{r_2}\right|. \tag{4.20}$$

This focal length is different for both lens orientations. The separation T between the two principal planes could be called the effective thickness and it is given by

$$T = \left|1 - \frac{F(P_1 + P_2)}{n}\right|t \approx (n \sim 1)\frac{t}{n}, \tag{4.21}$$

where P_1 and P_2 are the powers of both surfaces, defined by

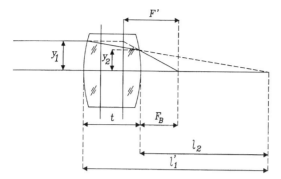

Figure 4.29 Light refraction in a thick lens.

$$P_1 = \frac{(n-1)}{r_1}; \qquad P_2 = \frac{(n-1)}{r_2}. \tag{4.22}$$

We see that the separations between the principal planes is nearly constant for any radii of curvature, and gives the effective focal length, as in Fig. 4.30.

System of Two Separated Thin Lenses

The effective focal length of a system of two separated thin lenses, as in Fig. 4.31, is given by

$$\frac{1}{f} = \frac{1}{f_1} + \frac{1}{f_2} - \frac{d}{f_1 f_2}, \tag{4.23}$$

where d is the distance between the lenses. An alternative common expression is

$$F = \frac{f_1 f_2}{f_1 + f_2 - d}. \tag{4.24}$$

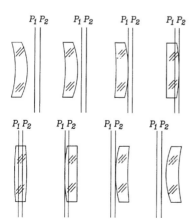

Figure 4.30 Principal planes' location in a thick lens.

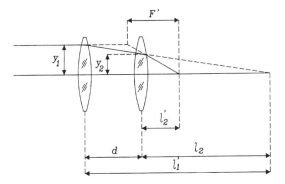

Figure 4.31 Light refraction in a system of two separated thin lenses.

As in thick lenses, the effective focal length is independent of the system orientation. The separation T between the two principal planes is

$$T = [1 - F(P_1 + P_2]d, \tag{4.25}$$

where the lens powers P_1 and P_2 are defined by

$$P_1 = \frac{1}{f_1}; \qquad P_2 = \frac{1}{f_2}. \tag{4.26}$$

4.2.3 Aspheric Lenses

Optical aberrations, such as spherical aberration, coma, and astigmatism, can seriously affect the image quality of an optical system. To eliminate the aberrations, several optical components lenses or mirrors have to be used so that the aberration of one element is compensated by the opposite sign on the others. If we do not restrict ourselves to the use of spherical surfaces, but use some aspherical surfaces, a better aberration correction can be achieved with less lenses or mirrors.

Hence, in theory the use of aspherical surfaces is ideal. They are, in general, more difficult to manufacture than spherical surfaces. The result is that they are avoided if possible; but sometimes there is no other option but to use them. Let us now consider a few examples for the use of aspherical surfaces:

> The large mirror in astronomical telescopes are either paraboloids (Cassegrainian and Newtonian telescopes) or hyperboloids of revolution (Ritchey–Chretién, as the Hubble telescope).
> Schmidt astronomical cameras have at the front a strong aspheric glass plate to produce a system free of spherical aberration. Coma and astigmatism are also zero, because of the symmetry of the system around a common center of curvature.
> A high-power single lens free of spherical aberration is frequently needed: for example, in light condensers and indirect ophthalmoscopes. Such a lens is possible only with one aspheric surface.

A rotationally symmetric aspheric optical surface is described by

Table 4.2 Values of Conic Constants for Conic Surfaces

Type of conic	Conic constant value
Hyperboloid	$K < -1$
Paraboloid	$K = -1$
Ellipse rotated about its major axis (prolate spheroid or ellipsoid)	$-1 < K < 0$
Sphere	$K = 0$
Ellipse rotated about its minor axis (oblate spheroid)	$K > 0$

$$z = \frac{cS^2}{1 + \sqrt{1 - (K + 1)c^2 S^2}} + A_4 S_4 + A_6 S^6 + A_8 S^8 + A_{10} S^{10}, \tag{4.27}$$

where K is the conic contrast, related to the eccentricity e by $K = -e^2$. The constants A_4, A_6, A_8, and A_{10} are called apsheric deformation constants.

The conic constant defines the type of conic, according to the Table 4.2. It is wasy to see that the conic constant is not defined for a flat surface.

4.2.4 Fresnel Lenses

A Fresnel lens was invented by Agoustine Fresnel and may be thought off as a thick plano-convex lens whose thickness has been reduced by breaking down the spherical face in annular concentric rings. The first Fresnel lens was employed in a lighthouse in France in 1822. The final thickness is approximately constant; therefore, the rings had different average slopes and also different widths. The widths decrease as the square of the semidiameter of the ring, as shown in Fig. 4.32.

In order to reduce the spherical aberration of a Fresnel lens the slope of each ring is controlled in order to produce an effective aspherical surface. In a properly designed and constructed lens, the on axis transverse aberration of all rings is zero; however, its phase difference is not necessarily zero, but random. Thus, they are not diffraction limited by the whole aperture but only by the central region.

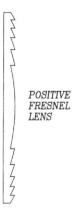

POSITIVE
FRESNEL
LENS

Figure 4.32 A Fresnel lens.

Fresnel lenses are made by hot pressing an acrylic sheet (Miller *et al.* [14], and Boettner and Barnett [15]). Thus, a nickel master must first be produced.

4.3 GRADIENT INDEX OPTICS

A gradient index (GRIN) optical component is one where the refractive index is not constant but varies within the transparent material (Machand [16]). In nature the gradient index appears on the hot air above roads, creating the familiar mirage. In optical instruments gradient index lenses are very useful as will be shown here.

The variation in the index of refraction in a lens can be:

(a) In the direction of the optical axis. This is called an axial gradient.
(b) Perpendicular to the optical axis, with the name of radial gradient.
(c) Symmetric about a point, which is the spherical gradient.

The spherical about gradient is rarely used in optical components mainly because they are difficult to fabricate and because they are equivalent to axial gradients.

Gradient index components are most often fabricated by an ion exchange process. They are made out of glass, polymers, zinc selenide/zinc sulfide and germanium (Moore [17]). Optical fibers with gradient index have been made by a chemical vapor deposition process.

4.3.1 Axial Gradient Index Lenses

Figure 4.33 shows a plano-convex lens with an axial gradient. If the lens is made with an homogeneous glass it is well known that a large amount of spherical aberration occurs, making the marginal rays converge at a point on the optical axis closer to the lens than the paraxial focus. A solution to correct this aberration is to decrease the refractive index near the edge of the lens. If a gradient is introduced, decreasing the refractive index in the direction of the optical axis, the average refractive index is lower at the edge of the lens than at its center. The refractive index along the optical axis can be represented by

$$n(z) = N_{00} + N_{01}z + N_{02}z^2 + \cdots, \tag{4.28}$$

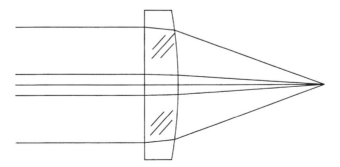

Figure 4.33 An axial gradient plano-convex lens.

where N_{00} is the refractive index at the vertex of the convex (first) lens surface. It can be shown (Moore [18]) that for a single plane convex lens, a good correction is obtained with linear approximation. The gradient should have a depth equal to the sagitta of the convex surface. With these conditions it can be proved that the refractive index change Δn along the optical axis is given by

$$\Delta n = \frac{0.15}{(f/N_{00})^2} \tag{4.29}$$

where $(f/\#)$ is the f-number of the lens. Thus, the necessary Δn for $\sin f/4$ lens is only 0.0094, while for an $f/1$ lens it is 0.15.

4.3.2 Radial Gradient Index Lenses

Radial index gradients are symmetric about the optical axis and can be represented by

$$n(r) = N_{00} + N_{10}r^2 + N_{20}r^4 + \cdots, \tag{4.30}$$

where r is the radial distance. A very thick plano-plano lens or rod, as shown in Fig. 4.34, refractor in a curved sinusoidal path all rays entering the lens. The wavelength L of this wavy path can be shown to be given by

$$L = 2\pi \left(-\frac{N_{00}}{2N_{20}}\right)^{1/2}. \tag{4.31}$$

Thus, if the rod has a length L, an object at the front surface can be imaged at the rear surface with unit magnification without any spherical aberration. This property is used in relay systems such as those in endoscopes or borescopes.

An interesting and useful application of gradient index optics rods is as an endoscopic relay, as described by Tomkinson *et al.* [19]. The great disadvantage is that endoscopes using these rods are rigid. In compensation, the great advantage is that they have a superior imaging performance in sharpness as well as in contrast.

4.4 SUMMARY

Optical elements made out of isotropic materials, homogeneous as well as inhomogeneous (gradient index), are the basic components of most optical systems.

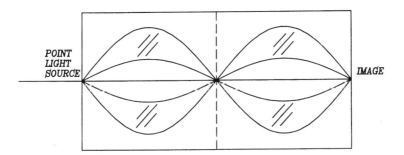

Figure 4.34 A radial gradient rod lens.

REFERENCES

1. Hopkins, R. E., "Mirror and Prism Systems," in *Military Standardization Handbook: Optical Design, MIL-HDBK* **141**, U.S. Defense Supply Agency, Washington, D.C., 1962.
2. Hopkins, R. E., "Mirror and Prism Systems," in *Applied Optics and Optical Engineering*, Kingslake, R. (ed.), **3**, Academic Press, San Diego, 1965.
3. Berkowitz, D. A., "Design of Plane Mirror Systems," *J. Opt. Soc. Am.*, **55**, 1464–1467 (1965).
4. Pegis, R. J. and M. M. Rao, "Analysis and Design of Plane Mirror Systems," *Appl. Opt.*, **2**, 1271–1274 (1963).
5. Walles, S. and R. E. Hopkins, "The Orientation of the Image Formed by a Series of Plane Mirrors," *Appl. Opt.*, **3**, 1447–1452 (1964).
6. Walther, A., "Comment on the Paper: Analysis and Design of Plane Mirror Sytems by Pegis and Rao," *Appl. Opt.*, **3**, 543–543 (1964).
7. Yoder, P. R. Jr., "Study of Light Deviation Errors in Triple Mirror and Tetrahedral Prism," *J. Opt. Soc. Am.*, **48**, 496–499 (1958).
8. Chandler, K. N., "On the Effects of Small Errors in Angles of Corner-Cube Reflectors," *J. Opt. Soc. Am.*, **50**, 203–206 (1960).
9. Eckhardt, H. D., "Simple Model of Corner Reflector Phenomena," *Appl. Opt.*, **10**, 1559–1566 (1971).
10. Friedman, Yl. and N. Schweltzer, "Classification of Stable Configurations of Plane Mirrors," *Appl. Opt.*, **37**, 7229–7234 (1998).
11. Schweltzer, N., Y. Friedman, and M. Skop, "Stability of Systems of Plane Reflecting Surfaces," *Appl. Opt.*, **37**, 5190–5192 (1998).
12. Pellin, P. and A. Broca, "Spectroscope a Déviation Fix," *J. Phys.*, **8**, 314–319 (1899).
13. Moosmüller, H., "Brewster's Angle Porro Prism: A Different Use for a Pellin-Broca Prism," *Opt. Phot. News*, **9**, 8140–8142 (1998).
14. Miller, O. E., J. H. McLeod, and W. T. Sherwood, "Thin Sheet Plastic Fresnel Lenses of High Aperture," *I. Opt. Soc. Am.*, **41**, 807–815 (1951).
15. Boettner, E. A. and N. E. Barnett, "Design and Construction of Fresnel Optics for Photoelectric Receivers," *J. Opt. Soc. Am.*, **41**, 849–857 (1951).
16. Marchand, E. W., *Gradient Index Optics*, Academic Press, New York, 1978.
17. Moore, D. T., "Gradient-Index Materials," in *CRC Handbook of Laser Science and Technology, Supplement 1: Lasers*, Weber, M. J. (ed.), CRC Press, New York, 1995, pp. 499–505.
18. Moore, D. T., "Design of a Single Element Gradient-Index Collimator," *J. Opt. Soc. Am.*, **67**, 1137–1137 (1977).
19. Tonkinson, T. H., J. L. Bentley, M. K. Crawford, C. J. Harkrider, D. T. Moore, and J. L. Ronke, "Rigid Endoscopic Relay Systems: A Comparative Study," *Appl. Opt.*, **35**, 6674–6683 (1996).

5

Reflective Optical Components

DANIEL MALACARA

Centro de Investigaciones en Optica, Leon, Mexico

5.1 MIRRORS

Concave or convex mirrors as well as lenses can be used in optical systems to form images. Like a lens, a mirror can introduce all five Seidel monochromatic aberrations. Only the two chromatic aberrations are not introduced. The possible variables are the mirror shape, aperture, pupil position, and image and object positions. In this chapter the main properties of these mirrors are described.

5.1.1 Spherical Mirror

Let us begin with the study of spherical mirrors (Malacara and Malacara [1]). A spherical mirror is free of spherical aberration only when the object and the image are both at the center of curvature or at the vertex of the mirror. If the object is at infinity, as shown in Fig. 5.1 ($l' = -f = r/2$), the image is at the focus and the longitudinal spherical aberration has the value

$$\text{SphL} = -\frac{y^2}{4r} = \frac{D^2}{32f} = -\frac{z}{2} \tag{5.1}$$

and, the transverse aberration is

$$\text{SphL} = -\frac{y^3}{2r^2} = \frac{D^3}{64f^2}, \tag{5.2}$$

where D is the diameter of the mirror, z is the sagitta of the surface and r is the radius of curvature. By integration of this expression, the wavefront aberration is given by

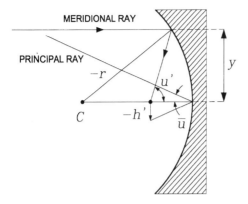

Figure 5.1   First-order parameters in a concave mirror.

$$W(y) = \frac{y^4}{4r^3}.$$ \hfill (5.3)

This wavefront distortion is twice the sagitta difference between a sphere and a paraboloid. This is to be expected, since the paraboloid is free of spherical aberration when the object is at infinity.

The value of the Petzval curvature, i.e., the field curvature in the absence of astigmatism, is

$$\text{Ptz} = \frac{h'^2}{r} = -\frac{h'^2}{2f}.$$ \hfill (5.4)

Thus, the Petzval surface is concentric with the mirror. In a spherical mirror, when the object and the image are at the center of curvature, the spherical aberration is zero. If the stop is at the center of curvature, the coma and astigmatism aberrations are also zero. Then, only the Petzval curvature exists and coincides with the field curvature.

The value of the sagittal coma aberration is a function of the stop position, given by

$$\text{Coma}_S = -\frac{y^2 h'^2 (\bar{l} - r)}{r^3}$$ \hfill (5.5)

where $\bar{l}$ is the distance from the mirror to the stop and h' is the image height. If the object is located at infinity, the spherical mirror would be free of coma with the stop at the mirror, only if the principal surface (the mirror surface itself) is a sphere with center of curvature at the focus, but obviously this is not the case. Then, when the stop is at the mirror ($\bar{l} = 0$), the value of the sagittal coma is

$$\text{Coma}_S = \frac{D^2 h'}{16f^2};$$ \hfill (5.6)

when the stop is at the center of curvature ($\bar{l} = r$), the value of the sagittal coma becomes zero, as mentioned before.

The longitudinal sagittal astigmatism in the spherical mirror is given by

$$AstL_S = -\frac{h'^2}{r}\left(\frac{\bar{l}-r}{r}\right)^2,$$ (5.7)

which may also be written as

$$AstL_S = -\left(\frac{\bar{l}-r}{r}\right)^2 Ptz.$$ (5.8)

As pointed out before, we can see that $AstL_S = Ptz$ when $\bar{l} = 0$ and the field is curved with the Petzval curvature, as shown in Fig. 5.2(a). In general it can be shown that the sagitta represented here by Best, the surface of best definition located between the sagittal and the tangential surfaces, is given by

$$Best = \left[1 - 2\left(\frac{\bar{l}-r}{r}\right)^2\right] \cong Ptz.$$ (5.9)

If the stop is located at

$$\frac{\bar{l}}{r} = \pm\frac{1}{\sqrt{3}} + 1 = 0.42 \text{ and } 1.58,$$ (5.10)

the tangential surface is flat, as shown in Fig. 5.2(b). When the stop is

$$\frac{\bar{l}}{r} = \pm\frac{1}{\sqrt{2}} + 1 = 0.29 \text{ and } 1.707,$$ (5.11)

the surface of best definition is a plane, as in Fig. 5.2(c). When the stop is at the mirror, as in Fig. 5.2(d), the sagittal surface is flat.

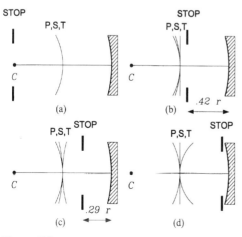

Figure 5.2 Anastigmatic surface for a spherical mirror with the stop at four different positions.

5.1.2 Paraboloidal Mirrors

A paraboloidal mirror is an aspherical surface whose conic constant is $K = -1$, as described in Chapter 4. The spherical aberration is absent if the object is located at infinity. However, if the object is at the center of curvature the spherical aberration appears. Now, let us examine each of the primary aberrations in a paraboloidal mirror. The exact expression for the longitudinal aberration of the normals to the mirror is given by

$$\text{SphL}_{\text{normals}} = f \tan^2 \varphi, \tag{5.12}$$

as illustrated in Fig. 5.3. When the object is at the center of curvature, the spherical aberration of the paraboloid is approximately twice the aberration of the normals. Thus, we may write

$$\text{SphL} = \frac{y^2}{r} = -\frac{D^2}{8f}. \tag{5.13}$$

For the spherical aberration of a spherical mirror with the object at infinity, the absolute values are different by a factor of four and are opposite in sign. In other words, the wavefront aberrations must have opposite signs and the same absolute values.

As for the sphere, if the object is located at infinity, the paraboloid would be free of coma with the stop at the mirror only if the principal surface is spherical with center of curvature at the focus; but again, this is not the case. The principal surface is the parabolic surface itself. Thus, the value of OSC (Malacara [1]) is given by

$$OSC = \frac{f_M}{f} - 1, \tag{5.14}$$

where f_M and f are the marginal and paraxial focal lengths, measured along the reflected rays, as shown in Fig. 5.3. For a paraboloid, we may show that

$$f_M = f - z, \tag{5.15}$$

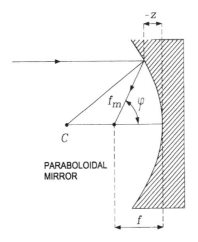

Figure 5.3 A paraboloidal mirror.

with the sagitta z given by

$$z = -\frac{D^2}{16f}.$$ (5.16)

This, the value of the sagittal coma can be shown to be

$$\text{Coma}_S = OSC \cdot h' = -\frac{zh'}{f} = \frac{D^2 h'}{16f^3}.$$ (5.17)

The coma aberrations are the same for spherical and parabolic mirrors when they stop at the mirror.

The astigmatism for a paraboloid and a sphere are related by

$$\text{AstL}_{S\,\text{parab}} = \text{AstL}_{S\,\text{sphere}} \left[1 - \left(\frac{i\,\bar{y}}{\bar{i}\,y} \right)^2 \right].$$ (5.18)

The astigmatism when the stop is at the mirror is equal to the astigmatism of a spherical mirror, which can also be written as

$$\text{AstL}_{S\,\text{total}} = \left[\frac{(\bar{l} - r)^2 - \bar{l}^2}{r^2} \right] \text{Ptz}.$$ (5.19)

We see that the surface of best definition is flat when $\bar{l}/r = 0.25$ and not 0.29, as in the case of the spherical mirror.

5.1.3 Ellipsoidal Mirrors

Ellipsoidal mirrors whose surface is generated by rotating an ellipse about its major axis produce an image free of spherical aberration when the image and object are located at each of their foci (Nielsen [2]). As described in Chapter 4, the conic constant for these ellipsoidal surfaces is in the range $-1 < K < 0$. If the major and minor semiaxes are a and b, respectively, as in Fig. 5.4, the separation between

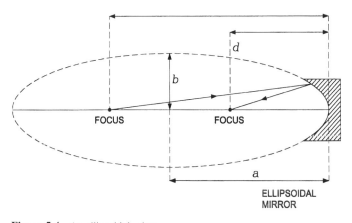

Figure 5.4 An ellipsoidal mirror.

the two foci of the ellipsoid is $a^2 - b^2$ and $K = -(1 - b^2/a^2)$. Thus, these two foci are at distances d_1 and d_2 from the vertex given by

$$d_1, d_2 = a \pm (a^2 - b^2)^{1/2} \tag{5.20}$$

which, in terms of the conic constant K, can be shown to be

$$d_1, d_2 = \frac{r}{(K+1)}(1 \pm \sqrt{-K}), \tag{5.21}$$

where r is the radius of curvature at the vertex of the ellipsoidal surface.

5.2 SOME MIRROR SYSTEMS

There are some optical systems that use only mirrors (Churilovskii and Goldis [3], Erdos [4]), such as the astronomical mirrors studied in Chapter 9 in this book. In this section, some other interesting systems are described.

5.2.1 Manguin Mirror

This mirror was invented in 1876 in France by Manguin as an alternative for the parabolic mirror used in search lights. It is made with a meniscus negative lens coated with a reflective film on the convex surface, as shown in Fig. 5.5. The radius of curvature of the concave refracting surface and the thickness are the variables used to correct the spherical aberration. A bonus advantage is that the coma aberration is less than half that of a parabolic mirror. This system has two more advantages; first, the surfaces are spherical not parabolic, making construction a lot easier; secondly the reflecting coating is on the back surface, avoiding air exposure and oxidation of the metal. A Manguin mirror made with crown glass (BK7) can be obtained with the following formulas

$$r_1 = 0.1540T + 1.0079F \tag{5.22}$$

and

$$r_2 = 0.8690 + 1.4977F, \tag{5.23}$$

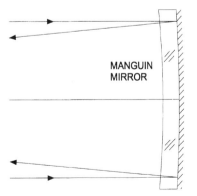

Figure 5.5 The Manguin mirror.

where T is the thickness, F is the effective focal length of the mirror, r_1 is the radius of curvature of this reflecting surface, and r_2 is the radius of curvature of the front surface.

5.2.2 Dyson System

Unit magnification systems are very useful for copying small structures or drawings, a typical application is in photolithography in the electronics industry. In general, these systems are symmetric, thus automatically eliminating coma, distortion, and magnification chromatic aberration. An example of these systems, illustrated in Fig. 5.6, was designed by Dyson [5].

The system is concentric. A marginal meridional ray on axis, leaving from the center of curvature, would not be refracted. Thus, spherical aberration and axial chromatic aberration are absent. The radius of curvature r_L of the lens is

$$r_L = \left(\frac{n-1}{n}\right) r_M, \qquad (5.24)$$

where r_M is the radius of curvature of the mirror, in order to make the Petzval sum zero. The primary astigmatism is also zero, since the spherical aberration contribution of both surfaces are zero. However, the high-order astigmatism appears not very far from the optical axis. Thus, all primary aberrations are corrected in this system.

It may be noted that since the principal ray is parallel to the optical axis in the object as well as in the image media, the system is both frontal and back telecentric.

5.2.3 Offner System

The Offner [6] system is another 1:1 magnification system, formed only by mirrors, as shown in Fig. 5.7. The system is concentric and with zero Petzval sum, as in the case of the Dyson system. This system may be also corrected for all primary aberrations, but since higher-order astigmatism is large in this configuration, actual Offner systems depart from this configuration. Primary and high-order astigmatism are balanced at a field zone to form a well-corrected ring where the sagittal and the tangential surfaces intersect.

5.3 REFLECTIVE COATINGS

Reflecting coatings can be of many different types (Hass [7]). The best option also depends on a large number of factors, such as the type of application, wavelength range, the environment around the optical system, cost limitations, etc. Thin films are described in detail in Chapter 23. However, here we describe some of the many

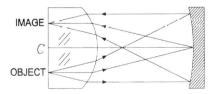

Figure 5.6 The Dyson system.

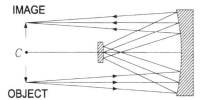

IMAGE

C

OBJECT

Figure 5.7 The Offner system.

thin film coatings used in reflective optical systems. The experimental procedures used to produce them are not described here.

5.3.1 Silver

An advantage of silver films is that they can be deposited by an inexpensive chemical procedure, without any vacuum deposition chamber (Strong [8]). Silver films have a good reflectance in the whole visible spectrum and in the near-ultraviolet, as illustrated in Fig. 5.8. A serious disadvantage is that silver oxidates and becomes black because of atmospheric humidity. This problem is solved when the silver is deposited on the back surface of the glass and then protected with a paint layer. Most non-optical common mirrors are made in this manner.

5.3.2 Aluminum

Aluminum coating has to be deposited in a vacuum deposition chamber by evaporation; this procedure makes these mirrors more expensive than the ones made with silver. The reflectivity of aluminum is worse than that of silver in the ultraviolet region, but better in the infrared region, as shown in Fig. 5.8.

The most important characteristic of aluminum is that although it oxidates quite rapidly, this oxide is transparent: it can be stained and easily scratched with dust. So, it is generally protected with a hard coating. Most optical mirrors are covered and protected with a layer of half a wave thick magnesium fluoride

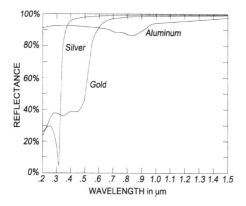

Figure 5.8 Reflectivity for aluminum, silver, and gold at different wavelengths.

(MgF) or silicon monoxide (SiO). The great advantage is that a protected mirror can be easily cleaned with isopropyl alcohol or acetone without damaging the mirror. Multilayer protective coatings are also used to enhance the reflectivity in the visible or in the ultraviolet regions.

5.3.3 Gold

Gold, like aluminum, has to be deposited by vacuum evaporation. As expected, it is more expensive than aluminum, but it has the great advantage of a good reflectivity in the infrared region, as shown in Fig. 5.8. Optical elements for use in the infrared frequently use gold. Gold mirrors can also be protected with a silicon monoxide overcoating.

5.3.4 Dielectric Films

Dielectric multilayer films are deposited by vacuum evaporation with a procedure that permits deposition of different kinds of materials with a precisely controlled thickness (see Chapter 23). Mirrors made in this manner are expensive. However, the great advantage is that the reflectivity at the desired wavelength range can be obtained as required. A common example are the mirrors used in gas lasers, with nearly 100% reflectivity at the proper wavelength. Unlike metal mirrors, they can reflect or transmit all of the incident light without absorbing anything; also, they can reflect some colors and transmit others.

An interesting application of these films is in the concave mirrors used in projector lamps, where the reflectivity has to be as high as possible in the visible region but as low as possible in the infrared region to avoid heating the film.

5.4 BEAMSPLITTERS

Beamsplitters are partially reflecting and transmitting mirrors. They can be made with an extremely thin film of metal or with a stack of thin films.

The reflectance and transmittance of thin metal films, as shown in Fig. 5.7, is almost flat over the visible wavelength range; therefore, the color of the reflected and transmitted light beam preserve their color. On the other hand, in a dielectric thin films beamsplitter the reflected and transmitted beams have complementary colors. They preserve their original colors only if the reflectivity is constant for all the visible spectrum. Another important difference is that metal films absorb a small part of the luminous energy.

Polarizing beamsplitters reflect a large percentage of the S polarized beam while transmitting the P polarized beam. Thus, if the incident beam is unpolarized, the reflected and the transmitted beams are linearly polarized, in orthogonal directions.

REFERENCES

1. Malacara, D. and Z. Malacara, *Handbook of Lens Design*, Marcel Dekker, New York, 1998.
2. Nielsen, J. R., "Aberrations in Ellipsoidal Mirrors Used in Infrared," *J. Opt. Soc. Am.*, **39**, 59–63 (1949).

3. Churilovskii, V. N. and K. I. Goldis, "An Apochromatic Catadioptric System Equivalent to a Parabolic Mirror," *Appl. Opt.*, **3**, 843–846 (1964).
4. Erdos, P., "Mirror Anastigmat with Two Concentric Spherical Surfaces," *J. Opt. Soc. Am.*, **49**, 877–886 (1956).
5. Dyson, J., "Unit Magnification System without Seidel Aberrations," *J. Opt. Soc. Am.*, **49**, 713 (1959).
6. Offner, A., "New Concepts in Projection Mask Aligners," *Opt. Eng.*, **14**, 131 (1975).
7. Hass, G., "Mirror Coatings," in *Applied Optics and Optical Engineering*, Academic Press, New York, 1965.
8. Strong, J., *Procedures in Experimental Physics*, Prentice-Hall, Englewood Cliffs, New Jersey, 1938.

6

Diffractive Optical Components

CRISTINA SOLANO

Centro de Investigaciones en Optica, Leon, Mexico

6.1 INTRODUCTION

An optical system can be thought of as a device that transforms input wavefronts into output wavefronts. The class of transformations that link the output to the input wavefronts in refractive–reflective optical systems is quite limited. For example it is not possible to design a refractive–reflective optical system for which the resultant image is three dimensional when the input wavefront is a collimated beam. This particular input–output transformation can be realized by using a hologram.

Holograms can also be made to have predefined optical transfer functions, in which case they are referred to as holographic optical elements (HOEs). The optical transfer function of an HOE is based on diffraction expressed by the diffraction equation,

$$\lambda = d(\sin\theta_1 + \sin\theta_2),\tag{6.1}$$

where θ_1 and θ_2 are the angles of incidence and diffraction, respectively, and d is the period of the diffraction grating.

The wavelength dependence of this grating will depend on its structure, as shown in Fig. 6.1. Consequently, HOEs are useful and sometimes indispensable components of optical systems when the source is monochromatic or when a wavelength-dependent system is desired.

The fundamental difference between a general hologram and an HOE is that the first one forms an image of some extended object that is recorded in holographic material in the form of a holographic interference pattern combining an object beam and a reference beam (Jannson *et al.*, 1994). This image is reconstructed using a beam with similar properties to the reference beam. On the other hand, an HOE is

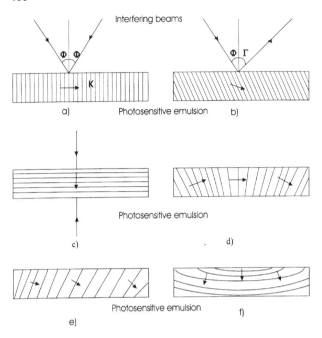

Figure 6.1 Some different fringes forms of the holographic optical elements (Jannson *et al.*, 1994).

recorded using a simple wavefront such as a spherical wave, a Gaussian, plane, elliptical, or any other elementary wave that satisfies, at least approximately, the eikonal equation (Sommerfeld, 1962).

In principle, given an arbitrary input wavefront, an HOE can be designed to transform into a desired output wavefront. In such a situation, the required HOE recording beams would most likely be produced by computer-generated holograms in conjunction with conventional refractive and reflective optical elements. Using the recent advances in micromachining and microelectronics techniques the fabrication of such structures with micron and submicron minimal details has become practical. Correspondingly, nonspectroscopic applications of gratings as diffractive optical elements (DOEs) have appeared. DOEs can not only replace reflective and refractive elements but in many cases can perform functions not even possible with conventional optics alone. The power of DOEs lies in their ability to synthesize arbitrary phase functions. They are used as components in novel devices, which were once considered too impractical but are now designed and fabricated. Complex microscopic patterns can be generated on the surface of many optical materials to improve the optical performance of existing designs as well as to make possible entirely new ones.

6.1.1 Holographic and Diffractive Optical Elements

In general, the term diffractive elements (DE) (or diffractive optical elements) refers to those that are based on the utilization of the wave nature of light. The HOEs and

DOEs are based on grating composition (Jannson *et al.*, 1994). Both are lens-like; and its main difference is mostly in their fabrication. An HOE is produced by holographic recording, while a DOE is usually fabricated by a photolithographic method.

For both kinds of diffractive elements (HOEs and DOEs) the grating effect is dominant and defines their function and limitations. In general, grating dispersion is much stronger than prism dispersion. Thus, chromatic (wavelength) dispersion strongly influences (and limits) the imaging properties of HOEs and DOEs. Moreover, almost all applications of HOEs and DOEs are the result of effectively controlling chromatic dispersion and chromatic aberrations.

Although DOEs typically have a periodic (grating) structure that is always located at their surface as a surface relief pattern, HOEs also have a periodic structure that is located either on the surface or within the volume of the holographic material (Fig. 6.2).

This categorization can be divided into several subsections:

- diffractive lenses: elements that perform functions similar to conventional refractive lenses, e.g., they form images
- kinoforms: diffractive elements whose phase modulation is introduced by a surface relief structure
- binary optics: kinoforms produced by lithography
- diffractive phase elements (DPEs): diffractive elements that introduce phase change.

Regardless of the name or method of fabrication, diffractive optics can be understood with just a few basic tools. The properties of diffractive optics that are shared with conventional elements (focal length, chromatic dispersion, aberration

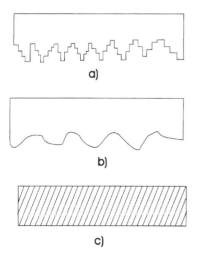

Figure 6.2 Illustration of diffractive optical elements (DOEs) and holographic optical elements (HOEs). (a) Multilevel DOE. (b) Surface (embossed) HOE. (c) Volume (Bragg) HOE; phase or amplitude modulation in the bulk of the material (Jannson *et al.*, 1994).

contributions, etc.) do not depend on the specific type of diffractive element. Given a phase function or, equivalently, a grating spatial frequency distribution, the influence of the diffractive element or an incident ray for a specified diffraction order is found via the grating equation. The specific type involved (kinoform, binary lens, HOE, etc.) only influences the diffraction efficiency. Because of this reason, though-out this chapter the general term of diffractive elements (DEs) will be used.

One factor that has stimulated much of the recent interest in diffractive optics has been the increased optical performance of such optical elements. This allows the fabrication of optical elements that are smaller, lighter, and cheaper to fabricate, are more rugged and have superior performance than the coventional optical components they often replace. In addition, the design capabilities for binary optics now available can make possible the design and manufacture of components having optical properties never before produced.

6.2 DESIGN ISSUES

Diffractive optical elements introduce a controlled phase delay into an optical wave-front by changing the optical path length, either by variating the surface height, or by modulating the refractive index as a function of position. Because of their unique characteristics, diffractive optical elements (DEs) present a new and very useful tool in optical design.

Innovative diffractive components have been applied in a number of new systems, such as laser-based systems, in which unusual features of these elements are utilized. Due to the great number of parameters that can be used to define a diffractive component, the efficient handling of this degree of freedom presents a technical challenge.

Probably the best way to describe the design procedure was presented by Joseph Mait (1995) who divided it into three basic stages: analysis, synthesis, and implementation.

(a) Analysis. There are two important points. First, it is necessary to understand the physics of the image formation required by the proposed diffractive element (DE) that will determine the method to be used (Fig. 6.3). Among the methods available are scalar or vector diffraction theory, geometrical, Fresnel and Fourier transform, convolution, correlation, and rigorous theory. The choice of method depends on the required diffraction properties of the DE and will affect the complexity of the design algorithm and the definition and value of the measured performance.

Secondly, it is important to take into account the fabrication model (i.e., the degree of linearity of the material and the possible errors in its fabrication) in which the data generated by computer will be recorded.

(b) Synthesis. Identifing the appropriate scheme that represents mathematically the underlying physical problem, the appropriate optimization techniques, design metrics (i.e., diffraction efficiency, quantization or reconstruction error, modulation transfer function, aberrations, undesired light, glare, etc.) and their degree of free-dom.

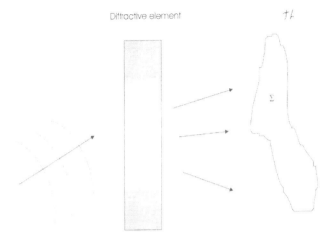

Diffractive element

Σ

Incident wave

Figure 6.3 Basic geometry of diffractive optics (transmission mode). The incident wave is diffracted by the diffractive element (DE). The resultant amplitude distribution Σ must satisfy the requirements of the specific design.

Among the proposed procedures for optimization of the problem are the quantization, steepest descent, simulated annealing, and iterative Fourier algorithm, which in general can be classified as direct and indirect approaches.

In the direct method the performance of the primary metric is optimized. This method, although simple, can be time consuming. For indirect approaches, the optimization of an alternate metric for solving the design problem is necessary.

(c) Implementation. This step considers the execution of the design, and the fabrication and testing of the resulting diffracting element. It is an iterative procedure. The DE performance can be improved by using the data collected during the testing, or by introducing more data on the material response into the design.

6.2.1 Modeling Theories

The theoretical basis for modeling diffractive optics can be divided into three regimes: geometrical optics, scalar, and vector diffraction. The main features of each regime are described below.

6.2.1.1 Geometrical Optics

In this case, rays are used to describe the propagation of the diffracted wavefront but neglecting its amplitude variations; i.e., geometrical optics can predict the direction of diffraction orders but not their relative intensities. Despite these limitations, if a diffractive element is used in an application which is normally designed by tracing rays, then ray tracing coupled with simple efficiency estimates will usually be sufficient. The majority of these applications are conventional systems (imaging systems, collimating or focusing optics, laser relay systems, etc.). In such systems the diffractive element corrects residual aberrations (chromatic, monochromatic, or thermal) or replaces a conventional optic (e.g., a Fresnel lens replacing a refractive lens). In most of these cases, the diffractive optic is blazed for a single order and can be though as a generalized grating – one in which the period varies across the element.

Two methods are used: the grating equation and the Sweat model. In the grating equation (Welford, 1986), the diffraction of an incident wave is calculated on the grating point by point, and the propagation of the diffracted wave through the rest of the optical system. This diffracted beam is calculated with the grating equation, which is the diffractive counterpart to Snell's law or the reflection law. Deviation of the observed wavefront from a spherical shape constitutes the system's aberrations. This approach is very useful for the analysis of these aberrations. Usually a thin hologram has been assumed, with an amplitude transmission during reconstruction that is proportional to the intensity during the recording process. Other assumptions lead to additional diffraction orders and different amplitude distributions. For a given diffraction order, the imaging characteristics are the same, regardless of the diffracting structure assumed.

Sweat (1977; 1979) showed that a diffracting element is mathematically equivalent to a thin refracting lens in which the refractive index approaches infinity and the lens' curvatures converge to the substrate of the diffracting lens. Snell's law is then used to trace rays through this refractive equivalent of the diffractive element. As the index approaches infinity, the Sweat model approaches the grating equation. Almost all commercially available ray tracing software can handle either of these models.

6.2.1.2 Scalar Diffraction Theory

This approach must be used when the variations amplitude are not negligible, if the value of the system's diffraction efficiency cannot be separated from the rest of the design, or if the diffraction element cannot be approximated by a generalized grating. There are two fundamental assumptions usually involved in the application of scalar diffraction theory to the design and analysis of DEs. The first is that the optical field just past the DE can be described by a simple transmission function. In this case, the thin element approximation and, often, the paraxial approximation are used, together with the treatment of the electromagnetic wave as a scalar phenomenon. This ensures that the design problems are comparatively simple to solve. The second assumption is the choice of propagation method to transform this field to the plane of interest. On this point, it is possible to use the mathematical formulation of Fourier optics when the image is in the far field elements or Fresnel optics when it is in the near field.

Scalar diffraction theory is a simple tool for the analysis of diffractive optical elements. In this case, the diffractive optic is modeled as an infinitely thin phase plate and the light's propagation is calculated using the appropriate scalar diffraction theory.

Using the scalar approach, the design of optical diffracting elements for parallel optical processing systems has become possible. There are now highly efficient space invariant spot array generators that provide a signal in the Fourier transform plane of a lens, and space variant lens arrays which provide a signal in the device focal plane. Diffractive beam samplers and beam shapers are also in use. Some detailed designs can be found in the work of Mait (1995).

Common optimization techniques include phase retrieval (Fienup, 1981), non-linear optimization, and search methods or linear systems analysis.

6.2.1.2.1 The Resonant Domain

In recent years more attention has been paid to elements that push against the validity of the scalar approximations or violate them completely. These work in the so-called resonance domain, which is characterized by diffracting elements' structure, with size w, that lie within the illuminating wavelength (λ) range

$$\frac{\lambda}{10} < w < 10\lambda. \tag{6.2}$$

The strongest resonance effects are produced when the size of the elemental features approach the wavelength of the illuminating light. In this case, polarization effects and multiple scattering must be taken into account by using the electromagnetic theory rigorously. Such DE when working at optical wavelengths can be fabricated using techniques such as direct-write electron beam lithography.

In the resonant domain the diffraction efficiency changes significantly with a slight change of grating parameters such as period, depth, and refractive index.

This principle can be used to form different DEs such as a diffractive optic beam deflector for high-power laser systems, where laser-induced damage limits the usefulness of conventional elements. Reflection gratings operating as a polarization-sensitive beamsplitter has also been proposed (Lightbody *et al.*, 1995). Some numerical simulation uses a single diffraction grating in the resonant domain for pulse compression (Ichikawa, 1999).

6.2.1.3 Vector Diffraction Theory

The scalar theory fails when the output of the DE to be used is in its near field and when the minimum size of the elemental features is on the order of the illumination wavelength. Diffraction analysis of these situations requires a vector solution of Maxwell's equations that avoids the approximations present in scalar theories.

It has become possible to fabricate computer-synthesized diffractive elements where the size of the elemental features are as small as a fraction of a wavelength due to the progress in fabrication methods that are well known within integrated circuits technology (Wei *et al.*, 1994). This has been pushed toward compact small-size and low-cost elements by industral requirements. An additional requirement is to incorporate several sophisticated functions into a single component.

A relatively simple method for finding an exact solution of the Maxwell's equations for the electromagnetic diffraction is by grating structures. It has been used successfully and accurately to analyze holographic and surface-relief grating structures, transmission and reflection planar dielectric/absorption holographic gratings, dielectric/metallic surfaces relief gratings, multiplexed holographic gratings, etc. (Moharam *et al.*, 1994; Maystre, 1989).

In this theory, the surface-relief periodic grating is approximated by a stack of lamellar gratings. The electromagnetic fields of each layer of the stack are decomposed into spatial harmonics having the same periodicity as the grating. These spatial harmonics are determined by solving a set of coupled wave equations, one for each layer. The electromagnetic fields within each layer are matched to those in the two adjacent layers and to the fields associated with the backward and forward propagating waves or evanescent waves in two exterior regions. The amplitudes of the diffracted fields are obtained by solving this system of equations.

Rigorous methods have led to several approaches: the integral, differential, or variational (Noponen and Saarinen, 1996; Mirotznik *et al.*, 1996); analytic continuation (Bruno and Reitich, 1995); and variational methods and others (Prather, 1995). The integral approach covers a range of methods based on the solution of integral equations. The differential methods use a formally opposite approach of solving first- or second-order differential equations. In some methods the wave equations are solved by numerical integration through the grating. Recently a method has been proposed for the calculation or the diffraction efficiency that includes the response of photosensitive materials that have a nonuniform thickness variation or erosion of the emulsion surface due to the developing process (Kamiya, 1998).

Another method used to model all diffractive effects rigorously is to solve Maxwell's equations by using the finite element method (FEM) that is based on a variational formulation of the scalar wave equation. The FEM is a tool in areas such as geophysics, acoustics, aerodynamics, astrophysics, laser function, fluid dynamics and electromagnetics, as well as to model complex structures. With this method the analysis of complicated material structures can be calculated (Lichtenberg and Gallagher, 1994).

Some hybrid integral–variational methods have also been studied (Mirotznik *et al.*, 1996; Cotter *et al.*, 1995). In this case an FEM is used to solve the Helmholtz equation for the interior of a DE. A boundary element method, a Green's function approach, is used to determine the field exterior to the DE. These two methods are coupled at the surface of the DE by field continuity conditions. This work has been applied to the design of a subwavelength diffractive lens in which the phase is continuous and the analysis of diffraction from inhomogeneous scatterers in an unbounded region.

There is a similar method for the design of periodic subwavelength diffracting elements (Noponen *et al.*, 1995; Zhou and Drabik, 1995) that begins with an initial structure derived from scalar theory that uses simulated annealing to improve its performance. However, the infinitely periodic nature of the structures allows the rigorous coupled wave theory (RCW) developed by Moharam and Gaylord (1981) to be used for the diffraction model. These methods are flexible and have been applied to surface-relief gratings and to gradient-index structures.

6.2.1.3.1 *Polarizing Components*

Vector theory allows the analysis of the polarization properties of surface-relief gratings and diffracted beams whenever cross-coupling between the polarization states takes place. This is shown in the design and tolerance of components for magneto-optical heads (Haggans and Kostuk, 1991). Another example is polarizing beamsplitter (PBS) that combines the form birefringence of a spatial frequency grating, with the resonant refractivity of a multilayer structure (Tyan *et al.*, 1996). The results demonstrate very high extinction ratios (1,000,000:1) when PBS is operated at the designed wavelength and angle of incidence, and good average extinction ratios (from 800:1 to 50:1) when the PBS is operated for waves of 20° angular bandwidth, with wavelength ranging from 1300 nm to 1500 nm, combining features such as small size and negligible insertion losses. The design has been optimized using rigorous couple-wave analysis (RCWA).

6.2.1.4 Achromatic Diffractive Elements

Diffractive optical elements (DEs) operate at the wavelength for which they were designed. When operating at a different wavelength, chromatic aberrations arise. This is a characteristic feature of diffractive lenses. As described by Bennett (1976), a complete analysis of the chromatic aberrations of holograms must take into account the lateral displacement of an image point (lateral dispersion), and its longitudinal dispersion, the change of magnification, third- and higher-order chromatic aberrations, and amplitude variation across the reconstructed wavefronts from thick holograms.

The Abbe value of a diffractive lens v_{diff}, defined over the wavelength range from λ_{short} to λ_{long} (Buralli, 1994)

$$v_{diff} = \frac{\lambda_0}{(\lambda_{short} - \lambda_{long})} \tag{6.3}$$

will always be negative as $\lambda_{long} > \lambda_{short}$. The absolute value of v_{diff} is much smaller than the Abbe value for a conventional refractive lens. For this lens, aberration coefficients can be derived.

Since many potential DE applications require the simultaneous use of more than one wavelength, correction of chromatic aberration is essential.

The unique properties of the DE can be used to correct the aberration of the optical systems that consists of conventional optical elements and DEs by combining this with refractive elements to produce achromatic diffractive/refractive hybrid lenses for use in optical systems. Figure 6.4 shows some hybrid eyepiece designs. Many of these elements have been designed for use with spectral bands ranging from the visible to mid-wave infrared and long-wave infrared regions. These show that a DE is very effective in the correction of primary chromatic aberrations in the infrared region and of primary and secondary chromatic aberrations for visible optical systems. Generally, a DE can improve optical system performance while reducing cost and weight. One can reduce the number of lens elements by approximately one-third; additional benefits can include reducing the sensitivity of the system to rigid body misalignments.

The advantages offered by hybrid refractive–diffractive elements are particularly attractive in infrared systems where the material used is a significant proportion of the overall cost. Hybrid elements allow, for example, passive athermalization of a lens in a simple aluminum amount with a minimum number of elements in which two elements do the athermalization while the dispersive properties of a diffractive surface are used to achromatize the system. In order to realize their full effectivity, hybrid elements must include a conventional aspheric lens with a diffractive structure on one surface (McDowell *et al.*, 1994). Diamond turning permits the aspheric profile and diffractive features to be machined on the same surface in a single process.

Another method of achromatic DE design is that of Ford *et al.* (1996), where the DE acts differently for each of the two wavelengths. The phase-relief hologram can be transparent at one wavelength (λ) yet diffracting efficiently at another (λ') provided that the phase delay is an integral number of wavelengths at λ and a half-

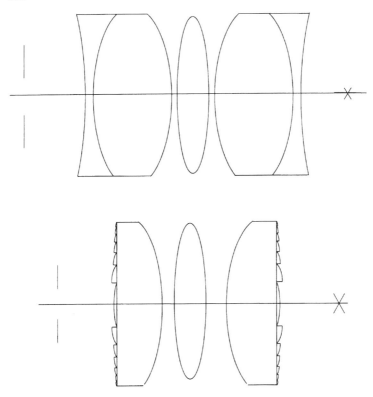

Figure 6.4 Eyepiece designs: (a) Erfle eyepiece and (b) an equivalent hybrid diffractive–refractive eyepiece (Missig and Morris, 1994).

integer number of wavelengths at λ'. In other words, there is an integral-multiple phase retardation of 2π to one wavelength until the suitable phase retardation of the secone wavelength is achieved (Fig. 6.5). In another method the DE is corrected for chromatic aberration designed by combining two aligned DEs made of different materials (Arieli *et al.*, 1998).

6.3 FABRICATION TECHNIQUES

The design of a diffractive optical element must include specifications for microstructure necessary to obtain the desired performance. With an appropriate fabrication technique, these microstructures will introduce a change in amplitude or phase that alters the incident wavefront.

A factor that has stimulated much of the recent interest in diffractive optics has been new manufacturing techniques that give the designer greater control over the phase function that introduces the diffracting element, resulting in a reasonably high diffraction efficiency. In fact, a scalar diffraction theory analysis indicates that a properly designed surface profile can have a first-order diffraction efficiency of 100% at the design wavelength.

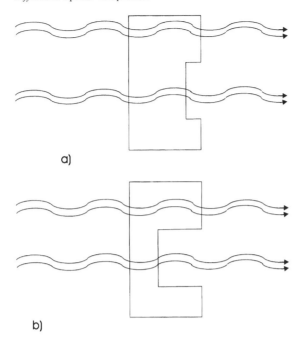

Figure 6.5 Effect of wavelength shift on (a) half-wave and (b) multiple-wave phase holograms. The path-length difference on wavelength shift is greater when the etch depth is optimum. With the correct etch depth, the phase delay at the second wavelength is zero, and there is no diffraction (Ford *et al.*, 1996).

In this section we discuss the main fabrication techniques. These are holographic recording, mask fabrication, and direct-writing techniques, as shown in Fig. 6.6.

6.3.1 Holographic Recording

Amplitude or phase modulation at high spatial frequencies can be obtained from a holographic recording. Off-axis diffractive optical elements have grating-like structures with submicron carrier frequency and diffraction efficiencies as high as 90%. The holographic recording process is rather complicated and is extremely sensitive to vibration, which can be avoided by using an active fringe stabilization system. With this technique, it is possible to obtain positioning errors below $\lambda/40$.

Probably one of the best-known materials is dichromated gelatine, which can be used to produce elements that introduce a phase-index modulation either in its bulk or on its surface. The advantages of this material are its high resolution, index modulation, diffraction efficiency, and low scattering. Further, factors such as humidity affect the holographic record in dichromated gelatine over time and hence the system is not stable unless it is properly sealed. This material is sensitive to the blue part of the spectrum, although there are some dyes that can be incorporated to make it sensitive to a different wavelength (Solano, 1987).

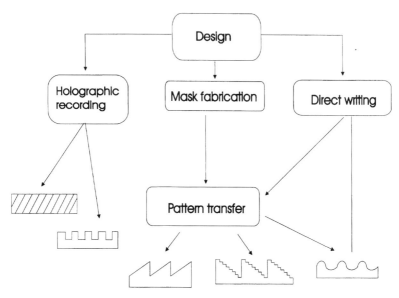

Figure 6.6 Main fabrication techniques of diffractive optical elements.

Other materials are photopolymers and are based on the photopolymeriza-tion of free-radical monomers such as acrylate esters (Lessard and Thompson, 1995). A relatively large number of free-radical monomers suitable for use in holographic recording have been developed. These allow the rapid polymerization of free-radical monomers with any of the common laser lines in the visible spec-trum. Problems with these materials include the inhibition of free-radical polymer-ization due to the presence of dissolved oxygen. To compensate, a high initial exposure to oxygen is required, which causes a significant volume contraction, distorting the recorded fringe pattern. Reprocity failure, reduced diffraction effi-ciency at low spatial frequencies, and time-consuming post-exposure fixing are limitations that are overcome in a photopolymer based on cationic ring-opening polymerization (Close *et al.*, 1969). Among those photopolymers with good stabi-lity and high index modulation are those made by Dupont (Chen and Chen, 1998), the laboratory made with poly(vinylalcohol) as a base and the ones containing acrylamide, and some dyes (Pascual *et al.*, 1999).

Surface-relief DE can be fabricated by holographic exposure in different mate-rials such as photoresists (Zaidi and Brueck, 1988), chalkogenide glasses (Tgran *et al.*, 1997), semiconductor-doped glasses (Amuk and Lawandy, 1997), and in liquid (Boiko *et al.*, 1991) and dry self-developing photopolymer materials (Calixto, 1987; Calixto and Paez, 1996; Neuman *et al.*, 1999), etc. Two types can be distinguished: those that approximate a staircase (Fresnel lens) and those based on diffractive optical elements (Fresnel zone plates, gratings, etc.).

It has been shown (Ehbets *et al.*, 1992) that almost any object intensity dis-tribution can be interferometrically recorded and transferred to a binary surface relief using a strongly nonlinear development. As a result, the sinusoidal interference pattern is then transformed into a rectangular-shaped relief grating.

6.3.2 Mask Fabrication: Binary Optical Elements

Binary optical elements are staircase approximations of kinoforms, which have multiple levels created by a photolithographic process, as shown in Fig. 6.7. The term binary optics comes from the binary mask lithography used to fabricate the multilevel structures.

The optical efficiency of a diffraction grating depends on the phase encoding technique. Binary optics is based in the creation of multilevel profiles, which requires exposure, development, and etching with different masks that are accurately aligned to each other. The number of phase levels realized through multiple binary masks depends on the specific encoding approach.

To explain the principle of these elements, assume that a blazed grating is to be written having the phase profile shown in Fig. 6.8(a), (Davis and Cottrell, 1994). Here the total phase shift over the grating is 2π radians and the period of the grating is defined as d. This grating would yield 100% diffraction efficiency into the first order. To fabricate this grating using binary optics techniques, masks are designed having increasingly finer resolutions of $d/2$, $d/4$, $d/8$, etc. Each mask is deposited sequentially onto a glass substrate. After the deposition of the first mask, the surface is etched in such a way that the phase difference between masked and unmasked areas is π radians, as shown in Fig. 6.8(b). However, the diffraction efficiency of this binary-phase-only mask is only 40.5%. To get higher diffraction efficiencies, increasingly finer masks are deposited one after the other and the substrate is etched in such a way as to produce increasingly smaller phase shifts. For the eight-phase level grating of Fig. 6.8(c), the diffraction efficiency reached 95%. However, to reach

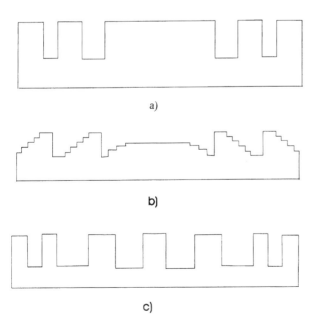

a)

b)

c)

Figure 6.7 Binary optics fabricated by binary semiconductor mask lithography: (a) Fresnel zone, (b) Fresnel lenslet, and (c) Dammann grating.

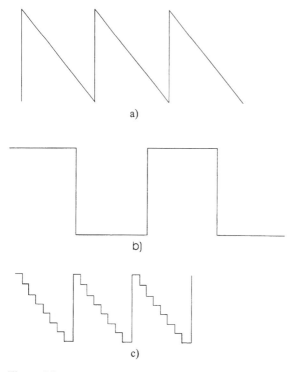

Figure 6.8 DE profiles: (a) phase profile for a grating with 100% diffraction efficiency; (b) binary phase grating profile; and (c) step phase grating profile (Davis and Cottrell, 1994).

higher diffraction efficiencies, the size of the elemental features must decrease in order to maintain the periodicity.

This staircase profile can be generated with masks or by using thin-film deposition (Beretta *et al.*, 1991).

These DE are constrained by spatial and phase quantization (Arrizon and Testorf, 1997). The complexity and quality of the reconstructed image determines the spatial complexity and phase resolution of the DE. The important issues in using masks are the alignment between the successive steps and the linewidth errors. This limits the fabrication of multilevel phase elements to three or four masks, corresponding to eight- or 16-phase levels.

Masks can be generated with electron-beam or laser beam lithography. These are amplitude elements that have to be transformed into surface-relief structures by exposure, chemical processing, and etching of the photoresist. These processes permit the fabrication of sawtooth, multilevel, or continuous profiles. For more rugged elements with high optical quality, the photoresist profiles are then transferred into a quartz substrate by standard techniques such as reactive ion etching.

With electron-beam lithography, one can write gratings with periods down to 100 nm, but beyond that they are limited by the proximity effect. Electron-beam lithography is a highly flexible means of generating arbitrary structures, even microlenses (Oppliger *et al.*, 1994). However, in the case of elementary feature sizes of the order of 50–100 nm, this approach is limited by the positioning accuracy during the writing process.

Binary and multilevel diffractive lenses with elementary feature sizes of the order of submicrometers have been produced on silicon and gallium phosphide wafers by using the CAD method, direct write electron-beam lithography, reactive ion etching, antireflection coating, and wafer dicing. Measurements indicate that it is possible to obtain aberration-free imaging and maximum diffraction efficiencies of 70% for lenses with numerical apertures (NAs) as high as 0.5. This technique has been applied to off-axis arrays for 18-channel parallel receiver modules (Haggans and Kostuk, 1991).

6.3.2.1 Photolithography

The interesting field of photolithography has developed as a result of the introduction of resist profiles to produce microoptical elements (Fresnel lenses, gratings, kinoforms, etc.). The thickness of the resist film that must be altered can be several micrometers thick to obtain the required profile depth of the optical element. The efficiency of those elements depend on the shape and quality of the resist profiles. Blazed and multilevel diffractive optical elements can reach a higher efficiency than binary optical elements.

Surface-relief lithography diffractive elements generated show promise for applications to magneto-optic heads for data storage due to their polarization selectivity, planar geometry, high diffractive efficiency, and manufacturability. Former applications of these elements had been limited due to the lack of information on their polarization properties.

The use of lithographic techniques opens the way to the development of optical elements that are economical, have high resolving power, and flexible design. These ideas are used in many systems at optical or near-infrared wavelengths.

6.3.2.2 Gray-Tone Masks

The gray-scale masks is an alternative approach to the multiple mask technique; it requires only one exposure and one etching process and yields a continuous profile. The gray levels are made by varying a number of transparent holes in a chromium mask that are so small that they are not resolved during the photolithographic step. Diffraction efficiencies reported are of the order of 75% for an element etched in fused silica, $\lambda = 633$ nm.

This process requires linearization of the photographic emulsion exposure as well as linearization of photoresist exposure: both are hard to reproduce (Däschner *et al.*, 1995).

6.3.3 Direct-Writing Techniques

High-intensity pulsed lasers can uniformly ablate material from the surface of a wide range of substrates (Braren *et al.*, 1993). Proper choice of laser wavelength allows a precise control of depth that can be applied in many materials that absorb in this region of the spectrum. These lasers have been used in lithographic processes. Direct-writing techniques yield higher phase resolutions (of the order of 64–128 phase levels) than photolithographic methods but at the expense of reduced spatial resolution.

Another alternative, however, is to use an excimer laser with an ultraviolet waveguide to etch diffractive structures directly into the substrate without masks or intermediate processing steps (Fig. 6.9) (Duignan, 1994). This technique can be applied to a large spectrum of substrate materials, such as glass, diamond, semiconductors, and polymers, and can also reduce time and cost to produce a diffractive element.

Figures 6.10 and 6.11 show the fabrication steps and a schematic of one of the systems: in this particular case, a He–Cd laser is used to fabricate the DE on a photoresist substrate (Gale *et al.*, 1994).

Direct writing in photoresist, with accurate control of the process parameters, enables one to fabricate a complex continuous relief microstructure with a single exposure and development operation, which has been shown to produce excellent results (Ehbets *et al.*, 1992). Because writing times can be relatively long (many hours for typical microstructures of $1\,cm^2$) a latent image decay must be compensated. A number of factors determine the fidelity of the developed microstructure. The dominant experimental errors in the writing process are surface structures of the coated and developed photoresist films, the profile of the focused laser spot, the accuracy of the exposure dose, the line straightness, and the accuracy of the interline distance of the raster scan on the substrate.

An example of elements fabricated by direct laser writing in photoresist (Gale *et al.*, 1993) is a fanout element and diffractive microlens with $NA = 0.5$, which has been produced with a diffraction efficiency of 60%.

6.3.4 Replication Techniques

The main attraction of micro-optical elements lies in the possibility of mass production using replication technology.

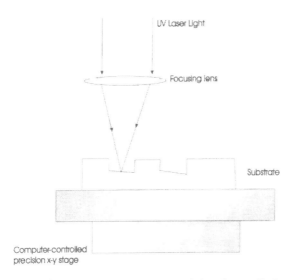

Figure 6.9   Direct-write photoablation of the substrate (Duignan, 1994).

Optical design

[Compute surface microrelief]

Exposure data
[Compute intensity data for exposure]

Fabrication of laser writing on the material

Development

Electroform shim

Replication

Figure 6.10 Fabrication steps for continuous-relief micro-optical elements (Gale *et al.*, 1994).

Replication technology is already well established for the production of diffractive foil, display holograms, and holographic security features in which the microrelief structures have a typical grating period of 0.5–1 μm with a maximum depth of about 1 μm. These are produced with a hot roller press applied to rolls of plastic up to 1 m wide and thousands of meters in length. (Kluepfel and Ross, 1991). For deeper microstructures, other replication techniques are required, such as hot-press, casting, or molding. In all cases it is necessary first to fabricate a metal shim, usually of nickel (Ni), by electroplating the surface of the microstructure. Figure 6.12 (Gale *et al.*, 1993) illustrates the steps involved in the fabrication of these shims. The recorded surface-relief microstructure in photoresist is first made conducting, either by the evaporation of a thin film of silver or gold of the order of 100 nm, or by using a commercial electronless Ni deposition bath. An Ni foil is then built up by electroplating this structure to a thickness of about 300 μm. Finally, the NI is separated from the resist/substrate and cleaned to give the master (first-generation) replication shim. This master shim can be used directly for replication, by hot-embossing or casting. It can also be supplied to a commercial shim facility for recombination to produce a large-area production shim.

The first-generation master can be used to generate multiple copies by electroplating further generations. The silver or nickel surface is first passivated by immer-

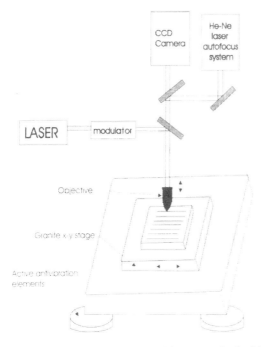

Figure 6.11   Schematic of laser-writing system for the fabrication of continuous relief micro-optical elements (Gale *et al.*, 1994).

sion in a dichromate solution or by O_2 plasma treatment, followed by further electroplating to form a copy that can readily be separated. In this way, numerous copies can be made from a single recorded microrelief.

Advances in the sol–gel process have made it possible to replicate fine-patterned surfaces in high-purity silica glass by a molding technique (Noguès and LaPaglia, 1989). The different types of diffractive optics that have already been replicated include binary grating, blazed grating, hybrid diffractive/refractive optical element, and plano kinoform. The requirements of the optics leads to an appropriate lens design, which then defines the design of the molds to be used to produce the optical components. To manufacture the mold, a tool that contains the required relief pattern must be fabricated. The mold is fabricated and used in the sol–gel process to produce prototype parts. Quality control then provides the necessary input to determine what, if any, changes are necessary in either mold or procedure for the final DE (Moreshead *et al.*, 1996).

There are three important advantages of the sol–gel replication process:

1. It provides a cost-effective way of producing optical elements with fine features. Although the mold surface is expensive, its cost can be amortized over a large volume of parts, thus making the unit cost relatively low.
2. The process can produce optical elements in silica glass, one of the best optical materials. The advantages of silica include a very high transmission over a broad wavelength range from 0.2 to 3.2 µm, excellent thermal sta-

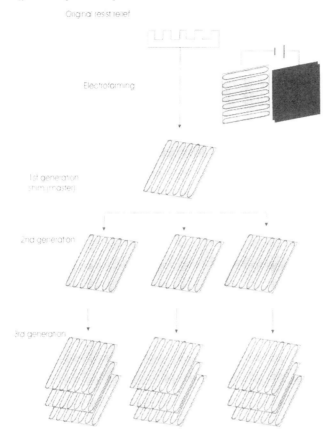

Figure 6.12 Fabrication of replication shims (Gale *et al.*, 1994).

bility, radiation hardness, and chemical treatment. Therefore, it can be used in harsh environments such as space, or for military uses and in high-power systems.

3. In the sol–gel replication process there is a substantial shrinkage that is controlled by adjusting the processing parameters. This shrinkage has been accurately quantified and has been found to be very uniform in all three dimensions, making it possible to fabricate parts with structures smaller than those made by other processing techniques. This reduces imperfections and tool marks by a factor of 2.5, reducing scattered light at the design wavelength.

Other techniques are compatible with the microfabrication techniques used in the semiconductor industry and require the generation of a gray-level mask such as that fabricated in high-energy beam sensitive (HEBS)-glass by means of a single electron-beam direct-write step (Wu, 1992). This mask was used in an optical contact aligner to print a multilevel DE in a single optical exposure. A chemically assisted ion-beam etching process has been used to transfer the DE structure from the resist into the substrate (Däschner *et al.*, 1995).

6.3.4.1 Plastic Optics

It is important to mention molded plastic DEs: they are low cost and can be mass-produced (Meyers, 1996). They can be produced in different shapes: rotationally symmetric, aspheric, and hybrid refractive/diffractive lenses. These are used in various applications, such as fixed focus and zoom camera lenses; camera viewfinders; diffractive achromatized laser-diode objectives; and asymmetric anamorphic diffractive concentrating and spectral filtering lenses for rangefinding and autofocus applications.

Planar micro-optical elements can be found in an increasing number of applications in optical systems and are expected to play a major role in the future. Typical elements for application at visible and infrared wavelengths are surface-relief microstructures with a maximum depth of about 5 μm. These can be mass-produced using current replication techniques such as hot-embossing, molding, and casting (Gale *et al.*, 1994).

6.4 DYNAMIC DIFFRACTIVE ELEMENTS

In recent years there has been a great deal of interest in active, dynamic diffractive elements that can be electrically programmed or switched. These elements can be divided into two classes. The first class uses an element-by-element addressing structure to produce diffracting patterns as spatial light modulators. The second class switches on a pre-patterned diffraction structure that has configured during fabrication. These devices could expand the range of application of the DE through the real-time control of an element's optical function. Both these devices have a large range of designs and methods for generating dynamically the phase or amplitude modulation of a spatial pattern in response to an electrical signal.

6.4.1 Programmable Diffraction Elements

Binary optics can be programmed to produce patterns with a large dynamic range. These have two functions. First, the spatial light modulator (SLM) serves as a programmable low-cost test for more complicated nonprogrammable binary optical elements. Secondly, the programmability of this system allows real-time image processing in which the optical element can be changed rapidly.

One way to obtain such elements is by using electro-optic material such as a liquid crystal (LC) layer. The LC materials exhibit a large field-induced birefringence effect, resulting in a local change in the index of refraction, polarization angle, or both. The main disadvantage is that the scale of the electrode patterns in these elements is larger than the microstructure needed for the diffractive elements. These elements show no diffraction effects except at their edges.

Therefore another allternative is to use the diffractive optical elements written onto an SLM (Parker, 1996). In this case, each phase region will be encoded onto an individual pixel element whose size is limited by the resolution of the SLM. These phase regions are limited by the operating physics of the SLM.

One type of SLM system used is the magneto-optic spatial light modulator (MOSLM). This binary modulator consists of a two-dimensional array of magneto-optic modulators that are fabricated monolithically on a nonmagnetic substrate. Each element of the array can be electronically addressed through an array of crossed electrodes (Psaltis *et al.*, 1984). By contrast, the phase-only nematic liquid crystal monitor can encode continuous phase levels of up to 2π radians (Davis, 1994). Assuming that a wide range of phases can be encoded, the diffraction efficiency can be increased by using a number of pixels to encode each period of the grating. However, the maximum number of pixels is limited by the size and resolution of the SLM. For this reason an increase in the optical efficiency of the grating is offset by a decrease in its resolving power. Similar problems exist in encoding Fresnel lenses using SLMs.

The SLM has been used in Fresnel lenses, magneto-optic spatial light modulators, optical interconnections, lens arrays, subdiffraction limited focusing lenses, derivative lenses, nondiffractive lenses, and Damman gratings.

6.4.2   Switched Optical Elements

Switched optical elements use transmitting or reflecting structures that incorporate a material that exhibits an index of refraction that can be varied electrically. When an electric field is applied to the resulting composite structure, a range of predetermined optical characteristics emerge that spatially modulate or redirect light in a controlled matter. The effect on an incident wavefront may be controlled by varying the applied electric field.

These devices are capable of producing diffraction effects when a drive signal is applied, or in some designs, when it is removed.

Typically, SLMs are restricted to relatively small pixel arrays on the order of 256×256 and with correspondingly low diffraction efficiency. Monolithic holograms, on the other hand, have extremely high resolution, high optical quality, and diffraction efficiency, with 1 million times the pixel density. Such elements can be used in devices that are significantly different, especially from SLMs, if the material is also of sufficiently high optical quality to permit series stacking (Sutherland *et al.*, 1994). Figure 6.13 shows a generic device made of stacks of switchable holograms.

Among the different approaches to these switching DE is the placing of electrodes over a layer of liquid crystals to respond to the localized fields with two-dimensional distribution birefringence. It is possible also to fill a surface-relief binary optical element or a sinusoidal relief grating etched on dichromated gelatin with a layer of liquid crystals (Sainov *et al.*, 1987; Stalder and Ehbets, 1994). Reported switching times ranged between 20 and 50 ms for an applied voltage of $20\,V_{rms}$. Some other work involves special materials such as LC selective polymerization, or fabrication of holographic gratings by photopolymerization of liquid crystalline monomers that can be switched with the application of an electric field (Zhang and Sponnsler, 1992).

Most of the reported approaches involve liquid crystals in one way or another although, in principle, semiconductor techniques could also be used (Domash *et al.*, 1994).

One of the most popular materials is the polymer-dispersed liquid crystal (PDLC) formed *in situ* by a single-step photopolymer reaction (Sutherland *et al.*, 1994). These materials are composites of LC droplets imbedded in a transparent

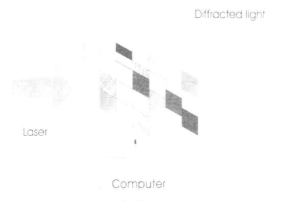

Figure 6.13 General active holographic interconnect device. Each plane contains an independent array of individually controlled patches of prerecorded holograms. By electrically addressing patches throughout the volume, many different diffractive functions can be programmed (Domash *et al.*, 1994).

polymer host whose refractive index falls between the LC ordinary and extraordinary indices. By modulating electrically the index match between LC droplets and polymer host, the characteristics of the volume holographic diffraction may be reduced. Fine-grained PDLCs have recently become available for electrically switchable holographic elements. The mechanism for the formation of the hologram grating is described by Sutherland, 1996. They have high diffraction efficiency, narrow angular selectivity, low voltage switching, and microsecond switching speed in elements with good optical quality as well as for the storage of the holographic image. Electro-optical read-out can be used with this new system material. Applications are for switchable notch filters for sensor application, reconfigurable optical interconnects in optical computing, fiber-optic switches, beam steering in laser radar, and tunable diffractive filters for color projection displays.

Dynamic-focus lenses that are controlled electrically are used in autofocusing devices for tracking in CD pickups, optical data storage components, and many other purposes. Some applications require continuous focusing; others call for switchable lenses with a discrete number of focal lengths. The basic concept is a diffractive lens material whose diffractive characteristics can be turned off by the application of an electric field. Using such a material, an electro-optic diffractive lens may be switched between two states – transparency (infinite focus) and finite focus.

A number of light-modulating SOE devices for display applications use structures that can be referred to as hybrid; i.e., structures that combine a fixed array of individually switched electrodes with a pre-patterned diffractive structure.

6.5 APPLICATIONS

6.5.1 Micro-optical Diffracting Elements

Micro-optical devices, such as diffractive and refractive microlenses have received considerable attention in the optics R&D community.

Technological advances make it possible to micromachine optical and diffractive devices with structures that have the same order of magnitude as the wavelength of the incident light. Devices that were once considered impractical because of the limitations of bulk optics are now designed and easily fabricated with advanced microelectronics technology.

Micro-optical elements can be refractive, such as hemispherical lenslet and lenslet arrays, diffractive as kinoforms, grating structures, etc., or a combination of both such as Fresnel microlenses. They can be continuous surface-relief micro structures (Gale *et al.*, 1993), binary or multilevel reliefs or made by direct laser writing (Ehbets *et al.*, 1992).

The ability to combine various optical functions, e.g., focusing and deflection, and the reduced thickness and weight of DE in comparison to their refractive counterparts essentially explain the concern of diffractive optics in micro-optics.

In processing optical materials two main classes of diffractive optics are higher power lasers and their periphery (such as the interconnection of a high power Nd:YAG laser with a fiber bundle) and the use of DE to shape the laser in order to provide the illumination beam required for the same application such as the production of diffuse illumination with high-power CO_2 lasers (Wyroski *et al.*, 1994).

On the other hand, such elements can be applied to holography for memory imaging, nondestructive testing in interferometry, wavefront shaping, and as spatial filters. They have been many practical applications, such as diffraction gratings to shape the phase and polarization of incident fields, reflectors for microwave resonant heating systems, microwave lenses, mode converters in integrated optical devices, for dispersion compensation and pulse shaping in ultrafast optics, etc. (Lichtenberg and Gallagher, 1994).

Technology for making binary optics is a broadly based diffractive optics using advanced submicrometer processing tools and micromachining processes to create novel optical devices. One potential role of binary optics is to integrate very large scale integration (VLSI) of microelectronic devices with micro-optical elements (Montamedi, 1994). Because small feature sizes and stringent process control have been two major considerations, attention has focused on microlithography during the past few years. The rapid growth of commercial devices, such as miniature compact disk heads, demands both higher accuracy and lower-cost microlenses' fabrication methods.

Binary optics microlenses arrays are typically fabricated from bulk material by multi-mask-level photoresist patterning and sequential reactive-ion etching to form multistep phase profiles that approximate a kinoform surface. To fabricate an efficient microlens' array, eight-phase-level zones are necessary. The main parameters involved in its design are the wavelength (λ), the microlens' diameter (d'), the focal length (f), $f_\# = f/d'$, and the smallest feature size or critical dimension (D). For a typical binary optic microlens with eight phase levels the D value is $D = (\lambda f_\#)/4$. The minimum value of VLSI is of the order of 0.5–1 µm. This limits the speed of binary optic microlenses designed for wavelengths (diode laser) from 0.632 µm to 0.850 µm to $f/6$ and $f/3$, respectively. Nevertheless, higher-speed microlenses can be fabricated for infrared applications.

As already mentioned, the diffraction efficiency of the light diffracted to the first-order focus increases with the number of phase levels. In practice, it decreases with the number of processing factors. Values of 90% have been obtained for eight-phase-level microlenses. The extent to which this is acceptable will depend on the application.

The surface relief of these diffractive microlenses has a planar structure of the order of the design wavelength. In a typical system this reduces the volume and weight of the optics relative to an all-refractive design.

Along with these developments in micro-optics technology is the development of micro-electro-mechanical (MEM) technology, which is based on micromachining methods for processing 3-D integrated circuits. MEM and micro-optics technologies have one critically important feature: both technologies are compatible with VSLI circuit processing. This feature means that the final device can be produced in volume at low cost. The standard VLSI process is generally confined to the surface of the wafer (Si or GaAs), extending only several micrometers under the surface. Multilayers of metal and dielectric are either deposited/grown on the surface or are etched into the surface.

Some micro-optical DE have been applied in optical choppers, optical switches, and scanners.

6.5.2 Optical Disk Memory Read–Write Heads

The optical head is an important component in optical disk storage. In it a laser beam is focused to a 1-μm diameter spot on the surface of the disk. The conventional optical head usually contains several optical elements such as a beamsplitter prism, a diffraction grating, a collimating lens, and a cylindrical lens, as shown in Fig. 6.14 (Huang *et al.*, 1994).

The disk moves under the optical axis of the head as it rotates. In this system it is necessary to detect and correct focus error to an accuracy of about ± 1 μm. This focus error is determined from the total intensity of the light reflected by the optical disk.

Some systems have been suggested for replacing each of the optical elements with a diffractive micro-optical element performing the three optical functions required for an optical head: splitting the beam, focusing, and tracking the error signals (Fig. 6.15) (Huang *et al.*, 1994).

6.5.3 Optical Interconnects

Diffractive optics will play an important role in optical interconnects and optical interconnecting networks necessary in high-parallel-throughput processing. Diffractive optical interconnect elements provide several advantages over conventional bulk elements such as spherical and cylindrical lenses (Herzig and Dändliker, 1993).

One of the most simple devices for fanning out signals in optical interconnecting systems is the diffraction grating. A basic fan-out arrangement, consisting of a diffraction grating and a collecting lens, is shown in Fig. 6.16. Diffraction by a periodic pattern, such as in a Damman grating, divides the incident wave into many beams that are then focused by the collimating lens onto the detector plane. The amount of splitted light is determined by the specific pattern of the grating. The light focused in the different orders illuminates photodetectors or fibers, depending on the application of the system. These systems can compensate for wavelength dispersion and distortion that occur in diffractive fan-out elements (Schwab *et al.*, 1994).

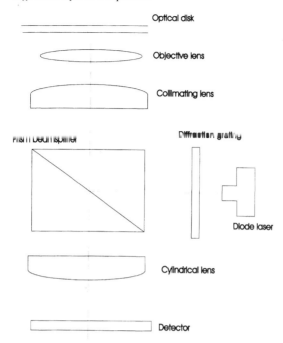

Figure 6.14 Configuration of a common optical head (Huang *et al.*, 1994).

The use of diffractive optics in interconnects is of considerable interest for several reasons. First, multiple diffractive optic elements can be cascaded on to planar substrates and more easily packaged with planar electronic substrates. To be effective, however, the diffractive optical system must separate and distribute optical signals in several dimensions. With diffractive optical interconnects for digital switching networks, current technology has the ability to form four-dimensional,

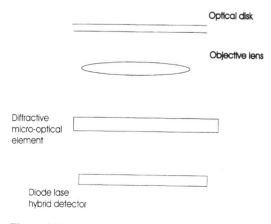

Figure 6.15 Optical head using a diffractive micro-optical element (Huang *et al.*, 1994).

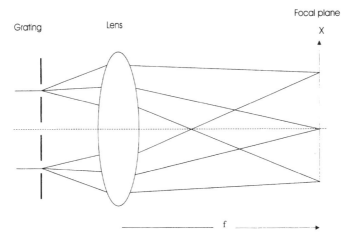

Figure 6.16 A grating is the simplest device for fanning out signals.

free-space optical interconnects with boundary conditions (Rakuljic *et al.*, 1992). The optics must be packaged with standard board substrates, and have alignment tolerances sufficient for board insertion, replacement, and changes of length caused by temperature variations.

Bidirectional information transfer is necessary at each information port. Parallel data transfer to increase information transfer rates is also important. It must be possible to broadcast greater data processing signal loads to multiple lateral and longitudinal locations.

In addition, the fabrication of microdiffractive optics using microlithographic and holography methods can produce many optical functions that can be used in both space variant and invariant systems. With these fabrication methods, units can be mass-produced, lowering overall system costs. Fabrication of diffractive optics uses computer-aided design (CAD) and microstructuring techniques. Reflection losses has been achieved using common techniques used in microelectronics technology such as ion-beam-sputter deposition. To reduce crosstalk and feedback, antireflection (AR) coatings or AR-structured surfaces have been suggested (Pawlowski and Kuhlow, 1994).

Diffractive optical elements (HOEs) have proven to be useful in optical interconnection and routing systems, especially where volume, weight, and design flexibility are important. Their characteristics can be increased by making them polarization-selective (Nieuborg *et al.*, 1997).

Finally, optical interconnect systems must be competitive in performance and cost with electrical interconnect methods. An example of a hybrid diffractive element design of a bidirectional interface is illustrated in Fig. 6.17 (Kostuk *et al.*, 1993).

6.5.4 Polarizing Elements

As mentioned earlier, another important application for diffractive elements is their ability to polarize light. Polarization-selective computer-generated holograms (CGH) or birefringent CGH (BCGH) have been found useful for image processing,

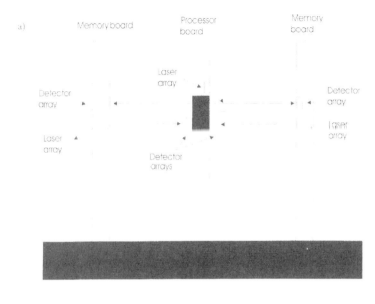

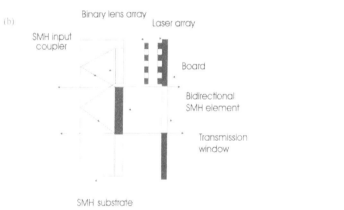

Figure 6.17 Schematic of (a) a single optical bus line connecting the processor board and (b) an expanded view showing the components on the central board transmitting to the adjacent boards through a bidirectional beamsplitter. This model uses a substrate-mode holographic (SMH) window element (Kostuk *et al.*, 1993).

photonic switching, and the packaging of optoelectronic devices and systems. With BCGH it is possible to perform two completely distinct functions using each of the two orthogonal polarizations.

Other applications are polarized beamsplitters (PBS) used for read–write magneto-optic disk heads (Ojima *et al.*, 1986), polarization-based imaging systems (Kinnstatter and Ojima, 1990; Kunstmann and Spitschan, 1990), and optical information processing such as free-space optical switching networks (McCormick *et al.*, 1992). These require the PBS to provide high extinction ratios, tolerate a wide angular bandwidth, a broad wavelength range of the incident light, and have a compact size for efficient packaging. For these applications tradtional birefringent structures such as the Wollaston prism or multilayer structures do not meet these requirements.

Diffractive optical elements (DE) have proven to be useful components in optical interconnection and routing systems, especially where volume, weight, and design flexibility are important. Their usefulness has been increased by making them polarization-selective using two wet etched anisotropic calcite substrates, joined at their etched surfaces and with their optical axes mutually perpendicular. The gap was filled with an index matching polymer (Nieuborg *et al.*, 1997). This element is less sensitive to fabrication errors. This method has been used to obtain elements that change the form of the emerging wavefront, depending on the polarization of the incident light, and has been applied in Fresnel lenses, gratings, and holograms that generate different images in their Fourier plane.

6.5.5 Holographic Memory Devices

An important characteristic of holographic memory is its ability to parallel input and record massive amounts of information into a memory. With this feature, memory devices can be created with high information quality. By information quality, we mean the product of the amount of recorded information and the retrieval rate.

The number of holograms that can be multiplexed in a given holographic system is primarily a function of the system's bandwidth in either temporal or spatial frequency, and the dynamic range of the material. One can record around 10 angle multiplexed holograms in a 38-μm thick film with diffraction efficiency of 10^{-3}. (Since it can typically work with holographic diffraction efficiencies on the order of 10^{-6}, we have sufficient dynamic range to record significantly more than 10 holograms.) The limitation in angular bandwidth can be alleviated with a thicker film, but scattering increases rapidly with thickness in these materials. Another method that has been previously used to increase the utilization of the available bandwidth of the system is fractal sampling grids (Mok, 1990), and peristrophic (consisting in turns) multiplexing has been used as a solution to the bandwidth limited capacity problem. With this method the hologram is physically rotated, with the axis of rotation perpendicular to the film's surface every time a new hologram is stored (Curtis and Psaltis, 1992).

6.5.6 Beam Shaping

In many applications one needs to reshape the laser beam intensity. The advantage of DE is that the beam energy is redistributed rather than blocked or removed, so that energy is preserved.

Some designs have been proposed using computer-generated holograms where the Gaussian beam has been converted into a ring distribution (Miler *et al.*, 1998), or using a two-element holographic system to obtain a flat distribution (Aleksoff *et al.*, 1991). Another proposed system is a Gaussian to top hat converter using a multilevel DE that can be fabricated with standard VLSI manufacturing equipment (Kosoburd *et al.*, 1994). This distribution is useful in material processing, laser radar, optical processing, etc. An interesting application is the collimation of high-power laser diodes (Goering *et al.*, 1999).

Free-space digital optical systems require optical power supplies that generate two-dimensional arrays of uniform intensity (Gray *et al.*, 1993). The resultant spot arrays are used to illuminate optoelectronic logic devices' arrays to optically encode and transfer information. The favored method for creating these regularly spaced beam arrays is to illuminate a computer-designed Fourier-plane hologram using a collimated laser source. These surface-relief gratings, also referred to as multiple beamsplitters, are designed using scalar diffraction theory by means of a computer optimization process that creates an array of beams of uniform intensity (Gale *et al.*, 1993). The quality of the hologram is measured by its diffraction efficiency in coupling light into a set of designated orders and the relative deviation of the beam intensities from their targeted values.

Other beam shapers include the Laguerre–Gaussian beam, which has a phase singularity that propagates along its axis (Miyamoto *et al.*, 1999). Work has also been done to convert a Gaussian-profile beam into a uniform-profile beam in a one-dimensional optical system as well as rotationally symmetric optical systems both for different fractional orders and different parameters of the beam (Zhang *et al.*, 1998). Another important application is the pseudo-nondiffracting beam DE, characterized by an intensity distribution that is almost constant axially over a finite axial region and a long propagation distance along the optical axis (Liu *et al.*, 1998) and the axicons. An axicon is an optical element that produces a constant light distribution over a long distance along the optical axis. A diffractive axicon with a discrete phase profile can be fabricated using lithographic fabrication techniques. Other elements can be fabricated with linear phase profiles (Lunitz and Jahns, 1999).

Another beam-shaping procedure is the projection pattern that can be applied to change the physical or chemical state of a surface with visible light or ultraviolet radiation. Important applications in industrial production processes are microlithography and laser material processing.

In conventional methods of pattern projection a real value (mostly binary) transmission mask pattern is reproduced on the target surface by imaging or shadow casting. This pattern is then formed by diffraction of the illuminating wave at the mask where the diffracted wave is transformed by propagation, either through a lens or through free space, to the target surface.

The use of DE allows us to add phase components to the mask, giving a complex transmission coefficient. This method is called phase masking. It can be used to improve the steepness of edges in projected patterns by reduction of the spatial bandwidth. Also, the mask may be located at some distance from the target surface or of its optical conjugate. The mask then contains the pattern to be projected in a coded form. When it is in the far field of the target surface, this code is essentially a Fourier transformation (Velzel *et al.*, 1994).

REFERENCES

Aleksoff, C. C., K. K. Ellis, and B. G. Neagle, "Holographic Convertion of Gaussian Beam to Near-Field Uniform Beam," *Opt. Eng.*, **30**, 537–543 (1991).

Amuk, A. Y. and N. M. Lawandy, "Direct Laser Writing of Diffractive Optics in Glass," *Opt. Lett.*, **22**, 1030–1032 (1997).

Arieli, Y., S. Noach, S. Ozeri, and N. Eisenberg, "Design of Diffractive Optical Elements for Multiple Wavelengths," *Appl. Opt.*, **37**, 6174–6177 (1998).

Arrizon, V. and M. Testorf, "Efficiency Limit of Spatially Quantized Fourier Array Illuminators," *Opt. Lett.*, **22**, 197–199 (1997).

Bennett, J., "Achromatic Combinations of Hologram Optical Elements," *Appl. Opt.*, **15**, 542–545 (1976).

Beretta, S., M. M. Cairoli, and M. Viardi, "Optimum Design of Phase Gratings for Diffractive Optical Elements Obtained by Thin-Film Deposition," *Proc. SPIE*, **1544**, 2–9 (1991).

Boiko, Y. B., V. M. Granchak, I. I. Dilung, V. S. Solojev, I. N. Sisakian, and V. A. Sojfer, "Relief Holograms Recording on Liquid Photopolymerizable Layers," in *Three-Dimensional Holography: Science, Culture, Education*, Jeong, T. H. and V. B. Markov (eds), *Proc. SPIE*, **1238**, 253–257 (1991).

Braren, B., J. J. Dubbowski, and D. P. Norton (eds), "Laser Ablation in Materials Processing: Fundamentals and Applications," *Material Research Society Proceedings*, **285**, 64–71 (1993).

Bruno, O. P. and F. Reitich, "Diffractive Optics in Nonlinear Media with Periodic Structure," *J. Opt. Soc. Am. A*, **12**, 3321–3325 (1995).

Buralli, D. A., "Using Diffractive Lenses in Optical Design," *OSA Technical Digest Diffractive Optics: Design, Fabrication, and Applications* **11**, 44–47 (1994).

Calixto, S., "Dry Polymer for Holographic Recording," *Appl. Opt.*, **26**, 3904–3910 (1987).

Calixto, S. and G. P. Paez, "Micromirrors and Microlenses Fabricated on Polymer Materials by Means of Infrared Radiation," *Appl. Opt.*, **35**, 6126–6130 (1996).

Chen, T. and Y. Chen, "Graded-Reflectivity Mirror Based on a Volume Phase Hologram in a Photopolymer Film," *Appl. Opt.*, **37**, 6603–6608 (1998).

Close, D. H., A. D. Jacobsob, J. D. Margerun, R. G. Brault, and F. J. McClung, *Appl. Phys. Lett.*, **14**, 159–160 (1969).

Cotter, N. P. K., T. W. Preist, and J. R. Sambles, "Scattering-Matrix Approach to Multilayer Diffraction," *J. Opt. Soc. Am. A*, **12**, 1097–1103 (1995).

Curtis, K. and D. Psaltis, "Recording of Multiple Holograms in Photopolymer Films," *Appl. Opt.*, **31**, 7425–7428 (1992).

Däschner, W., P. Long, M. Larson, and S. H. Lee, "Fabrication of Diffractive Optical Elements Using a Single Optical Exposure with a Gray Level Mask," *J. Vacuum Sci. & Techn. B*, **13**, 6–9 (1995).

Davis, J. A. and D. M. Cottrell, "Binary Optical Elements Using Spatial Light Modulators," *Proc. SPIE*, **2152**, 26, 226–236 (1994).

Domash, L. H., C. Gozewski, and A. Nelson, "Application of Switchable Polaroid Holograms," *Proc. SPIE*, **2152**, 13, 127–138 (1994).

Duignan, M. T., "Micromachining of Diffractive Optics with Excimer Lasers," *OSA Technical Digest Diffractive Optics: Design, Fabrication, and Applications*, **11**, 129–132 (1994).

Ehbets, P., H. P. Herzig, D. Prongué, and A. T. Gale, "High-Efficiency Continuous Surface-Relief Gratings for Two-Dimensional Array Generation," *Opt. Lett.*, **17**, 908–910 (1992).

Fienup, J. R., "Reconstruction and Synthesis Applications for an Iterative Algorithm," *Proc. SPIE*, **373**, 147–160 (1981).

Ford, J. E., F. Xu, and Y. Fainman, "Wavelength-Selective Planar Holograms," *Opt. Lett.*, **21**, 80–82 (1996).

Gale, M. T., M. Rossi, H. Schütz, P. Ehbets, H. P. Herzig, and D. Prongué, "Continuous-Relief Diffractive Optical Elements for Two-Dimensional Array Generation," *Appl. Opt.*, **32**, 2526–2533 (1993).

Gale, M. T., M. Rossi, J. Pedersen, and H. Schütz, "Fabrication of Continuous-Relief Micro-Optical elements by Direct Laser Writing in Photoresists," *Opt. Eng.*, **33**, 3556–3566 (1994).

Gray, S., F. Clube, and D. Struchen, "The Holographic Mask Aligner," *Holographic Systems, Components and Applications, Neuchâtel, CH Conference Publication No. 379* (Institution of Electrical Engineers, London, 265 (1993)).

Goering, R., B. Hoefer, A. Kraeplin, P. Schreiber, E. Kley, and V. Schmeisser, "Integration of High Power Laser Diodes with Microoptical Components in a Compact Pumping Source for Visible Fiber Laser," *Proc. SPIE*, **3631**, 191–197 (1999).

Haggans, C. W. and R. Kostuk, "Use of Rigorous Coupled-Wave Theory for Designing and Tolerancing Surface-Relief Diffractive Components for Magneto-Optical Heads," *Proc. SPIE*, **1499**, 293 296 (1991).

Herzig, H. P. and R. Dändliker, "Diffractive Components: Holographic Optical Elements," in *Perspectives for Parallel Interconnects*, Lalanne Ph. and P. Chavel (eds), Springer-Verlag, Berlin, 1993.

Huang, G., M. Wu, G. Jin, and Y. Yan, "Micro-Optic Element for Optical Disk Memory Read-Write Heads," *Proc. SPIE*, **2152**, 30, 261–265 (1994).

Ichikawa, H., "Analysis of Femtosecond-Order Pulses Diffracted by Periodic Structures," *JOSA A*, **16**, 299–304 (1999).

Jannson, T., T. M. Aye, and G. Savant, "Dispersion and Aberration Techniques in Diffractive Optics and Holography," *Proc. SPIE*, **2152**, 5, 44–70 (1994).

Kamiya, N., "Rigorous Couple-Wave Analysis for Practical Planar Dielectric Gratings: 1. Thickness-Change Holograms and Some Characteristics of Diffraction Efficiency," *Appl. Opt.*, **37**, 5843–5853 (1998).

Kamiya, N., "Rigorous Couple-Wave Analysis for Practical Planar Dielectric Gratings: 2. Diffraction by a Surface-Eroded Hologram Layer," *Appl. Opt.*, **37**, 5854–5863 (1998).

Kinnstatter, K. and K. Ojima, "Amplitude Detection for the Focus Error in Optical Disks Using a Birefringent Lens," *Appl. Opt.*, **29**, 4408–4413 (1990).

Kluepfel, B. and F. Ross (eds), *Holography Market Place*, Ross Books, Berkeley, CA, 1991.

Kosoburd, T., M. Akivis, Y. Malkin, B. Kobrin, and S. Kedmi, "Beam Shaping with Multilevel Elements," *Proc. SPIE*, **2152**, 48, 214–224 (1994).

Kostuk, R. K., J. H. Yeh, and M. Fink, "Distributed Optical Data Bus for Board-Levels Interconnects," *Appl. Opt.*, **32**, 5010–5021 (1993).

Kunstmann, P. and H. J. Spitschan, "General Complex Amplitude Addition in a Polarization Interferometer in the Detection of Pattern Differences," *Opt. Commun.*, **4**, 166–172 (1971).

Lessard, R. A. and B. J. Thompson (eds), "Selected Papers on Photopolymers," *Proc. SPIE*, Milestone Series, **MS114** (1995).

Litchtenberg, B. and N. C. Gallagher, "Numerical Modeling of Diffractive Devices Using the Finite Element Method," *Opt. Eng.*, **33**, 3518–3526 (1994).

Lightbody, M. T. M., B. Layet, M. R. Taghizadeh, and T. Bett, "Design of Novel Resonance Domain Diffractive Optical Element," *Proc. SPIE*, **2404**, 96–107 (1995).

Liu, B., B. Domg, and B. Gu., "Implementation of Pseudo-Nondiffracting Beams by Use of Diffractive Phase Elements," *Appl. Opt.*, **37**, 8219–8223 (1998).

Lunitz, B. and J. Jahns, "Computer Generated Diffractive Axicons," *European Optical Society*, **22**, 26–27 (1999).

McCormick, F. B., F. A. P. Tooley, T. J. Cloonan *et al.*, "Experimental Investigation of a Free-Space Optical Switching Network by Using Symmetric Self-Electro-Optic-Effective Devices," *Appl. Opt.*, **31**, 5431 5446 (1992).

McDowell, A. J., P. B. Conway, A. C. Cox, R. Parker, C. W. Slinger, and A. P. Wood, "An Investigation of the Defects Introduced when Diamond Turning Hybrid Components for Use in Infrared Optical Ssytem," *OSA Technical Digest Diffractive Optics: Design, Fabrication, and Applications*, **11**, 99–102 (1994).

Mait, J. N., "Understanding Diffractive Optic Design in the Scalar Domain," *J. Opt. Soc. Am. A*, **12**, 2145–2150 (1995).

Maystre D., "Rigorous Vector Theories of Diffraction Gratings," in *Progress in Optics*, Wolf, E. (ed.), North-Holland, Amsterdam, **XXI**, 1 (1984).

Meyers, M. M., "Diffractive Optics at Eastman Kodak Company," *Proc. SPIE*, **2689**, 31, 228–254 (1996).

Miler, M., I. Koudela, and I. Aubrecht, "Zone Areas of Diffractive Phase Elements Producing Focal Annuli," *Proc. SPIE*, **3320**, 210–213 (1998).

Mirotznik, M. S., D. W. Prather, and J. Mait, "Hybrid Finite Element-Bounder Element Method for Vector Modeling Diffractive Optical Elements," *Proc. SPIE*, **2689**, 2, 2–13 (1996).

Missig, M. D. and G. M. Morris, "Diffractive Optics Applied to Eyepiece Design," *OSA Technical Digest Diffractive Optics: Design, Fabrication, and Applications*, **11**, 57–60 (1994).

Miyamoto, Y., M. Masuda, A. Wada, and M. Takeda, "Electron-Beam Lithography Fabrication of Phase Holograms to Generate Laguerre–Gaussian Beams," *Proc. SPIE*, **3740**, 232 (1999).

Moharam, M. G. and T. K. Gaylord, "Rigorous Coupled-Wave Analysis of Surface-Gratings with Arbitrary Profiles," *J. Opt. Soc. Am.*, **71**, 1573–1574 (1981).

Moharam, M. G., D. A. Pommet, and E. B. Grann, "Implementation of the Rigorous Coupled-Wave Technique; Stability, Efficiency, and Covergence," *OSA Technical Digest Diffractive Optics: Design, Fabrication, and Applications*, **11**, 4–7 (1994).

Mok, F. H., "Angle-Multiplexed Storage of 5000 Holograms in Lithium Niobate," *Opt. Lett.*, **18**, 915–918 (1990).

Montamedi, M. E., "Micro-Opto-Electro-Mechanical Systems," *Opt. Eng.*, **33**, 3505–3517 (1994).

Montiel, F. and M. Neviere, "Differential Theory of Gratings: Extension to Deep Gratings of Arbitrary Profile and Permittivity through the R-Matrix Propagation Algorithm," *J. Opt. Soc. Am. A*, **11**, 3241 (1994).

Moreshead, W., J. L. Nogues, L. Howell, and B. F. Zhu, "Replication of Diffractive Optics in Silica Glass," *Proc. SPIE*, **2689**, 16, 142–152 (1996).

Neuman, J., K. S. Wieking, and K. Detlef, "Direct Laser Writing Surface Relief in Dry, Self-Developing Photopolymer Films," *Appl. Opt.*, **38**, 5418–5421 (1999).

Nieuborg, N., A. G. Kirk, B. Morlion, H. Thienpont, and L. P. Veretennicoff, "Highly Polarization-Selective Diffractive Optical Elements in Calcite with an Index-Matching Gap Material," *Proc. SPIE*, **3010**, 123–126 (1997).

Noguès, J. L. and A. J. LaPaglia, "Processing Properties and Applications of Sol–Gel Silica Optics," *Proc. SPIE*, **1168** (1989).

Noponen, E. and J. Saarinen, "Rigorous Synthesis of Diffractive Optical Elements," *Proc. SPIE*, **2689**, 6, 54–65 (1996).

Noponen, E., J. Turunen, and F. Wyrowski, "Synthesis of Paraxial Domain Diffractive Elements by Rigorous Electromagentic Theory," *J. Opt. Soc. Am. A*, **12**, 1128 (1995).

Ojima, M., A. Saito, T. Kaku *et al.*, "Compact Magnetooptical Disk for Coded Data Storage," *Appl. Opt.*, **25**, 483–489 (1986).

Oppliger, Y., P. Sixt, J. M. Stauffer, J. M. Mayor, P. Regnault, and G. Voirin, "One-step Shaping Using a Gray-Tone Cmask for Optical and Microelectronics Applications," *Microelectron. Eng.*, **23**, 449–452 (1994).

Parker, W. P., "Commercial Applications of Diffractive Switched Optical Elements (SOE's)," *Proc. SPIE*, **2689**, 27, 195–209 (1996).

Pascual, I., A. Marquez, A. Belendez, A. Fimia, J. Campos, and M. J. Yzuel, "Fabrication of Computer-Generated Phase Holograms Using Photopolymers as Holographic Recoding Material," *Proc. SPIE*, **3633**, 302–305 (1999).

Pawlowski, E. and B. Kuhlow, "Antireflection-Coated Diffractive Optical Elements Fabricated by Thin-Film Deposition," *Opt. Eng.*, **33**, 3537–3546 (1994).

Prather, D. W., M. S. Mirotznik, and J. N. Mait, "Boundary Element Method for Vector Modelling Diffractive Optical Elements," *Proc. SPIE*, **2404**, 28–39 (1995).

Psaltis, D., E. G. Paek, and S. S. Venkatesh, "Optical Image Correlation with a Binary Spatial Light Modulator," *Opt. Eng.*, **23**, 698–704 (1984).

Rakuljic, G. A., V. Leyva, and A. Yariv, "Optical Data Storage by Using Orthogonal Wavelength-Multiplexed Volume Holograms," *Opt. Lett.*, **15**, 1471–1473 (1992).

Sainov, S., M. Mazakova, M. Pantcheva, and D. Tonchev, *Mol. Cryst. Liq. Cryst.*, **152**, 609–612 (1987).

Schwab, M., N. Lindlein, J. Schwider, Y. Amitai, A. A. Friesem, and S. Reinhorn, "Achromatic Diffractive Fan-out Systems," *Proc. SPIE*, **2152**, 12, 173–184 (1994).

Solano, C., P. C. Roberge, R. A. Lessard, Methylene Blue senstized Gelatinas a Photodensitive medium for Conventional and Polarizing Holography, *"Appl. Opt.*, **26**, 1989-1977 (1987).

Sommerfeld, A., *Optics*, Vol. IV, Academic Press, New York, 1962.

Stalder, M. and P. Ehbets, "Electrically Switchable Diffractive Optical Element for Image Processing," *Opt. Lett.*, **19**, 1–3 (1994).

Sutherland, R. L., L. V. Natarajan, V. P. Tondiglia, T. J. Bunning, and W. W. Adams, "Development of photopolymer-liquid crystal composite materials for dynamic holo-gram applications, *Proc. SPIE*, **2152**, 303–312, 1994b.

Sutherland, R. L., L. V. Natarajan, V. P. Tondiglia, T. J. Bunning, and W. W. Adams, "Electrically Switchable Volume Gratings in Polymer-Dispersed Liquid Crystal," *Appl. Phys. Lett.*, **64**, 1074–1076 (1994).

Sweat, W. C., "Describing Holographic Optical Elements as Lenses," *J. Opt. Soc. Am.*, **67**, 803–808 (1977).

Sweat, W. C., "Mathematical Equivalence between a Holographic Optical Element and a Ultra-High Index Lens," *JOSA*, **69**, 486–487 (1979).

Tgran, V. G., J. F. Viens, A. Villeneuve, K. Richardson, and M. A. Duguay, "Photoinduced Self-Developing Relief Gratings in Thin Film Chalcogenide As_2s Si_3 Glasses," *J. Lightwave Technol.*, **15**, 1343–1345 (1997).

Tyan, R. C., P. C. Sun, and Y. Fainman, "Polarizing Beam Splitters Constructed of Form-Birefringent Multilayer Grating," *Proc. SPIE*, **2689**, 8, 82–89 (1996).

Velzel, C. H. F., F. Wyrowski, and H. Esdink, "Pattern Projection by Diffractive and Conventional Optics: a Comparison," *Proc. SPIE*, **2152**, 24, 206–211 (1994).

Wei, M., E. H. Anderson, and D. T. Attwood, "Fabrication of Ultrahigh Resolution Gratings for X-ray Spectroscopy," *OSA Technical Digest Diffractive Optics: Design, Fabrication, and Applications*, **11**, 91–94 (1994).

Welford, W. T., *Aberrations of Optical System*, Adam Hilger Ltd., Boston, 1996.

Wu, C., "Method of Making High Energy Beam Sensitive Glasses," U.S. Patent No. 5,078,771 (1992).

Wyroski, F., H. Esdonk, R. Zuidema, S. Wadmann, and G. Notenboom, "Use of Diffractive Optics in Material Processing," *Proc. SPIE*, **2152**, 16, 139–144 (1994).

Zaidi, S. H. and S. R. J. Brueck, "High Aspect-Ratio Holographic Photoresist Gratings," *Appl. Opt.*, **27**, 2999–3002 (1988).

Zhang, J. and M. B. Sponsler, "Switchable Liquid Crystalline Photopolymer Media for Holography," *J. Am. Chem. Soc.*, **14**, 1516–1520 (1992).

Zhang, Y., B. Gu, B. Dong, and G. Yang, "Design of Diffractive Phase Elements for Beam Shaping in the Fractional Fourier Transform Domain," *Proc. SPIE*, **3291**, 58–67 (1998).

Zhou, Z. and T. J. Drabik, "Optimized Binary, Phase-Only Diffractive Optical Elements with Sub-wavelength Features for 1.55 μm," *J. Opt. Soc. Am. A*, **12**, 1104–1112 (1995).

7

Some Lens Optical Devices

DANIEL MALACARA, JR.

Centro de Investigacionos en Optica, León, Mexico

7.1 INTRODUCTION

In this chapter some of the most important optical systems using lenses will be described. [1–4] However, telescopes and microscopes are not described here, since they are the subject of other chapters in this book. Since optical instruments cannot be studied without a previous background on the definitions of pupils, principal ray and skew and meridional rays, we will begin with a brief review of these concepts.

7.2 PRINCIPAL AND MERIDIONAL RAYS

In any optical system with several lenses two important meridional rays can be traced through the system, as shown in Fig. 7.1. [5] A meridional ray is in a common plane with the optical axis. A skew ray, on the other hand, is not.

(a) A ray from the object plane, on the optical axis, to the image plane, also on the optical axis, is called an on-axis meridional ray.

(b) A meridional ray from the edge (or any other off-axis point) on the object to the corresponding point on the image is called the chief or principal ray.

All the lenses in this optical system must have a minimum diameter to allow these two rays to pass through the whole system. The planes on which the on-axis meridional ray crosses the optical axis are conjugates to the object and image planes. Any thin lens at these positions does not affect the path of the on-axis meridional ray but affects the path of the principal ray. This lens is called a field lens. The diameter of the field lenses determines the diameter of the image (field). If a field lens limits the

191

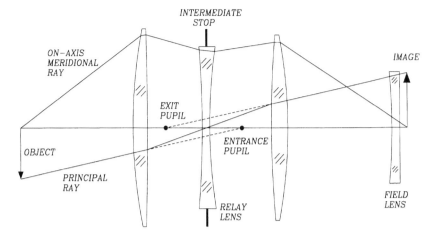

Figure 7.1   An optical system illustrating the concepts of stop, entrance pupil, and exit pupil.

field diameter more than desired by stopping a principal ray with a certain height, we have an effect called vignetting. The image does not have a sharp boundary, but its luminosity decreases very rapidly towards the outside of the field when maximum image height has been reached, i.e., when vignetting begins, because the effective pupil size decreases very fast, as shown in Fig. 7.2.

The planes on which the principal ray crosses the optical axis are said to be pupils or stop planes. At the first crossing of the ray, if it occurs before the first

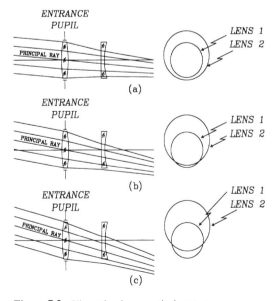

Figure 7.2   Vignetting in an optical system.

surface of the system, or at its straight extrapolation after leaving, the ray from the object, is the entrance pupil. At the last crossing of the ray, if it occurs after the last surface of the system, or at its straight extrapolation before arriving, the principal ray to the image, is the exit pupil. Any lens located at the pupil planes affects the path of the meridional ray but not the path of the principal ray. This lens is an imaging or relay lens. The diameter of the relay lenses determines the aperture of the system.

7.3 MAGNIFIERS

The most common and traditional use of a lens is as a magnifier. The stop of the system is the pupil of the observing eye. Since with good illumination this pupil is small (3–4 mm or even smaller), the aberrations that increase with the diameter of the pupil, such as spherical aberration, do not present any problem. On the other hand, the field of view is not small.; hence, field curvature, astigmatism, and distortion are the most important aberrations to be corrected.

The angular magnification M is defined as the ratio of the angular diameter β of the virtual image observed through the magnifier to the angular diameter α of the object as observed from a distance of 250 mm (defined as the minimum observing distance for young adults). If the lens is placed at a distance from the observed object so that the virtual image is at distance l' from the eye, the magnification M is given by

$$M = \frac{\beta}{\alpha} = \frac{250}{(l' + d)} \left(\frac{l'}{f} + 1 \right), \tag{7.1}$$

where d is the distance from the lens to the eye, f is the focal length, and all distances are in millimeters.

The maximum magnification, obtained when the virtual image is as close as 250 mm in front of the observing eye, and the lens is close to the eye $(d = 0)$ is

$$M = \frac{250}{f} + 1. \tag{7.2}$$

If the virtual image is placed at infinity to avoid the need for eye accommodation the magnification becomes independent of the distance d and has a value

$$M = \frac{250}{f}. \tag{7.3}$$

Here, we have to remember that eye accommodation occurs when the eye lens (crystalline) modifies its curvature to focus close objects.

We can thus see that for small focal lengths f, the magnification is nearly the same for all lens' positions with respect to the eye and the observed object, as long as the virtual image is not closer than 250 mm from the eye.

It has been shown [6] that if the magnifier is a single plano-convex lens, the optimum orientation to produce the best possible image with the minimum aberrations is

(a) with the plane on the eye's side if the lens is closer to the eye than to the object, as shown in Fig. 7.3(a), and

(b) with the plane on the side of the object if the lens is closer to the object than to the eye, as shown in Fig. 7.3(b).

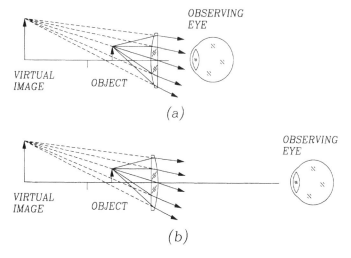

Figure 7.3 A simple magnifier (a) with the observing eye close to the lens and (b) with the observing eye far from the lens.

The disadvantage of the plano-convex magnifier is that if one does not know these rules one may use it with the wrong lens orientation. A safer magnifier configuration would be a double convex lens, but then the image is not the best. For this reason most high-quality magnifiers are symmetrical. To reduce the aberrations and have the best possible image, more complicated designs can be used, as illustrated in Fig. 7.4.

7.4 OPHTHALMIC LENSES

A human eye may have refractive defects that produce a defocused or aberrated image in the retina. The most common defects are:

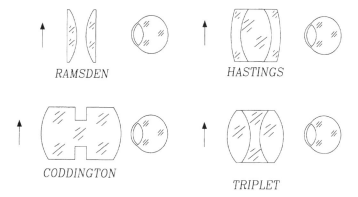

Figure 7.4 Some common magnifiers.

(a) Myopia, when the image of an object at infinity falls in front of the retina. Eye accommodation cannot compensate this defocusing error. A myopic eye sees in focus only close objects (see Fig. 7.5(b)).

(b) Hypermetropia, when the image of an object at infinity falls behind the retina. This error can be compensated in young people by eye accommodation. An hypermetropic eye feels the problem if it cannot accommodate either due to age or to the high magnitude of the defect (see Fig. 7.5(c)).

(c) Astigmatism, when the rays in two planes passing through two perpendicular diameters on the pupil of the eye have different focus positions along the optical axis (see Fig. 7.5(d)).

These refractive defects are corrected by means of a single lens in front of the eye. The geometry used to design an ophthalmic lens is shown in Fig. 7.6. The eye rotates in its skull socket to observe objects at different locations away from the lens' optical axis. Thus, its effective stop is not at the pupil of the eye but at the center of rotation of the eye.

The ophthalmic lens is not in contact with the eye, but at a distance d_v of about 14 mm in front of the cornea. An image magnification is produced because of the lens separation from the cornea. The focus of the ophthalmic lens has to be located at the point in space that is conjugate to the retina. This point is in front of the eye for myopic eyes and behind the eye for hypermetropic eyes.

Since the distance d_v is a fixed constant parameter, the important quantity in the ophthalmic lens is the back (or vertex) focal length F_v. The inverse of this vertex focal length is the vertex power

$$P_v = \frac{1}{F_v},$$ (7.4)

whyere P_v is in diopters if F_v is expressed in meters.

When an eye is corrected with an ophthalmic lens, the apparent image size changes. The magnification M produced by this lens is

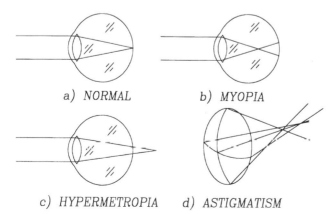

a) NORMAL   b) MYOPIA

c) HYPERMETROPIA   d) ASTIGMATISM

Figure 7.5 Geometry of vision refractive errors.

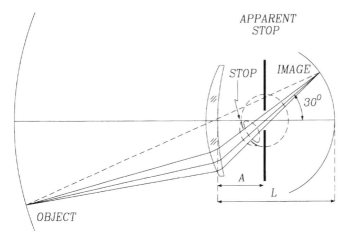

Figure 7.6 Geometry used to design eyeglasses.

$$M = \left(\frac{1}{1 - dP} - 1 \right) \times 100\%. \tag{7.5}$$

The eye with refractive defects has a different image size than the normal (emetropic) eye. Myopia is due to abnormal elongation image size of the eye globe. Hypermetropia is due to an abnormal shortening of the eye globe. Astigmatism arises when the curvature of the cornea (or one of the surfaces of the eye lens) has two different values in mutually perpendicular diameters, like in a toroid. The diameter of the eye globe, with a refractive defect of power P, is given by the empirical relation

$$D = \frac{P_v}{6} + 14.5, \tag{7.6}$$

where P_v is the vertex power in diopters of the required lens and D is in millimeters.

When designing an ophthalmic lens, the lens surface has to be spherical, concentric with the eye globe.

Since the pupil of the eye has a small diameter, the shape of the lens has to be chosen so that the field curvature and the off-axis astigmatism are minimized, with an equilibrium that can be selected by the designer. Two solutions are found, as shown in the Tscherning ellipses in Fig. 7.7. In these ellipses, we can observe the following:

(a) Two possible solutions exist: one is the Ostwald lens and the other is the more curved Wollaston lens.

(b) The solutions for zero-field curvature (off-axis power error) and for off-axis astigmatism are close to each another.

(c) There are no solutions for lenses with a vertex power larger than about 7–10 diopters. If an aspheric surface is introduced, the range of powers with solutions is greatly extended.

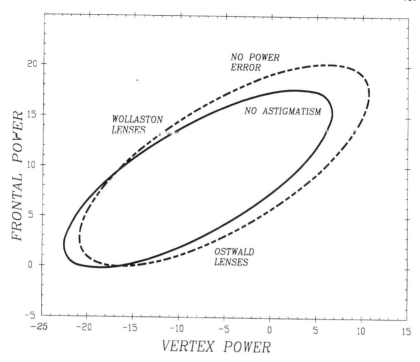

Figure 7.7 Ellipses for zero astigmatism and no power error in eyeglasses.

7.5 ACHROMATIC DOUBLETS AND COLLIMATORS

A single lens has axial chromatic aberration. Its spherical aberration can be minimized with the proper shape (bending), but it can never be zero. The advantage of a doublet made by joining together two lenses with different glasses is that a good correction of both axial chromatic and spherical aberrations can be achieved.

Considering a doublet of two thin lenses with focal lengths f_1, and f_2, the focal length of the combination for red light (C) and blue light (F) can be made the same to eliminate the axial chromatic aberration if

$$(n_{1C} - n_{1F})K_1 = (n_{2C} - n_{2F})K_2, \tag{7.7}$$

where K_i has been defined by the lens maker's equation:

$$\frac{1}{f_i} - (n_i - 1)\left(\frac{1}{r_{1i}} - \frac{1}{r_{2i}}\right) = (n_i - 1)K_i. \tag{7.8}$$

From Eq. (7.7) we can find

$$f_1 V_1 = -f_2 V_2 \tag{7.9}$$

where the Abbe number V_1 of the glass i has been defined as

$$V_i = \frac{(n_{iD} - 1)}{(n_{iC} - n_{iF})} \tag{7.10}$$

and from these expressions we finally obtain

$$f_1 = F\left(1 - \frac{V_2}{V_1}\right) \tag{7.11}$$

and

$$f_2 = F\left(1 - \frac{V_1}{V_2}\right). \tag{7.12}$$

We see that any two glasses with different values of the Abbe number V can be used to design an achromatic doublet. However, a small difference in the V values would produce lenses with low power, which are desirable. This means that large differences in the values of V are appropriate. Typically, the positive (convergent) lens is made with crown glass ($V > 50$) and the negative (divergent) lens with flint glass ($V < 50$). Of course, these formulas are thin-lenses approximations, but they produce a reasonable close solution to perform ray tracing in order to find an exact solution.

Another advantage of a doublet made with two different glasses is that the primary spherical aberration can be completely corrected if the proper shape of the lenses is used, as in Fig. 7.8; the design data are given in Table 7.1.

A disadvantage of the cemented doublet is that the primary spherical aberration is well corrected but not the high-order spherical aberration. A better correction is obtained if the two lens components are separated, as in Fig. 7.9; the design data are given in Table 7.2.

Achromatic lenses are used as telescope objectives or as collimators. When used as laser collimators, the off-axis aberrations are not as important as in the telescope objective.

7.6 AFOCAL SYSTEMS

An afocal system by definition has an infinite effective focal length. Thus, the image of an object at an infinite distance is also at an infinite distance. However, if the object is at a finite distance in front of the system a virtual image is formed, or vice

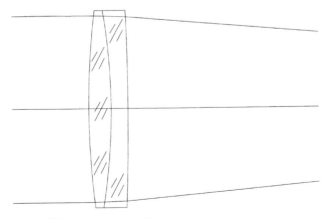

Figure 7.8 A cemented doublet.

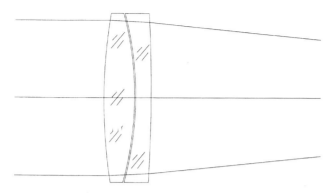

Figure 7.9 An air-spaced doublet.

Table 7.1 Achromatic Doublet (from Malacara and Malacara [6])

Radius of curvature (mm)	Diameter (mm)	Thickness (mm)	Material
12.8018	3.41	0.434	BaK-1
−9.0623	3.41	0.321	SF-8
−37.6563	3.41	19.631	Air
Focal ratio: 5.86632			
Effective focal length (mm):	20.0042		
Back focal length (mm):	19.6355		
Front focal length (mm):	−19.9043		

Table 7.2 Broken-Contact Aplanatic Achromatic Doublet (from Malacara and Malacara [6])

Radius of curvature (mm)	Diameter (mm)	Thickness (mm)	Material
58.393	20.0	4.0	BK7
−34.382	20.0	0.15	Air
−34.677	20.0	2.0	F2
−154.68			Air
Focal ratio: 4.99942			
Effective focal length (mm):	99.9884		
Back focal length (mm):	96.5450		
Front focal length (mm):	−99.3828		

versa. Let us consider, as in Fig. 7.10, a real object with height H equal to the semidiameter of the entrance pupil and the principal ray (dotted line) entering with an angle α. A marginal ray (solid line) enters parallel to the principal ray. Since the system is afocal and these two rays are parallel to each other, they will also be parallel to each other when they exit the system. Thus, we can write

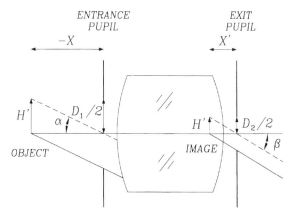

Figure 7.10 An afocal system forming a real image of a virtual object.

$$M = \frac{\tan \beta}{\tan \alpha} = \frac{D_2}{D_1} \frac{X}{X'} = \frac{1}{M} \frac{X}{X'}. \tag{7.13}$$

The distance X is positive if the object is after the entrance pupil (virtual object) and the distance X' is positive if the image is after the exit pupil (real image). These two quantities always have the same sign. The total distance L between the object and the image is given by

$$L = L_p - X\left(1 - \frac{1}{M^2}\right), \tag{7.14}$$

where L_p is the separation between the entrance and exit pupils. We notice then that the lateral magnification H'/H is a constant equal to the angular magnification of the system.

Another interesting property of afocal systems is that, if the object and image planes are fixed in space, the image can be focused by axially moving the afocal system, without modifying the lateral magnification.

These properties find applications in microlithography.

7.7 RELAY SYSTEMS AND PERISCOPES

A periscope is an afocal instrument designed to observe through a tube or long hole, as shown in Fig. 7.11.

An alternate series of imaging or relay lenses (RL) and field lenses (FL) form the system. The first imaging lens is at the entrance pupil and the last imaging lens is near the exit pupil, at the pupil of the observing eye. The field lenses are used to maintain the principal ray propagating to the system. Thin field lenses introduce almost no chromatic aberrations and distortion; hence, these are frequently thin single lenses. Imaging lenses, on the other hand, produce axial chromatic aberration. For this reason they are represented here by doublets.

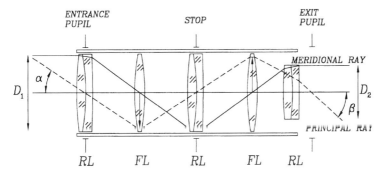

Figure 7.11 Optical schematics of a periscope.

A disadvantage of this system is that all lenses have positive power, producing a large Petzval sum. Thus, the only way of controlling the field curvature is by the introduction of a large amount of astigmatism.

It must be pointed out that the ideal field lenses are at the intermediate images positions: i.e., where the meridional ray crosses the optical axis. Then, the images will be located on the plane of the field lenses and any dirt or imperfection on these lenses will appear sharply focused on top of the image. This is not convenient. For this reason field lenses are axially displaced a little from the intermediate images positions.

Imaging lenses are located at the pupils' positions, where the principal ray crosses the optical axis. This location avoids the introduction of distortion by these lenses. However, the overall aberration balancing of the system may call for a different lens location.

As in any afocal system, if the object is located at infinity and the image is also located at infinity the meridional ray is parallel to the optical axis in the object space as well as in the image space. Under this condition, the ratio of the slope β of the principal ray in the image space to the slope α in the object space is the angular magnification of the system, given by

$$M = \frac{\tan \beta}{\tan \alpha} = \frac{D_1}{D_2} \tag{7.15}$$

as in any afocal system, where D_1 is the diameter of the entrance pupil and D_2 is the diameter of the exit pupil.

7.8 INDIRECT OPHTHALMOSCOPES AND FUNDUS CAMERA

These ophthalmic instruments are periscopic afocal systems, designed to observe the retina of the eye. They have the following important characteristics:

(a) The entrance pupil must have the same position and diameter as the eye pupil of the observed patient.
(b) To have a good observed field of the retina of the patient, the angular magnification should be below 1. Thus, the exit pupil should be larger than the entrance pupil.

A version of this system is illustrated in Fig. 7.12. The first lens in front of the eye is an imaging lens. The image of the retina is formed at the back focal plane of this lens. The height of the meridional ray (solid line) is the semidiameter of the pupil of the observed eye. This small aperture makes the on-axis aberations of this lens very small. This principal ray (dotted lines) arrives far from the center of the imaging lens, marking its diameter quite large if a good field angular diameter is desired. In order to form a good image of the pupil of the observed eye this imaging lens has to be aspheric.

At the image of the pupil of the observed eye a stop is located, which has three small windows. Two windows provide a stereoscopic view of the retina, sending the light from each window to a different observing eye. The third window is used to illuminate the retina of the observed eye.

The angular magnification M of the ophthalmoscope, as pointed out before, should be smaller than 1. Thus, the focal length of the eyepieces should be larger than the focal length of the aspheric lens (about five times larger).

A slightly different indirect ophthalmoscope is illustrated in Fig. 7.13, using a second imaging lens at the pupil of the observed eye, thus producing an erected image at the focal planes of the eyepieces. This is an achromatic lens, since the image of the pupil has been magnified. The aperture of this lens has a stop in contact with three small apertures: one for illumination and two for stereoscopic observation. The final image is observed with a pair of Huygens' eyepieces.

The angular magnification M of this ophthalmoscope is given by

$$M = \frac{\tan \beta}{\tan \alpha} = m\frac{f_a}{f_e}, \tag{7.16}$$

where f_a is the effective focal length of the aspheric lens, f_e is the effective focal length of the eyepiece, and m is the lateral magnification of the achromatic lens located at the image of the pupil of the observed eye.

To effectively use all the field width provided by the aspheric lens aperture, the tangent of the angular field semidiameter α_e of the eyepiece should be equal to the

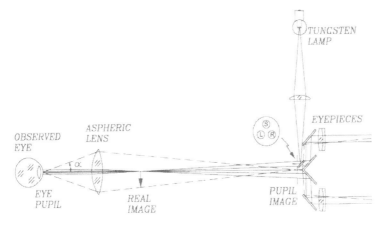

Figure 7.12   Portable stereoscopic indirect ophthalmoscope.

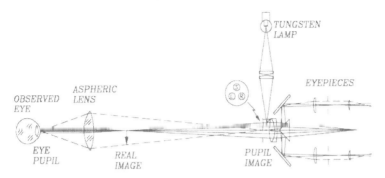

Figure 7.13 Schematics of a stereoscopic indirect ophthalmoscope.

tangent of the angular field semidiameter β of the aspheric lens multiplied by the angular magnification M, as follows:

$$\tan \beta = M \tan \alpha_e. \tag{7.17}$$

7.9 PROJECTORS

Projectors are designed to form the amplified real image of a generally flat object on a flat screen. There are several kinds of projectors, but their optical layout is basically the same.

7.9.1 Slide and Movie Projectors

The basic optical arrangemen in a slide projector is shown in Fig. 7.14. The light from the lamp must reach the image screen as much as possible, after illuminating the slide in the most homogeneous possible manner. If a lamp without reflector is used, a concave mirror with the center of curvature at the lamp is employed.

In order to achieve this, the illuminating system must concentrate the maximum possible light energy at the entrance pupil of the objective, after passing through the slide. The light must have a homogeneous distribution over the slide. There are two basic illuminating systems. The classical configuration is shown in Fig. 7.14, where the condenser can have several different configurations. The spherical mirror on the back of the lamp can be integrated in the lamp bulb.

Another more recent illumination configuration is shown in Fig. 7.15. The lamp has a paraboloidal or elliptical reflector with a large collecting solid angle. This lamp produces a more or less uniform distribution of the light on a plane about 142 mm from the rim of the reflector. The condensing lens is aspheric.

7.9.2. Overhead Projectors

An overhead projector is shown in Fig. 7.16. The light source is a lamp with a relatively small filament. The condenser is at the transparency plane, formed by a sandwich of two Fresnel lenses. This condenser forms, at the objective, an image of the small incandescent filament. Thus, the small image of the light source makes the

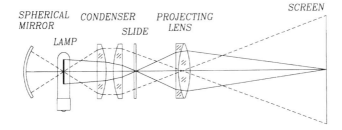

Figure 7.14 A classical slide projector.

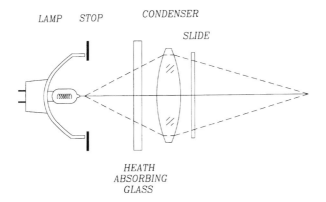

Figure 7.15 Illumination in a modern slide projector.

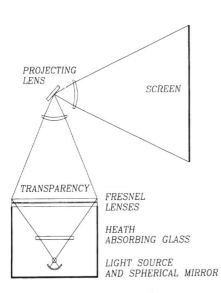

Figure 7.16 An overhead projector.

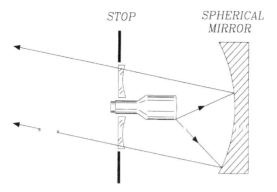

Figure 7.17 A television projector.

effective aperture of the objective also small. This minimizes the need for spherical aberration and axial chromatic aberration corrections. Off-axis aberrations are nearly corrected with an almost symmetrical configuration of the objective, formed by two identical meniscus lenses.

7.9.3 Television Projectors

A basic difference between slide projectors and projectors used for television is that the object and the light source are the same. Thus, the objective has to have a large collecting solid angle. However, this large aperture increases the requirements for aberration corrections. A common system uses an inverted Schmidt configuration, as shown in Fig. 7.17, but many others have been devised.

REFERENCES

1. Kingslake, R., "Basic Geometrical Optics," in *Applied Optics and Optical Engineering*, Vol. I, Kingslake, R. (ed.), Academic Press, San Diego, 1965.
2. Kingslake, R., "Lens Design," in *Applied Optics and Optical Engineering*, Vol. III, Kingslake, R. (ed.), Academic Press, San Diego, 1965.
3. Hopkins, R. R. and D. Malacara, "Optics and Optical Methods," in *Methods of Experimental Physics, Geometrical and Instrumental Optics*, Vol. 25, Malacara, D. (ed.), Academic Press, San Diego, 1988.
4. Welford, W. T., "Aplanatism and Isoplonatism," in *Progress in Optics*, Vol. XIII, Wolf, E. (ed.), North Holland, Amsterdam, 1976.
5. Hopkins, R. E., "Geometrical Optics," in *Methods of Experimental Physics, Geometrical and Instrumental Optics*, Vol. 25, Malacara, D. (ed.), Academic Press, San Diego, 1988.
6. Malacara, D. and Z. Malacara, *Handbook of Lens Design*, Marcel Dekker, New York, 1994.

8

Telescopes

GONZALO PAEZ and MARIJA STROJNIK
Centro de Investigaciones en Optica, León, Mexico

8.1 TRADITIONAL TRANSMISSIVE TELESCOPES

8.1.1 Introduction

According to the telescope's primary function or application, we divide them into three categories: the terrestrial telescope, the astronomical telescope, and the space telescope. While they all meet the same prerequisite of increasing the angular subtense of an object, each of them satisfies different additional requirements.

The layout of the basic telescope is shown in Fig. 8.1. It is used to view an object subtending the angle α at the viewer. By using the correctly curved optical surfaces of the two lenses that comprise the telescope, the rays from this distant object are incident into the observer's eye with a much larger angle of incidence, β. The function of a telescope is most easily understood with reference to Fig. 8.2. The radiation coming from a distant object subtends an angle α when viewed with an unaided (naked) eye. When seen through the telescope, the object is magnified transversely to subtend an angle β at the viewer.

The function of a telescope is exactly the same as that of a microscope, but for a small difference. Even though they both magnify the angular extent of an object that subtends a small angle at the human eye, in the case of a telescope, the small angle subtended by the object arises as a consequence of the (very) large, or infinite, object distance. In a microscope, on the other hand, the small angle is caused by the very small size (height) of the object and the minimum accommodation distance of the human eye.

So, we can summarize that we design and build the instruments, such as microscopes and telescopes, so that we adapt the world around us to our limited visual opto-neural system. Here we remember that the distribution of the rods and

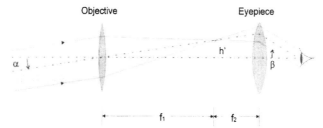

Figure 8.1 Basic (Keplerian) telescope with two converging lenses separated by the sum of their focal distances. The eyepiece lens, located closer to the viewer, is a strongly convergent lens. The collimated rays incident at an angle α are incident into the observer's eye on the right with a higher angle, β.

cones on the retina defines the limiting resolution of a normal human eye. The accommodation distance of the lens defines the minimum comfortable viewing distance of the human eye. The design procedures for classical telescopes have been described in the literature by many authors (see, for example, [1–3]).

8.1.2 Resolution

The resolution of a telescope may be defined in a number of ways. When a telescope with a circular aperture is diffraction-limited, the image of a point source is not a point, but rather a bright disk surrounded by a series of bright rings of much smaller and decreasing amplitude. Figure 8.3 shows the normalized intensity pattern obtained in the focal plane of a traditional telescope with a circular aperture. Theoretically, there is zero intensity in the first minimum, from the central peak, defining the first dark ring. About 84% of the optical energy is inside the first dark ring. According to the diffraction theory of imaging, the resulting intensity distribution is proportional to the square of the Fourier transform of a disk aperture, i.e., a Bessel function of the order 1 over its argument $[J_1(\rho)/\rho]$, in the case of a circular aperture; see [4, 5] for more examples. The bright spot inside the first dark ring generated by the diffraction pattern is called the Airy disk. The insert in Fig. 8.3 shows the corresponding gray-scale intensity, assuming a low exposure as expected for most applications of the telescopes. The first bright ring is barely visible.

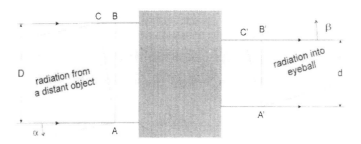

Figure 8.2 The function of a telescope is to increase the angle that the distant object subtends at the observer.

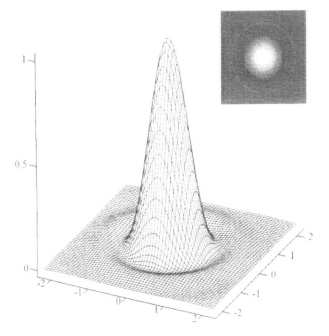

Figure 8.3 The normalized intensity pattern obtained in the focal plane of a traditional telescope with circular aperture. Theoretically, there is zero intensity in the first minimum, from the central peak, defining the first dark ring, or the radius of the resolution spot. The inset shows the corresponding gray-scale intensity, assuming a low exposure as expected for most applications of the telescopes. The first bright ring is barely visible.

The radius of the bright disk is defined by the radius of the first zero intensity ring,

$$r = 1.22\lambda f/D \text{ [m]}. \tag{8.1}$$

Here λ is wavelength [m], f is the focal length of the imaging system [m], and D is the diameter of the aperture stop [m]. The quantities are illustrated in Fig. 8.4 for a general optical system with the equivalent focal distance f and the stop or aperture of diameter D. Angle α is defined as $D/2f$.

The larger the aperture, the higher the angular resolution of the telescope. For the same magnitude of the figuring and the alignment errors, and the aberrations, the diameter of the diffraction spot increases also with wavelength. Viewed from a different perspective, the same telescope may give a diffraction-limited performance in the infrared (IR) spectral region, but not in the visible. A telescope is said to be diffraction-limited when the size of the diffraction spot determines the size of the image of a point source. Even though the stars come in different sizes, they are considered point sources due to their large (object) distance. Their image size depends on the telescope resolution or the seeing radius, depending on the telescope diameter (see Sec. 8.2.4.2), rather than the geometrical dimensions of the stars. Astronomers identify the stars by the shape and the size prescribed by the diffraction spot diameter; see, for example, [6, 7].

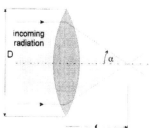

Figure 8.4 Any optical system, including a telescope, may be replaced by an equivalent optical system with limiting stop of diameter D and focal length f.

According to the Rayleigh resolution criterion, two point sources may be just distinguishable if the zero in the image-intensity distribution of the first one coincides with the peak of the second one. This means that, in the image space, they are separated by the distance given in Eq. (8.1). Figure 8.5 shows the normalized intensity patterns of two barely separated point objects obtained in the focal plane of a traditional telescope with circular aperture. Theoretically, the intensity minimum between the two peaks reaches 0.735. The first dark ring is barely seen, surrounding both spots. The inset shows the corresponding gray-scale intensity distribution, assuming a low exposure as expected for most applications of the telescopes. The separate low exposure images are clearly identifiable as arising from two individual point sources. The first bright ring is nearly invisible.

8.1.3 Geometrical Optics

In addition to diffraction effects, any optical instrument, including the telescope, also suffers from performance degradation due to aberrations, fabrication imperfections, and alignment errors. All these effects contribute to the spreading of the image point into a bright fuzzy disk. Aberrations are the variations by which an optical system forms an image of a point source that is spread out into a spot, even in the absence of diffraction effects. They are caused by the shape of the optical component and are a phenomenon purely of a geometrical optics. In the paraxial optics approximation, imaging is seen as merely a conformal transformation: the image of a point is a point, the image of a straight line is a straight line. Strictly speaking, spreading of the point into a spot is a consequence of aberrations, while deformation of a line into a curve is interpreted as a deformation. The true third-order aberrations are spherical aberration, astigmatism, and coma. A telescope, or for that matter, any optical system corrected for spherical aberration and coma, is referred to as aplanatic to the third order. When an optical system is additionally free of astigmatism, it is called aplanatic and anastigmatic. The shape of the surface on which the best image is located, the so-called Petzval curvature, is also controllable within third-order aberration theory.

Additionally, there is also a tilt and defocus which may be compensated for by the proper image displacement and image plane orientation.

When the spot size due to the effects of aberrations is smaller than the disk inside the first dark ring of the diffraction pattern, the resolution is still given by Eq.

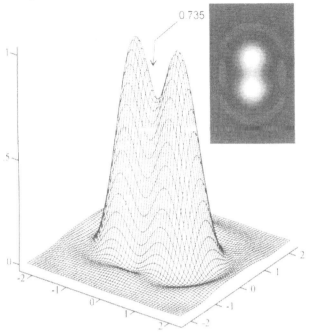

Figure 8.5 The normalized intensity patterns of two barely separated point objects obtained in the focal plane of a traditional telescope with a circular aperture. Theoretically, the intensity minimum between the two peaks reaches 0.735. The first dark ring is barely seen, surrounding both spots. The inset shows the corresponding gray-scale intensity distribution, assuming a low exposure as expected for most applications of the telescopes. The separate low exposure images are clearly identifiable as arising from two individual point sources. The first bright ring is nearly invisible.

(8.1). When this condition is met, the optical system is said to be limited in its performance by the diffraction effects; or put more succinctly, it is a diffraction-limited optical system.

The resolution of a moderately aberrated optical system is defined by the radius of the spot within which 90% of rays, originating at a point source, cross the image surface. This is often obtained visually, upon the review of the spot diagram. Figure 8.6 shows the spot diagrams for three image heights of a diffraction-limited Ritchey–Chretien telescope configuration with a two-element field corrector (see Sec. 8.2.2.3) [8]. The squares indicate the CCD pixel size, whose side has been chosen to be equal to the resolution in Eq. (8.1).

Figure 8.7 shows the intensity as a function of radial distance for an aberrated and diffracted traditional telescope in the best focus location. The parameter is the wavefront error due to the spherical aberration at the aperture edge, increasing from zero to 0.8λ in increments of 0.2λ. The central peak (Fig. 8.7(top)) decreases as the aberration is increased, while the intensity peak of the annulus increases. For large distances, the intensity patterns become similar, even though the similarity onset depends on the amount of the aberration. In the lower part, the same information is shown on a logarithmic scale.

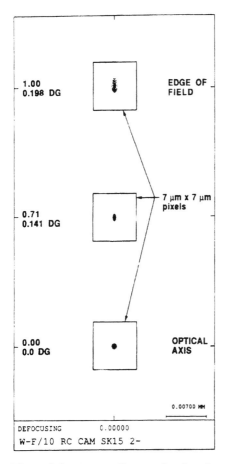

Figure 8.6 The spot diagrams for three image heights of a diffraction-limited Ritchey–Chretien telescope configuration with a two-element field corrector (from Scholl [8]).

Not surprisingly, in the paraxial focus, the effects of the aberrations rapidly overshadow the diffraction effects. The first few dark rings are drowned in the bright spot arising due to the aberrations. This is illustrated in Figure 8.8, where no intensity zero is observed for low-order dark rings. The number of dark rings that disappear is related to the amount of aberrations.

8.1.4 Modulation Transfer Function

The modern ray trace programs tend to summarize the performance of an optical system in terms of its capacity to image faithfully the individual spatial frequency components of the object. This presentation of results assumes that the object intensity is decomposed into its Fourier components. Then, the optical system may be considered as a black box which progressively decreases the modulation of the spatial frequencies and, maybe, modifies its phases. This interpretation is known

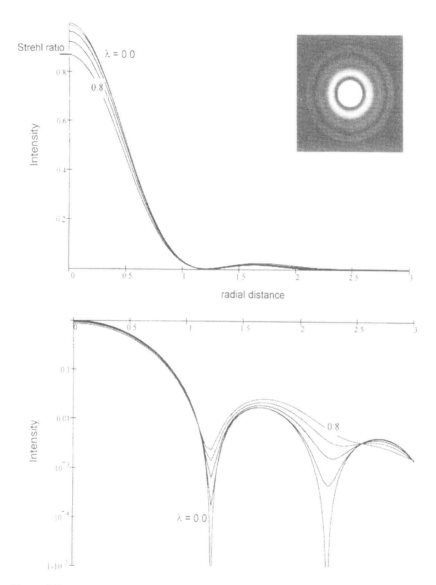

Figure 8.7 The intensity as a function of radial distance for an aberrated and diffracted basic telescope (see Fig 8.1) in the best focus location. The parameter is the wavefront error due to the spherical aberration at the aperture edge, increasing from zero to 0.8λ in increments of 0.2λ.

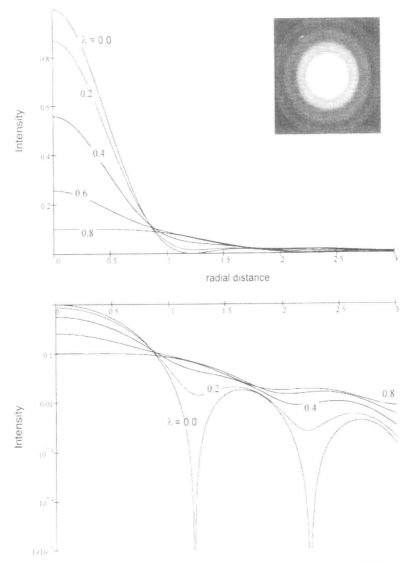

Figure 8.8 The intensity as a function of radial distance for an aberrated and diffracted basic telescope in the paraxial focus location. The parameter is the wavefront error due to the spherical aberration at the aperture edge, increasing from zero to 0.8λ in increments of 0.2λ.

as an optical transfer function (OTF), and its magnitude as a modulation transfer function (MTF), if the phase is not of interest. The significance of the modulation transfer function is illustrated in Fig. 8.9.

Figure 8.10 shows the best (theoretical) modulation transfer function as a function of position in the image plane of a traditional telescope with a circular aperture. Figure 8.10(a) shows the three-dimensional view to illustrate its depen-

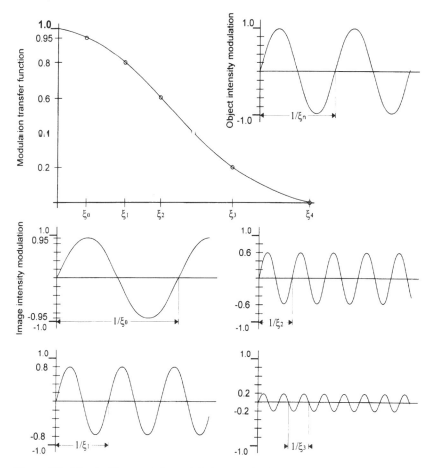

Figure 8.9 The significance of the modulation transfer function.

dence on two spatial frequency coordinates; Fig. (b) shows one cross section of this rotationally symmetric function. There exists one minimum spatial frequency in the object space whose amplitude is zero in the image space. This spatial frequency is referred to as the cutoff frequency, because the optical system images no spatial frequencies larger than this one. For an optical system with a circular aperture it is $D/f\lambda$.

The effects of the aberrations, component misalignment, and the imperfect surface figuring result in an MF that does not achieve the values of the theoretical MTF. This behavior may be seen in Fig. 8.11, showing the OTF as a function of the normalized spatial frequency for the case of 0.2–0.8λ of spherical aberration at the edge of the aperture. The theoretical MTF is shown for comparison. For an MTF of 0.4 as sometimes required for high-resolution imaging, we see that the incremental increase of spherical aberration of 0.2λ decreases the corresponding maximum spatial frequency by about 15%. For smaller values of MTF, the situation is only worsened. Assuming that one is looking for a specific spatial frequency, let us say

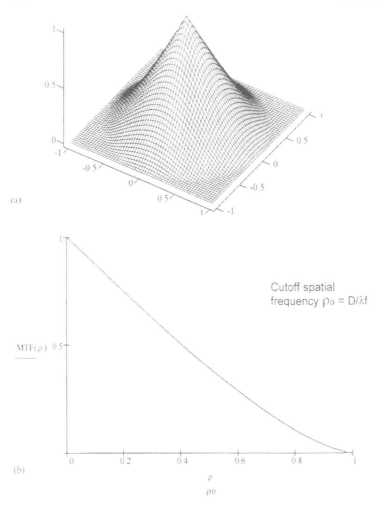

(a)

(b)

MTF(ρ)

Cutoff spatial
frequency ρₒ = D/λf

$\frac{\rho}{\rho_0}$

Figure 8.10 The best (theoretical) modulation transfer function as a function of position in the image plane of a traditional telescope with a circular aperture: (a) the three-dimensional view to illustrate the dependence on two coordinates; (b) one cross section of this rotationally symmetric function. No object frequencies are imaged larger than the system cutoff frequency.

at 0.5, the increment of spherical aberration of 0.2λ results in about 20% reduction in performance. The negative OTF arising from 0.8λ of spherical aberration corresponds to the phase reversal in addition to an insignificant amount of modulation. The MTF, the aberration effects, and image evaluation are described by Smith [9].

The publicized case of the spherical aberration on the primary mirror of the Hubble telescope has been estimated to be about 0.5λ.

The performance of a designed (potential) telescope for the imaging of individual spatial frequencies is best assessed when it is presented in the form of its MTF. Its predicted shape is compared with the optimum theoretically possible for the same

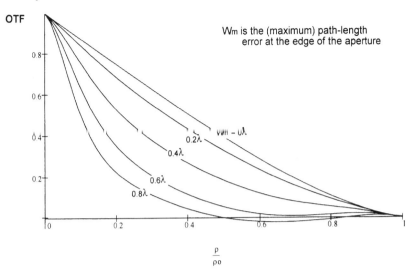

Figure 8.11 The optical transfer function (OTF) as a function of the normalized spatial frequency for the cases of 0–0.8λ of spherical aberration at the edge of the aperture in increments of 0.2λ.

aperture diameter. The MTF of a high-resolution telescope designed to survey the Martian surface with the engineering goal of constructing topographic maps (for requirements, see [10]) is shown in Fig. 8.12. The nearly perfect theoretically achievable MTF is obtained for the three field positions. The MTF value at the design spatial frequency of 71.7 cycles/mm is about 0.47. Fig. 8.12(b) shows that very small MTF degradation of a few percent is encountered for the range of flat CCD displacements out of the best image surface.

Generally, spatial frequencies in the image space with reduced amplitude will not contribute well to the faithful conformal imaging by the optical system. Their magnitudes may also be surpassed by the amplitudes of other sources of optical and electronic noise. Thus, there exists a minimum value of MTF such that the spatial frequencies larger than it are not recoverable from the noise background. This value of MTF_{min} depends on the amount of post-processing and/image enhancement, and on the degree and quality of the quantitative results that will be extracted from the measured intensity distribution in the images. Therefore, the MTF in Fig. 8.12 is shown only for the values of the MTF larger than 0.3 due to the requirement for the recovery of quantitative data.

8.1.5 Refractory or Transmissive Telescopes

The simplest and most common telescope used in the refractory mode is the basic, Keplerian telescope, shown in Fig. 8.1. Two positive lenses with focal distances f_1 and f_2 are combined in such a way that their focal points coincide in the common focal point between them. The magnification of this system is

$$M = -f_1/f_2 = D_1/D_2. \tag{8.2}$$

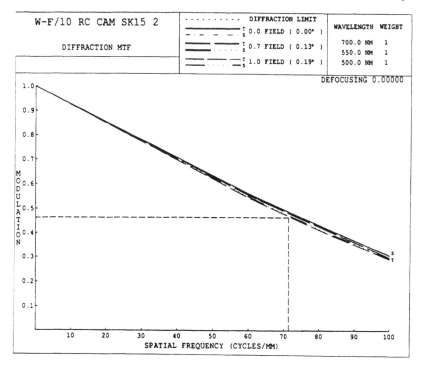

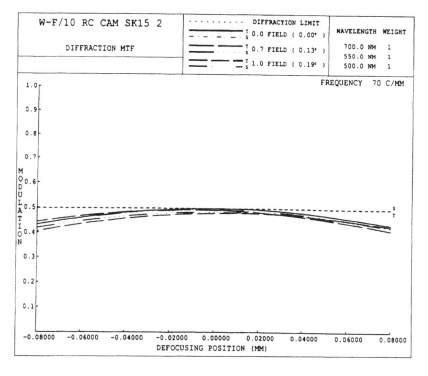

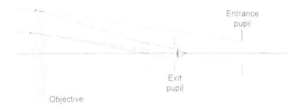

Figure 8.13 A Galilean telescope configuration.

This means that the diameter of the first lens has to be much larger than the second one, and that the ratio of their focal lengths is equal to the required angular magnification.

A Keplerian telescope may also be used as an image erector, or as a beam expander in the relatively low-power laser system. Its advantage of internal focus allows the incorporation of a pinhole for beam centering and spatial cleaning. However, it is not recommended for high-power laser applications due to excessive air heating in the focal volume that may introduce thermal lensing effects, air breakdown, or even generate the second-order electric field effects.

The need for telescope compactness favors a Galilean telescope configuration, depicted in Fig. 8.13. Here, the objective is a strong positive lens, while the eyepiece is a negative lens, placed in such a way that the focal points once again coincide, this time in the image space. The total telescope length is reduced by twice the focal distance of the eyepiece. Also, there is no real focus, making it an ideal candidate as a beam expander for high-power laser applications.

The simplest Galilean telescope is an example of an afocal system, both in the object and in the image space, and is shown in Fig. 8.14. The object space is all the space where the object may be located on the left of the telescope (optical system), assuming the convention that the light is incident from the left as is done in this monogram. The image space is all the space where the image may be found, on the right of the telescope, within the convention stated. An afocal system is one that has an effective focal distance equal to infinity. The effective focal distance is the distance between the stop and the point where the ray incident parallel to the optical axis in the object space crosses the optical axis in the image space. In an afocal system this point is at infinity.

8.1.6 Terrestrial Telescope

The terrestrial telescope is used by hunters and explorers for outdoor applications or as an 'opera glass' to observe new clothes of fine ladies in the audience. For those purposes, it is absolutely necessary that the image be erect, color-corrected, and a

---

Figure 8.12 The modulation transfer function (MTF) of a high-resolution telescope designed to survey the Martian surface with a long-term goal of constructing topographic maps. (a) The nearly perfect theoretically achievable MTF is obtained for the three field positions. The MTF with the values larger than 0.3 is shown only due to the requirements for the recovery of quantitative data. (b) A very small MTF degradation of a few percent is encountered for the range of displacements out of the best image surface. (From Scholl [8]).

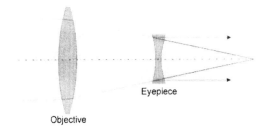

Eyepiece

Objective

Figure 8.14   A Galilean telescope in afocal configuration.

wide half-field of view is highly desirable. The terrestrial telescopes may be used in the visible spectral region and also in the very near-infrared for night-vision, with the night-glow as the source of illumination. The object distance for these telescopes is on the order of 50 m to 1000 m. For some applications like a vision scope, a telephoto arrangement is also desirable. For large distances, they tend to perform like surveillance cameras. They may have half-fields of view up to about 35 degrees.

The first lens, the one close to the object, is called the objective for its proximity to the object. The second lens, the one close to the observer, is referred to as the eyepiece because it is designed for direct viewing in the terrestrial systems. As the eyeball requires clearance, facility for the rotation around the eyeball axis, and a relatively large half-field of view, the eyepiece is designed for these specifications. Eye relief is the distance between the last surface of the eyepiece and the first surface of the human eyeball. It has to be long enough to accommodate (long) eyelashes and/or spectacle lenses for those with astigmatism. The eyepiece is often a compound lens incorporating a field lens. The Huygens, Ramsden, and Kellner eyepieces represent a successive improvement in the order of increasing degree of complexity, larger half-field of view (20 degrees), and consequently, the user's comfort. A review of a number of eyepieces and objectives is described by Smith [9].

The terrestrial telescope incorporates an image erector system. A lens relay system makes the system unnecessarily long, although a prism pair is considered heavier and more compact.

Either Keplerian or Galilean telescopes may be modified to meet the objectives of a terrestrial telescope. In practice, both the objective and the eyepiece are thick lenses, or a combination of lenses in order to minimize the chromatic aberrations and other aberrations, which is annoying to a human observer especially for large-field angles.

8.2   REFLECTING TELESCOPES

8.2.1   Increasing the Light-Collecting Area

An object is seen or detected by the telescope only if enough energy is collected to produce the requisite signal-to-noise ratio at the detector. This may be a photographic plate and, for many current applications, a CCD (charge-coupled device) which has the additional advantage of easy readout and storage for further processing, reusability without the need for refurbishment, and high degree of linearity of response.

The location of the best image produced by the telescope is on a curved surface, requiring image field correctors to accommodate the plane surface of the CCD or other semiconductor (solid state) detectors [11]. The CCD is placed in a plane such that the pixel distance from the surface of the best focus is minimized for all pixels. This is illustrated in Fig. 8.15 for the best image surface displaced only slightly from the CCD surface, due to the incorporation of a two-element corrector system.

The photographic plate still remains the medium of choice for precise spectro-scopic earth-based measurements, because its resolution, environmental stability, and a faithfulness of recording have not been surpassed by any other (electronic) recording medium.

For a great majority of astronomical applications, it is desirable to have a large-diameter primary mirror to collect as much radiation as possible from faint and distant celestial sources. With a large diameter for the primary lens, a number of fabrication and operational issues arise. First, these kinds of telescopes are no longer easily transportable; second, they require stability and a controlled environment, such as that offered by a removable dome in most observatories.

There are basically three significant, fundamental challenges when incorporat-ing a large-diameter primary lens into a transmissive telescope. The first one is due to fabrication of two sides of an optical surface, while the second one is in mounting the

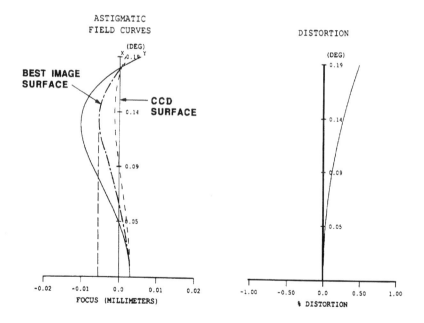

Figure 8.15 The CCD surface is placed in a plane such that the pixel distances from the surface of the best focus are minimized for all pixels. The best image surface is displaced only slightly from the CCD surface, due to the incorporation of a two-element corrector system (from Scholl [8]).

large primary lens in the appropriate structure. It is much easier to support the back of a single reflecting surface while it lays on the ground. This applies equally well to the fabrication stage as to the operational stage. Finally, large glass blanks are hard to produce.

Mirrors have other advantages over lenses. The transmissive components generally suffer from chromatic aberration, making them well corrected only over narrow spectral intervals. Broadband reflective coatings are now available over large spectral intervals with high reflectivity, permitting the use of a single design for different spectral regions. Also, metals perform well as reflectors, whether uncoated like aluminum, or gold, or coated. Finally, a mirror has less aberrations than a lens of the same power.

All-reflective optical systems are also known as *catoptric*. Many well-corrected reflective systems incorporate a transmissive lens or a corrector plate for further correction. When an optical system incorporates reflective as well as refractive components it is known as a *catadioptric* system.

8.2.2 Catadioptric Telescope

Mirror telescopes may be divided into one-, two-, and three-mirror systems. Flat beam turning mirrors are not considered in the mirror count, as they have no power. Many of them additionally incorporate transmissive components for specific aberration or as field correctors (catadioptric). An excellent corrector to the Ritchey–Chretien design has been described by Rosin [12], with the corrector placement in the hole of the primary mirror. The Schmidt camera is probably the best-known system where the aberrations are corrected with a large transmissive element in the incoming light. The field corrector is a lens (combination) used to flatten the surface of the best image location, as commented on for Fig. 8.15.

8.2.2.1 One-Mirror Telescope

The simplest one-mirror system is a parabolic telescope, shown in Fig. 8.16(a), consisting of a simple parabola. In this telescope or a simple collimator, the focal plane work is performed on axis, indirectly obstructing the incoming beam. When the stop is the front focal point, this telescope has no spherical aberration and astigmatism, with the image on a spherical surface.

In a Newtonian telescope, shown in Fig. 8.16(b), a small 45-degree flat mirror is placed in the beam path, deflecting the light out of the incoming beam to the work space. The great disadvantage of the reflecting telescopes is seen clearly in the case of the Newtonian telescope. Not all the area of the primary mirror is available to collect the radiation. The part of the beam taken up by the beam-deflecting 45-degree mirror is reflected back to the object space. Also, its field of view is limited.

By using only the part of the parabola above the focal plane, the need for the beam-turning mirror disappears. This configuration is known as the Herschelian telescope, seen in Fig. 8.16(c), and represents one of the first concepts in off-axis configurations now so popular for stray-light sensitive applications (see Sec. 8.2.6).

The Herschelian telescope and the simple parabola are the simplest collimator or telescope systems, depending on the direction of the incident light.

The elementary implementation of the Schmidt telescope incorporates a spherical primary with a stop and a corrector plate at the center of the curvature

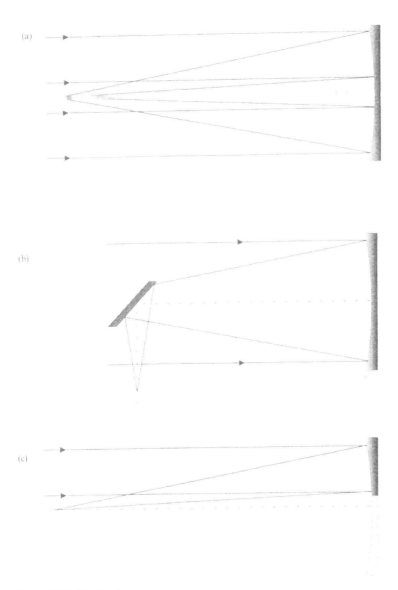

Figure 8.16 The development of a simple one-mirror telescope from (a) a simple parabola, to (b) a Newtonian telescope with an on-axis beam obstruction, to (c) a Herschelian telescope with an off-axis optical configuration that takes full advantage of the aperture of the primary mirror.

of the mirror. The freedom from coma and astigmatism results from monocentric construction with the stop at the mirror's center of curvature. For the object at infinity, the image is free of spherical aberration, coma, astigmatism, and distortion. The focal surface has radius equal to focal length. The design of the corrector plates has recently been discussed thoroughly by Rosete-Aguilar and Maxwell [13].

The fabrication of the corrector plate in the Schmidt telescope is complex, because it is a plate with one flat surface and the other a second-order polynomial. This telescope is popular because its inventor came up with an ingenious way of fabricating the corrective surface. It is one of the first cases of stress polishing, also used much later on for Keck segments. The mirror was polished under vacuum conditions, and then released to assume its correct form. This surface has been simplified in the Bouwers camera, on the basis of the work Schwarzchild had carried out on concentric surfaces [14]. In this case, the stop is at the center of curvature, but the corrector plate is a meniscus lens just before the focal point, so the rays transverse it only once. Its surfaces are concentric with the mirror surface, eliminating all off-axis aberrations. Maksutov designed his camera about the same time, with the added improvement that the radii of curvature of the meniscus lens are related to correct the axial chromatic aberration.

8.2.2.2 Optical Performance of Telescopes with Central Obstruction

The reduction of the effective light-collecting area of the telescope, i.e., the beam incident on the primary mirror, is the consequence of the introduction of a central obscuration. The light-collection area becomes

$$A = \pi(R^2 - r^2) = \pi R^2(1 - \varepsilon^2)\,[\text{m}^2] \tag{8.3}$$

Here R is the radius of the primary, and r is the radius of the central obstruction, usually the secondary assembly, that includes the mirror, the mounts, and any baffling and support structures that interfere with the passage of the incident beam. The obscuration ratio $\varepsilon = r/R$ is a parameter with a value of zero for no obscuration.

The intensity of the image of a point source at infinity in the telescope with the central obscuration is given as a function of radial distance in Fig. 8.17. The obscuration ratio ε is varied from 0 to 0.8 in increments of 0.2. The curve corresponding to $\varepsilon = 0$ has been shown in Fig. 8.3: the curves here are given in a logarithmic scale in order to see the trends clearly. As the obscuration radius increases, the first zero of the intensity pattern moves to shorter radial distances, thus increasing the resolution of the optical system [6]. This effect has been utilized also in microscopy to achieve so-called *ultra-resolution*. Also, the successive zeros in the intensity pattern are moved to shorter radial distances from the central peak.

In Fig. 8.17(a) where the intensities of all curves are normalized at the origin, we see that the increase in the beam obstruction ratio results in the increase in the intensity height of the first bright annulus. The relative intensity peak of the first annulus with respect to the first peak decreases with the increasing value of the obscuration ratio ε. In Fig. 8.17(b) we observe, that the actual peak of the intensity in the first annulus achieves the highest value for some intermediate value of ε, at about 0.5, corresponding to the area ratio of 25%, the value considered very desirable in telescope designs.

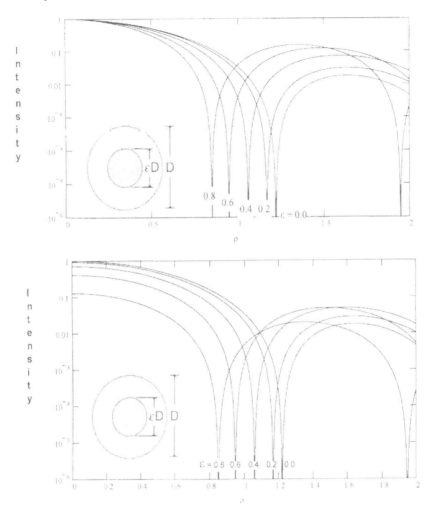

Figure 8.17 The intensity of the image of a point source at infinity in the telescope with a central obscuration as a function of radial distance with the obscuration ratio ε as a parameter, varying from 0 (no obscuration) to 0.8 in increments of 0.2.

Figure 8.18 gives the integrated intensity as a function of radial position for the obscuration ratio ε as a parameter, varying from 0 (no obscuration) to 0.8 in increments of 0.2. We identify the dark rings when the integrated intensity does not change with increasing radial distance, i.e., the curves come momentarily to a plateau. For all curves, the higher curve always corresponds to the case of lesser obscuration. These curves are normalized to the size of the light-collecting area: they all converge asymptotically to 1 (become 1 at infinity). Here we can read off the curve that 84% of the energy is indeed contained within the first dark ring. The radially integrated energy is a particularly convenient way of quantifying an aperture configuration when the spot sizes have radial symmetry.

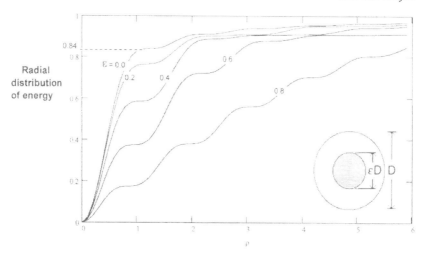

Figure 8.18 The integrated intensity as a function of radial position, with the obscuration ratio ε as a parameter, varying from 0 (no obscuration) to 0.8 in increments of 0.2. The compactness of the central spot may be recognized from the slope of this curve. There is 84% of energy within the first dark ring in an optical system without the obscuration.

While the compactness of the optical spot (the image of a point source) is measured most effectively from the intensity distributions in the focal plane, the MTF often provides the important complimentary information about the ability of the telescope to image a specific spatial frequency.

Thus, in Fig. 8.19 we show the MTF as a fucntion of radial spatial frequency coordinate, with the obscuration ratio ε as a parameter, varying from 0 (no obscuration) to 0.8 in increments of 0.2. Within the theory of imaging, this figure is equivalent to Fig. 8.17. However, the performance degradation upon the inclusion of the central obscuration is potentially made much more obvious in this presentation of results. It is clear that for the obscuration ratio of 0.5 (area obscuration of 0.25), the imaging of the intermediate spatial frequencies (0.25–0.55) is decreased by approximately 40% with respect to the highest theoretically achievable value, with the MTF of 0.25. For many imaging applications, this value is already considered too low for image reconstruction from within the noise frequencies.

8.2.2.3 Two-Mirror Telescopes

The evolution of two-mirror telescopes may be most easily appreciated by examining Figs. 8.20 and 8.21, where four different telescope configurations are shown with the constraint of the same aperture size, obstruction area, and the f-number. Figure 8.20 shows (a) Newtonian, (b) Cassegrain, and (c) Gregorian telescopes. In Fig. 8.21, a Schwarzschild configuration is shown separately due to its overall size, compared with a Cassegrain telescope.

Strictly speaking, the simplest on-axis two-mirror telescope is a Newtonian, using just a beam-turning mirror to bring the focal plane out of the incoming beam. The two basic refractive telescopes may also be implemented as two-mirror systems.

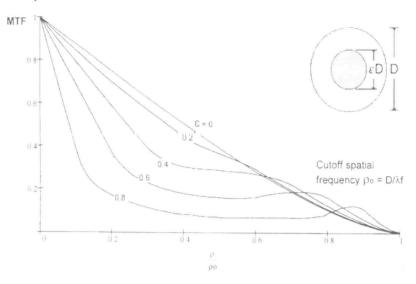

Figure 8.19 Modulation transfer function (MTF) as a function of radial spatial frequency coordinate, with the obscuration ratio ε as a parameter, varying from 0 (no obscuration) to 0.8 in increments of 0.2.

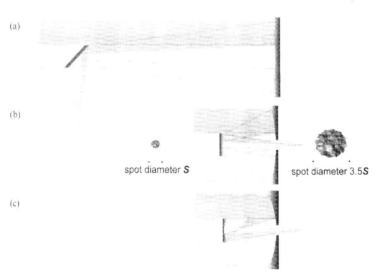

Figure 8.20 Comparison of (a) Newtonian, (b) Cassegrain, and (c) Gregorian telescopes under the conditions of the same f/number (F/#) and effective focal length. The image formed by the secondary mirror is magnified compared to that of the primary mirror.

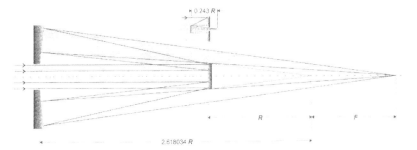

Figure 8.21 A Schwarzschild configuration representing aberration-free design on a flat field is appreciably larger than a Ritchey–Chretien telescope with the same F/# and effective focal length.

Figure 8.20(b) shows the case when the secondary mirror is divergent, resulting in an appreciably shorter instrument length and a smaller central obscuration, generally adopted for astronomical telescopes. This layout is known as a Cassegrain telescope, when the primary mirror is a paraboloid and the secondary mirror is a hyperboloid, a configuration necessary for the coincidence of the focal lengths behind the secondary mirror. The image formed by the primary mirror alone is smaller than the final image formed by the Cassegrain telescope because the secondary mirror actually magnifies it. When the secondary mirror is concave, there is a central focus between the mirrors and the system length is somewhat larger than the sum of the focal lengths. This layout is illustrated in Fig. 8.20(c), having the advantage that the final image just outside the primary mirror is erect (head-up). This layout may be used for terrestrial applications. This configuration is known as a Gregorian telescope, when the primary mirror is an on-axis section of a paraboloid, and the secondary mirror is an on-axis ellipsoid. In accordance with the established terminology, we will use shorter terms and leave out *on-axis*, when not explicitly referring to an off-axis configuration, and *section* even though this is always implied.

Table 8.1 lists the possible combinations of mirrors in the telescope systems employing the corrector plate, and the aberrations that remain in the image. The corrector plate may be used in the incoming beam only for small telescopes where weight is not a problem. In other configurations, it is placed just before the image. This refractory component has low power, so that its surface may be shaped to

Table 8.1 Primary/Secondary Mirror Combinations in Telescopes

Schmidt plate	Primary	Secondary	Spherical aberration	Coma	Astigmatism	Field curvature
Aspheric	Spherical	Spherical	0	Yes	Yes	Yes
Aspheric[a]	Spherical	Hyperbolic	0	0	Yes	Yes
Aspheric[b]	Oblate spheroid	Oblate spheroid	0	0	0	Yes

[a] Classical form of the shortened Schmidt–Cassegrain.
[b] Ritchey–Chretien form.

Table 8.2 Different Forms of Cassegrain Telescopes

	Primary mirror	Secondary mirror	Limiting aberrations
Classical	Parabolic	Hyperbolic	Coma Astigmatism Field curvature
Ritchey–Chretien	Hyperbolic	Hyperbolic	Astigmatism Field curvature
Dall–Kirkham	Elliptical	Spherical	Coma Astigmatism Field curvature

correct for the aberrations generated by the primary and the secondary mirrors. The Questar optical system uses a Schmidt-type corrector in the entering beam to correct the spherical aberration of two spherical mirrors.

Table 8.2 lists the different forms of Cassagrain telescopes. The Ritchey–Chretien configuration (see Fig. 8.22) is used nowadays for nearly all large telescopes for astronomical applications, including a space application to survey the Martian surface from orbit prior to proposed landing. An excellent review of these telescope configurations has been given by Wynne [15]. Its residual aberration is astigmatism, correctable with the introduction of an additional component. The field curvature cannot be corrected as it is the consequence of the curvature of individual components: however, it may be decreased by the introduction of a small field correcting element, shown in Fig. 8.23.

Figure 8.21 shows the famous Schwarzschild layout, which appears much like an inverted Cassagrain. This system is free of spherical aberration, coma, and astigmatism, because the primary and the secondary mirrors are concentric, when the stop is at the center of curvature. The radius of the image surface is equal to the focal length. This telescope becomes very large in comparison with other two-mirror layouts when the light-collecting aperture and the F/# are the same.

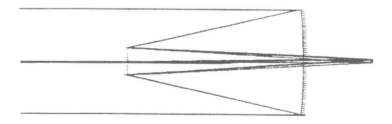

455. M

W–F/10 RC CAM SK15 2-   SCALE 0.055

Figure 8.22 Ritchey–Chretien layout of a 1-m diameter F-10 telescope proposed for the survey of the Martian surface with resolution of 0.25 m from a stable orbit (from Scholl [8]).

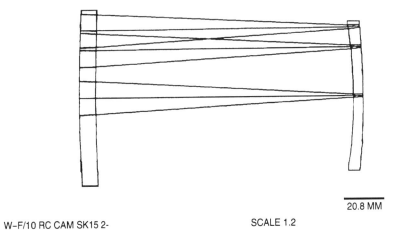

20.8 MM

W–F/10 RC CAM SK15 2-   SCALE 1.2

Figure 8.23 A two-element field corrector used to flatten the field of the telescope shown in Fig. 8.22 and to permit good-quality imaging in the infrared spectral region in addition to that in the visible (from Scholl [8]).

An interesting review article points out how a secondary mirror might be fabricated from the section of primary mirror that is cut out. Blakley [16] has proposed to call it Cesarian in honor of the first famous baby boy born by this method, while the mother was sacrificed for the good of the Empire. The idea of using both parts of the blank seems to favor the joint survival of mother and the child, so maybe Cosbian telescope might be a more appropriate term for it.

Afocal mirror combinations work very well, often including a beam-turning mirror to bring the focal plane instruments out of the main beam. A Mersenne telescope includes a parabola–parabola configuration. A very good review of afocal telescopes was presented by Purryayev [17] and Wetherell [18].

8.2.2.4 Unobscured Catadioptric Telescopes

There are three ways of implementing a mirror system without an obscuration. In the first, simplest one, the object, the image, and the mechanical axes of the parent components from which the off-axis segments are generated are on the same optical axis (see, for example [19]). The Herschelian telescope (see Fig. 8.16(c)) shows that an off-axis segment may be cut out of an on-axis component, which is traditionally referred to as the parent component. Its optical axis (the mechanical axis of the parent component) then does not coincide with the physical center of the off-axis component: the beam incident parallel to the optical axis reflects into a focal point outside the incident beam path. Here, the object and the image remain on the optical axis, while the stop displaces the beam position on the optical element to an off-axis location. The term *off-axis* refers to the fact that the on-axis part of the parent component is not used for the imaging.

In Figs. 8.20 and 8.21, the part of the incident beam above the optical axis has been shaded. This shows how most common telescopes can be incorporated as off-axis configurations without any beam obstruction. It is interesting that the Gregorian configuration results in a better use of the parent component, as a larger off-axis section may be used than in the other two-mirror configurations. Dials [20] describes the design of the high-resolution dynamics limb sounder (HIRDLS) instrument, which includes an off-axis Gregorian telescope developed to meet 0.7 arcseconds pointing and the 1% radiometric accuracy requirements.

When implementing reflecting systems with finite conjugates (image and object distances), the use of a z-configuration is favored. The field-dependent aberration coma is canceled when imaging with two off-axis identical sections of a parabola with parallel optical axis above and below the collimated beam. This is recommended for source reimaging in the monochromators or for the placement of a wind tunnel in the collimated light, in a *schlieren* system. The latter is illustrated in Fig. 8.24. The less-compact layout is acceptable as is often set up in the laboratory where space is not at a premium.

The optical axis of the parent component may also be broken, as in the second incorporation of the unobstructed optical systems [21]. The optical component is tilted. Another way of incorporating off-axis configurtions is when the optical axis is completely broken at each component, such as when three mirrors are employed for further aberration cancellation and field flattening. A combination of field-bias and aperture offset is also employed, as the third way [22].

8.2.2.5 Three-Mirror Telescopes

The number of aberrations that can be corrected in an optical system increases with the number of surfaces and their shapes. With three mirrors, the flat image surface is more easily achieved without using any refractory components. Generally, using an extra surface allows the correction of one more aberration than in a well-designed two-mirror system. The three-mirror systems nowadays are designed in an off-axis configuration.

Some three-mirror systems achieve the necessary performance by incorporating spherical mirrors. These systems, in general, tend to be optomechanically quite

Figure 8.24 Two off-axis parabolas may be used to generate a collimated beam, needed in schlieren systems. The asymmetrical z-configuration results in a smaller amount of coma critical in the monochromators.

complex; however the simplicity of fabrication, testing, and alignment of spherical (primary) mirrors makes them highly beneficial.

The first reference to the advantages of a three-mirror system in a telescope, depicted in Fig. 8.25, is made by Paul [23]. It is well known that two spherical surfaces may be replaced by an aspherical surface to achieve the same performance. He proposed to replace the hyperboloid in the classical Cassegrain with two spherical mirrors, with equal, but opposite radii of curvature. The first mirror, with a very large radius of curvature is the secondary. Its image would be very far away if it were not for the spherical tertiary located inside the hole of the primary mirror. This forms an aplanatic and anastigmatic image in the space between the primary and the secondary mirrors. The secondary mirror, whose diameter is equal to that of the tertiary mirror, is in the path of the incoming beam. There is little space for the focal plane instruments, making this a good candidate for off-axis configuration.

The idea of using a single primary mirror figured as two distinct optical surfaces started to become feasible when numerically controlled (diamond-turned) optical surfaces started to appear as a possibility. This may be particularly appropriate for metal mirrors and/or in the infrared. Thus, the delicate step of cutting a hole in the primary is avoided, no physical aperture is introduced, and the optical blank for the tertiary mirror comes integrated with the primary mirror [24]. A significant improvement on this idea follows with a design 20 years later [25], where the two reflections, that from the primary mirror and that from the tertiary mirror, are off the same mirror surface. Thus, the primary and the tertiary have the same form, a hyperboloid, and are on the same blank. As shown in Fig. 8.26, the focal plane in this aplanatic design with a central obscuration is conveniently located outside the telescope volume, behind the hyperboloidal secondary. This system could be considered an improved version of a Ritchey–Chretien telescope with decreased amount of aberrations.

Finally, three-mirror systems became popular with the advent of photolithography and laser beam handling systems. An unobscured configuration incorporating spherical mirrors with off-centered field and unit magnification is corrected for all five Seidel aberrations. The Offner system for microlithography consists of two

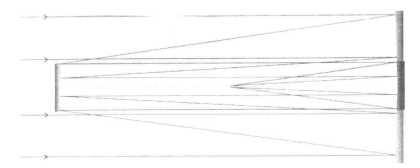

Figure 8.25 A three-mirror telescope proposed by Paul in which he replaces the hyperboloid in the classical Cassegrain with two spherical mirrors, with equal, but opposite radii of curvature. This telescope becomes an aplanatic and anastigmatic optical system. There is little space for the focal plane instruments, making this a good candidate for off-axis configuration.

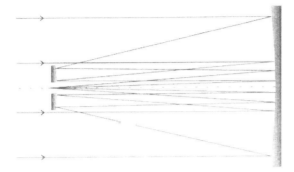

Figure 8.26 The primary and the tertiary form the same mirror surface, a hyperboloid. In this design with a central obscuration, the image is conveniently located outside the telescope volume, behind the hyperboloidal secondary. This system could be considered a version of a Ritchey–Chretien telescope with decreased amount of aberrations.

spherical mirrors which are approximately (exactly, in the third order version) concentric. The primary and tertiary are the same mirror, and the secondary has a radius equal to half that of the primary-tertiary. In the third order the aberrations are corrected by the concentric arrangement, but to balance out the higher order astigmatism the spacing must be changed by a very small, but critical amount, and of course the useful field is restricted to a ring or annulus, centered on the optical axis [26, 27]. Afocal three-mirror systems are used for (high energy) laser beam handling and conditioning systems. They incorporate two paraboloids and an ellipsoid or a hyperboloid, resulting only in the Petzval curvature.

All the optical layouts presented here represent results of exact ray tracing for comparison of the relative shapes and sizes of the telescopes. With the exception of the one to survey the Martian surface, designed using CODE V, the designs described here have been prepared using the lens design program, OpDes, written by the authors.

8.2.3 Astronomical Telescope

The primary function of the astronomical telescope is to collect as much energy as possible from a distant star, often feeding the energy to a secondary set of instruments. This requires the incorporation of a beam-turning mirror in a so-called Coudé of Nasmyth arrangement. These layouts generally require the incorporation of the type of mounts that follow the stars.

In the astronomical telescope the objective is known as the primary mirror, because it has the primary or critical function of collecting the light. The eyepiece becomes the secondary mirror, because it is really not needed as an eyepiece. Its secondary function, in terms of its importance, is to present the image in a suitable form for recording and data processing.

In astronomical applications, the objective lenses in the Galilean and Keplerian telescope configuration are replaced with the mirrors of the same power in order to have a large aperture without the associated weight and fabrication challenges.

In a mirror system completely equivalent to the refractive version, both the primary mirror and the secondary mirror are located on the optical axis: the secondary mirror obstructs the incident beam and projects the image inside the central section of the primary mirror. This procedure requires the construction of a physical hole in the primary mirror which adds to the risk in the fabrication and the amount of polishing.

Astronomical telescopes generally have an insignificant field of view compared with that of a terrestrial telescope. They tend to be designed for a specific spectral region, from ultraviolet (UV) to far-infrared. A most informative review of the requirements for the imaging in UV and some of the solutions to those is offered in the review article by Courtes [28]. The term 'camera' is sometimes used when referring to the telescope with wide-angle capabilities, such as Schmidt camera, with good half-field performance of up to more than 5 degrees. This term has arisen due to the similar requirements of the photographic camera [29].

A majority of the telescope configurations built in recent years take advantage of the excellent performance of the Ritchey–Chretien design. Its resolution is much higher than that actually achieved due to the atmospheric aberration.

The importance of the coupling of the telescope's optics and the instrument's optics is that the combined optical system has to be designed for the optimum performance of the overall system. This was considered when the Hubble telescope was identified as having a certain amount of spherical aberration [30–32]. The smaller imging camera was designed to compensate for this surface deformation, producing a much improved image in the camera image plane.

In space observational facilities, such as in the Hubble, ISO, SIRTF telescopes, the focal plane remains fixed relative to the instrument. However, its lack of accessibility means that the focal plane may be shared by several instruments concurrently. These might include a spectrometer, a monochromator, a photometer, a radiometer, a spectroradiometer, and a wide- and narrow-field imaging camera. A set of tertiary mirrors may have to be used to pick the beam into the individual instruments off the main beam. This is illustrated for one instrument in Fig. 8.27 for the preliminary version of the SIRTF telescope [33].

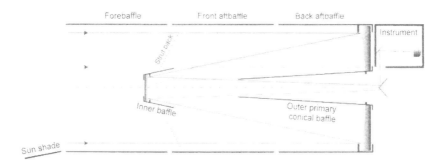

Figure 8.27 An astronomical telescope is used to collect the radiation and deliver it to the focal plane instruments. This is illustrated for one instrument and the pick-off mirror to deliver a portion of the beam [33].

More complex forms of spreading the detectors in the focal plane incorporate a pyramidal mirror, as was used in the Hubble telescope to enhance the number of resolution elements in the CCD by using four separate ones. A similar layout may also divide the available focal plane for a concurrent observation with several instruments.

The distant star is considered a point object whose image size is defined by the resolution of the telescope. In astronomical applications, we are much more interested in the angular position of the objects on the celestial sphere.

The half-angle corresponding to the diffraction blur radius is

$$\theta = r/f = 1.22\lambda/D \quad [sr], \tag{8.4}$$

where θ is the semi-angle of resolution [sr] and r is the radius of the first zero in the Bessel function [m]. This resolution criterion applies only to small-aperture telescopes, or space-based instruments, where the affects of the atmosphere do not destroy the local *seeing* conditions.

8.2.4 Atmospheric Effects

The atmosphere is a collection of gases that surround the earth's surface. It has two detrimental effects on the image quality of the telescope. The rays experience a different phase delay when incident on the mirror upon traversing the atmosphere having different amounts of air turbulence. The absorption characteristics of the gases that constitute the atmosphere do not transmit the radiation of all spectral intervals of interest to astronomers – in particular, the infrared and short wavelength UV regions.

8.2.4.1 Seeing

Earth-based telescopes have their resolution limited by turbulence in the upper layers of the atmosphere (stratosphere), referred to as 'seeing' at their specific sites. This diameter, referred to as "cell," is at most 26 cm maximum for the best observational sites under most favorable conditions. Lower values of 16–20 cm are more common. The seeing problems of large-diameter ground-based telescopes were discussed after the successful completion of multiple-mirror telescopes [34, 35].

For large-aperture earth-based telescopes, the air movement in the atmosphere provides the limiting resolution, especially in the visible region. Generally, the 'seeing' parameter of 10–20 cm is due to the stronger turbulence within the first several meters from the ground. The stratospheric layer limits the isoplanatic patch angle. There, different incident rays are subject to distinct conditions due to fluctuations in density, temperature, and humidity. In terms of resolving power, turbulence-induced wavefront distortions limit the telescope's aperture to an effective diameter of r_0, the coherence diameter of the atmosphere [36]. The seeing increases with the temperature stability the presence of large water mass with a high thermal constant, the absence of human population and its different forms of perturbation of the environment, and with the longer wavelengths. To improve the performance of the earth-based astronomical telescopes, active systems are employed to measure the aberrations introduced by the atmosphere in order to adjust the thin surface of the telescope primary mirror to compensate for this aberration. This active correction is referred to as 'active optics' and requires the employment of deformable optical surfaces figured with the selective application of actuators [37].

Because of the effective resolution limitation imposed by the atmosphere, the resolution of the earth-based astronomical telescope is not as important as its light-gathering capability. Fainter and more distant celestial objects may be detected by increasing the telescope diameter or by lengthening the signal-integration time.

8.2.4.2 Air Mass

Increased light-gathering capability is also achieved by prolonging the signal collection and integration time. One way of doing this in the surveillance cameras is by time delay and integration [29]. In the astronomical facility, this increased integration time is routinely achieved by employing tracking mounts in the observatory. Thus, the same object may be viewed for hours, limited by the air mass and the sky background light, by incorporating the alt-azimuth mounts to compensate for the earth's rotation. This observational mode is limited by the facility's physical construction and the different air mass between the early/late and optimum (zenith) observation periods. The air mass is the ratio of the thickness of the atmosphere that is traversed by a ray comming from an object in a zenith and that at an angle θ, see Fig. 8.28. In the case of star position B, the ray passes through the air thickness d, which is given by

$$d = h/\cos\theta \, [\text{km}]. \tag{8.5}$$

Here h is the height of the air column in the zenith. One air mass corresponds to the observation of the star in zenith. The air mass grows quite quickly with the angle. For example, when the angle to the line of view to the star is 60 degrees, the light traverses 2 air masses. Under this condition, the air turbulence is increased appreciably with respect to the single air mass; and furthermore, the image of the star is displaced due to refraction in the atmosphere.

8.2.4.3 Active and Adaptive Optics

The contribution of the atmosphere is to add a phase error to the light signal coming from an extra-atmospheric object. There are other sources of phase error: the fabrication error of an optical component, alignment of a system, and mount jitter. All

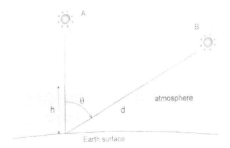

Figure 8.28 The air mass is the ratio of the apparent thickness of the atmosphere traversed by a ray coming from a star into the telescope at an angle θ divided by the thickness of the atmosphere experienced by a ray traversing the atmosphere from a point at a local zenith.

these phase errors add to the image deterioration. They are not known in advance, so a component or a system may not be built such as to compensate for their detrimental effects.

An adaptive optical component is an optical component that adjusts or adapts its performance to the specific conditions and the associated electro-optical system that makes it function this way. The most famous and the most used adaptive optics component is the lens inside the human eyeball. Its surfaces flatten in a young eye to focus from about 5 cm (with an effort) to infinity (5 m)

An active system uses a feedback loop to minimize the undesirable residual phase aberration. A phase measurement is performed to obtain the information about phase error experienced by the surface. The active system uses this information to send the instructions about the requisite changes on the surface shape of the deformable optical component.

The active optical system in the case of the human includes the complete visual-neuromotor system. When one cannot read the small print, the lymphatic neural system automatically sends an order to contract the muscles that control the lens shape (without the human even being aware of the process). A complex electromechanical system is used to change the shape of an optical component after the wavefront has been sensed as deformed and the degree of deformation has been quantified. Thus, a deformable component is just a subsystem of an active system.

The spherical aberration discovered on the Hubble telescope after it had been put in orbit might be an example of a phase error correctable with a deformable component. It is generally expected that such a large structure as a grand observatory would change its optomechanical characteristics after having been launched into space, having experienced the high accelerations and the associated forces, after adjusting to the gravity-free environment of space, and settling to the new temperature equilibrium distribution due to the absence of air. Indeed, the Hubble telescope incorporates a small number of actuators with a limited range to correct for the changes due to the different settled conditions of the space environment. However, this simple system was not sufficient to perform the requisite corrections. Due to the good fortune that the primary mirror was reimaged on the secondary mirrors on the wide-field and the planetary cameras (WFPC I), its surface shape was redesigned and adjusted on the second trial with the placement of WFPC II [39]. This is a form of an active system with a very long period between the identification of the phase error and the adjustment of optomechanical surfaces to correct it.

A study [40] was made of the feasibility of correcting with an adaptive optic component the spherical aberration on the primary mirror of a large-diameter telescope with the following parameters: primary mirror diameter 2.4 m, mirror obscuration diameter 0.1056 m, mirror radius of curvature 4.8 m, design wavelength 0.5 μm, suffering 6.5λ of spherical aberration.

In an adaptive mirror, the surface is deformed to produce an equal but opposite phase error to that which is to be canceled. The combined effect of the aberrated phase and the component intentionally deformed with the opposite algebraic sign is expected to be close to zero. Only those spatial frequencies in the phase that are smaller than the maximum spatial frequency in the actuator distribution may be corrected. Thus, some high frequencies are expected to remain in the corrected phase function. Figure 8.29 shows how the correction works. Each actuator in a rectangular array exerts a force on the thin surface of the deformable component.

SPATIAL DOMAIN   FREQUENCY DOMAIN

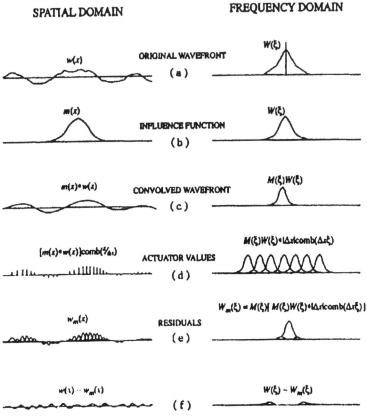

Figure 8.29   In an adaptive mirror, the surface is deformed to produce an equal but opposite phase error to that which is to be corrected in order to cancel the aberration. Its performance is best understood in spatial and spatial frequency domains. (a) The original aberrated wavefront is to be duplicated by applying a deformation to a thin faceplate using actuators. (b) The effect of an actuator force exerted at a specific position is to deform the surface in a shape similar to a Gaussian. (c) The amount of force applied at actuator locations reproduces the general shape of the wavefront error. (d) The specific value of the actuator amplitudes are calculated from the expected surface shape. (e) With the optimal choice of the width of the actuator influence function, controlled by the faceplate stiffness, the actuator separation may be set to determine the highest spatial frequency corrected in the phase. (f) The resulting wavefront error, given as the difference between the initial phase and the phase due to the surface deformation, contains only the high-frequency, low-amplitude phase components (after Scholl [40]).

Depending on the deformation characteristics of the thin faceplate, it is deformed in a controlled or uncontrolled fashion with some coupling between the neighboring actuators. A zonal model of the actuator phase–plate interaction has been used to model the possible range of phase-sheet deformations [42]. Obviously, a large-diameter mirror is more amenable to faithful shaping because more actuators and their control mechanisms can be fitted behind a large area. The physical space occupied by the actuators behind the mirrors makes them suitable correction candidates. When only low frequencies are corrected, a smaller secondary is the preferred choice.

Figure 8.30(a) shows the aberrated wavefront and the resulting intensity distribution in the telescope focal plane (see Fig. 8.30(b), where many intensity rings confirm the presence of a highly aberrated wavefront (see Fig. 8.8). Figure 8.30(c) shows the intensity slices along the optical axis, as illustrated in Fig. 8.31. The spot is spread along the optical axis. Using the deformable mirror model with the actuator spacing of 4.6 cm, the corrected spot (see Fig. 8.30 (right)) has a Strehl ratio of 0.95. When the number of actuators is reduced by a factor of 2 in each direction, the Strehl ratio reduces to 0.86. With the application of an adaptive component, the intensity distribution in the focal plane improves, showing only a single bright ring around the central bright spot. Also, the spot extent along the optical axis tightens appreciably. The corrected phase is seen to be nearly constant. Only a quantitative measure such as the Strehl ratio allows us to appreciate the incremental improvement of about 10% upon quadrupling the number of actuators.

The incorporation of a fully fledged deformable primary mirror into a space telescope would significantly improve its performance under most adverse conditions. The inclusion of the obligatory robust control system adds significant further complexities and increases the potential for risk of a single-point failure, which prohibit its implementation in applications to space astronomy at this time.

Even if an adaptive component in a space system were feasible, the correction implementation would not necessarily include a complete active system. The quality of the images and their review by the scientific community provide the wavefront sensing and evaluation to generate the control commands. This is basically a quasi-static correction system with phase error evaluation performed infrequently (once a month).

Atmosphere changes occur frequently, as a matter of fact constantly, as we can acertain when we observe the twinkling of a distant light at night. For this reason, many modern telescopes envision some active correction (see, for example, [43, 44]). Any correction of the atmospheric effects would have to be monitored and corrected with a frequency between 0.10 (for calm conditions) and 100 Hertz (for conditions of turbulence). Real-time compensation of atmospheric phase distortions is generally referred to as an active optics system [45, 46]. Such a temporally dependent system by necessity includes an active system of sensing and a wavefront correction feedback loop.

Only a fraction of the phase error may be corrected during each sampling to avoid instabilities and transients. The phase error has to be sampled at a frequency at least five times larger than its highest significant temporal frequency.

The resolution limitation on a large telescope imposed by the atmosphere is a generally accepted phenomenon. A great deal of effort has been expended in the last 40 years to understand its behavior, in order to be able to correct for it with an active optics system. Atmospheric aberration may be described by the Kolmogorov spectral distribution of the wavefront as a function of height above the earth surface, [47]

$$W^2(\rho, \lambda, n) = \frac{0.38}{\lambda^2 \cdot \rho^{11/3}} \int_{h_{min}}^{h_{max}} [C_n(h)]^2 \mathrm{d}h, \tag{8.6}$$

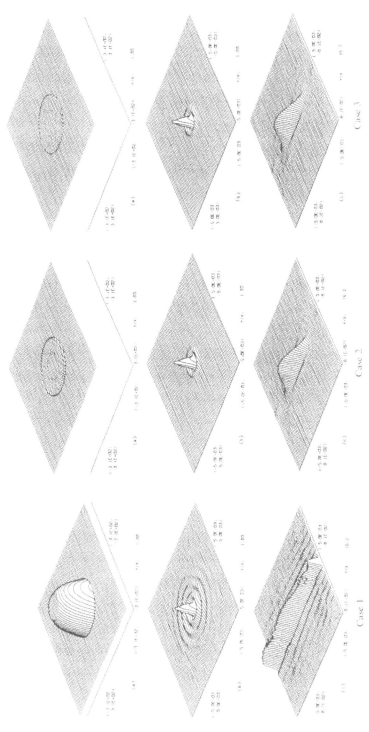

Figure 8.30 The simulation results of the correction of a wavefront with 6.5λ of spherical aberration with a deformable mirror demonstrate the high degree of correction possible with a deformable mirror of 2.4 m diameter. In each of the three cases, we show the wavefront shape (a) then the far-field image (b) and finally a transverse slice along the optical axis (c). Case 1: the energy in the point-spread function corresponding to this aberrated wavefront in the Cassegrain focus is spread out into the higher annuli (compare with Fig. 8.7). Case 2: the significant improvement in the wavefront is observed upon correction with actuators placed on a square grid, 9.2 cm apart, producing a rather compact spot with a Strehl ratio of 0.86. Case 3: the quadrupling of the number of actuators, the corrected phase exhibits a nearly perfectly flat profile, and the far-field pattern closely resembles the diffraction-limited shape of Fig. 8.3, with the Strehl ratio of 0.95 (after Scholl [40]).

Figure 8.31 The wavefront with a (compensated) aberration is incident on a perfect mirror, which forms the far-field image in the (Cassegrain) focal plane. Customarily, the point-spread function is shown as a function of the distance from the optical axis. Here the intensity slices are given as a function of one transverse coordinate for a number of positions along the optical axis (also known as inside, in, and outside focus positions). In Fig. 8.30(c), we show the intensity slices along the optical axis, in order to assess the three-dimensional extent of the bright spot (after Scholl [8]).

where C_n^2 is the refractive index structure of the atmosphere as a function of the height layer n, between h_{min} and h_{max}. Upon integration of this equation over all layers of the atmosphere, characterized by their respective refractive index structure, we can see clearly that the amount of atmospheric aberration is proportional to the inverse wavelength and the variation in the index of refraction of the air. These coefficients, of course, are dependent on the site in general, and on the specific climatic and environmental conditions, in particular.

In terms of the seeing parameter r_0 that is used to describe empirically the effect of the atmosphere on the phase error, this equation may also be written, when h_{min} and h_{max} include the integration over the whole atmosphere, [48]

$$W^2(\rho, r_0) = \frac{0.023}{(r_0)^{5/3} \cdot \rho^{11/3}} \cdot \frac{e^{-\rho^2 \cdot (L_i)^2}}{\left[1 + \frac{1}{(L_0 \cdot \rho)^2}\right]^{11/6}}, \tag{8.7}$$

where L_i and L_0 are the inner and outer scale, while ρ is the radial spatial frequency coordinate. On comparing Eqs (8.6) and (8.7) we see that the seeing radius increases with wavelength nearly linearly (exponent is 6/5).

$$r_0(\lambda) = \lambda^{1.2}[R_{o1}(\lambda_1)/\lambda_1^{1.2}]. \tag{8.8}$$

This equation may be rewritten for the visible region, for which the seeing parameter is usually known, between 0.1 and 0.2 m.

Using Eq. (8.4) for the angular resolution, we obtain

$$\theta(\lambda) = [0.61/\lambda^{0.2}][\lambda_1/R_{o1}(\lambda_1)^{1.2}] \text{ [rad]} \tag{8.9}$$

Thus, the angular resolution θ of earth-based telescopes without atmospheric correction increases with wavelength with an exponent of wavelength of -0.2. We recall that a small value for angular resolution means a better resolution. As the wavelength increases, θ in Eq. (8.9) decreases slowly. The angular resolution of earth-based telescopes as a function of wavelength is shown in Fig. 8.32. This figure incorporates two sets of parameters. The mirror diameter is varied from 1 m to 10.95 m (the diameter of the Keck telescope). Also the seeing radius is varied to include (in the visible) the range that correspond to good astronomical sites. The resolution limit corresponds to the higher of the curves for the applicable parameters. The region of the expected resolution values has been shaded. We observe

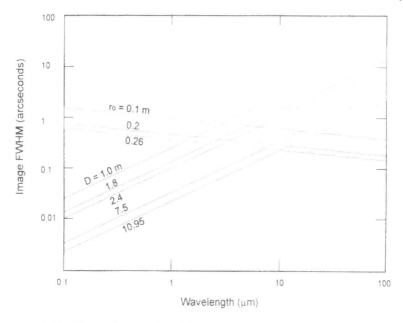

Figure 8.32   The angular resolution of the earth-based telescope as a function of wavelength with two sets of parameters: the mirror diameter varies from 1 m to 10.95 m; the seeing radius includes the range from 0.1 m to 0.26 m, corresponding to good astonomical sites in the visible.

that in the visible and the near-infrared the seeing limits the resolution of the telescope, independently of its diameter.

Thus, large-diameter telescopes are well utilized on the earth's surface only in the infrared. There, of course, the atmospheric spectral transmission as well as the Earth's emission generate a whole new set of problems.

The feasibility of correcting the atmospheric aberration using a deformable mirror has also been analyzed. A laser beam is first expanded to 0.5 m diameter, with a beam expander that introduces 0.2λ astigmatism at $0.48\,\mu m$; the effects of the atmospheric phase error are calculated, and the effects of the actuators are determined to neutralize the beam aberration, as shown in Fig. 8.33. With the actuator separation on a square grid of 0.04 m, we can see the sequence of intensity and phase at each of the steps illustrated in Fig. 8.34. The initial intensity and phase are depicted in Fig. 8.34(a), with the Strehl ratio of 1; the phase in Fig. 8.34(b) shows some astigmatism after the beam passes through the aberrated beam expander, now having the Strehl ratio of 0.924; the phase is severely, randomly, and unpredictably aberrated in Fig. 8.34(c), with a significant Strehl ratio degradation to 0.344; and, finally, the corrected phase in Fig. 8.34(d) exhibits a much improved Strehl ratio of 0.865. The residual phase error is characterized by the high-frequency components whose presence could not be eliminated by the given small number of actuators.

The phase error introduced by the atmosphere just before the mirror surface is equal to the deformation that needs to be placed on the mirror surface to produce the corrected wavefront.

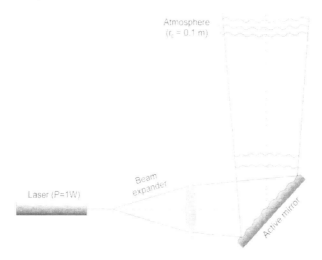

Figure 8.33 The optical model to assess the feasibility of correcting the atmospheric aberration with a deformable mirror includes the laser beam, first expanded to 0.5 m diameter, with a beam expander that introduces 0.2λ astigmatism at $0.48\,\mu m$; the phase error introduced by the atmosphere; and the actuator strengths, calculated so as to neutralize the beam aberration (after Scholl [45]).

In principle, a bright star could be used as a radiation source for determination of the atmospheric phase error to determine the requisite correction. When the atmosphere is rapidly changing, we need to use an active illumination system such as a laser beam directed to the upper atmospheric layers where the sodium atoms, so excited, produce fluorescence that acts as a bright point source. Due to the practical consideration of inadequate exposure, artificial sources have been developed using an active laser beam as illuminator. For example, the Strehl ratio has been improved from 0.025 to 0.64 on a 1.5 m telescope using an active system [49] and an artificial laser beacon. One of the disadvantages of this very effective tool is that it is insensitive to tilt.

8.2.5 Space Telescopes

With the advent of larger and more powerful rockets, it became feasible in the late 1960s to send a satellite into space with communications and scientific remote-sensing payloads. This was followed by telescope facilities, both in the UV and IR, because there the atmosphere is opaque, effectively making it impossible to obtain astronomical data with the ground facilities. The Far Infrared and Submillimeter Space Telescope (FIRST) is intended to open up for study [50] the entire submillimeter and far-infrared band (85–900 μm). When the seeing problem was first fully understood in the 1970s, a series of large observational telescope facilities incorporating large-diameter monolithic telescopes was initiated, including the Hubble telescope in the visible, followed by the Infrared Space Observatory (ISO) in the infrared. Remote sensing is another important area of application for space tele-

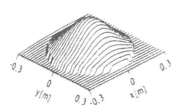

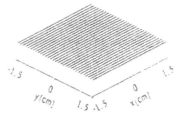

(a)

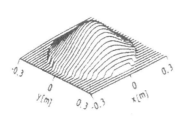

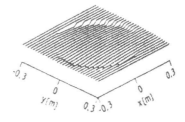

(b)

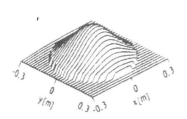

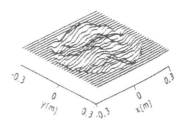

(c)

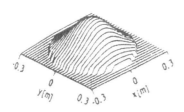

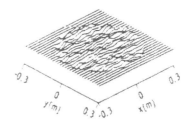

(d)

scopes: for example, in instruments used to monitor ozone depletion in the stratosphere [51].

The clear advantage of placing a telescope in space is that the detrimental effects of the atmosphere are avoided. The equally obvious disadvantage is that the telescope orbiting in space is not easily accessible for (routine) repairs and adjustments, as became significant in the case of the improperly formed primary mirror on the Hubble telescope, which ended up with approximately 0.5λ of spherical aberration [52,53]. Two instruments were redesigned to compensate for this aberration, the wide-field and the planetary cameras, and replaced in the observational facility several years later. In addition, COSTAR was added to allow the use of the nonimaging spectrographs and the faint object camera.

The IRAS was the first IR satellite to survey the skies, while the ISO was the first infrared astronomical observatory in space [54] operating at wavelengths from 2.5 to 200 μm. It was considered a success, having run longer than was originally planned due to its conservative design [55–58]. Its successful completion resulted in the implementation of a slightly larger facility, the Space Infrared Telescope (SIRTF), planned for launch in 2002.

8.2.6 Infrared Telescopes

Telescopes may also be classified according to the wavelength of observation into several important spectral intervals of observation: the visible, including blue and red; the UV; the near- and far-infrared, and the millimeter and radio waves. The infrared telescope facilities generally include the observations approaching a millimeter range, where the detection methods change, and the telescope primary mirror functions just as a light-collecting surface, even in space. Both the UV and the infrared regions are important to differentiate from the visible spectral regions, because of the different light transmission characteristics of the atmosphere and optical glasses in those two spectral regions. Additionally, the amount of stray light in the infrared becomes of uttermost importance because it generates the thermal noise.

Figure 8.34 The feasibility of correcting the atmospheric aberration using a deformable mirror has been confirmed with the Strehl ratio from 0.344 due to atmospheric aberration to the much improved Strehl ratio of 0.865 upon the application of the active optics system to control the surface shape of the deformable mirror. While the beam and the phase are shown at each significant step in the active optics system, the atmospheric aberration affects only the phase. The laser beam intensity scattering within the atmosphere is not considered. (a) The initial phase at the output of the laser is constant, with the Strehl ratio of 1. (b) A small amount of astigmatism (0.2λ) is introduced at the aberrated beam expander, reducing slightly the Strehl ratio to 0.924. (c) The phase is severely, randomly, and unpredictably aberrated due to its propagation through the atmosphere, with a significant Strehl ratio degradation to 0.344. (d) Finally, the corrected phase exhibits only small-amplitude, high-frequency ripples after reflection off the deformable mirror with the actuator separation on a square grid of 0.04 m, with a much improved Strehl ratio of 0.865. The residual phase error is characterized by the high-frequency components whose presence could not be eliminated by the given small number of actuators (after Scholl [45]).

The material–transmission characteristics of infrared glasses favor the implementation of the all-reflective configurations in this spectral region. First of all, there is an increased amount of absorption. Secondly, the index of refraction is larger, thus increasing the losses upon refraction at a glass–air interface. Such problems are eliminated if an all-reflective design is used. Additionally, the reflectivity of aluminum, silver, and gold increases in the red–infrared spectral region, making it even more advantageous to incorporate all-metal reflecting telescopes for the IR.

There is one more important issue to address when dealing with telescopes operating in the longer-wavelength regions. It is the significance of the thermal noise, generally addressed during the early design steps in order to eliminate the so-called stray light. The importance of the noise arising from sources outside the field of view whose radiation scatters inside the telescope barrel has been analyzed and identified as important already for assessing the performance of the Hubble telescope in the 1970s. A sun shade, similar to the one shown in Fig. 8.27, was used to prevent the sun rays from out of field of view from entering the telescope barrel. These potential noise sources include the sun, the moon, and the earth, as the radiators outside the field of view whose light may scatter inside the telescope tube. The out-of-field sources are even more detrimental in the IR, because the sun is a strong emitter in the IR Both the earth and the moon remain an emitter in IR even from the side not illuminated directly by the sun, due to their temperature. The temperature control of all parts of the telescope systems is of critical importance for the minimization of the internally generated stray light noise [59, 60]. For these two reasons, IR telescopes tend to be heavily baffled, as can also be seen in Fig. 8.27.

One detail that may be clearly appreciated in this layout with the exactly traced optical beam volume is that the central part of the secondary mirror is not used for imaging. As this part of the mirror is seen directly by the detector, it is obstructed by the small planar reflector. It misses its objective somewhat in the telescopes where the secondary mirror is nodded for the purpose of chopping, by looking at a different sky region.

The secondary mirror, seen directly by the detector, requires extra attention, resulting in a large obstruction to the incident radiation, much larger than the actual diameter of the secondary. With such an enlarged obscuration, the amount of light collected by an on-axis reflecting configuration is smaller, equal to $\pi(R^2 - r^2)$, where r is the radius of the obstruction, rather than that of the secondary mirror.

In the IR designs the secondary mirror tends to be very small, for two reasons. Due to its size it subtends a small angle at the detector in order to diminish the amount of stray light. Secondly, the small secondary mirror is easier to control to oscillate between the source and the reference patch of the sky using a fixed primary mirror. Of course, such mechanisms add to the bulkiness of the secondary assembly. The primary mirror has to be slightly oversized to accommodate the extreme position of the nodding secondary mirror. The designers of these telescopes have not come to an agreement whether it is better to have the stop on the primary or the secondary mirror.

The large infrared facilities, such as ISO and SIRTF, have been designed as on-axis configurations, due to improved mechanical stability of centered systems, even though the shade introduces some mechanical asymmetry. In the past decade, there has been an insurgence of small IR telescopes for dedicated missions of observing earth, seas, and shores. These revolutionary designs incorporate the off-axis tele-

Figure 8.35 A telescope designed with the objective of a decreased amount of stray light for an improved signal-to-noise ratio takes advantage of the off-axis configuration to employ fully the available light-collecting area of the primary mirror. To decrease the number of diffracting edges, a Gregorian layout incorporating the internal focus where a field stop may be located, and the Lyot stop for limiting the stray light (after Scholl [59]).

scope layouts with the Lyot stop configuration to minimize the amount of stray light on the detector plane.

In this approach, the whole instrument is designed with the single overriding purpose of maximizing the signal-to-noise ratio for detection of distant faint sources as in the case of the wide-field infrared explorer [61]. The same high signal-to-noise ratio may be achieved either with a large aperture for large light-collecting area or a small amount of noise. In the design depicted in Fig. 8.35, the telescope uses an off-axis reflective layout in the Gregorian configuration.

By choosing an off-axis reflecting mirror the whole primary is used as a light collection system. The secondary neither obstructs the radiation, nor does it modify the resolution (light distribution in the focal plane). The off-axis layout has the undesirable consequences of introducing field aberrations, such as astigmatism, coma, and the image is located on a surface with an increased amount of (Petzval) curvature.

8.3 FILLED-APERTURE TELESCOPE

The maximum diameter of constructed (Earth-based) astronomical telescopes is limited by the technological and engineering considerations of the day. First, there are fabrication issues, in terms of the uniformity of the blank, then its transport, optomechanical forming of the surface, and the component testing to meet the specifications.

In the 1960s the belief that the maximum feasible blank was limited to about 2.3 m led to the nearly simultaneous designs employing an arrangement of six mirrors, known as the multiple-mirror telescope (MMT); see for example [34, 62]. The advantage of a large light-collecting area was paid for by the difficult phasing of the different segments, aligning the segments so well with respect to each other to generate no phase error between them, as discussed by McCarthy [63]. This successful telescope configuration was followed by the first Keck mirror, with 356 hexagonal panels of 0.9-m side in a honeycomb arrangement; the first truly segmented mirror was little more than 10 m diameter (maximum diameter of 10.95 m). This telescope has been successfully fabricated, aligned. It has been used at visible as well as IR wavelengths from its first use [64–67]. This system incorporates an active control to align and maintain in alignment the large number of segments [68].

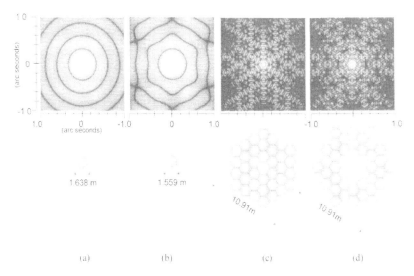

Figure 8.36 Diffraction patterns in a logarithmic scale illustrating the effect of the segments on the performance of the segmented array and the distribution of segments. (a) A single circular segment with diameter 1.628 m produces a set of circular rings, with the first-intensity zero at 0.33 arcseconds. (b) A single hexagonal segment with the diameter of the inscribed circle of 1.559 m, with the first zero ring a slightly-deformed circle, with resolution of 0.33 arcseconds. (c) Thirty-six hexagonal segments in a honeycomb configuration, with the central one missing, with the first zero ring at 0.055 arcseconds. (d) Thirty-six hexagonal segments in a honeycomb configuration, with the central seven segments missing, with the first zero ring at 0.045 arcseconds. The wavelength of 2.2 μm is taken for the resolution determination.

Figure 8.36 shows the diffraction patterns of two possible arrangements of 36 hexagonal segments with the diameter of the inscribed circle of 1.559 m. The diffraction pattern is shown on a logarithmic scale, such that intensity of 1 corresponds to gray level 255, and 10^{-6} and less intensity is black, with a gray level of zero. The wavelength of 2.2 μm is assumed for the resolution calculations. In both layouts, the aperture area is the same; thus, the light-collecting efficiency does not change.

In Fig. 8.36(c) we see a layout with a single central segment missing, the actual Keck configuration, while in Fig. 8.36(d) the first ring of mirrors is missing additionally. The six remaining mirrors are placed on the outside of the configuration shown in Fig. 8.36(c). There are three benefits realized with this modification. In both figures we observe the first zero ring that defines the resolution of the telescope, both corresponding to an hexagonal shape just like the outline of the aperture. Thus, the diameter of the first zero ring, representing the telescope resolution in Fig. 8.36(d) is smaller than that in Fig. 8.26(c). Additionally, the placement of six segments on the outside increases the diameter of the telescope, and therefore, the cutoff frequency is larger along these directions. If the segmented configuration is parabolic or hyperbolic in shape, then the curvature of the segment decreases with the segment distance from the apex.

In Fig. 8.36(c) and (d) we observe a strong dark ring at 0.33 arcseconds, which arises as a consequence of the diffraction pattern due to an individual segment, shown by itself in Fig. 8.36(b). Finally, we show for comparison the diffraction pattern of a circular aperture of diameter 1.638 m, producing a circular ring at the same location. While it is difficult to observe with the naked eye, we comment that the first zero ring in Fig. 8.36(b), (c), and (d) are in fact a hexagon, nearly circular in shape. Looking at the intensity distribution in Fig. 8.36(b), we therefore conclude that the diffraction pattern of a hexagon produces a near-circular hexagonal ring as its first zero, followed by a more clearly formed hexagons. When we search for the first hexagon with a circle inscribed in the diffraction pattern of 36 segment configurations, we find it has the radius of 0.055 arc sec, representing the true resolution of the segmented configurations.

In the 1970s, with the great success of radio astronomy [69], the successful techniques of aperture synthesis began to be studied in optics.

The funding was initially available for military research; however, diluted aperture optical systems turned out to be limited to a narrow field of view, precluding their use for surveillance [70]. Diluted apertures became of interest to the astronomical community for interferometric rather than imaging applications, for astrometric applications (exact star position), and for such exotic research as the detection of planets outside our solar system (see, for example, [71, 72]). It is believed that a single spatial frequency detected with an interferometric array could confirm the presence of such an ephemeral object as a dark, small planet orbiting a bright sun-like star.

The formation of the planets, solar systems, stars, galaxies, and the universe remains of great scientific and astronomical interest worldwide, even if the planet detection project has lost a bit of its appeal within the U.S. scientific community.

8.3.1 Increasing the Aperture Size

There are basically two ways of improving the radiometric sensitivity of a telescope. The first one that has been receiving a great deal of technological impetus, in parallel with the advances in the semiconductor technology due to the employment of similar materials, is the detector technology [73] and the focal plane architecture. The second one belongs to traditional optics: increasing the diameter of the light-collecting aperture to intercept more photons. Its challenges are often the limitations of the state of technology rather than the fundamental limits. This problem for the large-diameter telescopes was addressed at the successful completion of testing the multiple-mirror telescope by Woolf [74].

8.3.2 True Monolithic

When the Hubble telescope was first proposed, the 2.4-m diameter primary mirror was considered the largest mirror that could be figured and tested for use in space. Only a decade earlier, a 2-meter segment had been consider the best and the largest that could be fabricated. In order to build a larger telescope than that, a telescope had to be built of several segments of smaller size. The technology limitation of the day is not to be taken lightly: a size limitation of a furnace to produce a quality blank is a temporary technology limitation. A great advantage of the technology development programs is that with time, efforts, funds, and human ingenuity, such limitations may be overcome.

8.3.3 Monolithic in Appearance

Today, there exists a number of monolithic mirrors under development or already in a telescope with diameters of 7–8.5 m that started to be developed as light-weighted mirrors [75–83]. This development includes the replacement of the one in the housing originally prepared for the multiple-mirror telescope (see, for example, [84]). Actually, in most of these cases, it was human ingenuity rather than a breakthrough in technology and technology transfer from similar areas that brought about the improvements. Possibly the most interesting such case is the building of a primary from a set of hexagonal segments, whose surface is covered with a thin sheet of glass – technology transfer from adaptive optics [85].

8.3.4 Segmented for Any Size

The multiple-mirror telescope represented a novel way of achieving largeness: if one can combine six mirrors, why not 12, or 18 or even more: 36 in the case of Keck, and 127 for the extremely large telescope (ELT). Figure 8.37 shows the comparison of the primaries for the Hubble, the Keck, and the extremely large telescope. The anticipated growth of the segmented mirrors represents an extension of the theorem of natural numbers: if you can build two 2-m telescopes, then assume that you can build an N-mirror telescope, and see if you can build a $N + 1$ 2-meter telescope. Of course you can, and you will, because the astronomical community wants to intercept photons from even fainter sources, and they just need to count them. This has been the design philosophy of the Keck telescope as well as the Spanish telescope in the Canaries, described by Espinosa [86].

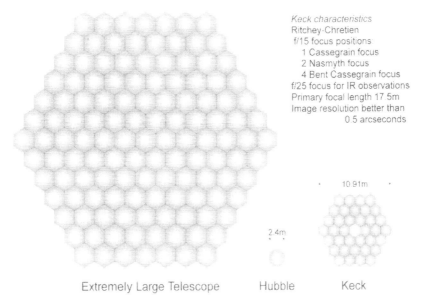

Keck characteristics
Ritchey-Chretien
f/15 focus positions
  1 Cassegrain focus
  2 Nasmyth focus
  4 Bent Cassegrain focus
f/25 focus for IR observations
Primary focal length 17.5m
Image resolution better than
  0.5 arcseconds

10.91m

2.4m

Extremely Large Telescope   Hubble   Keck

Figure 8.37 The primary apertures of Hubble; Keck, with 36 segments; and the extremely large telescope, with 127 segments.

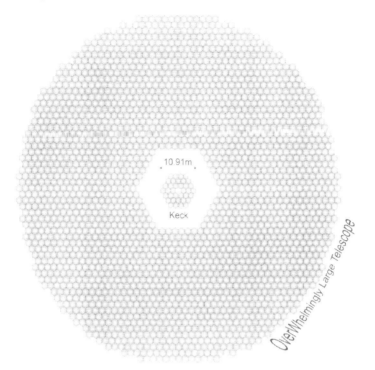

Figure 8.38 The primary apertures of the Keck telescope and the OWL, the overwhelmingly large telescope, incorporating 2000 spherical segments of 2.3-meter diameter.

Equally, the huge telescopes of the future will be constructed on the basis of the same principle. Consider the overwhelmingly large telescope (OWL) that incorporates 2000 segments of 2.3-m mirror diameter. In this telescope, the fabrication will have been much simplified: each segment is planned to be a sphere, fabricated at a rate of 1 per day. The primary will weight 1500 tons. The support structure is of the same size as the Eiffel Tower, employing 4000 tubes of 2 m diameter. The secondary is planned to be larger than the Keck. Figure 8.38 shows for comparison the apertures of the OWL and the Keck telescope.

At the same time that we learned that mozaicking a mirror surface is beneficial to light collection but not necessarily to the resolution, we also discovered the detrimental effects of the atmosphere. The appearance of segmented mirrors and the need for the phase combining them brought into focus the application of interferometry to the detection of a specific feature in the spatial frequency domain. The spatial frequency domain corresponds to the space in which a Fourier transform of a spatial field or intensity distribution is presented. Thus, it has been known for a long time that two separated apertures give information about one spatial frequency, that corresponding to the inverse separation between the apertures, along the line connecting the apertures. By changing the separation of the apertures, also referred to as the interferometer baseline, a continuous set of spatial frequencies could be sampled.

It turns out that the interferometry has been used since the 1950s in radio astronomy to synthesize the shape of radio sources, using a set of separated segments.

8.4 DILUTED-APERTURE TELESCOPE

8.4.1 Increasing the Resolution, But Not the Area

A segmented mirror is made of segments whose shapes combine to form a continuous surface with a small (2 mm) separation between segments. While the surface may be made arbitrarily large, the diffraction pattern is dominated by that of the individual segment, until the segments are correctly phased, as in the Keck telescopes. Then, the diffraction pattern is that of the full aperture (see Fig. 8.36c). So the practice of interferometry to measure a specific spatial frequency of a distant object through the visibility function, and the earlier experience of image synthesis with radio waves, led the researchers to ask themselves: why not separate the individual segments spatially, in order to sample different and higher spatial frequencies. The idea was to conserve the light-collecting area of the segments, and place the segments at such position as to collect information about the spatial frequencies of interest. This is the principle behind using an interferometric configuration to detect a planet around a nearby star. If a planet is there, its presence is confirmed by one spatial frequency corresponding to the star–planet separation. The separation of the segments may be arranged to look for this spatial frequency. Thus, we can think of a diluted-aperture array as a configuration where the light-collecting area is the same as in the segmented mirror, the diffraction pattern or resolution is that of an individual mirror, but the extent of specific spatial frequencies that may be sampled is increased in relation to the separation between individual apertures.

Conversely, we may achieve the same coverage in the spatial frequency domain by using a select set of apertures, but having a much smaller light-collecting area. One such distribution of small circular areas may replace the spatial frequency coverage even of the Keck telescope. The advantage of incorporating the diluted-aperture configuration is that a much lower area of segments needs to be constructed, still sampling the important, information-carrying spatial frequencies, related to the separation of the segments. The disadvantages of diluted-aperture configurations are the phasing of the now distant segments, which in radio astronomy was successfully overcome; the challenges of image reconstruction with the unusual optical transfer functions; and control, and construction of the system of telescopes. While it is clear that this is a complex task, there has been no fundamental limitation discovered to prevent its implementation; it has already been implemented in the radio portion of the electromagnetic spectrum. Rather, the most obvious limitation has been the application-bound one of the limited field of view, which makes it of limited interest to the defense and surveillance communities, its original proponents. The small field of view is not considered a limitation in astronomical application, where attempts are being made to measure stellar diameters and star–planet separations, all subtending very small angles.

The primary function of the diluted-aperture configuration is to detect specific spatial frequencies of interest rather than to form a faithful image of the object. Thus, those portions of the aperture that are involved in the imaging of the spatial frequencies of no interest may be eliminated. In Fig. 8.39, we compare the MTF of

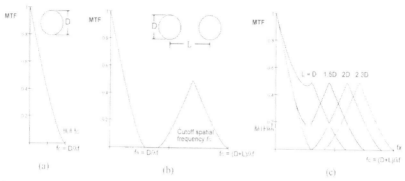

Figure 8.39　The modulation transfer function (MTF) of (a) the monolithic and (b, c) simplest diluted aperture composed of two apertures of the same diameter. Here f_c is the maximum spatial frequency imaged by the optical system; f_R is the maximum spatial frequency below which the MTF does not dip to zero; and f_F is the functional frequency, the maximum spatial frequency for which MTF remains higher than some MTF_{min}. As the latter depends on the specific application, the value of f_F depends also on the application rather than on the configuration of the diluted aperture. The f_c and f_R coincide in the case of a monolithic mirror and when the mirrors are close.

the simplest diluted aperture with that for filled one. One such ground-based configuration involves Keck I and Keck II [87] to do ground-based interferometry. Two apertures of diameter D and center-to-center separation L may produce the MTF that exhibits zero values, as shown in Fig. 8.39(b). It is assumed that these spatial frequencies have no information of interest. When the aperture separation L is sufficiently small, the two MTF peaks start to overlap, and all the intermediate spatial frequencies are also imaged with some modulation (see Fig. 8.39(c)). Here f_c is the maximum spatial frequency imaged by the monolithic optical system, as shown in Fig. 8.39(a); f_R is the maximum spatial frequency for which the MTF is higher than zero; and f_F is the functional frequency, i.e., the maximum spatial frequency for which MTF remains higher than some MTF_{min}. As the latter depends on the specific application, the value of the functional frequency f_F depends more on the application rather than on the configuration of the diluted aperture. For this reason, it is not a figure of merit for aperture optimization. The frequencies f_c and f_R coincide in the case of a monolithic mirror and when the mirror separation is small. Each of these frequencies is a significant figure of merit in different applications: high f_R and f_c are desirable for faithful imaging; high f_c is sought in interferometry, and high f_c, but not necessarily f_R, are required for imaging of select spatial frequencies.

Diluted imaging was analyzed in great depth for radiofrequencies. The first case of an optical system employing a diluted configuration for the visible and near-IR spectral regions was the multiple-mirror telescope, with six mirrors nearly in contact. An approximation to this layout is shown in the inset of Fig. 8.40, described as a six-aperture redundant configuration with a dilution factor of 1.5. A redundant configuration is the one where each spatial frequency is sampled more than once, a highly desirable feature for an imaging system. A dilution ratio is the ratio of the area of the aperture of the monolithic mirror with the same cutoff frequency to the combined area of the subapertures.

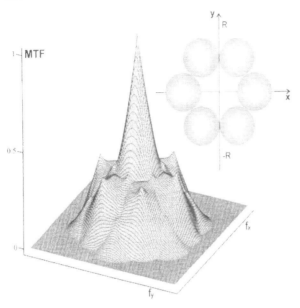

Figure 8.40 The modulation transfer function (MTF) of a six-aperture redundant configuration with a dilution factor of 1.5 illustrates the general features of the diluted aperture
systems. Due to a low dilution factor, the MTF for this system covers about the same spatial
frequencies as that of a monolithic aperture. The MTF exhibits a large plateau for moderate
values of the radial spatial frequency.

The MTF of the six circular apertures in contact, depicted in Fig. 8.40, illustrates the general features of diluted-aperture optical systems. Due to its relatively
low dilution ratio, the MTF has a nonzero value for nearly the same spatial frequencies as a monolithic mirror. However, the amplitudes of the spatial frequencies
are smaller: for the first 35% of the covered radial spatial frequencies, the MTF
decreases steeply to about 0.4; for the next 35% it is approxiamtely constant, at
about 0.4; in the last 30%, the MTF decreases at about the same rate as that of a
monolithic mirror. The MTF exhibits a large plateau for the moderate values of the
radial spatial frequency, much the same as that for a mirror with a central obscuration (see Fig. 8.19). The decrease of performance for intermediate radial spatial
frequencies may be compared also to the degradation due to aberrations, illustrated
in Fig. 8.11. A tradeoff of cost and performance is continually made to determine the
largest diameter aperture of a monolithic mirror afflicted with some aberrations vs.
the challenges of phasing the individual subapertures to obtain its optimum performance. Some of these issues are discussed in Hebden [88].

The interest of building diluted apertures has grown also in the Earth remote
sensing community. Figure 8.41 shows a configuration of a large mirror, flanked on
one side by several smaller ones, with the purpose of using the large mirror for
imaging up to its cutoff frequency and the smaller ones for the detection of a specific
phenomenon at a given spatial frequency and orientation. For illustrative purposes,
we present a potential redesign of this aperture, such that a complete coverage of the

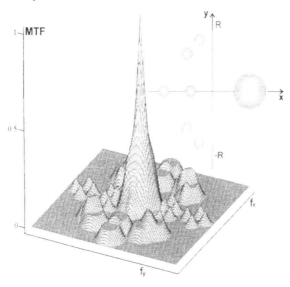

Figure 8.41   The polar stratospheric telescope (POST) unsymmetrical aperture configuration with a large mirror, flanked on one side by several smaller ones, with the purpose of using the large mirror for imaging up to its cutoff frequency and the smaller ones for the detection of a specific phenomenon at a given spatial frequency (after Ford [89]).

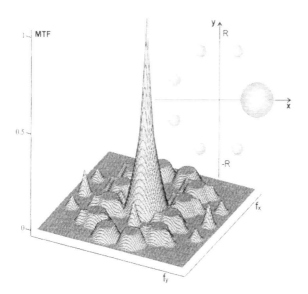

Figure 8.42   The diameters of the side mirrors in the POST layout are increased and their centers are redistributed in a nonredundant manner to obtain a complete, even if a low-amplitude, coverage of the spatial frequency plane up to the cutoff frequency.

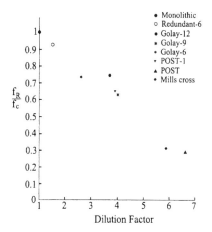

Figure 8.43 The increase in the dilution factor results in the decrease of the quality of imaging as measured by the maximum spatial frequency for which the modulation transfer function (MTF) is not zero, f_R, normalized to the cutoff frequency of the monolithic mirror f_c.

spatial frequency plane is achieved: the side mirror apertures are increased and their centers are redistributed in a nonredundant manner, as depicted in Fig. 8.42.

Figure 8.43 shows the general trend that the increase in the dilution factors results in the decrease of the quality of imaging as measured by the f_R (the maximum spatial frequency for which the MTF is not zero), normalized with respect to the cutoff frequency, f_c.

REFERENCES

1. Hopkins, R. E., *Military Standardization Handbook, Optical Design*, Washington, D. C., 1962, Chapters 10, 14, 15.
2. Kingslake, R., *Optical System Design*, Academic Press, New York, 1983.
3. Malacara, D. and Z. Malacara, *Handbook of Lens Design*, Marcel Dekker, New York, 1994.
4. Goodman, J. W., *Introduction to Fourier Optics*, McGraw-Hill, New York, 1968.
5. Born, M. and E. Wolf, *Principles of Optics*, Macmillan, New York, 1959.
6. Scholl, M. S., "Experimental Demonstration of a Star Field Identification Algorithm," *Opt. Lett.*, **18**(3), 412–404 (1993).
7. Scholl, M. S., "Experimental Verification of the Star Field Identification Algorithm in the Observatory Environment," *Opt. Eng.*, **34**(2), 384–390 (1996).
8. Scholl, M. S., "Apodization Effects Due to the Size of a Secondary Mirror in a Reflecting, On-Axis Telescope for Detection of Extra-Solar Planets," *Infrared Remote Sensing*, M. S. Scholl, ed., *Proc. SPIE*, **2019**, 407–412 (1993).
9. Smith, W. J., *Modern Optical Engineering. The Design of the Optical Systems*, McGraw-Hill, New York, 1990.
10. Scholl, M. S., Y. Wang, J. E. Randolph, and J. A. Ayon, "Site Certification Imaging Sensor for Mars Exploration," *Opt. Eng.*, **30**(5), 590–597 (1991).
11. Scholl, M. S., "Autonomous Star Field Identification Using Intelligent CCD-Based Cameras," *Opt. Eng.*, **33**(1), 134–139 (1994).

12. Rosin, S., "Ritchey–Chretien Corrector System," *Appl. Opt.*, **5**(4), 475–676 (1966).

13. Rosete-Aguilar, M. and J. Maxwell, "Design of Achromatic Corrector Plates for Schmidt Cameras and Related Optical Systems," *J. Modern Opt.*, **40**(7), 1395–1410 (1993).

14. Wilson, R., "The History of Optical Theory of Reflecting Telescopes and Implications for Future Projects; Computer Lens Design Workshop," C. Londoño and R. E. Fischer, eds, *Proc. SPIE*, **2871**, 784–786 (1995).

15. Wynne, C. G., "Ritchey–Chretien Telescopes and Extended Field Systems," *Astrophys. J.*, **152**, 676–691 (1968).

16. Blakley, R., "Cesarian Telescope Optical System," *Opt. Eng.*, **35**(11), 3338–3341 (1996).

17. Puryayev, D. T., "Afocal Two-Mirror System," *Opt. Eng.*, **32**(6), 1325–1329 (1993).

18. Wetherell, W. B., "All Reflecting Afocal Telescopes," in *Reflective Optics*, C. Londoño and R. E. Fischer, eds, *Proc. SPIE*, **751**, 126–134 (1987).

19. Scholl, M. S., "Recursive Exact Ray Trace Equations through the Foci of the Tilted Off-Axis Confocal Prolate Spheroids," *J. Modern Opt.*, **43**(8), 1583–1588 (1996).

20. Dials, M. A., J. C. Gille, J. J. Barnett, and J. G. Whitney, "A Description of the High Resolution Dynamics Limb Sounder (HIRDLS) Instrument," in *Infrared Spaceborne Remote Sensing VI*, M. S. Strojnik, and B. F. Andresen, eds, *Proc. SPIE*, **3437**, 84–91 (1998).

21. Páez Padilla, P. and M. Strojnik Scholl, "Recursive Relations for Ray-Tracing Through Three-Dimensional Reflective Confocal Prolate Spheroids," *Rev. Mex. de Fís.*, **43**(6), 875–886 (1997).

22. Cook, L. G., "The Last Three-Mirror Anstigmat?," in *Lens Design*, W. J. Smith, ed., SPIE Optical Engineering Press, Bellingham, WA 1992, pp. 310–324.

23. Paul, M., "Sistems Correcteurs pour Reflecteurs Astronomiques," *Revue D'Optique*, **5**, 169 (1935).

24. Rumsey, N. J., "A Compact Three-Reflection Astronomical Camera," *Optical Instruments and Techniques*, J. H. Dickson, ed., Oriel Press, Newcastle-upon-Tyne, 1969, p. 514.

25. Eisenberg, S. and E. T. Pearson, "Two-Mirror Three-Surface Telescope," in *Reflective Optics*, C. Londoño and R. E. Fischer, eds, *Proc. SPIE*, **751**, 126–134 (1987).

26. Offner, A., "Unit Power Imaging Catoptric Anastigmat," U.S. Patent #3,748,015 (July 24, 1973).

27. Korsch, D., *Reflective Optics*, Academic Press, New York (1991).

28. Courtes, G., P. Cruvellier, M. Detaille, and M. Saisse, "Some New Optical Designs for the Ultraviolet Bidimensional Detection of Astronomical Objects," in *Progress in Optics XX*, E. Wolf, ed., North-Holland, New York, 1983.

29. Scholl, M. S. and G. Paez Padilla, "Push–Broom Reconnaissance Camera with Time Expansion for a (Martian) Landing – Site Certification," *Opt. Eng.*, **36**(2), 566–573 (1997).

30. Roddier, C. and F. Roddier, "A Combined Approach to HST Wave-Front Distortion Analysis," in *Space Optics*, Vol. 19 of OSA 1991 Technical Digest Series (Optical Society of America, Washington, D.C., 1991), paper MB5-1, pp. 25–27.

31. Fienup, J. R., "Phase-Retrieval Algorithms for a Complicated Optical System," *Appl. Opt.*, **32**, 1737–1746 (1993).

32. Fienup, J. R., "HST Aberrations and Alignment Determined by the Phase Retrieval Algorithms," in *Space Optics*, Vol. 19 of OSA 1991 Technical Digest Series (Optical Society of America, Washington, D.C., 1991), paper MB3-1, pp. 19–21.

33. Scholl, M. S., "Stray Light Issues for Background-Limited Far-Infrared Telescope Operation," *Opt. Eng.*, **33**(3), 681–684 (1994).

34. Beckers, M. M., B. L. Ulich, and J. T. Williams, "Performance of the Multiple Mirror Telescope (MMT): I. MMT – The First of the Advanced Technology Telescopes," in

Advanced Technology Optical Telescopes, G. Burbidge and L. D. Barr, eds, *Proc. SPIE*, **332**, 2–8 (1982).

35. Woolf, N. J., J. R. P. Angel, J. Antebi, N. Carleton, and L. D. Barr, "Scaling the Multiple Mirror Telescope (MMT) to 15 meters – Similarities and Differences," in *Advanced Technology Optical Telescopes*, G. Burbidge and L. D. Barr, eds, *Proc. SPIE*, **332**, 79–88 (1982).

36. Fried, D. L., "Diffusion Analysis for the Propagation of Mutual Coherence," *J. Opt. Soc. Am.*, **58**, 961–969 (1968).

37. Merkle, F., K. Freischlad, and H.-L. Reischman, "Deformable Mirror with Combined Piezoelectric and Electrostatic Actuators," in *Advanced Technology Optical Telescopes*, G. Burbidge and L. D. Barr, eds, *Proc. SPIE*, **332**, 260–268 (1982).

38. Stock, J. and G. Keller, "Astronomical Seeing," in *Telescopes*, G. P. Kuiper and B. M. Middlehurst, eds, Univ. of Chicago Press, Chicago, 1960.

39. Page, N. A. and J. P. McGuire, Jr., "Design of Wide Field Planetary Camera 2 for Hubble Space Telescope," in *Space Optics*, Vol. 19 of OSA 1991 Technical Digest Series (Optical Society of America, Washington, D.C., 1991), paper MC4-1, pp. 38–40.

40. Scholl, M. S. and G. N. Lawrence, "Adaptive Optics for In-Orbit Aberration Correction – Feasibility Study," *Appl. Opt.*, **34**(31), 7295–7301 (1995).

41. Lawrence, G. N., "Optical System Analysis with Physical Optics Codes," in *Recent Trends in Optical System Design; Computer Lens Design Workshop*, C. Londoño and R. E. Fischer, eds, *Proc. SPIE*, **766**, 111–118 (1987).

42. Moore, K. T. and G. N. Lawrence, "Zonal Model of an Adaptive Mirror," *Appl. Opt.*, **29**, 4622–4628 (1990).

43. Hardy, J. H., "Active Optics – Don't Build a Telescope Without It," in *Advanced Technology Optical Telescopes*, G. Burbidge and L. D. Barr, eds, *Proc. SPIE*, **332**, 252–259 (1982).

44. Beckers, J. M., F. J. Roddier, P. R. Eisenhardt, L. E. Goad, and K.-L. Shu, "National Optical Astronomy Observatories (NOAO) Infrared Adaptive Optics Program I: General Description," in *Advanced Technology Optical Telescopes III*, L. D. Barr, ed., *Proc. SPIE*, **629**, 290–297 (1986).

45. Scholl, M. S. and G. N. Lawrence, "Diffraction Modeling of a Space Relay Experiment," *Opt. Eng.*, **29**(3), 271–278 (1990).

46. Massie, N. A., "Experiments with High Bandwidth Segmented Mirrors," in *Advanced Technology Optical Telescopes*, G. Burbidge and L. D. Barr, eds, *Proc. SPIE*, **332**, 377–381 (1982).

47. Roddier, F., "The Effects of Atmospheric Turbulence in Optical Astronomy," in *Progress in Optics XIX*, E. Wolf, ed., North-Holland, New York, 1981.

48. Lutomirski, R. F. and H. T. Yura, "Aperture Averaging Factor for a Fluctuating Light Signal," *J. Opt. Soc. Am*, **59**, 1247–1248 (1969).

49. Fugate, R. Q., "Laser Beacon Adaptive Optics – Boom or Bust," in *Current Trend in Optics*, Vol. 19, C. Dainty, ed., Academic Press, New York, 1994, pp. 289–304.

50. Poglitsch, A., "Far-Infrared Astronomy from Airborne and Spaceborne Platforms (KAO and FIRST)," in *Infrared Spaceborne Remote Sensing II*, M. S. Scholl, ed., *Proc. SPIE*, **2268**, 251–262 (1994).

51. Suzuki, M., A. Kuze, J. Tanii, A. Villemaire, F. Murcray, and Y. Kondo, "Feasibility Study on Solar Occultation with a Compct FTIR," in *Infrared Spaceborne Remote Sensing V*, M. S. Scholl and B. F. Andresen, eds, *Proc. SPIE*, **3122**, 2–15 (1997).

52. J. R. Fienup, J. C. Marron, T. J. Schulz and J. H. Seldin, "Hubble Space Telescope Characterized by Using Phase Retrieval Algorithms," *Appl. Opt.*, *32*, 1747–1768 (1993).

53. Burrows, C. and J. Krist, "Phase Retrieval Analysis of Pre- and Post-Repair Hubble Space Telescope Images," **34** (22), *Appl. Opt.*, (1995), p. 491–496.

54. Kessler, M. F., "Science with the Infrared Space Observatory," in *Infrared Spaceborne Remote Sensing*, M. S. Scholl, ed., *Proc. SPIE*, **2019**, 3–8 (1993).

55. Cesarsky, C. J., J. F. Bonnal, O. Boulade, *et al.*, "Development of ISOCAM, the Camera of the Infrared Space Observatory," in *Infrared Spaceborne Remote Sensing*, M. S. Scholl, ed., *Proc. SPIE*, **2019**, 36–47 (1993).

56. Clegg, P. C., "ISO Long-Wavelength Spectrometer," in *Infrared Spaceborne Remote Sensing*, M. S. Scholl, ed., *Proc. SPIE*, **2019**, 13–23 (1993).

57. Graauw, Th., D. Beintema, W. Luinge, *et al.*, "The ISO Short-Wavelength Spectrometer," in *Infrared Spaceborne Remote Sensing*, M. S. Scholl, ed., *Proc. SPIE*, **2019**, 24–27 (1993).

58. Cohen, R., T. Mast, and J. Nelson, "Performance of the W. M. Keck Telescope Active Mirror Control System," in *Advanced Technology Optical Telescopes V*, L. M. Stepp, ed., *Proc. SPIE*, **2199**, 105–116 (1994).

58. Lemke, D., F. Garzon, H. P. Gemünd, *et al.*, "ISOPHOT – Far-Infrared Imaging, Polarimetry and Spectrophotometry on ISO," in *Infrared Spaceborne Remote Sensing*, M. S. Scholl, ed., *Proc. SPIE*, **2019**, 28–33 (1993).

59. Scholl, M. S., "Design Parameters for a Two-Mirror Telescope for Stray-Light Sensitive Infrared Applications," *Infr. Phys. & Tech.*, **37**, 251–257 (1996).

60. Scholl, M. S. and G. Paez Padilla, "Image-Plane Incidence for a Baffled Infrared Telescope," *Infr. Phys. & Tech.*, **38**, 87–92 (1997).

61. Kemp, J. C. and R. W. Esplin, "Sensitivity Model for the Wide-field Infrared Explorer Mission," in *Infrared Spaceborne Remote Sensing III*, M. S. Scholl and B. F. Andresen, eds, *Proc. SPIE*, **2553**, 26–37 (1995).

62. Beckers, J. M. and J. T. Williams, "Performance of the Multiple Mirror Telescope (MMT): III. MMT – Seeing Experiments with the MMT," in *Advanced Technology Optical Telescopes*, G. Burbidge and L. D. Barr, eds, *Proc. SPIE*, **332**, 16–23 (1982).

63. McCarthy, D. W., P. A. Strittmatter, E. K. Hege, and F. J. Low, "Performance of the Multiple Mirror Telescope (MMT): VIII. MMT as an Optical-Infrared Interferometer and Phased Array," in *Advanced Technology Optical Telescopes*, G. Burbidge and L. D. Barr, eds, *Proc. SPIE*, **332**, 57–64 (1982).

64. Chanan, G., M. Troy, E. Sirko, "Phase Discontinuity Sensing: a Method for Phasing Segmented Mirrors in the Infrared", *Appl. Opt.*, **38**, (4) 704–713 (1999).

65. Chanan, G., C. Ohara, M. Troy, "Phasing the Mirror Segments of the Keck Telescopes II: the Narrow-Band Phasing Algorithm" *Appl. Opt.*, **39**, (25) 4706–4714 (2000).

66. Nelson, J. E. and P. Gilingham, "An Overview of the Performance of the W. M. Keck Observatory," in *Advanced Technology Optical Telescopes V*, L. M. Stepp, ed., *Proc. SPIE*, **2199**, 82–93 (1994).

67. Mast, T. S., J. E. Nelson, and W. J. Welch, "Effects of Primary Mirror Segmentation on Telescope Image Quality," in *Advanced Technology Optical Telescopes*, G. Burbidge and L. D. Barr, eds, *Proc. SPIE*, **332**, 123–134 (1982).

69. Cole, T. W., "Quasi-Optical Techniques of Radio Astronomy," in *Progress in Optics XV*, E. Wolf, ed., North Holland, New York, 1977.

70. Beckers, J. M., "Field of View Considerations for Telescope Arrays," in *Advanced Technology Optical Telescopes III*, L. D. Barr, ed., *Proc. SPIE*, **629**, 255–260 (1986).

71. Scholl, M. S., "Signal Detection by an Extra-Solar-System Planet Detected by a Rotating Rotationally-shearing Interferometer," *J. Opt. Soc. Am. A*, **13**(7), 1584–1592 (1996).

72. Strojnik Scholl, M. and G. Paez, "Cancellation of Star Light Generated by a Nearby Star–Planet System upon Detection with a Rotationally-Shearing Interferometer," *Infr. Phys. & Tech.*, **40**, 357–365 (1999).

73. Gillespie, A., T. Matsunaga, S. Rokugawa, and S. Hook, "Temperature and Emissivity Separation from Advanced Spaceborne Thermal Emission and Reflection Radiometer (ASTER) Images," in *Infrared Spaceborne Remote Sensing IV*, M. S. Scholl and B. F. Andresen, eds, *Proc. SPIE*, **2817**, 83–94 (1996).

74. Woolf, N. J., D. W. McCarthy, and J. R. P. Angel, "Image Shrienking in Sub-Arcsecond Seeing at the MMT and 2.3 m Telescopes (MMT): VIII. MMT as an Optical–Infrared Interferometer and Phased Array," in *Advanced Technology Optical Telescopes*, G. Burbidge and L. D. Barr, eds, *Proc. SPIE*, **332**, 50–56 (1982).

75. Angel, J. R. P. and J. M. Hill, "Manufacture of Large Glass Honeycomb Mirrors," in *Advanced Technology Optical Telescopes*, G. Burbidge and L. D. Barr, eds, *Proc. SPIE*, **332**, 298–306 (1982).

76. Anderson, D., R. E. Parks, Q. M. Hansen, and R. Melugin, "Gravity Deflections of Light-Weighted Mirrors," in *Advanced Technology Optical Telescopes*, G. Burbidge and L. D. Barr, eds, *Proc. SPIE*, **332**, 424–435 (1982).

77. Bely, P. Y., "A Ten-Meter Optical Telescope in Space," in *Advanced Technology Optical Telescopes III*, L. D. Barr, ed., *Proc. SPIE*, **629**, 188–195 (1986).

78. Enard, D., "The ESO Very Large Telescope Project," in *Advanced Technology Optical Telescopes III*, L. D. Barr, ed., *Proc. SPIE*, **629**, 221–226 (1986).

79. Kodaira, K. and S. Isobe, "Progress Report on the Technical Study of the Japanese Telescope Project," in *Advanced Technology Optical Telescopes III*, L. D. Barr, ed., *Proc. SPIE*, **629**, 234–238 (1986).

80. Pearson, E., L. Stepp, W-Y. Wong, *et al.*, "Planning the National New Technology Telescope (NNT): III. Primary Optics – Tests on a 1.8-m Borosilicate Glass Honeycomb Mirror," in *Advanced Technology Optical Telescopes III*, L. D. Barr, ed., *Proc. SPIE*, **69**, 91–101 (1986).

81. Siegmund, W. A., E. J. Mannery, J. Radochia, and P. E. Gillett, "Design of the Apache Point Observatory 3.5 m Telescope II. Deformation Analysis of the Primary Mirror," in *Advanced Technology Optical Telescopes III*, L. D. Barr, ed., *Proc. SPIE*, **629**, 377–389 (1986).

82. Mountain, M., R. Kurz, and J. Oschmann, "The GEMINI 8-m Telescope Project," in *Advanced Technology Optical Telescopes V*, L. M. Stepp, ed., *Proc. SPIE*, **2199**, 41–55 (1994).

83. Petrovsky, G. T., M. N. Tolstoy, S. V. Ljubarsky, Y. P. Khimich, and P. Robb, "A 2.7-Meter Diameter Silicon Carbide Primary Mirror for the SOFIA Telescope," in *Advanced Technology Optical Telescopes V*, L. M. Stepp, ed., *Proc. SPIE*, **2199**, 265–270 (1994).

84. Olbert, B., J. R. P. Angel, J. M. Hill, and S. F. Hinman, "Casting 6.5 Meter Mirrors for the MMT Conversion and Magellan," in *Advanced Technology Optical Telescopes V*, L. M. Stepp, ed., *Proc. SPIE*, **2199**, 144–155 (1994).

85. Miglietta, L., P. Gray, W. Gallieni, and C. Del Vecchio, "The Final Design of the Large Binocular Telescope Mirror Cells for the ORM," in *Optical Telescopes of Today and Tomorrow*, A. Ardeberg, ed., *Proc. SPIE*, **2871**, 301–313 (1996).

86. Espinosa, J. M. and P. Alvarez Martin, "Gran Telescopio Canarias: a 10-m Telescope for the ORM," in *Optical Telescopes of Today and Tomorrow*, A. Ardeberg, ed., *Proc. SPIE*, **2871**, 69–73 (1996).

87. Smith, G. M., "II Status Report," in *Optical Telescopes of Today and Tomorrow*, A. Ardeberg, ed., *Proc. SPIE*, **2871**, 10–14 (1996).

88. Hebden, J. C., E. K. Hege, and J. M. Beckers, "Use of Coherent MMT for Diffraction Limited Imaging," in *Advanced Technology Optical Telescopes III*, L. D. Barr, ed., *Proc. SPIE*, **629**, 42–49 (1986).
89. Ford, H., "A polar stratospheric telescope," in *Advanced Technology Optical Telescopes V*, L. Stepp, ed., *Proc. SPIE*, **2199**, 298–314 (1994).

ADDITIONAL REFERENCES ADDED IN PROOF

Scholl, M. S. and G. Paez Padilla, "Using the y, y-Bar Diagram to Control Stray Light Noise in IR systems," *Infr. Phys. & Tech.*, **38**, 25–30 (1997).
Tarenghi, M., "European Southern Observatory (ESO) 3.5 m in New Technology Telescope," in *Advanced Technology Optical Telescopes II*, L. D. Barr, ed., *Proc. SPIE*, **629**, 213–220 (1986).

9

Spectrometers

WILLIAM WOLFE
The University of Arizona, Tucson, Arizona

9.1 INTRODUCTION

A spectrum is a representation of a phenomenon in terms of its frequency of occurrence. One example is sunrise. At first thought we might conclude that the sun rises every 24 hours, so that the spectrum would be a one placed at a frequency of 1/24 cycle per hour. More careful consideration, of course, tells us that, as the seasons pass and at my latitude, the sun rises anywhere from about once every 12 hours to once every 16 hours. Therefore, for a period of a year, the spectrum would be ones starting at a frequency of 1 at 0.0625 to 0.0833 cycles per hour. There might also be several twos in there near the solstices when the sun has about the same period for several days. In the arctic, there would be a bunch of zeros in the summer and winter when the sun never sets or never rises. Another example, which is more or less constant for all of us, is that we receive from one to 20 advertisements in the mail every day, except on Sundays and official holidays. So this rate would vary from one to 20 per day and have a few, blessed zeros once in a while. In mathematics such a relationship is often called a spectral density, and the cumulative curve is called the spectrum, i.e., about one to 20 at the end of the first week, then more at the end of 2 weeks and still more by the end of the year. In this chapter, the spectral density will be called the spectrum for brevity. We will not have need of the cumulative spectrum.

A spectrometer is any device that measures a spectrum. In the example of the advertisements, you, the counter of the ads, would be the spectrometer. In an optical spectrometer, the emission, absorption, or fluorescence spectrum of a material is measured. Spectrometers come in many different forms, and most of them are discussed in this chapter. Their function can be based on any physical phenomenon that varies with optical wavelength or frequency.

Probably the earliest spectrum observed by man was the rainbow. Other observations were surely glories and perhaps the colors seen from oil slicks and other thin films. The first spectrometer may have been 'constructed' by Seneca during the first century AD, who observed that the colors that came from the edge of a piece of glass were similar to those of the rainbow. More likely it was the prism of Newton. He observed prismatic colors and, indeed, differentiated among different substances by their different *refrangibility*. [1]

The first observation of spectral lines was made by Thomas Melville in 1752. [1] He used a prism and small aperture to observe the different spectra produced by sal ammoniac, alum potash, and other substances. Wollaston (1766–1828) observed the spectrum of the blue base of a candle flame and William Swan saw what are now called the Swan bands in 1856. The first use of the spectrometer as a wave analyzer must belong to Fraunhofer. While measuring the refractive indices of various glasses in about 1815 (independent of the works cited above), he observed the sodium doublet. He then proceeded to observe the Fraunhofer lines of the sun, first with a slit and prism and then with a slit and a diffraction grating. This work was followed in 1835 by Herschel and many other luminaries of physics: Talbot, Foucault, Daguerre, Becquerel, Draper, Angstrom, Bohr, Kirchhoff, Bunsen, Stewart . . . . The critical step of identifying the molecular and atomic structure was the accomplishment of Bunsen and Kirchhoff, although others were involved, including Miller, Stokes, and Kelvin.

Spectroscopy was critical in most of the advances in atomic and molecular physics. The Bohr atom was defined and delineated by relating spectroscopic measurements to the motion of electrons. [2] The very advent of quantum physics was due to the exact representation of the spectrum of blackbody radiation. [3] Sommerfeld applied the theory of relativity to the motion of the perihelion of an electron in its orbit, [4] and Paschen and Meissner confirmed his theoretical predictions with spectroscopic measurements of the fine structure. [5]

Prism spectrometers were improved little by little by the introduction of better sources, better detectors, and better optics. Grating spectrometers had these improvements, but also were improved in very substantial ways by the improvements in grating manufacture. These advances were made by Wood, Harrison, Strong, and others. [6] Gratings were ruled by very precise mechanical devices. A major advance was the introduction of interferometric control. More modern devices now use holographic and replicated gratings. [7]

Spectroscopy is used today in many different applications. It is a standard technique for the identification of chemicals. There are even extensive catalogs of the spectra of many elements, molecules, and compounds to assist in this process. The exhaust gases of our vehicles are tested spectroscopically for environmental purposes, as are many smoke-stack emissions. The medical chemical laboratory uses colorimetry, a simple type of spectrometer for blood, urine, and other fluids analysis. Colorimetry, another simple form of spectroscopy, is used extensively in the garment and paper industries. (These are different types of colorimetry – the first is a transmission measurement at a single color; the second is a determination of chromaticity.)

9.2 SPECTROMETER DESCRIPTIONS

Spectrometers are described in terms of how well they resolve lines, how sensitive they are, whether there is a "jumble" of lines, and their different geometric and path configurations.

9.2.1 Spectral Lines

The concept of a *spectral line* probably arose when Fraunhofer used a slit and a prism. Where there was a local spectral region of higher intensity, it looked like a line, because it was the image of a slit. The plot of a spectrum as a function of the frequency can be low and flat until a local region of higher intensity occurs. If this is narrow enough it looks like a vertical line. *Line* is probably a nicer term than *spike* or *pip*, which could have been used. There is a story that has wonderful memories for me. When the animals were being named, the question arose: "Why did you call that a hippopotamus?" The answer was: "Because it looked like a hippopotamus." That applies here.

9.2.2 Spectral Variables

The spectra that are measured by spectrometers are described in terms of their *frequency* and their *wavelength* as independent variables and by *emission, absorption, fluorescence,* and *transmission* as dependent variables. Sometimes these are given as arbitrary output values and as digital counts.

The spectral variables include ν, the frequency in cycles per second or Hertz; σ, the wavenumber in cycles per centimeter; k, the radian wave number equal to $2\pi\sigma$ in radians per centimeter; and the wavelength λ, given variously in nanometers, micrometers, or angstroms. The wavenumber is also cited as kaysers, and angstroms are no longer on the "approved list" for a wavelength measure. Many authors use $\tilde{\nu}$, but this is awkward in typography and will not be used here. In summary and in equation form the spectral variables are related by

$$\sigma = \frac{k}{2\pi} = \frac{1}{\lambda} = \frac{\nu}{c}. \tag{9.1}$$

All of these have the same units, usually cm and cm/s. Almost always σ is given in reciprocal centimeters and λ in micrometers. In this case

$$\sigma = \frac{10000}{\lambda}. \tag{9.2}$$

9.2.3 Resolution and Resolving Power

Resolution is a measure of the fineness with which the width of a spectral line can be measured. One measure of this is the full width of the measured line at half the maximum value, the FWHM. This can be given in any of the spectral variables: $d\lambda$, $d\sigma$, dk, etc. Resolution can also be stipulated as a fraction of the wavelength: $d\lambda/\lambda$, $d\sigma/\sigma$, dk/k, etc. This is slightly awkward in that lower values are generally better: i.e., 0.01 is a higher resolution than 0.1! It is finer. *Resolving power* (RP) is a definition of resolution that avoids this little complication. It is the reciprocal of fractional resolution: $\lambda/d\lambda$, $\sigma/d\sigma$, etc. This is the same definition as the quality factor Q of an electrical circuit. Thus both RP and Q are used for the resolving power of a spectrometer, and some authors use $\mathscr{R}$ and call it resolvance. The fractional resolution and resolving power are equal no matter what the spectral variable:

$$Q = RP = \frac{\sigma}{d\sigma} = \frac{\lambda}{d\lambda} = \frac{k}{dk} = \frac{\nu}{d\nu} = \cdots. \tag{9.3}$$

However, the resolution $d\sigma$ is not equal to the resolution in wavelength, but, for σ in reciprocal centimeters and λ in micrometers, one finds

$$|d\lambda| = \left|\frac{\sigma}{\lambda}\,d\sigma\right| = \left|\frac{10000}{\lambda^2}\,d\sigma\right|. \tag{9.4}$$

Sometimes the base band is specified; it can be the full width at 1% of maximum or the full width to the first zeros of the spectral line. The shape factor is the ratio of the base band to the half width (FWHM).

9.2.4   Free Spectral Range and Finesse

In multiple-beam interference, which includes both Fabry–Perot interferometers and diffraction gratings, there can be an overlapping of orders. The *free spectral range* is the spectral interval between such overlaps. The *finesse* is the ratio of the free spectral range to the spectral slit width.

9.2.5   Throughput and Sensitivity

Whereas resolution is one important characteristic of a spectrometer, defining how well the spectral lines can be determined, throughput, is part of the determination of the sensitivity of the spectrometer. The signal-to-noise ratio of a spectrometer can be written as

$$\text{SNR} = \frac{D^* L_\lambda d\lambda Z}{\sqrt{A_d B}}, \tag{9.5}$$

where D^* is the specific detectivity, [8] L_λ is the spectral radiance, $d\lambda$ is the spectral bandwidth, A_d is the detector area, B is the electronic bandwidth, and Z is the throughput, which is defined by the following expression

$$Z = \frac{\tau A_e \cos\theta_f A_f \cos\theta_i}{f^2} = \tau A_f \cos\theta_i \Omega' = \tau A_f \cos\theta_i \frac{\pi}{4F^2} = \tau A_f \pi NA^2, \tag{9.6}$$

where τ is the transmission of the spectrometer, A_e is the area of the entrance pupil, A_f if the area of the field stop, Ω' is the projected solid angle, f is the focal length, F is the optical speed in terms of the F-number, and NA is the numerical aperture. These will be evaluated in later sections for specific instruments.

The specific detectivity is a function of the detector used, and the radiance is a function of the source used. The throughput is a property of the optical system and is invariant throughout the optics.

In classic treatments [9] the product of the resolving power and the throughput has been used as a figure of merit. While this is important and a reasonable figure, it is not the whole story. From the expression for the SNR, it is easy to see that square root of the detector area and noise bandwidth are important considerations. An extended figure of merit that incorporates the bandwidth and detector area will be used later to compare these different instruments.

9.3   FILTER SPECTROMETERS

In some ways these are the simplest of all spectrometers. There are, however, many different types of filters and filter designs. A filter spectrometer in concept is a device that consists of a number of filters, each of which passes a small portion of the spectrum. Filters are described by their transmission characteristics as *narrow band*, *wide band*, *cut on*, and *cut off*. These are, of course, general descriptions because *narrow* and *wide* are relative terms. Care must be exercised in reading about cut-on and cut-off filters. Usually a *cut-on* filter has zero or very low transmission at short wavelengths and then is transparent at longer wavelengths. The *cut-off* is the opposite. Sometimes, however, authors use the term *cut on* to mean a filter that has zero or low transmission at short *frequencies*. The former nomenclature will be used in this chapter.

9.3.1 Filter Wheels

It is easy to imagine the simple filter-wheel spectrometer as an optical system with a detector behind a large wheel that has a set of filters in it, as shown schematically in Fig. 9.1. The filter wheel is moved from position to position, thereby allowing different portions of the spectrum to fall on the detector. The resolution and resolving power depend upon the nature of the filter.

Filters can be classified as absorption filters, interference filters, Lyot filters, Christiansen filters, reststrahlen filters, and acousto-optical tunable filters.

Absorption filters [10] operate on the principle that absorption in any material is a function of wavelength. These filters are manufactured by a number of vendors, including Kodak, Corning, Zeiss, and Schott. [11] Their spectral bands are not regular (well behaved – neither symmetric nor of equal resolution in dλ or dσ or resolving power Q) because they are dependent upon the dyes that are usually used in a glass or plastic substrate. Semiconductors make fine cut-on filters. They absorb radiation of wavelengths short enough that the photons are energetic enough to cause an electronic transition. Then, at the critical wavelength, where the photons

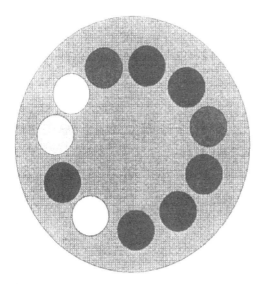

Figure 9.1 A filter wheel.

are no longer more energetic than the bandgap, there is a *sharp* transition to transmission. These can often be used in conjunction with interference filters as short-wave blockers. The transmissions of special crystals and glasses are tabulated in several publications. [12]

9.3.2 Christiansen Filters

These filters are based upon the fact that scattering of light is a function of wavelength. Solid particles are immersed and spread throughout a liquid. Light is scattered by the particles at every wavelength except the one for which the refractive index of the particles is the same as that of the liquid. Liquids are used because their refractive indices change faster than solids; some combinations of solids can certainly be used if the particles can be properly distributed in the volume. McAlister [14] made a set of Christiansen filters [13] with borosilicate glass spheres in different concentrations of carbon disulfide in benzene. They had center wavelengths from about 450 nm to 700 nm with a resolving power of about 10.

9.3.3 Reststrahlen Filters

Reststrahlen or residual ray filters are based on the variation of refractive index in the wavelength regime of an absorption band. In this region the reflectance varies from about 5 or 10% to almost 100%. By generating several reflections from such a material in this region, the high-reflection portion of the spectrum is passed through the system, while the low reflectivities are attenuated. Both the Christiansen and reststrahlen filters are cumbersome and little used as commercial spectrometers. They can be required for some parts of the spectrum, especially the infrared.

9.3.4 Lyot–Öhman Filters

Also called a polarization filter, filters of this type are based on the wavelength variation of the rotation of the plane of polarization by certain materials, quartz being one of the best known. They were developed independently by Lyot in France in 1933 and by Öhman in 1938 in Sweden. A linear polarizer precedes a rotation plate; another polarizer precedes a plate of twice the thickness, and so on, as shown in Fig. 9.2. If the rotation is 90 degrees, then the second plate does not pass the polarized light. To achieve 90-degree rotation, the plate must be 'half-wave', i.e., it

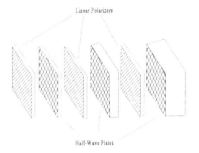

Figure 9.2   The Lyot–Öhman filter.

must have an optical path difference that is one half wavelength (in the medium), or $\Delta = nd\lambda/2$. Then wavelengths of all odd multiples of this will be blocked, and all even multiples, or full waves will be transmitted. The situation is the same for the second plate and its polarizers, but it is twice as thick. It therefore has twice the frequency of maxima. The successive transmissions and resultant are shown in Fig. 9.3. The passband is a function of wavelength because the path difference is a function of the refractive index and therefore of wavelength. Filters of this type have been made with half-widths of about 0.1 nm in the visible ($\Omega \sim 5000$) and peak transmissions of about 70%. [15]

9.3.5 Interference Filters

These are perhaps the most popular and useful filters in the arsenal [16]. They are composed of thin layers of transparent material in various combinations of different thicknesses and refractive indices. Of course the refractive index is a function of the wavelength, so the interference is a function of the wavelength. An analysis of the transmission of a single-layer filter can provide insight into the more complex designs. The expression for its transmission always includes the cosine of the incidence angle, a fact that can be critical to its use as a spectrometer. Figure 9.4 shows the transmission of the filter as a function of wavelength; Fig. 9.5 shows the filter transmission as a function of wavenumber and Fig. 9.6 shows a typical transmission as a function of angle. These together illustrate why most theoretical treatments of

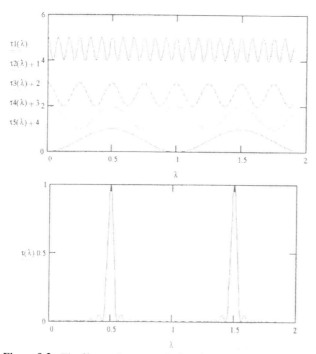

Figure 9.3 The filter actions: top, the interference of the components; bottom, the result.

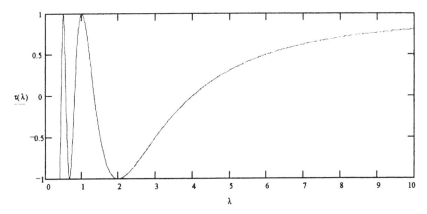

Figure 9.4   Filter transmission vs. wavelength.

interference filters are done on a frequency basis, i.e., the interference filter is a multiple-beam system with a finite free spectral range and transmission is a function of angle of incidence. The last fact implies that the passband can be altered by tipping the filter, and that the passband is broadened when a convergent beam is incident upon the filter.

Most bandpass interference filters are based on the Fabry–Perot interferometer in which the mirrors and the spacer are all thin films – with differing refractive index and thickness. Typically, one material is the high index and one other material is low index, and the thickness are all QWOT – quarter-wave optical thickness. These filters are also subject to the angle changes discussed above, although some special designs can reduce the effect. Details of the Fabry–Perot interferometer, which apply in large measure to the Fabry–Perot filter, are discussed later.

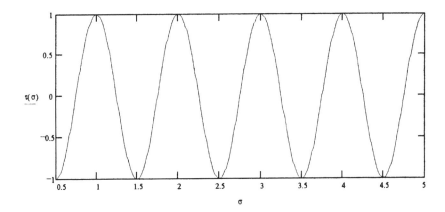

Figure 9.5   Filter transmission vs. frequency.

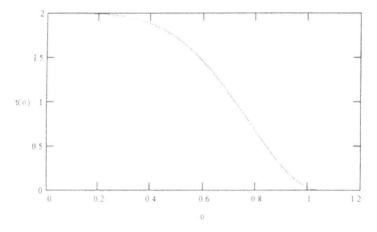

Figure 9.6 Filter transmission vs. angle.

9.3.6 Circular Variable Filters

Interference filters of the types described above can be made in a special way. The different layers can be deposited according to the prescription on a circular substrate – as the substrate is rotated at a variable rate. Therefore, each portion of the circle has layers of slightly different thickness. There is a circular gradient of layer thickness around the filter. Consequently, each different circumferential position of the filter has a different wavelength of maximum transmission. It is a circularly variable filter or CVF. These can be made to order, with different gradients, different resolving powers, and different spectral regions, but there are constraints on them. These filters are used in spectrometers of relatively low resolving power by rotating them in front of the optical system. A slit is used right behind the filter, and the optics then image the slit on the detector. As the filter rotates, different wavelength bands are incident upon the detector.

An Optical Coating Laboratories Inc. (OCLI) filter can serve as an example. It covers the region from 2.5 to 14.1 µm in three 86-degree arcs of radii 1.45 and 1.75 in. for bands from 2.5 to 4.3, 4.3 to 7.7, and 7.7 to 14.1 µm. The gradients are therefore 0.75, 1.42, and 2.67 µm per inch, or a slit 0.1-inch (0.25 cm) wide has a resolution of 0.075, 0.142, and 0.267 µm in each of the bands and the resolving powers are about 45. The resolution could be made finer with a finer slit, but the flux is reduced proportionately.

The throughput of a CVF spectrometer will include the slit area, the system focal length, and the detector area, as well as the filter transmission. The throughput will be determined by this aperture area, the size of the detector, and the focal ratio. For the 0.1-inch slit and the wheel cited above, and an F/1 lens, and a 25×75 µm detector, the throughput is 10^{-6} in^2 sr or 6.25×10^{-6} cm^2 sr. (I chose a small infrared detector or a big visible one.)

Since this is a plane, parallel plate in front of the optical system, narcissus (internal reflections) may be a problem. [17]

9.3.7 Linear Variable Filters

Gradient filters can also be made such that the gradient is linear; therefore, these are called linear variable filters or LVFs. These filters are generally meant for use with detector arrays. They are placed in close proximity with the arrays, thereby ensuring that each pixel in the array senses one component of the spectrum. This close proximity can also cause a certain amount of scattering that may give rise to cross-talk. The spectral range is 2:1 (an octave – do, re, mi, fa, so la, ti, do; do) with resolving power of 10–100 and peak transmission of about 70%. The infrared range of 1–20 μm is covered. The throughput can be larger, since a slit is not necessary.

With both CVFs and LVFs the manufacturer should be contacted and can surely make them to order with approximately these properties.

9.4 PRISM SPECTROMETERS

Prism spectrometers are probably the oldest type of spectrometer. First used by Fraunhofer to determine refractive indices of various materials for better designs of telescopes, they are still in use today and still provide good performance for some applications.

9.4.1 Throughput

The 'standard' configuration of a prism spectrometer is shown in Fig. 9.7. Light enters a slit with dimensions $h \times w$, and is collimated by the first lens. A collimated beam is incident upon the first face of the prism A_{prism}, where it is dispersed in a set of collimated beams that cover a range of angles corresponding to a range of wavelengths. These beams are collected by the camera lens and focused onto the exit slit. The light that passes through the exit slit is then focused onto an appropriate detector. The throughput is determined by the area of the slit (which is normal to the optical axis and is therefore equal to the projected area) and the projected area of the prism and the square of the focal length of the lens. The throughput is constant throughout. Since the prism is the critical element, the collimating lens is slightly larger, and the prism area is the aperture stop of the system. The camera lens must be a little larger than the collimating lens because of the dispersion. In a well-designed

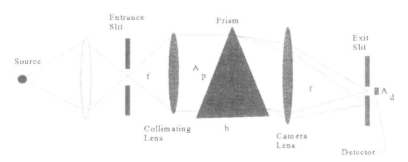

Figure 9.7 Prism spectrometer layout.

spectrometer the entrance and exit slits are equal in size, and either or both is the field stop. Thus, the throughput is

$$Z = \frac{A_{\text{slit}} A_{\text{prism}} \cos \theta}{f^2} = \frac{hw A_{\text{prism}} \cos \theta}{f^2}. \tag{9.7}$$

9.4.2 Prism Deviation

The collimated beams that pass through the prism can be represented nicely by rays. The geometry is shown in Fig. 9.8. The deviation δ is easily seen to be

$$\delta = \theta_1 - \theta_1' + \theta_2 - \theta_2' = \theta_1 + \theta_2 - (\theta_1' + \theta_2') = \theta_1 + \theta_2 - \alpha. \tag{9.8}$$

Minimum deviation can be found by taking the derivative and setting it to zero, as usual:

$$d\delta = d\theta_1 + d\theta_2 = 0 \tag{9.9}$$

The final expression is valid since two of the legs of the triangle formed by the normals and the ray in the prism are perpendicular to each other, so the third angle of the prism is α. Clearly, minimum deviation is obtained when the two angles are equal, but opposite in sign. The minimum deviation angle is $\delta = 2\theta_1 - \alpha$, and $\theta = \alpha/2$. This information can be used to obtain an expression used for finding the refractive index of such a prism. At the first surface, Snell's law is

$$n_1 \sin \theta_1 = n_2 \sin \theta_2, \tag{9.10}$$

$$\frac{n_2}{n_1} = \frac{\sin \theta_1}{\sin \theta_2} = \frac{\sin \dfrac{\delta + \alpha}{2}}{\sin \dfrac{\alpha}{2}}. \tag{9.11}$$

The angular magnification can be found by a reapplication of Snell's law. At the first surface, the relationship is

$$n_1 \sin \theta_1 = n_2 \sin \theta_1'. \tag{9.12}$$

Differentiation yields

$$n \cos \theta_1 d\theta_1 = n' \cos \theta_1' d\theta_1' \tag{9.13}$$

The same applies to the second surface, with 2's as subscripts. Division and a little algebra yield

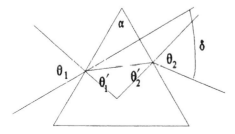

Figure 9.8 Ray geometry.

$$\frac{d\theta_2}{d\theta_1} = -\frac{\cos\theta_1 \cos\theta_2'}{\cos\theta_1' \cos\theta_2}. \tag{9.14}$$

From Eq. (9.9), at minimum deviation,

$$\frac{d\theta_2}{d\theta_1} = -1. \tag{9.15}$$

So, when the deviation is at a minimum, $\theta_1 = \theta_2$ and $\theta_1' = \theta_2'$. In this condition there is complete symmetry with respect to incident and exiting beams, and the beam is parallel to the base (of an isometric prism with base normal to the prism angle bisector). We note in passing that Eq. (9.7) provides the expression for the magnification generated by the prims, which is 1 for minimum deviation.

9.4.3 Dispersion

One of the classical techniques for measuring the refractive index of a prism was introduced by Fraunhofer. It can be written

$$n = \frac{n_2}{n_1} = \frac{\sin\theta_1}{\sin\theta_1'} = \frac{\sin(\alpha + \delta)/2}{\sin(\alpha/2)}, \tag{9.16}$$

which was obtained from the relationships above at minimum deviation. Then

$$\frac{d\delta}{dn} = \frac{2\sin\alpha/2}{\cos(\alpha + \delta)/2} = \frac{2\sin\alpha/2}{\sqrt{1 - n^2 \sin^2\dfrac{(\alpha + \delta)}{2}}} = \frac{2\sin\alpha/2}{\cos\theta_1}. \tag{9.17}$$

This expression can be used to find the angular dispersion of a prism of relative refractive index n, and the linear dispersion is $f d\delta$, where f is the focal length. The resolving power of the prism is given by

$$Q = \frac{\lambda}{d\lambda} = \frac{\lambda}{d\delta}\frac{d\delta}{d\lambda} = a\frac{d\delta}{d\lambda} = a\frac{d\delta}{dn}\frac{dn}{d\lambda}, \tag{9.18}$$

where a is the beam width. Then, substitution of $a = p\cos\theta_1$ and $b = 2p\sin\alpha/2$, gives

$$Q = a\frac{b/p}{a/p}\frac{dn}{d\lambda} = b\frac{dn}{d\lambda}, \tag{9.19}$$

where p is the length of the side of the (square) prism face and b is the base of the prism, or, more accurately, the maximum length of travel of the beam through the prism. The resolving power is just the 'base' times the dispersion of the prism material.

9.4.4 Some Mounting Arrangements

There are both refractive and reflective arrangements that have been invented or designed for prism spectrometers. A few apply to imaging prism spectrometers. Perhaps the most popular is the Littrow arrangement, which is illustrated in Fig. 9.9. Only the essential elements are shown. The light is retroreflected so that there are two passes through the prism. Somewhere in the optics in front of the prism there must be a beamsplitter or pick-off mirror to get to the exit slit. In the Wadsworth

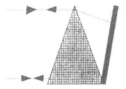

Figure 9.9 The Littrow mount.

arrangement the mirror is set parallel to the base of the prism, as shown in Fig. 9.10, in order to have an undeviated beam.

Symmetric and asymmetric mirror arrangements are shown in Fig. 9.11. The symmetric arrangement is shown as image 1, and the other as image 2. A little consideration leads to the conclusion that the asymmetric (right-hand) arrangement leads to balancing of off-axis aberrations – because the tilt in one direction helps to offset the aberrations that were introduced by the tilt in the other direction.

Figure 9.10 The Wadsworth mount.

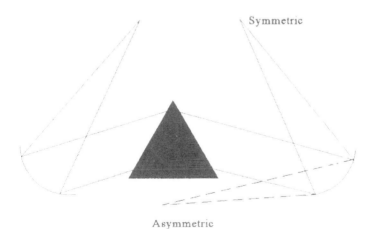

Figure 9.11 Symmetric and asymmetric mounts.

9.5 GRATING SPECTROMETERS

Grating spectrometers make use of the diffraction of light from a regularly spaced, ruled surface. They disperse the light by a combination of diffraction and interference rather than the refractive index variation with wavelength, as with a prism.

Probably, Joseph Fraunhofer [18] was (again) the first to use diffraction gratings (in 1823). Henry Rowland [19] later built extremely precise ruling engines that could make relatively large, about 10-inch, gratings of high resolving power. Rowland invented the concave grating. [20] Albert Michelson, America's first physics Nobel laureate, developed interferometric techniques for ruling gratings that were used and improved by John Strong and George Harrison. [21] It was R. W. Wood [22] who introduced the blaze, and Strong who ruled on aluminum films that had been coated on glass (for greater stability and precision than speculum). Although Michelson indicated in 1927 the possibility of generating, not ruling, gratings interferometrically, it took the advent of the laser to allow the development of holographic gratings.

There are three types of grating: the rule grating, the holographic grating, and replica gratings made from molds that have been ruled. The latter two are relatively cheap, and are used in most commercial instruments today because they are cheap, reproducible, and do not depend upon the transparency of a medium, as does a prism.

9.5.1 Diffraction Theory

It can be shown [23] that the expression for the irradiance pattern from a ruled grating is

$$E = E_0 \operatorname{sinc}^2(\pi w \sin\theta/\lambda) \left[\frac{\sin(N\pi s(\sin\theta_d + \sin\theta_i)/\lambda)}{\sin(\pi s \sin\theta/\lambda)} \right]^2 \tag{9.20}$$

where N is the number of rulings, s is the ruling spacing, θ_i is the angle of incidence, θ_d is the angle of diffraction, and λ is the wavelength of the light. This pattern consists of three terms: a constant, E_0, that depends upon the system setup; the source, optical speed, and transmission; a single-slit diffraction function and an interference function. The general form is shown in Fig. 9.12. If the rulings are sinusoidal rather than rectangular in cross section, the sinc function is replaced by a sine. (The sinc function is given by sinc $x = \sin x/x$.) It can be seen, based on Fourier transform theory, that this is the Fourier transform of the rectangle that is the full grating × the comb function that represents grooves. [24] This equation is sometimes written

$$E = E_0 \operatorname{sinc}^2\beta \left[\frac{\sin N\gamma}{\sin\gamma} \right]^2. \tag{9.21}$$

In this form it is emphasized that β is half the phase difference between the edges of a groove, and γ is half the phase difference between rulings. This theory does not take into account the generation of ghosts and other results of imperfect rulings.

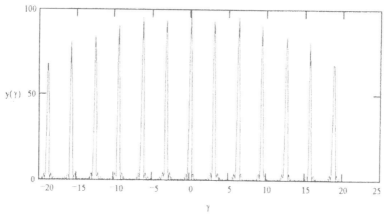

Figure 9.12 Multislit diffraction pattern.

9.5.2 Geometric Layout

The layout for a grating spectrometer is almost the same as for a prism. If the grating operates in transmission, then it simply replaces the prism and is straight through, as shown in Fig. 9.13. Configurations for plane and concave [25] reflective gratings are shown at the end of this section.

9.5.3 Resolution and Resolving Power

The equation for the position of the peaks may be obtained from the basic expression for the irradiance. The position of the peaks for each wavelength is given by

$$m\lambda = s(\sin\theta_i + \sin\theta_d), \tag{9.22}$$

where s is the line spacing, often called the grating spacing, θ_i is the angle of incidence and θ_d is the angle of diffraction. The order of interference is m and λ is the wavelength.

$$d\lambda = \frac{s\cos\theta_d d\theta_d}{m}. \tag{9.23}$$

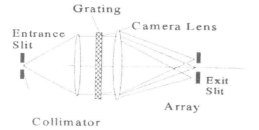

Figure 9.13 Transparent grating layout.

The resolution can be found by differentiation (where only the diffraction angle changes). Better (smaller) spectral resolution is obtained at higher orders, with smaller line spacing and at angles that approach 90 degrees. A grating has constant wavelength resolution as a function of angle in the image plane.

The resolving power is given by

$$Q = \frac{\lambda}{d\lambda} = mN. \tag{9.24}$$

The resolving power is just the order number × the number of lines in the grating. It is not possible to pick an arbitrarily large order or to get an unlimited number of lines. This is true partly because the grating is of finite size, and efficiency generally decreases with order number.

9.5.4 Throughput

The throughput of the grating is identical to that of a prism, as the product of the slit area and projected area of the grating divided by the focal length of the collimating lens. For this reason, it is desirable to have zero incidence angle when possible. This can be possible with a transmission grating, but is almost never available with a reflection grating. The entrance and exit slits should be identical.

9.5.5 Blazing Rulings

The efficiency may be increased by making the rulings triangular and using the slope to reflect the light to a particular order. The manufacturer will specify the blaze direction and the efficiency, typically about 60%. The rulings are still rectangular in cross section; they are slanted within the groove.

9.5.6 Grating Operation

The normal operation of the grating is the same as with a prism. The grating is rotated, and wavelength after wavelength passes the slit and its radiation is detected. There is, however, an additional complication. The unwanted orders must be filtered so that they do not taint the measurement.

The requirement to eliminate extraneous orders arises from the basic equation and indicates that lower diffraction orders are preferable. Equation (9.22) has geometry on the right-hand side. On the left-hand side is the product $m\lambda$, the order × the wavelength. So first order and 1 µm, for instance, is the same as second order and 0.5 µm, and third order and 0.33 µm, etc. It is preferable from one standpoint to operate at low orders, because it makes the separation of overlapping orders easier. The ratio of wavelength is given by

$$\frac{\lambda_2}{\lambda_1} = \frac{m+1}{m}. \tag{9.25}$$

The higher the order, the closer are the two wavelengths. The separation of overlapping orders is

$$\Delta\lambda = \frac{\lambda}{m}. \tag{9.26}$$

This is the free spectral range.

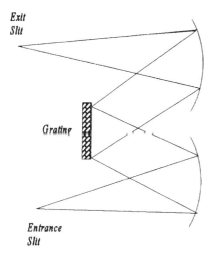

Exit
Slit

Grating

Entrance
Slit

Figure 9.14   Czerny–Turner mount.

9.5.7   Some Mounting Arrangements

Most gratings are reflective and need special mounting arrangements. These include Rowland, Eagle, and Paschen–Runge mountings for a concave grating, the Fastie Ebert, and Czerny–Turner mounts for plane gratings. Clearly, prism mounts, like the Littrow can be used for transparent gratings.

Both systems for use with plane gratings make use of the symmetry principle for the reduction of off-axis aberrations, but there is usually some residual astigmatism. The Czerny–Turner, Fig. 9.14 has more flexibility and uses two mirrors that can be adjusted independently. The Fastie–Ebert, Fig. 9.15, uses a larger, single spherical mirror.

The Rowland mounting was, of course, invented by Rowland, himself, the originator of the curved grating. It placed the curved grating and the detector

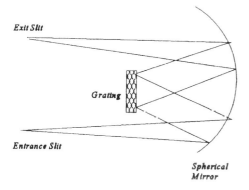

Exit Slit

Grating

Entrance Slit

Spherical
Mirror

Figure 9.15   The Fastie–Ebert mount.

(then a photographic plate) on the ends of a rigid bar, the length of which is equal to the radius of curvature of the grating. The ends of the bar rest in two perpendicular rails, as shown in Fig. 9.16. The entrance slit is placed at the intersection of the rails with the jaws perpendicular to the rail that carries the grating. By this arrangement, the slit, the centers of the grating, and the plate are all on a circle, the Rowland circle. This mounting is now obsolete as a result of various technological advances. The Abney mounting, now also obsolete, uses the same geometry, but the bar is rigid, while the slit rotates about a radius arm at the middle of the bar and itself rotates to keep the jaws perpendicular to the grating. Although this sounds more complicated, only the light (not heavy) slit arrangement moves.

The Eagle mounting, shown in Fig. 9.17, is similar to the prism Littrow mount in that the plate and slits are mounted close together on one end of a rigid bar. The concave grating is at the other end of the bar. The main advantage of the Eagle mount is its compactness. However, there are more adjustments to be made than with either the Abney or Rowland mounting arrangements.

The Wadsworth mounting is shown in Fig. 9.18. The most popular modern mounting is the Paschen–Runge mount, shown in Fig. 9.19.

9.6 THE FOURIER-TRANSFORM SPECTROMETER

This very important instrument had its origins in the interferometer introduced by Michelson in 1891 [26] for the examination of high-resolution spectra. It is more often used today in the form of a Twyman–Green interferometer, [27] which is essentially a Michelson interferometer used with collimated light. The Twyman–

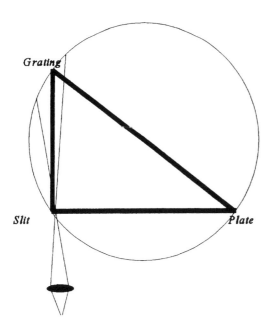

Figure 9.16 The Rowland mounting.

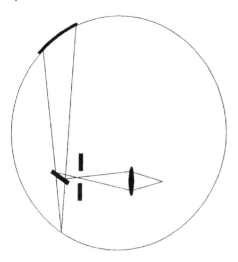

Figure 9.17 The Eagle mounting.

Green was introduced to test the figure and quality of optical components. It has more recently become the basis of many Fourier-transform spectrometers (FTSs). The use of interferometers as spectral analysis instruments was pioneered by Felgett, [28] Jacquinot, [29] and Strong [30].

9.6.1 Two-Beam Interference

When two monochromatic beams of light are superimposed, an interference pattern results. If Ψ_1 and Ψ_2 represent the complex electric fields of the two beams that are combined, the expression for the interference pattern in terms of the irradiance E is

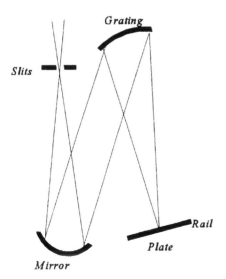

Figure 9.18 Wadsworth mounting.

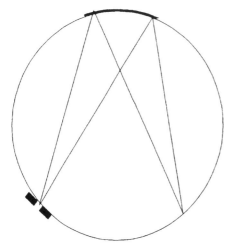

Figure 9.19   The Paschen–Runge mount.

$$E = \langle (\Psi^2) \rangle = \Psi \cdot \Psi^* = \Psi_1^2 + \Psi_2^2 + 2\Psi_1\Psi_2 \cos\left(\frac{2\pi nd \cos\theta}{\lambda}\right) \qquad (9.27)$$

The optical quantity that is sensed is the time average of the square of the electric field. The right-hand expression then consists of a dc or constant term plus the interference term.

9.6.2   Interference in the Michelson Interferometer [31]

The Michelson interferometer is shown schematically in Fig. 9.20. An extended source illuminates a beamsplitter, where the light is split into two separate beams.

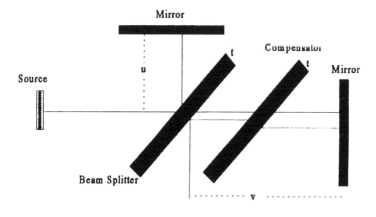

Figure 9.20   The Michelson interferometer.

The light is returned from each of the mirrors to the beamsplitter, which now acts as a beam combiner. The two beams then form an interference pattern at the focus of the lens.

There is a maximum in the pattern whenever the phase term is 0, or a multiple of π. This, in turn, can happen when $n\Delta$, the optical path difference is 0, or when $\cos\theta$ is 0 or a multiple of π. Thus the pattern is dependent upon both the effective separation of the two mirrors d and the angle off axis, θ.

The optical path of the on-axis beam that goes up, is $u + 2nt/\cos\theta' - 2t/\cos\theta$, where u is the (central) separation of the top mirror from the beamsplitter, n is the refractive index of the plate, θ' is the refracted angle in the beamsplitter and compensator plate, θ is the incidence angle, and t is their thickness. The path length of the horizontal beam is $2v + 2nt/\cos\theta' - 2t/\cos\theta$, since the beam goes through the beamsplitter twice for this configuration. (I have assumed that the bottom surface has an enhanced reflection while the bottom is coated for low reflection. It could be the other way around. It is troublesome if both surfaces have about the same reflection, because then there is multiple-beam interference generated in the beamsplitter.) This difference in the two path lengths is the reason that many instruments have a so-called compensation plate in one arm. The difference is just $2(u - v)$, but would be a function of refractive index (and therefore wavelength) if the compensator were not there, or if the top surface were the highly reflecting one. An off-axis beam is more complicated, as is shown in Fig. 9.21. There are more cosines, but the results are much the same. The differences are that the incidence and refraction angles are not the same as in the first case, and the paths in air are longer by the cosine. The result, with the compensator in place, is that the path difference is $2(u - v)/\cos\theta$. This path difference is part of the argument in the cosinusoidal interference term:

$$E = E_0[1 + \cos(2\pi n\Delta \cos\theta)]. \tag{9.28}$$

This gives the so-called bull's-eye pattern of a central ring with annuli that surround it of decreasing width, as shown in Fig. 9.22 in plan view and sectionally in Fig. 9.23.

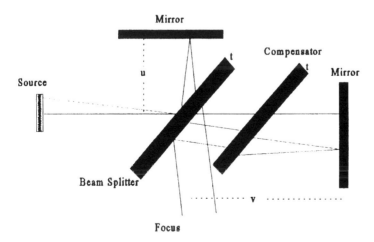

Figure 9.21 Off-axis beams in a Michelson.

Figure 9.22 Plan view of the bull's-eye pattern.

9.6.3 The Twyman–Green Interferometer

When the Michelson interferometer is illuminated with monochromatic collimated light, for instance by a lens that has a laser at its front focus, then the bull's-eye pattern disappears, and the pattern is a point. The flux density in this single on-axis spot varies as a function of the separation of the mirrors, the path difference. This is the basis of the Fourier-transform spectrometer.

9.6.4 The Fourier-Transform Spectrometer [32]

If the source of this on-axis interferometer is on axis, its flux density varies as the separation, in a manner that depends upon the optical path difference and the wavelength of the light in the usual way:

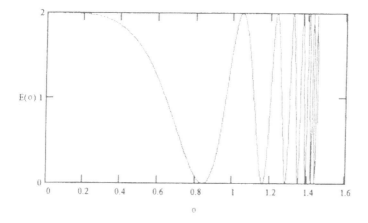

Figure 9.23 Sectional view of the bull's-eye pattern.

$$E + E_0\left[1 + \cos\left(\frac{2\pi\Delta}{\lambda}\right)\right].$$ (9.29)

The optical path difference, Δ, as shown above is twice the difference of the mirror separation from the beamsplitter. If the source is bichromatic, i.e., has two beams with two different frequencies and amplitudes, the interference pattern will be the sum of two cosines.

$$E = E_0\left[\frac{E_1}{E_0} + \cos\left(\frac{2\pi\Delta}{\lambda_1}\right) + \frac{E_2}{E_0}\cos\left(\frac{2\pi\Delta}{\lambda_2}\right)\right].$$ (9.30)

If there are many, the same is true, but with more components

$$E = \sum_i E_i\cos\left(\frac{2\pi\Delta}{\lambda_i}\right) \Rightarrow \int E(\lambda)\cos\left(\frac{2\pi\Delta}{\lambda}\right)d\lambda$$ (9.31)

and the integral represents a continuous summation in the usual calculus sense. The pattern obtained, as the mirror moves, is the sum of a collection of monochromatic interference patterns, each with its own amplitude. This is the interferogram. Its Fourier transform is the spectrum. If the interferogram is measured by recording the pattern of a given source as a function of the path difference (translation of one mirror), then the mathematical operation of Fourier transformation will yield the spectrum. We have just seen the essence of Fourier-transform spectroscopy, the so-called FTS technique.

Figures 9.24 and 9.25 further illustrate these relationships. Figure 9.24 shows five cosines plotted as a function of the path difference, which is assumed to be on axis and is just *nd*. The wavelengths of these cosines are 10–14 μm in steps of 1 μm. It can be seen how they all 'cooperate' in their addition at zero path difference, but as the path difference increases, their maxima and minima separate, and they get "muddled up." As the path difference increases, the interference pattern will decreases, but not monotonically. Figure 9.25 shows the square of the sum of

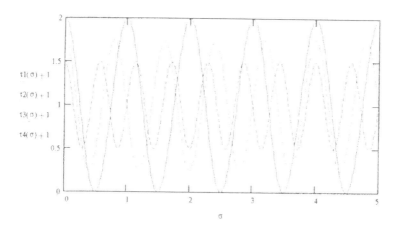

Figure 9.24 Waves in an FTS.

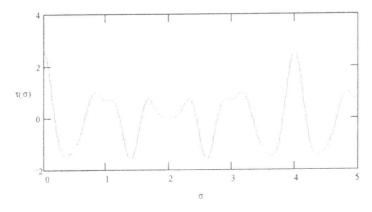

Figure 9.25 Superposition of waves in an FTS.

these five functions. It is the interferogram (within a constant and ignoring the constant term).

There is another way to look at this. For every position of the moving mirror the radiation that reaches the detector is a sum of components of different amplitudes and different frequencies, and for each frequency there is a different phase difference causing a different state of interference. This is the interferogram or interference pattern – a sum of sine waves with different amplitudes. The Fourier transform of this is the spectrum. The interferogram flux density can be written as

$$E(\delta) = E_0(1 + \cos 2\pi\sigma\delta), \tag{9.32}$$

where $E(\delta)$ is the incidence as a function of the path difference δ. For a nonmonochromatic beam it is

$$E(\delta) = \int_0^\infty E_0(1 + \cos 2\pi\sigma\delta)d\sigma, \tag{9.33}$$

where E_0 is the intensity measured at zero path difference between the two arms and $S(\sigma)$ is the spectrum of the light. Then, ignoring the dc or fixed background term, one has

$$E(\delta) = \int_0^\infty E_o \cos(2\pi\sigma\delta)d\sigma. \tag{9.34}$$

This is in the form of a Fourier cosine transform. The inverse transform provides the spectrum:

$$S(\sigma) = \int_0^\infty E_0 \cos(2\pi\sigma\delta)d\delta. \tag{9.35}$$

There is still another way to understand this relationship and process. The motion of the mirror generates the autocorrelation function of the incident light, and the power spectrum is the Fourier transform of the autocorrelation function. Usually the interferogram is recorded by reading the output from the detector as one arm of the interferometer is moved. Then, after the fact, the spectrum is calculated numerically by computer techniques.

The resolution can be found, integrating Eq. (9.22) from 0 to the full extent of the path difference δ_{max}. The result is a sinc of $2\pi\sigma\delta_{max}$. The first zero occurs when this argument equals π, and that gives the following condition:

$$\Delta\sigma = \frac{1}{2\delta_{max}} = \frac{\lambda^2}{20{,}000\delta_{max}}. \tag{9.36}$$

By the Shannon sampling theorem, one needs two samples for every cycle. The resolving power, of course, will be given by

$$Q = \frac{\sigma}{d\sigma} = \frac{\lambda}{d\lambda} = \frac{\sigma}{2\delta} = \frac{5000}{\lambda\delta}. \tag{9.37}$$

The relations for wavelength are obtained from those for frequency by the fact that $\lambda/d\lambda = \sigma/d\sigma$ and, because the wavelength is in μm and the wavenumber is in cm^{-1}, $\lambda = 10{,}000/\sigma$. The resolution in the frequency domain is independent of the frequency, but the resolution in wavelength terms is dependent upon the wavelength.

9.6.5 Throughput and Sensitivity [33]

The Michelson (or Twyman–Green) has both a throughput and a multiplex advantage. The detector is exposed to all the radiation in the total spectral band during the entire scan, whereas with the grating and prism spectrometers only the part of the spectral region within the resolution cell is incident upon the detector. The sensitivity calculations can be complicated. The following is a fairly simple argument that shows this advantage for the FTS. The expression for the interference on axis (assuming the beam flux densities are equal) is the average over a full scan of the moving mirror. This can be written

$$E = \frac{1}{\Delta}\int_\Delta\int_\sigma E(\sigma)[1 + \cos(2\pi\sigma nd)]d\sigma d\Delta = \int_\sigma E(\sigma)d\sigma. \tag{9.38}$$

The first term in the integral is a constant and represents the full density in the spectrum. The second term is a cosine that has many cycles in the full-path averaging and goes to zero. Of course, the optical efficiency of the beamsplitter, compensator plate, mirrors, and any foreoptics need to be taken into account.

The throughput is given by the usual expression,

$$Z = \frac{A_o\cos\theta A_d\cos\theta}{f^2} = A_o\cos\theta\Omega = A_o\Omega, \tag{9.39}$$

where A_o is the area of the entrance optics, and equal to the area of the focusing optics, A_d is the detector area, f is the focal length, and Ω is the solid angle the detector subtends at the optics. The optics are on axis, so the cosine is 1. The area of the detector can be at a maximum, the area of either the central circle of the bull's eye or any of the annuli or some combination. These can have a width that encompasses one full phase. The solid angle in general is given by

$$\Omega = 2\pi(\cos\theta_2 - \cos\theta_1), \tag{9.40}$$

where the θs represent the minimum and maximum angles of the annuli's radii. The central circle has a zero minimum angle. Since the phase is represented by

$2\pi\sigma\,\Delta\cos\theta$, $\cos\theta$ can be only as large as $1/\sigma\Delta$, which is the same as $2d\sigma/\sigma = 2/Q$. This leads to the relationship

$$Q\Omega = 4\pi, \tag{9.41}$$

which is one similar to that reported by Jacquinot [34]. This result is valid for the Michelson interferometer and is 2π for the Fabry–Perot instrument (which makes two passes of the spacing to generate interference), which is discussed next.

9.7 FABRY–PEROT SPECTROMETERS

The Fabry–Perot (FP) interferometer is a multiple-wave device. It is essentially a plane, parallel plate of material, that can be air or vacuum, between two partially reflecting plates, as shown in Fig. 9.26. (The rays are shown for a slightly off-axis point to emphasize the repeated reflections.) It can be in the form of two partially reflecting mirrors with a space between them, a plate of glass or other dielectric with reflective coatings on the faces, or layers of thin films that form a filter. As shown, there is walk off of the beams after many reflections. When used on axis, that is not the case. The figure also shows two separate partially reflective mirrors; it could be a single plate with two surfaces that perform the reflection functions. It is shown later that the field covered by an FP is indeed quite small so that walk off is not a problem, and the classic Airy function applies. Note, however, that in configurations that are used relatively far off axis, the light is attenuated; the summation of beams is not infinite, and modifications must be made.

The function that describes the transmission of the FP is the so-called Airy function: [35]

$$\tau(\sigma) = \frac{\tau_{\max}}{1 + R\sin^2\phi}, \tag{9.42}$$

where the maximum transmission τ_0 is given by

$$\tau_{\max} = \frac{\tau_1\tau_2}{(1 - \rho)^2}, \tag{9.43}$$

$$\phi = \pi\sigma\Delta - \frac{\varepsilon_1 + \varepsilon_2}{2} = \pi\sigma nd\cos\theta - \frac{\varepsilon_1 + \varepsilon_2}{2}, \tag{9.44}$$

where the ρ's are the reflectivities of the coating measured from the gap, the τ's are the transmittances of the coatings on

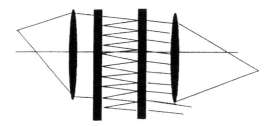

Figure 9.26 Schematic of a Fabry–Perot interferometer.

$$R = \frac{4\rho}{(1-\rho)^2},$$ (9.45)

$$\rho = \sqrt{\rho_1 \rho_2},$$ (9.46)

the plates, the εs are the phase shifts on reflection, n is the refractive index of the plate, and d is its thickness. The phase shifts serve only to shift maxima and minima; they do not affect resolution, resolving power, free spectral range, finesse, and throughput.

The pattern is shown in Fig. 9.27. The line width (FWHM) is given approximately, for small angles, by the following expression:

$$d\sigma = \frac{1-\rho}{2\sqrt{\rho}} \frac{1}{\pi\Delta}.$$ (9.47)

It can be found by setting the transmission expression to $\frac{1}{2}$ and solving for $\sin\phi$. Then, for small angles the angle is equal to the sine and the FWHM is twice the value found, yielding Eq. (9.47).

The free spectral range is the distance in frequency space between the maxima, i.e., when $F\sin\phi$ is zero. This occurs at integral multiples of π, for $\phi = m\pi$, or when $\sigma = m/\Delta$. Therefore, the free spectral range, the distance between the peaks, is $1/\Delta$. The resolving power is given by

$$Q = \frac{\sigma}{d\sigma} = \frac{\lambda}{d\lambda} = \sigma\pi\Delta \frac{\sqrt{\rho}}{1-\rho} = \pi\frac{\Delta}{\lambda} \frac{\sqrt{\rho}}{1-\rho}.$$ (9.48)

It is related to plate separation and the reflectivities, and is proportional to the number of waves in the separation. Assume that the reflectivity is 0.99, that the separation is 5 cm, and that the light is right in the middle of the visible, at 500 nm. Then, the resolving power is

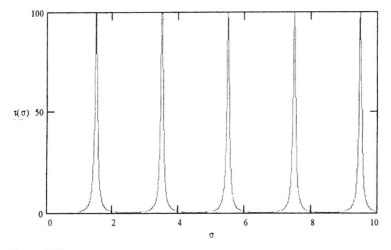

Figure 9.27 The Airy function.

$$Q = \frac{\sigma}{d\sigma} = \frac{\lambda}{d\lambda} = \pi \frac{\Delta}{\lambda} \frac{\sqrt{\rho}}{1-\rho} = 3.14 \frac{0.05}{0.5 \times 10^{-6}} \frac{\sqrt{0.99}}{0.01} = 3.14 \times 10^7 \times 99.5 \approx 3 \times 10^9.$$

(9.49)

This extremely high resolving power is the strength of the FP interferometer spectrometer. Most of it, 10^7, comes from the number of waves in the cavity, but having high reflectivity helps (by a factor of about 100) both the Q and the maximum transmission, which increases as the reflectivity of the plates increases! The transmission maximum occurs when the square of the sine is zero and minimum when it is 1. The maximum occurs when ϕ is an integral multiple of π, when σnd is a multiple of $1/2$, or when nd is an integral number of waves. The separation of the peaks, the free spectral range, is given by

$$\Delta\sigma = \frac{1}{\Delta} = \frac{1}{nd\cos\theta}.$$

(9.50)

The free spectral range decreases as the path separation increases (the order number increases), so high resolving power is obtained at the expense of free spectral range. For the example given above, the free spectral range is $0.2\,\mathrm{cm}^{-1}$ (the resolution is $2 \times 10^{-12}\,\mathrm{cm}^{-1}$). The finesse for collimated light is

$$\frac{\Delta\sigma}{Q} = \frac{1-\rho}{nd\cos\theta\sqrt{\rho m\pi}}.$$

(9.51)

The throughput is calculated based upon the area of the beam that traverses the plates and the size of the image. The image is a bull's eye pattern with alternating light and dark annuli, as with the Michelson, and it has the same expression. In fact the throughput of the Fabry–Perot is the same as that of the Michelson. Some workers [36] have used curved slits that match an arc of one of the annuli. In that case, the solid angle is $\phi(\cos\theta_2 - \cos\theta_1)$, with the angle ϕ replacing the full circumferential angle 2π.

The expressions given here are for an idealized instrument. Jacquinot provides information on the influence of various imperfections and provides some additional insights into the operation of the performance of Fabry–Perot interferometers.

Fabry–Perot interferometers are research tools, and are not available commercially.

9.8 LASER SPECTROMETERS

Laser spectrometers [37] are different from classical spectrometers in that there is no monochromator; the source is a tunable laser. The line width can be very narrow. These can be used for transmission, reflection, and related measurements, but cannot be used to measure emission spectra. The limitations are strictly those of the bandwidth and spectral range of tunable lasers. Some of the salient characteristics of lasers used in the visible and infrared are power, resolving power, and spectral range.

There are several different types of tunable lasers, all with different characteristics. [38] The most useful type seem to be semiconductor lasers [39] that can span the spectrum from about 350 nm to about 14 μm. This range must be accomplished with different materials, ranging from ZnCdS to HgCdTe, of various mixture ratios. Tuning is accomplished in these semiconductor lasers by varying either the drive current or the pressure. Powers range from microwatts to milliwatts; line widths as small as 10^{-6} cm^{-1} have been observed, but useful widths are somewhat larger. Note that this represents a resolving power of about 10^{10}!

The second most useful type is probably the family of dye lasers. [40] These cover the region from about 350 nm to 1 μm using several different dyes and solvents. Each combination can be tuned several hundred wavenumbers from its spectral peak with a concomitant reduction in power.

Color-center lasers, in which lasers like NdYAG pump certain alkali halides, cover the region from about 1–3 μm, again with several different pump and color-center lasers.

The fourth class is molecular lasers, such as carbon dioxide, carbon monoxide, deuterium fluoride, and carbon disulfide, which are used at relatively high pressures to broaden the vibrational–rotational bands of the gases themselves. These typically have much greater power, but somewhat lower resolving power.

These laser spectrometers have incredibly high resolving powers as a result of the narrowness of the laser lines. They can be used in special situations with relatively limited spectral ranges, with relatively low power and with the specter of mode hopping as a problem to be overcome and controlled.

A summary of the properties of the main tunable lasers that can be used for these spectrometers is given in Table 9.1.

9.9 UNUSUAL SPECTROMETERS

During the 1950s, largely as a result of the pioneering work of Jacquinot on the importance of throughput in spectrometers and that of Felgett on the value of multiplexing, many innovative designs arose. Some became commercial; others have died a natural death. The concepts are interesting.

The general treatments have been based on classical spectrometers like the prism and grating instruments described above. Therefore, these newer ones are described as having either a multiplex or a throughput advantage. The spirit of this section is to reveal these interesting concepts for those who might employ one or more of them.

9.9.1 Golay Multislit Spectrometer

Marcel Golay described the technique of using more than one entrance slit and more than one exit slit to gain a throughput and multiplex advantage. His first paper [41] describes what might be called dynamic multislit spectrometry. His example is repeated here; it is for an instrument with six slits. They are modulated by a chopper. Table 9.2 shows how each of the slits is blocked or unblocked by the chopper.

The convention is that 1 is complete transmission and 0 represents no transmission; it is a binary modulation of the slits. During the first half cycle, no radiation that passes through the first entrance slit gets through the first exit slit. During the second half cycle, all of the radiation passes through both. The same is true for the rest of the slits. This provides 100% modulation for corresponding entrance and exit slits. However, during the first half cycle, no light gets through the 1–4, 2–5, and 3–6 slit combinations and one-fourth gets through the others in each half cycle. There is,

Table 9.1 Properties of Spectrometer Lasers

Semiconductor lasers	Spectral range	Dye lasers	Spectral range	Molecular lasers	Spectral range	Color center lasers	Spectral range
ZnCdS	0.3–0.5	Stilbene	0.4–0.42	DF	3.8–4.2	NaF	1.0–1.2
CdSeS	0.5–0.7	Coumarin	0.42–0.55	CO	5.0–8.0	KF	1.2–1.6
GaAlAs	0.6–0.9	Rhodamine	0.55–0.7	CO_2	9.0–12.0	NaCl	1.4–1.8
GaAsP	0.6–0.9	Oxazine	0.7–0.8	CO_2	9.0–12.0	KCl	1.8–1.9
GaAs	0.8–0.9	DOTC	0.75–0.8			KCl:Na	2.2–2.4
GaAsSb	1.0–1.7	HIDC	0.8–0.9			KICl:Li	2.3–2.8
InAsP	1.0–3.2	IR-140	0.9–1.0			RbCl:Li	2.8–3.2
InGaAs	1.0–3.2						
InAsSb	3.0–5.5						
PbSSe	4.0–8.0						
HgCdTe	3.2–15						
PbSnTe	3.2–15						

Table 9.2 Dynamic Multislit Modulation Scheme

	Entrance slits		Exit slits	
Slit	First half cycle	Second half cycle	First half cycle	Second half cycle
1	0101	0101	1010	0101
2	0011	0011	1100	0011
3	0110	0110	1001	0110
4	0101	1010	1010	1010
5	0011	1100	1100	1100
6	0110	1001	1001	1001

therefore, strong modulation of the corresponding slits, but no modulation from the "off" combinations.

Static multislit spectroscopy uses some of the same ideas, but establishes two masks, an entrance and an exit mask. Again, using Golay's example, [42] assume that there is an optical system with two side-by-side entrance pupils. In each there is a set of eight slits, and there are corresponding exit pupils and slits. Table 9.3 shows the arrangement.

Light passes through the entire system only when 1's are directly above each other in each half. In this arrangement, there are four pairs that allow light through in the first pair of pupils and none in the second. Light of a nearby wavelength is represented by a shift of the entrance pupils. In the first set, only one pair will let light through for a shift of 1, and the same is true for the second set. This is also true for shifts of up to six, and no light gets through for shifts of more than six. These, of course, are not really shifts, but static blockages of unwanted spectra. The scheme can be adapted to more than one row and more slits.

9.9.2 Haddamard Spectrometer [42]

This device bears a very distinct similarity to the Golay devices just described. It consists of a collection of entrance slits and correlated exit slits. Decker [44] describes these as binary masks (100% transmitting and 0% transmitting) arranged in a pseudo-random pattern that do not depend upon any prior knowledge. They are an analog of frequency modulation and can be thought of as a set of n equations with n unknowns.

Table 9.3 Modulation Scheme for Static Multislit Spectroscopy

First entrance pupil	11001010
First exit pupil	11001010
Second entrance pupil	10000001
Second exit pupil	01111110

One fairly representative device operated in the 2.5–15 µm region, with a resolution of 3.5 cm^{-1}, from 666–4000 cm^{-1}. It used several gratings and order sorting that will not be described here. The effective throughput was 1.3 mm^2 sr.

9.9.3 Girard Grille [45]

This device is another version of a multislit spectrometer, although one must consider the word *slit* in a very general way. Rather than slits, the entrance and exit planes contain Fresnel zone plates. In other respects this system operates the same as the Golay static multislit spectrometer. At one specific wavelength, the light passes both grilles without attenuation. At other wavelengths, the dispersed light passes through the first zone plate on axis, but through the exit plate off axis. The autocorrelation function of these zone plates is peaked at the reference wavelength (zero displacement) and decays rapidly off axis. The Girard grille has high throughput, but does not have the multiplex advantage. It was once offered commercially by Huet in Paris, but does not seem to be in existence as a commercial instrument today. The instrument they built was a large (3 m × 2.3 m diameter with 2 m focal length mirrors) laboratory device that used a vacuum system and PbS, InSb, and extrinsic germanium detectors for operation from 1 to 35 µm.

9.9.4 Mock Interferometer

Although Mertz described this as a mock interferometer, a more descriptive appellation might be a *rotating multislit spectrometer*. Besides the usual optics, it consists of an entrance aperture, comprising a set of parallel bars, as shown in Fig. 9.28, which is a schematic of the arrangement. The exit slit is an image of the entrance slit. As these rotate, a modulation is imposed on the light that is proportional to the distance between the centers of the disks. Since the exit slit is an image of the entrance slit in the light dispersed from the prism, the modulation is a function of the frequency of

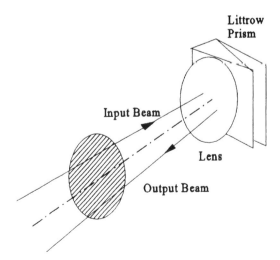

Figure 9.28 Mock interferometer.

the light. The detector measures a quantity proportional to the integral of the spectral flux × the cosine of the rotational frequency of the disks, i.e.,

$$V(t) = \mathcal{R} \int_{\sigma_1}^{\sigma_2} \Phi(\sigma)\cos^2(2\pi\sigma t + \phi_0)\mathrm{d}r \tag{9.52}$$

where $\mathcal{R}$ is the responsivity and Φ is the flux on the detector. This system has both the throughput and the multiplex advantage. Although Block engineering offered it commercially for some years, it no longer appears available as a commercial product.

9.9.5 SISAM

The acronym is from the French, *systemé interferential selection amplitude moduation* (well, it's almost English. Is this the Anglaise that some French despise?). The technique, developed by Connes [46] is to use a Michelson interferometer with gratings rather than mirrors. Motion of one of the gratings provides modulation of a given spectral line. The advantage here is throughput. Imagine that a monochromatic beam enters and that the (identical) gratings are set to reflect the beam directly back. Then at zero-path difference there is constructive interference of that beam. It can be modulated by rotating the compensation plate or by translating one of the gratings. In that way the radiation of that wavelength is modulated. Radiation of a nearby wavelength will also be modulated, but not as strongly, until the gratings are rotated (in the same direction) to obtain constructive interference for that wavelength.

It can be shown that the resolving power is $2\sigma\Delta$ without apodization and $\sigma\Delta$ with apodization (masking the gratings with a diamond shape); the throughput is the same as for a Michelson; the free spectral range is the same as that for a grating.

9.9.6 PEPSIOSIS [47]

No this is not some sort of stomach remedy; it stands for purely interferometric high-resolution scanning spectrometer. I confess that I do not understand the acronym. In its simplest form it is a Fabry–Perot with three etalons. More are described in the pertinent articles. Here it is enough to note that one can eliminate some of the maxima of one etalon by interference with the second etalon and thereby increase the finesse. The arguments and calculations are very similar to those involving the Lyot–Öhman filter.

9.10 ACOUSTO-OPTICAL FILTERS

Acousto-optics has provided us with a new method of performing spectral filtering. An acoustic wave can be set up in a crystal. This wave, which is an alternation of rare and dense portions of the medium, provides a diffraction grating in the crystal. The grating spacing can be adjusted by tuning the frequency of the acoustic wave. This is a tunable diffraction grating. It provides the basis for the acousto-optical tunable filter, the AOTF.

There are two types of acousto-optical tunable filters: collinear and noncollinear. In the collinear version, unpolarized light is polarized and propagated through a medium, usually a rectangular cylinder of the proper AO material. An

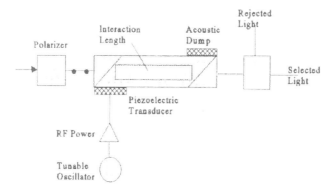

Figure 9.29   The collinear AOTF.

acoustic transducer is attached to the side of the cylinder and by way of a prism, as shown in Fig. 9.29, the acoustic waves propagate collinearly with the optical waves. There is energy coupling so that a new wave with a frequency that is the sum of the acoustic and optical frequencies is generated – as long as the phases match. The acoustic wave is reflected to a beam dump; the output light is passed through an analyzer in order to maintain only the phase-matched light. As a result of the two polarization operations, the maximum transmission is 0.25. In practice it will also be reduced by the several surfaces and by the efficiency of coupling. The collinear AOTF couples the energy from one polarization state to the other; the noncollinear system separates the beams in angle, as shown in Fig. 9.30.

The center wavelength of the passband λ_0 is given by

$$\lambda_0 = \frac{v \Delta n}{f},\tag{9.53}$$

where v is the acoustic velocity in the material, Δn is the difference between the ordinary and extraordinary refractive indices, and f is the acoustic frequency.

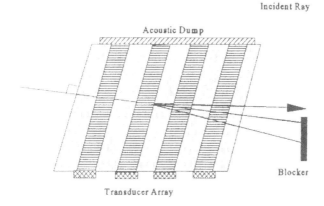

Figure 9.30   Noncollinear AOTF.

The spectral resolution is given by

$$d\lambda = \frac{0.9\lambda_0^2}{\zeta l \sin^2 \theta_i},$$ (9.54)

where l is the interaction length (the length in the material over which the acoustic waves and the optical waves are superimposed and interact), θ_i is the angle of incidence, and ζ is the dispersion constant, given by

$$\zeta = \Delta n - \lambda_0 \frac{\partial \Delta n}{\partial \lambda_0},$$ (9.55)

where Δn is the change in refractive index over the band and λ_0 is the center wavelength.

The solid angle of acceptance is given approximately by

$$\Omega = \frac{\pi n^2 \lambda_0}{l \Delta n}.$$ (9.56)

The more accurate representation includes both angles, and is given by

$$d\theta_1 = n\sqrt{\frac{\lambda_0}{dnlF_1}},$$ (9.57)

$$d\theta_2 = n\sqrt{\frac{\lambda_0}{dnlF_2}},$$ (9.58)

where

$$F_1 = 2\cos^2\theta_i - \sin^2\theta_i \qquad F_2 = 2\cos^2\theta_i + \sin^2\theta_i,$$ (9.59)

where θ_i is again the angle of incidence. The solid angle is obtained by integrating these two differential angles, and when multiplied by the projected area gives the throughput of the system, as usual.

The acoustic power required is given by

$$\tau_0 = \sin^2\left[\frac{\pi l}{\lambda_0}\sqrt{\frac{M_2 E_a}{2}}\right]$$ (9.60)

where τ_0 is the maximum transmission, E_a is the acoustic power density, and M_2 is the acoustic figure of merit.

In a noncollinear system, the separation of beams is given by

$$\Delta\theta_d = \Delta n \sin 2\theta_0,$$ (9.61)

as illustrated in Fig. 9.30.

One of the early AOTFs used in the visible [48] was TeO_2, operated from 450 to 700 nm with an acoustic frequency of 100–180 MHZ, a spectral band FWHM of 4 nm and an acoustic power of 0.12 W. Nearly 100% of the incident light is diffracted with an angle of about 6 degrees. The resolution is given by Eq. (9.54). The acceptance angles, by Eqs (9.57) and (9.58), are 0.016 and 0.015 rad (9.16 and 8.59 degrees and 2.4 msr). The required power is given as 120 mW. Finally, the deviation angle is about 6 degrees.

Table 9.4 Properties of the Westinghouse AOTF Imaging Spectrometer

Property	Value	Units
Spectral resolution	5	cm^{-1}
Resolving power	400–1000	
Spectral band	2–5	μm
Crystal length, area	3.5, 2.5 × 1.4	mm
Interaction length	2.6	cm
Acoustic input	0.86	$W\,cm^{-2}$ (80% efficiency)
Detector	128 × 128 InSb	$D^* = 4 \times 10^{11}$
Optics	TMA and reimager	
f, F, A	207.4, 3.27, 56.2 × 56.2	mm
SNR[a] vs. 400 K	13 dB @ 2 μm, 31 db @ 3 μm	for 1 frame/s
SNR[a] vs. 400 K	2 @ 2 μm, 126 @ 3 μm	for 0.01 frame/s

[a]The SNR data in db are given directly from the paper, where db = 10 log (ratio). A TMA is a Three-Mirror Anastigmat. [50]

Westinghouse built an AOTF System [49] for Eglin Air Force Base to measure the spectra of jet aircraft exhaust plumes. It was an imaging spectrometer. The properties are summarized in Table 9.4.

9.11 SPECTROMETER CONFIGURATIONS

Grating and prism spectrometers come in a variety of configurations: single and double pass, single and double beam, and double monochromator versions.

A classic spectrometer has a source, a monochromator, and a receiver. These configurations are different, according to how the monochromator is arranged. The simplest configuration is a single-pass, single-beam single monochromator. It is shown in Fig. 9.31. This configuration uses a source to illuminate the monochromator through the entrance slit, a prism with a Littrow mirror to disperse the light, and an off-axis paraboloidal mirror to focus and collimate the beam. The light of a single wavelength passes through the exit slit to a detector. Although the beam actually passes through the prism twice, it is considered a single-pass instrument. Figure 9.32

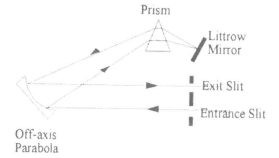

Figure 9.31 Single-pass Littrow spectrometer.

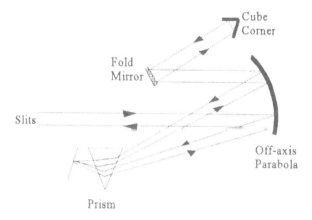

Figure 9.32 Double-pass Littrow spectrometer.

shows the double-pass version of this. The use of two passes, provides better resolution at the cost of a second prism and some additional components.

Double-beam spectrometers are used to obtain instrumental corrections for changes in atmospheric transmission, slit size variations, and spectral variations of the source and detector. They are nulling instruments, the most popular of which uses a comb-shaped attenuator to match the signals, and the position of the attenuator is monitored. A pair of choppers is used to switch from beam to beam in the nulling process, as shown in Fig. 9.33. With a single-beam instrument, a transmission measurement requires a measurement without a sample and then with, and the ratioing of the two. The double-beam does this instrumentally. The two choppers are synchronized so that light passes onto the detector alternately from the sample beam and the reference beam. In this way everything is the same, except for the comb-shaped attenuator in the reference beam. The position of the comb is mon-

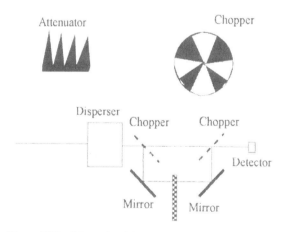

Figure 9.33 Schematic of the double-beam principle.

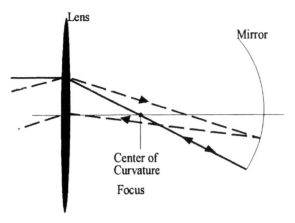

Figure 9.34   Cat's-eye reflector.

itored as it always seeks a zero difference between the two beams. Other attenuators can be used, but this one has the advantage of spectral neutrality. The top beam is the sample beam; the bottom beam is the reference beam.

Double-monochromator spectrometers use two monochromators in series to obtain greater immunity from scattered light.

Compound spectrometers may use a prism monochromator as a prefilter for a grating monochromator.

Interferometer spectrometers come in different configurations as well. The two-beam versions most often are in the form of a Michelson interferometer. However, it is critical that the moving mirror does not tip, tilt, or translate. One approach to eliminating tip and tilt is to use a cube corner as the moving mirror. This does not eliminate the possibility of translation. Another approach is the use of cat's-eye mirrors, as shown in Fig. 9.34. For Michelsons, in particular, knowledge of the position of the mirror is critical; most systems use a position-sensing laser system to record the position, and this information is used in the Fourier transformation. The transformation can be done accurately even if the velocity of translation is not constant.

Some have used interferometer configurations other than the Michelson to obtain the advantages of Fourier-transform spectrometers. These include the Mach–Zender, shown in Fig. 9.35, and the Sagnac. The Sagnac interferometer is one of a class of interferometers that are of the two-beam variety with beams that traverse the same path in opposite directions. There can be as few as three mirrors and as many as imagination and space will allow. The four-plate Sagnac is shown in Fig. 9.36 because it seems to be the easiest to illustrate and explain. The light enters the Sagnac at A where the beamsplitter sends it to B above and D to the right. The mirror D sends the light up to C, where it rejoins the beam that went from A to B to C. The two go through the output beamsplitter with a phase difference that is determined by the positions of the mirrors M_1 and M_2. That this is not multiple-beam interferometry can be seen by considering the lower beam that goes from A to D to C. It is reflected to B and to A, but then it is refracted out of consideration to E. Similarly, the ABCDA beam returns to the source. Similar considerations apply to

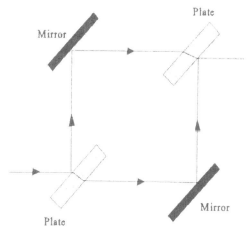

Figure 9.35   Mach–Zender interferometer.

these counter-rotational interferometers. Some of the history and performance are described by Hariharan, who also gives ample references. [51]

There are also several versions of multiple-wave interferometers. These include the double etalon and the use of spherical mirrors. [52] The spherical Fabry–Perot has the same transmission, resolving power, free spectral range, and contrast as the plane version, but the throughput is directly proportional to the resolving power rather than inversely proportional.

9.11.1   Spherical Fabry–Perot [53]

Consider two identical spherical mirrors, the upper parts of which are partially reflecting and the lower parts are completely reflecting, as shown in Fig. 9.37. They have centers of curvatures on the opposite mirror, as indicated by C_1 and

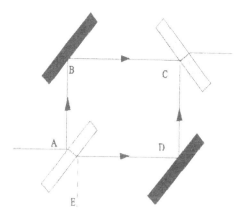

Figure 9.36   The Sagnac four-mirror interferometer.

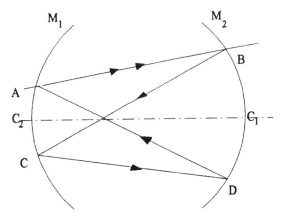

Figure 9.37   The spherical Fabry–Perot.

C_2. A ray, representing a collimated beam, enters the instrument at A and traverses to B, where it is partially reflected. It goes to C on the mirror part, where it is totally reflected towards C and back again to A. There it joins the incident ray with a phase shift that is exactly twice the mirror separation. This is why it has the properties of a Fabry–Perot with twice the etalon separation.

   The literature also contains descriptions of some designs for maintaining the position of the mirror while it translates. [54]

   Jackson describes a Fabry–Perot with a double etalon to increase the finesse. [55]

9.12   COMPARISONS

There are many similarities among the different types of spectrometers. They are presented here, and a summary of the salient properties is given in Tables 9.5 and 9.6. Table 9.5 lists the different types of spectrometers and the expressions for resolution, $d\sigma$ and $d\lambda$, resolving power Q, free spectral range $\Delta\sigma$, throughput Z, the required number of samples N_s during a spectral scan, and a normalized signal-to-noise ratio, SNR*. For purposes of this summary, the appropriate symbol definitions are reviewed here. The prism base is b; $dn/d\lambda$ is the prism material dispersion; m is the grating order number; N is the number of lines in the grating; Δ is the maximum total path difference; R is the reflection factor in the Fabry–Perot filter, i.e., $\sqrt{\rho}/(1-\rho)$; s is the grating spacing; h and w are the linear dimensions of the grating or prism slit; α and β are the angular measures; and the equivalent noise bandwidth B is given by $B = N_s/(2t_s)$.

   There have been other such comparisons, and it seems appropriate to compare their results as well. Jacquinot compared gratings, prisms and Fabry–Perots. He compared a grating to a prism that has the same base area as the grating area. The ratio of the flux outputs for the same resolving power was found to be

$$\frac{\Phi_{prism}}{\Phi_{grating}} = \frac{\lambda dn/d\lambda}{2\sin\theta} \rightarrow \lambda\frac{dn}{d\lambda}, \tag{9.62}$$

where the blaze angle has been taken as 30 degrees for the final expression. Since the dispersion is very low in the regions where the prism transmission is high, the practical advantage for the grating is about 100.

The comparison of the grating with the Fabry–Perot proceeded along the same lines, whereby

$$\frac{\Phi_{FP}}{\Phi_{grating}} = \frac{\tau_{FP}0.7\pi^2/2}{\tau_{grating}2\beta\sin\theta} = \frac{3.4}{\beta} \tag{9.63}$$

for equal resolving powers and areas, where the factor of 0.7 is the ratio of the effective resolving power to the theoretical value, β is the angular height of the slit, and θ is again taken as 30 degrees. Values for β typically run from 0.1 to 0.01; the Fabry–Perot is therefore from 34 to 340 times as good as a grating based on these assumptions and criteria. Any other interferometric device that does not have a slit will have an equivalent advantage over the grating and prism instruments. The evaluation criteria are valid, but they are limited. Other considerations may enter, including the size of the free spectral range, ease of scanning, and computation of spectra as well as the size, weight, availability, and convenience of the components and their implementation.

The comparisons change a little when the full sensitivity equation is used as part of the comparison. The equation that determines the signal-to-noise ratio, SNR, is repeated here for convenience,

$$SNR = \frac{D^*L_\lambda d\lambda Z}{\sqrt{A_d B}}, \tag{9.64}$$

where D^* is the specific detectivity, L_λ is the spectral radiance, $d\lambda$ is the spectral interval, Z is the throughput, A_d is the detector area, and B is the noise bandwidth. The values for the throughput derived above can be used in this expression. For prisms and gratings, the required sample number is $\Delta\lambda/d\lambda$, the total spectrum divided by the resolution. For a Fourier-transform spectrometer the required sample number is twice the number of minimum wavelengths in the maximum path difference, i.e., Δ/λ_{min}, and this can be related to the resolving power, $\sigma_{max}/d\sigma$. The other major difference is that the flux available to the prism and grating is the spectral radiance $L_\lambda \times$ the resolution, $d\lambda$, while for the Fourier-transform spectrometer it is the spectral radiance $\times$ the total spectral interval $\Delta\lambda$ (more precisely, the integral over the band of the spectral radiance). These equations can be reformulated in a variety of ways by the use of the expressions given in Table 9.5, which is a summary of the expressions that have been determined earlier.

The normalized SNR is found from the appropriate expressions for the SNR itself. In terms of the specific detectivity the SNR is

$$SNR - \frac{D^*I_\lambda d\lambda Z}{\sqrt{A_d B}} = \frac{D^*L_\sigma d\sigma Z}{\sqrt{A_d B}}. \tag{9.65}$$

Since the detectivity depends upon the detector selected, the spectral radiance depends upon the source used, and the sample time is a design variable, the normalized SNR takes these out of consideration. The normalized SNR is the SNR per unit specific detectivity, spectral radiance, and the square root of $2t_s$, i.e.,

Table 9.5 Spectrometer Expressions

Type	$d\sigma$	$d\lambda$	Q	$\Delta\sigma$	Z	N_s	SNR*
One-layer filter	$\dfrac{1}{\pi\Delta}$		$\pi\dfrac{\Delta}{\lambda}$		$\alpha\beta A_{\text{filter}}$	$\Delta\sigma/d\sigma = \Delta\lambda/d\lambda$	$\alpha\beta A_{\text{filter}}\sqrt{\dfrac{\Delta\lambda\, d\lambda}{A_d}}$
AOTF		$\dfrac{0.9\lambda_0^2}{\zeta l\sin^2\theta_i}$	$\dfrac{\zeta l\sin^2\theta_i}{0.9\lambda_0}$	$\cdot$	$\dfrac{\text{AOTF}\,\pi n^2\lambda}{l\Delta n}$	$\Delta\sigma/d\sigma = \Delta\lambda/d\lambda$	$\alpha\beta A_{\text{AOTF}}\sqrt{\dfrac{\Delta\lambda\, d\lambda}{A_d}}$
Prism		$\dfrac{\lambda}{b\,dn/d\lambda}$	$b\,dn/d\lambda$	∞	$\alpha\beta A_p\cos\theta$	$\Delta\sigma/d\sigma = \Delta\lambda/d\lambda$	$\alpha\beta A_p\cos\theta\sqrt{\dfrac{\Delta\lambda\, d\lambda}{A_d}}$
Grating		$\dfrac{s\cos\theta_d\, d\theta_d}{m}$	$mN = (\Delta/\lambda)(\sin\theta)$	λ_x/m	$\alpha\beta A_p\cos\theta$	$\Delta\sigma/d\sigma = \Delta\lambda/d\lambda$	$\alpha\beta A_g\cos\theta\sqrt{\dfrac{\Delta\lambda\, d\lambda}{A_d}}$
FTS	$\dfrac{1}{2\Delta}$		$2\Delta/\lambda$	∞	$\dfrac{4\pi A_{\text{FTS}}}{Q}$	$2\Delta/\lambda_n$	$\dfrac{A_{\text{FTS}}\phi\lambda_{\min}}{\sqrt{A_d}}\sqrt{\dfrac{2}{d\lambda\,\Delta}}$
FP	$\dfrac{R}{2\pi\Delta}$		$2\pi(\Delta/\lambda)2R$	$1/\Delta$	$\dfrac{2\pi A_{\text{FP}}}{Q}$		$\dfrac{A_{\text{FTS}}\phi\lambda_{\min}}{\sqrt{A_d}}\sqrt{\dfrac{2}{d\lambda\,\Delta}}$

$$\text{SNR}^* = \frac{\text{SNR}\sqrt{2t_s}}{D^*L_\sigma} = \frac{\text{SNR}\sqrt{2t_s}}{D^*L_\lambda}. \tag{9.66}$$

It can be expressed in terms of the radiance per unit wavelength or radiance per unit wavenumber as long as the spectral interval is expressed in the proper units.

Perhaps the first and most obvious generality is that grating, Fourier-transform spectrometer, and Fabry–Perot spectrometers, all of which are based on interference, have a resolving power that is proportional to the number of waves in the maximum path difference. The same is true for a Fabry–Perot filter. (This applies also to the MTF of a diffraction-limited imaging system. The cutoff frequency is $1/(\lambda F) = (1/f)(D/\lambda)$, where D is the optics diameter, f is the focal length, and F is the speed.)

Greenler has used the product of throughput and resolving power as a figure of merit for rating spectrometers. Using this lead, I define a figure of merit, FOM, as the normalized SNR × the resolving power. Then, using the information given above, the FOM is easily generated.

Table 9.6 shows some representative performance characteristics of the different types of spectrometers. Of course, a representation is only a guide and can only be approximate, i.e., representative. The spectral range $\Delta\lambda$ may need to be covered by more than one instrument. A second column indicates the range generally covered by one instrument, prism, grating, laser, etc. The filter spectral range depends upon materials, but generally covers the ultraviolet through the long-wave infrared and even the millimeter range. The individual range is usually one octave.

The resolving power is a function of construction; it can be made larger, but usually at the expense of maximum transmission. The AOTF range is covered by different materials. Each of them has a range that is somewhat less than the center wavelength λ_0. The aperture size today is typically less than 1 cm, thereby limiting the throughput. Prisms cover the range from the ultraviolet (with LiF) to the long-wave infrared, but are practically limited at about 80 mm (CsI). The limitation to a single prism is a result of having proper transmission and dispersion over a limited range. The resolving power is a function of the dispersion, typically from 0.001 to 0.01 μm in this range and the size of the prism. The (reflection) grating is limited in its single range by the groove spacing and the blaze efficiency as well as overlapping of orders. The octave is obtained only for use in the first order, where the resolving power is smaller. The Fourier-transform spectrometer has a spectral range that is

Table 9.6 Representative Spectrometer Properties

Type	$\Delta\lambda$ (μm)	$\Delta\lambda$ singly (μm)	Q	Aperture size (cm)
Filter	0.3–1000	octave	100	1–20
AOTF	0.5–14	$< \lambda_0$	1000	< 1
Prism	0.2 80	10	1000	1–20
Grating	1–1000	octave	10^6	1–40
FTS	1–1000	detector/splitter	10^6	10
FP	1–1000	detector/splitter	10^9	10
Laser	0.3–1.5	$< \lambda_0$	10^{12}	0.5

limited only by the transmission of the beamsplitter and the spectral sensitivity of the detector. Its resolving power is a function of the path difference and the minimum wavelength of the spectral region. The value given is for 1 μm and 1 m. This is somewhat extreme. The Fabry–Perot has the same spectral limitations as the FTS, but it has a higher resolving power as a result of multiple-wave interference. Laser spectrometers, like the AOTF, are limited to an individual spectral range of a little less than the center wavelength. Their resolving power is phenomenal. Size is not so important as the fact that the output is of very high radiance, very bright.

REFERENCES

1. Cajori, F., *A History of Physics*, Dover, 1962.
2. Bohr, N., *Philosophical Magazine*, **26**, 476, 857 (1913).
3. Planck, M., *Physikalische Abhandlungen und Vorträge*, Band III, Braunschweig (1958).
4. Sommerfeld, A., *Atombau und Spektrallinien*, 4th edn, 1924.
5. Millikan, R. A., *Proceedings of the American Philosophical Society*, **65**, 74 (1926).
6. Rowland, H. A., *Philosophical Magazine*, **23**, 257 (1887); Harrison, G. R., *Journal of the Optical Society of America*, **39**, 413 (1949).
7. Wallace, R. J., *Astrophysical Journal*, **22**, 123 (1905).
8. Dereniak, E. and D. Crowe, *Radiation Detectors*, Wiley, 1985.
9. Jacquinot, P., *Journal de physique et le radium*, **19**, 223 (1958); Greenler, R., *Journal of the Optical Society of America*; Jacquinot, P., *Journal of the Optical Society of America*, **44**, 536 (1954).
10. Dobrowolski, J., "Coatings and Filters," in Driscoll, W. and W. Vaughan, eds, *Handbook of Optics*, McGraw-Hill, 1978.
11. Strong, J., *Procedures in Experimental Physics*, Prentice Hall, New Jersey, 1938.
12. Wolfe, W., "Optical Materials," in Wolfe, W. and G. Zissis, eds, *The Infrared Handbook*; Wolfe, W., Volume 3, Chapter 1 "Optical Materials," in Acetta, J. and Schafer, D., eds, Intrafred and Electro-optical Systems Handbook; Tropf, W., Thomas M., and T. Harris, "Properties of Crystals and Glasses," in Bass, M.,,. Palmer, D., Van Stryland, E. and W. Wolfe, eds, *Handbook of Optics*, McGraw-Hill, 1995.
13. Christiansen, C., *Annalen der Physik und Chemie*, **23**, 298 (1884).
14. McAlister, E., *Smithsonian Miscellaneous Collection*, **93**, No. 7 (1935).
15. Strong, J., *Procedures in Applied Optics*, Dekker, 1988.
16. Dobrowolski, J., "Optical Properties of Films and Coatings," in Bass, M., Palmer, D., Van Stryland, E., and W Wolfe, eds, *Handbook of Optics*, McGraw-Hill, 1995.
17. Wolfe, W., *Introduction to Infrared System Design*, SPIE, 1996.
18. Fraunhofer, J., *Annalen der Physik*, **74**, 337 (1823).
19. Rowland, H. A., *Philosophical Magazine*, **13**, 469 (1882).
20. Ibid, **16**, 297 (1883).
21. Harrison, G. R., *Journal of the Optical Society of America*, **39**, 413 (1949).
22. Wood, R. W., *Nature*, **140**, 723 (1937).
23. Jenkins, F. A. and H. E. White, *Fundamentals of Optics*, 3rd edn, McGraw-Hill, 1957; and other basic texts.
24. Goodman, J. E., *Introduction to Fourier Optics*, McGraw-Hill, 1968.
25. Wood, R. W., *Physical Optics*, 3rd edn, Macmillan, 1934.
26. Michelson, A. A., *American Journal of Science*, **22**(3), 120 (1881).
27. Twyman, F. and A. Green, British Patent 103382 (1916).
28. Felgett, P. Thesis, University of Cambridge, 1951.
29. Jacquinot, P., and C. Dufour, *Journal of Research of CNRS*, **6**, 91 (1948).
30. Strong, J., *Concepts of Classical Optics*, Freeman, 1958.

31. Steel, W. H., *Interferometry*, Cambridge University Press.
32. Bell, R. J., *Introductory Fourier Transform Spectroscopy*, Academic Press, 1972.
33. Treffers, R. R., *Applied Optics*, **16**, 3103 (1977); Carli, B. and V. Natale, *Applied Optics*, **18**, 3954 (1979); Junttile, M., *Applied Optics*, **31**, 4106 (1993).
34. Jacquinot, P., *Journal of the Optical Society of America*, **44**, 536 (1954).
35. Anonymous, "Mil Handbook 141," *Military Standardization Handbook, Optical Design*, Defense Supply Agency, 1962.
36. Hirshberg, J. G., *Journal de physique et la de radium*, **19**, 256 (1958).
37. Hinkley, E. D., K. W. Nill, and F. A. Blum, "Infrared Spectroscopy with Tunable Lasers," in H. Walther, ed., *Laser Spectroscopy of Atoms and Molecules*, Springer-Verlag, 1974.
38. Weber, M. J., ed., *CRC Handbook of Laser Science and Technology*, CRC Press, 1986.
39. Koechner, W., *Solid State Laser Engineering*, Springer-Verlag, 1976.
40. Schaefer, F. P., "Principles of Dye Laser Operation," in F. P. Schaefer, ed., *Dye Lasers*, Springer-Verlag, 1973.
41. Golay, M., *Journal of the Optical Society of America*, **39**, 437 (1949).
42. Golay, M., "Static Multislit Spectrometry and its Application to the Panoramic Display of Infrared Spectra," *Journal of the Optical Society of America*, **41**, 468 (1951).
43. Harwit, M. and N. Sloane, *Haddamard Transform Optics*, Academic Press, 1979.
44. Decker, J., "Haddamard Transform Spectroscopy," *Industrial Research*, **2**, 61 (1973).
45. Girard, A., *Optica Acta*, **7**, 81 (1960); *Applied Optics*, **2**, 79 (1963).
46. Connes, P., *Journal de physique et la de radium*, **19**, 215 (1960).
47. Mack, J. E., D. P. McNutt, F. L. Roessler, and R. Chabbal, *Applied Optics*, **2**, 873 (1963); McNutt, D. P., *Journal of the Optical Society of America*, **55**, 288 (1965).
48. Bass, M., E. Van Stryland, D. Williams, and W. Wolfe, eds, *Handbook of Optics*, McGraw-Hill, New York, 1995, p. 12.30.
49. Taylor, L. H., D. R. Shure, S. A. Wutzke, P. L. Ulerich, G. D. Baldwin, and M. T. Myers, "Infrared Spectroradiometer Design Based on an Acousto-optical Filter," *Proc. SPIE*, Spring (1995).
50. Bass, M., D. Palmer, E. Van Stryland, and W. Wolfe, *The Handbook of Optics*, McGraw-Hill, 1995.
51. Hariharan, P., *Applied Optics*, **14**, 2319 (1975).
52. Connes, P., *Le journal de physique et la radium*, **19**, 262 (1958).
53. Connes, P., *Le journal de physique et le radium*, **19**, 262 (1958).
54. Chabbal, R. and M. Soulet, ibid, 274; Gorbert, J., ibid, 278; Roig, J., ibid, 284; Dupeyrat, R., ibid, 290; Peck, E. R., ibid, 397.
55. Jackson, D. A., ibid, 379.

APPENDIX: GLOSSARY

a	beam width
A_d	detector area
A_e	entrance pupil area
A_f	field stop area
A_{slit}	slit area
A_{prism}	prism face area
b	prism base
B	effective noise bandwidth
c	light speed in vacuo
d	distance
$d\lambda$	resolution

$d\sigma$	resolution
D^*	specific detectivity
E	irradiance
f	focal length
F	focal ratio, speed
$\mathscr{F}$	finesse
h	slit height
k	radian wavenumber
l	interaction length
L	radiance
L_λ	spectral radiance
m	order number
M_2	acoustic figure of merit
n	refractive index
N	number of rulings
NA	numerical aperture
p	length of prism side
Q	resolving power
R	$4\rho/(1-\rho)^2$
$\mathscr{R}$	responsivity
s	ruling spacing
SNR	signal-to-noise ratio
t	time
v	velocity
w	slit width
Z	throughput

Greek Symbols

α	prism angle
β	groove width phase
δ	deviation angle
ε	phase shift on reflection
ζ	dispersion constant
θ	general angle
$\Delta\lambda$	spectral range
Δ	path difference
$\Delta\sigma$	spectral range
γ	half spacing phase
λ	wavelength
ν	frequency
ρ	reflectance
σ	wavenumber
τ	transmittance
Φ	flux
Ψ	wave function
Ω	solid angle
Ω'	projected solid angle

Subscripts and Superscripts

d	diffraction
i	incident
d	detector
o	optics
1	first
2	second

10

Wavefront Slope Measurements in Optical Testing

ALEJANDRO CORNEJO-RODRIGUEZ and ALBERTO CORDERO-DAVILA

Centro de Investigaciones en Optica, León, Mexico

10.1 INTRODUCTION AND HISTORICAL REVIEW

In this chapter we analyze a group of methods for testing optical surfaces and systems whose main characteristics are the measurements of the rays' slopes at certain planes. Auxiliary optics are not required, which allows us to have a direct measurement of the wavefront under test that comes from the optical system or surface under test, which is not possible in other measuring methods.

Historically the first test that can be classified into this group here described, is the so-called knife edge or Foucault test (Foucault, 1858). Subsequently, other techniques were described in the first half of the last century, including the wire (Ritchey, 1904), Ronchi (Ronchi, 1923), Hartmann (Hartmann, 1900), and Platzeck–Gaviola tests (Platzeck, 1939). More recently, one technique derived from the Hartmann method is the Shack–Hartmann (Shack and Platt, 1971) test; two more of the tests were developed by Ichikawa (Ichikawa *et. al.*, 1988) and Roddier (Roddier *et al.*, 1988), both based on the theory of the irradiance transport equation (Teague, 1983). The Shack–Hartmann and Roddier tests are normally applied to wavefront testing of working astronomical telescopes, and mainly function under an adapative optics technique.

In general, each one of the tests mentioned has its own physical and mathematical description (Malacara, 1992). However, in this chapter, easier explanations of the methods are given, and a common step-by-step description of the older tests is presented. For the case of the Ronchi and Hartmann methods, a unified theory has also been developed (Cordero *et al.*, 1992). Even though in some cases geometrical and physical optics can be used to describe the tests, as mentioned, in this chapter geometrical optics is mainly used to explain the methods applied to measuring the slope rays of the wavefronts coming from optical surfaces or systems.

The presentation of each one of the tests contains comments on some practical aspects rather than the theoretical foundations of the methods. Of course, theoretical and practical aspects are considered elsewhere (Malacara, 1992; De Vany, 1967).

10.2 KNIFE TEST

It is assumed, in the first place, that the optical system under test is illuminated by a point source, or equivalently by a perfect spherical wavefront; hence, if a perfect spherical wavefront is leaving the exit pupil of an optical system, and you place the knife edge into the light beam, see Fig. 10.1a, then the border of the shade that you will see in any observation plane corresponds to a straight line, independently of the orientation and position of knife edge, and the wavelength of the light.

Figure 10.1b shows the observed pattern. From the geometrical point of view, the border between the dark and illuminated zones is defined by the plane that passes through the $\overline{BC}$ border of the knife for an image point A'. Despite the diffraction effects, the border will be a plane that, when it is intersected with any other observation plane, E'', a straight line $\overline{B''C'}$ is being defined and therefore, corresponds to the border of the shade that will be observed.

In the usual form of observation, see Fig. 10.1b, the camera lens is placed near the image point A' and you focus the plane of the exit pupil E. In this case, the virtual border (in similarity with the virtual images), is the observed shade and will be a straight line. A real image will be formed, with the aid of a camera lens, on the detector plane (charge-coupled device (CCD), photographic plate), or the retina of the eye in the case of a direct observation. It could also be recorded, in any further plane from the knife edge, without a lens. This last type of observation is very useful when the distance between the camera lens (or the eye) and the observation plane is very short and it is not achieved as an acceptable focus, or when the F# of the system under test is very small. In this last case, the pupil of the eye obstructs the passage of the rays, and only some parts of the pattern are observed with certain circular shape.

When the light rays do not converge toward a unique point, then the projected and observed shadow will not be a straight line; this means that the wavefront is not perfectly spherical. In such case, the border of the shadow is given by the intersection points of the rays that will pass exactly on the border of the knife: i.e., the shadow will not follow a straight line.

In this test the comparison between the experimental and the real patterns are very important; for this reason in the next section an algorithm will be presented, such that simulated Foucaultgrams can be obtained if skew rays can be traced through the optical system.

10.2.1 Foucaultgram Simulations

In order to simulate the Foucaultgrams, it is assumed that in the optical systems, see Fig. 10.2, there are several planes with their respective coordinates as follows: the entrance pupil plane (X, Y); the exit pupil plane (X_o, Y_o); the observation plane (X_{ob}, Y_{ob}), and the knife edge plane (T_X, T_Y).

If it is assumed that the knife edge is parallel to the X-axis, then all the points that belong to the semiplane of the knife edge could be expressed mathematically through the next inequality

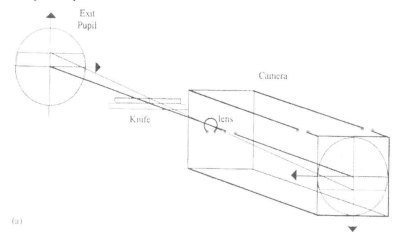

(a)

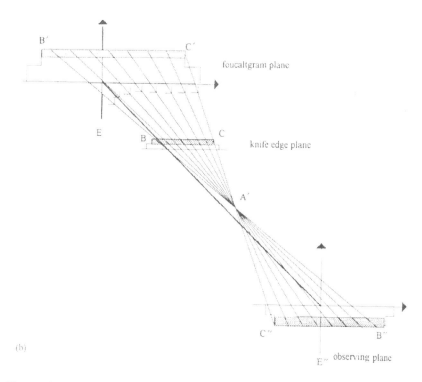

(b)

Figure 10.1 (a) Foucaultgram of a spherical wavefront. (b) Foucaultgram recorded by means of a camera focused at the exit pupil of the system under test.

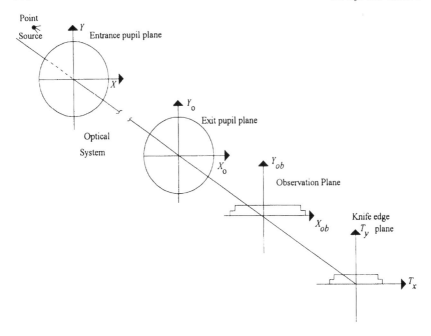

Figure 10.2 Planes used in the Foucaultgram simulations.

$$T_y - T_{yn} \leq 0 \tag{10.1}$$

where T_{yn} is the distance from the knife edge to the origin of the coordinate system. Equation (10.1) is equal to zero for the points that belong to the border of the knife; negative values are obtained for the points over the knife; while for the points that are not covered by the knife edge, values will be greater than zero. This property is important because it serves to distinguish, numerically, between a ray that falls on the knife and another which arrives to the transparent zone.

The Foucaultgram simulation has its basis in the previous idea and allows us to show that it is not necessary to trace all the rays but only a very limited number. However, several rays are traced from the point source to the knife plane, passing through the entrance pupil, on which the incident rays are located at equally spaced points and in one straight line parallel to the Y-axis, similar to the ray tracing in the optical design.

The important data for each traced ray are the pairs of coordinates (X, Y), (X_{ob}, Y_{ob}), and (T_X, T_Y). If T_y of this last pair satisfies the equality in Eq. (10.1), then the corresponding (X_{ob}, Y_{ob}), belongs to the border of the shade. However, when the T_y values are found which do not satisfy Eq. (10.1), then it is necessary to identify the successive values for pairs of the coordinates T_y, for which a change of sign according to the inequality (10.1) is obtained. Once a first interval is identified and satisfied, the bisection method is used until the limit of an established required precision. To have as many points as possible for the pattern, the previous procedure is repeated for a family of parallel straight lines within the entrance pupil.

When the edge is rotated an angle α, then it is applied a rotation to the coordinates (T_X, T_Y) for obtaining the new coordinate T_y^* which must be substituted in Eq. (10.1).

Two practical remarks are important. First, if we increase the distance between two consecutive points on the scanned line, then the time for obtaining the pattern diminishes; however we could not detect some points that would appear when the patterns are closed. Secondly, if you carry out the scan only on parallel straight lines to the axes, then it is difficult to locate some points of the pattern; this occurs mainly for the case when the straight lines of the scan are nearly parallel to the knife edge. To avoid this last problem, it is important to carry out two scans on perpendicular straight lines.

We develop the program FOURON to simulate Foucaultgrams and Ronchigrams shearing the same computer program for any section of a conic mirror and with the source at any position. In Figs. 10.3a and 10.3b we show some Foucaultgrams obtained with the program. We left out the simulation of Foucaultgrams for systems affected by Seidel aberrations, which can be calculated analytically (Ojeda, 1992).

In the development of this section it has been assumed that the illumination of the system under test is with a point source; however, it is possible to substitute it with a linear source, under the basic condition that the linear source is parallel to the border of the knife. In this case the patterns are almost identical for each point of the linear source. When the knife and the linear source are perpendicular to each other, astigmatic effects will appear that show tilts and displacements of the border of the shades; in this case, the patterns show a slow change of intensities in the border and therefore a more poor definition of the Foucaultgram. It is evident that the use of white light has no effects when the mirrors are tested, but for no corrected refractive systems the chromatic aberration becomes evident and appears as colored borders.

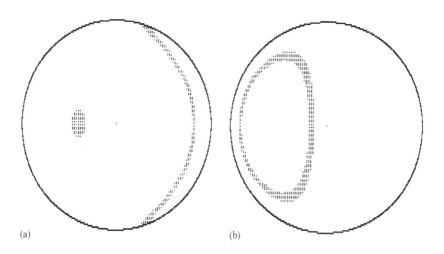

(a)   (b)

Figure 10.3 Foucaultgram simulated by using the FOURON program.

The Foucault test has been analyzed only from a simplified geometrical point of view, but it could also be analyzed using physical optics. For such an analysis, it is demonstrated that if an obstruction in the Fourier plane exists (image plane), then the observed intensity will be uniform. However, when the knife edges are in the Fourier plane, then some changes of intensity are observed which are proportional to the derivate of the amplitude of the field in the image plane (Ojeda, 1992).

10.3 WIRE TEST

The Foucault test is often used to analyze spherical wavefronts, and particularly to test spherical surfaces with the point source placed in the plane of the center of curvature of the surface. However, in testing aspherical mirrors, the Foucault test is not very sensitive, since details are lost. Ritchey (1904) developed the wire test, which is more sensitive as a zonal test and is analyzed in this section.

The wire test can be considered as an extension of the Foucault test, since a wire can be assumed to be a double-edged knife. Thus, a wiregram can be obtained from two simulated Foucaultgrams for which the two edge knifes are parallel, and at heights Y_a and Y_b from the optical axis. In the particular case of a perfect spherical wavefront, the zonal pattern will be formed by a straight band.

When the wire is located on the optical axis of an axisymmetrical mirror, at a distance L^* from the vertex of the mirror, then a dark ring can be observed on the mirror, of radius S^*, corresponding to the points from the zone S^* with radius of curvature at the distance L^*. Then the rays coming from zone S^* cross the wire at position L^*.

With this idea, Ritchey (1904) proposed to test parabolic surfaces by comparing the experimental and theoretical longitudinal spherical aberrations. He measured several values of L^* and S^* and compared them using theoretical calculus. The same idea is presented in what follows, applied to any conic mirror, but using the mathematical formulations of Sherwood (1958) and Malacara (1965). These authors demonstrated, independently that, if figure 4 is used, the equation of the reflected ray is given by

$$T = \frac{(l + L - 2z)\left[1 - \left(\dfrac{dz}{ds}\right)^2\right] + 2\left(\dfrac{dz}{ds}\right)\left[s - \dfrac{(l - z)(L - z)}{s}\right]}{\dfrac{(l - z)}{s}\left[1 - \left(\dfrac{dz}{ds}\right)^2\right] + 2\left(\dfrac{dz}{ds}\right)}, \qquad (10.2)$$

where T, the transverse aberration, is the height of the reflected ray at the distance L, l is the position of the point source, and z is the function that describes the saggita of the mirror, given by

$$z = \frac{cs^2}{1 + \sqrt{1 - (K + 1)c^2s^2}}. \qquad (10.3)$$

In this last equation s is the distance to the axis where the incident ray is reflected, c is the paraxial curvature, and K is the conic constant of the mirror

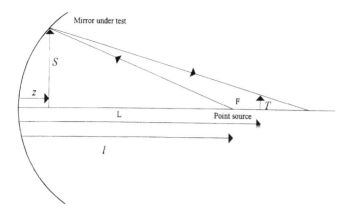

Figure 10.4 Geometry for the transverse aberration, T, of the reflected ray and in the wire test.

under test. From Eq. (10.2) it is evident that if we take an incident cone of rays with vertex in the point source and whose base is a circle of radius s on the mirror, then the reflected rays will also form a cone whose base is the same zone s, and whose vertex will be located in L^*, corresponding to the zero of Eq. (10.2) and given by

$$L^* = \frac{(l - 2z)\left[1 - \left(\dfrac{dz}{ds}\right)^2\right] + 2\left(\dfrac{dz}{ds}\right)\left[s - \dfrac{z(l - z)}{s}\right]}{2\left(\dfrac{dz}{ds}\right)\dfrac{(l - z)}{s} - \left[1 - \left(\dfrac{dz}{ds}\right)^2\right]}.$$ (10.4)

From Eq. (10.4) it is clear that for a given value of s we will have an only value of L^* and since s defines a circumference on the mirror, then, if the rays are obstructed precisely in L^*, then at least the rays coming from the circumference of radius s will be blocked. Besides, the point of the wire it will be observed as a straight line parallel to the wire. Figure 10.5 shows a pattern obtained with aid of the program FOURON.

The application of Ritchey's idea begins with the designing of a screen, for testing a parabolic mirror, that allows illumination of different zones as the paraxial, intermediate ($0.7071\ S_{max}$) and at the border of the mirror, see Fig. 10.6a. The screen is placed in front of the mirror and the wire is displaced along the axis to find the positions corresponding to the different ring zones of the screen on the mirror. In each case they become a series of measurements of L^* and they are compared with the theoretical calculus made with aid of the Eq. (10.4).

There are more versatile screens, see Fig. 10.6b; however, it is not possible to detect the asymmetrical defects (Ojeda, 1992). This type of test is highly recommended for use only in the first phases of fabrication of optical surfaces.

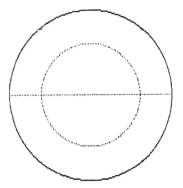

Figure 10.5 Wiregrams simulated with the program FOURON.

As it was pointed out, a linear source can be used when the wire and the linear source are parallel to each other. In this case, the advantages and disadvantages mentioned in the knife test are retained.

From the physical optics point of view, it could be possible to consider that the wire is a space filter located in the Fourier plane with the advantage, over the knife test, in that here the position and the thickness of the wire could be selected.

10.5 RONCHI TEST

As well as the wire test being considered as conceptually, like an extension of the knife test, the Ronchi test (1923) can be considered, see Fig. 10.7a, as an extension of the wire test, if we consider a Ronchi ruling as formed of several wires equidistant and parallel. In the Ronchi ruling the slits are alternated clear and dark, with the same width, and are assumed to be parallel to the X-axis. The extension for the Ronchi test also gives us global information of the surface, in this test the information is obtained at the same time from several wires and, therefore, Ronchigram fringes "cover" the exit pupil of the system under test.

From the wire test it is clear that if you are testing a spherical wavefront with a Ronchi ruling, a Ronchigram of parallel and equally spaced fringes will be observed

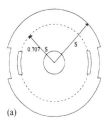
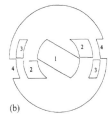

Figure 10.6 (a) Zonal screen and (b) Couder screen for the wire test.

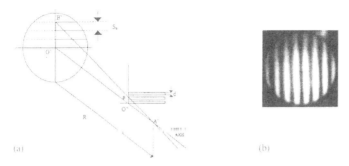

Figure 10.7 (a) Ronchigram for a spherical wavefront. A′ is the paraxial center of curvature; O″ is the ruling plane located along the optical axis; O′ is the exit pupil plane or surface under test plane. (b) Experimental Ronchigram for a spherical wavefront.

(see Fig. 10.7b). The separation, S_b, between two consecutive fringe borders in the Ronchigrams is equal to

$$S_b = \frac{R}{D_f}\, d, \tag{10.5}$$

where R is the radius of curvature of the wavefront, D_f is the separation, $A′O″$, between the ideal image point, A′, and the intersection point O″ of the ruling plane with the optical axis, and d is the distance between two consecutive borders in the Ronchi ruling, which is the half period of the ruling.

From Eq. (10.5) we obtain a well-known result in the application of the Ronchi test. If D_f is increased, i.e., the ruling moves away, before or after the image point A′, then the width of the observed fringes diminishes and you will see an increase in the number of fringes in the exit pupil plane. If D_f becomes zero, then S_b becomes infinite and, therefore, a field totally brilliant or dark is observed experimentally, depending upon whether the light arrives to a clear or dark slit on the Ronchi ruling. This result is important if you want to locate the image point of an optical system, A′; since in this case, when the ruling is placed at the plane that contains the image point, then the fringes will become, theoretically, infinitely wide. However, in practice, you cannot measure a width greater than the diameter of the exit pupil of the system under test; therefore, you will have an uncertainty in the localization of the image point A′.

For a spherical mirror, another well-known result of this test comes from Eq. (10.5), which is the dependence of the width of the fringes, S_b, with the width, d, of the ruling slits. In this case, if a ruling of greater frequency is used (a minor d), the frequency of the fringes will also increase; this means tht S_b will diminish.

10.4.1 Ronchigram Simulations

Up to now we have analyzed the Ronchigrams that could be obtained with spherical surface or wavefronts; in this section, an algorithm will be described to simulate Ronchigrams of optical systems by just doing ray tracing.

The procedure has its theoretical basis in the simulation of several wiregrams; in the Ronchigrams each wire corresponds to a dark slit of the grating. In practice, the simulation is based in the ray tracing for two neighboring rays (in the entrance pupil), through the optical system and by assuming that these two rays are neighbors also in the ruling plane. The rays at the entrance pupil for the positions (X, Y) and (X', Y') will correspond to the positions (T_X, T_Y) and (T_X', T_Y') at the Ronchi ruling (Fig. 10.7b). An important case for this pair of rays corresponds to when in the Ronchi ruling one ray falls in a dark zone and the other ray falls in an illuminated one. If the slits of the ruling are assumed parallel to the X-axis, then the borders can be described by the equation

$$T_y = md, \tag{10.6}$$

where $m = 0, \pm1, \pm2, \pm3 \ldots$. Then, in order to identify the existence of one border between two points (T_X, T_Y) and (T_X', T_Y'), it is possible to calculate

$$M = \text{int}\left(\frac{T_Y}{d}\right) \tag{10.7a}$$

and

$$M' = \text{int}\left(\frac{T_Y'}{d}\right), \tag{10.7b}$$

where [int] is the computer instruction by means of which the integer part of the quantity that appears between parentheses is calculated.

For two neighboring rays, two possibilities exist: the two integers are equal or different. In the first case, we have not passed over any border and in the second case, it is an indicative of that we have crossed some border. In this latter case, the separation between the neighboring rays in the entrance pupil can be diminished by using an intermediate point between (X, Y) and (X', Y'); then, the procedure is repeated in order to elect the new subinterval that we refine as in the bisection method. This procedure is followed until the desired limit is reached. As in the case of the Foucaultgrams, two scans are required: one parallel to the X-axis and other parallel to the Y-axis.

Using the previously described idea, the FOURON algorithm was developed to allow us to obtain simulated Ronchigrams for conic sections; centered or decentered. In Fig. 10.8 several Ronchigrams are simulated with the aid of the FOURON program are shown.

When the Ronchi ruling is rotated at an angle α, then the coordinates (T_X, T_Y) must be transformed through a rotation with the same angle α, in order to obtain a new coordinate T_Y^*, which is substituted in Eqs (10.7a) and (10.7b).

Despite the fact that the previous program has a general character, Malacara (1966) developed another option that can be used exclusively for systems with symmetry of revolution, with respect to the optical axis, and with the point source located on that axis. Under these conditions, the computer programming is easier and elegant, the calculus time is inferior to the mentioned FOURON program and the procedure is sufficient to simulate the majority of the Ronchigrams needed in an optical shop.

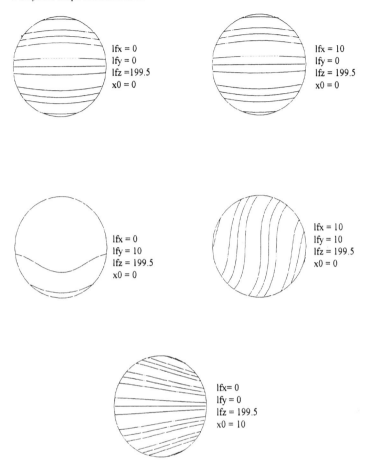

Figure 10.8 Ronchigrams simulated with the aid of the FOURON program. A parabolic mirror of 15 cm diameter and 200 cm radius of curvature was used. The position of the Ronchi ruling (LR = lfz), point source (lfx, lfy, lfz), and center of the conic section (X0), are indicated in each case.

The simplified procedure developed by Malacara (1966) can be interpreted as a ray tracing from a source, on axis, to a point located on the entrance pupil of the system (Fig. 10.9), to a certain distance, S, from the optical axis. Such a ray tracing represents all the traced rays over the circle with radius S, i.e., they are the traced rays that belong to a cone whose vertex is the point source and whose base is the circle with radius S, on the entrance pupil. Similarly, all the rays that left the system will belong to a cone of rays whose intersection points at the Ronchi ruling plane will define a circle whose points of intersection could be easily calculated. The idea of Malacara (1965) has been programmed in order to simulate Ronchigrams for any reflecting conic surface (ROMA), and a refractive systems with centered spherical surfaces (ROLEN).

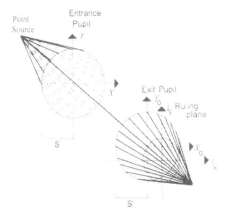

Figure 10.9 Geometry used in the simulations of the Ronchigrams for axis symmetrical systems.

10.4.2 Ronchigram Evaluations

As was mentioned previously, the aim of the application of these tests is to know quantitatively the wavefront deformations of any surface or, in general, an optical system. In order to know W (the wavefront deformations) using the Ronchi test, two crossed Ronchigrams are required; since the fringes of a Ronchigram can be interpreted as the level curves of only one of the components of the transverse aberration, $\vec{T}$, and for the transverse aberration components (T_X, T_Y), they are related with W by the equation

$$\vec{T} = -\vec{\nabla} W; \tag{10.8}$$

therefore, it is necessary to carry out an integration to evaluate W from the measurements of the transverse aberrations. Thus, if we have only one Ronchigram, and then only one partial derivative, then we cannot evaluate the line integral; given such situations, we need two crossed Ronchigrams in order to carry out the line integral.

In order to obtain two perpendicular Ronchigrams with only one grating, the first Ronchigram is obtained with the ruling oriented along one direction and, afterwards, the grating is rotated 90 degrees and then the second Ronchigram is recorded (Cornejo, 1992). In this case, it is not possible to simultaneously record the two patterns, and then precision is lost if there are temporal variations of W. A second option is achieved by amplitude division of the wavefront (by using a beamsplitter) into two similar channels, each one having a lens, a grating and a detector array. A third option (Meyers and Stahl, 1992; Cordero, et. al. 1998a) is by means of a squared grating, instead of the classical Ronchi ruling. This kind of grating could be considered like the intersection of two crossed Ronchi rulings; and then the obtained pattern could be considered like the intersection points of two crossed Ronchigrams, that in general, they are required in order to calculate W (Fig. 10.10).

For the Ronchigram evaluation (Fig. 10.11), with the slits of the Ronchi ruling parallel to the X-axis, N vectors (X_i, Y_i, M_{Xi}) are considered, where (X_i, Y_i) are the

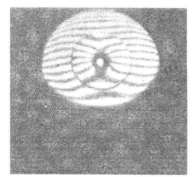
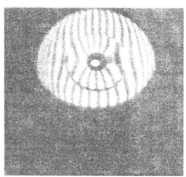

Figure 10.10 Two crossed Ronchigrams used for evaluating a mirror.

coordinates of the point associated to the maximums and/or minimums of intensity of the Ronchigram fringes, and M_{Xi} is the interference order associated to the fringe. In the next step, the least-square method is applied to the mentioned vectors in order to estimate the polynomial coefficients of degree $K - 1$, for the bidimensional polynomial transverse aberration component, T_X. In a similar way, the polynomial coefficients for T_Y are calculated for the crossed Ronchigram (Cornejo and Malacara, 1976). If it is assumed that W could also be written like another polynomial function but now of degree K, then the integration is carried out using the relationship among the coefficients of the polynomial expressions of W, T_X, and T_Y.

If a squared grating is used in the Ronchi test, the BIROEV program can be used for calculating the polynomial coefficients until degree 8. In this case (Fig. 10.10), N data vectors $(X_i, Y_i, M_{Xi}, M_{Yi})$ are taken into account, where the data for the point coordinates and the respective interference orders are considered for the application of the Ronchi test with a squared grating. The coordinates (X_i, Y_i) should be normalized to a unitary circle, in order to apply the Zernike polynomials.

An important aspect to start a pattern evaluation within a unitary circle is the knowledge of the center coordiantes and the radius of the exit pupil. Cordero *et al.*, (1993) have proposed an algorithm, BOFI that allows us to evaluate both the center coordinates and the radius of a pattern, starting with the border points that are supposed to be affected by Gaussian errors.

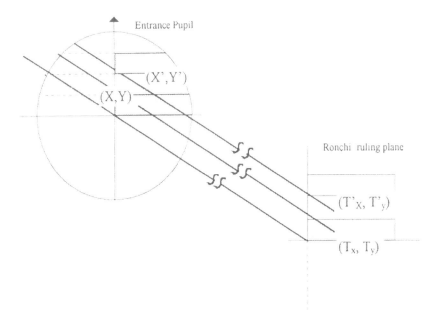

Figure 10.11   Ronchigram obtained with the square Ronchi ruling.

The Ronchigram evaluations using polynomial fittings have the disadvantage of introducing an overfitting through the election of the polynomial degree. Therefore, Cordero *et al.* (1994) have proposed a new algorithm by means of which the fitting is carried out assuming that the Gaussian errors are on the fringe coordinates. Another algorithm (Cordero *et al.*, 1998b), was developed in order to avoid the overfitting and the incorrect interpretation at the borders. In this case, numerical integration is used, as in the Hartmann test. Unfortunately, this procedure only could be used in the Ronchi test with the squared grating.

The Ronchigrams could be analyzed from the point of view of the physical optics predicting the intensities that one should observe (Malacara, 1990). In this last case, the Ronchigrams could also be considered as lateral shearing interferograms. The analysis can be simplified if it is assumed that only two of the diffracted orders interfere, and it is possible to drop the third derivative of W; hence, both interpretations, the geometrical and physical theory, coincide (Cornejo, 1992).

In the experimental setup of the Ronchi test it is common to use a point light source or a slit parallel to the ruling lines. Anderson and Porter (1992) suggested allowing the grating to extend over the lamp, instead of employing a slit source. In practice, a led (light emission diode) source for illuminating the grating from behind can be used; this simplifies the experimental array. More recently, Patorsky and Cornejo (1986) found that the setup can be further greatly simplified by illuminating the grating with daylight and setting a strip of aluminum foil just behind this part.

10.5 HARTMANN TEST

The exclusive use of Schlieren's tests in testing optical surfaces could be insufficient, as in the case of the Foucault test in that it is little sensitive to slow variations of the wavefront (Ghozeil, 1992). That is not an important problem in surfaces of small dimensions; however, it could be a severe drawback for large surfaces. On the other hand, with the wire test, as has been explained, only radial zones of the surface under test can be carried out.

It is evident that the interferometric tests are much more precise than the geometrical tests. However, interferometric tests could be a very expensive method for testing surfaces and/or optical systems of large aperture. However, the Hartmann test could prove to be an important and economical alternative for testing optics, mainly if large surfaces are under test.

The principal advantage of the Hartmann test is that it does not require the inclusion of wavefront compensators to convert the reflected wavefront into a spherical wavefront, which is required in the use of other methods. It is obvious that correctors can be a source of additional errors.

The basic hypothesis, in the evaluation of the Hartmann test, is that the slopes of the wavefront under test do not change abruptly, but rather in a slow manner; such an assumption is important because the surface or wavefront is sampled in a few zones, where the holes of the Hartmann screen are located, and continuous slow variations are considered.

In this test the Hartmann screen is a kind of filter, at a certain location plane, and usually is placed at the exit pupil of the system under test. The observation-registering plane is located near the image point of the point source illuminating the screen of the surface. If the coordinates of the centers of each hole in the Hartmann screen are well known and the positions of the centers of the dots in the Hartmanngram are measured, then the director cosines of the rays joining corresponding holes and dots, and then the slopes of the rays of the wavefront can be calculated. Alternatively, two Hartmanngrams recorded with the same Hartmann screen, at two different distances, could be used for evaluation in the Hartmann test, and to avoid the measurement of the distance from the vertex mirror to the observation or Hartmanngram plane; however, you should measure with precision the distance between the observed and registered Hartmanngram planes.

The evaluation of the wavefront is carried out by using Eq. (10.8), which gives us the relationship between the transverse aberration and the optical path difference, W. In testing a surface, you could suppose that the errors, h, are related to W by

$$W = 2h. \tag{10.9}$$

In Eq. (10.9) it is assumed normal incidence and, therefore, h is measured along the same direction.

One of the main advantages of the Hartmann method is that the evaluation of W can be done by using polynomial fittings or by means of numerical integration. Theoretically, it is possible to think that it is better to use polynomial fittings, because they avoid the errors due to the numerical integration methods. However, polynomial fittings usually introduce overfitting, which means artificiality in the surface under test or problems of interpretation in the borders.

When surfaces of large dimensions are tested, systematic errors are produced as a result of the laminar turbulence produced by local gradients of temperature, or mirror deformations due to the weight of the mirror, which can be seen as astigmatic aberrations of W.

10.5.1 Screen Design and Hartmanngram Evaluation

The holes distribution in the Hartmann screen has changed from the radial distribution used by Hartmann, to the helical one and, finally, to a squared-hole array that has been commonly employed for the testing of large primary mirrors of astronomical telescopes. The main disadvantage of the radial and helical screens is that sampling is not uniform, since the density of holes diminishes radially. Another important disadvantage of these screens is that concentric scratches, or other general defects, produced during the lapping working could not be detected with such a type of screen. Thus, in order to overcome those kinds of problems, a square screen is used, which produces a uniform sampling. Another advantage of this screen is that it could become rigid and then, with this structure, it could achieve high precision in the hole positions. This can be accomplished by making holes of largest size in the rigid structure and, later on, other holes, with the final size are made on small badges. Finally, the holes in the badges can be aligned and fixed with a higher precision in the rigid structure.

An analysis of the conditions that limit the applicability of the Hartmann test was made by Vitrichenko (1976) and Morales and Malacara (1983). Three important factors limiting the use of the Hartmann test are as follows:

- diffraction effects of the holes of the screen and its mechanical strength, fixed some limit for the diameters of the screen holes and the distance between their centers;
- the total numbers of holes is limited by the accuracy which must be developed to obtain results of a given reliability; and
- the adequacy of the description of the surface of an optical component is limited by the degree of smoothness of the surface.

The HASC program was developed in order to take into account these limitations and to calculate the optimum diameter of the holes and the minimum distance between two of them.

Some important practical recommendations when the Hartmann test is applied are:

- The screen must be centered accurately (Landgrave and Moya, 1986), since a decentering of the screens leads to an apparent presence of coma.
- The point light source used to illuminate must be centered properly to prevent the introduction of off-axis aberrations.
- The photographic plate or CCD array should be perpendicular to the optical axis.

The surface evaluation can be made by running the EVHAR program. The complete data are organized in three data sets:

- surface data (curvature radius, conic constant, and diameter);

- experimental array data (locations of the point source and the Hartmanngram plate); and
- coordinates and the number order of the dots of the Hartmanngram, corresponding to the order numbers of the holes in the Hartmanngram screen.

The processing of all the data sets is now analysed.

Once the Hartmanngram has been recorded, the location of the dots on the photographic plate (or CCD) must be measured to a high accuracy. The Hartmanngram coordinates (X_{dei}, Y_{dei}) can be measured by means of a microdensitometer or a measuring microscope having an X–Y traveling stage, or as is usual, if a CCD is used, the locations of the dots can be found after the Hartmanngram has been digitized and stored in a computer. For each dot, the coordinates are measured and a set of numbers (M_{Xi}, M_{Yi}) are assigned. They are related to the center of a hole in the Hartmann screen, with coordinates (X_i, Y_i), by means of the equation

$$(X_i, Y_i) = e(M_{Xi}, M_{Yi}), \tag{10.10}$$

where e is the distance between two nearest holes centers.

Starting with the N data vectors $(X_{dei}, Y_{dei}, M_{Xi}, M_{Yi})$, it is required to know the coordinates of each one of the dots of the Hartmanngram, with a coordinate system whose origin is at the intersection of the Hartmanngram plane with the optical axis of the system under test. An estimted origin could be found by supposing that the point source is located on the optical axis, and the surface under test is near to a axisymmetrical surface. In this case, the optical axis crosses the Hartmanngram plane at the point (X_{av}, Y_{av}), given by the average of all the coordinates of the dots in the Hartmanngram; and, therefore, the N vectors referred to the optical axis can be calculated from

$$(X_{oai}, Y_{oai}, M_{Xi}, M_{Yi}) = (X_{dei} - X_{av}, Y_{dei} - Y_{av}, M_{Xi}, M_{Yi}). \tag{10.11}$$

In the next step the ideal dot coordinates (X_{eri}, Y_{eri}) are calculated with the aid of an exact ray tracing. From these ideal coordinates the transverse aberration values can be calculated by means of

$$(X_{abi}, Y_{abi}, M_{Xi}, M_{Yi}) = (X_{oai}, X_{eri}, Y_{oai} - Y_{eri}, M_{Xi}, M_{Yi}); \tag{10.12}$$

the new N vectors describe the transverse aberration values of the wavefront that leaves the mirror or system under test. In this data, a defocus error can be present since the hypothetical location of the Hartmanngram plane in the exact ray tracing can be different from the actual Hartmanngram plane. The optimum focus can be found by subtraction of the linear term that can be calculated by means of two least-square fits applied to the data vectors given in Eq. (10.12). In this case the focus error data (X_{abi}^*, Y_{abi}^*), can be calculated as function of the hole centers with the equation

$$(X_{abi}^*, Y_{abi}^*) = (A + BX_i, A' + B'Y), \tag{10.13}$$

where A, B, A', and B' can be evaluated through two least-square independent fittings.

If we apply the two independent fits described above, we can eliminate some astigmatic terms in the linear dependence, which can be interpreted as a defocusing term, and therefore one can have an incorrect interpretation of the form of the surface. In order to avoid the previous problem only one least-square fitting should be carried out (Zverev et al., 1977).

With the new transverse aberration coordinates $(X_{abi} - X_{abi}^*, Y_{abi} - Y_{abi}^*)$ a numerical or polynomical integration is carried out to get the values of the optical path differences. Once these evaluations are concluded, we can do a new fitting with the terms to fourth degrees but special care must be taken for not subtracting them without the quantitative knowledge of the origin of each term; this last factor could be known after evaluating the expected flexions of the mirror and/or the decentering of the source of light.

10.6   NULL TESTS AND THE RONCHI AND HARTMANN TESTS

Even with the technological advances in the recording and evalutation of the patterns, important efforts have been done recently in order to design and construct modifications to the traditional tests in such a way that—although still not for spherical wavefronts—we have a direct interpretations as in the case of a pattern of right fringes in the Ronchi test or of a Hartmanngram whose dots are distributed on a squared array.

The previous idea has particular importance in testing aspherical conic sections and/or in the case of production in series since the criterion of acceptance/refusal is very direct. In order to achieve a simplified analysis, an additional optical system could be introduced and, with this, the original wavefront can be modified and converted into a spherical one, as in the case of the optical compensators, whose principal problem is the increment of the number of possible errors such as defects in their construction and/or assembling of their components. An unexpensive alternative for the Ronchi and Hartmann tests consists in the modification of the Ronchi ruling or the Hartmann screen. In the Ronchi test, this idea was developed qualitatively for conic surfaces by Pastor (1969), and with theory of third order by Popov (1972), Mobsby (1974); and with more precise solution by Malacara and Cornejo (1974). Finally, the idea was developed for any optical system and using exact ray tracing by Hopkins and Shagan (1977).

The null Hartmann test has been studied by Cordero *et al.* (1990). With an analysis based on a common treatment of both tests of Ronchi and Hartmann. Cordero *et al.* (1992) have demonstrated that if the centers of the holes in the Hartmann screen are distributed at the interesection points of two crossed Ronchigrams then the dots in the Hartmanngram will be located on the intersection points of the two crossed rulings. And for the Ronchi, if we desire to get a Ronchigram of right- and same-spaced fringes, then the lines of the null Ronchi ruling should contain the Hartmanngram points.

10.7   GRADIENT TESTS

In this section we present two techniques that solve the problem of finding the wavefront coming from a system or surface, mainly by sensing the irradiance for several different planes (Platzeck *et al.*), and at two planes (Teague, 1983), coming from the surface or wavefront under test.

10.7.1   Platzeck–Gaviola Test

A different and interesting approach to the testing of optical surfaces was presented by Platzeck and Gaviola in 1939. In such a technique, the different zones of an aspheric or conic surface are identified by means of the so-called caustic coordinates. These coordinates correspond to the center of curvature of each one of the zones in which the surface or system under test is divided. In Fig. 10.12a the meaning of caustic coordinates is illustrated (Cornejo, 1978); they physically correspond to the coordinates (x, y) of the paraxial center of curvature, zone a; as well as of the observed $b - b', c - c', \dots, l - l'$ zones, located at certain distances from the optical axis, and correspond to certain sagitta value Z_k. For each zone a spherical region is considered, with certain values for the radius of curvature, R, of the zone, and the corresponding localized center of curvature with coordinates (x, y). The values for the local radius of curvature can be obtained by means of the following equation, derived from calculus,

$$R = \frac{\left[1 + (\partial z/\partial s)^2\right]^{3/2}}{(\partial^2 z/\partial^2 s)}; \tag{10.14}$$

after obtaining the first and second derivatives of the sagitta z, then

$$R = \frac{1}{c}(1 - Kc^2 s^2)^{3/2}. \tag{10.15}$$

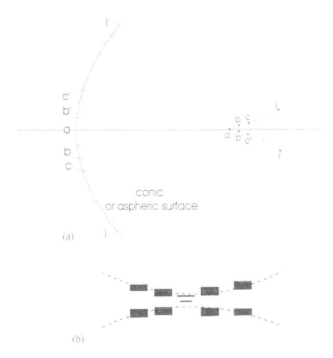

(a)

(b)

Figure 10.12 (a) Center of curvature for the different considered spherical zones of a general conic or aspheric surfaces corresponds to the paraxial center of curvature, taken as a reference point for the other zones. (b) Best focusing image for certain zones $a, b - b', c - c', \dots, l - l'$ of an aspherical or conic surface.

For the case of the conic constant $K = 0$, a sphere, the usual relation $R = 1/c$ is obtained. In order to obtain a set of equations for the caustic coordinates x, y, from Fig. 10.13, it is possible to write

$$\frac{y}{2(x + Kz)} = \frac{s}{(1/c) - (K + 1)^2} \tag{10.16}$$

and

$$R^2 = \left(s + \frac{y}{2}\right)^2 + (Hc + h - 2)^2. \tag{10.17}$$

By means of Eqs (10.15–10.17), equations for x and y can be obtained as

$$x = -Kz\{3 + cz(K + 1)[cz(K + 1) - 3]\} \tag{10.18}$$

and

$$y = -2scKz\left\{\frac{2 + cz(K + 1)[cz(K + 1) - 3]}{1 - cz(K + 1)}\right\}. \tag{10.19}$$

It is important to notice that the caustic coordinates (x, y) from Eqs (10.18) and (10.19) are measured taking as a origin the corresponding paraxial center of curvature for the zone a of the surface.

Experimentally what it is necessary to do is to register, in a photographic film or with the use of some other modern detector, the position of a wire or a slit that moves along the optical axis producing a sharp focusing image for the different

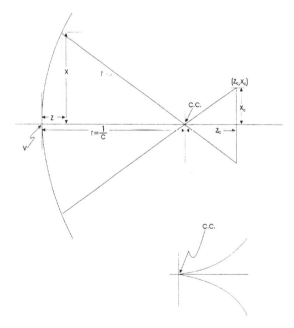

Figure 10.13 Caustic coordinates of an aspheric surface.

zones of the surface. A detailed procedure for this experiment can be read in Schroader (1953) paper, or in the same Platzeck and Gaviola (1939) work.

In Fig. 10.12b it can be seen how a complete set of focal points for the different zones of the surface under test are registered in a photographic plate.

10.7.2 Irradiance Transport Equation

The problem of finding the phase of a wavefront from irradiance measurements has been studied intensively by several authors in the last 25 years. Some of them have developed algorithms – for example, Gerchberg and Saxton (1972) and Teague (1982) – or established more comprehensive theories based in Helmholtz and irradiance transport equations, as those works by Teague (1983), Streibl (1984), Roddier *et al.* (1988), and Ichikawa *et al.* (1998); interesting review works are those by Fienup (1982) and, more recently, Campos (1985).

Before establishing a particular application of the solutions of the irradiance transport equations to the field of testing optical surfaces or systems, a general and brief review will be given about the development of obtaining information for the phase wavefront by means of intensity measurements. Finally, the presentation of the work by Teague (1983) will be described.

According to a paper by Gerchberg and Saxon, some of the first trials to obtain the wavefront phase from irradiance measurements were done by Hoppe, Schiske, and Erickson and Klug. In all these works they abandon the Gabor proposals to add a reference wave, in order to find the wavefront phase. In their first paper Gerchberg and Saxon (1971), they developed a method for the determination of the phase from intensity recordings in the imaging and diffraction planes. One of the most important characteristics of such a paper was that it was not limited to small phase deviations. In a second paper, the same authors, Gerchberg and Saxon (1972) improved the computing time, recognizing the wave relation in the imaging and diffraction planes by means of the Fourier transform.

Following the work of many authors to solve the retrieval of phase from irradiance measurements, a crucial problem was always the uniqueness of the solution [(see for example, Gonzalves, 1976, Devaney and Chidlaw (1978), Fienup (1982)]. A step forward to find the solution was given by Teague (1982), where for the first time he established that mathematically it is sufficient to retrieve the phase from irradiance data in two optical planes, and that the solution was deterministic. With this last result the previous problem of the uniqueness of the solution was finally solved. A year later Teague (1982) described an alternative method based on a Green's functions solution to the propagation equations of phase and irradiance; and Fienup made a comparison among the different algorithms to the retrieval of phase. More recently, Gureyev *et al.* (199x), and Salas found other possible solutions to the radiation transfer equation.

In the next paragraphs, we describe briefly how Teague (1983) derived the propagation equation for irradiance.

Starting from the Helmholtz equation

$$(\nabla^2 + k^2)u(x, y, z) = 0, \tag{10.20}$$

where $\nabla^2 = (\partial^2/\partial x^2) + (\partial^2/\partial y^2) + (\partial^2/\partial z^2)$, and $k = 2\pi/\lambda$, it will be assumed that any wave depending only of the position must obey Eq. (10.20). For a wave traveling in the z-positive direction and considering only the spatial component, then

$$\psi(x, y, z) = u(x, y, z)\exp(-ikz). \tag{10.21}$$

Substituting $\psi(x, y, z)$ of Eq. (10.21) into Eq. (10.20), we obtain

$$\left[\nabla_T^2 + \frac{\partial^2}{\partial z^2} - 2ik\left(\frac{\partial}{\partial z}\right)\right]u(x, y, z) = 0, \tag{10.22}$$

where $\nabla_T^2 = (\partial^2/\partial x^2) + (\partial^2/\partial y^2)$. Assuming that the amplitude u varies slowly along the Z-direction, implies that the term $\partial^2 u/\partial z^2$ can be dropped from Eq. (10.22), and the so-called paraxial wave equation is obtained:

$$\left[\nabla_T^2 - 2ik\left(\frac{\partial}{\partial z}\right)\right]u(x, y, z) = 0. \tag{10.23}$$

If now a Fresnel diffraction theory solution is proposed, without the term $\exp(ikz)$ for Eq. (10.23), then the so-called parabolic equation can be derived:

$$\frac{(\nabla_T^2)}{2k} - k - i\frac{\partial}{\partial z}u_F(x, y, z) = 0. \tag{10.24}$$

Following the Teague paper (1983), where $w_z(x, y, z)$ is normalized such that $|U_F(x, y, z)|^2 = I_F$, with I_F the irradiance at the point (x, y, z), and writing

$$U_F(x, y, z) = |I_F|^{1/2}\exp[i\phi(x, y, z)] \tag{10.25}$$

and after some algebraic manipulations and with $\phi = (2\pi/\lambda)W(x, y, z)$, where $W(x, y, z)$ is the wavefront, the irradiance transport equation is obtained as

$$\frac{\partial}{\partial^2}I + \nabla_T W \cdot \nabla_T I + I\nabla_T^2 W = 0, \tag{10.26}$$

or in a compact form

$$\nabla_T \cdot (I\nabla_T W) = \frac{\partial}{\partial z}I. \tag{10.27}$$

A similar result for the irradiance transport equation was obtained by Streibl, following a different and more simplified approach.

With the irradiance transport equation well established, Teague proved its validity and possible solution by means of the Green's functions, carrying out a numerical simulation. Streibl applied it for thin phase structures, obtaining mainly qualitative results. The first quantitative experimented results were obtained by Ichikawa, Lomhann, and Takeda (1988). In order to obtain experimental results, those last authors solved the ITE by the Fourier-transform method, and explained in detail the physical meaning of the different terms of Eq. (10.26). In their experiment, Ichikawa *et al.* used as a phase object a lens and, with the help of a grating and a CCD camera, the irradiance in two planes, separated by 0.7 mm, was registered and

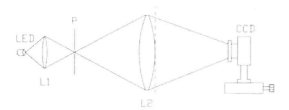

Figure 10.14 Experimental setup.

the wavefront phase was found with the experimental setup proposed (Fig. 10.14). Figure 10.15 shows the results obtained by Alonso *et al.*, using a Nodal bench, following Ichikawa *et al.*'s technique.

10.7.3 Roddier Method

For testing astronomical telescopes or for using with adaptive optics devices, Roddier developed a method to find the phase retrieval from wavefront irradiance measurements. In Fig. 10.16 is shown what Roddier called the curvature sensor,

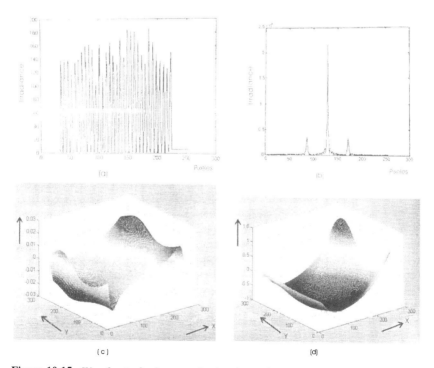

Figure 10.15 Wavefront of a lens tested using the method developed by Ichikawa *et al.* (a) One scan irradiance measurement. (b) Spatial frequency spectra of (a). (c) Derivative of the wavefront. (d) Wavefront shape. (Alonso *et al.*, 2000).

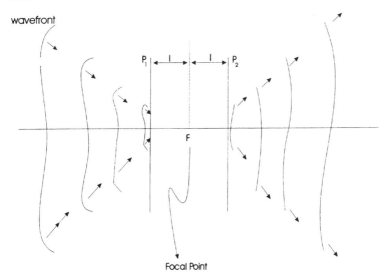

Figure 10.16 Roddier's method for measuring the irradiance at the two symmetrical planes, P_1 and P_2, from the focal point F.

because the aim was to obtain irradiance data at the two off-focus planes, P_1 and P_2, and obtain the phase of the wavefront coming from the optical system.

For his method, Roddier derived the next equation

$$\frac{I_1 - I_2}{I_1 + I_2} = \frac{f(f-1)}{l}\left[\frac{\partial}{\partial n} WS(r-a) - P\left(\frac{f}{l}\bar{r}\right)\nabla_1^2 W\right].$$ (10.28)

On the left side of Eq. (10.28) are the irradiance measurements at the two planes P_1 and P_2. On the right side of Eq. (10.28), f is the focal length of the system under test; l is the distance where the planes P_1 and P_2 are located symmetrically from the focal point F; and S is the circular Dirac distribution, representing the outward-pointing derivative at the edge of the pupil where it is different to zero.

Equation (10.28) can be derived in a more or less straightforward way from the ITE, Eq. (10.26), as has been explained by Roddier0′8. Assuming a plane wavefront at the pupil plane ($Z = 0$), then everywhere except in at the pupil border

$$\nabla I|_{20} = \begin{cases} 0 \\ -I|_{z=0}\hat{n}\partial(r-a). \end{cases}$$ (10.29)

Substituting Eq. (10.29) into Eq. (10.26), we obtain

$$\frac{\partial I}{\partial z}\bigg|_{z=0} = \left\{IS(r-a)\frac{\partial}{\partial n}W + P(\bar{r})I\nabla^2 W\right\}\bigg|_{z=0},$$ (10.30)

where $P(\vec{r})$ is the pupil transmittance and $\hat{n} \cdot \nabla W = (\partial W/\partial n)$.

The irradiance measurements at planes I_1, and I_2 can be written as

$$I_1 = I_{z=0} + \frac{\partial}{\partial z}I\big|_{z=0}\Delta Z_1$$

and

$$I_2 = I_{z=0} + \frac{\partial}{\partial z}I\big|_{z=0}\Delta Z_2; \tag{10.31}$$

an important condition for the Roddier method is that $|\Delta Z_1| = |\Delta Z_2| = \Delta Z$. With this last condition, substituting Eq. (10.30) into Eq. (10.31), the next expression can be derived:

$$\frac{I_1 - I_2}{I_1 + I_2} = \left\{ \partial(r - a)\frac{\partial}{\partial n}W - P(\bar{r})\nabla_T^2 W \right\}\Delta Z. \tag{10.32}$$

Using the thin lens equation, it can be proved that $\Delta Z \cong f(f - l)/l$; with this result Eq. (10.32) can be written as Eq. (10.28), which Roddier used for his technique. It is worth mentioning that it is advised that the planes P_1 and P_2 must be outside the caustic in order to avoid mixing the same rays and their irradiance.

Roddier developed an algorithm to find the solution for W of Roddier's equation (10.28), from the irradiance measurements I_1 and I_2. The Roddier algorithm solves the Poisson equation with Neumann boundary conditions, and the iterative method to find the solution is called overrelaxation.

REFERENCES

Cordero-Dávila, A., O. Cardona-Nuñez, and A. Cornejo-Rodríguez, "Null Hartmann and Ronchi–Hartmann Tests," *Appl. Opt.*, **29**, 4618 (1990).

Cordero-Dávila, A., O. Cardona-Nuñez, and A. Cornejo-Rodríguez, "The Ronchi and Hartmann Tests with the same Mathematical Theory," *Appl. Opt.*, **31**, 2370–2376 (1992).

Cordero-Dávila, A., O. Cardona-Nuñez, and A. Cornejo-Rodriguez, "Least-Squares Estimators for the Center and Radius of Circular Patterns," *Appl. Opt.*, **32**, 5683–5685 (1993).

Cordero-Dávila, A., O. Cardona-Nuñez, and A. Cornejo-Rodríguez, "Polynomial Fitting of Interferograms with Gaussian Errors on the Fringe Coordinates. I: Computer Simulations," *Appl. Opt.*, **33**, 7339–7342 (1994a).

Cordero-Dávila, A., O. Cardona-Nuñez, and A. Cornejo-Rodríguez, "Polynomial Fitting of Interferograms with Gaussian Errors on the Fringe Coordinates. II: Analytical Study," *Appl. Opt.*, **33**, 7343–7348 (1994b).

Cornejo-Rodríguez, A. and D. Malacara, "Wavefront Determination Using Ronchi and Hartmann Tests," *Bol. Int. Tonantzintla*, **2**, 127–129 (1976).

Cornejo-Rodríguez, A., "Ronchi Test," in *Optical Shop Testing*, D. Malacara, ed., Wiley, New York, 1992, pp. 321–365.

Foucault, L. M., "Description des Procédés Employés pour Reconnaitre la Configuration des Surfaces Optiques," *C. R. Academ. Sci. Paris*, **47**, 958 (1858); reprinted in *Classiques de la Science*, Vol. II, By Armand Colin.

Ghozeil, I., "Hartmann and Other Screen Tests," in *Optical Shop Testing*, D. Malacara, ed., Wiley, New York, 1992, pp. 367–396.

Hartmann, J., "Bemerkungen uber den Bau und die Justirung von Spoktrographen," *Zt., Instrumentenkd*, **20**, 47 (1900).

Hopkins, G. H. and R. H. Shagan, "Null Ronchi Gratings from Spot Diagram," *Appl. Opt.*, **16**, 2602–2603 (1977).Hartmann, J., "Bemerkungen uber den Bau und die Justirung von Spoktrographen," *Zt., Instrumentenkd*, **20**, 47 (1900).

Landgrave, J. E. A. and J. R. Moya, "Effect of a Small Centering Error of the Hartmann Screen on the Computed Wavefront Aberration," *Appl. Opt.*, **25**, 533 (1986).

Malacara, D., "Ronchi Test and Transversal Spherical Aberration," *Bol. Obs. Tonantzintla Tacubaya*, **4**, 73 (1966).

Malacara, D., "Geometrical Ronchi Test of Aspherical Mirrors," *Appl. Opt.*, **4**, 1371–1374 (1965).

Malacara, D., "Analysis of the Interferometric Ronchi Test," *Appl. Opt.*, **29**, 3633 (1990).

Malacara, D. and A. Cornejo, "Null Ronchi Test for Aspherical Surfaces," *Appl. Opt.*, **13**, 1778–1780 (1974).

Malacara, D., Ed., *Optical Shop Testing*, John Wiley & Sons, New York, 1st edn, 1979; 2nd edn, 1992.

Mobsby, E., "A Ronchi Null Test for Paraboloids," *Sky Telesc.*, **48**, 325–330 (1974).

Morales, A. and D. Malacara, "Geometrical Parameters in the Hartmann Test of Aspherical Mirrors," *Appl. Opt.*, **22**, 3957 (1983).

Ojeda-Castañeda, J., "Foucault, Wire, and Phase Modulation Test," in *Optical Shop Testing*, D. Malacara, ed., Wiley, New York, 1992, pp. 265–320.

Patorsky, K. and A. Cornejo-Rodríguez, "Ronchi Test with Daylight Illumination," *Appl. Opt.*, **25**, 2031–2032 (1986).

Platzeck, R. and E. Gaviola, "On the Errors of Testing a New Method for Surveying Optical Surfaces and Systems," *J. Opt. Soc. Am.*, **29**, 484 (1939).

Popov, G. M., "Methods of Calculation and Testing of Ritchey–Chrétien Systems," *Izv. Krym. Astrofiz, Obs.*, **45**, 188 (1972).

Ritchey, G. H., "On the Modern Reflecting Telescope and the Making and Testing of Optical Mirrors," *Smithson. Contrib. Knowl.*, **34**, 3 (1904).

Ronchi, V., "La Frange di Combinazione Nello Studio delle Superficie e dei Sistem Ottics," *Riv. Ottica Mecc. Precis*, **2**, 9 (1923a).

Ronchi, V., "Due Nuovi Metodi per lo Studio delle Superficie e dei Sistemi Occi," *Ann. Soc. Norm. Super. Pisa*, **15** (1923b).

Sherwood, A. A., "Ronchi Test Charts for Parabolic Mirrors," *J. Proc. R. Soc., New South Wales*, **43**, 19 (1959); reprinted in *Atti. Fond. Giorgio Ronchi Contrib. Ist. Naz. Ottica*, **15**, 340–346 (1960).

Vitrichenko, E. A., "Methods of Studying Astronomical Optics. Limitations of the Hartmann Method," *Sov. J. Opt. Technol.*, **20**, 3 (1976).

Zverev *et al.*, "Mathematical Principles of Hartmann Test of the Primary Mirror of the Large Azimuthal Telescope," *Sov. J. Opt. Technol.*, **44**, 78 (1977).

PROGRAMS

Cordero-Dávila, A. and S. Zárate-Vazquez, " 'ROMA'; Simulation of Ronchigrams for Centered Conic Mirrors and with the Source on the Optical Axis," F.C.F.M.-B.U.A.P. (1992).

Cordero-Dávila, A. and S. Zárate-Vazquez, " 'BIROEV'; Evaluation of Optical System with the Squared Grating," F.C.F.M.-B.U.A.P. (1993).

Cordero-Dávila, A. and S. Zárate-Vazquez, "'FOURON'; Simulation of Foucaultgrams and Ronchigrams for any conic section and with the source in any position," F.C.F.M.-B.U.A.P. (1995).

Luna-Aguilar, E. and A. Cordero-Dávila, " 'BOFI'; Least Square Estimators for the Center and Radius of Circular Pattern," UNAM-IAUNAM (1990).

Luna-Aguilar, E., " 'HASC'; Program for Designing the Hartmann Screens," UNAM-IAUNAM (1990).

Malacara-Hernández, D., A. Cornejo-Rodríguez, and A. Ma. Zárate-Rivera, " 'HAEV'; Evaluation of Optical Surfaces with the Hartmann Test," (1995).

Vazquez Montiel, S., A. Cordero-Dávila, and A. Cornejo-Rodríguez, " 'ROLEN'; Simulation of Ronchigrams for Refractive Systems of Centered Spherical Surfaces and with the Source on the Optical Axis," INAOE (1995).

11

Basic Interferometers

DANIEL MALACARA

Centro de Investigaciones en Optica, León, Mexico

11.1 INTRODUCTION

Interferometers have been described with detail in many textbooks (Malacara, 1992). They produce the interference of two or more light waves by superimposing them on a screen or the eye. If the relative phase of the light waves is different for different points on the screen, constructive and destructive interference appears at different points, forming interference fringes. Their uses and applications are extremely numerous. In this chapter only their basic configurations will be described.

To begin, let us consider the interference of two light waves, one having a flat wavefront (constant phase on a plane in space at a given time) and the other a distorted wavefront with deformations $W(x, y)$, as in Fig. 11.1. Thus, the amplitude $E_1(x, y)$ in the observing plane is given by the sum of the two waves, with amplitudes $A_1(x, y)$ and $A_2(x, y)$, given by

$$E_1(x, y) = A_1(x, y) \exp[ikW(x, y)] + A_2(x, y) \exp[ikx \sin \theta], \tag{11.1}$$

where $k = 2\pi/\lambda$ and θ is the angle between the wavefront. The irradiance function $I(x, y)$ may then be written as

$$E_1(x, y) \cdot E_1^*(x, y) = A_1^2(x, y)$$
$$+ A_2^2(x, y) + 2A_1(x, y)A_2(x, y) \cos k[x \sin \theta - W(x, y)], \tag{11.2}$$

where the symbol * denotes the complex conjugate. This function is plotted in Fig. 11.2. We see that the resultant amplitude becomes a maximum when the phase difference is a multiple of the wavelength and a minimum when the phase difference is an odd multiple of half the wavelength. These two conditions are constructive and destructive interference, respectively.

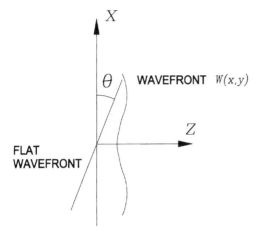

Figure 11.1 Two interfering wavefronts.

11.2 COHERENCE OF LIGHT SOURCES FOR INTERFEROMETERS

If light source has a single spectral line, we say that it is monochromatic. Then, it is formed by an infinite sinusoidal wavetrain or, equivalently, it has a long coherence length. On the other hand, a light source with several spectral lines or a continuous spectrum is nonmonochromatic. Then, its wavetrain or coherence length is short. A light source with a short wave train is said to be temporally incoherent and a monochromatic light source is temporally coherent.

The helium–neon laser has a large coherence length and monochromaticity. For this reason it is the most common light source in interferometry. However, this advantage can sometimes be a problem, because many undesired spurious fringes are formed. Great precautions must be taken to avoid this noise on top of the fringes.

With laser light sources extremely large, optical path differences (OPDs) can be introduced without appreciably losing fringe contrast. Although almost perfectly monochromatic, the light emitted by a gas laser consists of several spectral lines,

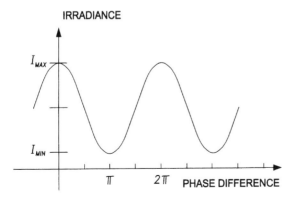

Figure 11.2 Irradiance as a function of phase difference for two-wave interference.

called longitudinal modes. They are equally spaced but very close together, with a frequency spacing $\Delta \nu$ given by

$$\Delta \nu = \frac{c}{2L}, \tag{11.3}$$

where L is the length of the laser cavity. If this laser cavity length L is modified due to thermal expansion or contraction or to mechanical vibrations, the spectral lines change their frequency, approximately preserving their separation, but with their intensities inside a Gaussian dotted envelope called a power gain curve, as shown in Fig. 11.3.

Helium–neon lasers with a single mode or frequency can be constructed, thus producing a perfectly monochromatic wavetrain. However, if special precautions are not taken, because of instabilities in the cavity length, the frequency may be unstable. Lasers with stable frequencies are commercially produced, making possible extremely large OPDs without reducing the fringe contrast. A laser with two longitudinal modes can also be frequency stabilized if desired, to avoid contrast changes. When only two longitudinal modes are present and they are orthogonally linearly polarized, one of them can be eliminated with a linear polarizer. This procedure greatly increases the temporal coherence of the laser.

With a multimode helium–neon laser the fringe visibility in an interferometer is a function of the OPD, as shown in Fig. 11.4. In order to have a good fringe contrast, the OPD has to be an integral multiple of $2L$.

A laser diode can also be used as a light source in interferometers. Creath (1985), and Ning *et al.* (1989) have described the coherence characteristics of laser diodes. Their coherence length is of the order of 1 mm, which is a great advantage in many applications, besides the common advantage of their low price and small size.

11.3 YOUNG'S DOUBLE SLIT

The typical interference experiment is the Young's double slit (Fig. 11.5). A line light source emits a cylindrical wavefront that illuminates both slits. The light is diffracted on each slit, producing cylindrical waves diverging from these slits. Any point on the screen is illuminated by the waves emerging from these two slits. Since the total paths from the point source to a point on the observing screen are different, the phases of the two waves are not the same. Then, constructive or destructive interference takes place at different points on the screen, forming interference fringes.

The amplitude E at the observing point D on the screen is

$$E = a e^{i\phi_1} + a e^{i\phi_2}, \tag{11.4}$$

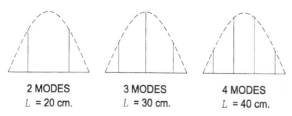

2 MODES	3 MODES	4 MODES
L = 20 cm.	L = 30 cm.	L = 40 cm.

Figure 11.3 Longitudinal modes in a He–Ne laser.

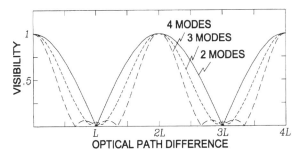

Figure 11.4 Contrast as a function of the optical path difference in a two interferometer.

where a is the amplitude at D due to each one of the two slits alone and ϕ_i are the phases whose difference is given by

$$\phi_2 - \phi_1 = k\text{OPD}, \tag{11.5}$$

where

$$\text{OPD} = (AC + CD) - (AB + BD). \tag{11.6}$$

From Eq. (11.2), the irradiance I at the point D is

$$I = EE^* = 2a^2(1 + \cos(k\text{OPD})). \tag{11.7}$$

The minima of the irradiance occurs when

$$\text{OPD} = m\lambda, \tag{11.8}$$

where m is an integer; thus

$$(AC + CD) - (AB + BD) = m\lambda, \tag{11.9}$$

which is the expression for a hyperbola. Thus, the bright fringes are located at hyperboloidal surfaces, as in Fig. 11.6. If a plane screen is placed a certain distance

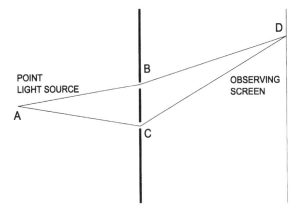

Figure 11.5 Interference in Young's double slit.

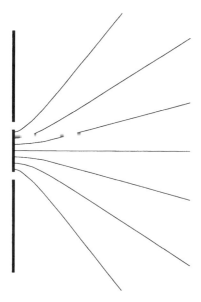

Figure 11.6 Locus of fringes in Young's double slit.

in front of the two slits, the fringes are straight and parallel with an increasing separation as they separate from the optical axis.

For a screen located at infinity, OPD is equal to CE − FB, as in Fig. 11.7. Then, Eq. (11.4) become

$$I = 2a^2[1 + \cos(kd(\sin\theta' - \sin\theta))], \tag{11.10}$$

where d is the slits' separation and θ is the angle of observation with respect to the optical axis. For small values of θ, these fringes are sinusoidal. The peaks of the irradiance (center of the bright fringes) are given by

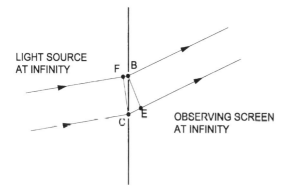

Figure 11.7 Young's experiment with light source and observing plane at infinite distances.

$$d \sin \theta = m\lambda, \tag{11.11}$$

so that their separation $\delta\theta$ has to be much larger than the eye resolution, which is about 1 arc minute ($d < 3500\lambda \sim 2\,\text{mm}$).

11.3.1 Coherence in Young's Interferometer

An ideal light source for many optics experiments is a point source with only one pure wavelength (color). However, in practice, most light sources are not a point but have a certain finite size and emit several wavelengths simultaneously. A point source is said to be spatially coherent if, when used to illuminate a system of two slits, interference fringes are produced. If an extended light source is used to illuminate the two slits and no interference fringes are observed, the extended light source is said to be spatially incoherent.

All proceeding theory for the two slits assumes that the light source is a point and also that it is monochromatic. Let us now consider the cases when the light source does not satisfy these conditions. If the light source has two spectral lines with different wavelengths, two different fringe patterns with different fringe separation will be superimposed on the observing screen, as shown in Fig. 11.8. The central maxima coincide but they are out of phase for points far from the optical axis. If the light soruce is white, the fringes will be visible only in the neighborhood of the optical axis.

Let us assume that the two slits are illuminated by two point light sources aligned in a perpendicular direction to the slits. Then, two identical fringe patterns are formed, but one displaced with respect to the other, as in Fig. 11.9. If the angular separation between the two light sources, as seen from the slits plane, is equal to half the angular separation between the fringe, the contrast is close to zero. With a single large pinhole the contrast is reduced. More details will be given in the next section, when studying the stellar Michelson interferometer.

11.3.2 Stellar Michelson Interferometer

Let us consider a double-slit interferometer with the light source and the observing screen at infinite distance from the slits' plane. If we have an extended light source, each element with apparent angular dimensions $d\theta_x d\theta_y$ will generate a fringe pattern with irradiance dI, given from Eq. (11.10) by

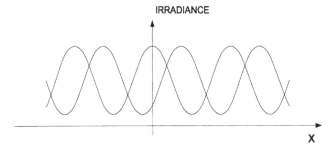

IRRADIANCE

X

Figure 11.8 Superposition of two diffraction patterns with the same frequency.

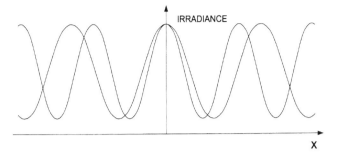

Figure 11.9 Superposition of two diffraction patterns with different frequencies.

$$dI = 1 + \cos[kd(\sin\theta_y' - \sin\theta_y))]\,d\theta_x d\theta_y, \tag{11.12}$$

which is a general expression; if we assume that the angular size of the light source is small and that the fringes are observed in the vicinity of the optical axis, we have

$$dI = [1 + \cos(kd(\theta_y' - \theta_y))]\,d\theta_x d\theta_y. \tag{11.13}$$

This is a valid expression for any shape of the light source. If we assume that it is square with angular dimensions equal to $2\alpha \times 2\alpha$, we have

$$I = \alpha^2 + 2\cos(kd\alpha_y')\sin(kd\theta_y'). \tag{11.14}$$

The fringe visibility V or contrast, defined by Michelson, is

$$V = \frac{I_{max} - I_{min}}{I_{max} + I_{min}}. \tag{11.15}$$

Thus, in this case, we have

$$V = \frac{\sin(kd\alpha_y')}{(kd\alpha_y')} = \text{sinc}(kd\alpha_y'). \tag{11.16}$$

We can see that the fringe visibility V is a function of the angular size of the light source, as shown in Fig. 11.10(a), with a maximum when it is a point source ($\alpha = 0$) or the slit separation d is extremely small. The first zero of this visibility occurs when $kd\alpha_y' = \pi$; that is, if $2d\alpha = \lambda$. Using this result, the apparent angular diameter of a square or rectangular light source can be found by forming the fringes with two slits with variable separation. The slits are separated until the fring visibility becomes zero.

If the light source is a circular disk, the visibility can be shown to be an Airy distribution:

$$V = \frac{J_1(kd\alpha_y')}{(kd\alpha_y')}, \tag{11.17}$$

which is plotted in Fig. 11.10(b). This interferometer has been used to measure the angular diameter of some stars.

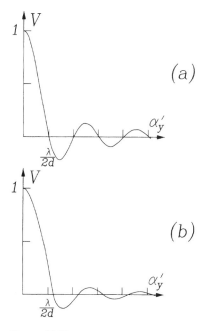

Figure 11.10 Contrast variation for (a) two slit light sources and (b) for two circular light sources.

11.4 MICHELSON INTERFEROMETER

A Michelson interferometer (Fig. 11.11) is a two-beam interferometer, illuminated with an extended light source. The beam of light from the light source is separated into two beams with smaller amplitudes, at the plane parallel glass plate (beamsplitter). After reflection on two flat mirrors, the beams are reflected back to the beamsplitter, where the two beams are recombined along a common path.

The observing eye (or camera) sees two virtual images of the extended light source, one on top of the other, but separated by a certain distance. The reason is that the two arms of the interferometer may have different lengths. Thus, the optical path difference is given by

$$\text{OPD} = 2[L_1 - L_2 - nT], \tag{11.18}$$

where T is the effective glass thickness traveled by the light rays on one path through the beamsplitter. On the other hand, from geometrical optics we can see that the virtual images of the extended light source are separated along the optical axis by a distance s, given by

$$s = 2\left(L_1 - L_2 - \frac{T}{n}\right); \tag{11.19}$$

we can see that these two expressions are different. Either the optical path difference or the two images separation can be made equal to zero.

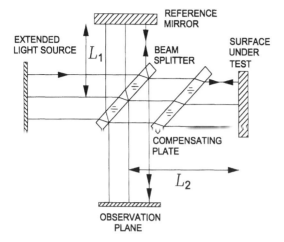

Figure 11.11 A Michelson interferometer.

To observe interference fringes with a nonmonochromatic or white light source the OPD must be zero for all wavelengths. This is possible only if the optical path is the same for the two interfering beams, at all wavelengths. We can see that the OPD can be made equal to zero by adjusting L_1 and L_2 for any desired wavelength but not for all the spectrum, unless T is zero or n does not depend on the wavelength. Only if a compensating glass plate is introduced, as in Fig. 11.11, can the OPD be made equal to zero for all wavelengths if $L_1 = L_2$. In this manner, white light fringes can be observed. In an uncompensated Michelson interferometer the optical path difference can also be written as

$$OPD = s - 2T\left(\frac{n-1}{n}\right),$$ (11.20)

but in a compensated interferometer the second term is not present.

If the light source is extended, but perfectly monochromatic, clearly spaced fringes can be observed if the two virtual images of the light sources are nearly at the same plane.

When the two virtual images of the light source are parallel to each other and the observing eye or camera is focused at infinity, circular equal inclination fringes will be observed, as in Fig. 11.13(a). We see in Fig. 11.12(a) that the OPD in a compensated interferometer is given by

$$OPD = s\cos\theta;$$ (11.21)

thus, the larger the images separation s, the greater the number of circular fringes in the pattern. The diameter of these fringes tends to infinity when the images separation becomes very small.

If there is a small angle between the two light source images, as in Fig. 11.12(b), the fringes appear curved, as if the center of the fringe had been shifted to one side, as in Fig. 11.13(b). In this case the fringes are not located at infinity, but close to the light source images. These are called localized fringes.

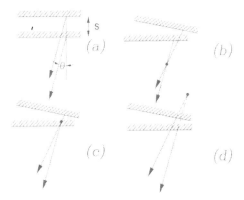

Figure 11.12 Images in space of two extended light sources in a Michelson interferometer for four different relative positions.

The fringes may appear to be in front of the two images of the light source, as in Fig. 11.12(b), between the two images, as in Fig. 11.12(c), or at the back of the two images, as in Fig. 11.12(d), depending on the relative position of these two images.

When the angle between the light sources is large, the optical path difference is nearly equal to the local separation between the two images of the light source. The fringes will be almost straight and parallel, as in Fig. 11.13(c). Their separation decreases when the angle increases. These are called equal thickness fringes.

A final remark about this interferometer, which is valid for all other amplitude division interferometers, is that there are two outputs with complementary interference patterns. In other words, a dark point in one of them corresponds to a bright point on the other and vice versa. The second interferogram is one with the wavefronts going back to the light source. These two interferogram patterns are exactly complementary if there are no energy losses in the system, as can be easily proved with Stokes' relations.

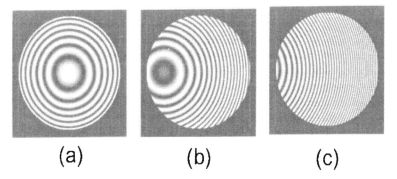

Figure 11.13 Fringe patterns in a Michelson interferometer: (a) equal inclination fringes, (b) localized fringes, and (c) equal thickness fringes.

11.5 FIZEAU INTERFEROMETER

The Fizeau interferometer, illustrated in Fig. 11.14, is quite similar to the Michelson interferometer described in the preceding section, producing the two interfering beams by means of an amplitude beamsplitter. Unlike in the Michelson interferometer, the illuminating light source is a monochromatic point, producing a spherical wavefront, which becomes flat after being collimated by a converging lens. This wavefront is reflected back on the partially reflecting front face of the beamsplitter plate. The transmitted beam goes to the optical element to be measured and is then reflected back to the beamsplitter.

The quality of many different optical elements can be evaluated with this interferometer: for example, a glass plate, which also serves here as the reference beamsplitter, as in Fig. 11.14. The optical path difference in this interferometer, when testing a single plane parallel plate, is given by

$$\text{OPD} = nt, \tag{11.22}$$

where t is the glass plate thickness. A field without interference fringes is produced when nt is a constant, but n and t cannot be determined, only its product.

In order to test a convex optical surface, the reference surface can be either flat or concave, as in Figs 11.15(a) and (b). The quality of a flat optical surface can be measured with the setup in Fig. 11.14. In this case the OPD is equal to $2d$. If we laterally displace the point light source by a small amount s, the refracted flat wavefront would be tilted at angle θ:

$$\theta = \frac{s}{f}, \tag{11.23}$$

where f is the effective focal length of the collimator. With this tilted flat wavefront, the optical path difference is given by

$$\text{OPD} = 2d \cos \theta. \tag{11.24}$$

The OPD with a small angle θ from the OPD on axis can be approximated by

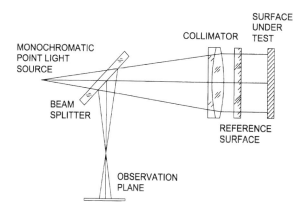

Figure 11.14   Fizeau interferometer.

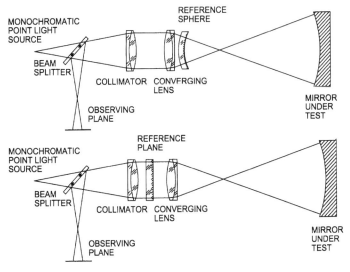

Figure 11.15 Fizeau interferometer to test a concave surface with (a) a reference sphere and (b) a reference plane.

$$\text{OPD} = 2d\left(1 - \frac{\theta^2}{2}\right) = 2d\left(1 - \frac{s^2}{2f}\right). \tag{11.25}$$

If a small extended light source with semidiameter s is used, the fringes have good contrast, as long as the condition

$$\Delta\text{OPD} = \frac{ds^2}{f} \le \frac{\lambda}{4} \tag{11.26}$$

is satisfied. The light source can increase its size s only if the air gap d is reduced.

If the collimator lens has spherical aberration, the collimated wavefront would not be flat. The maximum transverse aberration (TA) in this lens can be interpreted as the semidiameter s of the light source. Thus, the quality requirements for the collimator lens increase as the OPD is increased. When the OPD is zero, the collimator lens can have any value of spherical aberration without decreasing its precision.

11.5.1 Laser Fizeau and Shack Interferometers

An He–Ne gas laser can be used as a light source for a Fizeau interferometer with the great advantage that a large OPD can be used due to its light temporal coherence. This large OPD is highly necessary when testing concave optical surfaces with a long radius of curvature. However, the high coherence of the laser also brings some problems, such as undesired interference fringes from several optical surfaces in the system. With the introduction of a wedge in the beamsplitter and the use of pinholes acting as spatial filters, undesired reflections can be blocked out. Also, polarizing devices and antireflecting coating can be used for this purpose.

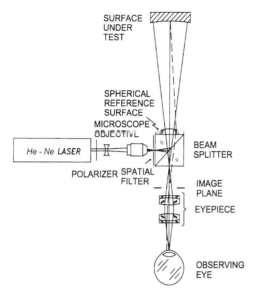

Figure 11.16 A Shack–Fizeau interferometer.

A Shack interferometer, as illustrated in Fig. 11.16, is an example of a Fizeau interferometer using an He–Ne laser.

The light from an He–Ne laser is focused on a spatial filter in contact with a nonpolarizing cube beamsplitter. Since the OPD is large, a high temporal coherence is needed. The gas laser is of such a length and characteristics that it contains two longitudinal modes with linear orthogonal polarizations. One of the two spectral lines is isolated by means of a polarizer.

The reference wavefront is reflected at the spherical convex surface of a plane convex lens cemented to the cube beamsplitter. This cube with the lens can be considered as a thick lens that forms a real image of the surface under test at the image plane; then, this image is visually observed with an eyepiece.

11.6   NEWTON INTERFEROMETER

The Newton interferometer can be considered as a Fizeau interferometer in which the air gap is greatly reduced to less than 1 mm, so that a large extended source can be used. This high tolerance in the magnitude of the angle θ also allows us to eliminate the need for the collimator, if a reasonably large observing distance is desired. Figure 11.17 shows a Newton interferometer, with a collimator, so that the effective observing distance is always infinite. The quality of this collimator does not need to be high. If desired, it can even be taken out, as long as the observing distance is not too short.

A Newton interferometer is frequently used in manufacturing processes, to test planes, concave spherical, or convex spherical optical surfaces by means of measuring test plates with the opposite curvature, placed on top of the surface under test.

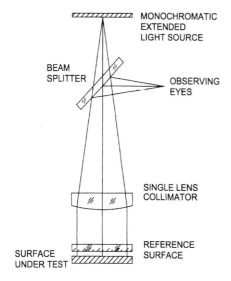

Figure 11.17 Newton interferometer.

11.7 TWYMAN–GREEN INTERFEROMETER

A Twyman–Green interferometer, designed by Twyman (1918) as a modification of the Michelson interferometer is shown in Fig. 11.18. The basic modification is to replace the extended light source by a point source and a collimator, as in the Fizeau interferometer. Thus, the wavefront is illuminated with a flat wavefront. Hence, the fringes in a Twyman–Green interferometer are of the equal-thickness type.

As in the Michelson interferometer, white light fringes are observed only if the instrument is compensated with a compensating plate. However, normally, a monochromatic light source is used, eliminating the need for the compensating plate.

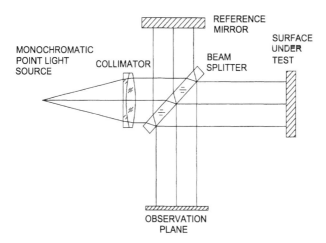

Figure 11.18 Twyman–Green interferometer.

The beamsplitter must have extremely flat surfaces and its material must be highly homogeneous. The best surface must be the reflecting one. The nonreflecting surface must not reflect any light, to avoid spurious interference fringes. Thus the nonreflecting face must be coated with an antireflection multilayer coating. Another possibility is to have an incidence angle on the beamsplitter with a magnitude equal to the Brewster angle and properly polarizing the incident light beam.

The size of the light source can be slightly increased to a small finite size if the optical path difference between the two interferometer arms is small, following the same principles used for the Fizeau interferometer.

A glass plate can be tested as in Fig. 11.19(a) or a convergent lens as in Fig. 11.19(b). When testing a glass plate, the optical path difference is given by

$$OPD = (n - 1)d. \tag{11.27}$$

When no fringes are present, we can conclude that $(n - 1)d$ is a constant, but not independent of n or d. If we compare this expression with the equivalent for the Fizeau interferometer (Eq. (11.22)), we see that n and t can be measured independently if both Fizeau and Twyman–Green interferometers are used.

A convex spherical mirror with its center of curvature at the focus of the lens is used to test convergent lenses, as in Fig. 11.20(a), or a concave spherical mirror can be used to test lenses with short focal lengths, as in Fig. 11.20(b). The small, flat mirror at the focus of the lens can also be employed. The small region being used on the flat mirror is so small that its surface does not need to be very accurate. However, the wavefront is rotated 180°, making the spatial coherence requirements higher and canceling odd aberrations like coma.

The Twyman–Green and Fizeau interferograms produce the same interferogram if the same aberration is present. The interferograms produced by the Seidel primary aberrations have been described by Kingslake (1925–1926) and their associated wavefront deformations can be expressed by

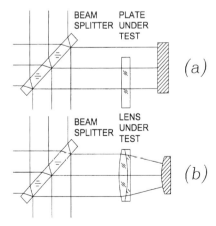

Figure 11.19 Testing (a) a glass plate and (b) a lens in a Twyman–Green interferometer.

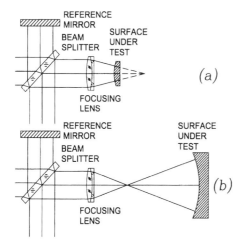

Figure 11.20 Testing (a) concave and (b) convex surfaces in a Twyman–Green interferometer.

$$W(x, y) = A(x^2 + y^2)^2 + By(x^2 + y^2) + C(x^2 - y^2)$$
$$+ D(x^2 + y^2) + Ex + Fy + G, \tag{11.28}$$

where

> A = spherical aberration coefficient
>
> B = coma coefficient
>
> C = astigmatism coefficient
>
> D = defocussing coefficient
>
> E = tilt about the y-axis coefficient (image displacement along the x-axis)
>
> F = tilt about the x-axis coefficient (image displacement along the y-axis)
>
> G = piston or constant term.

The interferograms produced by these Seidel primary aberrations are illustrated in Fig. 11.21. To determine these eight constants from measurements in the interferogram, the eight sampling points shown in Fig. 11.22 can be used.

11.7.1 Laser Twyman–Green Interferometer

Large astronomical mirrors can also be tested with a Twyman–Green unequal-path interferometer, as in Fig. 11.23, and described by Houston *et al.* (1967). However, there are important considerations to take into account because of the large OPD.

(a) As in the Fizeau interferometer, when the OPD is large, the collimator as well as the focusing lens must be almost perfect, producing a flat and a spherical wavefront, respectively.

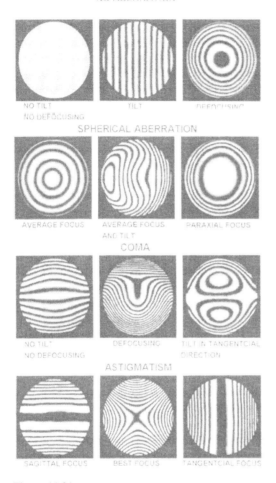

Figure 11.21   Interferograms of primary aberrations in a Twyman–Green interferometer.

(b)   The laser must have a large temporal coherence. Ideally, a single longitudinal mode has to be present.

(c)   The concave mirror under test must be well supported, in a vibration and atmospheric turbulence-free environment.

11.7.2   Mach–Zehnder Interferometer

The Twyman–Green or Michelson configurations are sometimes unfolded to produce the optical arrangement shown in Fig. 11.24. An important characteristic is that any sample located in the interferometer is traversed by the light beam only once. Another important feature is that since there are two beamsplitters, the interferometer is compensated if their thicknesses are exactly equal.

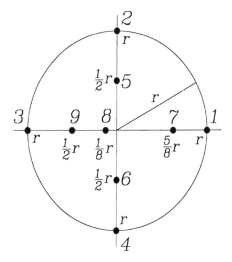

Figure 11.22 Points selected to evaluate the primary aberrations in a Twyman–Green interferogram.

11.8 COMMON PATH INTERFEROMETERS

In common path interferometers the two interfering wavefronts travel along the same path from the light source to the observing plane. The advantages are that the fringes are quite stable and also that the OPD is nearly zero, thus producing white light fringes.

There are many different types of common path interferometers. Here, a few of the most important will be described.

11.8.1 Burch and Murty Interferometers

The Burch interferometer, also called the scattering interferometer, is illustrated in Fig. 11.25(a). The real image of a small tungsten lamp is formed at the center of a

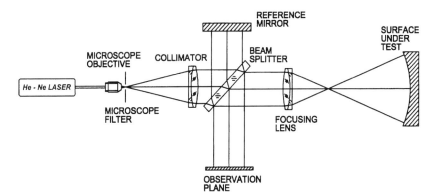

Figure 11.23 Unequal path Twyman–Green interferometer.

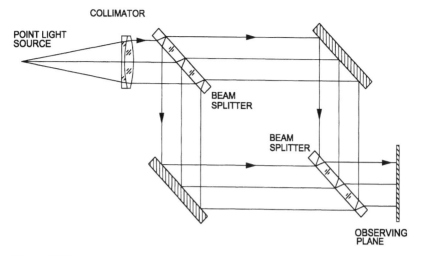

Figure 11.24 Mach–Zehnder interferometer.

concave surface under test. This light forming the image passes through a scattering glass plate SP_1 that can be made in several different manners, but the most common is with a half-polished glass surface. The light after the scattering plate can be considered as formed by two beams, one just transmitted undisturbed and another being scattered in a wide range of directions. The direct beam forms the image of the lamp on the central region of the concave surface and the diffracted one illuminates the whole surface of the mirror.

A second identical scattering plate SP_2 is located on the image of the plate SP_1 but rotated 180°. These scattering plates have to be identical point to point. This is the most critical condition, but one possible solution is to make a photographic copy

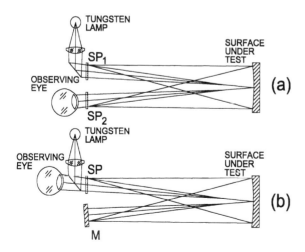

Figure 11.25 Scattering interferometer.

of the first plate SP$_1$. Both beams passing through the first scattering plate arrive at the second one. Here, the light beam not scattered on the first plate can go through the second plate, again, without being scattered. With this light, the observing eye sees a bright image of the lamp on the surface under test.

This direct beam from the SP$_1$ beam can also be scattered on the plate SP$_2$, producing many spherical wavefronts originating at each of the scattering points on the plates SP$_2$.

The scattered beam from SP$_1$ can also be considered to be formed by many spherical beams, with center on curvature on each scattering point in SP$_1$. Each of these spherical wavefronts illuminates the whole concave surface under test. If this surface is spherical, the reflected wavefront is also spherical and convergent to SP$_1$. However, if the concave surface is not spherical but contains deformations, the convergent wavefronts will also be deformed with twice the value present in the mirror. When these convergent deformed wavefronts pass through the plate SP$_2$ without being scattered, they interfere with the spherical wavefronts being produced there. The interference pattern is observed, projected over the concave surface.

To avoid the need for two identical scattering plates, a small flat mirror can be placed at the image of the scattering plate SP, as in Fig. 11.25(b). Then, the light goes back to the scattering plates after being reflected on this mirror and twice on the concave surface.

This interferometer is simpler to construct and is more insensitive to mechanical vibrations of the concave mirror. Another advantage is that the sensitivity to the deformations in the concave mirror is duplicated due to the double reflection here. There are two disadvantages: first, the concave surface has to be coated to increase its reflectivity because the double reflection here reduces the amount of light too much; secondly, the interferometer has no sensitivity to antisymmetric wavefront deformations. Thus, coma-like aberrations cannot be detected.

11.8.2 Point Diffraction Interferometer

In a point diffraction interferometer, first described by Linnik (1933) and later rediscovered independently by Smartt and Strong (1972), the aberrated wavefront passes through a specially designed plate, as in Fig. 11.26. This plate has a semitransparent small pinhole with a diameter equal to the Airy disk or smaller to that produced by a perfect spherical surface under test. The aberrated wavefront produces an image much greater than the pinhole size on this plate. The light from this aberrated wavefront goes through the semitransparent plate.

The small pinhole diffracts the light passing through it, producing a spherical wavefront with center of curvature at this pinhole. After the plate with the pinhole, the two wavefronts, one being aberrated and the second being spherical, produce the interferogram.

11.9 LATERAL SHEARING INTERFEROMETERS

Lateral shear interferometers produce two identical wavefronts, one laterally sheared with respect to the other, as shown in Fig. 11.27. The advantage is that a perfect

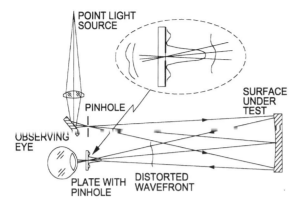

Figure 11.26 Point diffraction interferometer.

reference wavefront is not needed. The optical path difference in these interferometers can be written as

$$OPD = W(x, y) - W(x - S, y) + OPD_0, \tag{11.29}$$

where S is the lateral shear and OPD_0 is the optical path difference with two undistorted wavefronts. If this lateral shear is small compared with the aperture diameter, the smallest spatial wavelength of the Fourier components of the wavefront distortions is much smaller than S. Thus, we may obtain

$$OPD = S \frac{\partial W(x, y)}{\partial x}. \tag{11.30}$$

This interferometer can be quite simple, but a practical problem is that the interferogram represents the wavefront slopes in the shear direction, not the actual wavefront shape. Thus, to obtain the wavefront deformations a numerical integra-

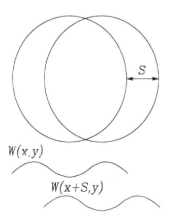

Figure 11.27 Wavefronts in a lateral shear interferometer.

tion of the slopes has to be performed; in addition, two laterally sheared interferograms in mutually perpendicular directions are needed.

If the shear is not small enough, the interferogram does not represent the wavefront slope. Then, to obtain the wavefront deformation, a different method has to be used. One of the possible procedures has been proposed by Saunders (1961) and is described in Fig. 11.28. To begin, let us assume that $W_1 = 0$. Then, we may write

$$
\begin{aligned}
W_1 &= 0 \\
W_2 &= \Delta W_1 + W_1 \\
W_3 &= \Delta W_2 + W_2 \\
&\cdots\cdots\cdots\cdots \\
W_n &= \Delta W_{n-1} - W_{n-1}
\end{aligned}
\tag{11.31}
$$

A disadvantage of this method is that the wavefront can be evaluated only at points separated by a constant distance S. Intermediate values have to be estimated by interpolation.

An extremely simple lateral shear interferometer was described by Murty (1964) and is shown in Fig. 11.29. The practical advantages of this instrument are its simplicity, low price, and fringe stability. The only disadvantage is that it is not compensated and, thus, it has to be illuminated by laser light.

The lateral shear interferograms for the Seidel primary aberrations may be obtained as follows. The interferogram for a defocused wavefront is given by

$$2DxS = m\lambda. \tag{11.32}$$

This is a system of straight, parallel, and equidistant fringes. These fringes are perpendicular to the lateral shear direction. When the defocusing is large, the spacing between the fringes is small. When there is no defocus, there are no fringes in the field.

For spherical aberration the interferogram is given by

$$4A(x^2 + y^2)xS = m\lambda. \tag{11.33}$$

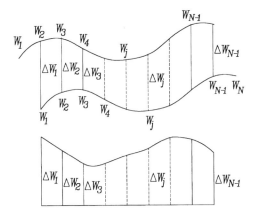

Figure 11.28 Saunders' method to find the wavefront in a lateral shear interferometer.

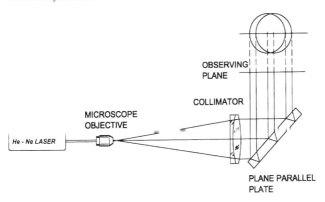

Figure 11.29 Murty's lateral shear interferometer.

If this aberration is combined with defocus we have

$$[4A(x^2 + y^2)x + 2Dx]S = m\lambda. \tag{11.34}$$

The interference fringes are cubic curves. When coma is present, the interferogram is given by

$$2BxyS = m\lambda \tag{11.35}$$

if the lateral shear is S in the *sagittal* direction. When the lateral shear is T in the tangential y-direction, the fringes are described by

$$B(x^2 + 3y^2)T = m\lambda. \tag{11.36}$$

For astigmatism, if the lateral shear is S in the sagittal x-direction, the fringes are described by

$$(2Dx + 2Cx)S = m\lambda \tag{11.37}$$

and, for lateral shear T in the tangential y-direction, we have

$$(2Dy - 2Cy)T = m\lambda. \tag{11.38}$$

Then, the fringes are straight and parallel as for defocus, but with a different separation in both interferograms. Typical interferograms for the Seidel primary aberrations are illustrated in Fig. 11.30.

The well-known and venerable Ronchi test, illustrated in Fig. 11.31, can be considered as a geometrical test but also as a lateral shear interferometer. In the geometrical model, the fringes are the projected shadows of the Ronchi ruling dark lines. However, the interferometric model assumes that several laterally sheared wavefronts are produced by diffraction. Thus, the Ronchi test can be considered as a multiple wavefront lateral shear interferometer.

11.10 TALBOT INTERFEROMETER AND MOIRÉ DEFLECTOMETRY

Projecting the shadow of a Ronchi ruling with a collimated beam of light, as in Fig. 11.32, the shadows of the dark and clear lines are not clearly defined due to diffrac-

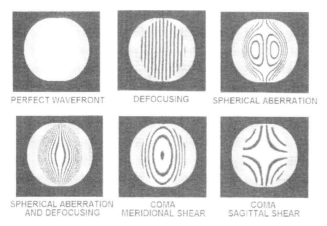

Figure 11.30 Interferograms in a lateral shear interferometer.

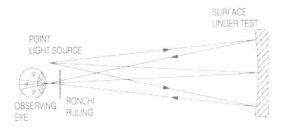

Figure 11.31 Ronchi test.

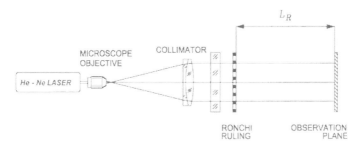

Figure 11.32 Observation of the Talbot effect.

tion. Sharp and well-defined shadows are obtained for extremely short distances from the ruling to the observing screen. When the observing distance is gradually increased, the fringe sharpness decreases, until, at a certain distance, the fringes completely disappear. However, as discovered by Talbot (1836), by further increasing the observing distance the fringes become sharp again and then disappear in a sinusoidal manner. With negative contrast a clear fringe appears where there should be a dark fringe and vice versa. Talbot was not able to explain this phenomenon but it was later explained by Rayleigh (1981). The period of this contrast variation is called the Rayleigh distance L_r, which can be expressed by

$$L_r = \frac{2d^2}{\lambda},\tag{11.39}$$

where d is the spatial period (lines separation) of the ruling and λ is the wavelength of the light.

When an aberrated glass plate is placed in the collimated light beam, the observed projected fringes will also be distorted, instead of straight and parallel.

A simple interpretation is analogous to the Ronchi test, with both the geometrical and the interferometric models. The geometrical model interprets the fringe deformation as due to the different local wavefront slopes producing different illumination directions, as in Fig. 11.33(a); then, the method is frequently known as deflectometry (Glatt and Kafri, 1988). The interferometric model interprets the fringes as due to the interference between multiple diffracted and laterally sheared wavefronts, as illustrated in Fig. 11.33(b). Talbot interferometry and their multiple applications have been described by many authors: for example, by Patorski (1988) and Takeda *et al.* (1984).

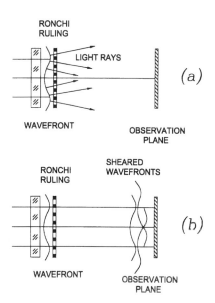

Figure 11.33 Interferometric and geometrical interpretations of Talbot interferometry.

The fringes being produced have a high spatial frequency and, thus, the linear carrier has to be removed by a Moiré effect with another identical Ronchi ruling at the observation plane.

11.11 FOUCAULT TEST AND SCHLIEREN TECHNIQUES

Leon Foucault proposed an extremely simple method to evaluate the shape of concave optical surfaces. A point light source is located slightly off axis, near the center of curvature. If the optical surface is perfectly spherical, a point image will be formed by the reflected light also near the center of curvature, as shown in Fig. 11.34.

A knife edge then cuts the converging reflected beam of light. Let us consider three possible planes for the knife edge:

- If the knife is inside of focus, the project shadow of the knife will be projected on the optical surface on the same side as the knife
- If the knife is outside of focus, the shadow will be on the opposite side to the knife
- If the knife is at the image plane, nearly all light is intercepted with even a small movement of the knife.

If the wavefront is not spherical, the shadow of the knife will create a light pattern on the mirror where the darkness or lightness will be directly proportional to the wavefront slope in the direction perpendicular to the knife edge. The intuitive impression is a picture of the wavefront topography. With this test even small amounts of wavefront deformations of a fraction of the wavefront can be detected.

If a transparent fluid or gas is placed in the light optical path between the lens or mirror producing a spherical wavefront and the knife edge, a good sensitivity to the refractive index gradients in the direction perpendicular to the knife edge is obtained. For example, any air turbulence can thus be detected and measured. This is the working principle of the Schlieren techniques used in atmospheric turbulence studies.

11.12 MULTIPLE REFLECTION INTERFEROMETERS

A typical example of a multiple reflection interferometer is the Fabry–Perot interferometer illustrated in Fig. 11.35. An extended light source optically placed at

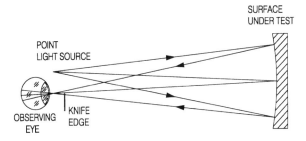

Figure 11.34 Foucault test.

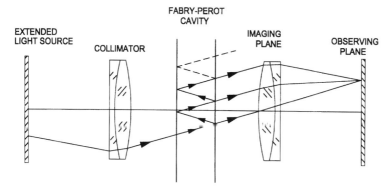

Figure 11.35 Fabry–Perot interferometer.

infinity by means of a collimator illuminates the interferometer, formed by a pair of plane and parallel interfaces. These two interfaces can be the two highly reflecting (coated) faces of a single plane parallel plate or two reflecting plane faces oriented front-to-front of a pair of glass plates. Then, the observed plane is optically placed at infinity by a focusing lens. Here, circular fringes will be observed.

As shown in Fig. 11.36, a ray emitted from the extended light source follows a path with multiple reflections. Then if the amplitude of this ray is a, the resultant transmitted amplitude $E_T(\phi)$ at a point on the observing screen located at an infinite distance is

$$E_T(\phi) = at_1t_2 + at_1t_2r_1r_2e^{i\phi} + at_1t_2r_1^2r_2^2e^{2i\phi} + at_1t_2r_1^3r_2^3e^{3i\phi} + \cdots \quad (11.40)$$

thus, obtaining

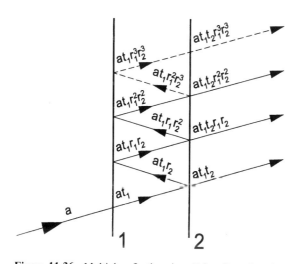

Figure 11.36 Multiple reflections in a Fabry–Perot interferometer.

$$E_T(\phi) = \frac{at_1t_2}{1 - r_1r_2e^{i\phi}}. \tag{11.41}$$

Assuming now that the two faces are equally reflecting faces and dielectric (nonabsorbing), we can consider the Stokes' relations to apply as follows:

$$r^2 + tt' = 1, \tag{11.42}$$
$$r = -r',$$

where t and r are for a ray traveling towards the interface from vacuum and t' and r' are for a ray traveling to the interface inside the glass. Thus, for this case we can write

$$E_T(\phi = a\frac{1 - r^2}{1 - r^2e^{i\phi}}. \tag{11.43}$$

The irradiance $I_T(\phi)$ of the transmitted interference pattern as a function of the phase difference ϕ between two consecutive rays is then given by the square of the amplitude of this complex amplitude:

$$I_T(\phi) = I_0\frac{1}{1 + \dfrac{4r^2}{(1 - r^2)^2}\sin^2\left(\dfrac{\phi}{2}\right)}, \tag{11.44}$$

where $I_0 = a^2$ is the irradiance of the incident light beam. This irradiance is plotted in Fig. 11.37 for several values of the reflectivity r of the faces. The interesting result is that the fringes become very narrow for high values of this reflectivity; then the position and shape of each fringe can be measured with a high precision.

As in any amplitude-division interferometer without energy losses, there are two complementary interference patterns. The sum of the energy in the reflected and the transmitted patterns must be equal to the incident energy. Thus, we can write

$$I_0 = I_T(\phi) + I_R(\phi), \tag{11.45}$$

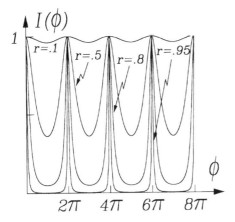

Figure 11.37 Irradiance as a function of the phase difference ϕ in a Fabry–Perot interferometer.

where the reflected irradiance is

$$I_R(\phi) = I_0 \frac{\dfrac{4r^2}{(1-r^2)^2}\sin^2\left(\dfrac{\phi}{2}\right)}{1+\dfrac{4r^2}{(1-r^2)^2}\sin^2\left(\dfrac{\phi}{2}\right)}. \tag{11.46}$$

There are a large number of multiple reflection interferometers whose principle is based on this narrowing of the fringes.

11.12.1 Cyclic Multiple Reflection Interferometers

Cyclic multiple reflection interferometers have been described by Garcia-Márquez *et al.* (1997), and are shown in Fig. 11.38. The amplitude $E(\phi)$ at the output can be found with a similar method to that used for the Fabry–Perot interferometer, obtaining

$$E(\phi) = a\frac{r+\sigma e^{i\phi}}{1+\sigma r e^{i\phi}}, \tag{11.47}$$

where σ is the coefficient for the energy loss of the system for one cyclic travel around the system ($\sigma = 1$ if no energy and $\sigma = 0$ if all the energy is lost). This coefficient can be the transmittance or absorbance of one of the mirrors when it is not 100% reflective. The irradiance $I(\phi)$ as a function of the phase difference ϕ between two consecutive passes through the system is given by

$$I(\phi) = I_0 \frac{(r+\sigma)^2 - 4\sigma r\sin^2\left(\dfrac{\phi}{2}\right)}{(1+\sigma r)^2 - 4\sigma r\sin^2\left(\dfrac{\phi}{2}\right)}, \tag{11.48}$$

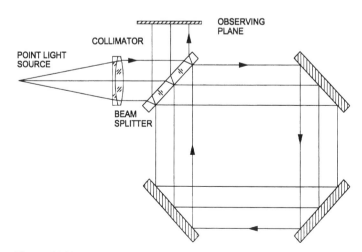

Figure 11.38 Cyclic multiple reflections interferometer.

where $I_0 = a^2$ is the incident irradiance. It is interesting to consider three particular cases:

(a) When the coefficient $\sigma = 1$, i.e., if no energy is lost, then $I(\phi) = a^2$. In other words, all energy arriving to the interferometer is in the output. An interesting consequence is that, then, no fringes can be observed.

(b) When the coefficient $\sigma \neq 1$ because the mirror M_2 is a semitransparent mirror with reflectance r, then $\sigma = -r$ and the irradiance $I(\phi)$ becomes equal to the irradiance $I_R(\phi)$ in the Fabry–Perot interferometer. In this case, there is a transmitted interference pattern in the semitransparent mirror, which acts as a second beamsplitter.

(c) When the coefficient $\sigma = 1$ due to energy absorption in the mirrors or because an absorbing material is introduced in the interferometer, then there is only one output in the interferometer, but it contains visible fringes. The complementary interference pattern is hidden as absorption.

11.13 CYCLIC INTERFEROMETERS

The basic arrangements for a cyclic interferometer are either square, as in Fig. 11.39, or triangular, as in Fig. 11.40. The two interfering wavefronts travel in opposite directions around the square or triangular path.

In both the square and the triangular configurations, the two interfering wavefronts keep their relative orientations on the output beam. Also, it can be observed in Figs 11.39 and 11.40 that these interferometers are compensated, so that their optical path difference OPD $= 0$ for all wavelengths. Thus, white light illumination can be used.

Any transparent object or sample located inside the interferometer will be traversed twice, in opposite directions, by the light beams. In the square configuration these beams pass through the sample with a reversal orientation (not 180°

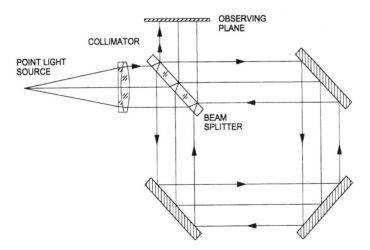

Figure 11.39 Square cyclic interferometer.

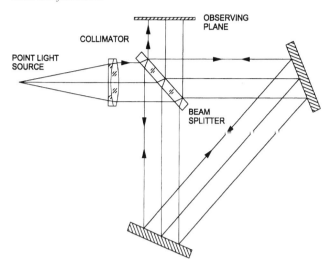

Figure 11.40 Triangular cyclic interferometer.

rotation), making it sensitive to antisymmetric aberrations, like coma or tilt in one of the mirrors.

In the triangular configuration the two interfering beams have an even number of reflections (two and four) going from the light source to the observing plane. Therefore, tilt fringes cannot be introduced by tilting any of the two mirrors. However, any tilt or displacement of the mirrors in a perpendicular direction to their surfaces produces a relative lateral shear of the two interfering beams on the observing plane. This property has been used to make lateral shear interferometers.

By introducing a telescopic afocal system in the interferometer a radial shear interferometer can also be made using this configuration.

11.14 SAGNAC INTERFEROMETER

The Sagnac interferometer (Sagnac 1913) is a cyclic interferometer with a typical square configuration, as in Fig. 11.39; it can also be made circular, with a coiled optical fiber, but the working principle is the same. The Sagnac interferometer was used as an optical gyroscope, to sense slow rotations.

Figure 11.41 shows the working principle. The beamsplitter A and the mirrors B, C, and D form the interferometer. The whole interferometer system rotates, including the light source and observer. Then, for a single travel of the light around the cyclic path in opposite directions, the beamsplitter and the mirrors have consecutive positions, labeled with subscripts 1, 2, 3, 4, and finally 5. When there is no rotation, the path length s from one mirror to the next is the same for both beams:

$$s = \sqrt{2}\,r,\tag{11.49}$$

where r is half the diagonal of the square arrangement. However, when the system is rotating, it can be observed in this figure that these paths have different lengths for the two beams, given by

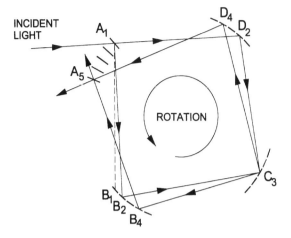

Figure 11.41   Sagnac interferometer.

$$s = \sqrt{2}\, r\left(1 \pm \frac{\theta}{2}\right), \tag{11.50}$$

where θ is the angle rotated between two consecutive positions in Fig. 11.41. Thus the OPD at the output for both interfering beams is

$$OPD = 2\sqrt{2}\, r\, \theta$$

hence it can be shown that

$$\frac{OPD}{\lambda} = \frac{4wr^2}{c\lambda} = \frac{4wA}{c\lambda}$$

where A is the area of the interferometer square.

This small shift in the fringes is constant given a fixed speed of rotation. If the interferometer plane is turned up side down, the fringe shifts in the opposite direction. This optical gyroscope has been used to detect and measure the earth's rotation.

REFERENCES

Cornejo, A., "Ronchi Test," in *Optical Shop Testing*, D. Malacara, ed., John Wiley and Sons, New York, 1992.

Creath, K., "Interferometric Investigation of a Laser Diode" *Appl. Opt.*, **24**, 1291–1293 (1985).

Fabry, C. and A. Perot, "Sur les Franges des Lames Minces Argentées et Leur Application a la Measure de Petites Epaisseurs d'air," *Ann. Chim. Phys.*, **12**, 459 (1897).

Foucault, L. M., "Description des Procédés Employés pou Reconnaitre la Configuration des Surfaces Optiques," *C. R. Acad. Sci. Paris*, **47**, 958 (1858).

Glatt, I. and O. Kafri, "Moiré Deflectometry – Ray Tracing Interferometry," *Opt. and Lasers in Eng.*, **8**, 277–320 (1988).

Houston, J. B. Jr., C. J. Buccini, and P. K. O'Neill, "A Laser Unequal Path Interferometer for the Optical Shop," *Appl. Opt.*, **6**, 1237–1242 (1967).

Kingslake, R., "The Interferometer Patterns Due to the Primary Aberrations," *Trans. Opt. Soc.*, **27**, 94 (1925–1926).

Linnik, W., "Simple Interferometer to Test Optical Systems," *Comptes Rendus del'Académie des Sciences d l'U.R.S.S.*, **1**, 208 (1933).

Malacara, D., ed., *Optical Shop Testing*, 2nd edn, John Wiley and Sons, New York, 1992.

Murty, M. V. R. K., "The use of a Single Plane Parallel Plate as a Lateral Shearing Interferometer with a Visible Gas Laser Source," *Appl. Opt.*, **3**, 331–351 (1964).

Ning, Y., K. T. V. Grattan, B. T. Meggitt, and A. W. Palmer, "Characteristics of Laser Diodes for Interferometric Use," *Appl. Opt.*, **28**, 3037–3061 (1989).

Patorski, K., "Moiré Methods in Interferometry," *Opt. and Lasers in Eng.*, **8**, 147–170 (1988).

Rayleigh, Lord, *Philos. Mag.*, **11**, 196 (1881).

Sagnac, G., "L'ether Lumineux Demontré por L'Effet due Vent Relatif d'ether Dans un Interferometre en Rotation Uniforme," *Comptes Rendus Academie Science Paris*, **157**, 361–362 (1913).

Saunders, J. B., "Measurement of Wavefronts without a Reference Standard: The Wavefront Shearing Interferometer," *J. Res. Nat. Bur. Stand.*, **65B**, 239 (1961).

Smartt, R. N. and J. Strong, "Point Diffraction Interferometer" (abstract only), *J. Opt. Soc. Am.*, **62**, 737 (1972).

Talbot, W. H. F., "Facts Relating to Optical Science," *Phil. Mag.*, **9**, 401 (1836).

Takeda, M. and S. Kobayashi, "Lateral Aberration Measurements with a Digital Talbot Interferometer," *Appl. Opt.*, **23**, 1760–1764 (1984).

Twyman, F., "Interferometers for the Experimental Study of Optical Systems from the Point of View of the Wave Theory," *Philos. Mag.*, Ser. **6**, 35, 49 (1918).

12

Modern Fringe Pattern Analysis in Interferometry

MANUEL SERVIN
Centro de Investigaciones en Optica, León, Mexico

MALGORZATA KUJAWINSKA
Institute of Precise and Optical Instruments Technical University, Warsaw, Poland

12.1 INTRODUCTION

Optical methods of testing with the output in the form of an interferogram, or more general fringe pattern, have been used since the early 1800s. However, the routine quantitative interpretation of the information coded into a fringe pattern was not practical in the absence of computers. In the late 1970s the advances in video CCD (charge-coupled device) cameras and image-processing technology coupled with the development of the inexpensive but powerful desktop computer provided the means for the birth and rapid development of automatic fringe pattern analysis. This caused a major resurgence of interest in interferometric metrology and related disciplines and formed an excellent basis for their industrial, medical, civil, and aeronautical engineering applications.

12.2 THE FRINGE PATTERN

12.2.1 Information Content

A fringe pattern can be considered as a sinusoidal signal fluctuation in two dimensional space (Fig. 12.1):

$$I(x, y) = a(x, y) + b(x, y) \cos \phi(x, y) + n(x, y) \tag{12.1}$$

or

$$I(x, y) = a(x, y)[1 + V(x, y) \cos \phi(x, y)] + n(x, y), \tag{12.2}$$

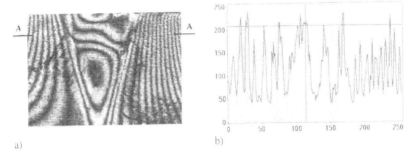

a)   b)

Figure 12.1 A fringe pattern as a sinusoidal signal fluctuation in space: (a) its cross section A–A and (b) its intensity distribution.

where $a(x, y)$, $b(x, y)$, $V(x, y)$ are background, local contrast, and fringe visibility functions, respectively, and $\phi(x, y)$ is the phase function obtained when an interferometer, Moiré system, or other device produces a continuous map which is an analogue of the physical quantity being measured (shape, displacement, deformation, strain, temperature, etc.).

Fringe pattern analysis (fringe analysis for short) (Robinson and Reid, 1993) refers to full reconversion to the original feature represented by a fringe pattern. In this process the only measurable quantity is intensity $I(x, y)$. The unknown phase $\phi(x, y)$ should be extracted from Eqs (12.1) or (12.2), although it is screened by two other functions $a(x, y)$ and $b(x, y)$; that is, $I(x, y)$ depends periodically on the phase, which causes additional problems:

- Due to periodicity, the phase is only determined mod 2π (2π phase ambiguity),
- Due to even character of the cosine function $\cos\phi = \cos(-\phi)$, the sign of ϕ cannot, in principle, be extracted from a single measurement of $I(x, y)$ without a priori knowledge (sign ambiguity), and
- in all practical cases some noise $n(x, y)$ is introduced in an additive and/or multiplicative way.

Additionally, the fringe pattern may suffer from a number of distortions degrading its quality and, additionally, screening the phase information (Schwider, 1990; Creath, 1991).

The background and contrast functions contain the intensities of interfering (superposed) fields and the various disturbances. Generally one can say that $a(x, y)$ contains all additive contributions – i.e., varying illumination and changing object reflectivity, time-dependent electronic noise due to electronic components of the image-capturing processing, diffraction of dust particles in the optical paths – while $b(x, y)$ comprises all multiplicative influences, including the ratio between the reference and object beams, speckle decorrelation and contrast variations caused by speckles.

For computer-aided quantitative evaluation the fringe pattern is usually recorded by a CCD camera and stored in the computer memory in a digital format, i.e., the recorded intensity is digitized into an array of $M \times N$ image points (pixels) and quantized into G discrete gray values. The numbers M and N set an upper bound to the density of the fringe pattern to be recorded. The sampling theorem demands more than two detection points per fringe; however, due to dealing with finite-sized detector elements, charge leakage to neighboring pixels, and noise in the fringe pattern, one has to supply at least 3–5 pixels per fringe period to yield a reliable phase estimation.

When intensity frames are acquired, the analog video signal is converted to a digital signal of discrete levels. In practice, a quantization into 8 bits corresponding to 256 gray values or into 10 bits, giving 1024 values, are the most common. Usually, 8 bits for reliable evaluation of fringe patterns are sufficient. The quantization error is affected by the modulation depth of the signal, as the effective number of quantization levels equals the modulation of the signal × the number of quantization levels.

12.2.2 Fringe Pattern Preprocessing and Design

A fringe pattern obtained as the output of a measuring system may be modified by the optoelectronic–mechanical hardware (sensors and actuators) and software (virtual sensors and actuators) of the system (Fig. 12.2) (Kujawinska and Kosinski, 1997). These modifications refer to both phase and amplitude (intensity) of the signal produced in space and time, so that the most general form of the fringe pattern is given by

$$I(x, y, t) = a_0(x, y) + \sum_{m=1}^{\infty} a_m(x, y) \cos m2\pi \left[f_{0x}x + f_{0y}y + v_0 t + \alpha(t) + \phi(x, y) \right]$$
$$+ n(x, y),$$

$$(12.3)$$

where $a_m(x, y)$ is the amplitude of mth harmonic of the signal, f_{0x}, f_{0y} are the fundamental spatial frequencies, v_0 is the fundamental temporal frequency, and α is the phase shift value. The measurand is coded in the phase $\phi(x, y)$; (x, y) and t represent

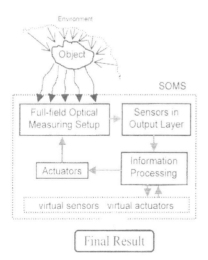

Figure 12.2 The scheme of a smart optical measuring system.

the space and time coordinates of the signal, respectively, and $n(x, y)$ represents additionally extracted random high-frequency noise.

Assuming a purely sinusoidal signal, Eq. (12.3) becomes

$$I(x, y, t) = a(x, y) + b(x, y)\cos m2\pi\left[f_{0x}x + f_{0y}y + v_0 t + \alpha(t) + \phi(x, y)\right] \\ + n(x, y), \tag{12.4}$$

where $a(x, y) = a_0(x, y)$ and $b(x, y) = a_1(x, y)$.

The fringe pattern has to be modified by hardware actuators in the measuring system in order to fulfil the demands of an a priori selected analysis method and resistance to environmental conditions. The sensors within the system enable it to determine and control the fringe pattern features $(f_{0x}, f_{0y}, v_0, \alpha, a(x, y), b(x, y))$ and in this way allow operational parameters of the actuators to be set. Table 12.1 shows the most commonly used actuators in modern research and commercial measuring systems. Special attention is paid to the new possibilities connected with the application of laser diodes (Kozlowska and Kujawinska, 1997), fiber optics, spatial light modulators (LCD, DMD) (Efron, 1989), and piezoelectric transducer (PZT) micropositioning devices. These devices not only allow one to design properly an output fringe pattern but also help to stabilize fringes in the presence of vibrations (Jones, 1994; Yamaguchi *et al.*, 1996; Olszak and Patorski, 1997).

However, in a given technical measurement, there is always a certain limit to which the appearance of a fringe pattern can be controlled. Real images are often noisy, low contrast, and with signficant variations of background. The fringe pattern analysis method described in the next sections should be designed to handle these problems. However, some general purpose image preprocessing techniques are often used to improve the original data prior to fringe analysis. Two main groups of operations are applied (Joenathan and Khorana, 1992; Van der Heijden, 1994).

- Arithmetic (pixel-to-pixel) operations including normalization, gamma correction, adding/subtractions, and multiplication/divisions performed directly on the images. These operations lead to production of a fringe pattern which looks better to human perception and is based on manipulations of the histogram of a digital image.
- Filtering operations which may be performed alternatively: directly on the image by convolution with a local operator or in Fourier space by multiplying the image spectrum with a filter window. In general, high-pass filtering weakens the influence of nonhomogeneous background $a(x, y)$, while a low-pass filter removes high-frequency noise $n(x, y)$.

After correctly performed hardware modifications of the features of fringe patterns and their software preprocessing, the images are ready for further analysis.

12.2.3 Classification of the Analysis Methods

The main task which has to be performed by fringe pattern analysis methods is to compute the phase $\phi(x, y)$ from the measured intensity values (Fig. 12.3). This means that an inverse problem has to be solved with all its difficulties:

- The regularization problem (an ill-posed problem due to unknown a, b, ϕ)

- The sign ambiguity problem
- The 2π phase ambiguity problem.

The first two difficulties may be overcome by two alternative approaches:

- Intensity methods in which we work passively on an image intensity distribution captured by a detector. These include fringe extrema localization methods (skeletoning, fringe tracking) and regularization methods.
- Phase methods for which we actively modify fringe pattern(s) in order to provide additional information to solve the sign ambiguity problem. These include:
  - temporal heterodyning (introducing running fringes) (Towers *et al.*, 1991);
  - spatial heterodyning (Fourier-transform method (Takeda *et al.*, 1982), PLL (Servin and Rodriguez-Vera, 1993), and spatial carrier phase shifting method (Servin and Cuevas, 1995);
  - temporal (Bruning *et al.*, 1974) and spatial phase shifting (Shough *et al.*, 1990), which are discrete versions of the above methods, where the time or spatially varying interferogram is sampled over a single period.

The third difficulty, the 2π phase ambiguity, coming from the sinusoidal nature of the signal is common to fringe pattern analysis methods (Fig. 12.3). The only method which measures nearly directly absolute phase with no 2π ambiguity is temporal heterodyning (Towers *et al.*, 1991). The other fringe pattern analysis methods determine absolute phase $\phi(x, y)$ by

- Fringe numbering and phase extrapolation (Robinson and Reid, 1993),
- Phase unwrapping (Huntley, 1994a; Takeda, 1996),
- Hierarchical unwrapping (Osten *et al.*, 1996), and
- Regularized phase-tracking techniques (Servin *et al.*, 1997b).

These procedures finalize the fringe measurement stage, which reduces a fringe pattern to a continuous phase map. To solve a particular engineering problem, the stage of phase scaling has to be implemented. It converts the phase map into the physical quantity to be measured in the form which enables further information processing and implementation system, finite element modeling, and machine vision systems (Schreiber *et al.*, 1996; Van der Heijden, 1994). This stage is specific application-oriented and is developing rapidly due to the implementation of fringe measurement to a vast range of different types of interferometers, Moiré, and fringe projection systems and due to the increased quality of the phase data obtained.

12.3 SMOOTHING TECHNIQUES

12.3.1 Introduction

It is very common that the fringe pattern, as captured by the video digitizing device, contains excessive noise. Generally speaking, fringe patterns contain mostly a low-frequency signal along with a degrading white noise (multiplicative or additive); therefore, a low-pass filtering (smoothing) of the fringe pattern may remove a substantial amount of this noise, making the demodulation process more reliable. We

Table 12.1 Actuators in Optical Measuring Systems

Actuator	Controllable		Influence on FP[a] features								Reference
	Signals	Parameters	f_{0x},f_{0y}	v_0	α	a	b	N_R	m		
Laser diode	Injection current Temperature	Output power Wavelength Degree of coherence	X	X	X		X	X	X		Kozlowska 1997 Takeda 1992
Noncoherent light source	Current Voltage	Output power				X	X	X			Van der Heijden 1994 Joenathan 1992
Fiber optics with accessories variable couplers	Depending on accessories	Optical path difference			X		X				Olszak 1997 Takeda 1992 Jones 1994
PZT	Voltage	Displacement Angle	X		X						Van der Heijden 1994 Ai 1987 Efron 1989
SLM (LCD, DMD)	Voltage	Intensity of each pixel or line	X		X		X	X	X		Efron 1989 Frankowski 1997 Patorski 1993
Polarizing optics + M[b]	Voltage	Angle Displacement	X		X		X				Robinson 1993 Shagam 1978 Asundi 1991

Element	Driver	Modulated quantity						References
Gratings + M	Voltage	Displacement	×					Robinson 1993
		Angle			×	×		Shagam 1978
								Huntley 1994
Optical wedges + M parallel plates	Voltage	Displacement	×					Robinson 1993
		Angle		×	×			Schwider 1990
								Shagam 1978
Acousto-optic devices	Voltage	Wavelength			×			Robinson 1993
		Angle				×		Shagam 1978
		Degree of coherence					×	
Ground glass + M	Voltage	Degree of coherence				×	×	Robinson 1993
								Schwider 1990
								Kujawinska 1997
CCD cameras	Voltage	Amplification			×	×	×	Jenathan 1992
Frame-grabber	Digital signal	Offset				×	×	Yamaguchi 1996
		γ correction						
		Exposure time					×	

[a] FP, fringe pattern.
[b] M, constant or alternate current motor.

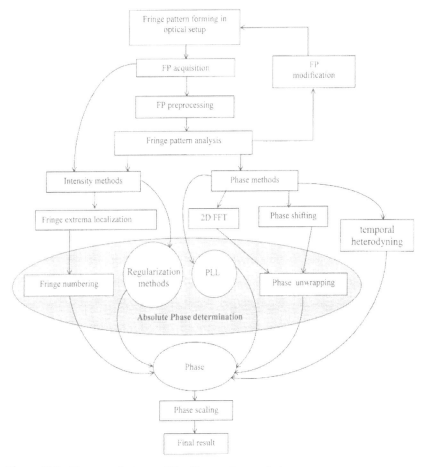

Figure 12.3   The general scheme of the fringe pattern analysis process.

are going to discuss two basic and commonly used low-pass filters, the averaging convolution window and the regularized low-pass filters.

12.3.2   Convolution Methods

The convolution averaging window is by far the most used low-pass filter in fringe analysis. The discrete impulse response of this filter may be represented by

$$h(x, y) = \frac{1}{9} \begin{pmatrix} 1 & 1 & 1 \\ 1 & 1 & 1 \\ 1 & 1 & 1 \end{pmatrix};$$ (12.5)

the frequency response of this convolution matrix is

$$H(\omega_x, \omega_y) = (1/9)[1 + 2\cos\omega_x + 2\cos\omega_y + 2\cos\sqrt{2}(\omega_x + \omega_y)$$
$$+ 2\cos\sqrt{2}(\omega_x - \omega_y)] \tag{12.6}$$

where ω_x, ω_y are the angular frequency in the x- and y-direction, respectively. This convolution filter may be used several times to decrease the bandpass frequency. Using a low-pass convolution filter several times changes the shape of the filter as well as its low-pass frequency. The frequency response of a series of identical low-pass filters will approach a Gaussian-shaped response, as can be seen in Fig. 12.4, which shows how rapidly the frequency response's shape of the 3×3 averaging filter changes as it is convolved with itself 1, 2, and 3 times.

12.3.3 Regularization Methods

The main disadvantage of convolution filters as applied to fringe pattern processing is their undesired effect at the edges of the interferogram. This distortion arises because at the boundary of the fringe pattern, convolution filters mix the background illumination with that of the fringe pattern, raising an estimated phase error in that zone. This undesired edge distortion may be so important that some

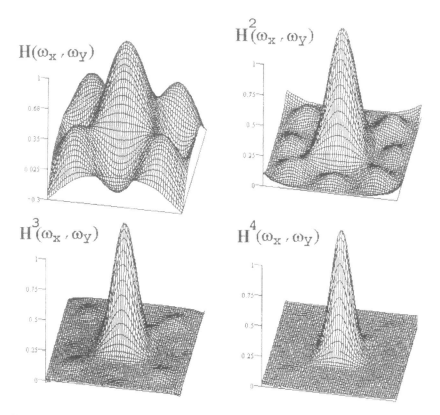

Figure 12.4 Frequency response of a 3×3 averaging window. As the number of convolution increases, the frequency response tends to a Gaussian shape.

people shrink the interferogram's area to avoid those unreliable pixels near the edge. Also, convolution filters cannot preserve fast gray-level variations while removing substantive amounts of noise. For these reasons, it is more convenient to formulate the smoothing problem in the ways described below.

Classical Regularization

The filtering problem may be stated as follows (Marroquin, 1993): find a smooth (or piecewise smooth) function $f(.)$ defined on a two-dimensional field L, given observations $g(.)$ that may be modeled by

$$g(x, y) = f(x, y) + n(x, y), (x, y) \in S, \tag{12.7}$$

where $n(.)$ is a noise field (for example a Gaussian white noise) and S is the subset of L, where observations have good signal-to-noise ratio.

The filtering problem may be seen as an optimizing problem in which one has a compromise between obtaining a smooth filtered field $f(x, y)$, while keeping a good fidelity to the observed data $g(x, y)$. In the continuous domain, a common mathematical form for the smoothing problem may be stated as the field $f(x, y)$ which minimizes the following cost or energy functional

$$U[f(x, y)] = \iint\limits_{(x,y)\in S} \left\{ [f(x, y) - g(x, y)]^2 + \lambda \left(\frac{\partial f(x, y)}{\partial x}\right)^2 + \lambda \left(\frac{\partial f(x, y)}{\partial y}\right)^2 \right\}.$$
$$\tag{12.8}$$

As the above equation shows, the first term is a measure of the fidelity between the smoothed field $f(x, y)$ and the observed data $g(x, y)$ in a least-squares sense. The second term (the regularizer) penalizes the departure from smoothness of the filtered field $f(x, y)$. The first-order regularizer is also known as a membrane regularizer. That is because the cost functional to be minimized corresponds to the mechanical energy of a two-dimensional membrane $f(x, y)$ attached by linear springs to the observations $g(x, y)$. The parameter λ measures the stiffness of the membrane model. A high stiffness value will lead to a smoother filtered field $f(x, y)$.

One may also smooth $f(x, y)$ using higher-order regularizers such as the second-order or thin-plate regularizer. In the continuous domain, the energy functional to be minimized for the filtered field $f(x, y)$ may be stated as

$$U[f(x, y)] = \iint\limits_{(x,y)\in S} \left\{ \frac{1}{\lambda}[f(x, y) - g(x, y)]^2 \right.$$
$$\left. + \left(\frac{\partial^2 f(x, y)}{\partial x^2}\right)^2 + \left(\frac{\partial^2 f(x, y)}{\partial y^2}\right)^2 + \left(\frac{\partial^2 f(x, y)}{\partial y\, \partial x}\right)^2 \right\}. \tag{12.9}$$

In this case the smoothed field $f(x, y)$ corresponds to the height of a metallic thin plate attached to the observations $g(x, y)$ by linear springs. Again, the parameter λ measures the stiffness of the thin-plate model or conversely (as in the last equation) the looseness of the linear spring connecting the thin plate (filtered field) to the observed data.

To optimize the cost functionals shown above using a digital computer one needs first to discretize the cost functional. Therefore, the functions $f(x, y)$ and

$g(x, y)$ are now defined on the nodes of a regular lattice L, so the integrals become sums over the domain of interest, i.e.,

$$U[f(x, y)] = \sum_{(x,y) \in S} \sum \{[f(x, y) - g(x, y)]^2 + \lambda R_1[f(x, y)]\}, \tag{12.10}$$

where S is the subset of L where observations are available. The discrete version of the first-order regularizer $R_1[f(x, y)]$ may be approximated by

$$R_1[f(x, y)] = [f(x, y) - f(x - 1, y)]^2 + [f(x, y) - f(x, y - 1)]^2 \tag{12.11}$$

and the second-order regularizer $R_2[f(x, y)]$ may be approximated by

$$\begin{aligned} R_2[f(x, y)] = &[f(x + 1, y) - 2f(x, y) + f(x - 1, y)]^2 + [f(x, y + 1) - 2f(x, y) \\ &+ f(x, y - 1)]^2 + [f(x + 1, y + 1) - f(x - 1, y - 1) \\ &+ f(x - 1, y + 1) - f(x + 1, y - 1)]^2. \end{aligned} \tag{12.12}$$

By considering only the first two terms of the second-order regularizer one may reduce significantly the computational load of the filtering process, while preserving a thin-plate-like behavior.

A simple way to optimize the discrete cost functions stated in this section is by gradient descent, i.e.,

$$f^{k+1}(x, y) = f^k(x, y) - \eta \frac{\partial U[f(x, y)]}{\partial f(x, y)}, \tag{12.13}$$

where k is the iteration number and $\eta \approx 0.1$ is the step size of the gradient search. Although this is a simple optimizing technique it is a slow procedure specially for high-order regularizers. One may use instead conjugate gradient methods to speed up the optimizing process.

Let us point out a possible practical way of implementing in a digital computer the derivative of the cost function $U[f(x, y)]$ using a irregularly shaped domain S. Let us define an indicator function $m(x, y)$ in the lattice L having $N \times M$ nodes. The indicator function $m(x, y) = 1$ if the pixel at (x, y) is inside S (valid observations) and 0 otherwise. Using this indicator field, the filtering problem with a first-order regularizer may be rewritten as

$$U[f(x, y)] = \sum_{x=0}^{N-1} \sum_{y=0}^{M-1} \{[f(x, y) - g(x, y)]^2 m(x, y) + \lambda R[f(x, y)]\}, \tag{12.14}$$

where

$$\begin{aligned} R_1[f(x, y)] = &[f(x, y) - f(x - 1, y)]^2 m(x, y) m(x - 1, y) \\ &+ [f(x, y) - f(x, y - 1)]^2 m(x, y) m(x, y - 1); \end{aligned} \tag{12.15}$$

then the derivative may be found as

$$\frac{\partial U[f(x, y)]}{\partial f(x, y)} = [f(x, y) - g(x, y)]m(x, y) + \lambda[f(x, y)$$

$$-f(x - 1, y)]m(x, y)m(x - 1, y) + \lambda[f(x = 1, y)$$
$$-f(x, y)]m(x + 1, y)m(x, y)$$
$$+ \lambda[f(x, y) - f(x, y - 1)]m(x, y)m(x, y - 1) + \lambda[f(x, y + 1)$$
$$-f(x, y)]m(x, y + 1)m(x, y).$$

$$(12.16)$$

As we can see, only the difference terms lying completely within the region of valid fringe data marked by $m(x, y)$ survive. In other words, the indicator field $m(x, y)$ is the function that actually decouples valid fringe data from its surrounding background.

Finally, one may consider regularization filters as being more robust than convolution filters in the following sense:

- Unlike convolution filters, the data outside the filtering area S do not affect the filtering process inside S; i.e., no cross-talking occurs between the filtered field inside S and its surrounding data in L. In other words, the edge effect of regularized filters at the boundary of S is minimized. This is specially important when dealing with irregular-shaped regions.
- They tolerate missing observations due to the capacity of these filters to interpolate over regions of missing data with a well-defined behavior. The interpolating behavior of the filter is given by the form of the regularization term.
- By modifying the potentials in the cost function one may obtain many different types of filters, such as quadrature bandpass filters (QFs), which are very important in fringe analysis as phase demodulators.

Frequency Response of Low-Pass Regularized Filters

The filtered field $f(x, y)$ that minimizes the cost functions seen in the previous section smooths out the observation field $g(x, y)$. To have a quantitative idea of the amount of smoothing one may find the frequency response of the regularizer. To see the frequency response of the first-order low-pass filter (Marroquin *et al.*, 1997a, 1997b) consider an infinite two-dimensional lattice. Setting the gradient of the cost function to 0 one obtains the following set of linear equations:

$$f(x, y) - g(x, y) + \lambda[-f(x - 1, y) + 2f(x, y) - f(x + 1, y) - f(x, y - 1)$$
$$+ 2f(x, y) - f(x, y + 1)] = 0$$

$$(12.17)$$

and taking the discrete Fourier transform of the last equation one obtains

$$G(\omega_x, \omega_y) = F(\omega_x, \omega_y)[1 + 2\lambda(2 - \cos\omega_x - \cos\omega_y)]; \qquad (12.18)$$

this leads to the following transfer function:

$$H(\omega_x, \omega_y) = \frac{F(\omega_x\omega_y)}{G(\omega_x, \omega_y)} = \frac{1}{1 + 2\lambda[2 - \cos(\omega_x) - \cos(\omega_y)]}, \qquad (12.19)$$

which represents a low-pass filter with a bandwidth controlled by the parameter λ.

12.4 TEMPORAL PHASE-MEASURING METHODS

12.4.1 Introduction

Phase-stepping interferometry (PSI) is the most common technique used to detect the modulating phase of interferograms. Given the high popularity of this method, there have been many researches contributing in this area of interferometry (Shagam, 1978; Morgan, 1982; Schwider *et al.*, 1983; Greivenkamp, 1984; Ai and Wyatt, 1987; Asundi and Yung 1991; Creath, 1991; Joenathan and Khorana, 1992; Robinson and Reid, 1993; Yamaguchi *et al.*, 1996). The PSI method was first introduced by Bruning *et al.* (1974) for testing optical components using a video CCD array to map, over a large number of points, the optical wavefront under analysis. In the PSI technique (Greivenkamp and Bruning, 1992) an interference pattern is phase stepped under computer control and spatially digitized over several phase steps. The digitized intensity values may then be linearly combined to detect the optical phase at every pixel in the interferogram. Since then, a number of algorithms have been used to recover the wavefront phase of interference patterns, for which the emphasis is laid on using a small number of phase-shifted samples. The measured intensity of a CCD detector may be written as

$$I(x, y, n\alpha) = a(x, y) + b(x, y)\cos[\phi(x, y) + n\alpha], \qquad n = -N, \ldots, 0, \ldots, N,$$
(12.20)

where $\phi(x, y)$ is the phase to be determined, $a(x, y)$ is the slowly varying background illumination, and $b(x, y)$ is the contrast of the interference fringes. The parameter α is the phase step among the interferograms obtained by linearly varying the path difference between the test and reference beams.

All the phase-shifting algorithms proposed work fine whenever the following ideal conditions are met:

- The light-intensity range should be within the linear range of the CCD.
- The phase-shifted interferograms are taken at exactly the right phase shift.
- The device used to move the reference beam moves linearly in a purely piston fashion.
- All the mechanical perturbation of the interferometer (vibrations, air turbulence) are small during the capture of the digital interferograms.

If the above requirements are fulfilled, then one may expect a reliable phase determination using any PSI interferometry method. Nevertheless, sometimes one or several conditions mentioned above are not met. In this case, we need a robust temporal quadrature filter. To evaluate the merits of different PSI formulas, one may analyze the frequency response of these PSI formulas; this was first done by Freischlad and Koliopoulos (1990).

As mentioned, the first important contribution to the understanding of the different flavors of phase shifting formulas interpreted as linear quadrature filters was given by Freischlad and Koliopoulos (1990). These researchers investigated in the Fourier domain several commonly used formulas in phase shifting interferometry. Using this frequency domain analysis one is able to classify and decide the best algorithm available in the literature for the experimental setup and/or interferometric data at hand. If no one of the available phase shifting formulas meet your experimental requirements one may also synthesize the one that best fit your particular needs. Such procedures were proposed by Surrel (1996), Tang (1996), and Servin *et al.* (1997a).

12.4.2 General Theory of PSI

The phase estimation by phase stepping is achieved by convolving in the time axis a discrete temporal quadrature filter with several phase-shifted interferograms. Here we will only consider PSI formulas involving equally and uniformly spaced phase steps. Suppose that we have the following equally spaced phase-stepped interferograms:

$$I(x, y, t) = \sum_{n=-N}^{N} \{a(x, y) + b(x, y) \cos[\phi(x, y) + t]\} \delta(t - n\alpha). \qquad (12.21)$$

Parameter α is the phase (time) step among the interferograms. Figure 12.5 shows four phase-shifted frinte patterns. The phase step among these fringe patterns is $\pi/2$.

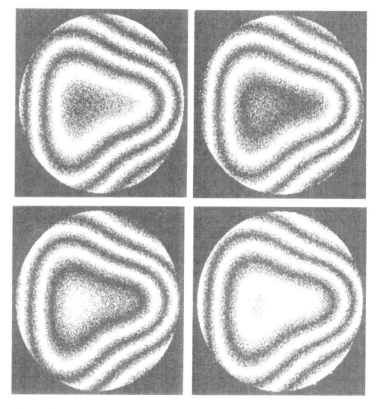

Figure 12.5 This figure shows four phase-shifted interferograms. The phase shift among them is $\pi/4$ radians.

To estimate the phase $\phi(x, y)$ of the above space-temporal signal $I(x, y, t)$ we need to filter it using a one-sided bandpass convolution filter tuned at the temporal frequency of the signal ($\omega = 1.0$). One-sided bandpass filters are normally called quadrature filters and are widely used in fringe analysis as phase estimators. A quadrature filter is a complex filter having the property that its output angle gives the searched phase. For example, consider the following discrete complex linear filter

$$h(t) = \sum_{n=-N}^{N} hr(n)\delta(t - n\alpha) + i \sum_{n=-N}^{N} hi(n)\delta(t - n\alpha), \tag{12.22}$$

where i is the square root of -1 and α is the sampling period of this discrete filter. This filter may be seen as being obtained from sampling a continuous filter $h(t)$ every α seconds. Quadrature filters are complex filters having a symmetric real component and an antisymmetric imaginary component, i.e.,

$$hr(n) = hr(-n), \qquad hi(n) = -hi(-n), \qquad hi(0) = 0; \tag{12.23}$$

taking the Fourier transform of the filter, the quadrature filter will have a frequency response given by

$$H(\omega) = \mathscr{F}\{h(t)\} = hr(0) + 2\sum_{n=1}^{N} hr(n)\cos(n\alpha\omega) + 2i\sum_{n=1}^{N} hi(n)\sin(n\alpha\omega). \tag{12.24}$$

The coefficients of this filter must be chosen in such a way to obtain a bandpass and one-sided frequency response. Filtering a sequence of $2N + 1$ interferograms $I(x, y, t)$ using the complex discrete filter shown above one obtains

$$g(x, y, t) = I(x, y, t) * h(t), \tag{12.25}$$

where the symbol * denotes a one-dimensional discrete convolution and $g(x, y, t)$ is the output sequence of complex-valued images. The convolution with the complex kernel of size $2N + 1$ temporal samples applied to a sequence of $2N + 1$ discrete temporal images generates $4N + 2$ complex images $g(x, y, n\alpha)$, but we are only interested in the phase of the complex signal located at the origin $g(x, y, 0)$ of this output sequence. The angle of the complex signal $g(x, y, 0)$ is the one that gives the most reliable phase determination. This is given by

$$\begin{aligned} g(x, y, 0) &= \sum_{n=-N}^{N} I(x, y, n\alpha)h(-n) \\ &= \sum_{n=-N}^{N} I(x, y, n\alpha)hr(n) - i\sum_{n=-N}^{N} I(x, y, n\alpha)hi(n), \end{aligned} \tag{12.26}$$

where we have used the symmetry relations of the complex filter. From the $2N + 1$ phase-shifted fringe patterns we have obtained two in quadrature fringe patterns, so the searched phase will be given by

$$\phi(x, y) = -\arctan\left(\frac{\sum_{n=-N}^{N} I(x, y, n\alpha)\, hi(n)}{\sum_{n=-N}^{N} I(x, y, n\alpha)\, hr(n)} \right). \tag{12.27}$$

The general theory of PSI using quadrature linear filters finishes here. But we have still not given any way for calculating the coefficients for the quadrature filter. Some examples of temporal quadrature linear filters used in phase stepping are presented in the next sections.

Given the limited number of phase-shifted interferograms that one usually obtains (from 3 to 20 typically), we have to make some comments about such a limited quadrature filter:

1. It is convenient that the quadrature linear filters used in PSI have the same number of coefficients as the temporal signal being filtered in order to use all the available information.
2. One needs to know the exact amount of phase step introduced; otherwise, a linear miscalibration would result. This linear phase error is called detuning; thus, it is convenient whenever possible to use a quadrature filter robust to detuning.
3. Sometimes we may have amplitude distortion in the frame-grabbed interferogram due to oversaturation or undersaturation of the CCD array used to obtain the video signal. In that case, harmonic signals are generated with a consequent degradation in the estimated phase. Therefore, it is convenient to also have a good rejection to the harmonics of the signal being filtered.

In the following sections we analyze several possible weights for the quadrature filter to obtain different types of phase-stepping formulas.

12.4.3 Performance of Some Commonly Used Phase-Shifting Methods

Before the frequency analysis of the linear quadrature filters used in PSI by Freischlad and Koliopoulos (1990), there was no systematic way to compare the different flavors of PSI formulas. In this section we are going to assign values to the quadrature filter coefficients and evaluate in the frequency domain some of the most used phase-detecting algorithms. The basic criteria to evaluate the performance of a PSI formula are

1. Robustness to detuning, that is linear phase miscalibration.
2. Harmonic signal rejection.
3. Highest possible noise rejection, i.e., small bandwidth of the quadrature filter.

Three-Step PSI

Given that in a fringe pattern one normally has three unknowns then, three interferograms is the practical minimum to make a PSI measurement. Let us begin with the form of the space-temporal signal:

$$I(x, y, t) = \sum_{n=-1}^{1} I(x, y, n\alpha)$$

$$= \sum_{n=-1}^{1} \{a(x, y) + b(x, y) \cos[\phi(x, y) + t]\} \delta(t - n\alpha),$$

(12.28)

where α is the phase step among the fringe patterns. In this equation we have three unknowns, namely $a(.)$, $b(.)$, and $\phi(.)$. Solving algebraically for $\phi(.)$ we obtain (Greivenkamp and Bruning, 1992)

$$\phi(x, y) = \arctan\left(\frac{(1 - \cos(\alpha))[I(-\alpha) - I(\alpha)]}{\sin(\alpha)[2I(0) - I(-\alpha) - I(\alpha)]}\right); \tag{12.29}$$

for clarity purposes the (x, y) dependance has been dropped from the irradiances. From here we may see that the linear quadrature filter may be written as

$$h(t) = \sin(\alpha)[2\delta(t) - \delta(t + \alpha) - \delta(t + \alpha)] + i(1 - \cos\alpha)[\delta(t - \alpha) - \delta(t + \alpha)]. \tag{12.30}$$

This quadrature filter has the following frequency response:

$$H(\omega) = Hr(\omega) + iHi(\omega) = (\sin\alpha)[2 - 2\cos(\alpha\omega)] + i2(1 - \cos\alpha)\sin(\alpha\omega). \tag{12.31}$$

In Fig. 12.6(a) the frequency response of the real and imaginary parts of this quadrature filter is shown: although they are always in quadrature, their magnitude is the same only for the nominal expected temporal frequency. In other words, we can only obtain a valid phase estimation if the real and imaginary parts of the complex filter

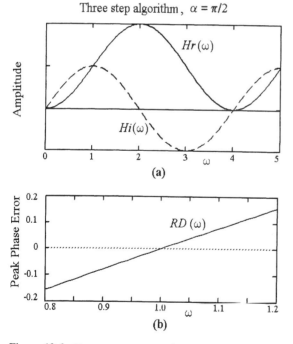

Figure 12.6 Frequency response of a three-step algorithm: the phase step (α) among the samples is $\pi/2$ radians. (a) Amplitude response of the real and imaginary parts of the quadrature filter used to estimate the phase. (b) Maximum phase error due to detuning. This PSI technique is very sensitive to detuning.

are in quadrature and have the same magnitude. This condition is fulfilled only for the temporal frequency equal to one. If for some reason the phase steps were not taken at exactly the required phase step, a detuning error will occur. The robustness to detuning $RD(\omega)$ of the three-step PSI algorithm is quite bad; it is given by

$$RD(\omega) = \text{atan}\left(\frac{Hi(\omega)}{Hr(\omega)}\right) - \text{atan}\left(\frac{Hi(1.0)}{Hr(1.0)}\right). \tag{12.32}$$

As seen from Fig. 12.6(b) the three-step phase-shifting algorithm is very weak to detuning. If you are not absolutely sure of having the exact phase steps, you must move to another quadrature filter. Finally, as Fig. 12.6(a) shows, the three sample quadrature filter has rejection to the fourth harmonic of the signal in the frequency range shown.

Four-Step PSI

Another widely used formula for phase stepping is the so-called four-frame method (Wyant, 1982). In this case the four recorded interferograms are given by

$$I(x, y, t) = \sum_{n=0}^{3}\left\{a(x, y) + b(x, y)\cos[\phi(x, y) + t]\right\}\delta(t - n\alpha), \alpha = \frac{\pi}{2}. \tag{12.33}$$

Solving for $\phi(x, y)$, the searched phase is given by

$$\phi(x, y) = \text{atan}\left(\frac{I(3\alpha) - I(\alpha)}{I(0) - I(2\alpha)}\right); \tag{12.34}$$

again, the spatial dependance of the fringe patterns has been removed for notation clarity. So the quadrature filter is given by

$$h(t) = \delta(t - 3\alpha) - \delta(t - \alpha) + i[\delta(t) - \delta(t - 2\alpha)]. \tag{12.35}$$

The frequency response of this four-frame quadrature filter is

$$H(\omega) = \mathscr{F}\{h(t)\} = [\exp(-i3\alpha\omega) - \exp(-i\alpha\omega)] + i[1 - \exp(-i2\alpha\omega)] \tag{12.36}$$

or, equivalently,

$$H(\omega) = \mathscr{F}\{h(t)\} = -2\sin(\alpha\omega)\exp(2\alpha\omega) + i2\sin(\alpha\omega)\exp(\alpha\omega). \tag{12.37}$$

This expression does not seem to represent a quadrature filter because both the amplitude of the real part and the amplitude of the imaginary part are the same (not in quadrature as expected) except at the angular frequency $\omega = 1.0$ where they are in quadrature. The main disadvantage of the mathematical representation of this filter is that its frequency response does not give an idea of how robust this filter is to detuning. To transform this filter into a filter having both components always in quadrature but with different amplitudes we may "shift" or more precisely "rotate" (in the sense described in Malacara *et al.*, 1998) the filter by $\pi/4$ radians. Therefore, the transformed-rotated filter may be written as

$$h'(t) = \delta(t + 3\alpha') - \delta(t + \alpha') - \delta(t - \alpha') + \delta(t - 3\alpha') + i[\delta(t + 3\alpha') + \delta(t + \alpha')$$
$$- \delta(t - \alpha') - \delta(t - 3\alpha')], \qquad \alpha' = \frac{\pi}{4}, \tag{12.38}$$

which has the following frequency response:

$$\frac{H'(\omega)}{2} = \cos(3\alpha'\omega) - \cos(\alpha'\omega) + i[\sin(3\alpha'\omega) + \sin(\alpha'\omega)].\qquad(12.39)$$

Now the filter looks more like a quadrature filter. Even though this quadrature filter has a very different mathematical form with respect to the one given before, it has the same properties regarding its phase-detection capabilities. The frequency response of the filter is shown in Fig. 12.7(a), and the robustness to detuning in Fig. 12.7(b). As can be seen from Fig. 12.7, this quadrature filter is quite sensitive to detuning; it is insensitive to the second, fourth, sixth, . . . harmonics of the signal.

Symmetrical Five-Step PSI

A very popular algorithm is the so-called Schwider–Hariharan five-step algorithm (Schwider *et al.*, 1983; Hariharan *et al.*, 1987). The five-phase stepped interferograms may be written as

$$I(x, y, t) = \sum_{n=-2}^{2} \{a(x, y) + b(x, y) \cos[\phi(x, y) + t]\}\delta(t - n\alpha), \qquad \alpha = \frac{\pi}{2}.$$

$$(12.40)$$

The formula to obtain the phase is

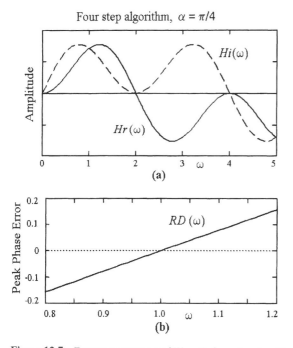

Figure 12.7 Frequency response of Wyant's four-step algorithm – the phase step among the samples is $\pi/4$ radians. (a) Amplitude response of the real and imaginary parts of the quadrature filter. (b) Phase error due to detuning.

$$\phi(x, y) = \operatorname{atan}\left(\frac{2[I(-\alpha) - I(\alpha)]}{I(-2\alpha) - 2I(0) + I(2\alpha)}\right), \tag{12.41}$$

where, for notation simplicity, the space dependance of the fringe pattern has been omitted. The quadrature filter associated with this formula is

$$h(t) = \delta(t + 2\alpha) - 2\delta(t) + \delta(t - 2\alpha) + i2[\delta(t + \alpha) - \delta(t - \alpha)]; \tag{12.42}$$

therefore, its frequency response is

$$H(\omega) = \mathscr{F}\{h(t)\} = Hr(\omega) + iHi(\omega) = 2\sin(\alpha\omega) + i[1 - \cos(2\alpha\omega)]. \tag{12.43}$$

The frequency response of the Schwider–Hariharan algorithm is shown in Fig. 12.8(a). This quadrature filter is insensitive to even harmonics of the signal. The real and imaginary components are always orthogonal and have the same magnitude (reliable phase estimation) around the frequency $\omega = 1.0$. This means that the symmetrical five-step PSI filter is robust to detuning, so it can tolerate small linear phase miscalibrations and still give a reliable phase estimation. The robustness to detuning $RD(\omega)$ may be found by

$$RD(\omega) = \operatorname{atan}\left(\frac{Hi(\omega)}{Hr(\omega)}\right) - \operatorname{atan}\left(\frac{Hi(1.0)}{Hr(1.0)}\right), \tag{12.44}$$

and it is graphically shown in Fig. 12.8(b).

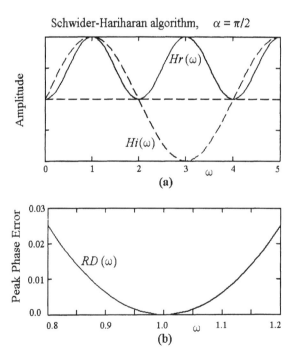

Figure 12.8 Frequency response of the five-step Schwider–Hariharan PSI algorithm using a phase step of $\pi/2$ radians. (a) Amplitude response of the real and imaginary parts of the quadrature filter. (b) Maximum phase error due to detuning.

Discussion

Before leaving this section, one cannot finish without mentioning some additional important works in temporal phase shifting, such as the ones made by Morgan (1982), Greivenkamp (1984), Groot (1995), Hibino *et al.* (1995), Schmith and Creath (1995a, 1996), and, finally, the recent work of Hibino and Yamaguchi (1998), which presents the following algorithm:

$$\tan\phi = \sqrt{3}\frac{5I(-3\alpha) - 6I(-2\alpha) - 17I(-\alpha) + 17I(\alpha) + 6I(2\alpha) - 5I(3\alpha)}{I(-3\alpha) - 26I(-2\alpha) - 25I(-\alpha) + 25I(\alpha) + 26I(2\alpha) + I(3\alpha)}.$$

(12.45)

There are still a lot more phase-stepping formulas. All of them have their particular merit. To find the best one for your needs you may review all the literature regardingt this or you may instead design your own PSI quadrature filter to fit your particular needs in terms of the number of phase-shifted interferograms, amount of phase shifting, detuning robustness, harmonic rejection, or noise removal. As it will be seen in the next section all these factors may be optimized to obtain the best possible linear quadrature filter given the number of phase shifted interferograms at hand.

Although many sources of error exist in taking several phase-shifted interferograms, the most common of these in the experimental implementation of phase-shifting interferometry is detuning. Having a detuning error means that the sampling of the temporal signal is uniform but does not occur at a phase spacing of the proper fraction of the signal period. Detuning is referred to as a linear phase shift miscalibration, in which, for example, the phase shift between the samples might be another quantity of degrees instead of the expected phase step. As seen above, the linear quadrature filter found by Schwider *et al.* (1983) and Hariharan *et al.* (1987) is robust to a small amount of detuning.

Regularizing methods can also be used when several phase-shifting interferograms are available (Marroquin *et al.*, 1998; Servin *et al.*, 1998b). This is revised in another section.

12.4.5 Synthesis of Quadrature Linear Filters for Phase Shifting

Given a number of phase-shifted fringe patterns with a constant phase shift α among them, there is an infinite number of quadrature filters (given by a different set of coefficients) that may estimate the desired phase. Up to now we have analyzed several quadrature filters and their properties in terms of harmonic rejection and phase detuning. Another more efficient method is to find the optimum quadrature filter coefficients given some desired properties in the frequency response of this filter (Servin *et al.*, 1997a).

Assume that we want to find the coefficients of the following linear quadrature filter in an optimum way:

$$h(t) = \sum_{n=-N}^{N} hr(n)\delta(t - n\alpha) + i \sum_{n=-N}^{N} hi(n)\delta(t - n\alpha),$$

(12.46)

where $hr(n)$ and $hi(n)$ are the filter coefficients. The filter given by the last equation is, in general, not in quadrature. To simplify our analysis assume that the real and imaginary parts of the filter are symmetrical and antisymmetrical, respectively. Then, the Fourier transform of the complex filter is

$$\mathscr{F}\{h(t)\} = Hr(\omega) + iHi(\omega) = hr(0) + 2\sum_{n=1}^{N} hr(n)\cos(n\alpha\omega) + 2i\sum_{n=1}^{N} hi(n)\sin(n\alpha\omega).$$

$$(12.47)$$

Given some predefined phase (time) steps α, we may define the required complex filter (that must be in quadrature at least in $\omega_0 = 1.0$) in terms of the coefficients $hr(n)$ and $hi(n)$ as the minimizer of the following cost function:

$$U = \lambda_0 Hr(0)^2 + \lambda_1 \int_{\omega=1.0-\Delta_1}^{1.0+\Delta_1} [Hr(\omega) - Hi(\omega)]^2 d\omega + \lambda_2 \int_{\omega=2.0-\Delta_2}^{2.0+\Delta_2}$$

$$[Hr(\omega)^2 + Hi(\omega)^2] + \cdots,$$

$$(12.48)$$

where λ_1, λ_2 are numbers that weight the different requirements that the filter must fulfill. The filter which minimizes this cost function will have several good properties. One of them is that the real part of the filter has minimum response at zero frequency (the imaginary part of $Hi(\omega)$ is already zero at the origin). The second requirement states that the complex filter must remain in quadrature at least in a frequency band equal to $2\Delta_1$ centered at the nominal or expected time-angular frequency (detuning robustness). The remaining terms deal with minimizing the filter response around some expected harmonic components that the temporal signal may have. Therefore the actual values needed for the complex filter given some predefined phase steps α may be obtained by optimizing the cost function U for the available free parameters $hr(n)$ and $hi(n)$. This may be accomplished by solving the following linear system of equations for the filter coefficients as

$$\frac{\partial U}{\partial hr(n)} = 0;$$

$$\frac{\partial U}{\partial hi(n)} = 0.$$

$$(12.49)$$

Having calculated our quadrature filter coefficients we may find the required phase information as

$$\phi(x, y) = \operatorname{atan}\left(\frac{\sum_{n=-N}^{N} hi(n)I(x, y, n\alpha)}{\sum_{n=-N}^{N} hr(n)I(x, y, n\alpha)}\right).$$

$$(12.50)$$

It must be remarked that having a summing notation running from $-N$ to N does not necessarily mean that we need $2N + 1$ interferogram samples; we may have, for example, four phase-centered samples.

As an example of this optimizing method, let us consider seven equally spaced sampling points with a phase interval of $\pi/2$ and optimize for detuning, using the following weights

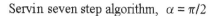

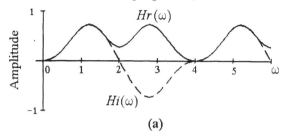

(a)

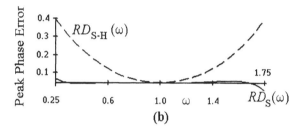

(b)

Figure 12.9 Frequency response of the seven-step Servin's PSI algorithm using a phase step of $\pi/2$ radians. (a) Amplitude response of the real and imaginary parts of the quadrature filter. (b) Maximum phase error due to detuning of the Servin's algorithm $RD_S(\omega)$ compared against the detuning error of the Schwider–Hariharan algorithm $RD_{S-H}(\omega)$.

$$\begin{aligned} \lambda_0 &= 1.0, \quad \lambda_1 = 1.0 \\ \lambda_2 &= 0.01 \\ \lambda_3 &= \lambda_4 \ldots = 0 \\ \Delta_1 &= 0.8 \\ \Delta_2 &= 0.1. \end{aligned} \tag{12.51}$$

With these parameters we define a quadrature filter with a large detuning robustness and some attenuation in the second harmonics. The solution of the linear system gives the following phase-estimation equation:

$$\tan\phi = \frac{I(-3\alpha) + 4.3I(-2\alpha) - 14I(-\alpha) + 14I(\alpha) - 4.3I(2\alpha) - I(3\alpha)}{1.5I(-3\alpha) - 6I(-2\alpha) - 4.5I(-\alpha) + 18I(0) - 4.5I(\alpha) - 6I(2\alpha) + 1.5I(3\alpha)}. \tag{12.52}$$

The frequency response of the real and imaginary parts of this filter is shown in Fig. 12.9(a). The detuning robustness is shown in Fig. 12.9(b) and it is compared against the Schwider–Hariharan PSI formula to have an idea of the superior detuning robustness.

12.5 SPATIAL PHASE-MEASURING METHODS

As seen before, PSI uses a sequence of phase-shifted interferograms to find the searched phase. The phase-estimating system is a one-dimensional quadrature filter tuned at the fundamental time frequency of the signal. Sometimes we cannot have several phase-shifted interferograms. In such cases one needs to deal with only one interferogram with either closed or open fringes (Ichioka and Inuiya, 1972; Macy, 1983; Mertz, 1983; Shough, 1990; Patorski, 1993). Open fringes may be obtained by introducing a large tilt in the reference beam of a two-path interferometer or by projecting a linear ruling in profilometry. A single carrier frequency interferogram (an open fringe interferogram) is much easier to demodulate than an interferogram which has closed fringes. Closed-fringe interferograms are difficult to demodulate due to the nonmonotonic variation of the phase field within the fringe pattern.

An open-fringe interferogram can always be written as

$$I(x, y) = a(x, y) + b(x, y)\cos[\omega_0 x + \phi(x, y)]. \tag{12.53}$$

The carrier frequency must be higher than the maximum frequency content of the phase in the carrier direction, i.e.,

$$\omega_0 > \frac{\partial \phi(x, y)}{\partial x}; \tag{12.54}$$

this condition ensures that the total phase of the interferogram will grow (or decrease) monotonically, so the slope of the total phase will always be positive (or negative).

Figure 12.10 shows a carrier frequency interferogram phase modulated by spherical aberration and defocusing.

12.5.1 The Fourier-Transform Method for Carrier Frequency Fringe Patterns

One may also use frequency domain techniques to estimate the modulating phase of fringe patterns. The phase estimation using the Fourier method is due to Takeda *et al.* (1982). This method is based in a bandpass quadrature filter in the frequency domain. The method works as follows: rewrite the carrier frequency interferogram given above as

$$I(x, y) = a(x, y) + c(x, y)\exp(i\omega_0 x) + c*(x, y)\exp(-i\omega_0 x), \tag{12.55}$$

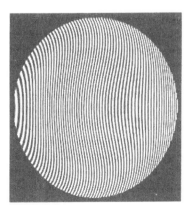

Figure 12.10 A typical carrier frequency interferogram.

with

$$c(x, y) = \frac{b(x, y)}{2} \exp[i\phi(x, y)]$$ (12.56)

and * denotes the complex conjugate. The Fourier transform of this signal is then

$$\mathcal{F}\{I(x, y)\} = A(\omega_x, \omega_y) + C(\omega_x + \omega_0, \omega_y) + C^*(\omega_x - \omega_0, \omega_y).$$ (12.57)

After this, using a quadrature bandpass filter, one keeps only one of the two $C(.)$ terms of the frequency spectrum; therefore,

$$C(\omega_x + \omega_0, \omega_y) = H(\omega_x + \omega_0, \omega_y)\mathcal{F}\{I(x, y)\},$$ (12.58)

where $H(\omega_x + \omega_0, \omega_y)$ represent a quadrature bandpass filter centered at $-\omega_0$ and with a bandwidth large enough to contain the spectrum of $C(\omega_x + \omega_0, \omega_y)$. Then, the following step is either to translate the information peak toward the origin to remove the carrier frequency ω_0 to obtain $C(\omega_x, \omega_y)$ or Fourier transform it to find directly the inverse Fourier transform of the filtered signal, so that one gets alternatively:

$$\mathcal{F}^{-1}\{C(\omega_x, \omega_y)\} = \frac{b(x, y)}{2} \exp[i\phi(x, y)],$$
$$\mathcal{F}^{-1}\{C(\omega_x - \omega_0, \omega_y)\} = \frac{b(x, y)}{2} \exp[i(\omega_0 x + \phi(x, y))],$$ (12.59)

so their respective phase is given by

$$\phi(x, y) = \mathrm{atan}\left(\frac{\mathrm{Im}[c(x, y)]}{\mathrm{Re}[c(x, y)]}\right),$$
$$\omega_0 x + \phi(x, y) = \mathrm{atan}\left(\frac{\mathrm{Im}[c(x, y)\exp(-i\omega_0 x)]}{\mathrm{Re}[c(x, y)\exp(-i\omega_0 x)]}\right).$$ (12.60)

In the second case the estimated phase $\phi(x, y)$ is usually obtained after a plane-fitting procedure performed on the total phase function. Of course, the estimated phase given above is wrapped because of the atan(.) function involved. Therefore, the last step in this process is to unwrap the phase. Figure 12.11(a) shows the frequency spectrum of the carrier frequency interferogram shown in Fig. 12.10. The detected phase using the phase information provided by keeping only one of the two side lobes of this spectrum is shown in Fig. 12.11(b). We can see how the recovered phase has some distortion at the boundary of the pupil. This phase distortion at the edge of the fringe pattern may be reduced by apodizing the fringe pattern intensity using, for example, a Hamming window (Malacara *et al.*, 1998).

While using the Fourier-transform method, which involves the global but pointwise operation on the interferogram spectrum, one has to be aware of the main sources of errors (Bone *et al.*, 1986; Roddier and Roddier, 1987; Takeda *et al.*, 1982; Kujawinska and Wójciak, 1991a,b):

- The errors associated with the use of fast Fourier transform (FFT): aliasing if the sampling frequency is too low; the picket fence effect, if the analyzed frequency includes a frequency which is not one of the discrete frequencies; and, the most significant error, leakage of the energy from one frequency into adjacent ones, due to fringe discontinuity or inappropriate truncation of the data.

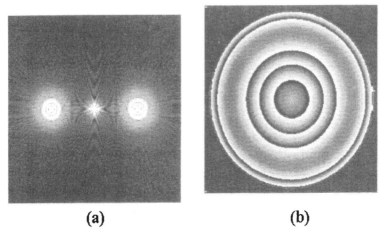

(a)   **(b)**

Figure 12.11   Phase detection using Takeda's Fourier technique. (a) Frequency spectrum of
the interferogram shown in Fig. 12.10. (b) Estimated phase of the quadrature filter.

- The errors due to incorrect filtering in the Fourier space, especially if a
  nonlinear recording of the fringe pattern has occurred.
- The influence of random noise and spurious fringes in the interferogram.

Here we refer to the most significant errors. The leakage of the energy can be reduced
significantly by apodizing the fringe pattern intensity using Haming, Hanning, bell,
or $\cos^4$ windows. The errors due to incorrect filtering are sometimes difficult to
avoid, especially if the assumption of $[\partial\phi(r)/\partial r]_{max} \ll f_0$ is not fulfilled and minimizing the filtering window is required due to noise in the image. Bone *et al.* (1986) have
shown that with an optimum filter window, the errors due to noise are approximately equal to the errors from information components lost from the filter window.

The Fourier transform method can also be modified by a technique proposed
by Kreis (1986). He transformed an interferogram without a spatial frequency carrier
added (but with a certain linear phase term intrinsic to the data) and obtained a
complex analytic signal by applying a filter function which covers nearly a half plane
of the Fourier space, which gives us the possibility of evaluating more complex fringe
patterns.

12.5.2   Spatial Carrier Phase Shifting Method

The spatial carrier phase shifting method (SCPI) is based on the use of the same
phase-stepping quadrature filters used in temporal PSI but in the space domain. So
the most simple quadrature filter to use is the three-step filter (Shough *et al.*, 1990).
This filter along the x-direction looks like

$$h(x) = hr(x) + ihi(x) = \sin(\omega_0)[2\delta(x) - \delta(x + \alpha) - \delta(x + \alpha)]$$
$$+ i(1 - \cos\omega_0)[\delta(x - \alpha) - \delta(x + \alpha)]. \tag{12.61}$$

When this filter is convolved with a carrier frequency fringe pattern given by

$$I(x, y) = a(x, y) + b(x, y)\cos[\omega_0 x + \phi(x, y)], \tag{12.62}$$

one obtains two fringe patterns in quadrature, i.e.,

$$g_1(x, y) + ig_2(x, y) = I(x, y) * hr(x) + iI(x, y) * hi(x). \tag{12.63}$$

So, the interesting phase may be estimated as

$$
\begin{aligned}
\phi(x, y) &= \operatorname{atan}\left(\frac{g_i(x, y)}{g_r(x, y)}\right) \\
&= \operatorname{atan}\left(\frac{1 - \cos(\omega_0)}{\sin(\omega_0)} \frac{I(x - 1, y) - I(x + 1, y)}{2I(x, y) - I(x - 1, y) - I(x + 1, y)}\right).
\end{aligned}
\tag{12.64}
$$

But, as we mentioned before, the three-step PSI filter has the disadvantage of being too sensitive to detuning. The detuning weakness of the three-step algorithm is evident when dealing with wide-band carrier-frequency interferograms. For that reason, one may use a more robust to detuning PSI algorithm, such as the five-step Schwider–Hariharan formula or the seven-step quadrature filter presented by Servin *et al.* (1995). The main inconvenience of using a larger-size convolution filter is that the first two or three pixels inside the boundary of the fringe pattern are not going to be phase estimated.

One possible solution is to stick with three samples but, instead of assuming a constant phase over three consecutive pixels, we will make a correction due to the instantaneous frequency variation within the three sample window. This was made by Ranson and Kokal (1986) and later, independently, by Servin and Cuevas (1995). The first step is to filter the interferogram $I(x, y)$ with a high-pass filter in order to eliminate the DC term $a(x, y)$. Now consider the following three consecutive pixels of the high-pass filtered fringe pattern $I'(x, y)$:

$$
\begin{aligned}
I'(x - 1, y) &= b(x - 1, y)\cos[\omega_0(x - 1) + \phi(x - 1, y)], \\
I'(x, y) &= b(x, y)\cos[\omega_0 x + \phi(x, y)], \tag{12.65} \\
I'(x + 1, y) &= b(x - 1, y)\cos[\omega_0(x + 1) + \phi(x + 1, y)].
\end{aligned}
$$

Assuming that the modulating function $b(x, y)$ remains constant over three consecutive pixels, and using a first-order approximation of $\phi(x, y)$ around the pixel at (x, y) along the x-axis, we obtain

$$
\begin{aligned}
I'(x - 1, y) &= b(x, y)\cos\left[\omega_0 x + \phi(x, y) - \omega_0 - \frac{\partial\phi(x, y)}{\partial x}\right], \\
I'(x, y) &= b(x, y)\cos[\omega_0 x + \phi(x, y)], \tag{12.66} \\
I'(x + 1, y) &= b(x, y)\cos\left[\omega_0 x + \phi(x, y) + \omega_0 + \frac{\partial\phi(x, y)}{\partial x}\right].
\end{aligned}
$$

In these equations we have three unknowns – namely, $b(x, y)$, $\phi(x, y)$, and $\partial\phi(x, y)/\partial x$ – so we may solve for $\phi(x, y)$ as

$$
\tan[\omega_0 x + \phi(x, y)] = \left(\frac{I'(x - 1, y) - I'(x + 1, y)}{\operatorname{sgn}[I'(x, y)]\sqrt{[2I'(x, y)]^2 - [I'(x - 1, y) + I'(x + 1, y)]^2}}\right),
\tag{12.67}
$$

where the function sgn[.] takes the sign of its argument. For a more detailed discussion of this method see Servin *et al.* (1995) or, in general, the n-point phase-shifting technique, as explained by Schmith and Creath (1995b) and Küchel (1997). A similar approach, with the assumption of constancy of the first derivative of phase within the convolution filter, is also used to the five-point algorithm (Pirga and Kujawinska, 1996), together with the concept of the two-directional spatial-carrier phase-shifting method (Pirga and Kujawinska, 1995). This last approach allows us to analyze multiplexed information coded into a fringe pattern.

12.5.3 Synchronous Spatial Phase Detection of Carrier Frequency Fringe Patterns

The method of synchronous spatial phase detection was first introduced in digital form by Womack (1984). To analyze this phase-demodulating system, consider as usual the following carrier frequency fringe pattern:

$$I(x, y) = a + b\cos[\omega_0 x + \phi(x, y)]. \tag{12.68}$$

The dependance on the spatial coordinates of $a(x, y)$ and $b(x, y)$ will be omitted in this section for notation clarity. This fringe pattern is now multiplied by the sine and cosine of the carrier phase as follows:

$$g_r(x, y) = I(x, y)\cos(\omega_0 x) = a\cos(\omega_0 x) + b\cos(\omega_0 x)\cos[\omega_0 x + \phi(x, y)],$$
$$g_i(x, y) = I(x, y)\sin(\omega_0 x) = a\sin(\omega_0 x) + b\sin(\omega_0 x)\cos[\omega_0 x + \phi(x, y)]; \tag{12.69}$$

this may be rewritten as

$$g_r(x, y) = a\cos(\omega_0 x) + \frac{b}{2}\cos[2\omega_0 x + \phi(x, y)] + \frac{b}{2}\cos[\phi(x, y)],$$
$$g_i(x, y) = a\sin(\omega_0 x) + \frac{b}{2}\sin[2\omega_0 x + \phi(x, y)] - \frac{b}{2}\sin[\phi(x, y)]. \tag{12.70}$$

To obtain the searched phase $\phi(x, y)$ we have to low-pass filter the signals $g_r(x, y)$ and $g_i(x, y)$ to eliminate the two first high-frequency terms. Finally to find the searched phase we need to find their ratio as

$$\phi(x, y) = \mathrm{atan}\left(\frac{g_i(x, y) * *h(x, y)}{g_i(x, y) * *h(x, y)}\right) = \mathrm{atan}\left(\frac{-(b/2)\sin[\phi(x, y)]}{(b/2)\cos[\phi(x, y)]}\right), \tag{12.71}$$

where $h(x, y)$ is a low-pass convolution low-pass filter.

12.5.4 Robust Quadrature Filters

A robust quadrature bandpass filter may be obtained by simply shifting in the frequency domain the regularizing potentials seen in Section 12.3.3 to the carrier frequency ω_{0x} of the fringe pattern (Marroquin *et al.*, 1997a). That is

$$U[f(x, y)] = \sum\sum_{(x,y)\in S}\{[f(x, y) - 2g(x, y)]^2 + \lambda R_1[f(x, y)]\}, \tag{12.72}$$

in which the first-order regularizer is now

$$R_1[f(x, y)] = [f(x, y) - f(x - 1, y)\exp(-\omega_{0x}x)]^2 + [f(x, y) - f(x, y - 1)]^2. \tag{12.73}$$

where we have shifted the first-order regularizer in the x-direction. The minimizer of this cost function given the observation field $2g(x, y)$ is a quadrature bandpass filter. To see this, let us find the frequency response of the filter that minimizes the above cost function; consider an infinite two-dimensional lattice. Setting the gradient of $U[f(x, y)]$ to zero, one obtains the following set of linear equations:

$$0 = f(x, y) - 2g(x, y) + \lambda[-f(x - 1, y)\exp(i\omega_{0x}) + 2f(x, y)$$
$$- f(x + 1, y)\exp(i\omega_{0x})] + \lambda[-f(x, y - 1) + 2f(x, y) - f(x, y + 1)],$$

(12.74)

and taking the discrete Fourier transform of this equation, one obtains

$$[1 + 2\lambda[2 - \cos(\omega_x - \omega_{ox}) - \cos(\omega_y)]F(\omega) = G(\omega),$$ (12.75)

which leads to the following transfer function $H(\omega)$:

$$H(\omega_x, \omega_y) = \frac{F(\omega_x, \omega_y)}{G(\omega_x, \omega_y)} = \frac{1}{1 + 2\lambda[2 - \cos(\omega_x - \omega_{ox}) - \cos(\omega_y)]},$$ (12.76)

which is a bandpass quadrature filter centered at the frequency ($\omega_x = \omega_{0x}, \omega_y = 0$) with a bandwidth controlled by the parameter λ. As this frequency response shows the form of the filter, it is exactly the same as the membrane low-pass filter studied in Section 12.3.3 but moved in the frequency domain to the coordinates (ω_{0x}, 0). So this filter may be used for estimating the phase of a carrier frequency interferogram.

An even better signal-to-noise ratio and edge-effect immunity may be obtained if one lets the tuning frequency vary in the two-dimensional space (i.e., $\omega_{0x} = \omega_x(x, y)$); that is,

$$R_f[f(x, y)] = [f(x, y) - f(x - 1, y)\exp(-i\omega_x(x, y))]^2 + [f(x, y) - f(x, y - 1)]^2.$$

(12.77)

Using this regularizer, one obtains an adaptive quadrature filter (Marroquin *et al.*, 1997b). In this case one must optimize not only for the filtered field $f(x, y)$ but also for the two-dimensional frequency field $\omega_x(x, y)$. Additionally, if we want the estimated frequency field $\omega_x(x, y)$ to be smooth, one needs also to use a regularizer for this field: for example, a first-order regularizer,

$$R_{\omega x}[\omega_x(x, y)] = [\omega_x(x, y) - \omega_x(x - 1, y)]^2 + [\omega_x(x, y) - \omega_x(x, y - 1)]^2.$$ (12.78)

The final cost function will have the following form:

$$U[f(x, y)] = \sum_{(x,y)\in S}\sum \{[f(x, y) - 2g(x, y)]^2 + \lambda_1 R_f[f(x, y)] + \lambda_2 R_{\omega x}[\omega_x(x, y)]\}.$$

(12.79)

Unfortunately, this cost function contain a nonlinear quadrature term (the $R_f[\omega_x(x, y)]$ term), so the use of fast convergence techniques such as conjugate gradient or transformed methods are precluded. One needs then to optimize this cost function following simple gradient search or Newtonian descent (Marroquin *et al.*, 1997b).

Finally, we may also optimize for the frequency in the y-direction $\omega_y(x, y)$. By estimating $\omega_x(x, y)$ and $\omega_y(x, y)$ altogether, it is possible to demodulate a single fringe pattern containing closed fringes (see Marroquin *et al.*, 1997b).

12.5.5 The Regularized Phase-Tracking Technique

The regularized phase-tracking (RPT) technique may be applied to almost every aspect of the fringe pattern processing. The RPT evolved from an early phase-locked loop (PLL) technique that was applied for the first time to fringe processing by Servin *et al.* (1993, 1994). In the RPT (Servin *et al.*, 1997b) technique, one assumes that locally the phase of the fringe pattern may be considered as spatially mono-chromatic, so its irradiance may be modeled as a cosinusoidal function phase modulated by a plane $p(.)$. Additionally, this phase plane $p(.)$ located at (x, y) must be close to the phase values $\phi_0(\epsilon, \eta)$ already detected in the neighborhood of the site (x, y).

Specifically, the proposed cost function to be minimized by the estimated phase $\phi_0(x, y)$ at each site (x, y), is

$$U(x, y) = \sum_{(\epsilon, \eta) \in (N_{x,y} \cap S)} \left\{ [I'(\epsilon, \eta) - \cos p(x, y, \epsilon, \eta)]^2 + \lambda[\phi_0(\epsilon, \eta) \right. \tag{12.80}$$
$$\left. -p(x, y, \epsilon, \eta)]^2 m(\epsilon, \eta) \right\}$$

and

$$p(x, y, \epsilon, \eta) = \phi_0(x, y) + \omega_x(x, y)(x - \epsilon) + \omega_y(x, y)(y - \eta), \tag{12.81}$$

where S is a two-dimensional lattice having valid fringe data (good amplitude modulation); $N_{x,y}$ is a neighborhood region around the coordinate (x, y) where the phase is being estimated; $m(x, y)$ is an indicator field which equals one if the site (x, y) has already been phase estimated, and zero otherwise. The fringe pattern $I'(\epsilon, \eta)$ is the high-pass filtered and amplitude normalized version of $I(x, y)$; this operation is performed in order to eliminate the low-frequency background $a(x, y)$ and to apprxoimate $b(x, y) \approx 1.0$. The functions $\omega_x(x, y)$ and $\omega_y(x, y)$ are the estimated local frequencies along the x- and y-directions, respectively. Finally, λ is the regularizing parameter that controls (along with the size of $N_{x,y}$) the smoothness of the detected phase.

The first term in Eq. (12.80) attempts to keep the local fringe model close to the observed irradiance in a least-squares' sense within the neighborhood $N_{x,y}$. The second term enforces the assumption of smoothness and continuity using only previously detected pixels $\phi_0(x, y)$ marked by $m(x, y)$. To demodulate a given fringe pattern we need to find the minimum of the cost function $U(x, y)$ with respect to the fields $\phi_0(x, y)$, $\omega_x(x, y)$, and $\omega_y(x, y)$. This may be achieved using the algorithm described in the next paragraph.

The first phase estimation on S is performed as follows: To start, the indicator function $m(x, y)$ is set to zero ($m(x, y) = 0$ in S). Then, one chooses a seed or starting point (x_0, y_0) inside S to begin the demodulation of the fringe pattern. The function $U(x_0, y_0)$ is then optimized with respect to $\phi_0(x_0, y_0)$, $\omega_x(x_0, y_0)$, $\omega_y(x_0, y_0)$; the visited site is marked as detected, i.e., we set $m(x_0, y_0) = 1$. Once the seed pixel is demodulated, the sequential phase demodulation proceeds as follows:

1. Choose the (x, y) pixel inside S (randomly or with a prescribed scanning order).
2. If $m(x, y) = 1$, return to the first statement.
   If $m(x, y) = 0$, then test if $m(x', y') = 1$ for any adjacent pixel (x', y').
   If no adjacent pixel has already been estimated, return to the first state-

ment.

If $m(x', y') = 1$ for an adjacent pixel, take $[\phi_0(x', y'), \omega_x(x', y'), \omega_y(x', y')]$ as initial condition to minimize $U(x, y)$ with respect to $[\phi_0(x, y), \omega_x(x, y), \omega_y(x, y)]$.

3. Set $m(x, y) = 1$.

4. Return to the first statement until all the pixels in S are estimated.

An intuitive way of considering the first iteration just presented is as a crystal growing (CG) process where new molecules are added to the bulk in that particular orientation which minimizes the local crystal energy given the geometrical orientation of the adjacent and previously positioned molecules. This demodulating strategy is capable of estimating the phase within any fringe pattern's boundary.

To optimize $U(x, y)$ at site (x, y) with respect to $(\phi_0, \omega_x, \omega_y)$, we may use simple gradient descent:

$$\phi_0^{k+1}(x, y) = \phi_0^k(x, y) - \tau \frac{\partial U(x, y)}{\partial \phi_0(x, y)},$$

$$\omega_x^{k+1}(x, y) = \omega_x^k(x, y) - \tau \frac{\partial U(x, y)}{\partial \omega_x(x, y)}, \tag{12.82}$$

$$\omega_y^{k+1}(x, y) = \omega_y^k(x, y) - \tau \frac{\partial U(x, y)}{\partial \omega_y(x, y)},$$

where τ is the step size and k is the iteration number. Only one or two iterations are normally needed (except for the demodulation of the starting seed point which may take about 20 iterations); this is because the initial conditions for the gradient search are taken from a neighborhood pixel already estimated. In this way the initial conditions are already very close to the stable point of the gradient search. It is important to remark that the two-dimensional RPT technique gives the estimated phase $\phi_0(x, y)$ already unwrapped, so no additional phase-unwrapping process is required.

The first global phase estimation in S (using the gradient search along with the CG algorithm) is usually very close to the actual modulating phase; if needed, one may perform additional global iterations to improve the phase-estimation process. Additional iterations may be performed using, again, Eq. (12.82), but now taking as initial conditions the last estimated values at the same site (x, y) (not the ones at a neighborhood site as done in the first global CG iteration). Note that for the additional iterations, the indicator function $m(x, y)$ in $U(x, y)$ is now everywhere equal to one; therefore, one may scan S in any desired order whenever all the sites are visited at each global iteration. In practice, only three or four additional global iterations are needed to reach a stable minimum of $U(x, y)$ at each site (x, y) in S.

Figure 12.12 shows the result of applying the RPT technique to the fringe pattern shown in Fig. 12.10. From Fig. 12.12 we can see that the estimated phase at the borders is given accurately.

As mentioned previously, the RPT technique may also be used to demodulate closed-fringe interferograms. Figure 12.13(a) shows a closed-fringe interferogram and Fig. 12.13(b) shows its estimated phase using the RPT technique. Some additional modifications (see Servin et al., 1997b) are needed to the RPT to make it more robust to noise when dealing with closed-fringe interferograms.

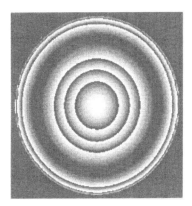

Figure 12.12 Estimated phase using the phase-tracking technique (RPT) applied to the carrier frequency interferogram shown in Fig. 12.10. As we can see, the interferogram's edge was properly estimated.

12.6 PHASE UNWRAPPING

Except for the RPT technique, all other interferometric methods give the detected phase wrapped (the modulo 2π of the true phase) due to the arc tangent function involved in the phase-estimation process. The relationship between the wrapped phase and the unwrapped phase may be stated as

$$\phi_{\mathrm{w}}(x, y) = \phi(x, y) + 2\pi k(x, y), \qquad (12.83)$$

where $\phi_{\mathrm{w}}(x, y)$ is the wrapped phase, $\phi(x, y)$ is the unwrapped phase, and $k(x, y)$ is an integer-valued correcting field. The unwrapping problem is trivial for phase maps calculated from good-quality fringe data; in such phase maps, the absolute phase

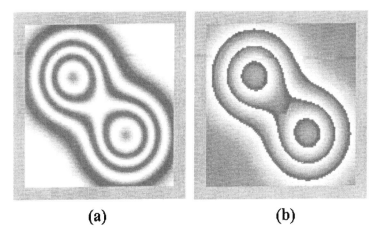

(a)   **(b)**

Figure 12.13 Estimated phase using the phase-tracking technique (RPT) applied to a closed-fringe pattern. (a) The fringe pattern. (b) Its estimated phase.

difference between consecutive phase samples in both the horizontal and vertical directions is less than π, except for the expected 2π discontinuities. Unwrapping is therefore a simple matter of adding or subtracting 2π offsets at each discontinuity (greater than π radians) encountered in the phase data or integrating wrapped phase differences (Itoh, 1982; Ghiglia *et al.*, 1987; Bone, 1991; Owner-Petersen, 1991; Huntley, 1994a, 1994b; Takeda, 1996).

Unwrapping becomes more difficult when the absolute phase difference between adjacent pixels at points other than discontinuities in the arctan() function is greater than π. These unexpected discontinuities may be introduced, for example, by high-frequency, high-amplitude noise, discontinuous phase jumps and regional undersampling in the fringe pattern, or a real physical discontinuity of the domain.

12.6.1 Unwrapping Using Least-Squares Integration of Gradient Phase

The least-squares technique was first introduced by Ghiglia *et al.* (1994) to unwrap inconsistent phase maps. To apply this method, start by estimating the wrapped phase gradient along the x- and y-direction, i.e.,

$$
\begin{aligned}
\phi_y(x, y) &= W[\phi_w(x, y) - \phi_w(x, y - 1)] \\
\phi_x(x, y) &= W[\phi_w(x, y) - \phi_w(x - 1, y)]
\end{aligned}
\tag{12.84}
$$

having an oversampled phase map with moderately low noise, the phase differences in Eq. (12.84) will be everywhere in the range $(-\pi, +\pi)$. In other words, the estimated gradient will be unwrapped. Now we may integrate the phase gradient in a consistent way by means of a least-squares integration. The integrated or searched continuous phase will be the one which minimizes the following cost function:

$$
U[\phi(x, y)] = \sum_{(x,y)\in S} \left\{ [\phi(x, y) - \phi(x - 1, y) - \phi_x(x, y)]^2 \right.
$$

$$
\left. + [\phi(x, y) - \phi(x, y - 1) - \phi_y(x, y)]^2 \right\}
\tag{12.85}
$$

The estimated unwrapped phase $\phi(x, y)$ may be found, for example, using simple gradient descent, as

$$
\phi^{k+1}(x, y) = \phi^k(x, y) - \tau \frac{\partial U}{\partial \phi(x, y)}
\tag{12.86}
$$

where k is the iteration number and τ is the convergence rate of the gradient search system. There are faster algorithms of obtaining the searched unwrapped phase among the techniques of conjugate gradient or the transform methods (Ghiglia *et al.*, 1994). Figure 12.14 shows the resulting unwrapped phase (Fig. 12.14(b)) applying the least-squares technique to a noiseless phase map (Fig. 12.14(a)).

We may also include regularizing potentials to the least-squares unwrapper in order to smooth out some phase noise and possibly interpolate over regions of missing data with a predefined behavior (Marroquin *et al.*, 1995).

When the phase map is too noisy, the fundamental basis of the least-squares integration technique may be broken; i.e., the wrapped phase difference may no longer be a good estimator of the gradient field due to a high amplitude noise. In

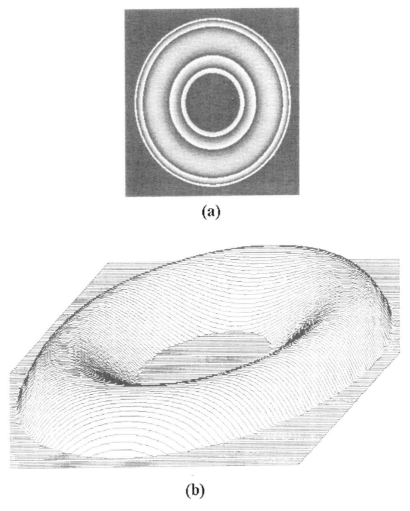

(a)

(b)

Figure 12.14 Phase unwrapping. (a) Wrapped phase. (b) Path-independent phase unwrapping using the least-squares integration technique.

this severe noise situation, the wrapped phase difference among neighborhood pixels is no longer less than π. As a consequence, a reduction of the dynamic range of the resulting unwrapped phase is obtained (Servin *et al.*, 1998a).

12.6.2 Unwrapping Using the Regularized Phase-Tracking Technique

The main motivation to apply the RPT method to phase unwrapping (Servin *et al.*, 1998a) is its superior robustness to noise with respect to the least-squares integration technique. The RPT technique as the least-squares integration of phase gradient is also robust to the edge effect at the boundary of the phase map.

The first step to unwrap a given phase map using the RPT technique is to put the wrapped phase into two-phase orthogonal fringe patterns. These fringe patterns may be obtained using the cosine and the sine of the map phase being unwrapped: i.e.,

$$
\begin{aligned}
I_C(x, y) &= \cos[\phi_w(x, y)], \\
I_S(x, y) &= \sin[\phi_w(x, y)],
\end{aligned}
\tag{12.87}
$$

where $\phi_w(x, y)$ is the phase map being unwrapped.

Now the problem of phase unwrapping may be treated as a demodulation of two phase-shifted fringe patterns using the RPT technique (Servin *et al.*, 1998a). Therefore, the cost function to be minimized by the unwrapped phase $\phi_0(x, y)$ at each site (x, y) is

$$
\begin{aligned}
U(x, y) = \sum_{(\epsilon, \eta) \in (N_{x,y} \cap S)} & \left\{ [I_C(\epsilon, \eta) - \cos p(x, y, \epsilon, \eta)]^2 + [I_S(\epsilon, \eta) - \sin p(x, y, \epsilon, \eta)]^2 \right. \\
& \left. \lambda[\phi_0(\epsilon, \eta) - p(x, y, \epsilon, \eta)]^2 m(\epsilon, \eta) \right\}
\end{aligned}
\tag{12.88}
$$

and

$$
p(x, y, \epsilon, \eta) = \phi_0(x, y) + \omega_x(x, y)(x - \epsilon) + \omega_y(x, y)(y - \eta),
\tag{12.89}
$$

where S is a two-dimensional lattice having valid fringe data (good amplitude modulation); $N_{x,y}$ is a neighborhood region around the coordinate (x, y) where the phase is being unwrapped; $m(x, y)$ is an indicator field which equals 1 if the site (x, y) has already been unwrapped, and 0 otherwise. The functions $\omega_x(x, y)$ and $\omega_y(x, y)$ are the estimated local frequencies along the x- and y-directions respectively. Finally, λ is the regularizing parameter which controls (along with the size of $N_{x,y}$) the smoothness of the detected-unwrapped phase.

The algorithm to optimize this cost function is the same as the one described in the RPT section 12.5.5, so we are not going into the details here.

12.6.3 Temporal Phase Unwrapping

This phase unwrapping technique was introduced by Huntley and Saldner in 1993 and it has been applied in optical metrology to measure deformation. The basic idea of this technique is to take several interferograms as the object is deformed; therefore, the number of deformation fringes within that object will grow due to the increasing applied force. If one wants to analyze each fringe pattern using the PSI technique, one needs to take at least three phase-shifted interferograms for each object's deformation to obtain its corresponding phase map. The sampling theorem must be fulfilled in the temporal space for every pixel in the fringe pattern; i.e., consecutive pixels should have a phase difference less than π in the time domain. The main advantage of this method is that the unwrapping of each pixel is an independent process from the unwrapping process of any other pixel in the temporal sequence of phase maps at hand.

As mentioned before, for each object's deformation the modulating phase of the object is estimated using the PSI technique. This gives us a temporal sequence of wrapped phases. The sequence of phase maps may be represented by

$$\phi_w(1, x, y), \phi_w(2, x, y), \phi_w(3, x, y), \ldots, \phi_w(N, x, y), \qquad (x, y) \in S, \qquad (12.90)$$

where S is the region of valid phase data and N is the total number of intermediate-phase maps. In order to fulfill the sampling theorems, the following condition must be fulfilled.

$$\left| W[\phi_w(i + 1, x, y) - \phi_w(i, x, y)] \right| < \pi, \qquad (1 \leq i \leq N - 1), \qquad (x, y) \in S$$
$$(12.91)$$

and $W[.]$ is the wrapping operator. The unwrapping process then proceeds according to

$$\phi(x, y) = \sum_{i=1}^{N-1} W[\phi_w(i + 1, x, y) - \phi_w(i, x, y)]. \qquad (12.92)$$

The distribution given by $\phi(x, y)$ gives the unwrapped phase difference between the initial state $\phi_w(1, x, y)$ and the final state $\phi_w(N, x, y)$.

Figure 12.15 shows a sequence of phase maps that may be unwrapped using the temporal phase unwrapping technique herein described. Each wrapped phase has less than half a fringe among them, as required by this technique. It must be pointed out that the unwrapped phase is going to be equal to the phase difference between the phase map shown in Fig. 12.15(a) and the phase map shown in Fig. 12.15(d).

12.7   EXTENDED RANGE FRINGE PATTERN ANALYSIS

Extended range interferometry allows us to measure larger numbers of aspheric wavefronts than standard two arms interferometers. The reason for this extended range is that using these techniques enables one to directly measure the gradient or curvature of the wavefront instead of the wavefront itself. If the wavefront being measured is smooth, then one needs less image pixels to represent its spatial variations. As we will see later, an exception to this is the sub-Nyquist interferometry, where the wavefront is taken in a direct way, with the disadvantage of needing a special-purpose CCD video array.

12.7.1   Phase Retrieval from Gradient Measurement Using Screen-Testing Methods

The screen-testing methods are used to detect the gradient of the wavefront under analysis. One normally uses a screen with holes or strips lying perpendicular to the propagation direction of the testing wavefront. Then, one collects the irradiance pattern produced by the shadow or the self-image (whenever possible) of the testing screen at some distance from it. If the testing wavefront is aberrated, then the shadow or self-image of the screen will be distorted with respect to the original screen. The phase difference between the screen's shadow and the screen is related to the gradient of the aberrated wavefront at the screen plane. Thus, to obtain the shape of the testing wavefront, one must use an integration procedure. The integration procedure that we employ is the least-squares solution, which has the advantage of being path-independent and robust to noise.

The two most used screens to test the wavefront aberration are the Ronchi ruling (Cornejo, 1992) and the Hartmann testing plate (Ghozeil, 1992). The Ronchi

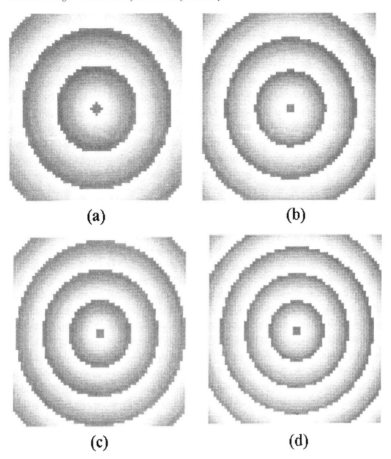

Figure 12.15 A sequence of wrapped phases suitable for being unwrapped by the method of temporal phase unwrapping. The unwrapped phase will be the difference between the first wrapped phase in (a) and the one in (d).

ruling, which is a linear grating, is inserted in a place perpendicular to the average direction of propagation of the wavefront being tested. The other frequently used screen is the Hartmann plate, which is a screen with a two-dimensional array of small circular holes evenly spaced in the screen. A linear grating such as a Ronchi ruling is only sensitive to the ray aberration perpendicular to the ruling strips. This means that, in general, we will need two shadow Ronchi images to fully determine the aberration of the wavefront. In contrast, only one Hartmann testing screen is needed to collect all the data regarding the wavefront aberration. Unfortunately, a Hartmanngram (the irradiance shadow of the Harmann screen at the testing plane) is more difficult to analyze than a Ronchigram. That is because a Ronchigram may be analyzed using robust and well-known carrier frequency interferometry.

The main advantage of using screen tests along with a CCD camera is to increase the measuring dynamic range of the tested wavefront: i.e., sensing the gradient of the testing wavefront instead of the wavefront itself allow us to increase the number of aberration waves that can be tested. This is why the most popular way of testing large optics, such as telescopes' primary mirrors, are screen tests. Thus, for a given number of pixels of a CCD camera, it is possible to measure more waves of aberration using screen tests than using a standard interferometer, which measures the aberration waves directly.

12.7.2 Wavefront Slope Analysis with Linear Gratings (Ronchi Test)

As mentioned earlier, a linear grating is easier to analyze using standard carrier fringe detecting procedures such as the Fourier method, the synchronous method, or the spatial phase shifting (SPSI) method. These techniques have already been discussed above. The Ronchi test has been a widely used technique and has been reported by several researchers in metrology (Yatagai, 1984; Omura and Yatagai, 1988; Wan and Lin, 1990; Fischer, 1992).

We may start with a simplified mathematical model for the transmittance of a linear grating (Ronchi rulings are normally made of binary transmittance):

$$T_x(x, y) = \frac{[1 + \cos(\omega_0 x)]}{2}, \tag{12.93}$$

where ω_0 is the angular spatial frequency of the Ronchi ruling. The linear ruling is then placed at the plane where the aberrated wavefront is being estimated. If a light detector is placed at a distance d from the Ronchi plate then, as a result of wavefront aberrations, we obtain a distorted irradiance pattern that will be given, approximately, by

$$I_\delta(x, y) = \frac{1}{2} + \frac{1}{2} \cos\left(\omega_0 x + \omega_0 d \frac{\partial W(x, y)}{\partial y}\right) \tag{12.94}$$

where $I_x(x, y)$ is the distorted shadow of the transmittance $T_x(x, y)$, and $W(x, y)$ represents the wavefront under test. As Eq. (12.94) shows, it is necessary to detect two orthogonal shadow patterns to completely describe the gradient field of the wavefront under test. The other linear ruling, located at the same testing plane but with its strip lines oriented in the y-direction, is

$$T_y(x, y) = \frac{[1 + \cos(\omega_0 y)]}{2}. \tag{12.95}$$

Thus, the distorted image of the Ronchi ruling at the collecting data plane is given by

$$I_y(x, y) = \frac{1}{2} + \frac{1}{2} \cos\left(\omega_0 y + \omega_0 d \frac{\partial W(x, y)}{\partial x}\right). \tag{12.96}$$

We may use any of the carrier fringe methods described in this chapter to demodulate these two Ronchigrams. Figure 12.16(a) shows a Ronchi ruling and Fig. 12.16(b) shows the same Ronchi ruling modulated by a wavefront containing aspheric aberration.

Once the detected and unwrapped phase of the ruling's shadows has been obtained, one needs to integrate the resulting gradient field. To integrate this phase gradient, one may use path-independent integration such as least-squares integration. The least-squares integration of the gradient field may be stated as the function which minimizes the following quadratic cost function:

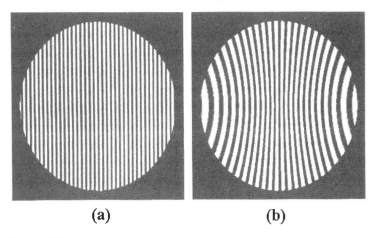

(a)   **(b)**

Figure 12.16 (a) Unmodulated Ronchi ruling. (b) Modulated Ronchi ruling.

$$U(\hat{W}) = \sum_{(x,y)\in S} \left[\hat{W}(x+1, y) - \hat{W}(x, y) - \frac{\partial W(x, y)}{\partial x} \right]^2 \qquad (12.97)$$
$$+ \sum_{(x,y)\in S} \left[\hat{W}(x, y+1) - \hat{W}(x, y) - \frac{\partial W(x, y)}{\partial y} \right]^2,$$

where the hat function $\hat{W}(x, y)$ is the estimated wavefront, and we have approximated the derivative of the searched phase along the x- and y-axis as first-order differences of the estimated wavefront. The least-squares estimator may then be obtained from $U(x, y)$ by simple gradient descent as

$$\hat{W}^{k+1}(x, y) = \hat{W}^k(x, y) - \tau \frac{\partial U(\hat{W})}{\partial \hat{W}(x, y)}, \qquad (12.98)$$

or using a faster algorithm such as conjugate gradient or transform methods (Fried, 1977; Hudgin, 1977; Noll, 1978; Hunt, 1979; Freischlad *et al.*, 1985, 1992; Takajo and Takahashi, 1988; Ghiglia and Romero, 1994).

12.7.3 Moiré Deflectometry

We may increase the sensitivity of the Ronchi test by placing the collecting data plane at the first self-image of the linear ruling. The first Talbot self-image for a collimated light beam appears at the so-called Rayleigh distance L_R, given by

$$L_R = \frac{2\,d^2}{\lambda}. \qquad (12.99)$$

The resulting deflectograms may be analyzed in the same way as the one described for the Ronchigrams.

12.7.4 Wavefront Analysis with Lateral Shearing Interferometry

Lateral shearing interferometry consists in obtaining a fringe by constructing an interfering pattern using two lateral displaced copies of the wavefront under analysis (Rimmer and Wyant, 1975; Hung, 1982; Yatagai and Kanou, 1984; Gasvik, 1987; Hardy and MacGovern, 1987; Welsh *et al.*, 1995). The mathematical form of the irradiance of a lateral sheared fringe pattern may be written as

$$
\begin{aligned}
I_x(x, y) &= \frac{1}{2} + \frac{1}{2}\cos\left\{\frac{2\pi}{\lambda}[W(x - \delta x, y) - W(x + \delta x, y)]\right\}, \\
I_x(x, y) &= \frac{1}{2} + \frac{1}{2}\cos\left[\frac{2\pi}{\lambda}\Delta_x W(x, y)\right],
\end{aligned}
\tag{12.100}
$$

where δx is half of the total lateral displacement. As the Eq. (12.100) shows, one also needs the orthogonally displaced shearogram to describe the wavefront under analysis completely. The orthogonal shearogram in the y-direction may be written as

$$
\begin{aligned}
I_y(x, y) &= \frac{1}{2} + \frac{1}{2}\cos\left\{\frac{2\pi}{\lambda}[W(x, y - \delta y) - W(x, y + \delta y)]\right\}, \\
I_y(x, y) &= \frac{1}{2} + \frac{1}{2}\cos\left[\frac{2\pi}{\lambda}\Delta_y W(x, y)\right].
\end{aligned}
\tag{12.101}
$$

These fringe patterns may be transformed into carrier frequency interferograms by introducing a large and known amount of defocusing to the testing wavefront (Mantravadi, 1992). Having linear carrier fringe patterns, one may proceed to their demodulation using standard techniques of fringe carrier analysis, as seen in this chapter. A shearing interferogram is shown in Fig. 12.17. This shearing interferogram corresponds to a defocused wavefront having a circular pupil and, as we can see from Fig. 12.17, interference fringes are only present in the common area of the two copies of the laterally displaced wavefront.

We may analyze in the frequency domain the modulating phase of a sheared interferogram (in the x-direction, for example) as the output of a linear filter:

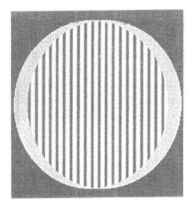

Figure 12.17 Lateral shearing of a circular pupil containing a defocused wavefront. Interference fringes form only at the superposition of both sheared pupils.

$$\mathcal{F}[\Delta_x W(x)] = \mathcal{F}[W(x - \delta x) - W(x + \delta x)]$$
$$= 2i \sin(\delta x\, \omega_x)\mathcal{F}[W(x)]. \tag{12.102}$$

As can be seen, the transfer function of the shearing operator in the frequency domain is a sinusoidal-shaped filter. As a consequence, the inverse filter (the one needed to obtain the searched wavefront) has poles in the frequency domain, so its use is not straightfoward. Instead of using a transformed method to recover the wavefront from the sheared data we feel it is easier to use a regularization approach in the space domain.

Assume that we have already estimated and unwrapped the interesting phase differences $\Delta_x W(x, y)$ and $\Delta_y W(x, y)$. Using this information, the least-squares wavefront reconstruction may be stated as the minimizer of the following cost function:

$$U(\hat{W}) = \sum_{(x,y)\in S_x} \left\{ \hat{W}(x - \delta x, y) - \hat{W}(x + \delta x, y) - \Delta_x W(x, y) \right\}^2$$
$$+ \sum_{(x,y)\in S_y} \left\{ \hat{W}(x, y - \delta y) - \hat{W}(x, y + \delta y) - \Delta_y W(x, y) \right\}^2 \tag{12.103}$$
$$U(\hat{W}) = \sum_{(x,y)\in S_x} U_x(x, y)^2 + \sum_{(x,y)\in S_y} U_y(x, y)^2,$$

where the "hat" function represents the estimated wavefront and S_x and S_y are two-dimensional lattices containing valid phase data in the x- and y-shearing directions (the common area of the two laterally displaced pupils). Unfortunately, the least-squares cost function stated above is not well posed, because the matrix that results from setting the gradient of U equal to 0 is not invertible (as seen previously, the inverse filter may contain poles in the frequency range of interest). Fortunately, we may regularize this inverse problem and find the expected smooth solution of the problem (Servin *et al.*, 1996b). As seen before, the regularizer may consist of a linear combination of squared magnitude of differences of the estimated wavefront within the domain of interest. In particular, one may use a second-order or thin-plate regularizer:

$$R_x(x, y) = \hat{W}(x - 1, y) - 2\hat{W}(x, y) + \hat{W}(x + 1, y),$$
$$R_y(x, y) = \hat{W}(x, y - 1) - 2\hat{W}(x, y) + \hat{W}(x, y + 1). \tag{12.104}$$

Therefore, the regularized cost function becomes

$$U(\hat{W}) = \sum_{(x,y)\in S_x} U_x(x, y)^2 + \sum_{(x,y)\in S_y} U_y(x, y)^2 + \lambda \sum_{(x,y)\in \mathrm{Pupil}} [R_x(x, y)^2 + R_y(x, y)^2], \tag{12.105}$$

where Pupil refers to the two-dimensional lattice inside the pupil of the wavefront being tested. The regularizing potentials discourage large changes in the estimated wavefront among neighboring pixels. As a consequence, the searched solution will be relatively smooth. The parameter λ controls the amount of smoothness of the estimated wavefront. It should be remarked that the use of regularizing potentials in this case is a must (even for noise-free observations) to yield a stable solution of the least-squares integration for lateral displacements greater than two pixels, as analyzed by Servin *et al.* (1996b).

The estimated wavefront may be calculated using simple gradient descent as

$$\hat{W}^{k+1}(x, y) = \hat{W}^k(x, y) - \tau \frac{\partial U(\hat{W})}{\partial \hat{W}(x, y)}, \tag{12.106}$$

where τ is the convergence rate. This optimizing method is not very fast. One normally uses faster algorithms such as conjugate gradient.

12.7.5 Wavefront Analysis with Hartmann Screens

The Hartmann test is a well-known technique for testing large optical components (Ghozeil, 1992; Welsh *et al.*, 1995). The Hartmann technique samples the wavefront under analysis using a screen of uniformly spaced holes situated at the pupil plane. The Hartmann screen may be expressed as

$$HS(x, y) = \sum_{n=-N/2}^{N/2} \sum_{m=-N/2}^{N/2} h(x - pn, y - pm), \tag{12.107}$$

where $HS(x, y)$ is the Hartmann screen and $h(x, y)$ are the small holes which are uniformly spaced in the Hartmann screen. Finally, p is the space among the holes of the screen. A typical Hartmann screen may be seen in Fig. 12.18, where the two-dimensional arrangement of holes is shown. The measuring wavefront must pass through these holes, and their shadow is recorded at a distance d from it. If we have wavefront aberrations higher than defocusing, the Hartmann screen's shadow will be geometrically distorted.

The collimated rays of light that pass through the screen holes are then captured by a photographic plate at some distance d from it. The uniformly spaced array of holes at the instrument's pupil is then distorted at the photographic plane by the aspherical aberrations of the wavefront under test. The screen deformations are then proportional to the slope of the aspheric wavefront: i.e.,

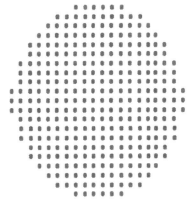

Figure 12.18 A typical Hartmann screen used in the Hartmann test.

$$H(x, y) = \left[\sum_{(n,m)=-N/2,}^{N/2} h'\left(x - pn - d\frac{\partial W(x, y)}{\partial x}, y - pm - d\frac{\partial W(x, y)}{\partial x}\right)\right] P(x, y),$$

(12.108)

where $H(x, y)$ is the Hartmanngram (the irradiance of the screen's shadow) obtained at a distance of d from the Hartmann screen. The function $h'(x, y)$ is the image of the screen's hole $h(x, y)$ as projected at the Hartmanngram plane. Finally, $P(x, y)$ is the pupil of the wavefront being tested. As seen in Eq. (12.108), only one Hartmanngram is needed to fully estimate the wavefront's gradient. The frequency content of the estimated wavefront will be limited by the sampling theorem to the hole's period p of the screen. A typical Hartmanngram may be seen in Fig. 12.19. This Hartmanngram corresponds to an aspheric wavefront having a strong spherical aberration component.

Traditionally, these Hartmanngrams (the distorted image of the screen at the observation plate's plane) are analyzed by measuring the centroid of the spots images $h'(x, y)$ generated by the screen holes $h(x, y)$. The deviation of these centroids from their uniformly spaced positions (unaberrated positions) are recorded. These deviations are proportional to the aberration's slope. The centroids' coordinates give a two-dimensional discrete field of the wavefront gradient that needs integration and interpolation over regions without data. Integration of the wavefront's gradient field is normally done by using the trapezoidal rule (Ghozeil, 1992). The trapezoidal rule is carried out following several independent integration paths and their outcomes averaged. In this way, one may approach a path-independent integration. Using this integration procedure, the wavefront is only known at the hole's position. Finally, a polynomial or spline wavefront fitting is necessary to estimate the wavefront's values at places other than the discrete points where the gradient data is collected. A two-dimensional polynomial for the wavefront's gradient may be proposed. This polynomial is then fitted by least squares to the slope data; it must contain every possible type of wavefront aberration, otherwise some unexpected features (specially at the

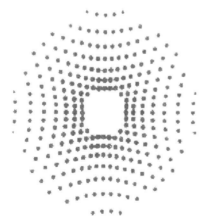

Figure 12.19 A distorted image of the Hartmann screen as seen inside the paraxial focus of a paraboloid under test.

edges) of the wavefront may be filtered out. On the other hand, if one uses a high-degree of polynomial (in order to ensure not to filter out any wavefront aberration), the estimated continuous wavefront may wildly oscillate in regions where no data are collected.

Recently, robust quadratic filters have been used to demodulate Hartmanngrams (Servin *et al.*, 1996d). Also the regularized phase tracker (RPT) has been used to demodulate the gradient information of the Hartmanngram. Using the RPT technique one is able to estimate the gradient field not only at the hole's positions but continuously over the whole pupil of the Hartmanngram (Servin *et al.*, 1999).

12.7.6 Wavefront Analysis by Curvature Sensing

Teague (1983), Streibl (1984), and Roddier (1990) have analyzed and demonstrated phase retrieval using the irradiance transport equation. Assuming a paraxial beam propagating along the z-axis, we may obtain the irradiance transport equation as (Teague, 1983; Streibl, 1984):

$$\frac{\partial I(x, y, z)}{\partial z} = -\nabla I(x, y, z) \cdot \nabla W(x, y, z) - I(x, y, z)\, \nabla^2 W(x, y, z) \qquad (12.109)$$

where $I(x, y, z)$ is the distribution of the illumination along the propagating beam, $W(x, y, z)$ is the wavefront surface at distance z from the origin, and ∇ is the $(\partial/\partial x, \partial/\partial y)$ operator. In the analysis of wavefronts using the transport equation there is no need for a codifying screen pupil, as in the case of the Ronchi or Hartmann test.

Following an interesting interpretation of the irradiance transport equation given by Ichikawa *et al.* (1988), one may note in the transport equation the following interpretation for each term:

- The first term $\nabla I \cdot \nabla W$ may be seen as the irradiance variation caused by a transverse shift of the inhomogeneous ($\nabla I \neq 0$) beam due to the local tilt of the wavefront whose normal (ray) direction is given by ∇W; this may be called a prism term.
- The second term $I\nabla^2 W$ may be interpreted as the irradiance variation caused by convergence or divergence of the beam whose local focal length is inversely proportional to $\nabla^2 W$; this may be called a "lens term."

Thus, the sum expresses the variation of the beam irradiance caused by the prism and lens effect as it propagates along the z-axis. Rewriting the transport equation as

$$-\frac{\partial I(x, y, z)}{\partial z} = \nabla \cdot [I(x, y, z)\, \nabla W(x, y, z)] \qquad (12.110)$$

and remarking that ∇W is the direction of the ray vector, we can easily see that the transport equation represents the law of light energy conservation, which is analogous to the law of mass or charge conservation, frequently expressed by

$$\frac{\partial \rho}{\partial t} = \mathrm{div}(\rho\, v), \qquad (12.111)$$

with ρ and v being the mass or charge density and the flow velocity, respectively.

The technique proposed by Roddier (1990) to use the transport equation in wavefront estimation is as follows. Let $P(x, y)$ be the transmittance of the pupil. That is, $P(x, y)$ equals 1 inside the pupil and 0 outside. Furthermore we may assume that the illumination at the pupil's plane is uniform and equal to I_0 inside $P(x, y)$. Hence $\nabla I = 0$ in $P(x, y)$ is everywhere 0 except at the pupil's edge, where it has the value

$$\nabla I = -I \, \mathbf{n} \, \delta_c \tag{12.112}$$

Here δ_c is a Dirac distribution around the pupil's edge and $\mathbf{n}$ is the unit vector perpendicular to the edge and pointing outward. Substituting the irradiance Laplacian in $P(x, y)$ into the irradiance transport equation, one obtains

$$\frac{\partial I(x, y, z)}{\partial z} = -I_0 \cdot \frac{\partial W(x, y, z)}{\partial n} \delta_c - I_0 P(x, y) \nabla^2 W(x, y, z) \tag{12.113}$$

where $\partial W / \partial n = \mathbf{n} \cdot \nabla W$ is the wavefront derivative in the outward direction perpendicular to the pupil's edge. Curvature sensing consists in taking the difference between the illumination observed in two close planes separated a distance $\pm \Delta z$ from the reference plane where the pupil $P(x, y)$ is located. Then we obtain the following two measurements as

$$
\begin{aligned}
I_1 &= I_0 + \frac{\partial I}{\partial z} \Delta z \\
I_2 &= I_0 - \frac{\partial I}{\partial z} \Delta z
\end{aligned}
\tag{12.114}
$$

Having these data, one may form the so-called sensor signal as

$$S = \frac{I_1 - I_2}{I_1 + I_2} = \frac{1}{I_0} \frac{\partial I}{\partial z} \Delta z \tag{12.115}$$

Substituting this into Eq. (12.113) yields

$$S = \left\{ \frac{\partial W(x, y)}{\partial n} \delta_c - P(x, y) \nabla^2 W(x, y) \right\} \Delta z \tag{12.116}$$

Solving this differential equation one is able to estimate the wavefront inside the pupil $P(x, y)$, knowing both the Laplacian of $W(x, y)$ inside $P(x, y)$ and $\partial W / \partial n$ along the pupil's edge as Neumann boundary conditions.

12.7.7 Sub-Nyquist Analysis

Testing of aspheric wavefronts is nowadays routinely achieved in the optical shop by the use of commercial interferometers. The testing of deep aspheres is limited by the aberrations of the interferometer's imaging optics as well as the spatial resolution of the CCD video camera used to gather the interferometric data. The CCD video arrays come typically with 256×256 or 512×512 image pixels. The number of CCD pixels limits the highest recordable frequency over the CCD array to π rad/pixel. This maximum recordable frequency is called the Nyquist limit of the sampling system. The detected phase map of an interferogram having frequencies higher than the Nyquist limit is said to be aliased and cannot be unwrapped using standard techniques such as the ones presented so far.

The main prior knowledge that is going to be used by us is that the expected wavefront is smooth (Greivankamp, 1987; Greivenkamp and Bruning, 1992). Then, one may introduce this prior knowledge into the unwrapping process. The main requirement to apply sub-Nyquist techniques is to have a CCD camera with detectors much smaller than the spatial separation among them. Another alternative is to use a mask with small holes over the CCD array to reduce the light-sensitive area of the CCD pixels. This requirement allows us to have a strong signal even for thin interferogram fringes. To obtain the undersampled phase map, one may use any well-known PSI techniques using phase-shifted undersampled interferograms.

The undersampled interferogram may be imaged directly over the CCD video array with the aid of an optical interferometer, as seen in Chapter 1. If the CCD sampling rate is Δx over the x-direction, and Δy over the y-direction and the diameter of the light-sensitive area of the CCD is d, we may write the mathematical expression for the sampling operation over the interferograms irradiance as

$$S[I(x, y)] = \left[I(x, y) ** \text{circ}\left(\frac{\rho}{d}\right)\right]\text{comb}\left(\frac{x}{\Delta x}, \frac{y}{\Delta y}\right), \rho = (x^2 + y^2)^{1/2}, \quad (12.117)$$

where the function $S[I(x, y)]$ is the sampling operator over the interferogram's irradiance. The symbol $(**)$ indicates a two-dimensional convolution. The $\text{circ}(\rho/d)$ is the circular size of the CCD detector. The comb function is an array of delta functions with the same spacing as the CCD pixels. The phase map of the subsampled interferogram may be obtained using, for example, three phase-shifted interferograms, as

$$I(x, y, t) = \sum_{n=-1}^{1} a_n(x, y) + b_n(x, y) \cos\left[\frac{2\pi}{\lambda}\phi(x, y) + t\right]\delta(t - n\alpha), \quad (12.118)$$

where the variable α is the amount of phase shift. Using well-known formulae we can find the subsampled wrapped phase by

$$\phi_w(x, y) = \tan^{-1}\left(\frac{1 - \cos(\alpha)}{\sin(\alpha)} \frac{S[I_1(x, y)] - S[I_3(x, y)]}{2S[I_1(x, y)] - S[I_2(x, y)] - S[I_3(x, y)]}\right). \quad (12.119)$$

As Eq. (12.119) shows, the obtained phase is a modulo 2π of the true undersampled phase due to the arc tangent function involved in the phase-detection process.

Now we may treat the problem of unwrapping undersampled phase maps due to smooth wavefronts: i.e., the only prior knowledge about the wavefront being analyzed is smoothness. This is far less restrictive than the null testing technique presented in the last section. Analysis of interferometric data beyond the Nyquist frequency was first proposed by Greivenkamp (1987), who assumed that the wavefront being tested is smooth up to the first or second derivative. Greivenkamp's approach to unwrap subsampled phase maps consists of adding multiples of 2π each time a discontinuity in the phase maps is found. The number of 2π values added is determined by the smoothness condition imposed on the wavefront in its first or second derivative along the unwrapping direction. Although Greivenkamp's approach is robust against noise, its weakness is that it is a path-dependent phase unwrapper.

In this section we present a method (Servin *et al.*, 1996a) which overcomes the path dependency of the Greivenkamp approach while preserving its noise robustness. In this case an estimation of the local wrapped curvature (or wrapped Laplacian) of the subsampled phase map $\phi_w(x, y)$ is used to unwrap the interesting deep aspheric wavefront. Once having the local wrapped curvature along the x- and y-directions, one may use least-squares integration to obtain the unwrapped continuous wavefront. The local wrapped curvature is obtained as

$$
\begin{aligned}
L_x(x, y) &= W[\phi_w(x - 1, y) - 2\phi_w(x, y) + \phi_w(x + 1, y)], \\
L_y(x, y) &= W[\phi_w(x, y - 1) - 2\phi_w(x, y) + \phi_w(x, y + 1)].
\end{aligned}
\tag{12.120}
$$

If the absolute value of the discrete wrapped Laplacian is less than π, its value will be nonwrapped. Then we may obtain the unwrapped phase $\psi(x, y)$ as the function which minimizes the following quadratic cost function (least squares):

$$
U[\phi(x, y)] = \sum_{(x, y) \in S} U_x(x, y)^2 + U_y(x, y)^2,
\tag{12.121}
$$

where S is a subset of a two-dimensional regular lattice of nodes having good amplitude modulation. The functions $U_x(x, y)$ and $U_y(x, y)$ are given by

$$
\begin{aligned}
U_x(x, y) &= L_x(x, y) - [\phi(x - 1, y) - 2\phi(x, y) + \phi(x + 1, y)], \\
U_y(x, y) &= L_y(x, y) - [\phi(x, y - 1) - 2\phi(x, y) + \phi(x, y + 1)].
\end{aligned}
\tag{12.122}
$$

The minimum of the cost function is obtained when its partial with respect to $\phi(x, y)$ equals zero. Therefore, the set of linear equations that must be solved is

$$
\begin{aligned}
\frac{\partial U[\phi(x, y)]}{\partial \phi(x, y)} &= U_x(x - 1, y) - 2U_x(x, y) + U_x(x + 1, y) + U_y(x, y - 1) \\
&\quad - 2U_y(x, y) + U_y(x, y + 1).
\end{aligned}
\tag{12.123}
$$

Several methods may be used to solve this system of linear equations; among others, there is the simple gradient descent:

$$
\phi^{k+1}(x, y) = \phi^k(x, y) - \eta \frac{\partial U}{\partial \phi(x, y)},
\tag{12.124}
$$

where the parameter η is the rate of convergence of the gradient search. The simple gradient descent is quite slow for this application; instead we may conjugate gradient or transformed techiques to speed up the computing time. Figure 12.20 shows a subsampled phase map along with its unwrapped version using the technique herein described.

Finally, if one has a good knowledge of the wavefront being tested up to a few wavelengths, one may use this information to obtain an oversampled phase map (Servin and Malacara, 1996c). This oversampled phase map is then the estimation error between what is the expected wavefront and the actual wavefront being tested. One also may reduce the number of aspherical aberration wavelengths by introducing a compensating hologram (Horman, 1965; Dörband and Tiziani, 1985).

12.8 APPLICABILITY OF FRINGE ANALYSIS METHODS

The success in implementation of optical full-field measuring methods into industrial, medical, and commercial areas depends on proper retrieval of a measurand coded in an arbitrary fringe pattern. This is the reason why such variety of tech-

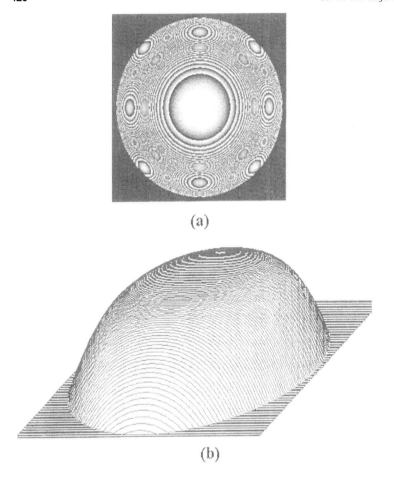

Figure 12.20 Undersampled phase unwrapping. (a) Undersampled phase map. (b) Unwrapped phase.

niques exist. Table 12.2 provides a comparison of the fringe pattern analysis methods, and indicates which techniques are most commonly used in commercial systems.

In order to fulfill the conditions of fast, automatic, accurate, and reliable analysis of fringe data the new solutions of phase-measuring methods focus on the following issues:

- Active approach to fringe-pattern forming.
- Active approach to design of phase analysis algorithms (phase shifting).
- Improving the methods.
- Given that many problems involved with fringe analysis are ill-posed, it is convenient to search for regularizers for the solution according to prior information available.

Table 12.2 Comparison of Features of Fringe Pattern (FPs) Analysis methods[a]

Method	Number of FPs per frame	Detector resolution requirements	Real time method for mod (2π)	Inherent image enhancement	Inherent phase interpolation	Automatic sign detection	Achievable accuracy	Experimental requirements	Complexity of processing	Dynamic events analysis	Commercial systems
Fringe extreme localization	1/1	$R_0^{\,b}$	No	No	No	No	Low	Low	Low	No	**
Regularization	1/1	R_0	No	Yes	Yes	Yes	Medium	Low	High	No	*
Temporal heterodyning	+	Single[c] detector and scanning	No	Yes	Yes	Yes	Very high	Very high	Hardware	No	*
Phase shifting:											
• temporal	min 3/3	R_0	Partly	No	No	Yes	High	High	Low	No	****
• spatial	3/1 3/3	$3R_0$	Yes	No	No	Yes	Medium	Medium	Low	Yes	
• carrier frequency	1/1	min $2R_0$	Yes	No	No	Yes	Medium	Low	Low	Yes	***
Fourier transform	1/1	min $2R_0$	Partly	Yes	Yes	Yes	High inside domain low at edges	Low	High	Yes	**
PLL	1/1	min $2R_0$	No	Yes	Yes	Yes	Medium	Low	Medium	No	
Space-domain processing	1/1	min $2R_0$	Yes	Yes	Yes	Yes	High	High	Hardware	Yes	

[a] The choice of the methods most often used is given.
[b] R_0 is detector resolution for the infinite fringe detection mode.
[c] The possibility of the use of an image dissector camera is not considered.

ACKNOWLEDGMENTS

Dr Manuel Servin wants to acknowledge the support of the CONACYT.

REFERENCES

Ai, C. and J. C. Wyant, "Effect of Piezoelectric Transducer Non-Linearity on Phase Shift Interferometry," *Appl. Opt.*, **26**, 1112–1116 (1987).

Asundi, A. and K. H. Yung, "Phase Shifting and Local Moiré," *JOSA A*, **8**, 1591–1600 (1991).

Bone, D. J., "Fourier Fringe Analysis: The Two Dimensional Phase Unwrapping Problem," *Appl. Opt.*, **30**, 3627–3632 (1991).

Bone, D. J., *et al.*, "Fringe Pattern Analysis Using a 2D Fourier Transform," *Appl. Opt.*, **25**, 1653–1660 (1986).

Bruning, J. H., D. R. Herriott, J. E. Gallagher, D. P. Rosenfel, A. D. White, and D. J. Brangaccio, "Digital Wavefront Measuring Interferometer for Testing Optical Surfaces and Lenses," *Appl. Opt.*, **13**, 2693–2703 (1974).

Cornejo, A., "The Ronchi Test," in *Optical Shop Testing*, Malacara, D., ed., John Wiley & Sons, New York, 1992.

Creath, K., "Phase Measuring Interferometry: Beware of These Errors," *Proc. SPIE*, **1559**, 313–220 (1991).

Dörband, B. and H. J. Tiziani, "Testing Aspheric Surfaces with Computer Generated Holograms: Analysis of Adjustment and Shape Errors," *Appl. Opt.*, **24**, 2604–2611 (1985).

Efron, U., *et al.*, "Special Issue on the Technology and Application of SLMs," *Appl. Opt.*, **28** (1989).

Fischer, D. J., "Vector Formulation for Ronchi Shear Surface Fitting," *Proc. SPIE*, **1755**, 228–238 (1992).

Freischlad, K., "Wavefront Integration from Difference Data," *Proc. SPIE*, **1755**, 212–218 (1992).

Freischlad, K. and C. L. Koliopoulos, "Wavefront Reconstruction from Noisy Slope or Difference Data Using the Discrete Fourier Transform," *Proc. SPIE*, **551**, 74–80 (1985).

Freischlad, K. and C. L. Koliopoulos, "Fourier Description of Digital Phase Measuring Interferometry," *J. Opt. Soc. Am. A*, **7**, 542–551 (1990).

Fried, D. L., "Least-Squares Fitting of a Wave-Front Distortion Estimate to an Array of Phase-Difference Measurements," *J. Opt. Soc. Am.*, **67**, 370–375 (1977).

Gasvik, K. J., *Optical Metrology*, John Wiley & Sons, New York, 1987.

Ghiglia, D. C. and L. A. Romero, "Robust Two Dimensional Weighted and Unweighted Phase Unwrapping That Uses Fast Transforms and Iterative Methods," *J. Opt. Soc. Am. A*, **11**, 107–117 (1994).

Ghiglia, D. C., G. A. Mastin, and L. A. Romero, "Cellular Automata Method for Phase Unwrapping," *J. Opt. Soc. Am.*, **4**, 267–280 (1987).

Ghozeil, I., "Hartmann and Other Screen Tests," in *Optical Shop Testing*, Malacara, D., ed., John Wiley & Sons, New York, 1992.

Greivenkamp, J. E., "Generalized Data Reduction for Heterodyne Interferometry," *Opt. Eng.*, **23**, 350–352 (1984).

Greivenkamp, J. E., "Sub-Nyquist Interferometry," *Appl. Opt.*, **26**, 5245–5258 (1987).

Greivenkamp, J. E. and J. H. Bruning, "Phase Shifting Interferometry," in *Optical Shop Testing*, Malacara, D., ed., John Wiley & Sons, New York, 1992, pp. 501–598.

Groot, P. D., "Derivation of Algorithms for Phase Shifting Interferometry Using the Concept of a Data-Sampling Windows," *Appl. Opt.*, **34**, 4723–4730 (1995).

Hardy, J. W. and A. J. MacGovern, "Shearing Interferometry: A Flexible Technique for Wavefront Measuring," *Proc. SPIE*, **816**, 180–195 (1987).

Hariharan, P., B. F. Areb, and T. Eyui, "Digital Phase-Stepping Interferometry: A Simple Error-Compensating Phase Calculation Algorithm," *Appl. Opt.*, **26**, 3899 (1987).

Hibino, K. and M. Yamaguchi, "Phase Determination Algorithms Compensating for Spatial Non-Uniform Phase Modulation in Phase Shifting Interferometry," *Proc. SPIE*, **3478**, 110–120 (1998).

Hibino, K., F. Oreb, and D. I. Farrant, "Phase Shifting for Nonsinusoidal Waveforms with Phase-Shifting Errors," *JOSA A*, **12**, 761–768 (1995).

Horman, M. H., "An Application of Wavefront Reconstruction to Interferometry," *Appl. Opt.*, **4**, 333–336 (1965).

Hudgin, R. H., "Wave-Front Reconstruction for Compensated Imaging," *J. Opt. Soc. Am.*, **67**, 375–378 (1977).

Hung, Y. Y., "Shearography: A New Optical Method for Strain Measurement and Nondestructive Testing," *Opt. Eng.*, **21**, 391–395 (1982).

Hunt, B. R., "Matrix Formulation of the Reconstruction of Phase Values from Phase Differences," *J. Opt. Soc. Am.*, **69**, 393–399 (1979).

Huntley, J. M., "Phase Unwrapping – Problems & Approaches," in *Proc. FASIG, Fringe Analysis '94*, York University, 391–393 (1994a).

Huntley, J. M., "New Methods for Unwrapping Noisy Phase Maps, *Proc. SPIE*, **2340**, 110–123 (1994b).

Huntley, J. M. and H. Saldner, "Temporal Phase Unwrapping Algorithm for Automated Interferogram Analysis," *Appl. Opt.*, **32**, 3047–3052 (1993).

Ichikawa, K., A. W. Lohmann, and M. Takeda, "Phase Retrieval Based on the Irradiance Transport Equation and the Fourier Transport Method: Experiments," *Appl. Opt.*, **27**, 3433–3436 (1988).

Ichioka, Y. and M. Inuiya, "Direct Phase Detecting System," *Appl. Opt.*, **11**, 1507–1514 (1972).

Itoh, K., "Analysis of the Phase Unwrapping Algorithm," *Appl. Opt.*, **21**, 2470 (1982).

Joenathan, C. and B. M. Khorana, "Phase Measuring Fiber Optic Electronic Speckle Pattern Interferometer: Phase Steps Calibration and Phase Drift Minimization," *Opt. Eng.*, **31**, 315–321 (1992).

Jones, J. D., "Engineering Applications of Optical Fiber Interferometers," *Proc. SPIE*, **2341**, 222–238 (1994).

Kreis, T., "Digital Holographic Interference-Phase Measurement Using the Fourier Transform Method," *JOSA A*, **3**, 847–855 (1986).

Kozlowska, A. and M. Kujawinska, "Grating Interferometry with a Semiconductor Light Source," *Appl. Opt.*, **36**, 8116–8120 (1997).

Kujawinska, M. and C. Kosinski, "Adaptability: Problem or Solution," in Jüptner, W. and W. Osten, eds, *Akademie Verlag Series in ptical Metrology*, **3**, 419–431 (1997).

Kujawinska, M. and J. Wójciak, "High Accuracy Fourier Transfer Fringe Pattern Analysis," *Opt. Lasers Eng.*, **14**, 325–329 (1991a).

Kujawinska, M. and J. Wójciak, "Spatial Carrier Phase Shifting Technique of Fringe Pattern Analysis," *Proc. SPIE*, **1508**, 61–67 (1991b).

Küchel, T., "Digital Holographic Interference-Phase Measurement Using the Fourier Transform Method," *JOSA A*, **3**, 847–855 (1986).

Macy, W., Jr., "Two-Dimensional Fringe Pattern Analysis," *Appl. Opt.*, **22**, 3898–3901 (1983).

Malacara, D., M. Servin, and Z. Malacara, *Optical Testing: Analysis of Interferograms*, Marcel Dekker, New York, 1998.

Mantravadi, M. V., "Lateral Shearing Interferometers," in *Optical Shop Testing*, Malacara, D., ed., John Wiley & Sons, New York, 1992.

Marroquin, J. L., "Deterministic Interactive Particle Models for Image Processing and Computer Graphics," *Computer and Vision, Graphics and Image Processing*, **55**, 408–417 (1993).

Marroquin, J. L. and M. Rivera, "Quadratic Regularization Functionals for Phase Unwrapping," *J. Opt. Soc. Am. A*, **12**, 2393–2400 (1995).

Marroquin, J. L., M. Servin, and R. Rodriguez-Vera, "Adaptive Quadrature Filters for Multi-Phase Stepping Images," *Opt. Lett.*, **24**, 238–240 (1998).

Marroquin, J. L., M. Servin, and J. E. Figueroa, "Robust Quadrature Filters," *JOSA A*, **14**, 779–791 (1997a).

Marroquin, J. L., M. Servin, and R. Rodriguez-Vera, "Adaptive Quadrature Filters and the Recovery of Phase from Fringe Pattern Images," *JOSA A*, **14**, 1742–1753 (1997b).

Mertz, L., "Real Time Fringe Pattern Analysis," *Appl. Opt.*, **22**, 1535–1539 (1983).

Morgan, C. J., "Least Squares Estimation in Phase Measurement Interferometry," *Opt. Lett.*, **7**, 368–370 (1982).

Noll, R. J., "Phase Estiamtes From Slope-Type Wave-Front Sensors," *J. Opt. Soc. Am.*, **68**, 139–140 (1978).

Olszak, A. and K. Patorski, "Modified Electronic Speckle Interferometer with Reduced Number of Elements for Vibration Analysis, *Opt. Comm.* **138**, 265–269 (1997).

Omura, K. and T. Yatagai, "Phase Measuring Ronchi Test," *Appl. Opt.*, **27**, 523–528 (1988).

Osten, W., Nadeburn, P. Andrä, "General Hierarchical Approach in Absolute Phase Measurement," *Proc. SPIE*, **2860**, 2–13 (1996).

Owner-Petersen, M., "Phase Unwrapping: A Comparison of Some Traditional Methods and a Presentation of a New Approach," *Proc. SPIE*, **1508**, 73–82 (1991).

Patorski, K., *Handbook of the Moiré Fringe Technique*, Elsevier, Amsterdam, 1993.

Pirga, M. and M. Kujawinksa, "Two Dimensional Spatial-Carrier Phase-Shifting Method for Analysis of Crossed and Closed Fringe Patterns," *Opt. Eng.*, **34**, 2459–2466 (1995).

Pirga, M. and M. Kujawinksa, "Errors in Two-Directional Spatial-Carrier Phase Shifting for Closed Fringe Pattern Analysis," *Proc. SPIE*, **2860**, 72–83 (1996).

Ransom, P. L. and J. B. Kokal, "Interferogram Analysis by a Modified Sinusoidal Fitting Technique," *Appl. Opt.*, **25**, 4199–4205 (1986).

Rimmer, M. P. and J. C. Wyant, "Evaluation of Large Aberrations Using a Lateral Shear Interferometer Having Variable Shear," *Appl. Opt.*, **14**, 142–150 (1975).

Robinson, D. W. and G. T. Reid, eds, *Interferogram Analysis: Digital Fringe Pattern Measurement Techniques*, Institute of Physics Publishing House, Bristol, 1993.

Roddier, F., "Wavefront Sensing and the Irradiance Transport Equation," *Appl. Opt.*, **29**, 1402–1403 (1990).

Roddier, C. and F. Roddier, "Interferogram Analysis Using Fourier Transform Techniques," *Appl. Opt.*, **26**, 1668–1673 (1987).

Schmith, J. and K. Creath, "Extended Averaging Technique for Derivation of Error Compensating Algorithms in Phase Shifting Interferometry," *Appl. Opt.*, **34**, 3610–3619 (1995a).

Schmith, J. and K. Creath, "Fast Calculation of Phase in Spatial *n*-Point Phase Shifting Technique," *Proc. SPIE*, **2544**, 102–111 (1995b).

Schmith, J. and K. Creath, "Window Function Influence on Phase Error in Phase-Shifting Algorithms," *Appl. Opt.*, **35**, 5642–5649 (1996).

Schreiber, W., G. Notni, P. Kühmstedt, J. Gerber, and R. Kowarschik, "Optical 3D Measurements of Objects with Technical Surfaces," in *Academic Verlag Series in Optical Metrology*, Jüptner, W. and W. Osten, eds, **2**, 1996, pp. 46–51.

Schwider, J., "Automated Evaluation Techniques in Interferometry," in *Progress in Optics, Vol. XXVIII*, Wolf, E., ed., Elsevier Science Publishers, Amsterdam, 1990.

Schwider, J., R. Burow, K. E. Elssner, J. Grzanna, R. Spolaczyk, and K. Merkel, "Digital Wavefront Interferometry: Some Systematic Error Sources," *Appl. Opt.*, **22**, 3421–3432 (1983).

Servin, M. and F. J. Cuevas, "A Novel Technique for Spatial-Phase-Shifting Interferometry," *J. Mod. Opt.*, **42**, 1853–1862 (1995).

Servin, M. and D. Malacara, "Path-Independent Phase Unwrapping of Subsampled Phase Maps," *Appl. Opt.*, **35**, 1643–1649 (1996a).

Servin, M. and D. Malacara, "Sub-Nyquist Interferometry Using a Computer Stored Reference," *J. Mod. Opt.*, **43**, 1723–1729 (1996c).

Servin, M. and R. Rodriguez-Vera, "Two Dimensional Phase Locked Loop Demodulation of Carrier Frequency Interferograms," *J. Mod. Opt.*, **40**, 2087–2094 (1993).

Servin, M., D. Malacara, and F. J. Cuevas, "Direct Phase Detection of Modulated Ronchi Rulings Using a Phase Locked Loop," *Opt. Eng.*, **33**, 1193–1199 (1994).

Servin, M., D. Malacara, and J. L. Marroquin, "Wave-Front Recovery from Two Orthogonal Sheared Interferograms," *Appl. Opt.*, **35**, 4343–4348 (1996b).

Servin, M., D. Malacara, and F. J. Cuevas, "New Technique for Ray Aberration Detection in Hartmanngrams Based on Regularized Band Pass Filters," *Opt. Eng.*, **35**, 1677–1683 (1996d).

Servin, M., D. Malacara, J. L. Marroquin, and F. J. Cuevas, "Complex Linear Filters for Phase Shifting with Very Low Detuning Sensitivity," *Journal of Mod. Opt.*, **44**, 1269–1278 (1997a).

Servin, M., J. L. Marroquin, and F. J. Cuevas, "Demodulation of a Single Interferogram by Use of a Two-Dimensional Regularized Phase-Tracking Technique," **36**, 4540–4548 (1997b).

Servin, M., F. J. Cuevas, D. Malacara, and J. L. Marroquin, "Phase Unwrapping through Demodulation using the RPT Technique," *Appl. Opt.*, **37**, 1917–1923 (1998a).

Servin, M., R. Rodriguez-Vera, J. L. Marroquin, and D. Malacara, "Phase Shifting Interferometry Using a Two Dimensional Regularized Phase-Tracking Technique," *Journ. of Mod. Opt.*, **45**, 1809–1820 (1998b).

Servin, M., F. Cuevas, D. Malacara, and J. L. Marroquin, "Direct Ray Aberration Estimation in Hartmanngrams Using a Regularized Phase Tracking System," *Appl. Opt.*, **38**, 2862–2869 (1999).

Shagam, R. N. and J. C. Wyant, "Optical Frequency Shifter for Heterodyne Interferometers," *Appl. Opt.*, **17**, 3034–3035 (1978).

Shough, D. H., O. Y. Kwon, and D. F. Leavy, "High Speed Interferometric Measurements of Aerodynamic Phenomena," *Proc. SPIE*, **1221**, 394–403 (1990).

Streibl, N., "Phase Imaging by the Transport Equation of Intensity," *Opt. Commun.*, **49**, 6–10 (1984).

Surrel, Y., "Design of Algorithms for Phase Measurement by Use of Phase Stepping," *Opt. Lett.*, **35**, 51–60 (1996).

Takajo, H. and Takahashi, T., "Least Squares Phase Estimation From Phase Differences," *J. Opt. Soc. Am. A*, **5**, 416–425 (1988).

Takeda, M. "Recent Progress in Phase-Unwrapping Techniques," *Proc. SPIE*, **2782**, 334–343 (1996).

Takeda, M. and M. Kitoh, "Spatio-Temporal Frequency-Multiplex Heterodyne Interferometry," *J. Opt. Soc. Am.*, **9**, 1607–1614 (1992).

Takeda, M., H. Ina, and S. Kobayashi, "Fourier Transform Methods of Fringe-Pattern Analysis for Computer-Based Topography and Interferometry," *J. Opt. Soc Am.*, **72**, 156–160 (1982).

Tang, S., "Generalized Generalized Algorithm for Phase Shifting Interferometry," *Proc. SPIE*, **2860**, 34–44 (1996).

Teague, M. R., "Deterministic Phase Retrieval: A Green's Function Solution," *J. Opt. Soc. Am.*, **73**, 1434–1441 (1983).

Towers, D. P., T. R. Judge, and P. J. Bryanston-Cross, "Automatic Interferogram Analysis Techniques Applied to Quasy-Heterodyne Holography and ESPI," *Opt. and Lasers Eng.*, **14**, 239–282 (1991).

Van der Heijden, F., *Image-Based Measurement Systems, Object Recognition and Parameter Estimation*, J. Wiley and Sons, Chichester, 1994.

Wan, D.-S. and D.-T. Lin, "Ronchi Test and a New Phase Reduction Algorithm," *Appl. Opt.*, **29**, 3255–3265 (1990).

Welsh, B. M., B. L. Ellerbroek, M. C. Roggemann, and T. L. Pennington, "Fundamental Performance Comparison of a Hartmann and a Shearing Interferometer Wave-Front Sensor," *Appl. Opt.*, **34**, 4186–4195 (1995).

Womack, K. H., "Interferometric Phase Measurement Using Spatial Synchronous Detection," *Opt. Eng.*, **23**, 391–395 (1984).

Wyant, J. C., "Interferometric Optical Metrology: Basic Principles and New Systems," *Laser Focus*, May, 65–71 (1982).

Yamaguchi, I., *et al.*, "Active Phase Shifting Interferometers for Shape and Deformation Measurements, *Opt. Eng.*, **35**, 2930–2937 (1996).

Yatagai, T., "Fringe Scanning Ronchi Test for Aspherical Surfaces," *Appl. Opt.*, **23**, 3676–3679 (1984).

Yatagai, T. and T. Kanou, "Aspherical Surface Testing with Shearing Interferometry Using Fringe Scanning Detection Method," *Opt. Eng.*, **23**, 357–360 (1984).

13

Optical Methods in Metrology: Point Methods

H. ZACARIAS MALACARA and RAMON RODRIGUEZ-VERA

Centro de Investigaciones en Optica, León, Mexico

13.1 INTRODUCTION

Light and optics have been used as the ultimate tools for metrology since olden days. An example of this is the way a ray beam is used as a reference for straightness, or in modern times, the definition of a given wavelength as a distance standard (Swyt, 1995). In this chapter, we provide an overview of some optical measuring methods and their applications to optical technology. We do not intend to cover all the optical methods in metrology. Several metrology techniques are described more extensively in other chapters in this book.

During the measuring process we need to adopt a common *measuring standard* from which all the references are made. The *SI measuring system* has a worldwide acceptance, and it is the one used in this chapter. Among the main characteristics of the SI system is the definition of a *primary standard* for every defined fundamental physical unit. A set of *derived units* is also defined from the primary units. The primary standard in the SI system is the *meter*. After several revised definitions (Baird and Howlett, 1963), the meter is now defined as the distance traveled by light in $1/299,792,458$ of a second (Swyt, 1995). Under this new definition, the meter is a derived unit from the time standard. To avoid the meter being a derived unit, it has been proposed to define the meter as "the length equal to $9,192,631,770/299,792,458$ wavelengths in the vacuum of the radiation corresponding to the transition between the two hyperfine levels of the ground state of the cesium 133 atom" (Giacomo, 1980; Goldman, 1980).

13.2 LINEAR DISTANCE MEASUREMENTS

Linear distance measurements are made through a direct comparison to a scale or a secondary standard. In other cases, an indirect measurement to the standard is done with known precision. Optics give us the flexibility and simplicity of both methods. We have an assortment of optical methods for distance measurements, depending on the scale of distances to be measured. Some representative methods are described next.

13.2.1 Large-Distance Optical Measurements

Large distances (about a human body and larger) can be measured by optical means to a high degree of precision. Direct distance comparison, distance and angle measurement, and time-of-flight methods are used.

13.2.1.1 Range Finders and Optical Radar

Range finders are devices used to measure distances in several ways. The simplest case of a range finder is called a stadia. A stadia is made from a telescope and a precision rotating mirror (Fig. 13.1). A bar with a known distance w is placed at the range to be measured R. A beamsplitting prism superimposes two images from the bar in the telescope. First, both images are brought into coincidence; then, opposite ends of the bar are put together by rotating the mirror at an angle θ from the coincidence point. This gives the angle subtense for the reference bar. The range R is then

$$R = \frac{W}{\theta},\tag{13.1}$$

where θ is small and expressed in radians. Another stadia technique is used by some theodolites that have a reticle for comparison against a graduated bar. If a bar with known length W is seen through a telescope with focal length f, the image of the bar on the reticle has a size i; then, the distance is

$$R = \left(\frac{f}{i}\right)W.\tag{13.2}$$

A range finder uses a baseline instead of a reference bar as the standard length. A basic range finder schematic is shown in Fig. 13.2. Two pentaprisms separated a

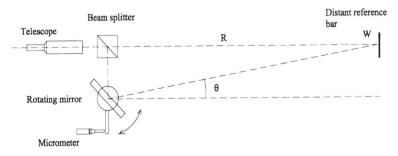

Figure 13.1 A stadia range finder.

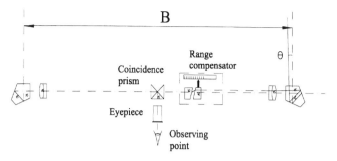

Figure 13.2 Range finder schematics.

distance B have each two identical telescope objectives. Both telescope images are brought to the same eyepiece by means of a beamsplitter or a coincidence prism. Originally, the instrument is built so that images from an infinite-distance located object are overlapped. A range compensator is a device inserted in one of the instrument branches to displace laterally one image on the focal plane (angularly on the object space), and brings into coincidence objects that are not at infinity. Figure 13.3 shows some devices used as range compensators. The baseline B viewed from the distant point subtends an angle θ and the range is then

$$R = \frac{B}{\theta}.$$ (13.3)

For a given error in the angle measurement, the range error is

$$\Delta R = -B\theta^{-2}\Delta\theta;$$ (13.4)

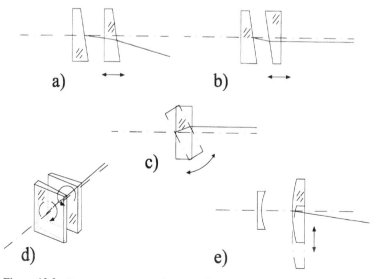

Figure 13.3 Range compensators for range finders.

also, from Eq. (13.3),

$$\Delta R = -\frac{R^2}{B}\Delta\theta, \tag{13.5}$$

To keep the range finder error low, the baseline must be as large as possible, but the error increases with the square of the distance. Since the eye has an angular acuity of about 10 arc seconds, the angular error is about 0.00005 radians (Smith, 1966). The image displacement can be measured using two CCD (charge-coupled device) cameras, one for each branch, and measuring the parallax automatically with a computer (Tocher, 1992). Another method correlates electronically the images from both telescopes (Schmidt, 1984).

Another method for distance measurement is the time of flight method, tested shortly after the laser invention (Buddenhagen *et al.*, 1961, Stitch *et al.*, 1961; Stitch, 1972). For a light beam traveling at a known speed, the distance can be measured by the time a light beam takes to go and return from the measuring point. For very large distance scales, optical radar is used, as in the moon distance determination (Faller and Wampler, 1970); but it has also been used in surveying instruments (Rüeger and Pascoe, 1989). While in RF radar the main problem was to obtain a fast rising pulse, the problem is well solved with Q-switched lasers. A block diagram for the moon ranging experiment is shown in Fig. 13.4.

A precise measurement of the time-of-flight method is done by measuring the phase of an amplitude-modulated beam. Several systems that use this method are described in the literature (Sona, 1972; Burnside, 1991). The light beam is amplitude modulated at a frequency ω, the output light beam has an amplitude $s_0 = A_0 \sin \omega t$, and the returning beam has an amplitude $s_r = A_r \sin \omega(t + \Delta t)$, where Δt is the time of flight for the light beam. The resulting phase difference between the returning and a local reference beam will be $\Delta\phi = \omega\Delta t$. Since the modulating signal is periodical, the returning and reference signal will be in phase for distances that are multiples of the modulating wavelength, or $\Delta\phi = n\pi$, n being an integer. The phase difference is equivalent to a distance of $x = c\Delta t = c\Delta\phi/\omega$; c is the speed of light in the medium.

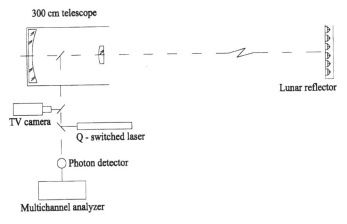

Figure 13.4 Moon ranging experiment.

For a returning beam in phase with the sending beam, the distance can be found from the modulating wavelength:

$$x = \frac{c\Delta\phi}{2\pi f} = n\lambda. \tag{13.6}$$

For any given distance, the distance in terms of the modulating wavelength is

$$D = n\lambda + \Delta\lambda. \tag{13.7}$$

Since the number of full wavelengths in Eq. (13.7) is unknown, we can use a wavelength longer than the range to be measured with the consequent loss in precision. An alternative is to use at least three close frequencies:

$$\begin{aligned} D &= n_1\lambda_1 + \Delta\lambda_1, \\ D &= n_2\lambda_2 + \Delta\lambda_2, \\ D &= n_3\lambda_3 + \Delta\lambda_3. \end{aligned} \tag{13.8}$$

To solve Eqs (13.8) we use three close frequencies, and assume the same n for the three measurements. Another solution could be made making λ_1 and λ_2 an exact multiple (about 1000) one from another, and solving for D and n. In another method, a system that resembles a phase-locked loop (PLL) is used (Takeshima, 1996). The system sweeps in frequency until locks, repeats for the next frequency $(n + 1)$, and two more frequencies. When the oscillator is locked, the phase difference is zero ($\Delta\lambda = 0$); we need only to determine D and n_1, n_2, and n_3.

Absolute distance laser ranging: For a laser distance measuring system, we assume a given refractive index n. The limiting factor will always be any change in the refractive index that is a function of the temperature, pressure, and moisture content. By measuring at two or more wavelengths, all the sources of uncertainty are removed. Systems have been designed for two wavelengths (Shipley and Bradsell, 1976; Dändliker *et al.*, 1998). A nonambiguity measuring range of 0.2 parts per million has been achieved.

13.2.1.2 Curvature and Focal Length

Optical manufacturing has several ways to measure a radius of curvature, including *templates, spherometers,* and *test plates*. Templates are rigid metal sheets with both concave and a convex curvature cut at opposite faces (Fig. 13.5). Templates are brought into direct contact with the sample. A minimal light space must be observed at the contact point. If the space between surfaces is small, the light between surfaces turns blue, due to diffraction. A template has the simplicity that it can be made in a mechanical shop with appropriate measuring tools, but it is also commercially available. For a more precise curvature measurement, a spherometer is used. Essentially, a spherometer measures the sagitta in a curved surface. Assume a bar spherometer, as shown in Fig. 13.6 (Cooke, 1964). For a leg separation y and a ball radius r, the radius of curvature R can be obtained from the sagitta z:

$$R = \frac{z}{2} + \frac{y^2}{2z} \pm r, \tag{13.9}$$

where the plus sign is used for concave surfaces and minus for convex surfaces.

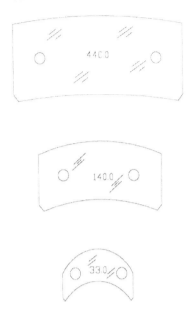

Figure 13.5 Templates.

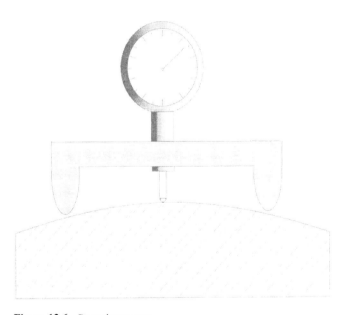

Figure 13.6 Bar spherometer.

The uncertainty in the measurement is found by differentiating the previous equation:

$$\Delta R = \frac{\Delta z}{2}\left(1 - \frac{y^2}{z^2}\right); \tag{13.10}$$

assuming no error in the determination of the parameters for the instrument, the leg's separation is perfectly known.

Several variants of the basic spherometer include the ring spherometer (Fig. 13.7), which is commonly used where a spherical surface is assumed. An interchangeable set of rings extends its use to improve the accuracy for a larger lens diameter. The Geneva gauge is a portable bar spherometer that reads the diopter power directly, assuming a nominal refractive index for ophthalmic lenses. Modern spherometers are now digital, and include a microprocessor to convert the sagitta distance to a radius of curvature, diopter power, or curvature. Spherometer precision and accuracy are analyzed by Jurek (1977).

Test plates (Malacara, 1992) are glass plates polished to a curvature opposite to the one we want to check. Curvature is tested by direct contact. A test plate is made for each cuvature to be tested. Test plates are appropriate for online curvature testing. Plane testing plates are common on the optical shop.

Several means have been devised for optical curvature measuring. For concave curvature measurements, probably the simplest method is the Foucault test. Analysis and applications for this test are covered in another chapter in this book. Another precise curvature radii's measurement device is the traveling microscope (Horne, 1972), shown in Fig. 13.8. A point source is produced at the front focus of a traveling microscope. Also, an illuminated reticle eyepiece could be used. In both cases, a sharp image of the point source or the reticle is sought when the surface coincides with the front focus. Then, the microscope is moved until a new image is found. The

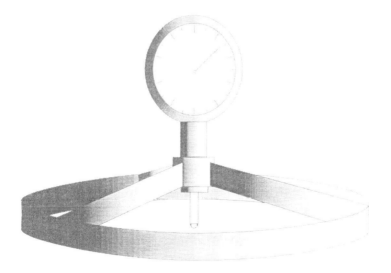

Figure 13.7 Ring spherometer.

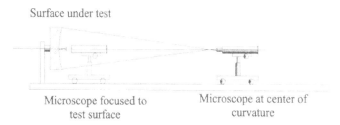

Surface under test

Microscope focused to   Microscope at center of
test surface   curvature

Figure 13.8   Traveling microscope for curvature measurements.

curvature radius is then the distance between these positions. Carnell and Welford (1971) describe a method using only one measurement. After focusing the microscope at the curvature center, a bar micrometer is inserted with one end touching the vertex of the surface. This method is also suitable for convex surfaces by inserting a well-corrected lens such that the conjugate focus is larger than the curvature radius under test. Convex curvatures can be measured using an autocollimator and an auxiliary lens (Boyd, 1969).

For a large curvature radius, an autocollimator and a pentaprism is used, as shown in Fig. 13.9. A shift in the reticle image is measured in the autocollimator. This is an indirect measure of the surface's slope. By scanning the surface with the prism, samples of the slope are obtained and by integration, the curvature is calculated. This system is appropriate for large curvature measurements and can be used both for concave and convex surfaces (Cooke, 1963).

Another commonly used method for curvature measurements is the confocal cavity method. The so-called optical cavity technique, described by Gerchman and Hunter (1979 and 1980), interferometrically measures the radii of curvature for long curvature concave surfaces. A Fizeau interferometer is formed, as shown in Fig. 13.10. A nth order confocal cavity is obtained where n is the number of times the optical path is folded. The radius of curvature is equal to $2n$ times the cavity length Z. The accuracy is about 0.1%.

13.2.2   Moiré Techniques in Medium Distances

When two periodic objects are superimposed, a well-known effect takes place; this is the *moiré effect* (Kafri and Glatt, 1990; Patorski, 1993; Post *et al.*, 1994). This effect

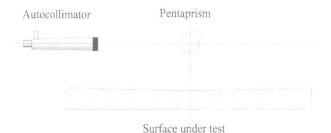

Autocollimator   Pentaprism

Surface under test

Figure 13.9   Curvature measurements with an autocollimator and a traveling pentaprism.

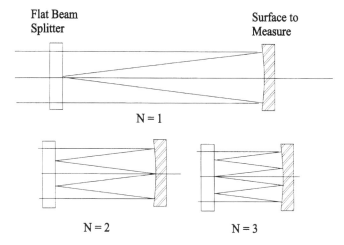

Flat Beam
Splitter

Surface to
Measure

N = 1

N = 2

N = 3

Figure 13.10 Confocal cavity arrangements for curvature measurements.

produces a low-frequency pattern of secondary fringes known as *moiré fringes*. In this manner, for example, in Fig. 13.11 we can see several overlapping periodic objects, showing moiré fringes. For measuring purposes, the periodic objects are usually constituted by gratings of alternating clear and dark lines. Figure 13.12(a) is a sample of a linear grid and Fig. 13.12(b) is the moiré pattern taking place by the overlapping of two such gratings. Ronchi gratings are a particular case of linear grids with a quadratic profile that can easily be reproduced.

From the geometric point of view (Nishijima and Oster, 1964), the moiré fringes are defined as a locus of points of two overlapping periodical objects. It is possible to determine the period p' and the angle φ of the moiré fringes knowing the periods p_1, p_2, and the angle θ among the lines of the gratings (Fig. 13.13). Then,

$$p' = \frac{p_1 p_2}{\sqrt{p_1^2 + p_2^2 - 2p_1 p_2 \cos \theta}}, \qquad (13.11)$$

and

$$\sin \varphi = \frac{p_2 \sin \theta}{\sqrt{p_1^2 + p_2^2 - 2p_1 p_2 \cos \theta}}. \qquad (13.12)$$

When $\theta = 0$, then $\varphi = 0$, and Eq. (13.11) becomes

$$p' = \frac{p_1 p_2}{|p_1 - p_2|}. \qquad (13.13)$$

Now, if $p_1 = p_2 = p$, then Eq. (13.11) transforms into

$$p' = \frac{p}{2 \sin \theta/2}. \qquad (13.14)$$

If $\theta \approx 0$, then $p' = p/\theta$ and $\varphi \approx 90°$. A quick analysis of Eq. (13.14) shows that when the angle between the gratings is large, the moiré pattern frequency increases

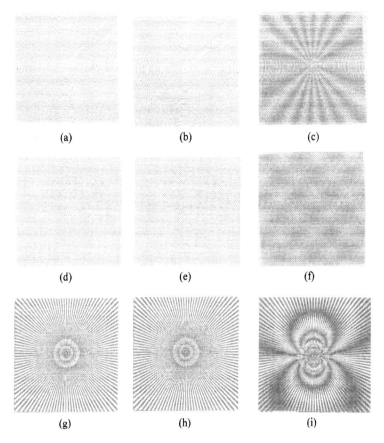

    (a)   (b)   (c)

    (d)   (e)   (f)

    (g)   (h)   (i)

Figure 13.11 Formation of moiré fringes with different periodic objects.

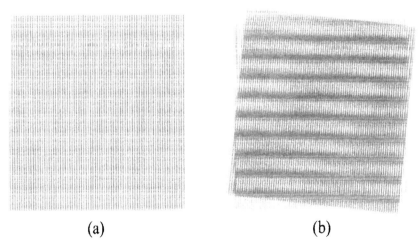

            (a)   (b)

Figure 13.12 (a) Linear grating; (b) moiré pattern formed by two linear gratings.

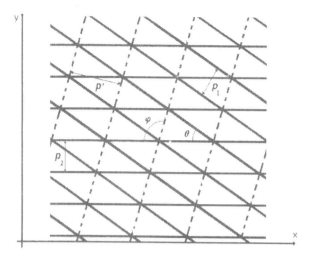

Figure 13.13 Diagram showing the intersection points of two superimposed linear gratings of periods p_1 and p_2; θ is the angle among the lines of the gratings. Dotted lines correspond to moiré fringes, which make an angle φ with the lines of the grating of period p_2.

(the period is reduced); otherwise, the frequency diminishes until the moiré fringes disappear.

One of the fundamental characteristics of a moiré pattern is that if one of the gratings is deformed and the other remains fixed, the moiré pattern is also deformed, as shown in the Fig. 13.14. Deformation of one of the gratings can arise because of the large size to be measured. For this reason, it is possible to call the deformed grating a *grating object*, while the one not deformed is the *reference grating*.

Figure 13.14 Moiré patterns formed by two linear gratings; one of them is lightly deformed.

A simple way to obtain a couple of linear gratings is to photocopy in a transparency the image of the Fig. 13.12(a). When superimposing these gratings the moiré effect is observed, similar to that of Fig. 13.12(b). Another characteristic of the moiré effect is that if we maintain a grating fixed and other displaced a small distance in the direction of its lines, the moiré pattern has a large displacement.

There are several ways of overlapping two gratings. The simplest, as already mentioned, is by contact. Another manner is to project the image of a grating by means of an optical system over another grating. A third form is by means of some logic or arithmetic operation in a digital way (Asundi and Yung, 1991; Rodriguez-Vera 1994). This last method can be carried out when the gratings are stored in a computer.

Several metrology techniques are based on the moiré effect. These techniques have been used in several applications of science and engineering. Some of the techniques will now be explained along with different measuring tasks.

13.2.2.1 Photoelectric Fringe Counting

Typical examples of distance meters based on the moiré effect are verniers and digital micrometers, and coordinate measuring machines (CMM). Their operation consists of the photoelectric detection of the moiré fringes' movement (Luxmoore, 1983). The basic elements of such a system are shown in the Fig. 13.15. This system consists of a light source, a collimating lens, two linear gratings, and a set of four photocells. The displacement is measured by counting moiré fringes when one of the gratings moves in the normal direction to its lines. The displacement of the moiré fringes will be analogous to the lateral displacement of the lines of the gratings. Using a photocell, the moiré fringes are detected during the movement (Watson, 1983). In practice, to determine the displacement sense, four photocells are required. These detectors are positioned at four points in a moiré fringe pattern, and they are spaced to a quarter of their period. Alternate count of phase steps and signs combines to be fed to an amplifier, in a pair of symmetrical signals in quadrature. The sensibility of the instruments based on this moiré technique can be up to 0.000250 inches (Farago, 1982).

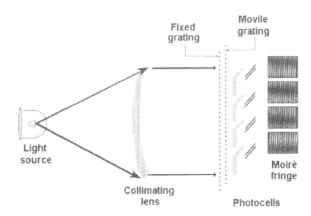

Figure 13.15 Opto-electronic arrangement of distance measurement based on the moiré effect.

13.2.2.2 The Talbot Effect

When a periodic object is illuminated with a spherical or plane wavefront, replicas of the object are obtained in very defined planes along the light propagation direction (Patorski, 1993). This phenomenon is known as the Talbot effect or self-imaging. In spherical illumination, the replication of the object is amplified and the distance between planes are not a constant. On the other hand, when the illumination of the periodic object is made by means of a collimated wave, the self-image planes are equispaced and well defined (Fig. 13.16). If the periodic object is a linear grating with a period p, the self-images are formed in planes, given by the equation (Malacara, 1974):

$$\Delta_n = \frac{np^2}{\lambda},\qquad(13.15)$$

where n is the n-order plane and λ is the illumination wavelength.

This effect is very useful for measuring the distance Δ between a grating and its Talbot image, forming a moiré pattern. In this manner, for example, Jutamulia *et al.* (1986) and Rodriguez-Vera *et al.* (1991a) used the Talbot effect to identify depths or separate planes in color. The color appears naturally by illuminating the grating with a white light source and by superimposing the self-image reflected from different planes of the scene into a second grating. Moiré fringes look at different colors depending on the plane position.

13.2.2.3 Liquid Level Measurement

As an extension of moiré techniques, the Talbot effect is used for liquid level and volume determination in containers (Silva and Rodriguez-Vera, 1996). This technique uses the reflected image from the liquid–air interface inside a container. The

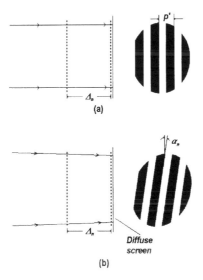

Figure 13.16 Moiré fringes formed by (a) collimated and (b) noncollimated light.

incident light to this interface comes from the image of a linear grating illuminated from a collimated monochromatic light source. By means of an appropriate optical system, the reflected image is formed onto a second grating, forming a sharp moiré pattern, when the second grating is at a distance equal to the Talbot plane. For a given longitudinal displacement in the liquid–air interface that corresponds to a level change, the moiré pattern becomes unsharp. Consequently, it is necessary to adjust the second grating mechanically to observe, again, a sharp moiré pattern. This linear mechanical adjustment reflects the level change in the container.

13.2.2.4 Focal Length Measurement

Moiré fringes and the Talbot effect have also been useful in the focal length measurement of an optical system. Two categories are used for these techniques: the first is based on the measurement of the moiré pattern's rotation (Nakano and Murata, 1985; Chang and Su, 1989; Su and Chang, 1990); the second is based on the beating of a moiré pattern (Glatt and Kafri, 1987b; Bernardo and Soares, 1988). In the first case, the focal length f is obtained by measuring the rotation angle α_n for the moiré fringes due to beam divergence changes to noncollimation the incident beam suffers at the first grating, as shown in Fig. 13.16. The moiré fringes are observed through a diffuse screen. The focal length will be calculated using the following equation (Nakano and Murata, 1985):

$$f = \frac{1}{\sin\theta\tan\alpha_n + \cos\theta - 1} \cdot \frac{np^2}{\lambda},\tag{13.16}$$

where θ is the angle between the lines of the gratings and p is the period of the gratings. A typical experimental setup used to make these arrangements is shown in Fig. 13.17.

The second case uses the beating of the moiré pattern produced when the lines of the grills are parallel ($\theta = 0$). In this case, the moiré fringes are caused by different periods on both grills, and the focal length is calculated by means of the equation

$$f = \frac{npp'}{\lambda}.\tag{13.17}$$

This last case is more difficult in practice, since the setup is more difficult to implement, and measuring the moiré pattern period p' is more time consuming than the angle α_n.

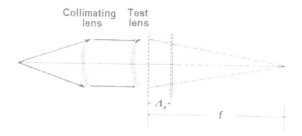

Figure 13.17 Experimental setup to determine the focal length of a lens.

An additional method that does not use collimated illumination has been reported (Sriram *et al.*, 1993). However, this method requires additional mechanical displacements to the gratings while the measurement is being performed.

Based also on moiré techniques and Talbot interferometry, systems have been built that not only measure the focal length of a lens but also its refractive power (Nakano *et al.*, 1990; Malacara Doblado, 1995). The strength of these techniques resides in the use of automatic digital processing of the moiré fringes.

13.2.2.5 Thickness Measurement

Thickness measurement is based on the very well-known method of projection moiré contouring (Hovanesian and Hung, 1971). Basically, it consists in projecting a grating over the object under study. The projected grating is deformed according to the topography of the object. By means of an optical system, the deformed grating is imaged on a similar reference grating, in such a way that a moiré pattern is obtained on the overlapping plane. Under this outline the moiré fringes represent contours or level curves of the surface object. This technique has been broadly used for measuring tasks in ways using a Talbot image as projection grating (Rodríguez-Vera *et al.*, 1991b), or determining form and deformation of engineering structures by means of digital grating superposition (Rodríguez-Vera and Servin, 1994).

A simple way to interpret the moiré fringes in this outline is through the contour interval (Dessus and Leblanc, 1973). For collimated illumination and far away observation, the contour interval (Rodriguez-Vera *et al.*, 1991b) is given by

$$\Delta z = \frac{p}{\sin \beta}, \tag{13.18}$$

where p is the projected grating period and β is the angle between illumination and detection optical axes, as shown in Fig. 13.18. Equation (13.18) displays height differences between a contour and the next one.

This projection outline can be simplified by analyzing the projected grating or a light line. Figures 13.19(a) and (b) show schematically a grating and a light line on a cube, respectively. Note that both projected grating and light line are deformed according to the topography of the surface. In this case we do not deal with closed contours or level lines, but with "grid contours" (Rodriguez-Vera *et al.*, 1992). These grid contours are similar to those of carrier frequency introduced in an interferom-

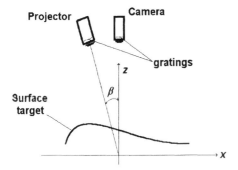

Figure 13.18 Scheme of the projection moiré contouring technique.

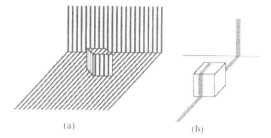

Figure 13.19 (a) Fringes produced by a projected linear grating, (b) projected light line.

eter. For the case of a step like the one shown in Fig. 13.20(a), the fringes are displaced. This displacement is a change of phase of the grid contour and, therefore, is related to the difference of height between the two planes. Figure 13.20(b) shows the three-dimensional plot of phase shift when applying the technique of phase-locked loop detection to determine the height between the planes that form the step (Rodriguez-Vera and Servin, 1994).

A limitation of this technique lies in the ambiguity that results when the differences of height are large enough so that projected fringes surpass the phase change by 2π. To solve this problem, the possibility of using a narrow light sheet to illuminate the object has been investigated recently (Muñoz Rodriguez *et al.*, 2000) as shown in Fig. 13.19(b). The object to be analyzed is placed on a servomechanism to be moved along an axis. During the movement, the object is illuminated with a sheet light beam (flat-shaped beam); the images are captured on a computer and processed. Figure 13.21 shows the object with the projected deformed light line formed by the incident light sheet and its topographical reconstruction.

13.2.3 Interferometric Methods in Small Distance Measurement

Fringe counting in an interferometer suggests an obvious application for the laser in short distance measurement. Interferometric methods are reviewed by Hariharan (1987). The high radiance and monochromaticity are the main useable characteristics in the laser. From the conceptually simple outline of a Michelson interferometer,

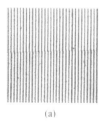

Figure 13.20 Projected linear grating on a step produced by two planes. (a) Grid contours; (b) difference of height between planes.

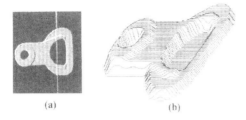

Figure 13.21 (a) Sheet of light projected on a metallic surface. (b) Three-dimensional reconstruction of the object of (a).

several improvements make it a more convenient and practical instrument. Some problems from the simple form are

(a) A light source must have a highly stable frequency to improve the precision. Since the light wavelength is used as a standard, as much as the source remains stable, the more precise the measurement will be. The original length standard was chosen as a discharge ^{86}Kr source. The frequency of this source is stable up to one part in 10^8, even better than some laser sources (Baird and Howlett, 1963). To improve the laser light stability, several means have been developed for frequency stabilization (McDuff, 1972). For a precise measurement device, frequency stabilization is essential.

(b) A light beam reflected back to the laser makes it unstable and the radiance will fluctuate with the moving-mirror displacement. For example, a Michelson interferometer forms two complementary interference patterns, one reflected back to the source. This reflected pattern eventually makes the laser unstable and the intensity will fluctuate as the optical path in one branch is changed. To avoid this problem, a nonreacting configuration must be used.

(c) Fringe counting in the interferometer is usually made electronically, which means the electronic counters should discriminate the fringe movement direction. To discriminate the direction movement, two detectors at a 90° phase difference are used.

(d) the fringe count across the aperature should be kept low and constant as the mirror is moved. This problem is easily solved by using retroreflectors instead of plane mirrors. Now, the phase shift across the full aperture is kept almost constant.

The so-called DC interferometer is shown in Fig. 13.22. The use of two corner cube retroreflectors is twofold: to have a noreacting configuration and to have a constant phase shift across the full aperture. For a nonabsorbing beamsplitter, the phase shift between both interference patterns will be 180°, but for an absorbing one, it can be adjusted to a 90° phase shift, making is possible to discriminate for the movement direction.

Another nonreacting interferometer for distance measuring is described by Minkowitz *et al.* (1967). A circularly polarized beam of light (Fig. 13.23) is obtained through a linearly polarized laser and a $\lambda/4$ phase plate. A first beamsplitter sepa-

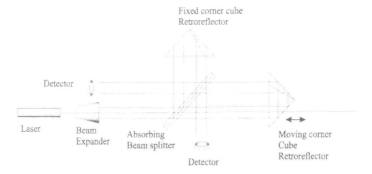

Figure 13.22   DC interferometer for distance measurement.

rates between the reference and measuring beam. Upon reflection, the reference beam changes its polarization while the measuring beam changes twice its polarization at both reflections on the moving corner cube reflector. Since both beams have opposite polarizations, the resulting beams are linearly polarized, with a polarization angle determined by the optical path difference. The polarization angle rotates 360° for every $\lambda/2$ path difference. Two polarizers are placed at each complementary output of the beamsplitter. Since the polarizers are rotated 90° each from the other, the irradiance varies sinusoidally with the corner cube displacement with a quadrature phase shift at the detectors.

A different approach is taken in an ac interferometer (Fig. 13.24) described by Burgwald and Kruger (1970) and Dukes and Gordon (1970) and commercially produced by Hewlett–Packard. A frequency stabilized He–Ne laser is Zeeman-split by a magnetic field and the optical beam has now two frequencies about 2 MHz apart. The two signals with frequencies f_1 and f_2 are circularly polarized and with opposite handedness. A $\lambda/4$ phase plate changes the signals f_1 and f_2 into two orthogonal linearly polarized beams, one in the vertical and the other horizontal plane. A

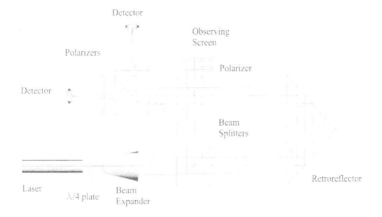

Figure 13.23   Minkowitz nonreacting interferometer.

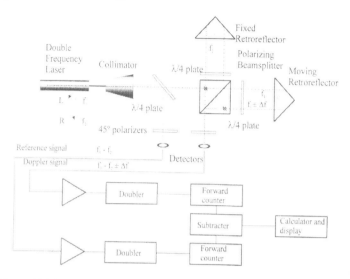

Figure 13.24 Hewlett–Packard distance measuring interferometer.

beamsplitter takes a sample of the beam. Since both signals are mixed, the resulting beam is linearly polarized, rotating at a frequency $f_1 - f_2$. Later, the polarizer makes a sinusoidally modulated beam at this frequency. Another portion of the beam is sent to a polarizing beamsplitter. One branch of the interferometer receives the f_1 component while the other receives the f_2 part of the signal. On each branch, a $\lambda/4$ phase plate converts the polarization to circularly polarized, but while one branch is left handed, the other is right handed. After two reflections at the corner cube reflectors, the handness is preserved but the $\lambda/4$ phase plate changes its polarization to a perpendicular one. Assuming a moving corner cube, the signal will be Doppler-shifted to a frequency $f_1 + \Delta f$. Both beams meet again at the polarizing beamsplitter and the outgoing beams are opposite-handed circularly polarized beams that add to form a slowly varying circularly polarized beam at a frequency $f_1 - f_2 + \Delta f$. The signals from the detectors are fed to a couple of digital counters, one increasing at a frequency $f_1 - f_2$ while the other at a frequency $f_1 - f_2 + \Delta f$. A digital subtracter obtains the accumulated pulse difference. For a stationary reflector, each counter increases at about two million counts per second, but the difference remains constant. Some advantages of this system lie in its relative insensibility to radiance variations from the laser.

13.2.3.1 Multiple-Wavelength Interferometry

At the best, all the measuring instruments can measure incremental fringes from an initial point. Absolute distance measurments can be made from a multiple-wavelength laser. The technique is similar to that described in Section 13.2.1 and is reviewed by Dändliker *et al.* (1998). For a double-pass interferometer, the distance is expressed as

$$D = N(\lambda/2) + \epsilon, \tag{13.19}$$

where N is an integer number and ϵ is a fractional excess less than half wavelength. For a multiple wavelength laser, the distance is

$$D = N_i[(\lambda_i)/2] + \epsilon_i, \tag{13.20}$$

Since N_i are integer constants, it is possible to know D from several wavelengths (λ_i's). Bourdet and Orszag (1979) use six wavelengths from a CO_2 laser for an absolute distance determination.

By using two wavelengths, a beating is obtained at the detector. This beating has a synthetic wavelength:

$$\Lambda = \frac{\lambda_1 \lambda_2}{\lambda_1 - \lambda_2}. \tag{13.21}$$

In practice, each wavelength has to be optically filtered for measurement. The absolute distance accuracy depends on the properties of the source. Both wavelengths must be known very accurately. To increase the nonambiguity range, multiple wavelengths are used by dispersive comb spectrum interferometry (Rovati *et al.*, 1998). If one laser is continuously tuned, a variable synthetic wavelength is used (Bechstein and Fuchs, 1998).

13.3 ANGULAR MEASUREMENTS

Angle measurements are done with traditional and interferometrical methods. For traditional methods, templates are used in the lower end for low-precision work while goniometers and autocollimators are used for high precision. The best precision is obtained with interferometric angle-measuring methods. Interferometric angle measurements have the additional advantage that they can be interfaced to automatic electronic systems.

13.3.1 Some Noninterferometric Methods

13.3.1.1 Divided Circles and Goniometers

For a rough scale, protractors and divided circles are used for angle determination. Although these devices are limited to a precision of about 30 min, modern electronic digital protractors and inclinometers can measure precise angles up to 0.5 min. At the optical shop, angle measurements can be made by means of a sine plate (Fig. 13.25). The sine plate can both support a piece of glass and measure its angle by itself or with a collimator. The angle is defined from the base plate's length and a calibrated plate inserted in one point. The angle is then calculated from the plate length and the calibrated plate length inserted in one leg to form a triangle. The angle is calculated assuming a rectangular triangle: hence its name of sine plate. With a good sine plate, an accuracy of 30 min can be achieved. A serrated table for angle measurement is described by Horne (1972) with an accuracy of 0.1 s.

Angle blocks are also used for comparison. These are available commercially to an accuracy of ±20 s. By reversing and combining the blocks, any angle between $1°$ and $90°$ can be obtained. Since angle blocks have precision flat polished faces, together with a goniometer, they can be used as an angle standard. A glass polygon used with an autocollimator can precisely calibrate goniometers or divided circles.

Figure 13.25 Sine plate.

A goniometer (Fig. 13.26) is a precision spectrometer table. Its divided circles are used for precise angle measurement. For angle measurement, the telescope reticle is illuminated and the reflected beam is observed. Prism or polygon angles are measured in a goniometer. The divided circle sets the precision for the instrument. The polygon under test is set at the table while the telescope is turned around until the reflected reticle is centered.

13.3.1.2 Autocollimators and Theodolites

An autocollimator is essentially a telescope with an illuminated reticle: for example, a Gauss or Abbe eyepiece (Fig. 13.27) at the focal plane of the objective. By placing a mirror at a distance from the autocollimator, the reflected beam puts an image from the reticle displaced from its original position at a distance

Figure 13.26 Goniometer.

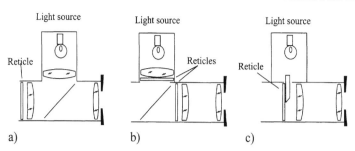

Figure 13.27   Illuminated eyepieces: (a) Gauss, (b) bright line, and (c) Abbe.

$$d = 2\alpha f, \tag{13.22}$$

where f is the collimator's focal distance and α the mirror tilt angle from perpendicularity. Since the angle varies with the focal length, the objective is critically corrected to maintain its accuracy. Precise focus adjustment can be achieved through Talbot interferometry (Kothiyal and Sirohi, 1987). Autocollimators are manufactured with corrected doublets, although some include a negative lens to form a telephoto system to get a more compact system. A complete description of autocollimators and applications can be found in Hume (1965).

Some autocollimators have a drum micrometer to measure the image displacement precisely, while others have an electronic position sensor to obtain the image centroid. Electronic autocollimators can go beyond diffraction limit, about an order of magnitude from visual systems. Micro-optic autocollimators use a microscope to observe the image position.

Autocollimators, besides measuring angles of a reflecting surface, can be used for parallelism in glass plates, or divided circles manufacturing (Horne, 1974). By slope integration, from an autocollimator, flatness measurements can be obtained for a machine tool bed or an optical surface (Young, 1967). In an autocollimator measurement, the reflecting surface must have high reflectivity and be very flat. When a curved surface is measured, this is equivalent of introducing another lens in the system with a change in the effective focal length (Young, 1967).

Theodolites are surveying instruments made from a telescope mounted in a precise altitude-azimuth mounting with a spirit level and a tree screw base for leveling. Besides giving a precise means of measuring both the elevation and the azimuth angles, a reticle with stadia markings permits distance measurements. Old instruments with engraved circles had an accuracy of 20 arcmin. Modern electronic theodolites contain digital encoders for angle measurements and electronic range finders. Angle accuracy is better than 20 arcsec. Errors derived from eccentricity and perpendicularity are removed by rotating both axes 180° and repeating the measurement.

Besides surveying, theodolites are used for angle measurement in the optical shop from reference points visually taken through the telescope for large baseline angles.

13.3.1.3 Angle Measurement in Prisms

Angle measurement in prisms is very important, since prisms are frequently used as angle standards. Optical prisms are made with 30°, 40°, and 90° angles. Optical means of producing these angles are easily obtained without the need for a standard.

A very important issue in prism measurement is the assumption of no pyramidal error. For a prism to be free of pyramidal error, surface normals for all faces must lie in a single plane (Fig. 13.28). Pyramidal error can be visually checked (Martin, 1924; Johnson, 1947; Taarev *et al*, 1985) for a set of three faces by examining both the reflected and refracted image from a straight line (Fig. 13.29). Under pyramidal error, the straight line appears broken. A far target can be used as a reference to measure the angle error.

Precise angle replication can be made by mounting the blank glass pieces in the same axis as the master prism (Twyman, 1957; DeVany, 1971). An autocollimator is directed to see the reflected beam from each face on the master prism (Fig. 13.30).

Precise 90° prisms can be tested either visually or with auxiliary instruments. By looking from the hypotenuse side, a retroreflected image (Johnson, 1947) from the pupil is seen that depends on the departure from the 90° angle, as shown in Fig. 13.31. This test can be improved, as shown by Malacara and Flores (1990) by using a target with a hole and a cross in the path, as shown in Fig. 13.32.

An autocollimator can be used to increase the sensitivity of the test (DeVany, 1968; Taarev, 1985). Two overlapped images are observed, with a separation $2N\alpha$, where α is the magnitude of the prism angle error and its sign is unknown. To determine the error of the sign, DeVany (1978) suggests defocusing the autocollimator inward. If the images tend to separate, then the angle in the prism is larger than 90°. Another means of determining the angle error is by introducing, between the autocollimator and the prism, a glass plate with a small wedge with a known orientation. The wedge is introduced to cover half the aperture. Polarized light can also be used as suggested by Ratajczyk and Bodner (1966).

Figure 13.28 Pyramidal error.

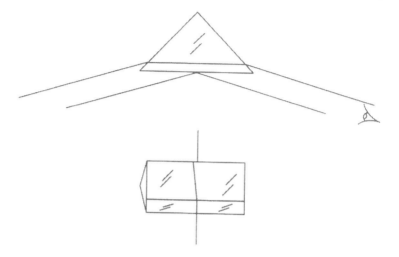

Figure 13.29 Pyramidal error check by reflected and refracted reference observation.

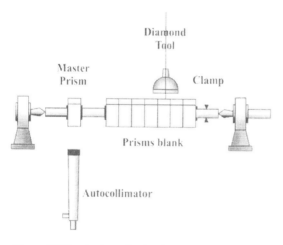

Figure 13.30 Angle replication in polygons with an autocollimator.

$\theta = 90^\circ$ $\theta < 90^\circ$ $\theta > 90^\circ$

Figure 13.31 Pupil's image in a rectangular prism with and without error.

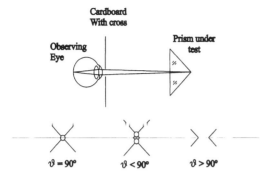

Figure 13.32 Prism angle error observation.

13.3.1.4 Level Measurement

Levels are optical instruments that define a horizontal line of sight. The traditional form is a telescope with a spirit level. Once the spirit level is adjusted, the telescope is aimed at a constant level line. For an original adjustment of the spirit level (Kingslake, 1983), a pair of surveying staves are set some distance apart (Fig. 13.33). The level is directed to a fixed point in one of the staves from two opposite directions and the corresponding point in the other staff is compared. Any difference in the reading is compensated by moving the telescope to a midpoint in the second staff; then the spirit level is fixed for the horizontal position.

An autoset level (Fig. 13.34) relies on a pendulum-loaded prism inside the telescope tube. A small tilt is compensated by the pendulum movement, although other mechanisms are also used (Ahrend, 1968). A typical precision for this autoset level lies within 1 arcsecond and works properly for a telescope angle within 15 arcmin (Young, 1967).

13.3.1.5 Parallelism Measurement

Rough parallelism measurements can be done with micrometers and thickness gauges, but they need a physical contact with a damage risk; besides, it is necessary to make several measurements to make a reliable test. Optically, parallelism is mea-

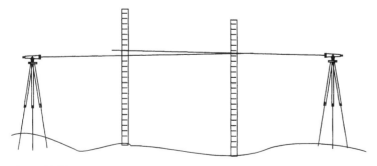

Figure 13.33 Level adjustment with two staves.

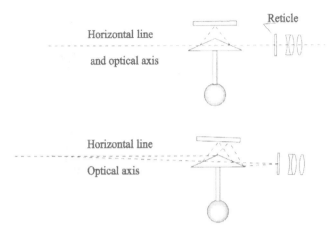

Figure 13.34   The autoset level.

sured with the versaility provided by an autocollimator. The reticle from the autocollimator is reflected back by both surfaces simultaneously in a transparent optical
plane. Any departure from parallelism is seen as two reticles. Under this setup, the
reflected image from the first surface is adjusted for perpendicularity and the angle
for the second surface is simply found from the Snell's law. Parallelism in opaque
surfaces can be measured after Murty and Shukla (1979) by using a low-power laser
incident in an optical wedged plate as shown in Fig. 13.35. The plate under test is
placed in a three point support base, then the reflected point is noted on a distant
screen. Next the plate is rotated 180° and the new position for the beam is recorded.
Let the separation for both points be d, and the screen distance from the plate is D;
then the wedge angle α is

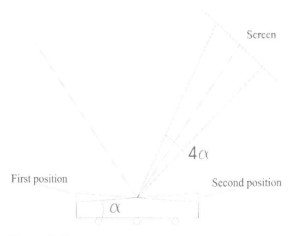

Figure 13.35   Parallelism measurement in an opaque plate.

$$\alpha = \frac{d}{4D}.$$

(13.23)

By using two opposed collimators in a single optical bench, they are adjusted to center the other collimator image. The plate to be checked for parallelism is inserted between both collimators and first adjusted for perpendicularity in one of the collimators while the other reads the amount of departure for the second surface (Tew, 1966). The latter technique could be used for nontransparent surfaces.

13.3.2 Moiré Methods in Level, Angle, Parallelism and Perpendicularity Measurements

The different optical techniques that follow have their foundation in the formation of moiré fringes (see Section 13.2.2). These techniques, in principle, will give the shadowy characteristics of angular measurement with applications toward the same angular measurement, parallelism or collimation, perpendicularity, alignment, slope, and curvature.

If a grating of period p_p is projected on a flat object, the size of the period changes if it is observed to a different angle from that of projection, as shown in Fig. 13.36. In this figure, it is supposed that the projection and observation systems are far from the surface, in such a way that the illumination and reflected beams are plane wavefronts. An angular change of the surface also produces a change in the period of the observed grating, as shown in Fig. 13.37. If the observed gratings are superimposed before and after the surface is tilted, a moiré pattern of period p', given by Eq. (13.14), will be observed. Then, we can find the relationship between the observed pattern's period and the angular displacement of the surface, knowing the period of the projected grating. The observed period p_1 before the surface is moved is given by (see Fig. 13.37):

$$p_1 = p_p \frac{\cos \delta}{\cos \beta},$$

(13.24)

where p_p is the period of the projected grating and δ and β are the angles to the z-axis of the optical axes of observation and projection systems, respectively. Figure 13.37

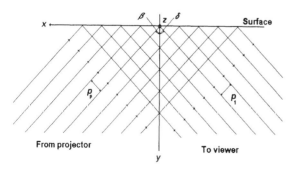

Figure 13.36 Diagram of a linear grating projected on a flat surface. The grating lines are perpendicular to the x–y plane.

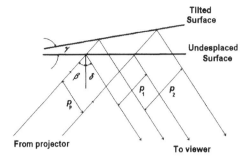

Figure 13.37 Parameters for calculating the moiré fringe period when the test surface is tilted.

shows all these parameters. The observed period p_2 after the surface is tilted is given by

$$p_2 = p_{\mathrm{p}} \frac{\cos(\delta - \gamma)}{\cos(\beta + \gamma)}, \tag{13.25}$$

where γ is the angular displacement that suffered the surface. Substituting Eqs (13.24) and (13.25) in (13.13), one has

$$p' = p_{\mathrm{p}} \left| \frac{\cos\delta\cos(\delta - \gamma)}{\cos(\beta + \gamma)\cos\delta - \cos\beta\cos(\delta - \gamma)} \right|. \tag{13.26}$$

For the particular case of a perpendicular observation to the surface, where $\delta = 0$, Eq. (13.26) is transformed into

$$\tan\gamma = \frac{p_{\mathrm{p}}}{p'\sin\beta}. \tag{13.27}$$

This last equation allows us to calculate the inclination that suffers a surface by means of the moiré method. The tilt angle γ that moves the surface from its original position is determined from the moiré pattern's period p', the projected grating period p_p, and the projection angle β.

13.3.2.1 Tilt Measurement

Moiré fringes are used in three methods to measure angular variations of a surface. One is based on the use of linear grating (Nakano and Murata, 1986; Nakano, 1987); another is the use of nonlinear gratings (Glatt and Kafri, 1987a; Ng and Chau, 1994; Ng, 1996); and, the third, in the projection of interference fringes (Dai *et al.*, 1995, 1997).

The interference fringe projection method is based on the detection of phase changes of the projected fringes when the object is tilted at a small angle (Dai *et al.*, 1995). The projected interference fringes on the object are reflected and they are detected in two points. The change in the phase difference between the two detected phase points is a function of the object rotation angle. The sensibility of the technique depends on the position of the two detection reference points in the fringe pattern. With this technique, angular variations up to 17 mrad/arcsec can be detected. The technique suffers from an ambiguity: the maximum detected phase change is 2π. In order to increase the sensibility of the method, projection of two interference fringe patterns with different period were employed (Dai *et al.*, 1997).

Techniques of angular measurement with nonlinear gratings can use two (Glatt and Kafri, 1987a; Ng, 1996) or a single circular grating (Ng and Chau, 1994). The basic technique consists of detecting the moiré pattern when the circular gratings are superimposed. For the case of a single circular grating, the shadow moiré technique is used (Takasaki, 1970). The grating is placed in front of a reference plane and illuminated. A projected shadow of the grating is formed on the reference plane and is superimposed with the physical grating, producing in this way the moiré pattern. The observed moiré pattern, with a camera located perpendicularly to the reference plane, is similar to that of the Fig. 13.11(c).

Nakano (1987) uses the Talbot effect to measure small angular variations of a surface. Figure 13.38 is an experimental diagram to carry out tilt measurements. By means of a collimated monochromatic beam a grating g_1 is illuminated. The wavefront passing through the grating is reflected on a surface M that makes an angle γ with respect to the optical axis. This reflected wavefront impinges on a second grating g_2. The grating g_2 is placed at a Talbot distance, with respect to g_1. So, in the observation plane, a moiré pattern is formed. If the angle between the lines of the gratings is θ, the moiré fringes appear inclined a quantity α_1, to the x'-axis. If the surface object is tilted an angle $\Delta\gamma$ from its original position, the Talbot self-image suffers a modification. This modification means that the moiré fringes lean at an angle α_2, to the same x'-axis. Then, the small angular variation $\Delta\gamma$ will be given by Nakano (1987).

The inclination angles of the moiré fringes are measured in a direct way or automatically, taking as reference the first position of the surface M. The accuracy of this technique depends on the method used to measure the inclination angles of the moiré fringes, α. Other error sources affecting the sensitivity of the method are the measurement of the Talbot distance and the angle between the grating lines, θ.

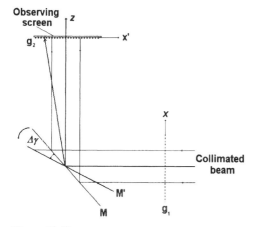

Figure 13.38 Experimental outline to measure the tilt angle $\Delta\gamma$ of the surface M.

13.3.2.2 Optical Collimation

A very simple technique for beam collimating is also based on the moiré and Talbot effects (Silva, 1971; Foueré and Malacara, 1974; Yokozeki *et al.*, 1975). The experimental arrangement consists of a couple of identical linear gratings, a laser, a beam expander, and the collimating lens, as shown in Fig. 13.39. The principle consists on adjusting the collimating lens mechanically to coincide its focal points with that of the expander lens. A way to corroborate this coincidence of points is by means of the moiré pattern. It is known, by Eq. (13.14), that when two identical linear gratings are superimposed, the moiré pattern's period becomes infinitely big when the angle among the lines of the gratings goes to zero. On the other hand, if the beam illuminating the first grating, in the outline of the Fig. 13.39, is not collimated, the self-image is amplified and the moiré pattern appears; even the angle among the grating lines becomes zero (see Eq. (13.13). So, the mechanical movement of the collimating lens (along the optical axis) and of the second grating must be adjusted until the moiré pattern disappears when $\theta = 0$. A similar outline to that previously described has been reported by Kothiyal *et al.* (1988) and Kothiyal and Sirohi (1987). Each one of the gratings is built with two different frequencies and dispositions of the lines, giving a greater sensitivity to the technique.

13.3.2.3 Optical Level and Optical Alignment

Under certain conditions an optical collimator, as the one described in previous sections, can serve as an optical level. This way, for example, it is possible to place one or several objects in a straight line. This line can be parallel or perpendicular to the optical axis of the system. In consequence, this system can serve as help for alignment, center, and measure perpendicular deviation. This idea of using the moiré as an optical measurement tool has resulted in instruments for precise optical levels and "aligners" (Palmer, 1969). Some of these use circular gratings (Patorski *et al.*, 1975) or combinations of linear and circular gratings (Reid, 1983).

13.3.2.4 Slope and Curvature Measurements

Measurement of a surface by means of the moiré effect is very well known (Takasaki, 1970; Hovanesian and Hung, 1971). The fundamental characteristic of this technique is the formation of contours or level curves, as mentioned in the section 13.2.2, Eq. (13.18). However, for some applications, it is useful to know contours of slopes or local surface curvature, mainly when it is required to know the field of mechanical strains on this surface. A useful technique to measure the slope and curvature of the surface is the reflection moiré (Rieder and Ritter, 1965; Kao and Chiang, 1982;

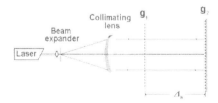

Figure 13.39 Experimental arrangement for optical collimating.

Asundi, 1994). This technique is based on projecting a linear grating on a mirror-like test surface. The projected grating is recorded photographically or electronically twice before and after subjecting the object to a flection. As a result, a moiré pattern is formed and interpreted as a slope map (derived) in the perpendicular direction to the lines of the projected grating.

Another possibility is to obtain a local curvature map on the plate (Ritter and Schettler-Koehler, 1983). To obtain this map, two slope moiré maps are superimposed, giving a second moiré pattern (moiré of moiré). The slope maps that will be superimposed to give the moiré of moiré are displaced a small quantity. This curvature map (second derived) is also perpendicular to the lines of the projected grating.

13.3.3 Interferometric Methods

Interferometric angle measurement seems an obvious task for a simple Michelson interferometer. A small tilt in one mirror produces a fringe pattern. Unfortunately, since the interferometer sensitivity is very high, this device could practically measure very small angles only. Interferometric angle measurement with both high precision and large range is a desired device. Angular measurements can be done (Sirohi and Kothiyal, 1990) with a distance-measuring interferometer, built as shown in Fig. 13.40. By tilting the retroreflectors assembly, the distance difference changes and the angle can be obtained from the equation:

$$\theta = \arcsin \frac{\Delta x}{L}, \tag{13.28}$$

where L is the mirror separation and Δx is the distance difference between retroreflectors. From this equation, it is evident that the angle precision depends on the angle, so the measurement range is also limited.

A laser Doppler displacement interferometer is used for a large angle precise measurement in a telescope. The setup is the same as the previously described system. With a single interferometer, the resolution is 0.01 arcsec. For a mirror separa-

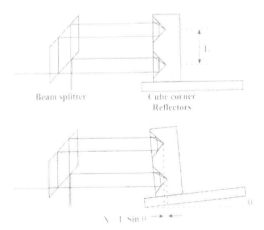

Figure 13.40 Interferometric angular measurement.

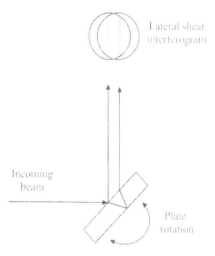

Figure 13.41 Lateral shear interferometer for angle measurement.

tion of 28 mm, the maximum angle is 4°. To cover the full 360° circle, a polygon and two measuring systems can be used (Ravensbergen *et al.*, 1995).

A Murty's shearing interferometer has been proposed by Malacara and Harris (1970) that is accurate within tenths of arcseconds. The basic setup is shown in Fig. 13.41. A collimated beam of light is reflected from the two faces of a plane parallel faces glass. The angle can be obtained from the fringe count as the plate rotates. This method can be used for any angle but only for a limited angle span.

Laser speckle interferometry has been suggested for angle measurements. The objected is illuminated with a laser and a defocused image is formed at H (Fig. 13.42). A double exposure of the object will form a fringe pattern that reflects the amount of rotation (Tiziani, 1972; Françon, 1979).

13.4   VELOCITY AND VIBRATION MEASUREMENT

Velocity measurement u involves two physical magnitudes to be determined: displacement Δx and time Δt. The measurement of time is essentially a process of counting. Any phenomenon that repeats periodically can be used as a time measurement; this measure consists on counting the repetitions. Of the many phenomenon of this type that occur in nature, the rotation of the Earth around its axis is adopted. This movement, when reflected in the apparent movement of the stars and the sun, is a basic unit that one has easily to reach.

Another way of measuring time is artificially, by means of two apparatuses: one that generates periodic events and another to count these events. Today, optical clocks operate by means of atomic transition measurements between energy levels of the cesium atom (Lee *et al.*, 1995; Teles *et al.*, 1999). Also, electronics can be used to implement instrumentation of time measurement. Perhaps the most convenient and widely utilized instruments for accurate measurement of time interval and frequency

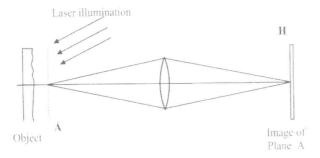

Laser illumination

H

Object

A

Image of
Plane A

Figure 13.42 Speckle angle measurement.

are based on piezoelectric crystal oscillators, which generate a voltage whose frequency is very stable (Bottom, 1982).

Distance measurement also has its historical ingredient, and has also been based on the observation of natural phenomena. For example, Eratosthenes (273 BC) tried to calculate the diameter of the Earth. This was based on his observations in Assuan that, in the summer solstice, the sun's rays fell perpendicularly and therefore a shadow was not cast, while at the same time, in Alexandria, there was a ray inclination of $7.2°$ in the shadows. Performing the appropriate calculations, this Greek astronomer obtained the Earth's diameter with an error of 400 km from the actual value (Eratosthenes measured 39,690 km; the actual diameter is 40,057 km). Currently, popular techniques for distance measurement use graduated rules, traced to the standard meter (Busch, 1989). However, as we have seen in the present chapter, there are several optical techniques that can be used to make this measurement with a very high precision.

This section describes some different optical methods for local displacement, velocity, acceleration, and vibration measurements.

13.4.1 Velocity Measurement Using Stroboscopic Lamps

Translation and rotational velocity can conveniently be measured using electronic stroboscopic lamps which emit light in a controlled and intermittent way. For the case of translation velocity measurement, the movement of the target is recorded on a photographic film when the camera shutter is opened and several shots of the strobe lamp are fired during that time. The measurement is made of the number of lamp shots time, and the serial position of the registered target. In order to measure the target displacement, a fixed rule is placed along the target path.

Sometimes, it is desirable to measure a value of average velocity of an object over a short distance or time interval, and velocity or time is not required in a continuous way. A useful basic optical method is to somehow generate a pulse of light when the object in movement passes through two points whose spacing is exactly known (Doebelin, 1990). If the velocity were constant, any spacing could be used: for large spacing, of course, one has a better accuracy. If the velocity is varying, the spacing Δx should be small enough so that the average velocity over Δx is not very different from the velocity at either end of Δx.

Rotational velocity can be measured using a strobe lamp firing the target and adjusting its shot frequency until the object is observed motionless. At this setting the lamp frequency and motion frequency are identical, and the numerical value can be read from lamp's calibrated dial for pulse repetition rate to an accuracy of about ±1% of the reading; or up to 0.01% in some units with crystal-controlled timebase (Doebelin, 1990).

13.4.2 The Laser Interferometer

Although the principle of the interference of light as a measuring tool is very old, it continues to be advantageous. Albert A. Michelson, in the 1890s, was the first to use the interferometer, that bears his name, to measure distance and position in a very precise way.

The first efforts to use interferometry for the study of mechanical vibrations dates back to the 1920s (Osterberg, 1932; Thorton and Kelly, 1956). The use of the interferometer replaced stroboscopic techniques in those experiments where measuring oscillation frequencies are higher than those in the range of stroboscopic lamps.

The advent of the laser in the 1960s popularized the interferometer as a useful optical instrument for measuring distances and displacements with high precision (Dukes and Gordon, 1970; Hariharan, 1987; Fischer *et al.*, 1995; Gouaux *et al.*, 1998), as well as accelerometer and vibrometer calibration (Ruiz Boullosa and Perez Lopez, 1990; Ueda and Umeda, 1995; Martens *et al.*, 1998) to high measuring velocities.

Figure 13.43 shows a Michelson interferometer. A laser beam is divided in two parts: beam 1 hits the moving (test) mirror directly and beam 2 impinges on the fixed (reference) mirror. Due to the optical path difference between the overlapping beams, light and dark fringes are produced on the observing screen. The motionless test mirror produces a static fringe pattern. Cycles of maxima and minima passing through a fixed point on the observation screen are detected when the mobile mirror

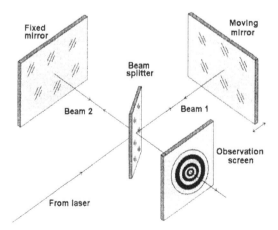

Figure 13.43 The laser interferometer.

moves. Fringe displacement is due to a phase change (and therefore, optical path change) of the beam 1 with respect to the fixed one. If we know the laser light wavelength to be, for example 0.5×10^{-6} m, then each 0.25×10^{-6} m of mirror movement corresponds to one complete cycle (light to dark to light) at the fixed point on the observation screen. It is possible to calculate the displacement of the mobile mirror between two positions by counting the number of cycles. Therefore, the mobile mirror velocity is determined by calculating the number of cycles in the unit of time.

Electronic fringe counting involves two uniform interferograms, one of them with an additional quarter wavelength optical path difference introduced between the interfering beams. Two detectors viewing these fields provide signals in quadrature, which are used to drive a bidirectional counter which gives the changes in the integral fringe order (Peck and Obetz, 1953). Nowadays, the two signals can also be processed in a microcomputer to give an accurate estimate of the fractional fringe order (Smythe and Moore, 1984; Martins *et al.*, 1998).

Another technique used to carry out fringe counting uses a laser emitting two frequencies, which avoids low-frequency laser noise (Dukes and Gordon, 1970). These kinds of interferometers have been widely used for industrial measurements over distances up to 60 m.

Additional techniques, where the phase of moving interference fringe patterns are measured, have been applied. These methods are heterodyne phase measurement and phase lock detection (Greivenkamp and Bruning, 1992; Dändliker *et al.*, 1995; Malacara *et al.*, 1998). In the heterodyne technique, also known as AC interferometry, the interferometer produces a continuous phase shift by introducing two different optical frequencies between both arms. With this approach the interferogram intensity is modulated at the difference frequency. Phase lock interferometry involves applying a small sinusoidal oscillation to the reference mirror (Johnson and Moore, 1977; Matthews *et al.*, 1986; Fischer *et al.*, 1995).

13.4.3 Laser Speckle Photography and Particle Image Velocimetry

Laser speckle photography (LSP) and particle image velocimetry (PIV) are two nearly related optical techniques for the measurement of in-plane two-dimensional displacement, rotation, and velocity (Siorhi, 1993, Raffel *et al.*, 1998). LSP is used primarily for the measurement of the movement of solid surfaces, while PIV is used in applications of fluid dynamics. In both cases, the principle of operation is based on photographic recording under light laser illumination.

In LSP, light scattered from a moving object illuminated by coherent laser light is double-exposure photographed with a known time delay between exposures, as shown in Fig. 13.44(a). In this way, locally identical but slightly shifted speckle patterns are recorded, which can be analyzed optically to find local displacement vectors at the surface of the moving object surface.

In PIV a double-pulsed laser sheet of light is used to illuminate a plane within a seeded flow, which is photographed to produce a double-exposure transparency as shown in Fig. 13.44(b). It is important to note, however, that, for practical seeding densities in PIV, the recorded image consists no longer of two overlapping speckle patterns but of discretely resolved particle–image pairs corresponding to both exposures.

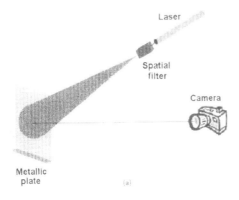

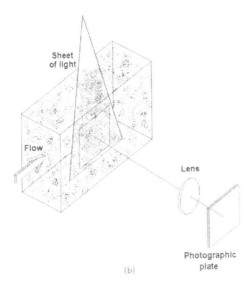

Figure 13.44 (a) Experimental recording of laser speckle photography and (b) particle image velocimetry.

In both cases, the double-exposure transparencies could be analyzed in two ways. The first method, a whole-field approach, consists of optically filtering the recorded transparency in a setup, as schematized in Fig. 13.45.

Supposing that the object shift is *s* along the *x*-axis; displacement fringe contours appear on the Fourier focal plane described by (Gasvik, 1987)

$$|\mathcal{F}(|U(x)|^2)|^2 \cos^2\left(\frac{s\omega}{2}\right),$$ (13.29)

where $\mathcal{F}\{U(x)\}$ means the Fourier transform of $U(x)$, the complex amplitude distribution of the object, and $\omega = 2\pi\xi/\lambda f$, with f being the focal length. Then, object's displacement can be measured by

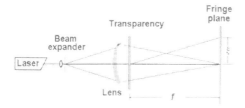

Figure 13.45 Fringe formation by the Fourier method.

$$s_0 = \frac{\lambda f}{M\xi},$$ (13.30)

where M is a magnification produced by the recorder camera objective, and ξ denotes the distance from zero diffraction order, parallel to the x-axis.

The second method, a point-by-point approach, consists of scanning the transparency by means of a narrow laser beam (Fig. 13.46). The laser beam diffracted by the speckles (or seed images in the PIV case) lying within the beam area gives rise to a diffraction halo. The halo is modulated by an equidistant system of fringes arising from the interference of two identical but displaced speckles, like the Young's interferometer. The directions of these fringes are perpendicular to the displacement direction. The magnitude of displacement, inversely proportional to the fringe spacing d, is given by

$$S = \frac{\lambda z}{Md},$$ (13.31)

where z is the separation between the transparency and the observation screen. Figure 13.47 shows examples of such a point-by-point approach obtained for different linear displacements of a metallic plate (Lopez-Ramirez, 1995). Young fringes were formed by displacing the metallic plate 120, 100, 80, 60, 40, and 20 µm, respectively. Note that the number of fringes increases with the displacement.

By using LSP, in-plane rotation of a metallic plate is measured. In this case, in-plane displacements are not uniform point to point, because different speckles move in different directions. Near to the rotation axis, speckles will move less than at the

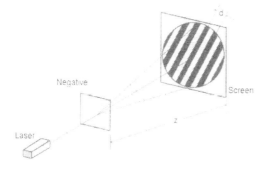

Figure 13.46 Young's fringe formation.

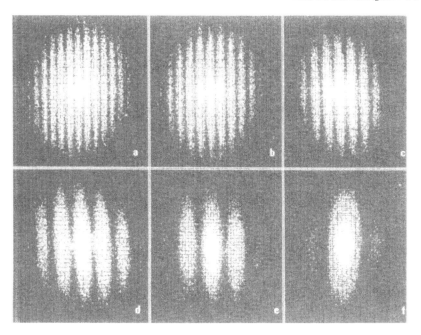

Figure 13.47 Young's experimental fringes for different displacements (reproduced by kind permission of J. Manuel Lopez-Ramirez).

ends. The recording step is made in such a way that the rotation and viewing optical axes coincide. By placing the transparency into to slide, movable in both horizontal and vertical directions, measurement on different points on the transparency were carried out. Knowing the distance r from rotation center to the sampling point and the displacement s, it is easy to know the rotation angle by means of the relationship

$$\alpha = \frac{Ms}{y},\tag{13.32}$$

By combining Eqs (13.32) and (13.31), we obtain

$$\alpha = \frac{Mz}{yd},\tag{13.33}$$

Figure 13.48(a) shows Young's fringe images for different points on the same plate. These points are localized, as shown in Fig. 13.48(b). Both direction and fringe separation can be measured directly on the observation screen. In order to check experimentally, three zones on the surface object are analyzed. Figure 13.49 shows such zones. Measurements from the center to each zone are measured directly on the photographic plate. In this case, the distance between transparence and observation screen is $z = 46$ cm, the wavelength 632.8 nm, and the optical recording system amplification 0.2222. Table 13.1 show figures of the measurements obtained by using Eq. (13.33).

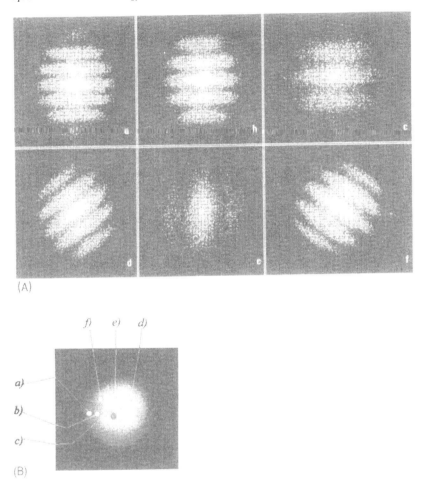

Figure 13.48 (A) Young's experimental fringes obtained for a plate under rotation from the points shown in (B) (reproduced by kind permission of J. Manuel Lopez-Ramirez).

Presented results have been made by manual evaluation of the Young's fringes; therefore, it is a time-consuming, tedious and impractical procedure. For this reason, considerable effort has been put into developing optical systems with digital image processing to automate the analysis procedure (Pickering and Halliwell, 1984; Kaufmann, 1993; Malacara *et al.*, 1998).

Other ways of reducing times of dynamic phenomena by means of PIV and LSP are using streak and CCD cameras instead of making photographic recordings (Grant *et al.*, 1998; Fomin *et al.*, 1999; Funes-Galanzi, 1998). Nowadays, experimental setups have been implemented by using two cameras to make three-dimensional measurements from stereoscopic images. Together with the modern recording instruments, computational techniques have been developed for a reliable and quick interpretation, such as the use of neural networks (Grant *et al.*, 1998).

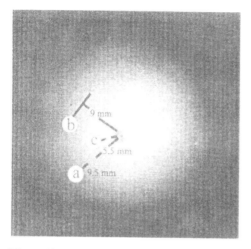

Figure 13.49 Zones analyzed on the transparency (reproduced by kind permission of J. Manuel Lopez-Ramirez).

13.4.4   Laser Doppler Velocimetry

13.4.4.1   Physical Principle

Laser Doppler velocimetry (LDV) also known as laser Doppler anemometry (LDA) is now a well-established optical nondestructive technique for local measurements of velocity. Although initially this technique, developed at the end of the 1960s, was applied in the field of fluid flows, it has now been extended to solid mechanics (Drain, 1980).

The physical principle of all LDVs lies in the detection of Doppler frequency shift of coherent light scattered from a moving object. The frequency shift is proportional to the component of its velocity along the bisector angle between illuminating and viewing directions. This frequency shift can be detected by beats produced either by the scattered light and a reference beam or by scattered light from two illuminating beams incident at different angles. An initial frequency offset can be used to distinguish between positive and negative movement direction (Durst *et al.*, 1976).

Figure 13.50(a) shows a schematic diagram of the effect when a particle moving with a velocity u scatters light in a direction k_2 from a laser beam, fundamentally, a single frequency traveling in a direction k_1, where k_1 and k_2 are wavenumber vectors. The light frequency shift, Δf, produced by the moving point is given by (Yeh and Cummins, 1964):

Table 13.1 Displacement Calculation and Angle for Rotational Movement

Zone	d (mm)	r (mm)	s (mm)	α (rad)
A	20.66	9.50	63.40	0.00148
B	21.60	9.00	60.64	0.00150
C	36.00	5.50	36.40	0.00147

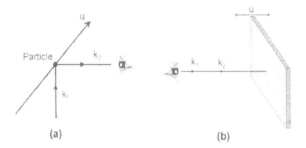

Figure 13.50 The Doppler effect.

$$\Delta f = (k_2 - k_1) \cdot u = K \cdot u, \tag{13.34}$$

where $K = k_2 - k_1$.

Figure 13.50(b) shows a geometry which is appropriate for solid surface velocity measurement where the laser beam is directed at a solid surface which acts like a dense collection of particle scatters. In this situation k_1 and k_2 are parallel so that the Doppler frequency shift measured corresponds to the surface velocity component in the direction of the incident beam and is given by

$$\Delta f = 2ku. \tag{13.35}$$

Measuring the frequency shift Δf, then the linear relationship with surface velocity to measure u can be used. Furthermore, tracking the changing Dopper frequency it is possible to have a means of time-resolved measurement. The scattered light, as shown in Figs 13.50(a) and (b) has a frequency which is typically 10^{15} Hz; that is too high to demodulate directly. Therefore, Δf should be measured electronically by mixing the scattered light with another frequency shifted *reference* beam so that the two signals are heterodyned on a photodetector face. This is shown schematically in Figs 13.51 and 13.52, where a beamsplitter has been used to mix the two beams. These figures also demonstrate the need to frequency preshift the reference beam. If the reference beam is not preshifted, as in Fig. 13.51 then, when the target surface moves through zero velocity, the Doppler signal disappears and cannot be tracked (Halliwell, 1990). Figure 13.52 shows the situation required where the target surface frequency modulates the carrier frequency which is provided by the constant frequency preshift (f_R) in the reference beam. Frequency tracking the changing Doppler frequency then provides a time-resolved voltage analogue to the surface velocity.

13.4.4.2 Frequency Shifting and Optical Setups

All LDVs work on the physical principle described above and differ only in the choice of optical geometry and the type of frequency-shifting device used. Just as frequency shifting is paramount for solid surface velocity measurement, it is also included as a standard item in commercially available LDV systems which are used for flow measurement. It is obviously necessary for measurements in highly oscillatory flows and in practice it is extremely useful to have a carrier frequency corre-

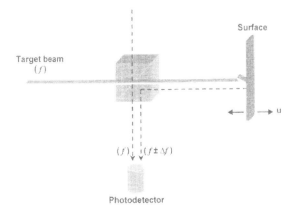

Figure 13.51   Reference beam mixing.

sponding to zero motion for alignment and calibration purposes. For instrumentation purposes there are six types of signal which can be used to modulate sensors. These are mechanical, thermal, electrical, magnetic, chemical, and radiant (Medlock, 1986). But, the most commonly used form of frequency shifting is the Bragg cell (Crosswy and Hornkohl, 1973). The incident laser beam passes through a medium in which acoustic waves are traveling and the small-scale density variations diffract the beam into several orders. Water or glass are amongst the media of choice and usually the first-order diffracted beam is used. In this way the frequency-shifted beam emerges at a slight angle to the incident beam, which requires compensation in some setups. Physical limitations often restrict the frequency shift provided by a Bragg cell to tens of megahertz, which is rather high for immediate frequency tracking demodulation. Consequently, two Bragg cells are often used which shift both target and reference beams by typically 40 MHz and 41 MHz, respectively. Subsequent heterodyning then provides a carrier frequency that is readily demodu-

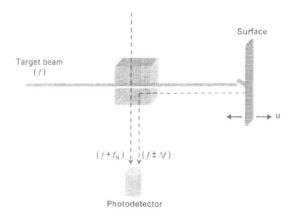

Figure 13.52   Preshifted reference beam mixing.

lated. An alternative scheme sometimes utilized is to preshift with one cell and to electronically downbeat the photodetector output prior to demodulation (Hurst *et al.*, 1981).

Figures 13.53 and 13.54 show two possible optical setups which can be used for solid surface velocity measurement and which incorporate Bragg cells for frequency shifting. Their compactness, electronic control, and freedom from mechanically moving parts make them popular as preshift devices in commercially available LDV systems for laboratory uses. Their presence does, however, add to the expense of the system, which then requires additional electronic and optical components.

Rotating a diffraction grating disk through an incident laser beam provides another common means of frequency shifting. Just as the small density variations in the Bragg cell diffracted the beam, in the diffraction grating case, the small periodic thickness variations perform the same task. Advantages over the Bragg cell are the smaller shifts obtained (~ 1 MHz) and the easy and close control of the latter through modification of the disk speed. Disadvantages are the mechanically moving parts and the inherent fragility of the disk itself, which is expensive to manufacture. Extra optical components are again needed to control the cross-sectional areas of the diffracted orders. A typical optical geometry for the measurement of solid surface vibration using a rotating diffraction grating as a frequency shifter, as shown in Fig. 13.55. Other frequency shifting devices that have been utilized consist of Kerr cells (Drain and Moss, 1972), Pockel's cells, and rotating scatter plates (Rizzo and Halliwell, 1978; Halliwell, 1979). In the case of an LDV using a CO_2 laser, a liquid-nitrogen-cooled mercury–cadmium–telluride (MCT) detector is used to detect a Doppler-shift signal (Churnside, 1984). The liquid nitrogen should be replenished

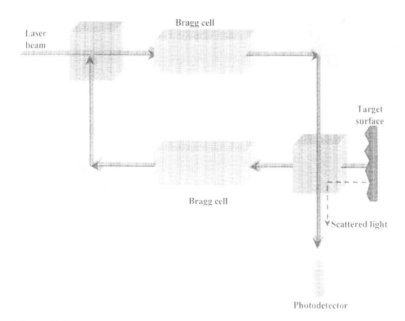

Figure 13.53 Geometry on axis.

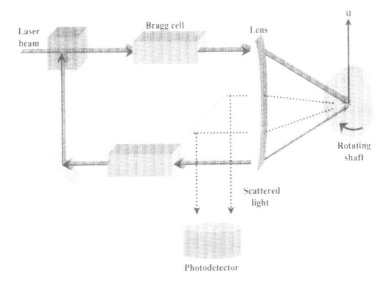

Figure 13.54   Torsional geometry.

in the MCT detector at a predetermined time interval. Lately, optogalvanic effect has been investigated with the purpose of detecting a Doppler-shift signal in the self-mixing-type CO_2 LDV without using an MCT detector (Choi *et al.*, 1997).

Another important variation of the LDVs uses stabilized double-frequency lasers (Doebelin, 1990; Müller and Chour, 1993). These designs provide portable measuring systems that are highly precise and easy to use. For example, in Fig. 13.56

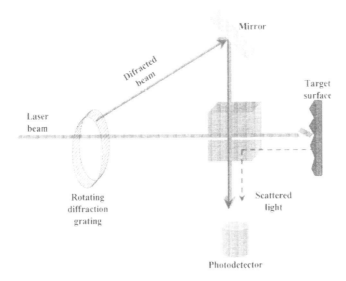

Figure 13.55   Frequency shifting by diffraction grating.

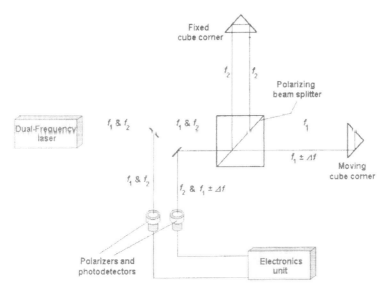

Figure 13.56 The dual-frequency method.

is shown a diagram of one such systems. A frequency-stabilized laser generates a beam composed by two emission modes with frequencies f_1 and f_2 and a known frequency difference $f_1 - f_2$. Beams of frequency f_1 and f_2 are polarized perpendicularly, in such a way that due to the polarizing beamsplitter, this only allows the beam of frequency f_1 to pass to the movable corner cube. The beam of frequency f_2 impinges directly on the fixed corner cube. When neither of the two retroreflectors moves, the two photodetectors detect both frequencies. When the mobile mirror moves, frequency f_1 suffers a change Δf as a result of the Doppler effect and this is related to the velocity.

Another particular geometry is shown in Fig. 13.57, in which in-plane surface velocity measurement is made. This experimental geometry is called *Doppler differential mode*, to differentiate it from the previous ones known as *reference beam mode*. In the Doppler differential mode, two symmetrical beams from the same laser are used for illuminating the moving point-object (or particle). The Doppler frequency shift is given by (Ready, 1978)

$$\Delta f = \frac{2u}{\lambda} \sin \frac{\theta}{2}, \tag{13.36}$$

where λ is the wavelength of the laser light and θ is the angle between the two illuminating beams

The operation of this dual-beam system is possible to visualize in terms of fringes applied in fluid dynamics. Where the two beams cross the region, the light waves interfere to form alternate regions of high and low intensity. If one particle traverses the fringe pattern, it will scatter more light when it crosses through regions of high intensity. Thus, the light received by the detector will show a varying elec-

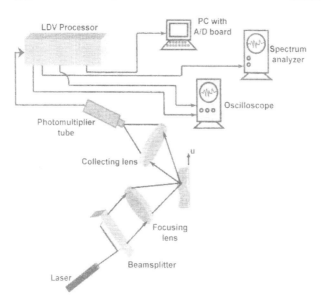

Figure 13.57   The differential geometry.

trical signal whose frequency is proportional to the rate at which the particle crosses the interference fringes.

13.4.4.3   Photodetector and Doppler Signal Processing

(a)   Photodetectors

A very important step in the measurement process of velocity is the manner of how the Doppler signal is detected and processed. Normally a photodetector and its associated electronics for use in LDVs convert incident optical energy into electrical energy. Since early days, *photosensors* have been used for fringe detection and are based on the photoelectric effect: i.e., an incident photon on a given material may remove an electron from its surface (cathode). These removed electrons are accelerated towards, and collected by, an anode. This mechanism involves vacuum and gas-filled phototubes. Electron–ion pairs are also produced as the primary electrons collide and atoms of the gas, so that electron multiplication occurs. One of the most sensitive sensors of this family is the *photomultiplier*. This same effect can be carried out *in solid-state sensors* in which the photoelectric effect is made internally (Watson, 1983). Phototubes are primarily sensitive to the energetic photons of the UV and blue regions, while solid-state detectors are sensitive over the red and IR regions of the spectrum.

There is a variety of different configurations and technologies used for the manufacture of solid-state sensors, including frame transfer and interline transfer charge-coupled devices (CCDs), charge injection devices (CIDs), and photodiode arrays. Solid-state sensors can be classified by their geometry as either area arrays or linear arrays. Area arrays are the most commonly used for image detection and permit the measurement of a two-dimensional section of a surface. The system spatial resolution is related to the number of pixels along each dimension of the sensor, and commonly available sensors have resolutions appropriate for television applications, typically about 500 × 500 pixels or less. Some newer sensors designed for machine vision applications or high-definition television, have dimensions of about 1000 × 1000 or even 2000 × 2000 pixels, but these sensors are expensive and are difficult to use. These kinds of sensors make slower frame rates and the amount of required computer memory and processing is quite high. Linear arrays, on the other hand, measure only one-dimensional trace across the part, but make up for this disadvantage by providing a higher spatial resolution along this line. Linear sensors containing over 7000 pixels are available today, so that measurements with extremely high spatial resolution can be obtained. The amount of data from these large linear arrays is small and easily handled when compared with the 100,000 or more pixels on even a low-resolution area array. Because of advances in semiconductor fabrication, we can expect to see the dimensions of available sensors continuing to grow.

Photoresistors are sensors that consist essentially of a film of material whose resistance is a function of incident illumination. This film can be either an intrinsic or extrinsic semiconductor material, but for the visible part of the spectrum, the chalcogenides, cadmium sulphide and cadmium selenide, are the more common. The size and shape of the active film determines both the dark resistance and the sensitivity. However, for fringe counting, a small strip of material upon which the fringes can be focused is most suitable, and this strip must be connected to a pair of electrodes. The resistance of the cell decreases as the illumination is increased, so that it is common to consider the inverse of this resistance or the conductance, as being the basic parameter, which, accounts for the alternative name of the sensor, the *photoconductive cell*. Electrically, therefore, it is resistance or conductance changes which must be detected. This implies that a current must be passed through the sensor, and either variations in this, or in associated voltage drops, should be measured.

If a single crystal of a semiconductor such as silicon or germanium is doped with both donor or acceptor impurities to form P and N regions, then the junction between these regions causes the crystal to exhibit *diode properties*. The commonly used photodiodes include the simple P–N junction diode, the P–I–N diode, and the avalanche diode (Sirohi and Chau, 1999). The operation region of a photodiode is limited to that of light polarization change. Incident light to the junction P–N will give an energy transfer as a result of the atomic structure, which originates a bigger inverse current level. The current returns to zero once the polarization of the light changes 90°, and so forth.

(b)   Doppler Signal Processing

The choice of a Doppler signal-processing method is dictated by the characteristics of the Doppler signal itself that is directly related to the particular measurement problem.* In fluid flows (water excepted), for example, it is usually necessary to seed the flow with scattering particles in order to detect sufficient intensities of Doppler-shifted light (Hinsch, 1993). Clearly for time-resolved measurements the ideal situation requires a continuous Doppler signal but unfortunately in practice the latter is

---

*For more details, a complete description of the various means of demodulating a Doppler signal is found in Chapters 6–9 of Hurst's book (Hurst *et al.*, 1981).

often intermittent due to changes in seeding particle density, which occur naturally. The three most popular methods of signal processing are frequency tracking, frequency counting, and photon correlation (although in the specialized literature others like burst counting, Fabry–Perot interferometry, and filterbank are found). Tracking requires a nearly continuous Doppler signal, while correlation has been developed to deal with situations where seeding is virtually absent. The increased ability to deal with intermittent signals (in what is really a statistical sampling problem) is usually indicative of the expense of the commercial processor concerned, and the relationship is not linear.

Fortunately in the case of solid-surface vibration measurements Doppler signals are continuous and frequency tracking demodulation is the treatment of choice. With this form of processing, a voltage-controlled oscillator (VCO) is used to track the incoming Doppler signal and is controlled via a feedback loop (Watson *et al.*, 1969; Lewin, 1994). Usually, a mixer at the input stage produces an "error" signal between the Doppler and VCO frequencies which is bandpass filtered and weighted before being integrated and used to control the oscillator to drive the error to a minimum. The feedback loop has an associated "slew rate" which limits the frequency response of the processor. With respect to the Doppler signal the tracker is really a low-pass filter which outputs the VCO volage as a time-resolved voltage analogue of the changing frequency. The frequency range of interest for vibration measurements (20 kHz in dc) is well within the range of this form of frequency demodulation. Some trackers carry sophisticated weighting networks that tailor the control of the VCO according to the signal-to-noise ratios of the incoming signal. A simple form of this network will hold the last value of Doppler frequency being tracked if the amplitude of the signal drops below a preset level. In this way the Doppler signal effectively "drops out", and careful consideration must be given to the statistic of what is essentially a sampled output especially when high-frequency information of the order of a drop-out period is required.

In several outdoors applications the sensitivity of the LDV suffers, along the free space path, from cochannel interference arising from spurious scattered light from rain, moisture, speckle, refractive index change, and dust, among others. New methods of demodulation signals from LDVs have been proposed. In order to minimize spurious scattering, an amplitude-locked loop (ALL) is combined with a phase-locked loop (PLL). The signal from the photodiode is down-converted to an intermediate frequency before being demodulated by a PLL to obtain baseband information, i.e., the vibration frequency of the mirror. The ALL is a high-bandwidth servo loop that is able to obtain extra information on the amplitude of the spikes. The incoming corrupted FM signal is directly connected to the ALL. The output of the ALL gives a fixed output FM signal, which is connected to the PLL input (Dussarrat *et al.*, 1998; Crickmore *et al.*, 1999).

Commercially available LDVs were originally designed for use in fluid-flow situations. Consequently, a great deal of research and development work has been directed toward solving the signal drop-out problem and other Doppler uncertainties produced by the finite size of the measurement volume. Since the early work in the middle to late 1970s manufacturers now appear to prefer frequency counting for the standard laboratory system. This represents a successful compromise between tracking and correlation. Modern electronics will allow very fast processing, so that a counter will provide a time-resolved analogue in a continuous signal situation while producing reliable data when seeding is sparse.

Compact LDVs have come to revolutionize the instrumentation of the velocity sensors. Optical systems of gradient index, together with diode lasers and optical fibers, have made the use of portable systems a reality. Fast response of modern photodetectors, as well as rapid computers for Doppler-shifted signal analysis and data reduction, has come to aid performance of such instruments (Jentink *et al.*, 1988; D'Emilia, 1994; Lewin, 1998). Basic investigation has also continued to be carried out, such as near-field investigations to determine three-dimensional velocity fields by combining differential and reference beam LDVs and the use of evanescent waves for determining flow velocity (Ross, 1997; Yamada, 1999).

13.4.5 Optical Vibrometers

Displacement sensors and transducers are used to modulate signals with the purpose of making measurements of dynamic magnitudes. In previous sections, a number of optical transducers have been analyzed, which have provided different forms of signal modulating to measure displacement and velocity. Now, in this section, a survey of punctual optical techniques to determine mechanical vibrations is given.

The measurement of vibration of a solid surface is usually achieved with an accelerometer or some other form of surface-contacting sensor (Harris, 1996). There are, however, many cases of engineering interest where this approach is either impossible or impractical, such as for lightweight, very hot, or rotating surfaces. Practical examples of these loudspeakers, engine exhausts, crankshafts, etc. Since the advent of the laser in the early 1960s, optical metrology has provided a means of obtaining remote measurements of vibration in situations which had been hitherto thought unobtainable. The first demonstration of the use of the laser as a remote velocity sensor was in the measurement of a fluid flow (Yeh and Cummins, 1964). The physical principles of the optical vibrometers have their roots in velocity and displacement measurement systems, like those described above. These concepts have already been treated in previous sections; however, they will be described again in an electronics context.

A variety of methods are at hand for utilizing the frequency (temporal) coherence, spatial coherence, or modulation capacity of laser light to measure the component's dynamics of a moving object. For general applications, the element under study is moving with displacements of normal and angular deflection, or tilt, with respect to some axis in the plane of the surface. Methods for detection for both types of motion will be described.

In general, the surface motion can be detected by observing its effect on the phase or frequency of a high-frequency subcarrier which has been amplitude modulated onto the optical carrier, or by observing the phase or frequency changes (Doppler shift) on the reflected optical carrier itself. In addition, the surface acts as a source of reflected light that changes its orientation in space with respect to the optical receiver; hence, the arrival angle of reflected light varies with target position and can be detected. All of the vibration measurement techniques that will be described fit into these general classifications. From the historical point of view, optical vibrometers will be described, starting with the first systems that appeared in scientific literature (based on subcarriers) and their evolution, up to the state of the art (scanning LDV).

When the subcarrier methods are used, the vibration-induced phase shift is proportional to the ratio of the vibration amplitude to the subcarrier wavelength. Obviously, then, it is desirable to use the highest possible subcarrier or modulation frequency at which efficient modulation and detection can be performed.

When the optical carrier is used, a photodetector alone is not adequate, as already mentioned above, to detect frequency changes as small as those produced in this application. Some type of optical interference is required to convert optical phase variations on the reflected signal into intensity variations (interference fringes) which the photosensor can detect. A reference beam from the laser transmitter may be used to produce the interference with the incoming signal beam collected by the receiver optics. In such a system, the reference beam is often called the *local oscillator beam*, using the terminology established for radiofrequency receivers. Demodulation using a reference beam is known as coherent detection. If the frequency of the reference or local oscillator beam is the same as the transmitted signal, the system is a coherent optical phase detector, or *homodyne* system (Deferrari *et al.*, 1967; Gilheany, 1971). When the frequency of the reference is shifted with respect to the transmitted wavelength, an electrical beat is produced by the square-law photodetector at the difference, or intermediate, frequency between the two beams. Such a system is called a *heterodyne* detector or coherent optical intermediate frequency system (Ohtsuka, 1982; Oshida *et al.*, 1983). Obviously, the coherent phase detector or homodyne is a special case of heterodyne detection with the intermediate frequency equal to zero.

13.4.5.1 Subcarrier Systems

Three different subcarrier systems are analyzed. Each of them uses an electro-optic amplitude modulator to vary the intensity of the transmitted laser beam at a microwave rate (Medlock, 1986). In practical devices, only a fraction, M, of the light intensity is modulated. These intensity variations represent a subcarrier envelope on the optical carrier; they can be detected by a photodetector, which has good enough high-frequency performance to respond to intensity variations at the subcarrier frequency. When the modulated light is reflected from the moving surface, the phase of the subcarrier envelope will vary with time, according to the expression:

$$\phi(z) = \frac{4\pi z_0}{\lambda_{\mathrm{m}}} \sin \omega_r t, \tag{13.37}$$

where z_0 is the zero-to-peak vibration amplitude, $\omega_r = 2\pi f_r$ is the vibration frequency, and λ_{m} is the microwave subcarrier wavelength. For a subcarrier frequency of 3 GHz, λ_{m} is 10 cm; thus, the peak of phase deviation for small vibrations is much less than 1 rad. The information on the vibration state of the surface appears on the reflected light in the subcarrier sidebands produced by the time-varying phase changes. For small peak-phase deviations, which are always of interest in determining the maximum sensitivity of a given system, only the first two sidebands are significant (the exact expression gives an infinite set of sidebands with Bessel function amplitudes, most of which are negligibly small). In that case, the reflected spectrum of light intensity is given by the approximation:

$$P = \frac{P_0}{2}$$
$$+ \frac{MP_0}{2} \sin \omega_m t \qquad (13.38)$$
$$+ \frac{MP_0 \pi \chi_0}{2\lambda_m} \sin(\omega_m + \omega_r)t + \frac{MP_0 \pi \chi_0}{2\lambda_m} \sin(\omega_m - \omega_r)t,$$

where ω_m is the microwave signal, M is the modulation index of the incident light, and P_0 is the reflected power. The first term of Eq. (13.38) is the unmodulated light; the second term, the subcarrier; and, third term, shows the sidebands. The necessary receiver bandwidth is simply the maximum range of vibration frequencies to be measured. Thus, any usable microwave system must be able to detect the sideband amplitudes in the presence of receiver noise integrated over the required bandwidth. It should be noted that Eq. (13.58) is for *optical* power; the demodulated power spectral components in the *electrical* circuits will be proportional to the squares of the individual terms in Eq. (13.38).

(a)   Direct Phase Detection System

This approach is the simplest subcarrier technique and is shown schematically in Fig. 13.58. The subcarrier is recovered after reflection by the photodetector, which might be a tube or solid state diode. The diode has the disadvantages of no multiplication and effective output impedances of only a few hundred ohms at best. However, solid-state quantum efficiencies can approach unity (Blumenthal, 1962).

The low-noise traveling-wave amplifier and tunnel diode limiter are required to suppress low-frequency amplitude fluctuations on the received signal due to surface

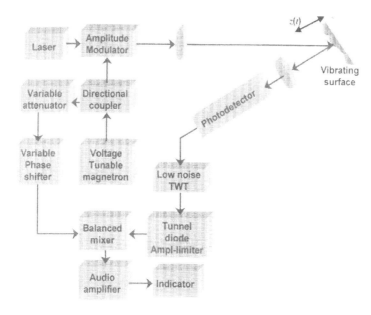

Figure 13.58   Microwave direct-phase detection system.

tilt, laser amplitude fluctuations, and other effects. The balanced mixer further rejects amplitude modulation on either of the microwave input signals.

The voltage-tunable magnetron exciter for the optical modulator is used. This might be tuned by a servo-control system, not shown in Fig. 13.58, to follow the drift in modulator resonant frequency during warm-up. Such a servo-system would include a sensor placed in or near the modulator cavity to sample the phase and amplitude of the cavity field. This signal would be processed by narrow-band electronics to derive the tuning voltage needed to make the oscillator track the cavity resonance.

The directional coupler, attenuator, and variable phase shifter provide the reference or local oscillator signal to the balanced mixer at the phase and amplitude for phase quadrature detection.

(b) Intermediate Frequency Detection System

This technique is illustrated schematically in Fig. 13.59. The receiver portion of this system differs from the one above because the local oscillator signal to the microwave mixer has been shifted in frequency. This is the purpose of the intermediate frequency (IF) generator and sideband filter. The microwave mixer produces a beat frequency when a signal is present. A second mixer, known as a phase detector, is used to demodulate the phase modulation produced on the IF by the surface motion. Thus the mixing down to audio is done in two steps. This improves the mixer noise figures, and with limiting in the IF amplifier it might be possible to operate the phototube directly into the first mixer. As pointed out in the previous section, the mixer noise is negligible anyway, so the additional complexity of this

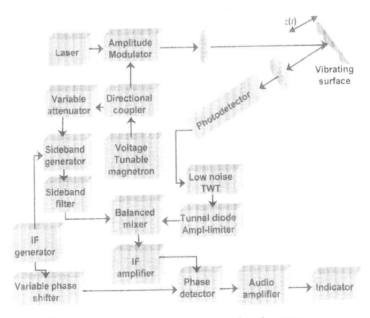

Figure 13.59 Microwave intermediate frequency detection system.

approach appears unnecessary. Sensitivity is the same as for the direct phase detector analyzed above.

(c) Double Modulation System

The fundamental limitation in performance of the systems above is imposed by the low quantum efficiency of the microwave phototube. A method which avoids this is illustrated in Fig. 13.60. Here, the phase demodulation is done not with a high-frequency detector but with a gated receiver and low-frequency detector. Gating of the microwave rate is accomplished by passing the reflected light back through the microwave optical modulator. A beamsplitter mirror or calcite prism and quarter-wave plate might be used to allow transmission and reception through the same optical system. The re-modulated signal in this case will have audiofrequency intensity variations corresponding to the subcarrier phase shifts (Doppler effect) produced by the vibration. Efficient, low-noise phototubes and diodes are now available at these frequencies.

With the external optical path adjusted so that the second modulation occurs $90°$ out of phase with the first one, the detected optical power is of the form:

$$P = P_r \left[1 + M^2 \frac{2\pi z_0}{\lambda_m} \sin(\omega, t) \right], \tag{13.39}$$

where P_r is the average power reaching the photodetector.

Several limitations were found in this system. It was not possible to detect the phase shift on the subcarrier because of the large spurious amplitude fluctuations due to surface tilt. This was true even when the vibrating mirror was placed at the focal point of a lens, an optical geometry that minimizes the angular sensitivity. Thus, any substantial amplitude change due to laser noise or reflected beam deflection will overcome the desired signal. Another limitation is that the subcarrier demodulation

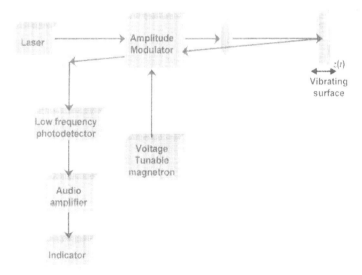

Figure 13.60 Microwave double modulation system.

to audio is done optically; thus, it is not possible to use limiting to remove the large amplitude variations before detection. For this reason it is doubtful that this system could be used to its theoretical limit of performance even if the modulator were improved to make M almost unity. If this problem could somehow be eliminated, the double-modulation system would realize the advantage of requiring no microwave receiver components and could make use of the best possible optical detectors.

13.4.6 Types of Laser Doppler Vibrometers

For overcoming the limitations imposed by subcarrier systems, coherent optical detection methods, based on the Doppler effect, began to be used. Combining the signal beam with a local oscillator beam that acts as a phase reference can coherently demodulate the vibration-induced phase shifts on the optical carrier. When both beams are properly aligned and are incident on an optical detector, the output current is proportional to the square of the total incident electric field. This current may be written as (Oshida *et al.*, 1983)

$$I(t) = I_r + I_s + 2\sqrt{I_r I_s} \cos(\omega_r - \omega_s)t, \tag{13.40}$$

where I_r is the local oscillator field only, I_s, is the direct field due to signal alone, and ω_r and ω_s are the frequencies of the local oscillator and the signal wave, respectively (see Fig. 13.56). The *instantaneous* frequency shifts from the local oscillator and target surface are, equivalently,

$$\omega_r = 2ku + \phi, \tag{13.41}$$

$$\omega_s = 2kz_0\omega_v \cos(\omega_v t), \tag{13.42}$$

where k is the wavenumber of the laser light and $(z_0 \cos \omega_v t)$ represents the target surface displacement of amplitude z_0 and frequency ω_v.

$$i(t) = A \cos\{2k[u - z_0\omega_v \cos(\omega_v t)]t + \epsilon(t)\}, \tag{13.43}$$

The function $\phi(t)$ represents a pseudo-random phase contribution due to the changing population of particulate scatters in the laser spot. Neglecting constant terms, we can write $A = (I_r I_s)^{1/2}$ and $\epsilon(t) = t\phi(t)$. The function $\epsilon(t)$ is *pseudo-random* since when using a reference beam oscillating to frequency shift the spatial distribution of scatters repeats after each revolution. These cause the frequency spectrum of the noise floor of the instrument to be a periodogram since the random amplitude modulation of $i(t)$ due to $\epsilon(t)$ repeats exactly after each oscillation period.

With reference to Eq. (13.43) a frequency tracking demodulator follows the frequency modulation of the carrier frequency $(2ku)$ to produce a voltage output which is an analogue of the changing surface velocity of amplitude $(z_0\omega_v)$.

13.4.6.1 Referenced (Out-of-Plane) Vibrometer

This LDV measures the vibrational component $z(t)$ which lies along the laser beam, already analyzed for velocity measuring, which is the most common type of LVD system. The system is a heterodyne interferometer, as shown in Fig. 13.61, which means that the signal and reference beams are frequency shifted relative to one another to allow the FM carrier generation. The two outputs of the interferometer provide complementary signals which, when differentially combined, generate a

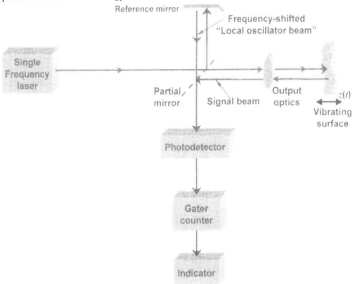

Figure 13.61 Out-of-plane system.

zero-centered FM carrier signal. The superimposed frequency (or phase) shifts are related to the surface position via the wavelength of the laser used as optical source, given by Eq. (13.40). This system has a resolution up to 10^{-15} m $(Hz)^{1/2}$ (Lewin, 1994).

Axial measurements can be obtained by approaching the same measurement point from three different directions.

13.4.6.2 Dual Beam (In-Plane) or Differential Vibrometer

The basic differential LDV vibrometer arrangement is shown in Fig. 13.57. The output from a laser is split into two beams of about equal intensity. A lens focuses the two beams together in a small spot on the face of a vibrating plate in direction $x(t)$, which is the target surface. Light scattered by the target surface is collected by a second lens and focused into a pinhole in front of a photodetector. The photodetector output is processed by an LDV counter in a similar manner to that of referenced vibrometer (Ross, 1997). By rotating the probe by 90°, $x(t)$ or $y(t)$ can be measured.

13.4.6.3 Scanning Vibrometer

An extension of the standard out-of-plane system, the scanning LDV uses computer-controlled deflection mirrors to direct the laser to a user-selected array of measurement points. The system automatically collects and processes vibration data at each point; scales the data in standard displacement, velocity, or acceleration engineering units; performs fast Fourier transform (FFT) or other operations; and displays full-field vibration pattern images and animated operational deflection shapes. The role of this system is found in its scanning system. Scanning mirrors can be moved by using voltage changes (Zeng *et al.*, 1994) or galvanometer-based (Li and Mitchell, 1994; Stafne *et al.*, 1998), requiring high-precision mechanical mounts. But once the scanning system is calibrated, the unit gives a haughty standard LDV.

13.4.6.4 Spot Projection Systems

Other optical vibrometers non-less important are spot projection-based or triangulation principle methods (Doebelin, 1990, pp. 283–286). If the moving surface is a diffuse reflector, it is possible to obtain information about some components of the motion by projecting one or more spots of laser light onto the surface and measuring the motion-induced effects on reflected light collected by an optical receiver. One method, in which the apparent motion of the spot is measured, does not make use of the spectral coherence of the laser. This system has been called the incoherent spot projection technique or triangulation-based technique. Another approach, in which two spots are projected and the interference between reflected waves from both of them is used, is known as coherent spot projection, because the laser coherence is utilized.

(a) Incoherent Spot Projection System

This system is illustrated in Fig. 13.62. The laser beam is projected to a small spot on the vibrating surface. A rectangular receiver aperature collects some of the reflected light. In the receiver focal plane, motion of the surface produces a lateral motion of the spot image. If a knife-edge stop is placed a short distance behind the image, where the beam has expanded to a rectangle, motion of the image affects the fraction of the light that passes the stop. In practice the stop would cover half the beam on average, and the distance behind focus would be adjusted to accommodate the largest expected image displacements in the linear range. The fraction of power

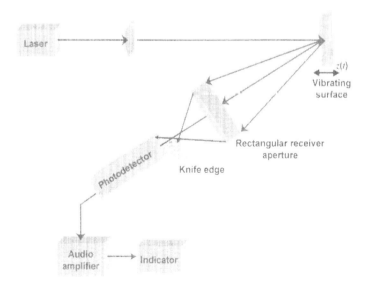

Figure 13.62 Incoherent spot projection.

passing the knife edge is measured by a photodetector whose current output is a linear analogue of the surface displacement along the axis of the transmitted beam.

It is interesting to notice that the sensitivity can approach that of the Doppler-shift-based systems if the maximum displacement is no greater than an optical wavelength, good optics are used, and the receiver aperature is large. In such a situation, the spatial coherence of the laser is fully utilized. Such a system is therefore not incoherent in the limit.

For large vibrations the system has many advantages and a few disadvantages. The main disadvantage is the need for careful alignment of the receiver and the knife edge. For scanning, the separate receiver and transmitter are troublesome. However, tolerance on the optical components is not severe for large maximum displacements, and other components are simple and reliable.

(b)   Coherent Spot Projection System

The above *incoherent* system measures normal displacement of the surface. The coherent system to be described measures angular tilt of the surface in the plane determined by the transmitter and receiver axes. Figure 13.63 illustrates the system schematically. Two small spots are projected by high-quality optics. They are separated on the surface by a distance approximately equal to a spot diameter. Then, at the receiver plane there will be interference fringes produced by reflected light from the two spots. If the spot separation is sufficiently small, the fringes can become large enough to fill a receiver aperature of a few inches. A tilt of the surface corresponding to a relative motion between the spots equal to 1/2 wavelength will move the fringe pattern laterally by a full spatial period. Power changes due to motion of the fringe pattern are detected in the receiver. The spot separation corresponding to a 3-inch

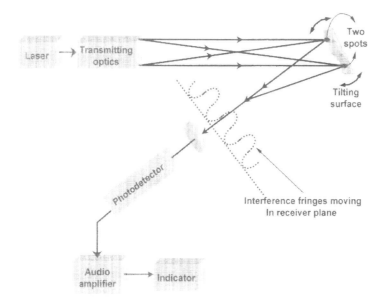

Figure 13.63   Coherent spot projection.

receiver at 3 feet is about 4×10^{-4} inches or 10^{-2} mm. The tilt associated with a half-wave relative motion is then approximately 30 mrad, or about 1.7°.

Some limitations of this technique are encountered. First of all, the diffuse surface causes self-interference in the reflected light from each spot. Thus, at any point in the receiver plane, the fields from the spots are likely to be far from equal. Consequently, the desired fringes have very low contrast over most of the plane, and must be detected in the presence of a large, strongly modulated background of random interference. The second problem is the inability to produce very small spots on a diffuse surface. Scattering among the rough elements near the illuminated region spreads the effective spot size considerably, thus reducing the maximum usable receiver aperture. Because of these problems, the receiver aperture needs to be carefully positioned in the fringe pattern for linear demodulation (a condition difficult to meet if the surface must be scanned).

13.4.7 Vibration Analysis by Moiré Techniques

The first works on moiré techniques that appeared for vibration measuring were based on the use of two gratings in contact (Aitchison *et al.*, 1959). Maximum and minimum of the moiré pattern in movement, due to vibration, is detected by means of photoelectric cells and transduced through an electronic system. Recently, the use of techniques such as fringe projection (Hovanesian and Hung, 1971; Dessus and Leblanc, 1973), reflection moiré (Theocaris, 1967; Asundi, 1994), shadow moiré (Dirckx *et al.*, 1986), moiré deflectometry (Kafri *et al.*, 1985), and holographic moiré (Sciammarella and Ahmadshahi, 1991) have been applied to vibration analysis. All these techniques have been used in a similar way to the time-average speckle technique (Jones and Wykes, 1983). On vibrating surfaces, antinodal positions will give continuously varying deflections, while nodes will produce zero deflection. A time-averaged photograph of a projected grating on to a vibrating object will produce areas of washed-out grating at the antinodes and a sharp grating image at the nodal positions. If a second grating is superimposed on the photograph to produce dark fringes at the nodes, the antinodes will appear brilliant, and one immediately has a contour map of the nodal positions.

ACKNOWLEDGMENT

R. Rodriguez-Vera wishes to thank Manuel Bujdud for drawing some of the figures.

REFERENCES

Ahrend, M., "Recent Photogrammetric and Geodetic Instruments," *Appl. Opt.*, **7**, 371–374 (1968).

Aitchison, T. W., J. W. Bruce, and D. S. Winning, "Vibration Amplitude Meter Using Moiré-Fringe Technique," *J. Sci. Instrum.*, **36**, September, 400–402 (1959).

Asundi, A., "Novel Techniques in Reflection Moiré," *Exp. Mech.*, **34**, September, 230–242 (1994).

Asundi, A. and H. Yung, "Phase-Shifting and Logical Moiré," *J. Opt. Soc. Am.*, **8**, 1591–1600 (1991).

Baird, K. M. and L. E. Howlett, "The Interntional Length Standard," *Appl. Opt.*, **2**, 455–463 (1963).

Bechstein, K. and W. Fuchs, "Absolute Interferometric Distance Measurements Applying a Variable Synthetic Wavelength," *J. Opt.*, **28**, 179–182 (1998).

Bernardo, L. M. and O. D. D. Soares, "Evaluation of the Focal Distance of a Lens by Talbot Interferometry," *Appl. Opt.*, **27**, 296–301 (1988).

Blumenthal, R. H., "Design of a Microwave-Frequency Light Modulator," *Proc. IRE 50*, April, 452 (1962).

Bosch, T. and M. Lescure, eds (1995), *Selected Papers on Laser Distance Measurement*, SPIE Milestone Series, Vol. 115, SPIE, Bellingham, Wa.

Bottom, V. E., *Introduction to Quartz Crystal Unit Design*, Van Nostrand, New York, 1982.

Bourdet, G. L. and A. G. Orszag, "Absolute Distance Measurements by CO_2 Laser Multiwavelength Interferometry," *Appl. Opt.*, **18**, 225–227 (1979).

Boyd, W. R., 1969, "The Use of a Collimator for Measuring Domes," *Appl. Opt.*, **8**, 792 (1969).

Buddenhagen, D. A., B. A. Lengyel, F. J. McClung, and G. F. Smith, *Proceedings of the IRE International Convention*, New York, 1961, Part 5 (Institute of Radio Engineers, New York), 1961, p. 285.

Burgwald, G. M. and W. P. Kruger, "An Instant-On Laser for Length Measurements," *Hewlett Packard Journal*, **21** (1970).

Burnside, C. D., *Electronic Distance Measurement*, 3rd edn, BSP Professional Books, London, 1991.

Busch, T., *Fundamentals of Dimensional Metrology*, 2nd edn, Delmar Publishers Inc., USA, 1989.

Carnell, K. H. and W. T. Welford, "A Method for Precision Spherometry of Concave Surfaces," *J. Phys.*, **E4**, 1060–1062 (1971).

Chang, Ch-W. and D-Ch. Su, "An Improved Technique of Measuring the Focal Length of a Lens," *Opt. Comm.*, **73**, 257–262 (1989).

Choi, J.-W., Y.-P. Kim, and Y.-M. Kim, "Optogalvanic Laser Doppler Velocimetry Using the Self-Mixing Effect of CO_2 Laser," *Rev. Sci. Instrum.*, **68**, 4623–4624 (1997).

Churnside, J. H., "Laser Doppler Velocimetry by Modulating a CO_2 Laser with Backscattered Light," *Appl. Opt.*, **23**, 61–66 (1984).

Cooke, F., "Optics Activities in Industry," *Appl. Opt.*, **2**, 328–329 (1963).

Cooke, F., "The Bar Spherometer," *Appl. Opt.*, **3**, 87–88 (1964).

Crickmore, R. I., S. H. Jack, D. B. Hann, and C. A. Greated, "Laser Doppler Anemometry and the Acousto-Optic Effect," *Opt. Laser Technol.*, **31**, 85–91 (1999).

Crosswy, F. L. and J. O. Hornkohl, *Rev. Sci. Instrum.*, **44**, 1324 (1973).

Dai, X., O. Sasaki, J. E. Greivenkamp, and T. Suzuki, "Measurement of Small Rotation Angles by Using a Parallel Interference Pattern," *Appl. Opt.*, **34**, 6380–6388 (1995).

Dai, X., O. Sasaki, J. E. Greivenkamp, and T. Suzuki, "High Accuracy, Wide Range, Rotation Angle Measurement by the Use of Two Parallel Interference Patterns," *Appl. Opt.*, **36**, 6190–6195 (1997).

Dändliker, R., K. Hug, J. Politch, and E. Zimmermann, "High-Accuracy Distance Measurements with Multiple-Wavelength Interferometry," *Opt. Eng.*, **34**, 2407–2412 (1995).

Dändliker, R., Y. Savadé, and E. Zimmermann, "Distance Measurement by Multiple-Wavelength Interferometry," *J. Opt.*, **29**, 105–114 (1998).

D'Emilia, G., "Evaluation of Measurement Characteristics of a Laser Doppler Vibrometer with Fiber Optic Components," *Proc. SPIE*, **2358**, 240–246 (1994).

Deferrari, H. A., R. A. Darby, and F. A. Andrews, "Vibrational Displacement and Mode-Shape Measurement by a Laser Interferometer," *J. Acoust. Soc. Am.*, **42**, 982–990 (1967).

Dessus, B. and M. Leblanc, "The 'Fringe Method' and Its Application to the Measurement of Deformations, Vibrations, Contour Lines and Differences," *Opto-Electronics*, **5**, 369–391 (1973).

DeVany, A. S., "Making and Testing Right Angle and Dove Prisms," *Appl. Opt.*, **7**, 1085–1087 (1968).

DeVany, A. S., "Reduplication of a Penta-Prism Angle Using Master Angle Prisms and Plano Interferometer," *Appl. Opt.*, **10**, 1371–1375 (1971).

DeVany, A. S., "Testing Glass Reflecting–Angles of Prisms," *Appl. Opt.*, **17**, 1661–1662 (1978).

Dirckx, J. J. J., W. F. Decraemer, and J. L. Janssens, "Real-Time Shadow Moiré Vibration Measurement: Method Featuring Simple Setup, High Sensitivity, and Exact Calibration," *Appl. Opt.*, **25**, 3785–3787 (1986).

Doebelin, E. O., *Measurement Systems, Application and Design*, 4th edn, McGraw-Hill International Editions, Singapore, 1990.

Drain, L. E. *The Laser Doppler Technique*, John Wiley & Sons, Chichester, 1980.

Drain, L. E., and B. C. Moss, "The Frequency Shifting of Laser Light by Electro-Optic Techniques," *Opto-Electronics*, **4**, 429 (1972).

Dukes, J. N. and G. B. Gordon, "A Two-Hundred-Foot Yardstick with Graduations Every Microinch," *Hewlett-Packard J.*, **21**, 2–8 (1970).

Durst, F., A. Mellin, and J. H. Whitelaw, *Principles and Practice of Laser-Doppler Interferometry*, Academic Press, London, 1976.

Dussarrat, O. J., D. F. Clark, and T. J. Moir, "A New Demodulation Process to Reduce Cochannel Interference for a Laser Vibrometer Sensing System," *Proc. SPIE*, **3411**, 2–13 (1998).

Faller, E. F. and E. J. Wampler, "The Lunar Laser Reflector," *Sci. Am.*, **223**(3), 38–47, March (1970).

Farago, F. T., *Handbook of Dimensional Measurement*, 2nd edn, Industrial Press Inc., New York, 1982.

Fischer, E., E. Dalhoff, S. Heim, U. Hofbauer, and H. J. Tiziani, "Absolute Interferometric Distance Measurement using a FM-Demodulation Technique," *Appl. Opt.*, **34**, 5589–5594 (1995).

Fomin, N., E. Laviskaja, W. Merzkirch, and D. Vitkin, "Speckle Photography Applied to Statistical Analysis of Turbulence," *Opt. Laser Technol.*, **31**, 13–22 (1999).

Foueré, J. C. and D. Malacara, "Focusing Errors in a Collimating Lens or Mirror: Use of a Moiré Technique," *Appl. Opt.*, **13**, 1322–1326 (1974).

Françon, M., *Laser Speckle and Applications in Optics*, Academic Press, New York, 1979.

Funes-Gallanzi, M., "High Accuracy Measurement of Unsteady Flows Using Digital Particle Image Velocimetry," *Opt. Laser Technol.*, **30**, 349–359 (1998).

Gasvik, K. J., *Optical Metrology*, John Wiley & Sons, New York, 1987, pp. 150–156.

Gerchman, M. C. and G. C. Hunter, "Differential Technique for Accurately Measuring the Radius of Curvature of Long Radius Concave Optical Surfaces," *Opt. Eng.*, **19**, 843–848 (1980).

Gerchman, M. C. and G. C. Hunter, "Differential Technique for Accurately Measuring the Radius of Curvature of Long Radius Concave Optical Surfaces," *Proc. SPIE*, **192**, 75–84 (1979).

Giacomo, P., "Metrology and Fundamental Constants," *Proc. Int. School of Phys. "Enrico Fermi," course 68*, North Holland, Amsterdam, 1980.

Gilheany, J. J., "Optical Homodyning, Theory and Experiments," *Am. J. Phys.*, **39**, May, 507–512 (1971).

Glatt, I. and O. Kafri, "Beam Direction Determination by Moiré Deflectometry Using Circular Gratings," *Appl. Opt.*, **26**, 4051–4053 (1987a).

Glatt, I. and O. Kafri, "Determination of the Focal Length of Non-Paraxial Lenses by Moiré Deflectometry," *Appl. Opt.*, **26**, 2507–2508 (1987b).

Goldman, D. T., "Proposed New Definitions of the Meter," *J. Opt. Soc. Am.*, **70**, 1640–1641 (1980).

Gouaux, F., N. Servagent, and T. Bosch, "Absolute Distance Measurement with an Optical Feedback Interferometer," *Appl. Opt.*, **37**, 6684–6689 (1998).

Grant, I., X. Pan, F. Romano, and X. Wang, "Neural-Network Method Applied to the Stereo Image Correspondence Problem in Three-Component Particle Image Velocimetry," *Appl. Opt.*, **37**, 3656–3663 (1998).

Greivenkamp, J. E. and J. H. Bruning, "Phase Shifting Interferometry," in *Optical Shop Testing*, Malacara, D., ed., 2nd edn, John Wiley & Sons, New York, 1992.

Halliwell, N. A., "Laser Doppler Measurement of Vibrating Surfaces: A Portable Instrument," *J. Sound and Vib.*, **62**, 312–315 (1979).

Halliwell, N. A., "Laser Properties and Laser Doppler Velocimetry," in *Vibration Measurement Using Laser Technology*, Course notes, Sira Communications Ltd, Loughborough University of Technology, 3–5 April 1990.

Hariharan, P., "Interferometric Metrology: Current Trends and Future Prospects," *Proc. SPIE*, **816**, 2–18 (1987).

Harris, C. M., ed., *Shock and Vibration Handbook*, McGraw-Hill, New York, 1996.

Hinsch, K. D., "Particle Image Velocimetry," in *Speckle Metrology*, Sirohi, R. S., ed., Marcel Dekker, New York, 1993.

Horne, D. F., *Optical Production Technology*, Adam Hilger, London, and Crane Russak, New York, 1972, Chapter XI.

Horne, D. F., *Dividing, Ruling and Mask Making*, Adam Hilger, London, 1974, Chapter VII.

Hovanesian, J. D. and Y. Y. Hung, "Moiré Contour-Sum, Contour-Difference, and Vibration Analysis of Arbitrary Objects," *Appl. Opt.*, **10**, 2734 (1971).

Hume, K. J., *Metrology with Autocollimators*, Hilger and Watts, London, 1965.

Hurst, F., A. Mellin, and J. H. Whitelaw, *Principles and Practice of Laser Doppler Anemometry*, 2nd edn, Academic Press, London, 1981.

Jentink, H. W., F. F. de Mul, H. E. Suichies, J. G. Aarnoudse, and J. Greve, "Small Laser Doppler Velocimeter Based on the Self-Mixing Effect in a Diode Laser," *Appl. Opt.*, **27**, 379–385 (1988).

Johnson, B. K., *Optics and Optical Instruments*, Dover, New York, 1947, Chapters II and VIII.

Johnson, G. W. and D. T. Moore, "Design and Construction of a Phase-Locked Interference Microscope," *Proc. SPIE*, **103**, 76–85 (1977).

Jones, R. and C. Wykes, *Holographic and Speckle Interferometry*, Cambridge University Press, London, 1983.

Jurek, B., *Optical Surfaces*, Elsevier Scientific Publ. Co., New York, 1977.

Jutamulia, S., T. W. Lin, and F. T. S. Yu, "Real-Time Color-Coding of Depth Using a White-Light Talbot Interferometer," *Opt. Comm.*, **58**, 78–82 (1986).

Kafri, O. and I. Glatt, *The Physics of Moiré Metrology*, John Wiley, New York, 1990.

Kafri, O., Y. B. Band, T. Chin, D. F. Heller, and J. C. Walling, "Real-Time Moiré Vibration Analysis of Diffusive Objects," *Appl. Opt.*, **24**, 240–242 (1985).

Kao, T. Y. and F. P. Chiang, "Family of Grating Techniques of Slope and Curvature Measurements for Static and Dynamic Flexure of Plates," *Opt. Eng.*, **21**, 721–742 (1982).

Kaufmann, G. H., "Automatic Fringe Analysis Procedures in Speckle Metrology," in *Speckle Metrology*, Sirohi, R. S., ed., Marcel Dekker, New York, 1993.

Kingslake, R., *Optical System Design*, Academic Press, New York, 1983, Chapter 13.

Kothiyal, M. P. and R. S. Sirohi, "Improved Collimation Testing Using Talbot Interferometry," *Appl. Opt.*, **26**, 4056–4057 (1987).

Kothiyal, M. P., R. S. Sirohi, and K. J. Rosenbruch, "Improved Techniques of Collimation Testing," *Opt. Laser Technol.*, **20**, 139–144 (1988).

Lee, W. D., J. H. Shirley, J. P. Lowe, and R. E. Drullinger, "The Accuracy Evaluation of NIST-7," *IEEE Trans. Instrum. Meas.*, **44**, 120–123 (1995).

Lewin, A., "The Implications of System 'Sensitivity' and 'Resolution' on an Ultrasonic Detecting LDV," *Proc. SPIE*, **2358**, 292–304 (1994).

Lewin, A. C., "Compact Laser Vibrometer for Industrial and Medical Applications," *Proc. SPIE*, **3411**, 61–67 (1998).

Li, W. X. and L. D. Mitchell, "Error Analysis and Improvements for Using Parallel-Shift Method to Test a Galvanometer-Based Laser Scanning System," *Proc. SPIE*, **2358**, 13–22 (1994).

Lopez-Ramirez, J. M., *Medición de Desplazamientos de Partículas Mediante Holografía y Moteado*, MSc thesis, CIO-Universidad de Guanajuato, Mexico, 1995.

Luxmoore, A. R., *Optical Transducers and Techniques in Engineering Measurement*, Applied Science Publishers, London, 1983.

McDuff, O. P., "Techniques of Gas Lasers," in *Laser Handbook, Vol. 1*, Arecchi, F. T. and E. O. Schulz-Dubois, eds, North-Holland, Amsterdam, 1972, pp. 631–702.

Malacara, D., "Some Properties of the Near Field of Diffraction Gratings," *Opt. Acta.*, **21**, 631–641 (1974).

Malacara, D. (ed.), *Optical Shop Testing*, 2nd edn, Wiley, New York, 1992.

Malacara Doblado, D., *Problems Associated to the Analysis of Interferograms and their Possible Applications*, PhD thesis, CIO-Mexico, 1995.

Malacara, D. and R. Flores, "A Simple Test for the 90 Degrees Angle in Prisms," *Proc. SPIE*, **1332**, 678 (1990).

Malacara, D. and O. Harris, "Interferometric Measurement of Angles," *Appl. Opt.*, **9**, 1630–1633 (1970).

Malacara, D., M. Servin, and Z. Malacara, *Interferogram Analysis for Optical Testing*, Marcel Dekker, New York, 1998.

Martens, von H-J., A. Täubner, W. I. Wabinsk, A. Link, and H-J. Schlaak, "Laser Interferometry-Tool and Object in Vibration and Shock Calibrations," *Proc. SPIE*, **3411**, 195–206 (1998).

Martin, L. C., *Optical Measuring Instruments*, Blackie and Sons Ltd., London, 1924.

Matthews, H. J., D. K. Hamilton, and C. J. R. Sheppard, "Surface Profiling by Phase-Locked Interferometry," *Appl. Opt.*, **25**, 2372–2374 (1986).

Medlock, R. S., "Review of Modulating Techniques for Fibre Optic Sensors," *J. Opt. Sensors*, **1**, 43–68 (1986).

Minkowitz, S. and W. A. Smith-Vanir, *J. Quantum Electronics*, **3**, 237 (1967).

Müller, J. and M. Chour, "Two-Frequency Laser Interferometric Path Measuring System for Extreme Velocities and High Accuracy's," *Int. J. Optoelectron.*, **8**, 647–654 (1993).

Muñoz Rodriguez, A., R. Rodriguez-Vera, and M. Servin, "Direct Object Shape Detection Based on Skeleton Extraction of a Light Line," *Opt Eng.* **39**, 2463–2471 (2000).

Murty, M. V. R. K. and R. P. Shukla, "Methods for Measurement of Parallelism of Optically Parallel Plates," *Opt. Eng.*, **18**, 352–353 (1979).

Nakano, Y., "Measurements of the Small Tilt-Angle Variation of an Object Surface Using Moiré Interferometry and Digital Image Processing," *Appl. Opt.*, **26**, 3911–3914 (1987).

Nakano, Y. and K. Murata, "Talbot Interferometry for Measuring the Focal Length of a Lens," *Appl. Opt.*, **24**, 3162–3166 (1985).

Nakano, Y. and K. Murata, "Talbot Interferometry for Measuring the Small Tilt Angle Variation of an Object Surface," *Appl. Opt.*, **25**, 2475–2477 (1986).

Nakano, Y., R. Ohmura, and K. Murata, "Refractive Power Mapping of Progressive Power Lenses Using Talbot Interferometry and Digital Image Processing," *Opt. Laser Technol.*, **22**, 195–198 (1990).

Ng, T. W., "Circular Grating Moiré Deflectometry Analysis by Zeroth and First Radial Fringe Order Angle Measurement," *Opt. Comm.*, **129**, 344–346 (1996).

Ng, T. W. and F. S. Chau, "Object Illumination Angle Measurement in Speckle Interferometry," *Appl. Opt.*, **33**, 5959–5965 (1994).

Nishijima, Y. and G. Oster, "Moiré Patterns: Their Application to Refractive Index and Refractive Index Gradient Measurements," *J. Opt. Soc. Am.*, **54**, 1–5 (1964).

Ohtsuka, Y., "Dynamic Measurement of Small Displacements by Laser Interferometry," *Trans. Inst. Meas. Control*, **4**, 115–124 (1982).

Oshida, Y., K. Iwata, and R. Nagata, "Optical Heterodyne Measurement of Vibration Phase," *Opt. Lasers Eng.*, **4**, 67–69 (1983).

Osterberg, H., "An Interferometer Method for Studying the Vibration of an Oscillating Quartz Plate," *J. Opt. Soc. Am.*, **22**, 19–35 (1932).

Palmer, C. H., "Differential Angle Measurements with Moiré Fringes," *Opt. Laser Technol.*, **1**(3), 150–152 (1969).

Patorski, K., *Handbook of the Moiré Fringe Technique*, Elsevier, Amsterdam, 1993.

Patorski, K., S. Yokozeki, and T. Suzuki, "Optical Alignment Using Fourier Imaging Phenomenon and Moiré Technique," *Opt. Laser Technol.*, **7**, 81–85 (1975).

Peck, E. R. and S. W. Obetz, "Wavelength or Length Measurement by Reversible Fringe Counting," *J. Opt. Soc. Am.*, **43**, 505–509 (1953).

Pickering, C. J. D. and N. A. Halliwell, "Laser Speckle Photography and Particle Image Velocimetry: Photographic Film Noise," *Appl. Opt.*, **23**, 2961–2969 (1984).

Post, D., B. Han, and P. Ifju, *High Sensitivity Moiré*, Springer-Verlag, New York, 1994.

Raffel, M., C. Willert, and J. Kompenhans, *Particle Image Velocimetry, a Practical Guide*, Springer-Verlag, Berlin, 1998.

Ratajczyk, F. and Z. Bodner, "An Autocollimation Measurement of the Right Angle Error with the Help of Polarized Light," *Appl. Opt.*, **5**, 755–758 (1966).

Ravensbergen, M., R. Merino, and C. P. Wang, "Encoders for the Altitude and Azimuth Axes of the VLT," *Proc. SPIE*, **2479**, 322–328 (1995).

Ready, J. F., *Industrial Applications of Lasers*, Academic Press, New York, 1978.

Reid, G. T., "A Moiré Fringe Alignment Aid," *Opt. Lasers Eng.*, **4**, 121–126 (1983).

Rieder, G. and R. Ritter, "Krummungsmessung an Belasteten Platten Nach dem Ligtenbergschen Moiré-Verfahren," *Forsch. Ing.-Wes.*, **31**, 33–44 (1965).

Ritter, R. and R. Schettler-Koehler, "Curvature Measurement by Moiré Effect," *Exp. Mech.*, **23**, 165–170 (1983).

Rizzo, J. E. and N. A. Halliwell, "Multicomponent Frequency Shifting Self-Aligning Laser Velocimeters," *Rev. Sci. Instrum.*, **49**, 1180–1185 (1978).

Rodriguez-Vera, R., "Three-Dimensional Gauging by Electronic Moiré Contouring," *Rev. Mex. Fis.*, **40**, 447–458 (1994).

Rodriguez-Vera, R. and M. Servin, "Phase Locked Loop Profilometry," *Opt. Laser Technol.*, **26**, 393–398 (1994).

Rodriguez-Vera, R., D. Kerr, and F. Mendoza-Santoyo, "Three-Dimensional Contouring of Diffuse Objects Using Talbot Interferometry," *Proc. SPIE*, **1553**, 55–65 (1991a).

Rodriguez-Vera, R., D. Kerr, and F. Mendoza-Santoyo, "3-D Contouring of Diffuse Objects by Talbot-Projected Fringes," *J. Mod. Opt.*, **38**, 1935–1945 (1991b).

Rodriguez-Vera, R., D. Kerr, and F. Mendoza-Santoyo, "Electronic Speckle Contouring," *J. Opt. Soc. Am.*, **A9**, 2000–2008 (1992).

Ross, M. M., "Combined Differential Reference Beam LDV for 3D Velocity Measurement," *Opt. Lasers Eng.*, **27**, 587 619 (1997).

Rovati, L., U. Minoni, M. Bonardi, and F. Docchio, "Absolute Distance Measurement Using Comb–Spectrum Interferometry," *J. Opt.*, **29**, 121–127 (1998).

Rüeger, J. M. and R. W. Pasco, "Performance of a Distomat Wild DI 3000 Distance Meter," Tech. Paper No. 19, *Aust. Survey Congress*, Hobart, 1989.

Ruiz Boullosa, R. and A. Perez Lopez, "Interferometro Láser y Conteo de Franjas Aplicado a la Calibración de Acelerómetros y Calibradores de Vibraciones," *Rev. Mex. Fis.*, **36**, 622–629 (1990).

Schmidt, H., US Patent # 4465366, 1984.

Sciammarella, C. A. and M. N. Ahmadshahi, "Nondestructive Evaluation of Turbine Blades Vibrating in Resonant Modes," *Proc. SPIE*, **1554B**, 743–753 (1991).

Shipley, G. and R. H. Bradsell, "Georan I, a Compact Two-Color EDM Instrument," *Survey Review*, XXIII (1976).

Silva, D. E., "A Simple Interferometric Method of Beam Collimation," *Appl. Opt.*, **10**, 1980–1982 (1971).

Silva, A. A. and R. Rodriguez-Vera, "Design of an Optical Level Using the Talbot Effect," *Proc. SPIE*, **2730**, 423–426 (1996).

Sirohi, R. S., ed., *Speckle Metrology*, Marcel Dekker, New York, 1993.

Sirohi, R. S. and F. S. Chau, *Optical Methods of Measurement, Wholefield Techniques*, Marcel Dekker, New York, 1999, Chapter 4.

Sirohi, R. S. and M. P. Kothiyal, *Optical Components, Systems and Measurement Techniques*, Marcel Dekker, New York, 1990.

Smith, W. J., *Modern Optical Engineering*, McGraw-Hill, New York, 1966.

Smythe, R. and R. Moore, "Instantaneous Phase Measuring Interferometry," *Opt. Eng.*, **23**, 361–364 (1984).

Sona, A., *Laser Handbook*, Vol. 2, Arecchi, F. and E. Schulz-Dubois, eds, New-Holland, Amsterdam, 1972.

Sriram, K. V., M. P. Kothiyal, and R. S. Sirohi, "Use of Non-Collimated Beam for Determining the Focal Length of a Lens by Talbot Interferometry," *J. Optics*, **22**, 61–66 (1993).

Stafne, M. A., L. D. Mitchell, and R. L. West, "Positional Callibration of Galvanometric Scanners Used in Laser Doppler Vibrometers," *Proc. SPIE*, **3411**, 210–223 (1998).

Stitch, M. L., "Laser Ranging," in *Laser Handbook*, Vol. 2, Arecchi, F. T. and E. O. Schulz-Dubois, eds, North-Holland, Amsterdam, 1972, pp. 1745–1804.

Stitch, M. L., E. J. Woodbury, and J. H. Morse, "Optical Ranging System Uses Laser Transmitters," *Electronics*, **34**, 51–53 (1961).

Su, D-Ch. and Ch.-W. Chang, "A New Technique for Measuring the Effective Focal Length of a Thick Lens or a Compound Lens," *Opt. Comm.*, **78**, 118–122 (1990).

Swyt, D. A., "The International Standard of Length," in *Coordinate Measuring Machines and Systems.*, Bosch, J. A., ed., Marcel Dekker, New York, 1995.

Taarev, A. M., "Testing the Angles of High-Precision Prisms by Means of an Autocollimator and a Mirror Unit," *Sov. J. Opt. Technol.*, **52**, 50–52 (1985).

Takasaki, H., "Moiré Topography," *Appl. Opt.*, **9**, 1457 (1970).

Takeshima, A., US Patent # 5534992, 1996.

Teles, F., D. V. Magalhaes, M. S. Santos, and V. S. Bagnato, "A Cesium-Beam Atomic Clock Optically Operated," *Proceedings of the International Symposium on Laser Metrology for Precision Measurement and Inspection*, Florianópolis, Brazil, October 13–15, 1999, pp. 3.37–3.45.

Tew, E. J., "Measurement Techniques Used in the Optics Workshop," *Appl. Opt.*, **5**, 695–700 (1966).

Theocaris, P. S., "Flexural Vibrations of Plates by the Moiré Method," *Brit. J. Appl. Phys.*, **18**, 513–519 (1967).

Thorton, B. S. and J. C. Kelly, "Multiple-Beam Interferometric Method of Studying Small Vibrations," *J. Opt. Soc. Am.*, **46**, 191–194 (1956).

Tiziani, H. J., "A Study of the Use of Laser Speckle to Measure Small Tilts of Optically Rough Surfaces Accurately," *Opt. Commun.*, **5**, 271–276 (1972).

Tocher, A. J., US Patent # 5483336, 1992.

Twyman, F., *Prisms and Lens Making.*, 2nd edn, Hilger and Watts, London, 1957.

Ueda, K. and A. Umeda, "Characterization of Shock Accelerometers Using Davies Bar and Laser Interferometer," *Exp. Mech.*, **35**, 216–223 (1995).

Watson, J., "Photodetectors and Electronics," in *Optical Transducers and Techniques in Engineering Measurement*, Luxmoore, A. R., ed., Applied Science Publishers, London, 1983, Chapter 1.

Watson, R. C., R. D. Lewis, and H. J. Watson, "Instruments for Motion Measurements Using Laser Doppler Heterodyning Techniques," *ISA Transactions*, **8**, 20–28 (1969).

Yamada, J., "Evanescent Wave Doppler Velocimetry for a Wall's Near Field," *Appl. Phys. Lett.*, **75**, 1805–1806 (1999).

Yeh, Y. and Z. Cummins, "Localized Fluid Flow Measurements with an He–Ne Laser Spectrometer," *Appl. Phys. Lett.*, **4**, 176–178 (1964).

Yokozeki, S., K. Patorski, and K. Ohnishi, "Collimating Method Using Fourier Imaging and Moiré Technique," *Opt. Comm.*, **14**, 401–405 (1975).

Young, A. W., "Optical Workshop Instruments," in *Applied Optics and Optical Engineering*, Kingslake, R., ed., Academic Press, New York, 1967, Chapter 7.

Zeng, X., A. L. Wicks, and L. D. Mitchell, "The Determination of the Position and Orientation of a Scanning Laser Vibrometer for a Laser-Based Mobility Measurement System," *Proc. SPIE*, **2358**, 81–92 (1994).

14

Optical Metrology of Diffuse Objects: Full-Field Methods

MALGORZATA KUJAWINSKA
Institute of Precise and Optical Instruments Technical University, Warsaw, Poland

DANIEL MALACARA
Centro de Investigaciones en Optica, León, Mexico

14.1 INTRODUCTION

Optical metrology of diffuse objects, as opposed to metrology of specularly reflecting optical surfaces (Malacara, 1992), is performed with some procedures that are described in this chapter. These techniques are extremely useful in engineering to measure solid bodies or structures that do not have a specularly reflecting surface. Optical metrology of diffuse objects focuses on gathering information about shape and displacement or deformation vector $\mathbf{r}(u, v, w)$ of three-dimensional solid state bodies or structures with nonspecularly reflecting surfaces.

In this chapter, the full-field coherent and non-coherent methods that are most frequently applied in engineering (Table 14.1) are considered. The specific feature of these methods is coding the measurand into a fringe pattern that is analyzed by one of the methods described in Chapter 12.

In Table 14.1 $w()$ is an out-of-plane displacement in the z-direction, and $u()$ and $v()$ are in-plane displacements in the x- and y-directions, respectively.

14.2 FRINGE PROJECTION

The shape of a solid three-dimensional body can be measured by projecting the image of a periodic structure or ruling over the body (Idesawa *et al.*, 1977; Takeda, 1982; Doty, 1983; Gåsvik, 1983; Kowarschik *et al.*, 2000) or by interference of two tilted plane or spherical wavefronts (Brooks and Heflinger, 1969). The fringes may be projected on the body by a lens or slide projector (Takasaki, 1970, 1973; Parker, 1978; Pirodda, 1982; Suganuma and Yoshisawa, 1991; Halioua *et al.*, 1983; Gåsvik, 1995).

Table 14.1 Review of the Methods Used in Optical Metrology of Diffuse Objects

Method	Features of object	Measurand (range)
Grid projection	Arbitrary shape, diffuse object	Shape, w (μm–cm)
Out-of-plane moiré		
• Projection	Arbitrary shape	Shape, w
• Shadow	diffuse object	(μm–cm)
Holographic interferometry		
• Classical	Arbitrary shape, arbitrary	u, v, w (nm–μm)
• Digital	surface including diffuse objects	Shape (μm–cm)
Speckle interferometry		
• ESPI in plane	Flat object, diffuse surface	v, v (nm–μm)
• ESPI out of plane	Arbitrary shape, diffuse surface	w (nm–μm) Shape (μm–cm)
Shearography		
• In plane	Flat object, diffuse surface	Derivatives of u and v
• Out of plane	Arbitrary shape, diffuse surface	derivative of w
Speckle photography		
• Conventional	Diffuse object, flat surface	u, v (μm–mm)
• Digital	often painted white	
In-plane moiré	Flat sample with grating attached	u, v
• Conventional	$f < 40$ lines/mm	(μm–mm)
• Photographic (high resolution)	$f < 300$ lines/mm	(μm–cm)
Grating interferometry (moiré)	Flat sample with high frequency grating $f < 3000$ lines/mm	u, v (nm–μm)

Depending on the topology of the surface the fringes are distorted and may be described by the equation:

$$I(x, y) = a(x, y) + b(x, y) \cos[2\pi f_0 x + \phi(x, y)], \qquad (14.1)$$

where $a(x, y)$ and $b(x, y)$ are background and local modulation functions, while f_0 is the frequency of projected fringes in a reference plane close to the object and $\phi(x, y)$ is the phase related to the height of the object.

The fringes are imaged on the observing plane by means of an optical system, photographic camera, or a charged-coupled device (CCD) camera. These fringes can be analyzed directly ($I(x, y)$) or preprocessed by one of the moiré techniques: projection moiré (Idesawa *et al.*, 1977) and shadow moiré (Pirodda, 1982). In projection moiré the distorted fringes are superimposed on a linear ruling with approximately the same frequency as the projected fringes. The reference ruling may be real or software generated on the computer analyzing the image (Asundi, 1991). In shadow moiré, a Ronchi ruling is located just in front of the object under study and is obliquely illuminated. The moiré fringes are formed as a result of beating between the distorted shadow grid and the grid itself.

Recently, fringe projection is most frequently used in engineering, medical, and multimedia applications. The process of measurement may be divided into two steps:

- To each image point of the object a phase $\phi(x, y)$ is assigned as the primary measuring quantity. The phase is calculated from the fringe pattern (Eq. (14.1)) by phase shifting or the Fourier-transform method (Patorski, 1993). In the case of steep slopes or step like objects the methods which allow us to avoid unwrapping are applied, namely the coded light or gray code technique or hierarchical absolute phase decoding (Osten *et al.*, 1996) which rely on using the combination of at least two projected patterns with different spatial frequencies.
- Based on the geometrical model of the image formation process the three-dimensional coordinates are determined using these phase values, and certain parameters of the measurement system have to be identified in advance.

The basis of the evaluation is the triangulation principle. A light point is projected onto the surface, which is observed under the so-called triangulation angle θ (Fig. 14.1). Using an optical system, this point is imaged on a light-sensitive sensor. Consequently, the measurement of the height h is reduced to the measurement of the lateral position Δx on the CCD chip. For the calculation of h the imaging geometry and the triangulation angle is needed. There are three basic configurations, as illustrated in Figs 14.2 and 14.3. In the first two cases (Fig. 14.2(a) and (b)) the optical axis of projection and observation systems intersect under an angle θ; however, the fringes are projected by the telecentric lens. If the observation point is located at a height l from the reference plane the contour surfaces are not planes (except the reference one) and the height h of a body is given by (Fig. 14.3(a))

$$h = \frac{l\Delta x}{l \tan \theta + x}. \tag{14.2}$$

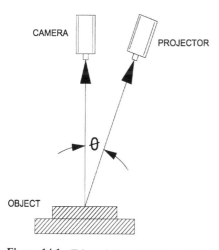

Figure 14.1 Triangulation principle used in fringe projection.

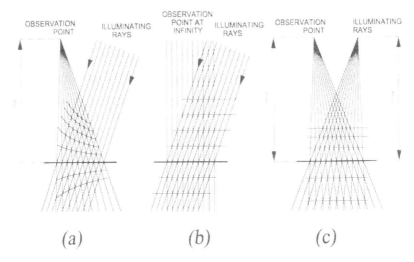

Figure 14.2 Three configurations to project a periodic structure over a solid body to measure its shape.

where the fringe deviation Δx is given by

$$\Delta x = \frac{d}{2\pi}\phi(x, y) \tag{14.3}$$

where d is the period of the projected fringes (in the reference plane) and x is measured perpendicularly to the projected fringes with an origin at the point where the axis of the projection intersects the observation lenses.

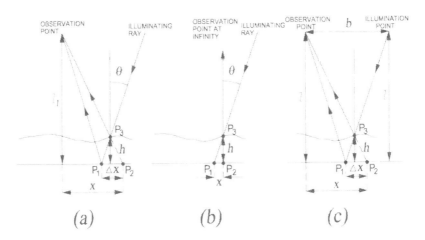

Figure 14.3 Geometries for the three basic configurations used to calculate the object height in the fringe projection.

Knowing the phase values $\phi(x, y)$ from fringe pattern analysis (Malacara *et al.*, 1998), the height of an object is calculated from

$$h = \frac{1}{2\pi} \left(\frac{ld}{l \tan \theta + x} \right) \phi(x, y). \tag{14.4}$$

The easiest scaling of the height of a three-dimensional object is given by a system with telecentric projection and detection (Figs 14.2(b) and 14.3(b)) and is given by

$$h = \frac{1}{2\pi} \left(\frac{ld}{\tan \theta} \right) \phi(x, y). \tag{14.5}$$

However, in this configuration the size of the measured object is restricted by the diameter of the telecentric optics.

The third configuration is based on the geometry, with mutual parallel optical axis of projecting and observation systems (Figs 14.2(c) and 14.3(c)). In this case, the height of the body is given by

$$h = \frac{1}{2\pi} \frac{ld}{b + \dfrac{d}{2\pi} \phi(x, y)} \phi(x, y). \tag{14.6}$$

If the geometry of the fringe projection system is not predetermined a calibration of the measurement volume is required (Sitnik and Kujawinska, 2000) or combining photogrametric and triangulation/phase-measuring approaches (Reich *et al.*, 2000). Modern shape measurement systems deliver data about an object's coordinates (x, y, z) in the form of a cloud of points measured and merged from different directions; these data are extensively used in CAD–CAM and rapid prototyping systems as well as in computer graphics and virtual reality environments.

14.3 HOLOGRAPHIC INTERFEROMETRY

Holographic interferometry (Ostrovsky *et al.*, 1980) is the most universal method of investigation of diffuse object. However, the bottleneck of holographic interferometry is the recording medium. Silver halides provide the best resolution with high sensitivity and good-quality holographic reconstruction, but need wet chemical processing. Photothermoplasts require special electronics, and are limited in size, resolution, and diffraction efficiency. The new solution is brought together with progress in high-resolution CCD cameras and fast computers. Below, the principles of both optical (conventional) and digital holographic interferometry are described.

14.3.1 Optical Holographic Interferometry

The basis for holographic interferometry is that a reconstructed hologram contains all the information (phase and amplitude) about the recorded object. If this holographic image is superimposed with the object wave from the same object, but slightly changed, the two waves will interfere (Kreis, 1996). This can be done by making the first exposure from the object in a reference state with an amplitude $E_1 \exp(i\phi_1(\mathbf{r}))$, and the second one – after some changes in the state of the object – with an amplitude $E_2 \exp(i\phi_2(\mathbf{r}))$ (Fig. 14.4(a)). It is called a double-exposure hologram. After the development of the photographic plate, the hologram is illuminated

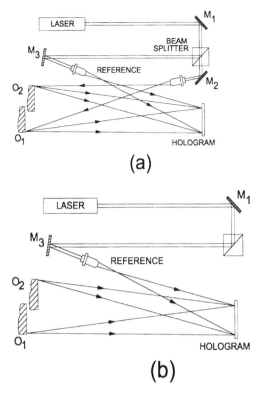

Figure 14.4 Optical arrangement in digital holographic interferometry with an off-axis reference beam. (a) Recording of a double-exposure hologram; (b) reconstruction of a double-exposure hologram.

with the reference wave, thus leading to the reconstruction of both states of the object at the same time in the original position (Fig. 14.4(b)). It is also possible to record the reference object state only, and monitor interference between the object wavefront and the wavefront reconstructed from the hologram (real-time holographic interferometry). If the change of the object is small enough, the two reconstructed waves interfere forming a fringe pattern given by

$$
\begin{aligned}
I = \ & E_1^2 + E_2^2 + E_1 E_2 \exp[i(\phi_1(\mathbf{r}) - \phi_2(\mathbf{r}))] + E_1 E_2 \exp[-(i(\phi_1(\mathbf{r}) - \phi_2(\mathbf{r})))] \\
= \ & I_1 + I_2 + 2 I_1 I_2 \cos[\phi_1(\mathbf{r}) - \phi_2(\mathbf{r})].
\end{aligned}
$$

$$(14.7)$$

The two phase terms can be combined to $\phi(x, y, z)$, which is calculated from the intensity distribution by one of the fringe pattern analysis algorithms. The phase difference between two object states is related to the optical path difference (OPD) by

$$
\phi(x, y) = \left(\frac{2\pi}{\lambda}\right) \text{OPD}.
$$

$$(14.8)$$

The optical path difference OPD is the geometrical one but projected onto the sensitivity vector $\mathbf{s} = \mathbf{p} - \mathbf{q}$ (Pryputniewicz, 1994), which is the vector given by the difference of the unity vector $\mathbf{q}$ from the illumination source to the point P with coordinates (x, y, z) located at the object and the unity vector $\mathbf{p}$ from P to the observation point (Fig. 14.5)

$$\text{OPD} = \mathbf{d} \cdot (\mathbf{p} - \mathbf{q}), \tag{14.9}$$

where $\mathbf{d}$ is the displacement vector from the point P with coordinates (x, y, z) to the shifted point with coordinates (x', y', z'). The sensitivity vector $\mathbf{s}$ is along the bisectrix for the angle between the illuminating ray and the ray traveling from the illuminated object to the observing point. Its maximum magnitude is 2 when the two light rays coincide.

Since the phase term is only the projection of the displacement vector onto the sensitivity vector, one needs three sensitivity vectors for the full determination of $\mathbf{d}$. They may be introduced by changing the observation direction or (most often) by changing the illumination direction.

A specific application of displacement measurement is the analysis of vibrating objects where the displacement of each point is

$$\mathbf{d}(P, t) = \mathbf{d}(P) \sin \omega t; \tag{14.10}$$

this task is performed by

- Stroboscopic holographic interferometry, which consists of recording a hologram by using a sequence of short pulses that are synchronized with the vibrating object (Hariharan *et al.*, 1987).
- Time-average holographic interferometry, in which the object is recorded holographically with a single exposure which is long compared with the period of vibration (Pryputniewicz, 1985). The resulting intensity in the reconstructed image is

$$I(P) = I_0(P) J_0^2 \left[\frac{2\pi}{\lambda} \mathbf{d} \cdot \mathbf{s} \right] \tag{14.11}$$

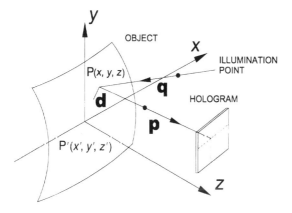

Figure 14.5 Optical path difference calculation in holographic interferometry.

where $\mathbf{s} = [\mathbf{p} - \mathbf{q}]$. Maximum intensity occurs at the nodes of the vibration modes and dark fringes appear when $(2\pi/\lambda)(\mathbf{d} \cdot \mathbf{s})$ equals the arguments of the zeros of the zero-order Bessel function of the first kind.

Besides the measurement of a map of object displacement, holographic interferometry enables us to measure the shape of an object. Holographic contouring requires recording two holograms of a static object with two different sensitivity vectors, introduced by changing alternatively:

- Laser wavelengths (from λ to λ') (Friesem and Levy, 1976):

$$\Delta z = \frac{\lambda\lambda'}{[(\lambda - \lambda')(1 + \cos\theta)]}, \tag{14.12}$$

where Δz is the depth difference.

- The directions of illumination (DeMattia and Fossati-Bellani, 1978). It is most frequently used with collimated beams which produce equidistant parallel contour surfaces with the depth distance

$$\Delta z = \frac{\lambda}{2\sin(\theta/2)} \tag{14.13}$$

- Refractive index of the medium surrounding an object (from n to n') (Tsuruta *et al.*, 1967):

$$\Delta z = \frac{\lambda}{2(n - n')} \tag{14.14}$$

14.3.2 Digital Holographic Interferometry

The principle used in digital holographic interferometry is basically the same used in conventional holography. The main difference is that instead of a holographic photographic plate, a CCD detector is used to record the image. The typical size of a CCD is about 7 mm with 1000×1000 pixels. Since the resolution is low compared with that of the holographic plate, the fringe separation must be much larger. Thus, if γ_{max} is the maximum angle between the reference beam and the object point with the largest angle from the light source as seen from the detector, and illustrated in Fig. 14.6, we should have

$$\sin\gamma_{max} \ll \frac{\lambda}{d}, \tag{14.15}$$

where d is the distance between two consecutive pixels on the CCD. This requirement can be satisfied if the object is small, of the order of a few millimeters and the light source producing the collimated reference beam is at least at a distance of about

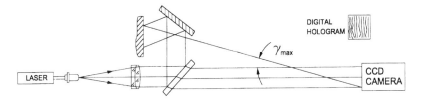

Figure 14.6 Optical arrangement in digital holographic interferometry.

1 m. These restrictions may be soon not valid if CCD cameras with the pixel size of 1 μm are available.

In digital holography (Yaroslavski and Merzlyakov, 1980; Schnars, 1994) the object reconstruction is performed not by an optical reconstruction as in classical holography but numerically in a computer, by performing a discrete finite Fresnel transformation. The phase on the object surface is thus computed. If two holograms are registered, one of them after a small displacement or deformation of the body, the phase difference can be obtained. This is called digital holographic interferometry.

In earlier digital holography, the off-axis setup just described was used, with zero order and the two conjugate images. This limitation was removed, allowing on-axis configuration, by using phase-shifting techniques with a piezoelectric translated in one of the mirrors. The distribution on complex amplitude, thus including the phase, at the hologram plane can be obtained (Yamaguchi, 1999).

The hologram is imaged on a CCD detector. The reconstruction plane is located very close to the object to be measured. The measured complex amplitude at the CCD detector is represented by $h(x, y)$. Then, the complex amplitude $E(\xi, \eta)$ on the reconstruction plane at distance z from the hologram is given by (Goodman, 1975a):

$$
\begin{aligned}
E(\xi, \eta) = \frac{iU_0}{\lambda z} \exp&\left[-\frac{i\pi}{\lambda z}(\xi^2 + \eta^2)\right] \int\int h(x, y) \exp\left[-\frac{i\pi}{\lambda z}(x^2 + y^2)\right] \\
&\exp\left[-\frac{i\pi}{\lambda z}(x\xi + y\eta)\right] d\xi \, d\eta;
\end{aligned}
\tag{14.16}
$$

hence, the phase on this plane is given by

$$
\phi(\xi, \eta) = \arctan\frac{\mathrm{Im}[E(\xi, \eta)]}{\mathrm{Re}[E(\xi, \eta)]}.
\tag{14.17}
$$

There are two methods to calculate the phase difference between the two body states. The general principles involved in these procedures are described in the block diagram in Fig. 14.7 (Kreis, 1996).

Using holographic interferometry, the displacement vector at any point of the body being examined can be detected and measured. It has sensitivity to in plane in any direction as well as out of plane.

14.4 ELECTRONIC SPECKLE PATTERN INTERFEROMETRY

Speckle pattern interferometry has been developed to study vibrations and deformations by many authors, especially by Macovski *et al.* (1971) in the United States and by Butters and Leendertz (1971) in England. Later, many other developments followed: for example, by Løkberg (1987) and by Jones and Wykes (1981, 1989). A review on this subject can be found in the book by Cloud (1995) and in the article by Ennos (1978). To understand the procedure, let us consider an extended nonspecular diffusing surface, illuminated with a single well-defined wavefront. In other words, the illuminating light beam must be spatially coherent. If a second diffusing surface is located in front of the illuminated surface, as illustrated in Fig. 14.8(a), each bright point in the illuminated surface contributes to illuminate the surface in front of it. Since the light arriving at point P comes from all points in the illuminated surface

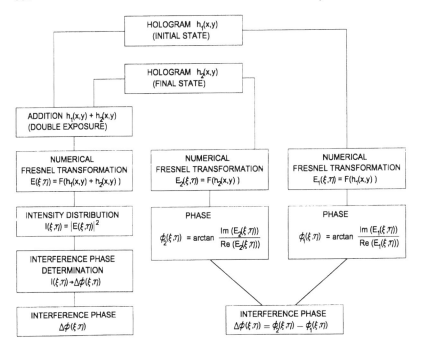

Figure 14.7 Procedures used to compute the object phase distribution in holographic interferometry.

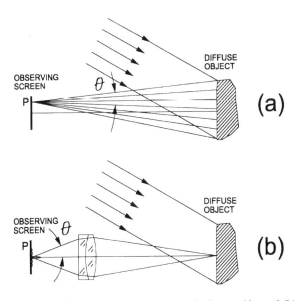

Figure 14.8 Speckle formation: (a) subjective speckles and (b) objective speckles.

and they are coherent to each other, multiple interference takes place. The second surface appears as if covered by many small and bright speckles. These speckes are real, as can be easily proved by substituting the second surface by a photographic plate. These are called objective speckles. Their average size d depends on the angular semidiameter θ of the illuminated surface as follows:

$$d = 1.22\lambda \sin \theta. \tag{14.18}$$

If we place a lens in front of the illuminated surface, as in Fig. 14.8(b), this lens forms the image of the illuminated surface on the plane of another diffusing surface. Each point of the illuminated surface is imaged as a small Airy disk with a diameter given by Eq. (14.18), where θ is the angular diameter of the imaging lens, as seen from the image point P. If F is the focal length of the lens, D its diameter, and m its magnification,

$$d = \frac{1.22\lambda D}{2F(m+1)}. \tag{14.19}$$

The interference between neighboring Airy disks produces speckles, which are called subjective speckles.

The speckle patterns are extremely complicated, but they depend only on two factors, i.e., the roughness configuration of the surface being illuminated and the phase distribution of the illuminating light beam. We can conclude that, given a diffusing surface, if the relative phase distribution for all points on the diffusing surface or on its image at the observing screen also changes, the speckle pattern structure is also modified. Furthermore, it can be seen that this relative phase distribution changes only if there are two interfering beams present. One of these two beams can go to illuminate the diffusing surface and the other directly to the observing screen, or both can go to the diffusing surface.

14.4.1   Optical Setups for Electronic Speckle Pattern Interferometry

To understand how the speckle pattern structure can change, let us consider some typical illumination configurations producing subjective speckle patterns. Any possible displacement along the coordinate axes x, y, z of the body being examined will be represented by u, v, w, where the coordinate z is along the perpendicular to the surface on the body.

(a)   Coaxial Arrangement

If the object is illuminated with a normally incident flat wavefront and the illuminated surface moves in its own plane, the speckle pattern also moves with this surface, but its structure remains unchanged. If this surface moves in the normal direction, the speckle structure also remains unchanged. However, if a reference wavefront traveling directly to the detector is added, out-of-plane movements could be detected. This setup is easily obtained if one of the mirrors in a Twyman–Green interferometer is replaced by the diffusing surface whose out-of-plane movement is to be detected.

(b) Asymmetrical Arrangement

When the illuminating wavefront has oblique incidence, as with normal illumination, it can be seen that neither of the two possible movements (in-plane or out-of-plane movements) of the illuminated surface change the speckle structure, unless there is a reference wavefront. A possible experimental arrangement with the reference beam going directly to the detector with normal incidence is illustrated in Fig. 14.9(a). Then, if the plane of incidence is the plane x, z, for a small in-plane displacement in the plane of incidence (along the x-axis) and a small out-of-the-plane displacement w (along the z-axis) the phase difference increases or decreases by an amount $\Delta\phi$ given by

$$\Delta\phi = ku\sin\theta + kw(1 + \cos\theta), \tag{14.20}$$

where $k = 2\pi/\lambda$ and θ is the angle of incidence for the oblique illuminating beam. The expression for a normally illuminating beam (coaxial arrangement) can be obtained with $\theta = 180°$, as

$$\Delta\phi = 2kw. \tag{14.21}$$

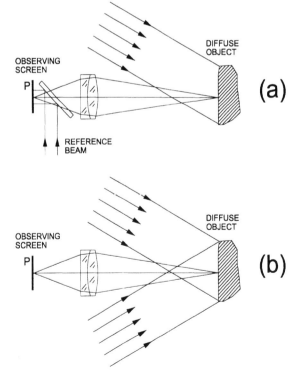

Figure 14.9 Schematic arrangements for out-of-plane movement detection in speckle interferometry: (a) out-of-plane and (b) in-plane movements.

The displacement values u and w can be separated only with two measurements using two different values of θ. When we have a combined displacement u and w, the movement cannot be detected if

$$\frac{u}{w} = \frac{1 + \cos\theta}{\sin\theta}. \tag{14.22}$$

An optical arrangement to perform electronic speckle pattern interferometry with a single beam with oblique incidence is illustrated in Fig. 14.10(a).

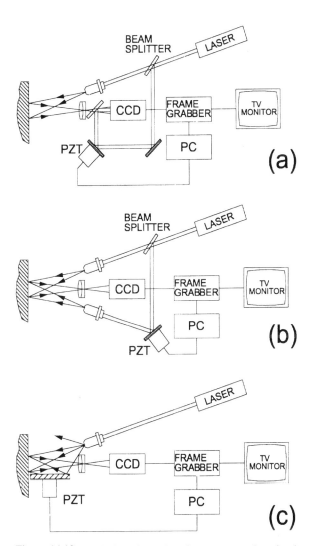

Figure 14.10 Optical configurations for movement detection in speckle interferometry. (a) In-plane and out-of-plane movements; (b) in-plane movement; and (c) in-plane movement.

(c) Symmetrical Arrangement

Let us assume that the two illuminating wavefronts have the same angle of incidence, but in opposite directions, in the plane of incidence x, z, as shown in Fig. 14.10(b). We can consider the observed speckle pattern to be formed by the interference of two speckle patterns, produced by each of the two illuminating light beams. These two speckle patterns have identical structure if observed independently, but their average phase changes linearly in opposite directions. This superposition has a different structure from any of its two components.

In-plane v movements in a direction perpendicular to the plane of incidence of the illuminating beams (y-axis) do not change the speckle structure. The phase difference between the two illuminating beams remains the same for all points in the surface. Thus, there is insensitivity for these movements.

With an in-plane movement u (x-axis direction), the phase increases for one of the beams and decreases for the other beam. The phase difference between the two beams increases in one direction and decreases in the opposite direction. Thus, the phase difference changes by an amount $\Delta\phi$, given by

$$\Delta\phi = 2ku\sin\theta, \tag{14.23}$$

where θ is the angle of incidence for the two illuminating beams. Thus, the larger the angle θ, the larger the sensitivity. It is important to notice that the introduced displacement should be smaller than the speckle size. An optical arrangement to perform electronic speckle pattern interferometry with two symmetric divergent beams with oblique incidence is illustrated in Fig. 14.10(b). The lack of collimation of the illuminating beams produces a nonconstant sensitivity over the illuminated area to in-plane displacements. Another configuration uses a flat mirror on one side of the object and thus produces the two illuminating beams with only one, as in Fig. 14.10(c). An arrangement with two collimated symmetrical beams to provide constant sensitivity over the illuminated area is shown in Fig. 14.11, with the disadvantage that the measured area is relatively small.

Another possible experimental setup using fiber optics is shown in Fig. 14.12. The beam is divided into two beams of equal intensity by a directional coupler. The light in one of the two arms of the interferometer passes through a phase shifter, which is formed by a PTZ cylinder with some fiber loops wrapped around. The phase shifted is used to perform phase-shifting interferometry controlled by a computer.

In speckle interferometry, two similar speckle patterns must be superimposed on each other by an additive or subtractive procedure. One speckle pattern is taken before and the other after a certain object movement has taken place. The correlation between these two patterns appear as moiré fringes. These fringes represent the phase difference between the two positions or shapes of the object on which the speckle pattern is formed. Each one of the two superimposed speckle patterns is made by the interference of the speckle pattern of the diffuse surface being measured with a reference beam. As pointed out by Jones and Wykes (1981), speckle interferometers can be classified into two categories:

- *Class 1*, where the reference beam is a single wavefront, generally plane or spherical, as in the arrangement just described.

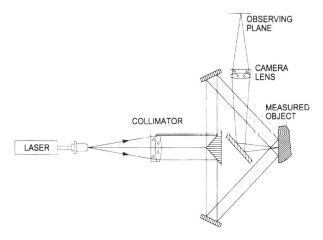

Figure 14.11 Optical configuration for in-plane movement detection with uniform sensitivity.

- *Class 2*, where both the object beam and the reference beams are speckle patterns, as in a Michelson interferometer with both mirrors replaced by diffuse objects.

In speckle photographic interferometry (Butters and Leendertz, 1971) a photograph of the previous speckle pattern is superimposed on the new speckle pattern, after the diffusing screen has been distorted or bended. In another method, two photographs may be placed on top of each other. In electronic speckle pattern interferometry (ESPI) (Ennos, 1978) two television images are electronically super-

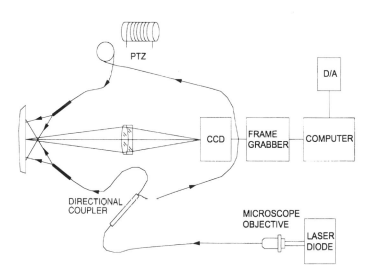

Figure 14.12 Configuration for in-plane movement detection using fiber optics.

imposed. An important requirement is that the speckle size must be greater than the size of each detector element.

14.4.2 Fringe Formation by Video Signal Subtraction or Addition

Let us consider the interference of two speckle patterns with irradiances I_0 and I_R and with random phases ψ_0 and ψ_R. Both these irradiances and these phases change very rapidly from point to point, due to the speckle nature. The interference of these two speckle patterns produce another speckle pattern with irradiance I_1, given by

$$I_1 = I_0 + I_R + 2\sqrt{I_0 I_R} \cos \psi, \tag{14.24}$$

where $\psi = \psi_0 - \psi_R$ is a random phase. Now, an additional smooth (not random) phase $\Delta\phi$ is added to one of the two interfering speckle patterns by a screen displacement or deformation. The new phase is $\psi + \Delta\phi$, and, then, the irradiance in the interference pattern is

$$I_2 = I_0 + I_R + 2\sqrt{I_0 I_R} \cos(\psi + \Delta\phi). \tag{14.25}$$

These two speckle patterns produced by interference are quite similar, with relatively small differences introduced by the phase $\Delta\phi$. A high-pass spatial filter is also applied to each of the two speckle patterns to remove low-frequency noise and variation in mean speckle irradiance. When these patterns are combined, their correlation becomes evident as a moiré pattern. There are two ways to produce moiré fringes between these two speckle patterns. One is by subtracting these two irradiances, obtaining

$$I_1 - I_2 = 2\sqrt{I_0 I_R} \sin\left(\psi + \frac{\Delta\phi}{2}\right) \sin\left(\frac{\Delta\phi}{2}\right). \tag{14.26}$$

This irradiance has positive as well as negative values. Thus, the absolute value must be taken

$$I_1 - I_2 = 2\sqrt{I_0 I_R} \left|\sin\left(\psi + \frac{\Delta\phi}{2}\right)\right| \left|\sin\left(\frac{\Delta\phi}{2}\right)\right|. \tag{14.27}$$

Then, a low-pass filter is applied to eliminate the high-frequency components produced by the speckle, obtaining a brightness B on the screen, given by

$$B_- = 2K\sqrt{I_0 I_R} \left|\sin\left(\frac{\Delta\phi}{2}\right)\right| \tag{14.28}$$

This subtraction method is used for the analysis of static events, as it requires the separate storing of two images.

The second method to form the fringes is by adding the irradiances of the two speckle patterns, as follows

$$I_1 + I_2 = 2(I_0 + I_R) + 4\sqrt{I_0 I_R} \cos\left(\psi + \frac{\Delta\phi}{2}\right) \cos\left(\frac{\Delta\phi}{2}\right). \tag{14.29}$$

This irradiance has a constant average value given by

$$\langle I_1 + I_2 \rangle = 2\langle I_0 \rangle + 2\langle I_R \rangle \tag{14.30}$$

but a variable contrast produced by the phase $\Delta\phi$. The variance σ of this irradiance for many points in the vicinity of a given point in the pattern is given by

$$\sigma^2 = \langle I^2 \rangle - \langle I \rangle^2, \tag{14.31}$$

and the standard deviation σ is

$$\sigma = \sqrt{\langle I^2 \rangle - \langle I \rangle^2}, \tag{14.32}$$

where $\langle I \rangle$ is the mean value of the irradiance values in the vicinity of the point being considered. This standard deviation of the irradiance over many points in the vicinity of a point has the advantage of eliminating the DC bias and at the same time rectifies the signal and applies a low-pass filter to eliminate the speckle. This standard deviation is given by

$$\sigma^2 = 4\sigma_R^2 + 4\sigma_0^2 + 8\langle I_R I_0 \rangle \cos^2\left(\frac{\Delta\phi}{2}\right). \tag{14.33}$$

On the other hand, assuming a Poissonian probability for each of the two speckle patterns, as shown by Goodman (1975b), it is possible to show that the standard deviation for each of these two speckle patterns is

$$\sigma_R = \left[\langle I_R^2 \rangle - \langle I_R \rangle^2\right]^{1/2} = \langle I_R \rangle \tag{14.34}$$

and

$$\sigma_0 = \left[\langle I_0^2 \rangle - \langle I_0 \rangle^2\right]^{1/2} = \langle I_0 \rangle. \tag{14.35}$$

Thus, the brightness B_+ on the screen which is proportional to this standard deviation of the irradiance, is given by

$$B_+ = 4K\left[\langle I_R \rangle^2 + \langle I_0 \rangle^2 + 2\langle I_R I_0 \rangle \cos^2\left(\frac{\Delta\phi}{2}\right)\right]^2. \tag{14.36}$$

The two speckle patterns to be correlated in order to produce the fringes are added together at the camera CCD detector. The two patterns do not need to be simultaneous because of the characteristic time persistence of the detector of about 0.1 s. This property permits the observation of dynamic or transient events or modal analysis of membranes, with a pulsed laser, using two consecutive pulses.

Areas in the fringe pattern with a large correlation have the maximum contrast. However, in the addition process the contrast is lower than in the subtraction method. The great advantage is that storage space for the two images is not needed. This problem of low fringe contrast can be solved by a computer processing of the image with different spatial filters (Alcalá-Ochoa et al., 1997).

Figure 14.13 shows two speckle images, before and after the object was modified. The effect of adding and subtracting the image is also illustrated. Figure 14.14 shows a block diagram with the basic steps followed in electronic speckle interferometry. A great advantage in this method is its relatively high speed with low environmental requirements. A disadvantage is its high noise content.

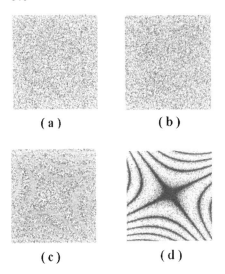

Figure 14.13 Speckle images. (a) First frame, (b) second frame, (c) subtraction and (d) addition.

14.4.3 Real-Time Vibration Measurement

When the out-of-plane vibrations of an object, like a membrane or loud-speaker diaphragm, are to be measured by speckle interferometry, the system is operated in time average. Commonly an addition of the frames is performed. At any time t, the irradiance in the speckle image is given by

$$I(t) = I_0 + I_R + 2\sqrt{I_0 I_R} \cos\left(\Delta\phi + \frac{4\pi}{\lambda} a_0 \sin \omega t\right), \tag{14.37}$$

where $a_0 \sin \omega t$ represents the position of the vibrating membrane at the time t. This irradiance is averaged over a time τ, obtaining

$$I_\tau = I_0 + I_R + \frac{2}{\tau}\sqrt{I_0 I_R} \int_0^\tau \cos\left(\Delta\phi + \frac{4\pi}{\lambda} a_0 \sin \omega t\right) d\tau. \tag{14.38}$$

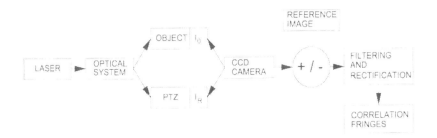

Figure 14.14 Speckle interferometry procedure to obtain the fringes.

If we take the time average for several cycles, we can assume that $2\pi/\omega \ll \tau$. Then, it is possible to show that

$$I_\tau = I_0 + I_R + 2\sqrt{I_0 I_R}\, J_0^2\left(\frac{4\pi}{\lambda}\right)\cos\phi, \tag{14.39}$$

where J_0 is the zero-order Bessel function.

Figure 14.15 shows an interferogram of the vibrating modes of a square diaphragm.

Vibration analysis with an in-plane sensitive arrangement using pulsed phase-stepped electronic speckle pattern interferometry has also been performed (Mendoza-Santoyo *et al.*, 1991).

14.5 ELECTRONIC SPECKLE PATTERN SHEARING INTERFEROMETRY

Electronic speckle pattern shearing interferometry (Sirohi, 1993) also called shearography was introduced by Hung (1982) and later further developed by many

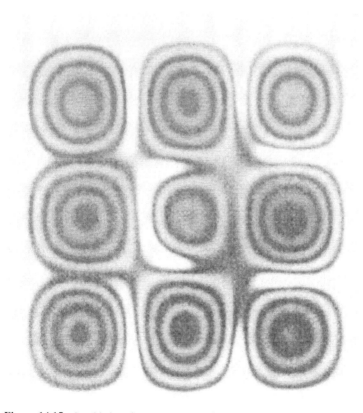

Figure 14.15 Speckle interferometry pattern for a square vibrating membrane.

researchers (Sirohi, 1984a, 1984b; Owner-Petersen, 1991. In this method two different points, laterally separated by a small distance called the shear, are added together at the same point on the detector.

Shearography provides information about out-of-plane deformations but in a direct manner about the slopes (derivatives) of these deformations. As in conventional lateral shearing interferometry, the fringes are the locus of the points with the same slope in the direction of the shear. Thus, an object tilt can not introduce any fringes. These properties make these interferograms more difficult to interpret and analyze. The deformations can be obtained only after numerical integration of the slopes in two perpendicular directions. This requires the generation of two shearograms with lateral shear in two orthogonal directions. However, in the theory of small deflections of a thin plate the strains are directly related to the second derivative of the deflection. Thus, to obtain the strains in shearography, only one derivative is necessary (Toyooka *et al.*, 1989). As described by Sirohi (1984b) and Ganesan *et al.* (1988), speckle shear interferometry can also be performed with radial or rotational shear, but the most common is lateral shear. If desired, as in conventional lateral shearing interferometry, a linear carrier can be introduced by defocusing, as described by Templeton and Hung (1989).

The principle is basically the same as in normal speckle pattern interferometry. As in conventional speckle interferometry, two images with a relative lateral shear are recorded, one before and one after the body deformations due to external applied forces. Then, the irradiances of these two images are added or subtracted. The result after low-pass filtering to remove the speckle noise is a fringe pattern with information about the body deformations. Using an oblique illumination beam with an angle of incidence θ, the phase differences with small lateral shear Δx for the sheared speckle patterns before deformation of the sample are

$$\Delta\phi = k(\Delta x)\sin\theta \tag{14.40}$$

and after deformation

$$\Delta\phi = k(\Delta x + \Delta u)\sin\theta + k(\Delta v)(1 + \cos\theta), \tag{14.41}$$

where Δx is the lateral shear and u and w are small local in-plane (along x-axis) and out-of-plane displacements. Thus, the change in the phase difference after these small local displacements is

$$\Delta\phi = k(\Delta u)\sin\theta + k(\Delta v)(1 + \cos\theta), \tag{14.42}$$

which can also be written as

$$\Delta\phi = k(\Delta x)\left[\frac{\partial u}{\partial x}\sin\theta + \frac{\partial w}{\partial x}(1 + \cos\theta)\right]. \tag{14.43}$$

If both types of displacement u and w are present and only the slope of w is desired, normal illumination can be used. However, if the slope of u is desired, two different measurements with different angle illuminations have to be used. Figure 14.16 shows three possible arrangements to produce this superposition of two laterally sheared speckle images on the detector. In the first case, a Michelson interferometer configuration is used, with a tilt of one of the mirrors. In the second example, the lens aperture is divided into two parts, one of them covered with a glass wedge. Finally, the third example shows a system with a Wollaston prism. Two

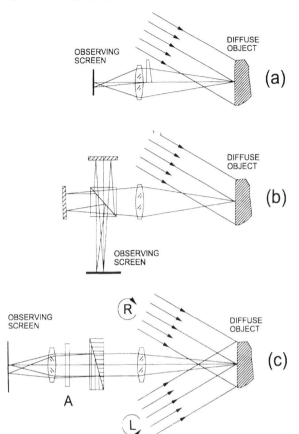

Figure 14.16 Configuration for speckle shearing interferometry. (a) With a glass wedge covering half of the lens aperture, (b) with a Michelson interferometer configuration with a tilted mirror and (c) with a Wollaston prism.

symmetrically oriented beams are used, with opposite angles of incidence, producing a system which is insensitive to out-of-plane displacements w. This system is sensitive only to in-plane displacements u. The change in the phase difference $\Delta\phi$ due to these in-plane displacements is

$$\Delta\phi = k\frac{\partial u}{\partial x}(\Delta x)\sin\theta. \tag{14.44}$$

Since the two illuminating beams are circularly polarized with opposite senses, the analyzer should be oriented at $+45°$ or $-45°$, where two complementary patterns are formed.

The setups normally used are simple and stable, making them ideal for in-situ measurements even with rough environmental conditions. In conclusion, electronic pattern shearing interferometry or shearography has many advantages, such as good vibration isolation and fringes of acceptable contrast, however most often it is used for qualitative, but quantitative evaluation.

14.6   IN-PLANE GRID AND MOIRÉ METHODS

Moiré methods, also called moiré interferometry, were developed by Post (1965, 1982, 1986), Abbiss *et al.* (1976), and many other researchers. As in electronic speckle pattern interferometry, small deformations or displacements of a solid body can be measured (Olszak *et al.*, 1996). In grating or moiré interferometry a specially prepared diffraction grating is attached to a nearly flat surface in a solid body, as shown in Fig. 14.17. Depending on the specimen material and the requested frequency of the specimen grating, the surface of an object is prepared by:

- Gluing printed patterns, photographic prints or stencilled paper patterns (frequencies up to 100 lines/mm),
- By replicating in epoxy a relief-type master grating or by exposing interfero-metric fringes in a photoresist layer spined at the specimen surface (frequencies up to 4000 lines/mm).

When the specimen is deformed by external forces, the attached grating is displaced and deformed as well. Frequencies up to 4000 lines/mm have been used. The grating is illuminated with two oblique and symmetrically oriented collimated light beams, with an angle of incidence θ given by

$$\sin \theta = \frac{\lambda}{d},\tag{14.45}$$

where d is the period of the grating attached to the solid body.

Thus, two conjugated orders of diffraction (+1 from illuminating beam A and −1 from illuminating beam B, as shown in Fig. 14.18) emerge from the grating, traveling along the normal to the grating. These two wavefronts produce an inter-ference pattern. The illuminated surface is then imaged on a CCD camera to register and analyze the fringe pattern.

The system is sensitive mainly to in-plane displacements along the incidence plane of the illuminating beams. The observed fringes represent the locus of the point on the object with a constant equal displacement. The displacement u of point on the fringe with order n, relative to the point on the fringe with order zero, is given by

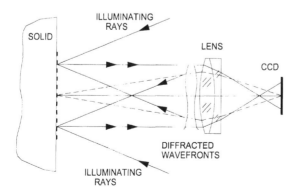

Figure 14.17   Grating interferometry configuration.

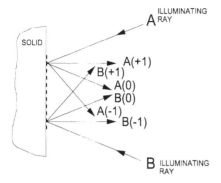

Figure 14.18　Diffracted beams in grating interferometry.

$$u = \frac{n}{f} = nd, \tag{14.46}$$

where f is the grating frequency and d is the period.

If there are changes in the spatial frequency of the grating attached to the solid body, due to stresses, expansions, contractions, or bendings, the two diffracted beams become distorted.

Moiré interferometry can be explained in the preceding manner with a physical model, as the interference of two wavefronts. However, it can also be explained as a geometrical moiré effect.

14.6.1　Basis of Grid Method

Whenever two slightly different periodic structures are superimposed, a beating between the two structures is observed in the form of another periodic structure with a lower spatial frequency, called a moiré pattern. As in Fig. 14.19, the two illuminating wavefronts interfere with each other, projecting a fringe pattern on the

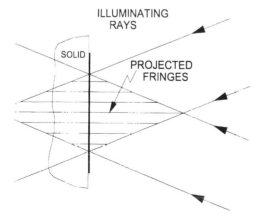

Figure 14.19　Projection of a virtual grating in grating interferometry.

grating. Thus, we can consider the grating to be illuminated by a periodic linear structure, with a period d equal to that of the grating. The observed fringe pattern can thus be interpreted as a moiré of the two superimposed structures.

Moiré techniques were used in metrology by Sciammarella (1982), Patorski (1988), and many other authors, with different configurations. Moiré patterns are frequently analyzed and interpreted from a geometrical point of view.

In moiré interferometry, the image of a grating produced by the interference of the two illuminating beams is projected over the grating attached to the solid body being examined. The two illuminating flat wavefronts produce interference fringes in space as dark and clear walls, parallel to each other and perpendicular to the plane of incidence and to the illuminated surface. Thus, a fringe pattern with straight lines, called a virtual grating, is projected on the body. The superposition of this virtual grating with the real grating on the body produces the moiré fringe pattern. This superposition can be interpreted as a multiplication of the irradiance on the virtual grating by the reflectivity on the real grating.

14.6.2 Conventional and Photographic Moiré Patterns

In the general case the superposition of the two periodic structures to produce the moiré patterns can be performed by multiplication, addition, or subtraction.

The multiplication can be implemented, for example, by superimposing the slides of two images or as in moiré interferometry (Post, 1971). The irradiance transmission of the combination is equal to the product of the two transmittances in the case of two slides. The contrast in the moiré image is smaller than in each of the two images. In addition or subtraction the contrast in the moiré pattern is higher.

In moiré interferometry the fringes are due to the superposition of the grating attached to the surface to be analyzed, which is illuminated by the projected virtual grating. Let us consider the fixed grating as a periodic structure, which is phase modulated (distorted) and whose amplitude reflectance $R(x, y)$, assuming the maximum contrast, may be described by

$$R(x, y) = 1 + \cos\left[\frac{2\pi}{d}(x + u(x, y))\right], \tag{14.47}$$

where d is the grating period and $u(x, y)$ represents the local displacement of the grating at the point (x, y). Let us now illuminate this distorted grating to be measured by a projected virtual reference grating with an amplitude $E_r(x, y)$ given by

$$E_r(x, y) = A(x, y) \cos\left(\frac{2\pi}{d} x\right). \tag{14.48}$$

There is no DC term, since this is a virtual grating without zero order. The observed amplitude pattern is the product of these two functions. Thus, the observed amplitude $E(x, y)$ in the moiré pattern is

$$E(x, y) = A(x, y)\left[1 + \cos\left(\frac{2\pi}{d}(x + u(x, y))\right)\right] \times \left[\cos\left(\frac{2\pi}{d} x\right)\right], \tag{14.49}$$

from which we may obtain

$$E(x, y) = A(x, y)\cos\left(\frac{2\pi}{d}x\right) + \frac{A(x, y)}{2}\cos\left(\frac{2\pi}{d}(2x - u(x, y))\right)$$
$$+ \frac{A(x, y)}{2}\cos\left(\frac{2\pi}{d}u(x, y)\right). \tag{14.50}$$

The first term represents the two zero-order beams. The second term represents the +1 order beam from illuminating beam A and the −1 order beam from the illuminating beam B. Finally, the last term represents the first-order beam from illuminating beam B and the negative first-order beam from the illuminating beam A. The last term squared is equal to Eq. (14.44), as we should expect. Figure 14.20 shows an interferogram obtained with moiré interferometry. Sometimes, instead of a projected virtual grating, a real reference grating placed over the body grating has been used (Post, 1982).

As in electronic speckle pattern shearing interferometry, moiré interferometry can be modified as proposed by Patorski *et al.* (1987) to produce fringes with information about the object slopes, for strain analysis.

14.6.3 Moiré Interferometry

As pointed out before, a local change in the frequency of the grating produces a wavefront distortion in the diffracted beams, by changing the local wavefront slope. Since the order of diffraction of the two diffracted interfering beams is of apposite sign, they are conjugate. In other words, their wavefront deformations $W(x, y)$

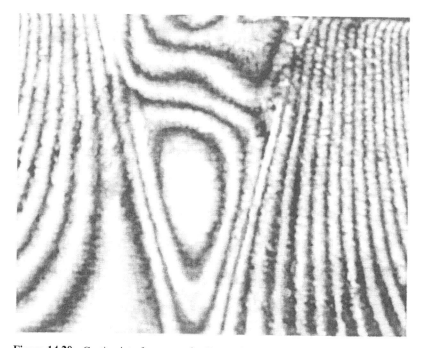

Figure 14.20   Grating interferogram of a distorted object.

produced by the local displacements $u(x, y)$ have opposite signs. As a result, the sensitivity to in-plane deformations of the grating lines is doubled.

Any lateral displacement (in plane) $u(x, y)$ of the grating in a perpendicular direction to the grating lines produces a relative phase shift on the two diffracted beams, shifting the interference fringes. The amplitude of the $+1$ order of diffraction beam, produced by the illuminating beam A, with maximum amplitude $A(x, y)$ is

$$E_{+1}^A = A(x, y) \exp\left\{i\left[\frac{2\pi}{d}u(x, y) + \frac{2\pi}{\lambda}w(x, y)\right]\right\} \tag{14.51}$$

and the amplitude of the -1 order of diffraction beam, produced by the illuminating beam B, with maximum amplitude $B(x, y)$ is

$$E_{-1}^B = B(x, y) \exp\left\{-i\left[\frac{2\pi}{d}u(x, y) - \frac{2\pi}{\lambda}w(x, y)\right]\right\} \tag{14.52}$$

where $k = 2\pi/\lambda$. Thus, the irradiance in the interferogram is

$$I(x, y) = [A(x, y)]^2 + [B(x, y)]^2 + [A(x, y)][B(x, y)]\cos\left[\frac{4\pi}{d}u(x, y)\right]. \tag{14.53}$$

The phase shifts produced by an out-of-plane displacement w are of the same magnitude and sign for both beams, thus keeping the interference fringe pattern unchanged.

If there is a local change in the slope of the grating in the perpendicular direction to the lines, the two diffracted beams also tilt by different amounts, as illustrated in Fig. 14.21. If the local grating tilt is θ, the angle $\delta\beta$ between the two diffracted wavefronts is

$$\delta\beta = \frac{\lambda\alpha^2}{d} = \alpha^2 \sin\theta, \tag{14.54}$$

where d is the grating period. This effect provides sensitivity to slope changes in the object being measured (McKelvie and Patorski, 1988), and should be taken into consideration if significant out-of-plane displacements are present.

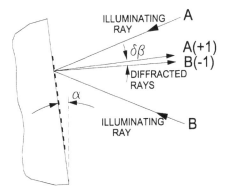

Figure 14.21   Grating interferometry with a tilted object.

With special experimental setups, two independent interferograms can be obtained from the two diffracted complementary wavefronts. This permits us to obtain the information not only about in-plane displacements but also about out-of-plane displacements, as shown by Basehore and Post (1982).

14.6.4 Multiple-Channel Grating Interferometric Systems

Phase-shifting techniques for the analysis of interferometer patterns require at least three interference patterns taken at three different phase-difference values. In the most common procedures the three pictures are registered in a CCD camera in a sequential manner. The problem is that in an unstable environment with vibrations and mechanical instabilities this is not possible. It is then necessary to generate the three phase-spaced fringe patterns simultaneously. This is called a multichannel interferometer (Kwon and Shough, 1985; Kujawinska, 1987; Kujawinska *et al.*, 1991; Kujawinska, 1993). Three types of interferometers are specially useful when measuring stresses and deformations of diffuse objects with speckle or moiré interferometry. An example is the three-channel phase-stepped system for moiré interferometry (Kujawinska *et al.*, 1991), illustrated in Fig. 14.22.

This moiré interferometer is similar to the one described in Fig. 14.9(c). A collimated beam of circularly polarized light illuminates the system. The light reflected on the flat mirror changes its sense of circular polarization. Thus, the body under examination is illuminated by two symmetrically oriented beams of circularly polarized light as in the speckle interferometer in Fig. 14.16(c). The key characteristic is the presence of the high-frequency grating that splits the interfering beams into three. Again, as in the system in Fig. 14.16(c), an analyzer selects only one plane of polarization, producing the interference pattern. Since the two interfering beams are circularly polarized in opposite directions, a rotation of the analyzer changes the phase difference between these beams. Thus, using three analyzers with different orientations, three interference patterns with different phase differences are produced.

If a light beam goes through a diffraction grating, several diffracted beams traveling in different directions are produced. A lateral displacement of the grating in

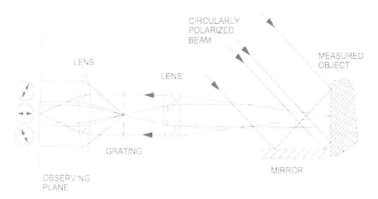

Figure 14.22   Three-channel system for moiré interferometry.

the direction perpendicular to the grating lines changes the relative phases of the diffracted beams. This effect has been used to construct several multichannel systems, as described by Kujawinska (1987, 1993).

REFERENCES

Alcalá, O. N., F. Mendoza-Santoyo, A. J. Moore, and C. Pérez-López, "Contrast Enhancement of Electronic Speckle Pattern Interferometry Addition Fringes," *Appl. Opt.*, **36**, 2783–2787 (1997).

Asundi, A., "Projection Moire Using PSALM," *Proc. SPIE*, **1554B**, 254–265 (1991).

Basehore, M. L. and D. Post, "Displacement Field (U, W) Obtained Simultaneously by Moiré Interferometry," *Appl. Opt.*, **21**, 2558–2562 (1982).

Brooks, R. E. and L. O. Heflinger, "Moire Gauging Using Optical Interference Fringes," *Appl. Opt.*, **8**, 935–939 (1969).

Butters, J. N. and J. A. Leendertz, "Speckle Pattern and Holographic Techniques in Engineering Metrology," *Opt. Laser Tech.*, **3**, 26–30 (1971).

Cloud, G., *Optical Methods of Engineering Analysis*, Cambridge University Press, Cambridge, 1995.

DeMattia, P. and V. Fossati-Bellani, "Holographic Contouring by Displacing the Object and the Illuminating Beams," *Opt. Comm.*, **26**, 17–21 (1978).

Doty, J. L., "Projection Moiré for Remote Contour Analysis," *J. Opt. Soc. Am.*, **76**, 366–372 (1983).

Ennos, A. E., "Speckle Interferometry," in *Progress in Optics*, Vol. 16, Wolf, E., ed., North-Holland, Amsterdam, 1978, pp. 231–288.

Friesem, A. A. and U. Levy, "Fringe Formation in Two-Wavelength Contour Holography," *Appl. Opt.*, **15**, 3009–3020 (1976).

Ganesan, A. R., D. K. Sharma, and M. P. Kothiyal, "Universal Digital Speckle Interferometer," *Appl. Opt.*, **27**, 4731–4734 (1988).

Gåsvik, K. J., "Moire Technique by Means of Digital Image Processing," *Appl. Opt.*, **22**, 3543–3548 (1983).

Gåsvik, K. J., *Optical Metrology*, 2nd edn, John Wiley and Sons, New York, 1995.

Goodman, J. W., *Introduction to Fourier Optics*, McGraw-Hill, New York, 1975, Chapter 4.

Goodman, J. W., *Laser Speckle and Related Phenomena*, Dainty, J. C., ed., Springer-Verlag, Berlin, 1975, Chapter 2.

Halioua, M., R. S. Krishnamurthy, H. Liu, and F. P. Chiang, "Projection Moiré with Moving Gratings for Automated 3-D Topography," *Appl. Opt.*, **22**, 805–855 (1983).

Hariharan, P., B. Oreb, and C. H. Freund, "Stroboscopic Holographic Interferometry Measurement of the Vector Components of a Vibration," *Appl. Opt.*, **26**, 3899–3903 (1987).

Hung, Y. Y., "Shearography: A New Method for Strain Measurement and Non Destructive Testing," *Opt. Eng.*, **21**, 391–395 (1982).

Idesawa, M., T. Yatagai, and T. Soma, "Scanning Moiré Method and Automatic Measurement of 3-D Shapes," *Appl. Opt.*, **16**, 2152–2162 (1977).

Jones, R. and C. Wykes, "General Parameters for the Design and Optimization of Electronic Speckle Pattern Interferometers," *Optica Acta*, **28**, 949–972 (1981).

Jones, R. and C. Wykes, *Holographic and Speckle Interferometry*, Cambridge University Press, Cambridge (1989).

Kowarschik, R., P. Kühmstedt, J. Gerber, and G. Notni, "Adaptive Optical Three-Dimensional Measurement with Structured Light," *Opt. Eng.*, **39**, 150–158 (2000).

Kreis, T., *Holographic Interferometry*, Akademie Verlag, Berlin, 1996.

Kujawinska, M. and Salbut, L., "Recent development in instrumentation of automated grating interferometry, *Appl Opt.*, **25**, 211–232 (1995).

Kujawinska, M., L. Salbut, and K. Patorski, "Three-Channel Phase Stepped System for Moiré Interferometry," *Appl. Opt.*, **30**, 1633–1635 (1991).

Kujawinska, M., "Spatial Phase Measuring Methods," in *Interferogram Analysis*, Robinson, D. W. and G. T. Reid, eds, Institute of Physics Publishing, Bristol and Philadelphia, 1993.

Løkberg, O. J., "The Present and Future Importance of ESPI," *Proc. SPIE*, **746**, 86–97 (1987).

McKelvie, J. and K. Patorski, "Influence of the Slopes of the Specimen Grating Surface on Out-of-Plane Displacements Measured by Moiré Interferometry," *Appl. Opt.*, **27**, 4603–4606 (1988).

Macovski, A., S. D. Ramsey, and L. F. Shaefer, "Time-Lapse Interferometry and Contouring Using Television Systems," *Appl. Opt.*, **10**, 2722–2727 (1971).

Malacara, D., ed., *Optical Shop Testing*, 2nd edn, John Wiley and Sons, New York, 1992.

Malacara, D., M. Servin, and Z. Malacara, *Interferogram Analysis for Optical Testing*, Marcel Dekker, New York, 1998.

Mendoza-Santoyo, F., M. C. Schellabear, and J. R. Tyrer, "Whole Field In-Plane Vibration Analysis Using Pulsed Phase-Stepped ESPI," *Appl. Opt.*, **30**, 717–721 (1991).

Osten, W., W. Nadeborn, and T. Andra, "General Approach in Absolute Phase Measurement," *Proc. SPIE*, **2860**, 2–13 (1996).

Ostrovsky, Y. I., M. M. Butusov, and G. V. Ostrovskaya, *Interferometry by Holography*, Springer-Verlag, Berlin, 1980, p. 142.

Owner-Petersen, M., "Digital Speckle Pattern Shearing Interferometry: Limitations and Prospects," *Appl. Opt.*, **30**, 2730–2738 (1991).

Parker, R. J., "Surface Topography of Non Optical Surfaces by Oblique Projection of Fringes from Diffraction Gratings," *Opt. Acta.*, **25**, 793–799 (1978).

Patorski, K., "Moiré Methods in Interferometry," *Opt. and Lasers in Eng.*, **8**, 147–170 (1988).

Patorski, K., *Handbook of the Moire Fringe Technique*, Elsevier, Amsterdam, London, 1993.

Pirodda, L., "Shadow and Projection Moiré Techniques for Absolute and Relative Mapping of Surface shapes," *Opt. Eng.*, **21**, 640–649 (1982).

Post, D., "The Moiré Grid-Analyzer Method for Strain Analysis," *Exp. Mech.*, **5**, 368–377 (1965).

Post, D., "Developments in Moiré Interferometry," *Opt. Eng.*, **21**, 458–467 (1982).

Post, D., Han, B. and Ifju, "High Sensitivity Moiré," Springer-Verlag, Berlin, 1994.

Pryputniewicz, R. J., "Time-Average Holography in Vibration Analysis," *Opt. Eng.*, **24**, 843–848 (1985).

Pryputniewicz, R. J., "Quantitative Determination of Displacements and Trains from Holograms," in *Holographic Interferometry*, Rastogi, P. K., ed., Springer Series in Optical Sciences, Vol. 68 (1994).

Reich, C., R. Ritter, and J. Thesing, "3-D Shape Measurement of Complex Objects by Combining Photogrametry and Fringe Projection," *Opt. Laser Tech.*, **3**, 224–231 (2000).

Schnars, U., "Direct Phase Determination in Hologram Interferometry with Use of Digitally Recording Holograms," *J. Opt. Soc. Am. A*, **11**, 2011–2015 (1994).

Schreiber, W. and G. Notni, "Theory and Arrangements of Self-Calibrating Whole Body 3D Measurement System Using Fringe Projection Technique," *Opt. Eng.*, **39**, 159–169 (2000).

Sciammarella, C. A., "The Moiré Method. A Review," *Exp. Mech.*, **22**, 418–433 (1982).

Sirohi, R. S., "Speckle Shear Interferometry – A Review," *J. Optics (India)*, **13**, 95–113 (1984a).

Sirohi, R. S., "Speckle Shear Interferometry," *Opt. Las. Tech.*, **84**, 251–254 (1984b).

Sirohi, R. S., *Speckle Metrology*, Marcel Dekker, New York, 1993.

Sitnik, R. and M. Kujawinska, "Opto-Numerical Methods of Data Acquisition for Computer Graphics and Animation System," *Proc. SPIE*, **3958**, 36–45 (2000).

Suganuma, M. and T. Yoshisawa, "Three Dimensional Shape Analysis by Use of a Projected Grating Image," *Opt. Eng.*, **30**, 1529–1533 (1991).

Takasaki, H., "Moiré Topography," *Appl. Opt.*, **9**, 1467–1472 (1970).

Takasaki, H., "Moiré Topography," *Appl. Opt.*, **12**, 845–850 (1973).

Takeda, M., "Fringe Formula for Projection Type Moiré Topography," *Opt. Las. Eng.*, **3**, 45–52 (1982).

Templeton, D. W. and Y. Y. Hung, "Shearographic Fringe Carrier Method for Data Reduction Computerization," *Opt. Eng.*, **28**, 30–34 (1989).

Toyooka, S., H. Nishida, and J. Takesaki, "Automatic Analysis of Holographic and Shearographic Fringes to Measure Flexural Strains in Plates," *Opt. Eng.*, **28**, 55–60 (1989).

Tsuruta, T., N. Shiotake, J. Tsujiuchi, and K. Matsuda, "Holographic Generation of Contour Map of Diffusely Reflecting Surface by Using Immersion Method," *Jap. J. Appl. Phys.*, **6**, 66 (1967).

Yamaguchi, I., "Phase-Shifting Digital Holography with Applications to Microscopy and Interferometry," *Proc. SPIE*, **3749**, 434–435 (1999).

Yaraslavskii, L. P. and N. S. Merzlyakov, *Methods of Digital Holography*, Consultants Bureau, New York, 1980.

15

Holography

CHANDRA S. VIKRAM

The University of Alabama in Huntsville, Huntsville, Alabama

15.1 INTRODUCTION

In common photography, a two-dimensional projection of the object intensity distribution is stored on the photosensitive medium. The information about the depth or the phase is lost. This loss of information can be explained in a very simple way. A photograph cannot reveal the object's distance from the camera or the photographer. Holography provides a powerful solution. Although a hologram is generally a two-dimensional intensity distribution on a photosensitive medium, with suitable illumination by a so-called *reconstruction beam* it creates a three-dimensional scene which in many respects is a replica of the original object. A hologram is an interference pattern between the light representing the object (called object beam) and another light, called the reference beam. Thus, the reference beam acts as a coding device about the object information, including depth or phase. Illuminated by a reconstruction beam which may be an exact replica of the original reference beam, the decoding occurs in the form of an erect three-dimensional image in space. Thus, a hologram is a complex grating. Likewise, a common grating can be called a simple hologram.

The history of holography can be traced to X-ray crystallography, which led to the Bragg X-ray microscope in 1942 (Bragg, 1942) where the image of the crystal structure can be reconstructed. Starting in 1948, Gabor (1948, 1949, 1951) introduced the idea of holographic imaging for increasing the resolution in electron microscopy. In early experiments, the object was opaque lines on a clear transparency. A collimated monochromatic beam illuminated the transparency and the subsequent diffraction pattern was stored on a photographic plate. The diffraction pattern is the interference between directly transmitted light (the reference wave) and the one diffracted by the lines. Once the processed transparency is illuminated by a monochromatic light, directly transmitted light and two diffracted waves (one corresponding to the original object wave and the other conjugate) are erect; the way a diffraction grating yields direct and ±1 order waves. These historical developments did not yield good-quality images. Two images (twin-image problem) and the direct light were superimposed. Although several techniques were proposed and demonstrated for better-quality images, the interest in optical holography declined. However, with the development of the laser, the off-axis reference beam concept introduced by Leith and Upatnieks (1962, 1963, 1964) solved the twin-image problem. Now one may divide a beam into two parts – one to illuminate the object and the other to send directly with an angle to the hologram recording plane. As a result of the high degree of coherence, the object and reference beams will still interfere to form the complex interference pattern, or hologram. Upon reconstruction, different diffracted waves are angularly separated.

Parallel to these developments, Denisyuk (1962, 1963, 1965) reported holograms where the object and reference beams reached the hologram plane from opposite directions. Such holograms are formed basically by placing the photosensitive medium between the light source and a diffusely reflecting object. Besides simplicity of the recording procedure, such holograms can be viewed by white light, as only a narrow wavelength region is reflected back in the reconstruction process.

For very small objects such as aerosols, a special type of holography called far-field holography or Fraunhofer holography evolved around the pioneering work of Thompson (1963, 1964). Due to individual objects in the far-field, the twin images are not a problem. Early applications on high-resolution imaging of aerosols were followed by many others, and enthusiasm still persists.

Once high-quality images were formed as stated above, the subject went through an explosive growth period in imaging itself as well as diverse applications such as information storage, information processing, and interferometry. These areas of application are still limited by imagination only, and developments are continuously being reported in literature. Web bookseller Amazon.com lists well over 200 book titles on holography. Fairly complete descriptions on optical holography are available: for example, in references by Collier (1971), Smith (1975), Caulfield (1979), Syms (1990), and Hariharan (1996). The bibliography of Hariharan (1996) lists 42 books.

Developments in different aspects continue. Abramson (1998) describes the holodiagram as a powerful nonmathematical tool. An up-to-date summary on electron holography is also available (Tonomura, 1999).

15.2 TYPES OF HOLOGRAMS

There are several (Caulfield and Lu, 1970) possible ways of classifying holograms based on geometry, application type, object type, and type of recording media, often with overlapping characteristics. In this section we describe several common types. However, the off-axis approach forming Fresnel holograms is the most general and can explain or help an understanding of the other types.

15.2.1 Off-Axis Holography

A general principle of off-axis holography can be described by Fig. 15.1. A quasi-monochromatic light such as that from a laser is divided into two parts. The division can be either by wave front or amplitude.

Lenses, mirrors, spatial filters, optical fibers, etc., are used to obtain clean beams of desired cross sections. One part, often called the object beam, illuminates the object [a typical point on the object is O]. The scattered light (the case of a diffusively reflecting object is presented) reaches the recording (x, y) plane. Let us represent the object beam light amplitude distribution at the hologram plane as $o(x, y) \exp(i\varphi_o)$. $o(x, y)$ is real and positive and called the absolute amplitude, whereas φ_o represents the local phase at the point. The other beam, called the reference beam, represented as a divergent beam in the figure, originates from the point R and directly illuminates the recording plane. Like the object beam, the light amplitude distribution of the reference beam at the recording plane can be represented by $r(x, y) \exp(i\varphi_R)$. The definition of the reference beam is very general, although a practical case of a divergent beam is shown. However, the point R is generally far away from the recording plane, so that $r(x, y)$ is practically a constant over the recording plane. In another common case of a collimated reference beam $(z_R = -\infty)$, $r(x, y)$ is a constant anyway. Let us also assume that the optical path difference between the two beams at every point on the recording plane is small or

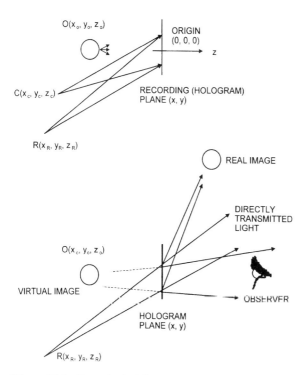

Figure 15.1 General principle of off-axis holographic recording and reconstruction.

well within the coherence length of the laser. In that case, two beams will interfere, yielding the irradiance distribution:

$$I(x, y) = \left| o(x, y)\exp(i\varphi_o) + r(x, y)\exp(i\varphi_R) \right|^2$$
$$= o^2(x, y) + r^2(x, y) + 2o(x, y)r(x, y)\cos(\varphi_o - \varphi_R). \tag{15.1}$$

We have assumed that the object and reference beams at the hologram plane have identical polarization states. Practically, using a vertically polarized laser, that is not a problem. The experimental setup on the horizontal laboratory table, nondepolarizing optical components and objects generally retain the vertical polarization direction on the recording plane as well. Nevertheless, in the case of seriously depolarizing objects, a polarizer may be used before the recording plane. Otherwise, the holographic fringes given by Eq. 15.1 may become partially or completely modulated or washed out. On the other hand, by proper manipulation of the recording and reconstruction beam geometry and polarization state, the original object beam polarization can be reconstructed.

Suppose the amplitude transmittance of the processed hologram (recording time t) is linearly related to the intensity in the interference pattern. Thus, the amplitude transmittance of the hologram can be represented as

$$t(x, y) = \tau_0 + \beta t I(x, y), \tag{15.2}$$

where τ_0 is the average transmittance and β is the slope of the exposure–transmittance curve. For common negative hologram, β is a negative quantity. For practical realization of the linear relationship given by Eq. (15.2), it is necessary that the amplitude of the harmonic term in Eq. (15.1) is small compared with the background. To assure this, a reference-to-object beam intensity ratio $r^2(x, y)/o^2(x, y)$ between 3 and 10 is a commonly accepted standard.

When the processed hologram is illuminated by the reconstructed beam (shown originating at point C in Fig. 15.1) whose complex amplitude distribution at the hologram plane is $c(x, y)\exp(i\varphi_c)$, the transmitted amplitude becomes

$$a(x, y) = t(x, y)c(x, y)\exp(i\varphi_c)$$
$$= a_1(x, y) + a_2(x, y) + a_3(x, y), \tag{15.3}$$

where

$$a_1(x, y) = [\tau_0 + \beta t o^2(x, y) + \beta t r^2(x, y)]c(x, y)\exp(i\varphi_c), \tag{15.4}$$
$$a_2(x, y) = \beta t r(x, y)c(x, y)o(x, y)\exp[i(-\varphi_R + \varphi_c)]\exp(i\varphi_o), \tag{15.5}$$

and

$$a_3(x, y) = \beta t r(x, y)c(x, y)o(x, y)\exp[i(\varphi_R + \varphi_c)]\exp(-i\varphi_o). \tag{15.6}$$

Under the practical assumption that τ_0, β, t, $r(x, y)$, $c(x, y)$ are either constant or slow varying over the hologram plane and also $r^2(x, y) \gg o^2(x, y)$, $a_1(x, y)$ is the directly transmitted reconstruction beam; $a_2(x, y)$ is proportional to the original object wave and, in the case of identical reference and reconstruction beams $[\varphi_R = \varphi_c]$, truly represents the original object wave except with some intensity modulation. To an observer, the image would appear in the original object position. Obviously, the image is virtual. The third term, $a_3(x, y)$, is also an image but with reversed phase. The real image is pseudoscopic, i.e., appears to be turned inside out.

Although we have represented a point object, a diffusing object consists of a large number of scattering points or centers. In that case, the object beam at the recording plane can be represented by a sum of beams from all the points. During the reconstruction, all the point images and hence the entire image is erect.

The reconstruction beam wavelength and geometry is general and can be different from the original reference beam. Consequently the images may be altered from the original position as described by holographic magnification relationships (Caulfield and Lu, 1970; Collier, 1971; Smith, 1975; Caulfield, 1979; Syms, 1990). The reconstruction beam geometry is often used to manipulate the desired image type (say real), minimize aberrations, etc. (Collier, 1971; Smith, 1975; Caulfield, 1979; Syms, 1990).

Equation (15.2) is the outcome of common thin absorption process. Nevertheless, such (amplitude or absorption) holograms can be bleached to make them transparent but with thickness and/or refractive index varying proportionally to $I(x, y)$ to be classified as phase holograms. Likewise, the recording media itself can be phase kind, such as a photopolymer or photoresist. In those cases, the amplitude transmittance of the processed hologram becomes $\exp|\tau_0' + \beta' t I(x, y)|$, where τ_0' and β' are new constants. In general, besides positive and negative first orders, all other orders are present. However, since higher orders are angularly separated from the desired one, the bleaching process is advantageous from the diffraction efficiency point of view. Diffraction efficiency is the fraction of the reconstructed image intensity to the reconstruction beam intensity. In the case of the amplitude hologram described by Eq. (15.2), even with equal object and reference beam intensity ratio, the optimum transmittance to cover the entire range between 0 and 1 is $1/2 + 1/4[\exp(\varphi_o - \varphi_R) + \exp(-\varphi_o + \varphi_R)]$. The maximum efficiency of a particular image term is $(1/4)^2$ or about 6.3% only. For the linear recording, one has to increase the reference-to-object beam intensity ratio typically between 3 and 10 further, reducing the modulation and hence the diffraction efficiency. The theoretical maximum diffraction efficiency of a thin phase transmission hologram dramatically increases to about 34% (Caulfield and Lu, 1970; Collier, 1971; Smith, 1975; Caulfield, 1979; Hariharan, 1996). At this stage it is appropriate to introduce the concept of thick holograms. If the thickness of the recording medium is significantly large compared with the wavelength of the laser or the hologram fringe spacing, the holographic fringes are no longer two-dimensional or surface phenomena. They will be volume gratings throughout the depth of the material. The reconstruction will no longer be governed by Eq. (15.2) but by Bragg's law (Caulfield and Lu, 1970; Collier, 1971; Smith, 1975; Caulfield, 1979; Syms, 1990; Hariharan, 1996). The holographic reconstruction becomes wavelength- and orientation-selective. So, even if white light is used for the reconstruction, a particular color region image is reconstructed. The maximum theoretical reconstruction efficiency of a thick phase hologram is 100%. An experimental efficiency close to 100% is also commonly achieved. Another advantage of volume holograms is in multiplexing. Localized holograms are not only on the surface but throughout the volume, thus significantly increasing the number of holograms being stored.

Besides the object beam, the reference beam can be of a general form such as a speckle pattern. Local reference beam holography (Caulfield and Lu, 1970) derives the reference beam from a focused spot on the object or a tiny mirror attached to the object. Thus, a successful hologram storage of a moving object is possible. Zel'dovich *et al.* (1995) describe holography with speckle waves in some detail.

The recording medium must be able to store the fine structure associated with the hologram. The fringe spacing, in the case of a two-dimensional recording plane inclined at an angle ϕ with respect to the bisector of a two plane waves (say object and reference waves of wavelength λ and half angle θ between them) is $\lambda/(2\sin\theta\sin\phi)$. Object and reference beams should be separated by some angle so that different reconstructed waves are angularly separated. At a practical θ, say $30°$, and He–Ne laser wavelength $0.6328\,\mu m$, the minimum possible fringe spacing (at $\phi = 90°$) becomes $0.6328\,\mu m$. The frequency of 1580 lines/mm cannot be resolved by ordinary photographic emulsions. However, extremely fine-grain silver halide materials, commercially known as holographic materials, are available. These are generally described in holographic books (Caulfield and Lu, 1970; Collier, 1971; Smith, 1975; Caulfield, 1979; Hariharan, 1996). Besides the fine-grain silver halide emulsions, dichromated gelatin, photoresists, photopolymers, photocromics, photothermoplastics, photorefractive crystals, spectral hole-burning materials, and even bacteriohodopsin films are used. These materials have unique characteristics, such as fast-in-place dry processing, real-time capability, etc. Several volumes are devoted to holographic recording materials (Smith, 1977; Bjelkhagen, 1995, 1996; Lessard and Manivannan, 1995). Nevertheless, recording material research is still one of the most active areas in holography. Unique concepts, such as silver halide sensitized gelatin (Neipp *et al.*, 1999) and photopolymer-filled nanoporous glass (Schnoes *et al.*, 1999) have always been of interest to holographic materials research. So are critical characterization of available materials (Kostuk, 1999), ultrafast holography with instantaneous display (Ramanujam *et al.*, 1999), and materials dispersed in liquid crystals (Cipparrone *et al.*, 1998).

The hologram recording material is generally but not necessarily plane. Cylindrical film has been used for 360° view holography (Jeong *et al.*, 1966; Jeong, 1967). For ballistic applications, half-cylindrical films have been used for behind-armor-debris analysis (Hough *et al.*, 1991; Anderson *et al.*, 1997).

15.2.2 Fraunhofer Holograms

The object can be very small compared with its distance from the recording medium and the hologram itself. In that case, the object beam at the hologram plane becomes a Fraunhofer diffraction pattern. As such, the general formulation of off-axis or Fresnel holography is still valid. The hologram of such a small object can still be stored in an off-axis manner. However, if the object and reference beams are along the same direction, the twin-image contributions become negligible, as it is now a Fraunhofer diffraction pattern (Thompson, 1963, 1964) from twice the original object distance. Such holography led to the enormous development of in-line Fraunhofer holography (Vikram, 1992, 1997; Thompson, 1996). As shown in Fig. 15.2, a volume containing small dynamic objects (bubbles, particles, droplets, etc.) is illuminated by a suitably pulsed laser (collimated shown but not necessary) and the hologram stored. It is assumed that a large portion of the beam that is uninterrupted by the objects acts as the reference beam. Now, when reconstructed, the directly transmitted light and the real and virtual images (of the original object cross sections in the case of opaque objects), are reconstructed. These images are generally magni-

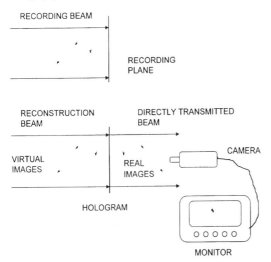

Figure 15.2 Recording and reconstruction processes of in-line Fraunhofer holography.

fied using a closed-circuit television (CCTV) system. Although a particular plane in the image space is in focus at a time, by moving the hologram or the camera on a x–y–z translation the entire volume can be covered. As stated earlier, when studying the real images, the contribution of virtual images is negligible due to far-field (twice the original object distance from the hologram). Such holography is very common for studying dynamic micro-objects in a volume. Processes are, for example, cavitation, combustion, ice crystal formation, fog studies, high-speed ballistic events, accelerometry in space microgravity conditions, droplet growth and coarsening phenomena, and holographic particle velocimetry (PIV) (Vikram, 1992, 1997; Thompson, 1996).

15.2.3 Fourier-Transform Holograms

These holograms, basically, are interference patterns between the Fourier transform of the object and reference sources. Such holograms are generally employed as spatial filters in pattern recognition. One basic requirement is that the reference source and subject are coplanar. Thus, such holograms are restricted to planar subjects or transparencies. One of the original arrangements of such holography is shown in Fig. 15.3. Subject transparency and source point are Fourier transformed on the recording plane using a lens of focal length f. Upon reconstruction the inverse Fourier-transform process yields the zero order, the original and conjugate (inverted) images. The generated images remain static even if the hologram is translated. Other geometrical arrangements and even lensless Fourier transform are possible (Collier, 1971).

15.2.4 Image Plane Holograms

In this type of hologram, the object is imaged near the recording plane using a lens, as shown in Fig. 15.4. Upon reconstruction, obviously part of the image may be real

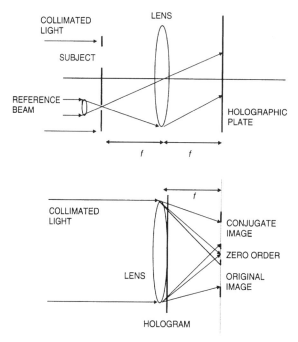

Figure 15.3 A typical Fourier-transform hologram recording and reconstruction arrangement.

and part imaginary. The view angle of the image is limited by the lens aperture. One special characteristic of such a hologram is that larger sources (such as bright, extended ones) with little spatial coherence can be used to view the image. The relaxation of the spatial coherence is ideal at the hologram plane and image degradation occurs as the distance increases. Thus, the object (image) depth should still be as small as possible.

Similarly, the bandwidth allowance of the reconstruction source may be large and even a white light source may be used.

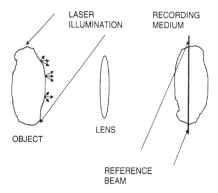

Figure 15.4 Image plane holographic recording arrangement.

15.2.5 Pinhole Holograms

This type of hologram (Xu *et al.*, 1989) of multiplex holography has the distinct advantage of selective reconstruction with the help of spatial light modulators and has applications in optical computing and page-oriented holographic memory.

As shown in Fig. 15.5, a pinhole is placed between the object and the recording plane. During the reconstruction with a conjugate beam, the entire object system including the image of the pinhole will be erected. Thus, in the pinhole plane the area except the pinhole is not containing the object information. Consequently, different holograms for different objects can be stored by changing the pinhole location. All the images will be simultaneously reconstructed. However, only a group of images will be reconstructed if the pinhole plane is masked to allow only corresponding locations to transmit. The masking can be done mechanically, or more conveniently using a spatial light modulator connected to a computer. Nondiffusing objects such as amplitude masks can also be multiplexed this way by illuminating them by a converging beam from a lens.

15.2.6 Color Holograms

There are several detailed discussions on color holograms (Collier, 1971; Smith, 1975; Caulfield, 1979; Syms, 1990; Hariharan, 1983, 1996). Basically, superimposed images with three primary colors are to be simultaneously reconstructed to have the desired color effect. Let us start our discussion with thin holograms. If the holograms with different wavelengths are superimposed (multiple exposure) on the same recording emulsion, then at the time of reconstruction (with all the wavelengths simultaneously) each wavelength will reconstruct several unwanted images along with the

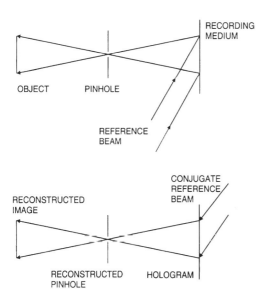

Figure 15.5 Schematic diagram of pinhole holography recording and reconstruction processes.

desired one. The reason is the presence of holograms corresponding to other wave-lengths. This type of cross-talk makes such color holography impractical.

Each image can be collected by a lens and spatially filtered to remove unwanted images. Besides unwanted system problems, the object depth has to be limited because spatial filtering works best for planar objects.

Spatial multiplexing is another option. Separate holograms are recorded on different portions of the hologram. Reconstruction of all the colors simultaneously is done with the aid of proper color filters to illuminate desired sections with particular colors. The process yields high-quality images. However, separate holograms must be small for the viewer to see all the colors simultaneously.

Thicker or volume holograms act as a filter and solve the multiple image problems as stated above. Rather than hologram formation on the surface, volume gratings are formed. These gratings, due to the Bragg effect, have wavelength selec-tivity. Thus, each color will be reconstructed without unwanted images. However, care must be taken to avoid emulsion shrinkage effects for ideal color reproduction.

15.2.7 Rainbow Holograms

The problem of disturbing superimposition of different images can be partly solved by this approach. Although the hologram is stored by monochromatic or laser source, the reconstruction can be viewed by a white light sources. Loss of vertical parallax is accepted for the gain in the color reconstruction by white light.

The classical two-step recording process (Fig. 15.6) pioneered by Benton (1969) involves first storing a primary hologram in the usual fashion with a laser source. This hologram is illuminated by the original laser source to reconstruct images. Near the reconstructed real image the hologram plate (for rainbow holo-graphy) is kept to store the second hologram using a convergent reference beam. However, a long slit near the primary hologram is used. Since images can be reconstructed using small hologram apertures, this does not pose a problem except a minor loss in resolution.

When reconstructed by a conjugate divergent source (may be white light now) real slit images are reconstructed and different color images can be viewed through these slits.

The different slit locations are due to holographic magnification which is wavelength dependent. Significant developments have occurred following the work of Benton (1969). Improved processes, interferometric applications, display applica-tions, etc., followed (Yu, 1983) and are continuing (Taketomi and Kubota, 1998; Zhang *et al.*, 1988; Guan *et al.*, 1999). The approach of Guan *et al.* (1999) does not use a physical slit. A synthetic slit is created by a short movement of the object.

15.2.8 Computer-Generated Holograms

A laser-illuminated subject can be described by a finite but large number of light scattering centers. Geometrical and physical parameters can then be used to compute the amplitude and phase of the light reaching hologram points, first due to individual centers and then cumulatively, to obtain the object beam parameters. Similarly, the reference beam parameters can be obtained in a more straightforward manner. The hologram amplitude transmittance thus obtained can be used to plot the hologram via a suitable display or storage device. The hologram thus stored will be able to

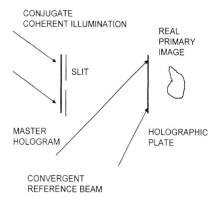

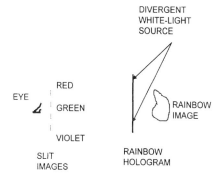

Figure 15.6 Rainbow holographic recording from master hologram and subsequent reconstruction of the rainbow image.

reconstruct the image of the object form used for the original computation. Several detailed descriptions on such holograms are available (Lee, 1978, 1992; Yoraslavskii and Merzlyakov, 1980; Bryngdahal and Wyrowski, 1983; Casasent, 1985; Soifer and Golub, 1994; Yaroslavsky and Eden, 1996). A reprints collection by Lee (1992) is classified based on theory and design, fabrication, performance, and applications.

There are several applications of such holograms. Complex objects (still having mathematical descriptions) to be fabricated usually do not have a master for comparisons or testing. A computer hologram can be used to reconstruct the desired image and then compare it with the fabricated object interferometrically. The complex objects could be, for example, aspheric lenses or mirrors, and lenses with multiple focal points. Creath and Wyant (1992) have summarized the use of computer-generated holograms in optical testing.

Some established application areas are complex spatial filtering, laser beam mode selection, weighted optical interconnections in neural networks, complex optical information processing, and display holography. New applications or refinements are constantly appearing. These are, for example in security applications (Yoshikawa *et al.*, 1998), generating diffuser for color mixing (Chao *et al.*, 1998), generating high-density intensity patterns (Takaki and Hojo, 1999), fabricating high-frequency gratings (Suleski *et al.*, 1999), waveguide grating coupling with tailored spectral response (Li *et al.*, 1999), pattern enhancement with photorefractive crystal (Nakagawa *et al.*, 1999), optical particle trapping (Reicherter *et al.*, 1999), color image generation (Suh, 1999) and auto design (Noaker, 1999). Conventional calculation of the hologram is time consuming. A recent process can gain the speed to a factor of 60–90 times (Ritter *et al.*, 1999).

15.3 APPLICATIONS

15.3.1 Holographic Interferometry

We have seen (Eq. (15.5)) that an exact replica of the original object wave can be reconstructed. Let us, for example, assume that half of the hologram exposure was provided with the object in a static position and the remaining in a different static position, say slightly stressed position to create surface strains. The object beam in the stressed position can be represented by $\varphi_o + \Delta\varphi_o$ where $\Delta\varphi_o$ corresponds to the optical phase change due to the strain. Now, upon reconstruction, both the images will be simultaneously present. The reconstructed amplitude will be proportional to $\exp(i\varphi_o) + \exp(i\varphi_o + i\Delta\varphi_o)$. In other words, the image irradiance will be modulated by $\cos^2(\Delta\varphi_o/2)$. As $\Delta\varphi_o$ varies over the image, the modulation may reach maxima and minima forming interference fringes. These fringes can then yield local strain, as $\Delta\varphi_o$ is known in terms of geometrical parameters and the wavelength of light used. Generally, $\Delta\varphi_o$ is directly proportional to the strain and inversely proportional to the wavelength.

The above description is called double-exposure holographic interferometry. A single exposure while the object is changing, say vibrating, can also be stored. Such holograms are termed as *time-average*. Upon reconstruction the time-averaged effect causes a fringe pattern over the image, such as the one governed by a Bessel function yielding local vibration amplitude.

One may also store a hologram and replace it in the original recording position for the image reconstruction. If the original object is also present and illuminated by the original object beam, one may observe interference between the object and reconstructed image beams. Live interference fringes can be observed as the object is moving, vibrating, etc. The fringes will yield the strain or other parameter in real time. Such holography is called real-time holography. An example is presented in Fig. 15.7. A hologram of a common ceramic coffee cup was stored on a thermoplastic plate using a diode-pumped, frequency-doubled neodymium vanadate laser operating at 532 nm. To view real-time or live interference fringes, the cup was filled with warm water to a point just above the intersection of the bottom of the handle and the body. Changing strains created by the thermal process resulted in the changing interference pattern, of which one frame is shown in the figure.

The above description is also valid for phase objects when the object beam passes to such sections in a combustion chamber, wind tunnel, etc.

The objects do not have to move to create the phase difference. Different wavelengths, changing illumination angles, etc., can yield the desired effects related to object contours.

Object, reference, and reconstruction beams in any combination can be temporally modulated, frequency translated, phase modulated, etc., for enhanced sensi-

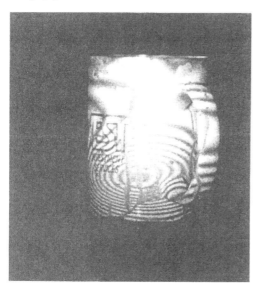

Figure 15.7 A fringe pattern from a real-time observation of heating of a coffee cup by warm water. (Picture courtesy of Dr. Martin J. Pechersky, Savannah River Technology Center, Aiken, South Carolina.)

tivity and advanced quantitative applications. Advanced quantitative tools such as phase-shifting interferometry and heterodyne interferometry have also been used for quantitative analysis.

Being such a general tool for deformation analysis for objects in their natural form, holographic interferometry has been a popular subject, as evident by a large number of books and extended discussions (Erf, 1974; Vest, 1979; Von Balley, 1979; Schumann and Dubas, 1979; Schumman et al., 1985; Jones and Wykes, 1989; Takemoto, 1989; Robillard and Caulfield, 1990; Ostrovsky, 1991; Conley and Robillard, 1992; Robinson and Reid, 1993; Rastogi, 1994b, 1997; Kreis, 1996; Sirohi and Hinsch, 1998). Applications have been in engineering, biology, and geophysics, as well as in commercial areas. Study areas cover aspects such as heat and mass transfer, vibration, deformation, stress, flaw detection, surface topography, etc. With TV holography (Jones and Wykes, 1989; Sirohi, 1993, 1996; Rastogi, 1994, 1997; Kreis, 1996; Sirohi and Hinsch, 1998; Meinlschmidt et al., 1996), the conventional emulsion is replaced by solid-state camera making the interferometric analysis convenient and fast.

New applications are constantly appearing. As reported by Gaughan (1999), the classical flaw-detection tool (localized abrupt strain changes in the region of flaw against an appropriate loading) of holographic interferometry has an application in breast cancer detection. A double-exposure hologram under different ambient conditions yields a surface-change profile. Nonuniformities result in nonuniform changes as asymmetries in the pattern. The results so far are encouraging and a powerful early detection tool is possible.

The phase change in the reconstructed image yielding fringes has been the main source of information for subsequent analysis. However, surface microstructure may also change between two exposures or observation times. These changes will partially to completely decorrelate the usual fringes. The measurement of the loss of the correlation (or fringe contrast) is also a very powerful tool in studies on erosion, corrosion, and mechanical wear (Vikram, 1996) including mechanical contacts (Ostrovsky, 1991). Finally, there have been unique approaches to perform holographic interferometry between two different surfaces (Rastogi, 1994a).

15.3.2 Hologram Copying and Embossing

It is often desirable to have several identical holograms. Although one may store several holograms to begin with, mass copying from a single master (Caulfield, 1979; Syms, 1990; Saxby, 1991; Hariharan, 1996) has economic and technical advantages. In an obvious approach, the reconstructed image from the master can be used as the object for the new or copied hologram. The recording geometry and parameters can now be altered to change the copy to a desired one, say to reflection, transmission, virtual, real, image-plane, enhanced reconstruction efficiency, different type of emulsion, with different wavelength, etc.

Contact printing of the original hologram is another optical option. However, for diffraction effects to be negligible, the separation between the original and copy must not be more than Δ^2/λ, where Δ is the fringe spacing and λ is the wavelength used. Usually the hologram frequency is very high, so the condition is difficult to meet; still, satisfactory copies can be made even for considerable hologram-copy separation by using a highly coherent source for the copying. However, three (directly transmitted and $\pm$ orders) waves from the original holograms will interfere at the copy plane, creating unwanted image-doubling effects.

Although these optical methods are useful, the most common mass production technique is embossing. Beside now common applications like those seen on credit cards, book covers, greetings cards, etc., one recent application is embossing on chocolate or hard candy (Peach, 1997). An embossing die is first made through a complicated process. However, once the die is made, duplication for mass production of holograms is a rather easy and cheap process. These holograms are now common, for example on bank cards.

First, a master hologram is made on a photoresist medium. Assuming a positive (soluble on exposure to light) photoresist is coated on glass, the resulting processed record will be a relief on the substrate. This type of phase hologram is formed due to thickness variation of the coating. Commonly, 488 nm wavelength Ar^+ laser light is used due to sensitivity requirements of the photoresist.

The next step is to chemically or vacuum deposit a layer of metal on the master or photoresist hologram. Finally, nickel is grown on the master by electroforming. The electroformed nickel is mounted on an embossing die or roller.

Now holograms can be hot rolled continuously on plastic materials. These relief holograms are usually coated with aluminum to make them reflective and also for surface protection.

Such holograms can be safely and securely applied to credit cards or other surfaces by hot-foil blocking. Being rainbow holograms, they can generate images in ordinary white light. However, for serious security applications more than one channel with different colors can be made to make counterfeiting very difficult.

15.3.3 Acoustical and Microwave Holography (Kock, 1981)

Acoustical holograms are formed by sound wave interference pattern. The pattern itself can be converted into a light wave pattern, say, by scanning the sound field and displaying the intensity variations and then storing into a photosensitive emulsion. After proper size reduction, the optical image can be veiwed by laser light.

One unique way employs a liquid–air interface to create stationary ripples to display the sound wave interference pattern. The advantage is real-time optical reconstruction. Short-wavelength coherent ultrasonic waves are used. The reference wave is directly sent to the interface, whereas the object wave, after passing through the object, is focused near the surface using an acoustic lens. The interference creates sound pressure variations and, hence, stationary ripples. Optical reconstruction will display absorption characteristics of the object, such as the presence of small breast tumors.

Likewise, two sets of coherent microwaves can be generated by a klystron and allowed to interefere. With a microwave-receiving antenna and suitable scanning mechanism, a photographic record of the microwave field or the hologram can be stored.

The most important contribution of microwave holography is synthetic aperture radar. An aircraft moving in a straight path emits highly coherent microwave pulses for the terrain illumination. The reflected signals from each of the points along the flight path are combined with the reference signal. Transformation to a light pattern and photographic record or hologram follows. Even if the aircraft has traveled a large distance, the signals being coherent, the processed effective antenna size is the long distance covered. This large effective aperture yields high-resolution pictures, even through clouds where microwave transmission is high.

15.3.4 Holographic Optical Elements

Holograms can effectively perform many of the operations of conventional optical elements like lenses and mirrors. For example, a hologram of a point source can be used to perform focusing operations. The mass fabrication eliminates grinding, polishing, coating, etc., associated with conventional processes. Holographic optical elements (HOE) are flat so they are lightweight and take less space even if multiple elements are stacked for complex operations such as multifocus capability. Volume holographic elements (VOHE) can also be fabricated on thick media. However, the term HOE is generally used and VOHE is a thick HOE.

Although not a substitute for all well-designed and established optical elements in cameras, precision telescopes, precision mirrors, etc., HOE have been found to be very useful in many applications. The principle of HOE is simple. Suppose a hologram is formed by interference between light coming from two point sources. Illuminating by either, the other will be reconstructed. Thus, the hologram works like a lens for the focusing operation. Real, virtual, transmittive (like a lens), reflective (like a mirror), on- or off-axis, etc., can be realized by holographic magnification relationships. In fact, a simple grating formed by two interfering waves is an HOE.

Hologram motion will shift the image point. Thus, a scan can easily be obtained, such as is common in supermarket bar code scanners. Beiser (1988) provides a detailed description on holographic scanning.

The holographic optical head in compact disk players performs multifunction beamsplitting, focusing, and tracking-error detection simultaneously.

Displays such as helmet-mounted, night-vision applications, selective rejection of certain wavelengths, etc., have found military applications. Selective rejection of a certain wavelength or wavelengths by a holographic filter is based on volume holographic recording. Such holograms have peak reflectance at a desired wavelength governed by Bragg's law. The notch filters can be used for laser eye protection while maintaining high overall transmittance for the view. In Raman spectroscopy, the Rayleigh line can be suppressed while allowing the useful Stokes' region to dramatically enhance the signal-to-noise ratio. For reduced solar heating, the near-infrared region can be rejected, while maintaining high transmittance in the visible region.

Laser wavelength selection is another application of dye lasers. The cavity may contain a single holographic grating for the wavelength selection, different gratings for different wavelength selection, or several superimposed for simultaneous multiple-wavelength emission.

The hologram of a point made with a single wavelength, if illuminated by a group of wavelengths simultaneously, will form different color image (like slit images in rainbow holography) points on different or shifted locations. Thus, a demultiplexing operation is performed.

Multiple elements can also be obtained by holographic multiplexing. These can be used in selective interconnects, etc.

These, as well as other applications such as in Fabry–Perot etalons and solar astronomy, have been described by Syms (1990) and Hariharan (1996). A collection of reprints on the subject is available (Stone and Thompson, 1991). The reprints are classified based on properties, types of elements, and applications. At this stage, it should be noted that these elements may be fabricated by the traditional holographic process as well as computer-generated. The applications are very diverse, such as in multiple imaging, laser machining, data processing, helmet-mounted display, sterography, fiber-optic coupling, multifocusing, optical computation, speckle interferometry, telescopy, compact disk applications, wide-field imaging, optical testing, and phase correction. A review of applications in speckle metrology is also available (Shakhar, 1993). Several critical spects and applications are also reported (Sirohi, 1990) and are continuing (Yadav and Ananda Rao, 1997; Petrov and Lau, 1998; Moustafa *et al.*, 1999).

Even residual aberration of readout lens in in-line holography can be corrected by another holographic optical element or an auxillary off-axis hologram (Ozkul and Anthore, 1991). Such phase conjugation (Barnhart *et al.*, 1994) or playback geometry to compensate for aberrations has been known (McGehee and Wuerker, 1985).

Novel applications of HOE are continuing in correcting large spherical aberrations in fast optics for high-resolution imaging, lithography, and communications (Andersen *et al.*, 1997; Andersen and Knize, 1998; Bains, 1999). For example, microscope objectives with high resolution require low working distances. A large aperture would allow larger working distances but with multielement corrected optics it becomes bulky and costly. In the HOE approach (Fig. 15.8), a

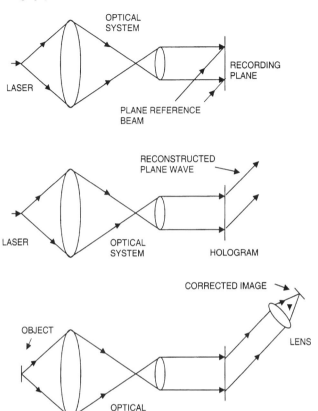

Figure 15.8 Diagrams representing holographic correction of aberration of an optical system.

hologram of spatially filtered light is passed through the microscope objective (say a large Fresnel lens) and collimated by an imaging lens is used as an object beam. A reference beam is collimated. If the reconstruction is performed by the spatially filtered light using the original system, the real original reference beam will be reconstructed. The aberrations caused by the original recording arrangement will be canceled out. If the spatial filter plane is replaced by an object, the aberration corrected images can be obtained. In summary, with a large and not critically corrected objective allowing a large objective–object distance, high-resolution images can be obtained.

Some other recent applications of optical elements are in optical interconnections (Chcbcn *et al.*, 1998), automotive windshield displays (Chao and Chi, 1998a), nondegenerate four-wave mixing (Gurley *et al.*, 1998), color filtering (Chao and Chi, 1998b), polarization selective holograms as photonic delay lines (Madamopoulos and Riza, 1998), laser mode selection (Yarovoy, 1998), grating for dual-wavelength operation (Lepage and McCarthy, 1998), variable beamsplitting (Blesa *et al.*, 1999), switching between single mode fibers (Wolffer *et al.*, 1999), collimating optics (Miler *et al.*, 1999), wideband phase-array antennas (Fu *et al.*, 1999), confocal microscopy (Barbastathis *et al.*, 1999), and spatial light modulation (Wu *et al.*, 1999).

15.3.5 Display Holograms

Being three-dimensional, reconstruction of images, displays, or pictorial applications have long been an area of interest. Using pulsed lasers, holograms of human subjects were successfully stored (Collier, 1971). Nevertheless, conventional laser recording and reconstruction have limitations for the purpose. A limited angle of view is one, while low image illuminance, etc., further limits the role of a conventional approach in this regard. There have been innovative approaches to solve these problems.

To solve limited view angles, holograms using several plates or cylindrical films can be recorded simultaneously to cover different views (Hariharan, 1996; Jeong *et al.*, 1966; Jeong, 1967). For a 360° view, the cylindrical film surrounds the object. A spatially filtered divergent laser beam (Fig. 15.9) illuminates the system. Portions reaching the film directly act as the reference beam, whereas those scattered by the object act as the object beam. Upon reconstruction, one may move around the cylindrical hologram to constantly view the corresponding side. These holograms are bulky, due to the need to surround the film on a rigid support such as a glass cylinder. A compromise is to use flat holograms with capability to view front and back of the image by viewing from two sides of the hologram. A sequence of storing front and back sides of the object finally yields a single hologram with this capability (Hariharan, 1996). Needless to mention, if stored on a thick volume material, such holograms can be viewed with ordinary white light. Another approach to simulate a 360° view on a flat hologram is to store a slit hologram using a mask. Multiple images at different slit locations are stored while rotating the object on a turntable. When the entire hologram is reconstructed, the eye movement across the hologram will yield constantly changing views.

A hologram of an ordinary photographic transparency can be holographically stored and reconstructed. Using a movable slit mask, holograms of the transparencies of the object taken from different angles can be multiplexed. Upon reconstruction, sterographic or acceptable three-dimensional views can be obtained. Such holograms, being non-laser source for the original storage, find applications in displays of X-ray and ultrasound images (Hariharan, 1996). The volume multiplexing has found applications in computed axial tomography (CAT), scanning tunnelling microscopy (STM), confocal scanning optical microscopy, etc. (Syms, 1990). White light stereography from a movie film is also possible (Hariharan, 1996). Commercial systems capable of making stereograms in a few minutes are underway (Bains, 1998).

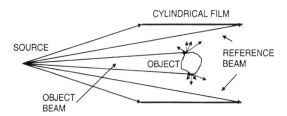

Figure 15.9   Diagram illustrating 360° view holography around a cylindrical film.

For display applications the volume occupied by the reconstruction beam of ordinary off-axis holography is disadvantageous. In the edge-lit holography, the reference and reconstruction beam enter from a side parallel to the plate. Thus, the reconstruction setup is in a plane rather in a volume. Compact portable displays can thus be obtained.

As we have seen in Section 15.2.7, bright ordinary white light sources can be used for the reconstruction in rainbow holography. This is very advantageous for the display applications. Color holography (Section 15.2.6) as well as image plane holography (Section 15.2.4) have primarily been developed for display applications. Color holography for display applications critically depends on high-quality practical recording materials, with considerable progress made in the recent past (Kaplan, 1998).

There is significant current commercial interest in holographic display technology (Wilson, 1998), including auto design (Suh, 1999) and holographic three-dimensional photocameras (Levin *et al.*, 1998). Specific tools such as pixelated holograms for three-dimensional display (Chen and Chatterjee, 1998) and improved holography on liquid-crystal displays (Mishina *et al.*, 1999) are constantly being investigated.

15.3.6 Information Storage and Processing

Holography have long been applied to information storage and processing. Synthesis of complex filters to perform numerous operations such as image enhancement, image restoration, optical pattern recognition, imaging through random media, high-density data storage, etc., have long been of active interest (Goodman, 1966; Stroke, 1966; Casasent, 1981, 1985; Yu, 1983) and still are being continued and applied (Collings, 1988; Mikaelian, 1994; Yu and Jutamula, 1996). In this section we describe some primary concepts.

A common area of interest is associative storage property and applications to pattern recognition, processing, coding, etc. Associative storage is a basic property of holography. Either (object or reference) of the original beams can illuminate the hologram to reconstruct the other. A common way is to store a Fourier-transform hologram because several operations can be performed in the Fourier plane, including manipulations by a spatial light modulator (SLM). Let us consider the object transparency O (Fig. 15.10) centered on the optical axis and a reference point R (point not necessary but convenient for describing the process) in the back focal plane of a Fourier-transform lens. They are illuminated by the same laser source, and in the front focal plane the hologram is recorded. The processed hologram is replaced for the reconstruction. If illuminated by the original reference point and reconstruction inverse Fourier transformed, the image will be obtained at O' in the focal plane of the lens as shown. As such, the outcome is obvious from the holographic point of view. Now, suppose instead of using the original reference point, the object or the illuminated transparency is used for the reconstruction. In that case, a sharp reconstructed image point will appear at R' corresponding to the original point at R. Now if the original transparency is changed, the hologram illumination beam may be randomly different and the intensity of the point at R' will sharply decline. This characteristic is very useful in pattern recognition. Several useful aspects have been demonstrated in the past, including variations in the architecture. One is for coding applications. The reference source can be diffused,

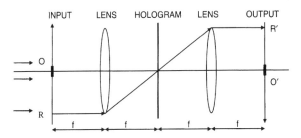

Figure 15.10 Diagram demonstrating associative storage property of holographic storage using the Fourier-transform method.

such as that derived by focusing a laser beam on a ground glass. However, different coded sources may be obtained just by moving the glass. The perfect reconstruction will be obtained with the ground glass in the original position; otherwise, the image will disappear or be very noisy. Thus, even if the hologram is available, the reconstruction is possible only if the code (diffuser and its orientation, position, etc.) is known.

At the reconstruction stage, processing or modifications in the Fourier (hologram) plane can yield a modified picture at O'. For example, if the original transparency was recorded with an imperfect optical system, the image can still be corrected. By creating filters using the imperfect system and placing in the hologram plane, the image at O' can be as if it was stored with a perfect system.

More complex operations for storing holographic arrays, particularly for page-oriented holographic memory, have been performed.

Optical information in compact form in a microfilm is limited by the film resolution beyond which information is lost. Holographic information storage is also related to the recording medium resolution, but the process itself provides serious advantages. On the same medium or emulsion, a multiplexed storage (say by different reference beams) yields many superimposed holograms. With a thick emulsion, a large number of holograms can be stored even with volumetric multiplexing. The information coding (for readout only the desired hologram at a time) is performed by wavelength, polarization, reference beam direction, orientation, etc. In 1993 as many as 5000 holograms were recorded in a single crystal of Fe:LiNbO$_3$ (Mok, 1993). Since then, there has been considerable progress in high-density holographic data storage materials and processes (Burr *et al.*, 1998; Dhar *et al.*, 1988, 1999; Guenther *et al.*, 1998; Adibi *et al.*, 1999; An *et al.*, 1999; Cheng *et al.*, 1999; Markov *et al.*, 1999; Zhou *et al.*, 1999).

Such capacity resulted in archival storage of documents and maps, videotape and videodisk systems, digital audio storage, page-oriented storage systems, credit and identity card verification, and other security applications (Hariharan, 1996).

Optical interconnection is another area associated with high-density storage. Using a single ordinary or computer-generated hologram, a light beam can reconstruct in several output beams (a diffraction grating is a simple form of such a hologram). Likewise, several optical beams can be combined into one. Unlike electrical connections, the optical approach yields minimum propagation delays and space requirements. Using such a capability is an active area of research with, applications to switching, optical neural networks, etc.

15.4 OTHER DEVELOPMENTS

Although we have covered many common aspects of holography, the discussion is far from complete. There have been constant critical re-evaluations, innovations, and new applications. We briefly describe some of them here. Lebrun *et al.* (1999) applied optical wavelet transform to perform the reconstruction process. Their particular interest is to reduce the noise associated with conventional reconstruction when the object is glass fibers being drawn in a turbulent flame. Wallace (1999a) reported a recent development of contact printing or "rubber stamp" of features down to 20 nm. Obviously, such technology has implications in the mass fabrication of holograms as well. Spinning disk holography (Wallace, 1999b) is another recent development for storing movies and high-speed events.

Gated holography (Abramson, 1981; Shih, 1999; Shih *et al.*, 1999) is another important development. Gating can be performed by several methods, such as short-coherence, Fourier synthesis, and spectral decomposition (Shih *et al.*, 1999). For example, if a short pulse is used as a light source, the object light arriving at different times will not be stored. The different light may be, for example, due to diffusion in the object media. Thus, one can avoid unwanted diffusion effects, such as while imaging through highly scattering media.

Digital recording and numerical reconstruction (Pomarico *et al.*, 1995) has long been of interest, as reported by Yamaguchi and Zhang (1998). The hologram is captured by a video camera and the image is reconstructed by mathematical means. As a result of significant progress in storage and processing technology, the interest in various aspects and applications has been growing (Marquardt and Richter, 1998; Nilsson and Carisson, 1998; Zhang and Yamaguchi, 1998; Cuche *et al.*, 1999; Pedrini *et al.*, 1999; Takaki and Ohzu, 1999).

Optical scanning holography (Poon *et al.*, 1996) stores holographic information using heterodyning. The output, being in the form of electrical current rather than light intensity in common holographic recording, has several unique features of information storage, transmission, and processing. New applications such as fluorescence imaging (Schilling *et al.*, 1997; Indebetouw *et al.*, 1998) and three-dimensional image recognition (Poon and Kim, 1999) are continuing.

Critical aberration aspects of underfluid objects have also been of significant interest (Coupland and Halliwell, 1997; Fang and Hobson, 1998).

Obviously, new concepts and applications are limited by imagination only. For example, Hariharan (1998) finds similarities between Lippmann color photography and holography. Maripov and Shamshiev (1999) described "the superhologram" possessing properties of Gabor's, Leith and Upatnieks's, Denisyuk's, Benton's, and others.

ACKNOWLEDGMENTS

I am very grateful to Professor Brian J. Thompson for encouragement and going through an earlier version of the manuscript with very valuable comments. I also express my gratitude to Dr. Martin J. Pechersky for kindly providing a beautiful picture (Fig. 15.7) for use in the chapter.

REFERENCES

Abramson, N. H., *The Making and Evaluation of Holograms*, Academic Press, London, 1981.

Abramson, N. H., *Light in Flight of the Holodiagram*, SPIE – The International Society for Optical Engineering, Bellingham, 1998.

Adibi, A., K. Buse, and D. Psaltis, "Multiplexing Holograms in LiNbO$_3$:Fe:Mn Crystals," *Opt. Lett.*, **24**, 652–654 (1999).

An, X., D. Pasaltis, and G. W. Burr, "Thermal Fixing of 10,000 Holograms in LiNbO$_3$:Fe," *Appl. Opt.*, **38**, 386–393 (1999).

Andersen, G. and R. J. Knize, "Holographically Corrected Microscope with a Large Working Distance," *Appl. Opt.*, **37**, 1849–1853 (1998).

Andersen, G., J. Munch, and P. Veitch, "Compact Holographic Correction of Aberrated Telescopes," *Appl. Opt.*, **36**, 1427–1432 (1997).

Anderson, C., J. Gordon, D. Watts, and J. Marsh, "Measurement of Behind-Armor Debris Using Cylindrical Holograms," *Opt. Eng.*, **36**, 40–46 (1997).

Bains, S., "MIT Media Laboratory Spins Off Holographic Hard Copy," *OE Reports*, **172**, 3 (1998).

Bains, S., "Holographically Corrected Microscopes May Have Applications to Lithography," *OE Reports*, **185**, 9 (1999).

Barbastathis, G., M. Balberg, and D. J. Brady, "Confocal Microscopy with a Volume Hologram Filter," *Opt. Lett.*, **24**, 811–813 (1999).

Barnhart, D. H., R. J. Adrian, and G. C. Papen, "Phase-Conjugate Holographic System for High Resolution Particle-Image Velocimetry," *Appl. Opt.*, **33**, 7159–7170 (1994).

Beiser, L., *Holographic Scanning*, John Wiley, New York, 1988.

Benton, S. A. "Hologram Reconstructions with Extended Incoherent Sources," *J. Opt. Soc. Am.*, **59**, 1545–1546 (1969).

Bjelkhagen, H. I., *Silver Halide Materials for Holography & Their Processing*, 2nd edn, Springer-Verlag, Berlin, 1995.

Bjelkhagen, H. I., *Selected Papers on Holographic Recording Materials*, SPIE – The International Society for Optical Engineering, Bellingham, 1996.

Blesa, A., I. Arias, J. Atencia, and M. Quintanilla, "Holographic Volume Dual-Element as an Optical Variable Beamsplitter," *J. Mod. Opt.*, **46**, 155–165 (1999).

Bragg, W. L., "X-Ray Microscope," *Nature*, **149**, 470–471 (1942).

Bryngdahal, O. and F. Wyrowski, "Digital Holography-Computer-Generated Holograms," in *Progress in Optics*, Vol. XXVIII, Wolf, E., ed., North-Holland, Amsterdam, 1983, pp. 1–86.

Burr, G. W., G. Barking, H. Goufal, J. A. Hoffnagle, C. M. Jefferson, and M. A. Neifeld, "Gray-Scale Data Pages for Digital Holographic Data Storage," *Opt. Lett.*, **23**, 1218–1220 (1998).

Casasent, D., ed., *Optical Data Processing*, Springer-Verlag, Berlin, 1981.

Casasent, D., "Computer Generated Holograms in Pattern Recognition: a Review," *Opt. Eng.*, **24**, 724–730 (1985).

Caulfield, H. J., ed., *Handbook of Optical Holography*, Academic Press, New York, 1979.

Caulfield, H. J. and S. Lu, *The Applications of Holograpny*, John Wiley, New York, 1970.

Chao, W. and S. Chi, "Diffraction Properties of Windshield Laminated Photopolymer Holograms," *J. Opt.*, **29**, 95–103 (1998a).

Chao, W. and S. Chi, "Novel Holographic Colour Filters with Double-Transmission Holograms," *J. Opt.*, **29**, 259–264 (1998b).

Chao, W., S. Chi, C. Y. Wu, and C. J. Kuo, "Computer-Generated Holographic Diffuser for Color Mixing," *Opt. Commun.*, **151**, 21–24 (1998).

Cheben, P., M. L. Calvo, T. Belenguer, and A. Nunez, "Substrate Mode Hologram for Optical Interconnects," *Opt. Commun.*, **148**, 18–22 (1998).

Chen, S.-T. and M. R. Chatterjee, "Implementation of a Spatially Multiplexed Pixelated Three-Dimensional Display by Use of a Holographic Optical Element Array," *Appl. Opt.*, **37**, 7504–7503 (1998).

Cheng, Y.-S., W. H. Su, and R. C. Chang, "Disk-Type Multiplex Holography," *Appl. Opt.*, **38**, 3093–3100 (1999).

Cipparrone, G., A. Mazzulla, F. P. Nicoletta, L. Lucchetti, and F. Simoni, "Holographic Grating Formation in Dye Doped Polymer Dispersed Liquid Crystals," *Opt. Commun.*, **150**, 297–304 (1998).

Collier, R. J., C. B. Burckhardt, and L. H. Lin, *Optical Holography*, Academic Press, New York, 1971.

Collings, N., *Optical Pattern Recognition Using Holographic Techniques*, Addison-Wesley, Workingham, 1988.

Conley, E. and J. Robillard, eds, *Industrial Applications for Optical Data Processing and Holography*, CRC Press, Boca Raton, 1992.

Coupland, J. M. and N. A. Halliwell, "Holographic Displacement Measurements in Fluid and Solid Mechanics: Immunity to Aberrations by Optical Correlation Processing," *Proc. R. Soc. Lond. A*, **453**, 1053–1066 (1997).

Creath, K. and C. Wyant, "Holographic and Speckle Tests," in *Optical Shop Testing*, 2nd edn, Malacara, D., ed., John Wiley, New York, 1992, pp. 603–612.

Cuche, E., F. Bevilacqua, and C. Depeursinge, "Digital Holography for Quantitative Phase-Contrast Imaging," *Opt. Lett.*, **24**, 291–293 (1999).

Denisyuk, Yu. N., "Photographic Reconstruction of the Optical Properties of an Object in Its Own Scattered Radiation Field," *Soviet Physics-Doklady*, **7**, 543–545 (1962).

Denisyuk, Yu. N., "On the Reproduction of the Optical Properties of an Object by the Wave Field of Its Scattered Radiation," *Optics and Spectroscopy*, **15**, 279–284 (1963).

Denisyuk, Yu. N., "On the Reproduction of the Optical Properties of an Object by Its Scattered Radiation II," *Optics and Spectroscopy*, **18**, 152–157 (1965).

Dhar, L., K. Curtis, M. Tackitt *et al.*, "Holographic Storage of Multiple High-Capacity Digital Data Pages in Thick Photopolymer systems," *Opt. Lett.*, **23**, 1710–1712 (1988).

Dhar, L., A. Hale, H. E. Katz, M. L. Schilling, M. G. Schnoes, and F. C. Schilling, "Recording Media that Exhibit High Dynamic Range for Digital Holographic Data Storage," *Opt. Lett.*, **24**, 487–489 (1999).

Erf, R. K., ed., *Holographic Non-Destructive Testing*, Academic Press, New York, 1974.

Fang, X. and P. R. Hobson, "Effect of Spherical Aberration on Real-Image Fidelity From Replayed In-Line Holograms of Underwater Objects," *Appl. Opt.*, **37**, 3206–3214 (1998).

Fu, Z., C. Zhou, and R. T. Chen, "Waveguide-Hologram-Based Wavelength-Multiplexed Pseudoanalog True-Time-Delay Module for Wideband Phased-Array Antennas," *Appl. Opt.*, **38**, 3053–3059 (1999).

Gabor, D., "A New Microscope Principle," *Nature*, **161**, 777–778 (1948)

Gabor, D., "Microscopy by Reconstructed Wavefronts," *Proc. Roy. Soc. A*, **197**, 454–487 (1949).

Gabor, D., "Microscopy by Reconstructed Wavefronts II," *Proc. Phys. Soc. (London) B*, **64**, 449–469 (1951).

Gaughan, R., "Tire-Inspection Method Prompts Cancer Detection Study," *Biophotonics Int.*, **6**(3), 53–55 (May/June, 1999).

Goodman, J. W., *Introduction to Fourier Optics*, McGraw-Hill, New York, 1968.

Guan, C., Z. Wang, and D. Li, "Rainbow Holography with Large Depth," *Appl. Opt.*, **38**, 3726–3727 (1999).

Guenther, H., R. Macfarlane, Y. Furukawa, K. Kitamura, and R. Neurgaonkar, "Two-Color Holography in Reduced Near-Stoichiometric Lithium Niobate," *Appl. Opt.*, **37**, 7611–7623 (1998).

Gurley, K. S., G. P. Andersen, R. J. Knize, W. R. White, and M. D. Johnson, "Wavelength Dependence of Wavefront Correction Using Holography in a Saturable Absorber," *Opt. Commun.*, **152**, 16–18 (1998).

Hariharan, P., "Colour Holography," in *Progress in Optics*, Vol. XX, Wolf, E., ed., North-Holland, Amsterdam, 1983, pp. 365–324.

Hariharan, P., *Optical Holography*, 2nd edn, Cambridge University Press, Cambridge, 1996.

Hariharan, P., "Lippmann Photography or Lippmann Holography?" *J. Mod. Opt.*, **45**, 1759–1762 (1998).

Hough, G. R., D. M. Gustafson, and W. R. Thursby, Jr., "Enhanced Holographic Recording Capabilities for Dynamic Applications," *Proc. SPIE*, **1346**, 194–199 (1991).

Indebetouw, G., T. Kim, T. C. Poon, and B. W. Schilling, "Three-Dimensional Location of Fluorescent Inhomogeneities in Turbid Media by Scanning Heterodyne Holography," *Opt. Lett.*, **23**, 135–137 (1998).

Jeong, T. H., "Cylindrical Holography and Some Proposed Applications," *J. Opt. Soc. Am.*, **57**, 1396–1398 (1967).

Jeong, T. H., P. Rudolf, and A. Luckett, "360° Holography," *J. Opt. Soc. Am.*, **56**, 1263–1264 (1966).

Jones, R. and C. Wykes, *Holographic and Speckle Interferometry*, 2nd edn, Cambridge University Press, Cambridge, 1989.

Kaplan, H., "New Materials Make Holography More Attractive," *Photonics Spectra*, **32**(3), 59–60 (March 1998).

Kock, W. E., "Lasers and Holography: An Introduction to Coherent Optics, Second Enlarged Edition," Dover Publications, New York, 1981.

Kostuk, R., "Dynamic Hologram Recording Characteristics in DuPont photopolymers," *Appl. Opt.*, **38**, 1357–1363 (1999).

Kreis, T., *Holographic Interferometry*, Akademie Verlag, Berlin, 1996.

Lebrun, D., S. Belaid, and C. Ozkul, "Hologram Reconstruction by Use of Optical Wavelet Transform," *Appl. Opt.*, **38**, 3730–3734 (1999).

Lee, S. H., *Selected Papers on Computer-Generated Holograms and Diffractive Optics*, SPIE – The International Society for Optical Engineering, Bellingham, 1992.

Lee, W. H., "Computer-Generated Holograms: Techniques and Applications," in *Progress in Optics*, Vol. XVI, Wolf, E., ed., North-Holland, Amsterdam, 1978, pp. 119–232.

Leith, E. N. and J. Upatnieks, "Reconstructed Wavefronts and Communication Theory," *J. Opt. Soc. Am.*, **52**, 1123–1130 (1962).

Leith, E. N. and J. Upatnieks, "Wavefront Reconstruction with Continuous-Tone Objects," *J. Opt. Soc. Am.*, **53**, 1377–1381 (1963).

Leith, E. N. and J. Upatnieks, "Wavefront Reconstruction with Diffused Illumination and Three-Dimensional Objects," *J. Opt. Soc. Am.*, **54**, 1295–1301 (1964).

Lepage, J.-F. and N. McCarthy, "Apodizing Holographic Gratings for Dual-Wavelength Operation of Broad-Area Semiconductor Lasers," *Appl. Opt.*, **37**, 8420–8424 (1998).

Lessard, R. A. and G. Manivannan, *Selected Papers on Photopolymers*, SPIE – The International Society for Optical Engineering, Bellingham, 1995.

Levin, G. G., T. V. Bulygin, and G. N. Vishnyakov, "Holographic 3D Photocamera," *OE Reports*, **171**, 12 (1998).

Li, M., B. S. Luo, C. P. Grover, Y. Feng, and H. C. Liu, "Waveguide Grating Coupler with a Tailored Spectral Response Based on a Computer-Generated Waveguide Hologram," *Opt. Lett.*, **24**, 655–657 (1999).

McGehee, J. H. and R. F. Wuerker, "Holographic Microscopy through Tubes and Ducts," *Proc. SPIE*, **523**, 70–74 (1985).

Madamopoulos, N. and N. A. Riza, "Polarization Selective Hologram-Based Photonic Delay Lines," *Opt. Commun.*, **157**, 225–237 (1998).

Maripov, A. and T. Shamshiev, "The Superhologram," *J. Opt. A: Pure Appl. Opt.*, **1**, 354–358 (1999),

Markov, V., J. Millerd, J. Trolinger, M. Norrie, J. Downie, and D. Timucin, "Multilayer Volume Optical Memory," *Opt. Lett.*, **24**, 265–267 (1999).

Marquardt, E. and Richter, "Digital Image Holography," *Opt. Eng.*, **37**, 1514–1519 (1998).

Meinlschmidt, P., K. D. Hinsch, and R. S. Sirohi, eds, *Selected Papers on Electronic Speckle Pattern Interferometry Principles and Practice*, SPIE–The International Society for Optical Engineering, Bellingham, 1996.

Mikaelian, A. L., *Optical Methods for Information Technologies*, Allerton Press, New York, 1994.

Miler, M., I. Koudela, and I. Aubrecht, "Holographic Diffractive Collimators Based on Recording with Homocentric Diverging Beams for Diode Lasers," *Appl. Opt.*, **38**, 3019–3024 (1999).

Mishina, T., F. Okano, and I. Yuyama, "Time-Alternating Method Based on Single-Sideband Holography with Half-Zone-Plate Processing for the Enlargement of Viewing Zones," *Appl. Opt.*, 3703–3713 (1999).

Mok, F. H., "Angle-Multiplexed Storage of 5000 Holograms in Lithium Niobate," *Opt. Lett.*, **18**, 915–917 (1993).

Moustafa, N. A., J. Kornis, and Z. Fuzessy, "Comparative Measurement by Phase-Shifting Digital Speckle Interferometry Using Holographically Generated Reference Wave," *Opt. Eng.*, **38**, 1241–1245 (1999).

Nakagawa, K., S. Iguchi, and T. Minemoto, "Enhancement of Selected Diffraction Pattern in a Computer-Generated Hologram with Photorefractive Crystal," *Opt. Eng.*, **38**, 947–952 (1999).

Neipp, C., I. Pascual, and A. Belendez, "Silver Halide Sensitized Gelatin Derived From BB-640 Holographic Emulsion," *Appl. Opt.*, **38**, 1348–1356 (1999).

Nilsson, B. and T. E. Carisson, "Direct Three-Dimensional Shape Measurement by Digital Light-in-Flight Holography," *Appl. Opt.*, **37**, 7954–7959 (1998).

Noaker, P. M., "Holography Drives Ford Car Design," *Laser Focus World* (February, 1999), pp. 14–16.

Ostrovsky, Yu, I., *Holographic Interferometry in Experimental Mechanics*, Springer-Verlag, Berlin, 1991.

Ozkul, C. and N. Anthore, "Residual Aberration Correction in Far Field in-Line Holography Using an Auxillary Off-Axis Hologram," *Appl. Opt.*, **30**, 372–373 (1991).

Peach, L. A., "Tasteful Holograms Adorn Candy," *Laser Focus World*, **35**(5), 50 (May, 1997).

Pedrini, G., P. Froning, H. J. Tiziani, and F. M. Santoyo, "Shape Measurement of Microscopic Structures Using Digital Holograms," *Opt. Commun.*, **164**, 257–268 (1999).

Petrov, V. and B. Lau, "Electronic Speckle Pattern Interferometry with Thin Beam Illumination of Miniature Reflection and Transmission Speckling Elements for In-Plane Deformation Measurements," *Opt. Eng.*, **37**, 2410–2415 (1998).

Pomarico, J., U. Schnars, H. J. Hartmann, and W. Juptner, "Digital Recording and Numerical Reconstruction of Holograms: a New Method for Displaying Light in Flight," *Appl. Opt.*, **34**, 8095–8099 (1995).

Poon, T.-C. and T. Kim, "Optical Image Recognition of Three-Dimensional Objects," *Appl. Opt.*, **38**, 370–381 (1999).

Poon, T.-C., M. H. Wu, K. Shinoda, and Y. Suzuki, "Optical Scanning Holography," *Proc. IEEE*, **84**, 753–764 (1996).

Ramanujam, P. S., M. Pedersen, and S. Hvilsted, "Instant Holography," *Appl. Phys. Lett.*, **74**, 3227–3229 (1999).

Rastogi, P. K., "Holographic Comparison between Waves Scattered from Two Physically Distinct Rough Surfaces," *Opt. Eng.*, **33**, 3484–3485 (1994a).

Rastogi, P. K., ed., *Holographic Interferometry*, Springer-Verlag, Berlin, 1994b.

Rastogi, P. K., "Holographic Interferometry – An Important Tool in Nondestructive Measurement and Testing," in *Optical Measurement Techniques and Applications*, Rastogi, P. K., ed., Artech House, Boston, 1997.

Reicherter, M., T. Haist, E. U. Wagemann, and H. J. Tiziani, "Optical Particle Trapping with Computer-Generated Holograms Written on a Liquid-Crystal Display," *Opt. Lett.*, **24**, 608–610 (1999).

Ritter, A., J. Bottger, O. Deussen, M. Konig, and T. Strothotte, "Hardware-Based Rendering of Full-Parallax Synthetic Holograms," *Appl. Opt.*, **38**, 1364–1369 (1999).

Robillard, J. and H. J. Caulfield, eds, *Industrial Applications of Holography*, Oxford University Press, New York, 1990.

Robinson, D. W. and D. W. Reid, eds, *Interferogram Analysis: Digital Processing Techniques for Fringe Pattern Measurement*, IOP, London, 1993.

Saxby, G., *Manual of Practical Holography*, Focal Press, Oxford, 1991.

Schilling, B. W., T. C. Poon, G. Indebetouw, *et al.*, "Three Dimensional Holographic Fluorescence Microscopy," *Opt. Lett.*, **22**, 1506–1508 (1997).

Schnoes, M. G., M. L. Schilling, S. S. Patel, and P. Wiltzius, *Opt. Lett.*, **24**, 658–660 (1999).

Schumann, W. and M. Dubas, *Holographic Interferometry*, Springer-Verlag, Berlin, 1979.

Schumann, W., J. P. Zurcher, and D. Cuche, *Holography and Deformation Analysis*, Springer-Verlag, Berlin, 1985.

Shakhar, C., "Speckle Metrology Using Hololenses," in *Speckle Metrology*, Sirohi, R. S., ed., Marcel Dekker, New York, 1993, pp. 473–506.

Shih, M. P., "Spectral Holography for Imaging through Scattering Media," *Appl. Opt.*, **38**, 743–750 (1999).

Shih, M. P., H. S. Chen, and E. N. Leith, "Spectral Holography for Coherence-Gated Imaging," *Opt. Lett.*, **24**, 52–54 (1999).

Sirohi, R. S., ed., *Holography and Speckle Phenomena and Their Industrial Applications*, World Scientific, Singapore, 1990.

Sirohi, R. S., ed., *Speckle Metrology*, Marcel Dekker, New York, 1993.

Sirohi, R. S., ed., *Selected Papers on Speckle Metrology*, SPIE – The International Society for Optical Engineering, Bellingham, 1996.

Sirohi, R. S. and K. D. Hinsch, *Selected Papers on Holographic Interferometry Principles and Techniques*, SPIE – The International Society for Optical Engineering, Bellingham, 1998.

Smith, H. M., *Principles of Holography*, 2nd edn, Wiley-Interscience, New York, 1975.

Smith, H. M., *Holographic Recording Materials*, Springer-Verlag, Berlin, 1977.

Soifer, V. and M. Golub, *Laser Beam Mode Selection by Computer-Generated Holograms*, CRC Press, Boca Raton, 1994.

Srtoke, G. W., *An Introduction to Coherent Optics and Holography*, Academic Press, New York, 1966.

Stone, T. W. and B. J. Thompson, eds, *Selected Papers on Holographic and Diffractive Lenses and Mirrors*, SPIE – The International Society for Optical Engineering, Bellingham, 1991.

Suh, H. H., "Color-Image Generation by Use of Binary-Phase Hologram," *Opt. Lett.*, **24**, 661–663 (1999).

Suleski, T. J., B. Baggett, W. F. Dalaney, C. Koehler, and E. G. Johnson, "Fabrication of High-Spatial-Frequency Gratings through Computer-Generated Near-Field Holography," *Opt. lett.*, **24**, 602–604 (1999).

Syms, R. R. A., *Practical Volume Holography*, Clarendon Press, Oxford, 1990.

Takaki, Y. and J. Hojo, "Computer-Generated Holograms to Produce High-Density Intensity Patterns," *Appl. Opt.*, **38**, 2189–2195 (1999).

Takaki, Y. and H. Ohzu, "Fast Numerical Reconstruction Technique for High-Resolution Holography," *Appl. Opt.*, **38**, 2204–2211 (1999).

Takemoto, S., ed., *Laser Holography in Geophysics*, Ellis Horwood, Chichester, 1989.

Taketomi, Y. and T. Kubota, "Reflection Hologram for Reconstructing Deep Images," *Opt. Lett.*, **22**, 1725–1727 (1997); errata, **23**, 231 (1998).

Thompson, B. J., "Fraunhofer Diffraction Patterns of Opaque Objects with Coherent Background," *J. Opt. Soc. Am.*, **53**, 1350 (1963).

Thompson, B. J., "Diffraction of Opaque and Transparent Particles," *J. Soc. Photo-Optical Instrumentation Engineers*, **2**, 43–46 (1964).

Thompson, B. J., "Holographic Methods," in *Liquid- and Surface-Borne Particle Measurement Handbook*, Knapp, J. Z., Barber, T. A., and Lieberman A., eds, Marcel Dekker, New York, 1996, pp. 197–233.

Tonomura, A., *Electron Holography*, second revised edn, Springer-Verlag, New York, 1999.

Vest, C. M., *Holographic Interferometry*, John Wiley, New York, 1979.

Vikram, C. S., *Particle Field Holography*, Cambridge University Press, Cambridge, 1992.

Vikram, C. S., "Holographic Metrology of Micro-Objects in a Dynamic Volume," in *Optical Measurement Techniques and Applications*, Rastogi, P. K., ed., Artech House, Boston, 1997, pp. 277–304.

Vikram, C. S., "Holography of Erosion, Corrosion, and Mechanical Wear: Possible Role of Phase-Shifting Interferometry," *Opt. Eng.*, **35**, 1795–1796 (1996); see also references cited therein.

Von Balley, G., ed., *Holography in Medicine & Biology*, Springer-Verlag, Berlin, 1979.

Wallace, J., "Rubber Stamp Prints Microscopic Features," *Laser Focus World*, **35**(5), 17–18 (May, 1999a).

Wallace, J., "Spinning Disk Records Holographic Movies," *Laser Focus World*, **35**(8), 16–18 (August, 1999b).

Wilson, A., "New Technologies Vie with Holographic Techniques for 3-D Displays," *Vision Systems Design*, **3**(5), 50–56 (May 1998).

Wolffer, N., B. Vinouze, and P. Gravey, "Holographic Switching Between Single Mode Fibers Based on Electrically Addressed Nematic Liquid Crystal Gratings with High Deflection Accuracy," *Opt. Commun.*, **160**, 42–46 (1999).

Wu, P, D. V. G. L. N. Rao, B. R. Kimball, M. Nakashima, and B. S. DeCristofano, "Spatial Light Modulation with an Azobenzene-Doped Polymer by Use of Diphotonic Holography," *Opt. Lett.*, **24**, 841–843 (1999).

Xu, S., G. Menders, S. Hart, and J. C. Dainty, "Pinhole Hologram and Its Applications," *Opt. Lett.*, **14**, 107–109 (1989).

Yadav, H. L. and S. Ananda Rao, "Design and Analysis of Hololens to Obtain Appreciable Illumination Over the Entire Image of an Extended Object for Its Use in Speckle Metrology," *J. Opt.*, **28**, 181–185 (1997).

Yamaguchi, I. and T. Zhang, "Phase Shifting Digital Holography," *Opt. Lett.*, **22**, 1268–1270 (1997).

Yaroslavskii, L. P. and N. S. Merzlyakov, *Methods of Digital Holography*, Consultants Bureau, New York, 1980.

Yaroslavsky, L. and M. Eden, *Fundamentals of Digital Optics: Digital Signal Processing in Optics and Holography*, Birkhauser, Boston, 1996.

Yarovoy, V. V., "Non-Diffractive Model of Capturing a Mode of Ring Laser with Holographic Mirror by a Single Speckle-Wave," *Opt. Commun.*, **158**, 351–359 (1998).

Yoshikawa, N., M. Itoh, and T. Yatagi, "Binary Computer-Generated Holograms for Security Applications from a Synthetic Double-Exposure Method by Electron-Beam Lithography," *Opt. Lett.*, **23**, 1483–1485 (1998).

Yu, F. T. S., *Optical Information Processing*, John Wiley, New York, 1983.

Yu, F. T. S. and S. Jutamula, eds, *Optical Storage and Retrieval: Memory, Neural Networks and Fractals*, Marcel Dekker, New York, 1996.

Zel'dovic, B. Ya, A. V. Mamaev, and V. V. Shkunov, *Speckle-Wave Interactions in Application to Holography and Nonlinear Optics*, CRC Press, Boca Raton, 1995.

Zhang, T. and I. Yamaguchi, "Three-Dimensional Microscopy with Phase-Shifting Digital Holography," *Opt. Lett.*, **23**, 1221–1223 (1998).

Zhang, X., S. Liu, H. Lai, and P. Chen, "One-Step Two-Dimensional Rainbow-Holographic Technique with a Nonlaser Light Source," *Opt. Lett.*, **23**, 1055–1056 (1988); errata, **23**, 1408 (1988).

Zhou, C., S. Stankovic, C. Denz, and T. Tschudi, "Phase Codes of Talbot Array Illumination for Encoding Holographic Multiplexing Storage," *Opt. Commun.*, **161**, 209–211 (1999).

16

Fourier Optics and Image Processing

FRANCIS T. S. YU

The Pennsylvania State University, University Park, Pennsylvania

16.1 FOURIER OPTICS AND IMAGE PROCESSING

The discovery of intensive lasers has enabled us to build more efficient optical systems for communication and signal processing. Most of the optical processing architectures to date have confined themselves to the cases of complete coherence or complete incoherence. However, a continuous transition between these two extremes is possible. Added to the recent development of spatial light modulators, this has brought optical signal processing to a new height. Much attention has been focused on high-speed and high-data-rate optical processing and computing.

In this chapter we shall discuss the basic principles of Fourier optics as applied to image processing and computing.

16.1.1 Fourier Transformation by Optics

To understand the basic concept of Fourier optics, we shall begin our discussion with the Fresnel–Kirchhoff theory. Let us start from the Huygens' principle, in which the complex amplitude observed at the point p$'$ of a coordinate system $\sigma(\alpha, \beta)$, due to a monochromatic light field located in another coordinate system $\rho(x, y)$, as shown in Fig. 16.1, can be calculated by assuming that each point of light source is an infinitesimal spherical radiator. Thus, the complex light amplitude $h_l(\alpha, \beta; k)$ contributed by a point p in the (x, y) coordinate system can be considered to be that from an unpolarized monochromatic point source, such that

$$h_l = -\frac{i}{\lambda r} \exp[i(kr - \omega t)], \tag{16.1}$$

where λ, k, and ω are the wavelengths, wave number, and angular frequency, respectively, of the point source, and r is the distance between the point source and the point of observation.

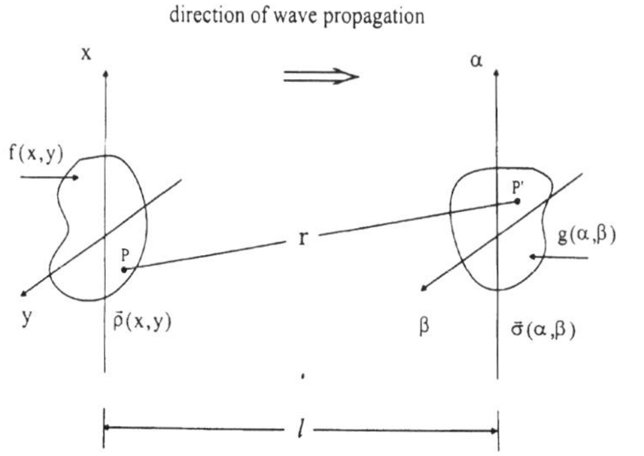

Figure 16.1   Fresnel–Kirchhoff theory.

If the separation l of the two-coordinate systems is assumed to be large compared with the regions of interest in the (x, y) and (α, β) coordinate systems, r can be approximated by

$$r = l + \frac{(\alpha - x)^2}{2l} + \frac{(\beta - y)^2}{2l}, \tag{16.2}$$

which is known as a *paraxial approximation*. By the substituting into Eq. (16.1), we have

$$h_l(\sigma - \rho; k) \cong -\frac{i}{\lambda l} \exp\left\{ ik\left[l + \frac{(\alpha - x)^2}{2l} + \frac{(\beta - y)^2}{2l} \right] \right\}, \tag{16.3}$$

which is known as the *spatial impulse response*, where the time-dependent exponent has been dropped for convenience. Thus, we see that the complex amplitude produced at the (α, β) coordinate system by the monochromatic radiating surface $f(x, y)$ can be written as

$$g(\alpha, \beta) = \iint_{x,y} f(x, y) h_l(\sigma - \rho; k)\,dxdy, \tag{16.4}$$

which is the well-known *Kirchhoff's integral*. In view of the proceeding equation, we see that the Kirchhoff's integral is in fact the convolution integral, by which can be written

$$g(\alpha, \beta) = f(x, y) * h_l(x, y), \tag{16.5}$$

where the asterisk denotes the convolution operation.

$$h_l(x, y) = C \exp\left[i\frac{k}{2l}(x^2 + y^2) \right] \tag{16.6}$$

and $C = (i/\lambda l) \exp(ikl)$ is a complex constant. Consequently, Eq. (16.5) can be represented by a block box system diagram, as shown in Fig. 16.2. In other words, the complex wave field distributed over the (α, β) coordinate system plane can be evaluated by the *convolution integral* of Eq. (16.4).

Figure 16.2 Linear system representation.

It is well known that a two-dimensional Fourier transformation can be obtained with a positive lens. Fourier-transform operations usually require complicated electronic spectrum analyzers or digital computers. However, this complicated transform can be performed extremely simply with a coherent optical system.

To perform Fourier transformations in optics, it is required that a positive lens is inserted in a monochromatic wave field of Fig. 16.1. The action of the lens can convert a spherical wavefront into a plane wave. The lens must induce a phase transformation, such as

$$T(x, y) = C \exp\left[-i\frac{\pi}{\lambda f}(x^2 + y^2)\right], \tag{16.7}$$

where C is an arbitrary complex constant.

Let us now show the Fourier transformation by a lens, as illustrated in Fig. 16.3, in which a monochromatic wave field at input plane (ξ, η) is $f(\xi, \eta)$. Then, by applying the Fresnel–Kirchhoff theory of Eq. (16.5), the complex light distribution at (α, β) can be written as

$$g(\alpha, \beta) = C\{[f(\xi, \eta) * hl(\xi, \eta)]T(x, y)\} * h_f(x, y), \tag{16.8}$$

where C is a proportionality complex constant, $h_l(\xi, \eta)$ and $h_f(x, y)$ are the corresponding spatial impulse responses, and $T(x, y)$ is the phase transform of the lens, as given in Eq. (16.7).

By a straightforward, tedious evaluation, we can show that

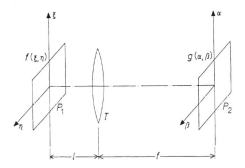

Figure 16.3 Fourier transformation by a lens.

$$g(\alpha, \beta) = C_1 \exp\left[-i\frac{k}{2f}\frac{1-v}{v}(\alpha^2 + \beta^2)\right] \int\int f(\xi, \eta) \exp\left[-i\frac{k}{f}(\alpha\xi + \beta\eta)\right]d\xi d\eta,$$

(16.9)

which is essentially the Fourier transform of $f(\xi, \eta)$ with a quadratic phase factor. If the signal plane is placed at the front focal plane of the lens, that is $l = f$, the quadratic phase factor vanishes, which leaves an exact Fourier transformation,

$$G(p, q) = C_1 \int\int f(\xi, n) \exp[-i(p\xi + qn)]d\xi d\eta,$$

(16.10)

where $p = k\alpha/f$ and $q = k\beta/f$ are the angular spatial frequency coordinates.

16.1.2 Coherent and Incoherent Processing

Let a hypothetical optical processing system be as shown in Fig. 16.4. Assume that the light emitted by the source Σ is monochromatic, and $u(x, y)$ is the complex light distribution at the input signal plane due to the incremental source $d\Sigma$. If the complex amplitude transmittance of the input plane is $f(x, y)$, the complex light field immediately behind the signal plane is $u(x, y)f(x, y)$. We assume the optical system in the black box is linearly spatially invariant with a spatial impulse response of $h(x, y)$, the output complex light field, due to $d\Sigma$, can be calculated by

$$g(\alpha, \beta) = [u(x, y)f(x, y)] * h(x, y),$$

(16.11)

which can be written as

$$dI(\alpha, \beta) = g(\alpha, \beta)g^*(\alpha, \beta)d\Sigma,$$

(16.12)

where the superscript asterisk denotes the complex conjugate. The overall output density distribution is therefore

$$I(\alpha, \beta) = \int\int |g(\alpha, \beta)|^2 d\Sigma,$$

which can be written in the following convolution integral:

$$I(\alpha, \beta) = \int\int\int\int \Gamma(x, y; x', y')h(\alpha - x, \beta - y)h^*(\alpha - x', \beta - y')$$
$$\cdot f(x, y)f^*(x', y')dxdydx'dy',$$

(16.13)

where

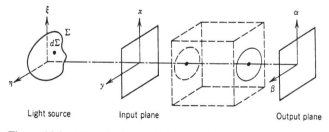

Figure 16.4 A hypothetical optical processing system.

$$\Gamma(x, y : x', y') = \int\int_{\Sigma} u(x, y)u^*(x'y')d\Sigma$$

is the *spatial coherence function*, also known as the *mutual intensity function*, at input plane (x, y).

By choosing two arbitrary points Q_1 and Q_2 at the input plane, and if r_1 and r_2 are the respective distances from Q_1 and Q_2 to $d\Sigma$, the complex light distrubances at Q_1 and Q_2 due to $d\Sigma$ can be written as

$$u_1(x, y) = \frac{[I(\xi, \eta)]^{-1/2}}{r_1} \exp(ikr_1) \tag{16.14}$$

and

$$u_2(x', y') = \frac{[I(\xi, \eta)]^{-1/2}}{r_1} \exp(ikr_2), \tag{16.15}$$

where $I(\xi, \eta)$ is the intensity distribution of the light source. By substituting Eqs (16.14) and (16.15) into Eq. (16.13), we have

$$\Gamma(x, y; x', y') = \int\int_{\Sigma} \frac{I(\xi, \eta)}{r_1 r_2} \exp[ik(r_1 - r_2)]d\Sigma. \tag{16.16}$$

In the paraxial case, $r_1 - r_2$ may be approximated by

$$r_1 - r_2 \simeq \frac{1}{r}[\xi(x - x') + \eta(y - y')],$$

where r is the separation between the source plane and the signal plane. Then, Eq. (16.16) can be reduced to

$$\Gamma(x, y; x', y') = \frac{1}{r^2}\int\int_{\Sigma} I(\xi, \eta) \exp\left\{i\frac{k}{r}[\xi(x - x') + \eta(y - y')]\right\}d\xi d\eta \tag{16.17}$$

which is known as the *Van Cittert–Zernike theorem*. Notice that Eq. (16.17) forms an inverse Fourier transform of the source intensity distribution.

Now one of the two extreme cases is by letting the light source become infinitely large: for example, $I(\xi, \eta) \simeq K$; then Eq. (16.17) becomes

$$\Gamma(x, y; x', y') = K_1\delta(x - x', y - y'), \tag{16.18}$$

which describes a completely *incoherent illumination*, where K_1 is an appropriate constant.

On the other hand, if the light source is vanishingly small, i.e., $I(\xi, \eta) \simeq K$, $\delta(\xi, \eta)$, Eq. (16.17) becomes

$$\Gamma(x, y; x', y') = K_2, \tag{16.19}$$

which describes a completely *coherent illumination*, where K_2 is a proportionality constant. In other words, a monochromatic point source describes a strictly coherent processing regime, while an extended source describes a strictly incoherent system. Furthermore, an extended monochromatic source is also known as a *spatially incoherent* source.

By referring to the completely incoherent illumination, we have $\Gamma(x, y; x', y') = K_1\delta(x - x', y - y')$, the intensity distribution at the output plane, can be shown

$$I(\alpha, \beta) = \int\int |h(\alpha - x, \beta - y)|^2 |f(x, y)|^2 \mathrm{d}x\mathrm{d}y, \tag{16.20}$$

in which we see that the output intensity distribution is the convolution of the input signal intensity with respect to the intensity impulse response. In other words, for the *completely incoherent illumination*, the optical signal processing system is linear in *intensity*, i.e.,

$$I(\alpha, \beta) = |h(x, y)|^2 * |f(x, y)|^2, \tag{16.21}$$

where the asterisk denotes the convolution operation. On the other hand, for the completely coherent illumination, i.e., $\Gamma(x, y; x', y') = K_2$, the output intensity distribution can be shown as

$$\begin{aligned} I(\alpha, \beta) = g(\alpha, \beta)g^*(\alpha, \beta) &= \int\int h(\alpha - x, \beta - y)f(x, y)\mathrm{d}x\mathrm{d}y \\ &\cdot \int\int h^*(\alpha - x', \beta - y')f(x', y')\mathrm{d}x'\mathrm{d}y' \end{aligned} \tag{16.22}$$

when

$$g(\alpha, \beta) = \int\int h(\alpha - x, \beta - y)f(x, y)\mathrm{d}x\mathrm{d}y, \tag{16.23}$$

for which we can see that the optical signal processing system is linear in *complex amplitude*. In other words, a coherent optical processor is capable of processing the information in complex amplitudes.

16.1.3 Fourier and Spatial Domain Processing

The roots of optical signal processing can be traced back to Abbe's work in 1873, which led to the discovery of spatial filtering. However, optical pattern recognition was not appreciated until the complex spatial filtering of Van der Lugt in 1964. Since then, techniques, architectures, and algorithms have been developed to construct efficient optical signal processors. The objective of this section is to discuss the optical architectures and techniques as applied to image processing. Basically, there are two frequently used signal-processing architectures in Fourier optics: namely, the Vander Lugt correlator (VLC) and the joint transform correlator (JTC). Nevertheless, image processing can be accomplished either by *Fourier domain filtering* or *spatial domain filtering* or both. Processors that use Fourier domain filtering are known as VLCs and the spatial-domain filtering is often used for JTC. The basic distinctions between them are that VLC depends on a Fourier-domain filter, whereas JTC depends on a spatial domain filter. In other words, the complex spatial filtering of Vander Lugt is input scene *independent*, whereas the joint transform filtering is input scene *dependent*. The reason is that once the Fourier domain spatial filter is synthesized, the structure of the filter will not be altered by the input scene (e.g., for multitarget detection or background noise). Thus the performance of Fourier domain filtering is independent of the input scene. On the other hand, the joint transform power spectrum would be affected by the input noise and multitarget detection, for which the performance in the JTC is input scene dependent.

It is apparent that a pure optical correlator has drawbacks which make certain tasks difficult or impossible to perform. The first problem is that optical systems are difficult to program, in the sense of a general-purpose digital computer. A second problem is that accuracy is difficult to achieve in Fourier optics. A third problem is that optical systems cannot easily be used to make decisions. Even a simple decision cannot be performed optically without the intervention of electronics. However, many deficiencies of optics happen to be the strong points of the electronic counterpart. For example, accuracy, controllability, and programmability are the obvious traits of digital computers. Thus, by combining an optical system with its electronic counterpart is rather natural to have a better processor, as shown in Figs 16.5 and 16.6, in which spatial light modulators (SLMs) are used for input-object and spatial filter devices. One of the important aspects of these hybrid-optical architectures is that decision making can be done by the computer.

We shall now consider the VLC (depicted in Fig. 16.5), which we assume is illuminated by a coherent plane wave. If an input object $f(x, y)$ is displayed at the input plane, the output complex light can be shown as

$$g(\alpha, \beta) = Kf(x, y) * h(x, y), \tag{16.24}$$

where $*$ denotes the convolution operation, K is a proportionality constant, and $h(x, y)$ is the spatial impulse response of the Fourier domain filter, which can be generated on the SLM. We note that a Fourier domain filter can be described by a complex-amplitude transmittance such as

$$H(p, q) = |H(p, q)| \exp[i\phi(p, q)], \tag{16.25}$$

for which the physically realizable conditions are imposed by

$$|H(p, q)| \leq 1 \tag{16.26}$$

and

$$0 \leq \phi(p, q) \leq 2\pi, \tag{16.27}$$

where (p, q) is the angular spatial frequency coordinator system. We stress that the complex transmittance imposed by the physical constraints can be represented by a

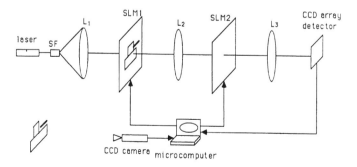

Figure 16.5 Hybrid optical Vander Lugt correlator (VLC). SLMs, spatial light modulators; CCDs, charge-coupled detectors; SF, spatial filter; Ls, lenses.

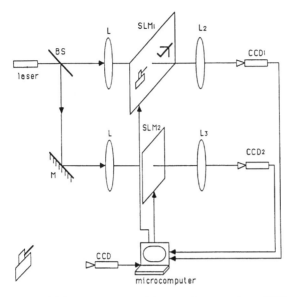

Figure 16.6 Hybrid optical joint transform correlator (JTC). SLMs, spatial light modulators; CCDs, charge-coupled detectors; BS, beamsplitter; M, mirror; Ls, lenses.

set of points within or on a unit circle in a complex plane shown in Fig. 16.7. In other words, a physically realizable filter can only be synthesized within or on the unit circle of the complex plane.

Let us now illustrate a complex signal detection using the VLC. We assume that a Fourier-domain matched filter is generated at the spatial frequency plane, as given by

$$H(p, q) = K\{1 + |F(p, q)|^2 + 2|F(p, q)|\cos[\alpha_0 p + \phi(p, q)]\}, \qquad (16.28)$$

which is a positive real function subject to the physical constraints of Eqs (16.26) and (16.27), where α_0 is the spatial carrier frequency. It is straightforward to show that the output complex light distribution is given by

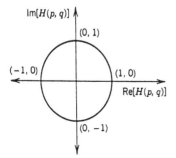

Figure 16.7 Complex amplitude transmittance.

$$g(\alpha, \beta) = K[f(x, y) + f(x, y) * f(x, y) * f^*(-x, -y) + f(x, y) * f(x + x_0, y)$$
$$+ f(x, y) * f^*(-x + \alpha_0, -y)],$$

$$(16.29)$$

in which the first and second terms represent the zero-order diffraction; the third and fourth terms are the convolution and cross-correlation terms, which are diffracted in the neighborhood of $\alpha = -\alpha_0$, and $\alpha = \alpha_0$ respectively. The zero-order and convolution terms are of no particular interest here; it is the cross-correlation term that is used in signal detection.

Now, if the input signal is assumed to be embedded in an additive white Gaussian noise n, that is,

$$f'(x, y) = f(x, y) + n(x, y), \tag{16.30}$$

then the correlation term in Eq. (16.29) would be

$$R(\alpha, \beta) = K[f(x, y) + n(x, y)] * f^*(-x + \alpha_0, -y). \tag{16.31}$$

Since the cross-correlation between $n(x, y)$ and $f^*(-x + \alpha_0, -y)$ can be shown to be approximately equal to zero, Eq. (16.31) reduces to

$$R(\alpha, \beta) = f(x, y) * f^*(-x + \alpha_0, -y), \tag{16.32}$$

which is the autocorrelation distribution of $f(x, y)$.

To ensure that the zero-order and the first-order diffraction terms will not overlap, the spatial carrier frequency α_0 is required that

$$\alpha_0 > l_f + \tfrac{3}{2}l_s, \tag{16.33}$$

where l_f and l_s are the spatial lengths in the x-direction of the input object transparency and the detecting signal $f(x, y)$, respectively. To show that this is true, we consider the length of the various output terms of $g(\alpha, \beta)$, as illustrated in Fig. 16.8.

Since lengths of the first, second, third, and fourth terms of Eq. (16.29) are l_f, $2l_s + l_f$, and $l_f + l_s$, respectively, to achieve complete separation the spatial carrier frequency α_0 must satisfy the inequality of Eq. (16.33).

Complex spatial filtering can also be performed by JTC (shown in Fig. 16.6) in which the input object displayed on the SLMs are given by

$$f_1(x - \alpha_0, y) + f_2(x + \alpha_0, y), \tag{16.34}$$

where $2\alpha_0$ is the main separation of the input objects. The corresponding joint transform power spectrum (JTPS), as detected by charge-coupled detector (CCD1), is given by

$$I(p, q) = |F_1(p, q)|^2 + |F_2(p, q)|^2 + 2|F_1(p, q)|^2|F_2(p, q)|^2$$
$$\cdot \cos[2\alpha_0 p - \phi_1(p, q) + \phi_2(p, q)], \tag{16.35}$$

where

$$F_1(p, q) = |F_1(p, q)|e^{i\phi_1(p, q)},$$

and

$$F_2(p, q) = |F_2(p, q)|e^{i\phi_2(p, q)}.$$

If the JTPS is displayed on SLM2, the output complex light distribution would be

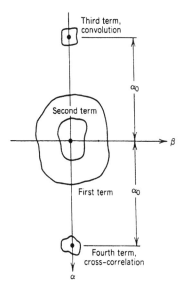

Figure 16.8   Sketch of the output diffraction.

$$g(\alpha, \beta) = f_1(x, y) \otimes f_1^*(x, y) + f_2(x, y) \otimes f_2^*(x, y) + f_1(x, y)$$
$$\otimes f_2^*(x - 2\alpha_0, y) + f_1^*(x, y) \otimes f_2(x + 2\alpha_0, y), \tag{16.36}$$

where $\otimes$ denotes the correlation operation. The first two terms represent overlapping correlation functions $f_1(x, y)$ and $f_2(x, y)$, which are diffracted at the origin of the output plane. The last two terms are the two cross-correlation terms, which are diffracted around $\alpha = 2\alpha_0$ and $\alpha = -2\alpha_0$, respectively.

To avoid the correlation term overlapping with the zero diffraction, it is required that the separation between the input object should be

$$2\alpha_0 \geq 2\chi, \tag{16.37}$$

where χ is the width of the input object.

For complex target detection, we assume that target $f_1(x - \alpha_0, y)$ is embedded in an additive white Gaussian noise, i.e., $f_1(x - \alpha_0, y) + n(x - \alpha_0, y)$ and $f_2(x - \alpha_0, y)$ is replaced by $f_1(x - \alpha_0, y)$. Then it can be shown that the JTPS is given by

$$I(p, q) = 2|F_1|^2 + |N|^2 + F_1 N^* + N F_1^* + (F_1 F_1^* + N F_1^*)e^{-2\alpha_0 p}$$
$$+ (F_1 F_1^* F_1 N^*)e^{i2\alpha_0 p}. \tag{16.38}$$

Since the noise is assumed to be additive and Gaussian distributed with zero mean, we note that

$$\int\int f_1(x, y) n(\alpha + x, \beta + y) dx dy = 0.$$

Thus, the output complex light field is

$$g(\alpha, \beta) = 2f_1(x, y) \otimes f_1^*(x, y) + n(x, y) \otimes n^*(x, y) + f_1(x, y) \otimes f_1^*(x - 2\alpha_0, y) \\ + f_1(x, y) \otimes f_1^*(x + 2\alpha_0, y),$$

$$(16.39)$$

in which the autocorrelation terms are diffracted at $\alpha_0 = 2\alpha_0$ and $\alpha_0 = -2\alpha_0$, respectively.

16.1.4 Exploitation of Coherence

The use of a coherent source enables optical processing to be carried out in complex amplitude processing, which offers a myriad of applications. However, coherent optical processing also suffers from coherent artifact noise, which limits its processing capabilities. To alleviate these limitations, we discuss methods to exploit the coherence contents from an incoherent source for complex amplitude processing. Since all physical sources are neither strictly coherent nor strictly incoherent, it is possible to extract their inherent coherence contents for coherent processing.

Let us begin with the exploitation of spatial coherence from an extended incoherent source. By referring to the conventional optical processor shown in Fig. 16.9, the *spatial coherence* function at the input plane can be written as

$$\Gamma(x_2 - x_2') = \iint \gamma(x_1) \exp\left[i2\pi \frac{x_1}{\lambda_f}(x_2 - x_2')\right] dx_1, \qquad (16.40)$$

which is the well-known Van Citter–Zernike theorem, where $\gamma(x_1)$ is the extended source, f is the focal length of the collimated lens, and λ is the wavelength of the extended source. Thus, we see that the spatial coherence at the input plane and the source-encoding intensity transmittance form a Fourier-transform pair, given by

$$\gamma(x_1) = \mathscr{F}[\Gamma(x_2 - x_2')], \quad \text{and} \quad \Gamma(x_2 - x_2') = \mathscr{F}^{-1}[\gamma(x_1)], \qquad (16.41)$$

where $\mathscr{F}$ denotes the Fourier-transform operation. In other words, if a specific spatial coherence requirement is needed for information processing, a source-encoding can be performed. The source-encoding $\gamma(x_1)$ can consist of apertures of different shapes or slits, but it should be a positive real function that satisfies the following physically realizable constraint:

$$0 \le \gamma(x_1) \le 1. \qquad (16.42)$$

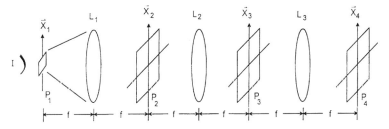

Figure 16.9 Incoherent source optical processor: I, incoherent source; L_1, collimating lens; L_2 and L_3, achromatic transformation lenses; P_1, source encoding mask; P_2, input plane; P_3, Fourier plane, and P_4, output plane.

For the exploitation of *temporal coherence*, we note that the Fourier spectrum is linearly proportional to the wavelength of the light source. It is apparently not capable of (or inefficient at) using a broadband source for complex amplitude processing. To do so, a narrow-spectral-band (i.e., temporally coherent) source is needed. In other words, the spectral spread of the input object should be confined within a small fraction fringe spacing of the spatial filter, which is given by

$$\frac{p_m f \Delta\lambda}{2\pi} \ll d, \tag{16.43}$$

where d is the fringe spacing of the spatial filter, p_m is the upper angular spatial frequency content of the input object, f is the focal length of the transform lens, and $\Delta\lambda$ is the spectral bandwidth of the light source. In order to have a higher temporal coherence requirement, the spectral width of the light source should satisfy the following constraint:

$$\frac{\Delta\lambda}{\lambda} \ll \frac{\pi}{h p_m}, \tag{16.44}$$

where λ is the center wavelength of the light source, $2h$ is the size of the input object transparency, and $2h = (\lambda f)/d$.

There are ways to exploit the temporal coherence content from a broadband source. One of the simplest methods is by dispersing the Fourier spectrum, which can be obtained by placing a spatial sampling grating at the input plane P_2. For example, if the input object is sampled by a phase grating, as given by

$$f(x_2)T(x_2) = f(x_2)\exp(ip_0 x_2), \tag{16.45}$$

then the corresponding Fourier transform would be

$$F(p, q) = F\left(x_3 - \frac{\lambda f}{2\pi}p_0\right), \tag{16.46}$$

in which we see that $F(p, q)$ is smeared into rainbow colors at the Fourier plane. Thus, a high temporal coherence Fourier spectrum within a narrow-spectral-band filter can be obtained, as given by

$$\frac{\Delta\lambda}{\lambda} \cong \frac{4p_m}{p_0} \ll 1. \tag{16.47}$$

Since the spectral content of the input object is dispersed in rainbow colors, as illustrated in Fig. 16.10(a), it is possible to synthesize a set of narrow-spectral-band filters to accommodate the dispersion.

On the other hand, if the spatial filtering is a 1D operation, it is possible to construct a fan-shaped spatial filter to cover the entire smeared Fourier spectrum, as illustrated in Fig. 16.10(b). Thus, we see that a high degree of temporally coherent filtering can be carried out by a simple white light source. Needless to say, the (broadband) spatial filters can be synthesized by computer-generated techniques.

In the preceding, we have shown that spatial and temporal coherence can be exploited by spatial encoding and spectral dispersion of an incoherent source. We have shown that complex amplitude processing can be carried out with either a set of 2-D narrow-spectral-band filters or with a 1-D fan-shaped broadband filter. Let us first consider that a set of narrow-spectral-band filters is being used, as given by

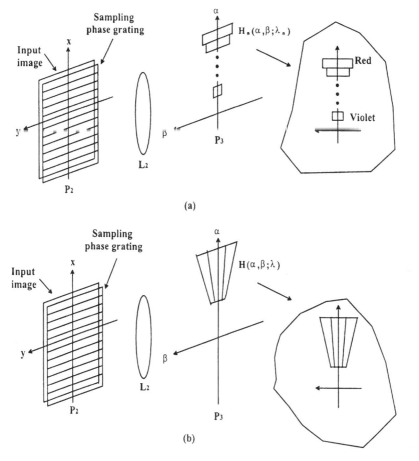

Figure 16.10 Broad spectral filtering: (a) using a set of narrow-spectral-band filters, (b) using a 1-D fan-shaped broadband filter.

$$H_n(p_n, q_n; \lambda_n), \qquad \text{for } n = 1, 2, \ldots, N, \tag{16.48}$$

where (p_n, q_n) represents the angular frequency coordinates and λ_n is the center wavelength of the narrow-width filter; then it can be shown that the output light intensity would be the incoherent superposition of the filtered signals, as given by

$$I(x, y) \cong \sum_{n=1}^{N} \Delta\lambda_n \left| f(x, y; \lambda_n) * h(x, y; \lambda_n) \right|^2, \tag{16.49}$$

where $*$ denotes the convolution operation, $f(x, y; \lambda_n)$ represents the input signal illuminated by λ_n, $\Delta\lambda_n$ is the narrow spectral width of the narrow-spectral-band filter, and $h(x, y; \lambda_n)$ is the spatial impulse response of $H_n(p_n, q_n; \lambda_n)$, that is,

$$h_n(p_n, q_n; \lambda_n) = \mathscr{F}^{-1}[H(p_n, q_n; \lambda_n)]. \tag{16.50}$$

Thus we see that by exploiting the spatial and temporal coherence, an incoherent source processor can be made to process the information in complex amplitude as a coherent processor. Since the output intensity distribution is the sum of mutually incoherent image irradiances, the annoying coherent artifact noise can be avoided.

On the other hand, if the signal processing is a 1-D operation, then the information processing can be carried out with a 1-D fan-shaped broadband filter. Then the output intensity distribution can be shown as

$$I(x, y) = \int_{\Delta\lambda} \left| f(x, y; \lambda) * h(x, y; \lambda_n) \right|^2 d\lambda, \tag{16.51}$$

where the integral is over the entire spectral band of the light source. Again, we see that the output irradiance is essentially obtained by incoherent superposition of the entire spectral band image irradiances, by which the coherent artifact noise can be avoided. Since one can utilize a conventional white light source, the processor can indeed be used to process polychromatic images. The advantages of exploiting the incoherent source for coherent processing are that, it enables the information to be processed in complex amplitude as a coherent processor and it is capable of suppressing the coherent artifact noise as an incoherent processor.

16.2 IMAGE PROCESSING

16.2.1 Coherent Processing

In the preceding sections, we have seen that by simple insertion of spatial filters, a wide variety of image processings can be performed by coherent processors (i.e., VLC or JTC). In this section we shall describe a couple of image processings by coherent light.

One of the interesting applications of coherent optical image processing is the restoration of blurred photographic images. The Fourier spectrum of a blurred (or distorted) image can be written as

$$G(p) = S(p)D(p), \tag{16.52}$$

where $G(p)$ is the distorted-image, $S(p)$ is the nondistorted image, $D(p)$ is the distorting function, and p is the angular spatial frequency. Then the inverse filter transfer function for the restoration is

$$H(p) = \frac{1}{D(p)}. \tag{16.53}$$

However, the inverse filter function is generally not physically realizable. If we would accept some restoration errors, then an approximated restoration filter may be physically realized. For example, let the transmission function of a linear smeared point image be

$$f(\xi) = \begin{cases} 1, & -1/2\Delta\xi \le \xi \le 1/2\Delta\xi, \\ 0, & \text{otherwise} \end{cases} \tag{16.54}$$

where $\Delta\xi$ is the smear length. If the preceding smeared transparency is displayed on the input SLM1 of the VLC shown in Fig. 16.5 the complex light field at the Fourier plane is given by

$$F(p) = \Delta\xi \frac{\sin(p\Delta\xi/2)}{p\Delta\xi/2}, \qquad (16.55)$$

which is plotted in Fig. 16.11. In principle, an inverse filter as given by

$$H(p) = \frac{p\Delta\xi/2}{\sin(p\Delta\xi/2)} \qquad (16.56)$$

should be used for the image restoration. However, it is trivial to see that it is not physically realizable, since it has infinite poles. If one is willing to sacrifice some degree of restoration, an approximate filter may be realized as follows:

An approximated inverse filter can be physically constructed by combining an *amplitude filter* with a *phase filter*, as shown in Figs 16.12 and 16.13, respectively, by which the combined transfer function is given by

$$H(p) = A(p)e^{i\phi(p)}. \qquad (16.57)$$

Thus by displaying the preceding inverse filter in SLM2, the restored Fourier spectrum is given as given by

$$F_1(p) = F(p)H(p). \qquad (16.58)$$

By denoting T_m as the minimum transmittance of the amplitude filter, then the restored Fourier spectrum is the shaded area of Fig. 16.11. Let us define the degree of image restoration as given by

$$\vartheta(T_m)(\%) = \frac{1}{T_m\Delta p}\int_{\Delta p}\frac{F(p)H(p)}{\Delta\xi}dp \times 100, \qquad (16.59)$$

where Δp is the spectral bandwidth of restoration. A plot as a function of T_m is shown in Fig. 16.14. We see that high degree of restoration can be achieved as T_m approaches zero. However, at the same time the restored Fourier spectrum is also vanishing small, for which no restored image can be observed. Although the inverse filter in principle can be computer generated, we will use a holographic phase filter for the demonstration, as given by

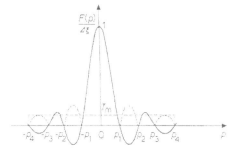

Figure 16.11 The solid curve represents the Fourier spectrum of a linear smeared point image. The shaded area represents the restored Fourier spectrum.

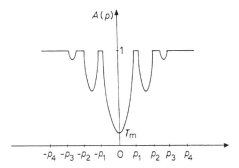

Figure 16.12 Amplitude filter.

$$T(p) = \tfrac{1}{2}\big\{1 + \cos[\phi(p) + \alpha_0 p]\big\}, \tag{16.60}$$

where α_0 is an arbitrarily chosen constant, and

$$\phi(p) = \begin{cases} \pi, & p_n \leq p \leq p_{n+1}, \quad n = \pm 1, \pm 3, \pm 5, \ldots \\ 0, & \text{otherwise.} \end{cases} \tag{16.61}$$

Thus, by combining with the amplitude filter we have

$$H_1(p) = A(p)T(p) = \tfrac{1}{2}A(p) + \tfrac{1}{4}[H(p)\exp(i\alpha_0 p) + H^*(p)\exp(-i\alpha_0 p)]. \tag{16.62}$$

If this complex filter $H_1(p)$ is inserted in the Fourier domain, then the transmitted Fourier spectrum would be

$$F_2(p) = \tfrac{1}{2}F(p)A(p) + \tfrac{1}{4}[F(p)H(p)\exp(i\alpha_0 p) + F(p)H^*(p)\exp(-i\alpha_0 p)], \tag{16.63}$$

in which we see that the second and third terms are the restored Fourier spectra, which will be diffracted around $\alpha = \alpha_0$ and $\alpha = -\alpha_0$, respectively, at the output plane. It is interesting to show that the effects of restoration due to amplitude filter alone, phase filter alone, and the combination of both, as plotted in Fig. 16.15. We see that by phase filtering alone it offers a significant effect of restoration as compared with the one using the amplitude filter. To conclude this section, a restored image using this technique is shown in Fig. 16.16. We see that the smeared image can indeed be restored with a coherent processor, but the stored image is also contaminated with coherent noise, as shown in Fig. 16.16(b).

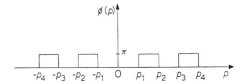

Figure 16.13 Phase filter.

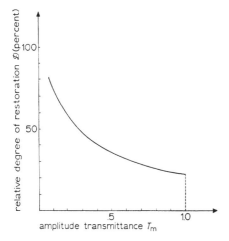

Figure 16.14 Relative degree of restoration as a function of T_m.

Another coherent image processing we shall demonstrate is the image subtraction. Image subtraction can be achieved by using a diffraction grating technique, as shown in Fig. 16.17. By displaying two input images on the input SLM1,

$$f_1(x - h, y) + f_2(x + h, y). \tag{16.64}$$

The corresponding Fourier spectrum can be written as

$$F(p, q) = F_1(p, q)e^{-ihp} + F_2(p, q)e^{ihp}, \tag{16.65}$$

where $2h$ is the main separation between the two input images. If a bipolar grating, given by

$$H(p) = \sin hp, \tag{16.66}$$

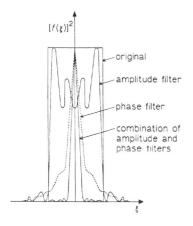

Figure 16.15 Restoration due to amplitude, phase, and complex filters.

(a)   (b)

Figure 16.16 Image restoration: (a) smeared image; (b) restored image with coherent light.

is displayed on SLM2 at the Fourier domain, it can be shown that the output light-intensity distribution is given by

$$I(x, y) = \int f_1(x - 2h, y)|^2 + |f_1(x, y) - f_2(x, y)|^2 + |f_1(x + 2h, y)|^2, \qquad (16.67)$$

in which we see that the subtracted image (i.e., $f_1 - f_2$) is diffracted around the origin at the output plane. Figure 16.18 shows an experimental result obtained with the coherent processor. Again, we see that the subtracted image is contaminated by severe coherent artifact noise.

Mention must be made that JTC can also be used as an image processor. Instead of using the Fourier domain filter, JTC uses the spatial domain filter. For example, an image $f(x, y)$ and a spatial filter domain $h(x, y)$ are displayed at this input plane of the JTC shown in Fig. 16.2, as given by

$$f(x - h, y) + h(-x + h, -y). \qquad (16.68)$$

It can be shown that output complex light distribution is

$$\begin{aligned} g(\alpha, \beta) = &f(x, y) \otimes f(x, y) + h(-x, -y) \otimes (-x, -y) + f(x, y) * h(x - 2h, y) \\ &+ f(x, y) * h(x + 2h, y), \end{aligned}$$

$$(16.69)$$

in which two processed images (i.e., the convolution terms) are diffracted around $x = \pm 2h$. Thus, we see that JTC can indeed be used as an image processor, for which it uses a spatial domain filter.

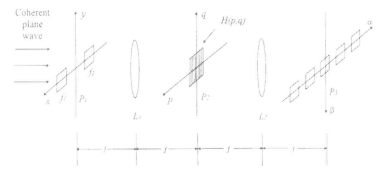

Figure 16.17 Optical setup for image subtraction.

(a)

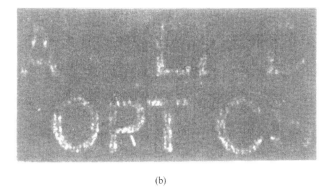

(b)

Figure 16.18 (a) Input images. (b) Subtracted image with coherent light.

16.2.2 Processing with Incoherent Light

In Section 16.1.2 we have shown that the incoherent processor is only capable of processing the image in terms of intensity and it is, however, not possible to process the information in complex amplitude. It is for this reason that makes coherent processors are more attractive for optical signal processing. Nevertheless, we have shown that the complex amplitude processing can be exploited from an incoherent source, as described in Section 16.1.4. In this section, we demonstrate a couple of examples that complex amplitude image processing can indeed be exploited from an incoherent light processor.

Let us now consider the image deblurring under the incoherent illumination. Since smeared image deblurring is a 1-D processing operation, inverse filtering takes place with respect to the smeared length of the blurred object. Thus, the required spatial coherence depends on the smeared length instead of the entire input plane. If we assume that a spatial coherence function is given by

$$\Gamma(x_2 - x_2') = \sin c\left\{\frac{\pi}{\Delta x_2}(x_2 - x_2')\right\}, \tag{16.70}$$

then the source-encoding function can be shown as

$$\gamma(x_1) = \text{rect}\left(\frac{x_1}{w}\right), \tag{16.71}$$

where Δx_2 is the smeared length, $w = (f\lambda)/(\Delta x_2)$ is the slit width of the encoding aperture as shown in Fig. 16.19(a), and

$$\text{rect}\left(\frac{x_1}{w}\right) = \begin{cases} 1, & -\dfrac{w}{3} \le x_1 \le \dfrac{w}{2} \\ 0, & \text{otherwise.} \end{cases} \tag{16.72}$$

As for the temporal coherence requirement, a sampling phase grating is used to disperse the Fourier spectrum in the Fourier plane. Let us consider the temporal coherence requirement for a 2-D image in the Fourier domain. A high degree of temporal coherence can be achieved by using a higher sampling frequency. We assume that the Fourier spectrum dispersion is along the x-axis. Since the smeared image deblurring is a 1-D processing, a *fan-shape* broadband spatial filter to accommodate the smeared Fourier spectrum can be utilized. Therefore, the sampling frequency of the input phase grating can be determined by

$$p_0 \ge \frac{4\lambda p_m}{\Delta\lambda}, \tag{16.73}$$

in which λ and $\Delta\lambda$ are the center wavelength and the spectral bandwidth of the light source, respectively, and p_m is the x-axis spatial frequency limit of the blurred image.

Figure 16.20(a) shows a set of blurred letters (OPTICS) due to linear motion. By inserting this blurred transparency in an incoherent source processor of Fig. 16.9 a set of deblurred letters is obtained, as shown in Fig. 16.20(b). Thus we see that by properly exploiting the coherence contents, complex amplitude processing can indeed be obtained from an incoherent source. Since the deblurred image is obtained by incoherent integration (or superposition) of the broadband source, the coherent

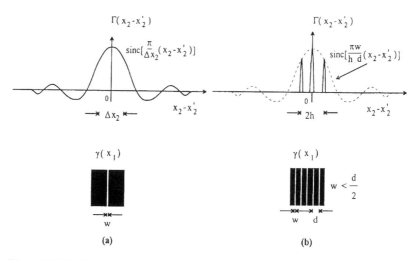

Figure 16.19 Source encoding and spatial coherence Γ, spatial coherence function γ, and source-encoding mask (a) for image deblurring and (b) for image subtraction.

Figure 16.20 (a) Smeared image. (b) Restored image with incoherent light.

artifact can be suppressed. As compared with the one obtained with coherent illu-mination of Fig. 16.16(b), we see that the coherence noise has been substantially reduced, as shown in Fig. 16.20(b).

Let us now consider an image subtraction processing with incoherent light. Since the spatial coherence depends on the corresponding point-pair of the images to be subtracted, a strictly broad spatial coherence function is not required. Instead a point-pair spatial coherence function is needed. To ensure the physically reliability of the source-encoding function, we let the point-pair spatial coherence function be given by

$$\Gamma(x_2 - x_2') = \frac{\sin[N(\pi/h)(x_2 - x_2')]}{N \sin[(\pi/h)(x_2 - x_2')]} \operatorname{sin} c\left[\frac{\pi w}{hd}(x_2 - x_2')\right], \tag{16.74}$$

where $2h$ is the main separation of the two input image transparencies. As $N \gg 1$ and $w \ll d$, Eq. (16.74) converges to a sequence of narrow pulses located at $(x_2 - x_2') = nh$, where n is a positive integer, as shown in Fig. 16.19(b). Thus, a high degree of coherence between the corresponding point-pair can be obtained. By Fourier transforming Eq. (16.74), the source-encoding function can be shown as

$$\gamma(x_1) = \sum_{n=1}^{N} \operatorname{rect}\left(\frac{x_1 - nd}{w}\right), \tag{16.75}$$

where w is the slit width, and $d = (\lambda f)/h$ is the separation between the slits. By plotting the preceding equation, the source-encoding mask is represented by N equally spaced narrow slits, as shown in Fig. 16.19(b).

Since the image subtraction is a 1-D processing operation, the spatial filter should be a *fan-shaped* broadband sinusoidal grating, as given by

$$G = \frac{1}{2}\left[1 + \sin\left(\frac{2\pi xh}{\lambda f}\right)\right]. \tag{16.76}$$

Figure 16.21(a) shows two input image transparencies inserted at the input plane of the incoherent source processor of Fig. 16.9. The output subtracted image obtained is shown in Fig. 16.21(b) in which we see that the coherent artifact noise is sup-pressed.

To conclude this section we note that the broadband (white-light) source con-tains all the visible wavelengths; the aforementioned incoherent source processor can be utilized to process color (or polychromatic) images

16.3 NEURAL NETWORKS

Electronic computers can solve some classes of computational problems thousands of times faster and more accurately than the human brain. However, for cognitive

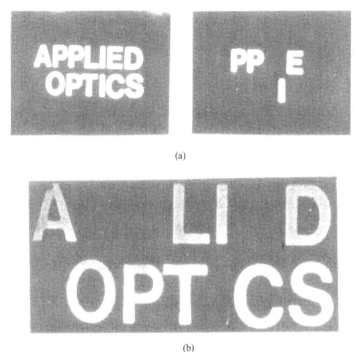

(a)

(b)

Figure 16.21   (a) Input images. (b) Subtracted image with incoherent light.

tasks, such as pattern recognition, understanding and speaking a language, etc., the human brain is much more efficient. In fact, these tasks are still beyond the reach of modern electronic computers.

A neural network consists of a collection of processing elements, called *neurons*. Each neuron has many input signals, but only one output signal which is fanned out to many pathways connected to other neurons. These pathways interconnect with other neurons to form a network. The operation of a neuron is determined by a transfer function that defines the neuron's output as a function of the input signals. Every connection entering a neuron has an adaptive coefficient called a *weight* assigned to it. The weight determines the interconnection strength between neurons, and they can be changed through a learning rule that modifies the weights in response to input signals and the transfer function. The learning rule allows the response of the neuron to change with time, depending on the nature of the input signals. This means that the network *adapts* itself to the environment and organizes the information within itself, which is a type of learning.

16.3.1   Optical Neural Net Architectures

Generally speaking, a one-layer neural network of N neurons has N^2 interconnections. The transfer function of a neuron can be described by a nonlinear relationship such as a step function, making the output of a neuron either 0 or 1 (binary), or a sigmoid function, which gives rise to analog values. The state of the ith neuron in the network can be represented by a *retrieval equation*, given by

$$u_i = f\left\{\sum_{j=1}^{N} T_{ij}u_j - \theta_i\right\}, \tag{16.77}$$

where u_i is the activation potential of the ith neuron, T_{ij} is the *interconnection weight* matrix (IWM) (associative memory) between the jth neuron and the ith neuron, θ_i is a phase bias, and f is a nonlinear processing operator. In view of the summation within the retrieval equation, it is essentially a matrix-vector outer-product operation, which can be optically implemented.

Light beams propagating in space will not interfere with each other and optical systems generally have large space-bandwidth products. These are the traits of optics that prompted the optical implementation of neural networks (NNs). An optical NN using a liquid-crystal television (LCTV) SLM is shown in Fig. 16.22, in which the lenslet array is used for the interconnection between the IWM and the input pattern. The transmitted light field after LCTV2 is collected by an imaging lens, focusing at the lenslet array and imaging onto a CCD array detector. The array of detected signals is sent to a thresholding circuit and the final pattern can be viewed at the monitor, and it can be sent back for the next iteration. The data flow is primarily controlled by the microcomputer, by which the LCTV-based neural net just described is indeed an *adaptive* optical NN.

16.3.2 Hopfield Neural Network

One of the most frequently used neural network models is the *Hopfield model*, which allows the desired output pattern to be retrieved from a distorted or partial input pattern. The model utilizes an associative memory retrieval process equivalent to an iterative thresholded matrix–vector outer product expression, as given by

$$\begin{aligned} V_i &\to 1 \quad \text{if } \sum_{j=1}^{N} T_{ij}V_j \quad \geq 0, \\ V_i &\to 1 \qquad\qquad\qquad < 0, \end{aligned} \tag{16.78}$$

where V_i and V_j are binary output and binary input patterns, respectively, and the associative memory operation is written as

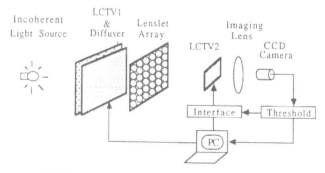

Figure 16.22 An optical hybrid neural network. CCD, charge-coupled detectors; LCTVs, liquid-crystal televisions.

$$T_{ij} = \begin{cases} \sum_{m=1}^{N}(2V_i^m - 1)(2V_j^m - 1), & \text{for } i \neq j \\ 0, & \text{for } i = j \end{cases} \qquad (16.79)$$

where V_i^m and V_j^m are ith and jth elements of the mth binary vectory.

The Hopfield model depends on the outer-product operation for construction of the associated memory, which severely limits storage capacity, and often causes failure in retrieving similar patterns. To overcome these shortcomings, neural network models, such as back propagation, orthogonal projection, multilevel recognition, interpattern association, moment invariants, and others have been used. One of the important aspects of neural computing must be the ability to retrieve distorted and partial inputs. To illustrate partial input retrieval a set of letters shown in Fig. 16.23(a) were stored in a Hopfield neural network. The positive and negative parts of the memory matrix are given in Fig. 16.23(b) and (c), respectively. If a partial image of A is presented to the Hopfield net, a reconstructed image of A converges by iteration, and is shown in Fig. 16.23(d). Thus, we see that the Hopfield neural network can indeed retrieve partial patterns.

16.3.3 Interpattern Association Neural Network

Although the Hopfield neural network can retrieve erroneous or partial patterns, the construction of the Hopfield neural network is through *intrapattern association*, which ignores the association among the stored exemplars. In other words,

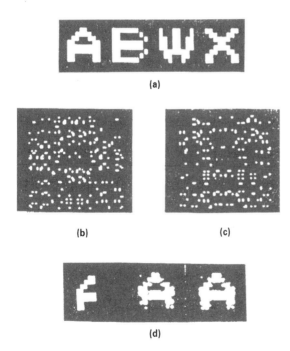

(a)

(b)   (c)

(d)

Figure 16.23 Results from a Hopfield model: (a) training set; (b) and (c) positive and negative IWMs, respectively; and (d) retrieved image.

Hopfield would have a limited storage capacity and it is not effective or even incapable of retrieving similar patterns. One of the alternative approaches is called the *interpattern association* (IPA) neural network. By using simple logic operations, an IPA neural network can be constructed. For example, consider three overlapping patterns given in the Venn diagram shown in Fig. 16.24, where the common and the special subspaces are defined. If one uses the following logic operations, then an IPA neural net can be constructed:

$$I = A \wedge \overline{(B \vee C)},$$

$$II = B \wedge \overline{(A \vee C)}, \qquad V = (B \wedge C) \wedge \overline{A},$$

$$III = C \wedge \overline{(A \vee B)}, \qquad VI = (C \wedge A) \wedge \overline{B},$$

$$IV = (A \wedge B) \wedge \overline{C}, \qquad VII = (A \wedge B \wedge C) \wedge \overline{\phi}.$$

$$(16.80)$$

If the interconnection weights are assigned 1, −1, and 0, for excitory, inhibitory, and null interconnections, then a tristate IPA neural net can be constructed. For instance, in Fig. 16.25(a), pixel 1 is the common pixel among patterns A, B, and C, pixel 2 is the common pixel between A and B, pixel 3 is the common pixel between A and C, whereas pixel 4 is the special pixel, which is also an exlusive pixel with respect to pixel 2. Applying the preceding logic operations, a tristate *neural network* can be constructed as shown in Fig. 16.25(b), and the corresponding IPA IWM is shown in Fig. 16.25(c).

By using the letters B, P, and R as the training set for constructing the IPA IWM shown in Fig. 16.26(a), the positive and negative parts of the IWM are shown in Fig. 16.26(b) and (c). If a noisy pattern of B, (SNR = 7 dB), is presented to the IPA neural network, a retrieved pattern of B is obtained, as shown in Fig. 16.26(e). Although the stored examples B, P, and R are very similar, the retrieved pattern can indeed be extracted by using the IPA model.

For comparison of the IPA and the Hopfield models, we have used an 8 × 8 neuron optical NN for the tests. The training set is the 26 letters lined up in sequence based on their similarities. Figure 16.27 shows the error rate as a function of the number of stored letters. In view of this plot, we see that the Hopfield

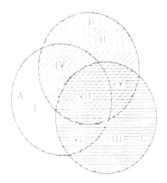

Figure 16.24 Common and special subspaces.

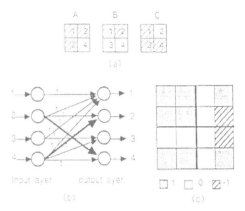

Figure 16.25 Construction of IPA neural network: (a) three reference patterns; (b) one-layer neural net; and (c) IWM.

model becomes unstable to about 4 patterns, whereas the IPA model is quite stable even for 10% input noise, which can retrieve 12 stored letters effectively. As for the noiseless input, the IPA model can in fact produce correct results for all 26 stored letters.

Pattern translation can be accomplished using the heteroassociation IPA. Using similar logic operations among input–output (translation) patterns, a hetero-associative IWM can be constructed. For example, an input–output (translation)

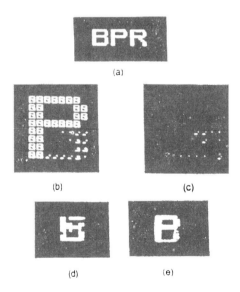

Figure 16.26 IPA neural network: (a) input training set; (b) and (c) positive and negative IWMs; (d) noisy input SNR = 7 dB; and (e) retrieved image.

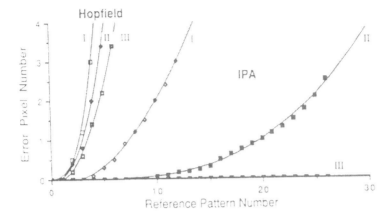

Figure 16.27 Comparison of the IPA and the Hopfield models. Note I, 10% noise level; II, 5% noise level; III, no noise.

training set is given in Fig. 16.28(a). Using the logic operations, a heteroassociation neural net can be constructed, as shown in Fig. 16.28(b), while Fig. 16.28(c) is its IWM. To illustrate the optical implementation, an input–output training set is shown in Fig. 16.29(a). The positive and negative parts of the heteroassociation IWMs are depicted in Fig. 16.29(b). If a partial Arabic numeral 4 is presented to the optical neural net, a translated Chinese numeral is obtained, as shown in Fig. 16.29(c). Thus, the heteroassociation neural net can indeed translate Arabic numerals into Chinese numerals.

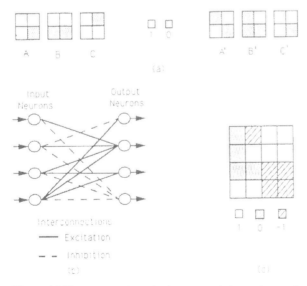

Figure 16.28 Construction of a heteroassociation IPA neural network: (a) input–output training sets; (b) a heteroassociation neural net; and (c) heteroassociation IWM.

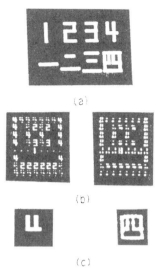

(a)

(b)

(c)

Figure 16.29 Pattern translation: (a) Arabic–Chinese training set; (b) heteroassociation IWM (positive and negative parts); and (c) partial Arabic numeral to the translated Chinese numeral.

16.4 WAVELET TRANSFORM PROCESSING

A major question concerning this section may be asked: What is a wavelet? Why is it interesting for solving signal-processing problems? These are crucial remarks that a signal analyst would want to know. The answer to these questions may be summarized as: wavelet representation is a versatile technique having, very much, physical and mathematical insights with great potential applications to signal and image processing. In other words, wavelets can be viewed as a new basis signals and images representation, which can be used for signal analysis and image synthesis.

16.4.1 Wavelet Transform

For nonstationary signal processing, the natural way to obtain joint time–frequency resolution of a signal is to take the Fourier transform of the signal within a time window function. This transform is known as *short-time Fourier transform* (STFT), where the size of the window is assumed invariant. However, if the size of the window changes as the analyzing frequency changes, then the transform is known as a *wavelet transform* (WT). The expression of the STFT can be written as

$$\text{STFT}(\tau, \omega) = \int x(t)h^*(t - \tau)\exp(-i\omega t)dt, \tag{16.81}$$

where $h(t)$ is an analyzing window function, ω is the analyzing frequency, and τ is an arbitrary time shift. Notice that if $h(t)$ is a Gaussian function, then the transform is also known as the *Gabor transform*. the STFT has been widely used in signal processing, such as time-varying signal analysis and filtering, spectral estimation, signal compression, and others. Usually the STFT offers very good performances for signals having uniform energy distribution within an analyzing window. Thus, the selection of the analyzing window size is critically important for achieving an optimum joint time–frequency resolution. However, the apparent drawback of STFT must be the invariant size of the analyzing window. To ovecome this shortcoming, the WT can be used:

$$WT(\tau, a) = \frac{1}{\sqrt{a}} \int x(t)\psi^*\left(\frac{t-\tau}{a}\right)dt, \tag{16.82}$$

where a is a scaling factor and $\psi(t)$ is called the *mother wavelet*. We note that the shape of $\psi(t)$ shrinks as the scaling factor a decreases, while it dilates as a increases. The shrunken and dilated wavelets are also known as the *daughter wavelets*. Thus, to have a better time resolution, a narrower WT window should be used for higher frequency content. In principle, the WT suffers from the same time–frequency resolution limitation as the STFT; that is, the time resolution and the frequency resolution cannot be resolved simultaneously, as imposed by the following inequity:

$$\Delta t \Delta\omega \geq 2\pi, \tag{16.83}$$

where Δt and $\Delta\omega$ are defined as

$$\Delta t = \frac{\int |h(t)|dt}{|h(0)|}, \text{ and } \Delta\omega = \frac{\int |H(\omega)|d\omega}{|H(0)|}. \tag{16.84}$$

Since window functions having a smaller resolution cell are preferred, the Gaussian window function is the best in the sense of meeting the lower bound of the inequality. However, the Gaussian function lacks either the biorthogonality or the orthogonality, which is the constraint of window functions for perfect signal (or image) reconstruction. We note that perfect reconstruction is one of the objectives for using the STFT and the WT: for example, as applied to nonlinear filtering, image compression, and image synthesis.

Let us begin with the basic definitions of the semicontinuous WT, which is given by

$$WT(\tau, n) = a_0^n \int X(t)\psi^*(a_0^n(t-\tau))dt, \tag{16.85}$$

and its Fourier domain representation is written as

$$WT(\tau, n) = \frac{1}{2\pi} \int X(\omega)\psi^*(a_0^{-n}\omega)\exp(i\omega\tau)d\omega, \tag{16.86}$$

where $a_0 > 1$ is a scaling factor (i.e., $a = a_0^{-n}$), n is an integer, $\psi(t)$ is the mother wavelet, and $\psi(\omega)$ is its Fourier transform. Equation (16.85) is somewhat different from Eq. (16.82), where $1/\sqrt{a}$ (i.e., $a_0^{n/2}$) is used instead of a_0^n (i.e., $1/a$). We note that this modification simplifies the optical implementation of the WT, as will be shown later. Similar to the STFT, the WT can be regarded as a multifilter system in the Fourier domain by which a signal can be decomposed into different spectral bands. Although WT uses narrower and wider band filters to analyze the lower and the higher frequency components, the operation is essentially similar to that of STFT.

To meet the admissibility of WT, $\psi(t)$ has to be a bandpass filter; however, for signals having rich low-frequency contents, a scaling function $\varphi(t)$ to preserve the low-frequency spectrum is needed. The scaling transform of the signal is defined as

$$\mathrm{ST}(\tau) = \int x(t)\phi^*(t - \tau)\mathrm{d}t. \tag{16.87}$$

Thus, the inverse operation of the WT can be written as

$$x(t) = \int \mathrm{ST}(\tau)s_\phi(t - \tau)\mathrm{d}\tau + \sum_{n=-\infty}^{\infty} \int \mathrm{WT}(\tau, n)s_\psi(a_0^n(t - \tau))a_0^n\mathrm{d}\tau, \tag{16.88}$$

and its Fourier domain representation can be shown as

$$x(t) = \frac{1}{2\pi} \int F\{\mathrm{ST}(\tau)\}S_\psi(\omega)\exp(i\omega t)\mathrm{d}\omega + \frac{1}{2\pi} \sum_{n=-\infty}^{\infty} \int F\{\mathrm{WT}(\tau, n)\}$$

$$\tau S_\psi(a_0^{-n}\omega)\exp(i\omega t)\mathrm{d}\omega, \tag{16.89}$$

where $s_\phi(t)$ and $s_\psi(t)$ are the synthesis scaling function and wavelet function, respectively, and $S_\phi(\omega)$ and $S_\psi(\omega)$ are the corresponding Fourier transforms.

If the WT is used for signal or image synthesis, for a perfect reconstruction $\varphi(t)$ and $\psi(t)$ must satisfy the following biorthogonality and orthogonality constraints:

$$\Phi^*(\omega) + \sum_{n=-\infty}^{+\infty} \psi^*(a_0^{-n}\omega) = C, \quad \text{for } s_\phi(t) = \delta(t), \quad \text{and } s_\psi(t) = \delta(t), \tag{16.90}$$

$$|\Phi^*(\omega)|^2 + \sum_{n=-\infty}^{+\infty} |\psi^*(a_0^{-n}\omega)|^2 = C, \quad \text{for } s_\phi(t) = \phi(t), \quad \text{and } s_\psi(t) = \psi(t). \tag{16.91}$$

Similar to STFT, $\psi(\omega)$ that satisfies the biorthogonality constraints is given by

$$\psi(\omega) = \begin{cases} 0, & \omega < \omega_0, \\ \sin^2\left[\dfrac{\pi}{2}v\dfrac{\omega - \omega_0}{\omega_0(a_0 - 1)}\right] & \omega_0 \le \omega \le a_0\omega_0, \\ \cos^2\left[\dfrac{\pi}{2}v\dfrac{\omega - a_0\omega_0}{\omega_0 a_0(a_0 - 1)}\right] & a_0\omega_0 \le \omega \le a_0^2\omega_0, \\ 0, & \omega > a_0^2\omega_0, \end{cases} \tag{16.92}$$

for which the scaling function $\Phi(\omega)$ can be shown as

$$\Phi(\omega) = \cos^2\left[\frac{\pi}{2}v\frac{|\omega|}{(a_0 + 1)\omega_0/2}\right], \quad |\omega| \le (a_0 + 1)\omega_0/2, \tag{16.93}$$

where the function $v(\cdot)$ has the same definition as the STFT shown in Fig. 16.30. Thus, we see that $\psi(\omega)$ forms a biorthogonal wavelet, and the squared-root $\sqrt{\psi(\omega)}$ is known as the *Mayer's wavelet*, which is orthogonal in terms of the constraint of Eq. (16.92).

Figure 16.30 shows a set of Fourier domain wavelets, given by

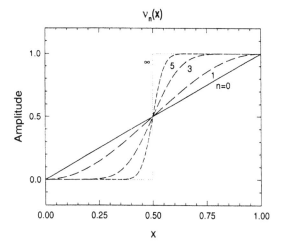

Figure 16.30 Function v as n increases.

$$\psi_1(\omega) = \begin{cases} 1, & 1.5 \le \omega \le 3, \\ 0, & \text{others} \end{cases}$$

$$\psi_2(\omega) = \begin{cases} \sin[(\omega - 1)\pi/2], & 1 \le \omega \le 2 \\ \cos[(\omega - 2)\pi/4], & 2 \le \omega \le 4 \\ 0, & \text{others} \end{cases}$$

$$\psi_3(\omega) = \begin{cases} \sin[(\omega - 1)\pi/2], & 1 \le \omega \le 2 \\ \cos[(\omega - 2)\pi/4], & 2 \le \omega \le 4 \\ 0, & \text{others} \end{cases}$$

$$\psi_4(\omega) = \begin{cases} \exp[-\pi(\omega - 2)^2], & \omega \le 2 \\ \exp[-\pi(\omega - 2)^2/4], & \omega \ge 2, \end{cases}$$

(16.94)

where we assume $a_0 = 2$. By plotting the real parts, as shown in Fig. 16.31, we see that the biorthogonal window $\psi_3(t)$ is an excellent approximation to the Gaussian function $\psi_4(t)$, both in the Fourier and the time domains. Therefore, the wavelet $\psi_3(t)$ has the advantages of having the smaller joint time–frequency resolution and biorthogonality, which simplifies the inversion of the WT. The function $\psi_2(t)$ can be used as an orthogonal wavelet, which has a relatively good joint time–frequency resolution. Nevertheless, wavelet $\psi_1(t)$ is often used, since its Fourier transform is a rectangular form, which is rather convenient for the application in Fourier domain processing.

Although our discussions are limited to the wavelets which have similar forms to the window functions, in practice the WT offers more solutions than the STFT: namely, one can select the wavelets for specific applications.

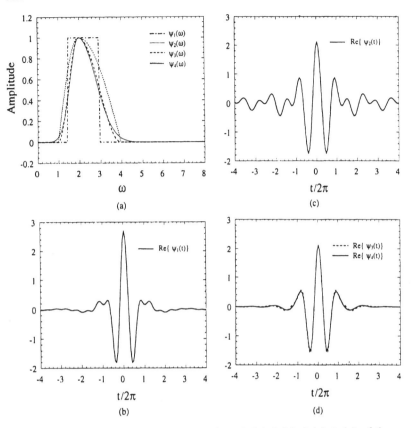

Figure 16.31 (a) Fourier domain representations, $\Psi_1(\omega)$, $\Psi_2(\omega)$, $\Psi_3(\omega)$, $\Psi_4(\omega)$, of the complex-valued wavelets (b) $\Psi_1(t)$, (c) $\Psi_2(t)$, (d) $\Psi_3(t)$ and $\Psi_4(t)$.

16.4.2 Optical Implementations

A couple of possible optical implementations for 1-D WT processing are shown in Fig. 16.32. The architecture of Fig. 16.32(a) is ued for biorthogonal WT, in which we assume that the synthesis function $s(t) = \delta(t)$. For example, let an input signal be displayed on a spatial light modulator (SLM) at P1 and a set of filter banks are placed at the Fourier plane P2. Then WT signals can be obtained in the back focal plane P3. Thus, the reconstructed signal can be obtained at P4, by summing all the WT signals diffracted from the filter banks. We notice that real-time processing can be realized by simply inserting an SLM filter at P3. Let us assume that the desired filter is $F(\tau, n)$; then the reconstructed signal would be

$$x'(t) = ST(\tau) + \Sigma \, WT(\tau, n) F(\tau, n). \tag{16.95}$$

Figure 16.32(b) shows the implementation of the orthogonal WT, in which we assume that the orthogonal wavelets $s(t) = \psi(t)$. Notice that the optical configuration is rather similar to that of Fig. 16.32(a) except the inverse operation. By virtue of the reciprocity principle, the inverse operation can be accomplished by placing a

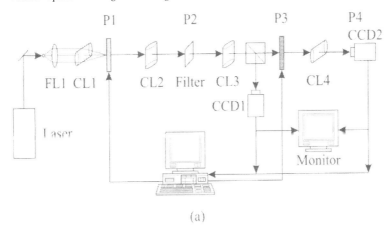

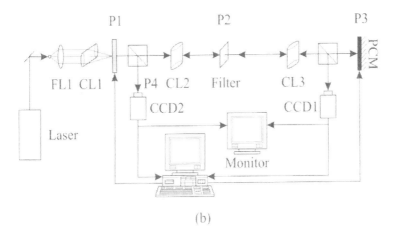

Figure 16.32 Optical implementations for WT: (a) for biorthogonal windows, i.e., $s(t) - \delta(t)$; (b) for orthogonal windows, i.e., $S(t) = \Psi(t)$.

phase conjugate mirror (PCM) behind plane P3. The major advantage of the PCM must be the self-alignment, for which the filter alignment can be avoided. As the return phase conjugate signal, $WT^*(\tau, n)$, arrives at plane P2, it is subjected to the Fourier transformation. By inserting a desired filter at plane P3, we see that real-time processing can indeed be obtained. We note that the filter at the plane P3 would be proportional to $F(\tau, n)^{1/2}$, since the signal has gone through the filter twice. Thus, the reconstructed signal at plane P4 would be

$$
\begin{aligned}
x_r^*(t) = &\int F\{ST^*(\tau)F(\tau)\}\Phi^*(\omega)\exp(i\omega t)d\omega \\
&+ \sum_n \int F\{WT^*(\tau, n)\}_\tau \psi^*(d_0^n \omega \exp(i\omega t)d\omega,
\end{aligned}
\tag{16.96}
$$

which can be observed with a CCD camera. Although the preceding discussion is limited to 1-D signal processing, the basic concepts can be easily extended to 2-D image processing.

For comparison, a set of Fourier domain filter banks, for complex-valued STFT and WT, are shown in Fig. 16.33(a) and (b) respectively. Note that the filter banks for WT are constant Q-filters, by which the bandwidth varies. We have omitted the filters for negative frequency components since the scalegram from the negative frequency components is the same as that from the positive components. A test signal, that includes a chirp and a transient shown at the bottom of Fig. 16.33(c) and (d), is given by

$$x(t) = \sin\left[\frac{\pi}{256}(t - 127)^2\right] + \cos\left[\frac{3\pi}{4}(t - 27)\right]\exp\left[\frac{|(t - 127)|}{2}\right], \quad 0 \le t < 256.$$

$$(16.97)$$

The STFT spectrogram and the WT scalegram are shown at the upper portion of these figures. Although STFT offers a relatively good time–frequency resolution for the chirp signal, it gives rise to a weaker response for the transient which is located at the center. On the other hand, the WT provides a higher time resolution for the

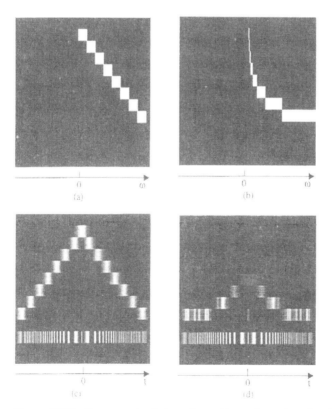

Figure 16.33 Computer simulations, for a chirp and transient signal: (a) filter banks for STFT; (b) filter banks for WT; (c) spectrogram; and (d) scalegram.

transient signal, but offers relatively poorer performance for the uniformly distributed chirp signal.

16.5 COMPUTING WITH OPTICS

Reaffirmation of optics parallelism and the development of picosecond and femtosecond optical switches have thrust optics into the nontraditional area of digital computing. The motivation primarily arises from the need for higher performance general-purpose computers. However, computers with parallelism require complicated interconnections which is difficult to achieve by using wires or microcircuits. Since both parallelism and space interconnection are the inherent traits of optics, it is reasonable to look into the development of a general-purpose optical computer. We shall, in this section restrict ourselves to discussing a few topics where computing can be performed conveniently by optics.

16.5.1 Logic-Based Computing

All optical detectors are sensitive to light intensity: they can be used to represent binary numbers 0 and 1, with dark and bright states. Since 1 (bright) cannot be physically generated from 0s (dark), there are some difficulties that would occur when a logical 1 has to be the output from 0s (e.g., NOR, XNOR, NAND). Nevertheless, shadow casting method can solve these problems, by simply initially encoding 1 and 0 in a dual-rail form.

The shadow-casting logic essentially performs all sixteen Boolean logic functions based on the combination of the NOT, AND, and OR. For example, 1 and 0 are encoded with four cells, as shown in Fig. 16.34, in which the spatially encoded formats A and B are placed in contact, which is equivalent to an AND operation, as shown in Fig. 16.34(b). On the other hand, if the uncoded 1 and 0 are represented by transparent and opaque cells, they provide four AND operations, i.e., AB, A$\overline{\text{B}}$, $\overline{\text{A}}$B,

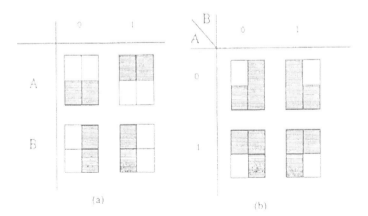

Figure 16.34 (a) Encoded input patterns. (b) Product of the input patterns for the shadow-casting logic array processor.

and $\overline{AB}$, as shown in Fig. 16.34(b). This superposed format is the input-coded image of the optical logic array processor, and is set in the input plane. Four spatially distributed light-emitting diodes (LEDs) are employed to illuminate the encoded input. The shadows from each LED will be cast onto an output screen, as shown in Fig. 16.35. A decoding mask is needed to extract only the true output. The shadow casting is essentially a selective OR operation among AB, $A\overline{B}$, $\overline{A}B$, and $\overline{AB}$. If the on–off states of the LEDs are denoted by α, β, γ, δ (where on is 1, and off is 0), the shadow-casting output can be expressed as follows:

$$G = \alpha(AB) + \beta(A\overline{B}) + \gamma(\overline{A}B) + \delta(\overline{AB}), \tag{16.98}$$

which is the intensity at the overlapping cell. The complete combination for generating the 16 Boolean functions is given in Table 16.1. A schematic diagram of a hybrid-optical logic array processor is depicted in Fig. 16.36, in which the endoded inputs A and B are displayed on LCTV SLM1 and SLM2, respectively, and the SLM3 is employed to determine the values of α, β, γ, and δ. However, the OR operation is performed electronically rather than optically. This has the advantage that no space is needed for shadow-casting NOR. The problems with this system are (1) the encoding operation would slow down the whole process, (2) if the OR operation is performed in parallel, a large number of electronic OR gates and wire intercommunications are required, and if it is performed sequentially, a longer processing time is required. Since all the optic logic processors use encoded inputs, if the coding is done by electronics, the overall processing speed will be substantially reduced. On the other hand, the optical output is also an encoded pattern that requries a decoding mask to obtain only the true output. Although the decoding process is parallel, and thus takes no significant processing time, the decoding mask does change the format of the output. A noncoded shadow-casting logic array that is free from these difficulties is shown in Fig. 16.37. An electronically addressed SLM, such as an LCTV, can be used to write an image format. The negation can be done by rotating the LCTV's analyzer (by 90°) without altering the addressing electronics or software processing. The electronic signal of input A is split into four individual SLMs. Two of them display input A, and the other two display $\overline{A}$. Input B follows a similar procedure to generate two Bs and two $\overline{B}$s. The products of AB, $A\overline{B}$, $\overline{A}B$, and $\overline{AB}$ are straightforwardly obtained by putting two corresponding SLMs up to each other. Finally, beamsplitters combine the AB, $A\overline{B}$, $\overline{A}B$, and $\overline{AB}$. The logic functions are

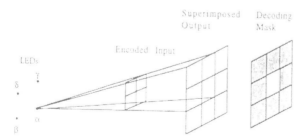

Figure 16.35 A shadow-casting logic array processor. The shadows of the encoded input generated by four laser-emitting diodes (LEDs) are superimposed.

Table 16.1 Generation of Sixteen Boolean Functions

	α	β	Y	δ
F0	0	0	0	0
F1	1	0	0	0
F2	0	1	0	0
F3	1	1	0	0
F4	0	0	1	0
F5	1	0	1	0
F6	0	1	1	0
F7	1	1	1	0
F8	0	0	0	1
F9	1	0	0	1
F10	0	1	0	1
F11	1	1	0	1
F12	0	0	1	1
F13	1	0	1	1
F14	0	1	1	1
F15	1	1	1	1

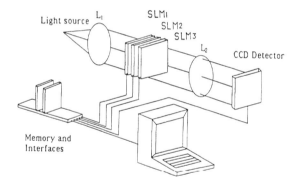

Figure 16.36 Shadow-casting logic processor using cascaded spatial light modulators (SLMs).

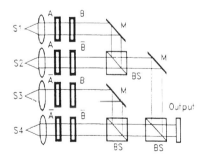

Figure 16.37 A noncoded shadow-casting logic array processor: S, light source; BS, beamsplitter; M, mirror.

controlled by the on–off state (α, β, γ, and δ) of the illuminating light (S1, S2, S3, and S4), as illustrated in Table 16.1.

The implementation of these logic functions using LCTVs is straightfoward, since no electronic signal modification or software data manipulation is required to encode inputs A and B. Although it seems that more SLMs are needed (four times as many), there is no increase in the space–bandwidth product of the system. In the original shadow-casting logic array, four pixels are used to represent a binary number 1 or 0, while a binary number can be represented by only one pixel in the noncoded system. The use of four sets of individual SLMs is to utilize fully the capability of the LCTV to form a negative image format by simply rotating the analyzer by 90°. This method can eliminate the bottleneck of the shadow-casting optical parallel logic array, which is introduced by the coding process.

16.5.2 Matrix–Vector and Matrix–Matrix Processors

The optical matrix–vector multiplier can be implemented as shown in Fig. 16.38. The elements of the vector are entered in parallel by controlling the intensities of N light-emitting diodes (LEDs). Spherical lens L_1 and cylindrical lens L_2 combine to image the LED array horizontally onto the matrix mask M, which consists of $N \times N$ elements. The combination of a cylindrical lens L_3 and a spherical lens L_4 is to collect all the light from a given row and bring it into focus on one detector element that measures the value of one output vector element. Thus, we see that it is essentially the matrix–vector multiplication operation. Furthermore, this configuration can be further simplified by fabricating line-shape LEDs and a line-shape detector array, as depicted in Fig. 16.39. Note that the LED array can be replaced by an SLM with a uniform illumination.

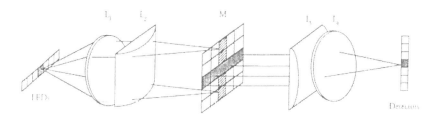

Figure 16.38 Schematic diagram of an optical matrix–vector multiplier.

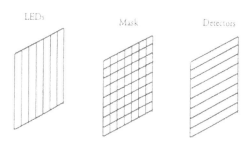

Figure 16.39 Optical matrix–vector multiplier using line-shape LEDs and detectors.

Many important problems can be solved by the iterative multiplication of a vector by a matrix, which includes finding eigenvalues and eigenvectors, solving simultaneous linear equations, computing the discrete Fourier transform, and implementation of neural networks. For example, a neural network consists of several layers of neurons in which two successive neuron layers are connected by an interconnection net. If the neuron structure in a layer is represented by a vector, then the interconnect can be represented by a matrix.

For example, the Hopfield neural network (see Section 16.3.2) uses an associative memory retrieval process, which is essentially a matrix–vector multiplier, as given by

$$
\begin{aligned}
V_i &\rightarrow 1 \text{ if } \sum T_{ij} V_i \geq 0, \\
V_i &\rightarrow 1 \text{ if } \sum T_{ij} V_i < 0,
\end{aligned}
\tag{16.99}
$$

where V_i and V_j are the output and the input binary vectors, respectively, and T_{ij} is the interconnect matrix.

If the matrix is binary, the matrix–vector multiplier becomes a *crossbar switch*. The crossbar switch is a general switching device that can connect any N inputs to any N outputs; this is called *global interconnect*. Crossbar switches are usually not implemented in electronic computers because they would require N^2 individual switches; however, they are used in telephone exchanges. On the other hand, an optical crossbar interconnected signal processor would be very useful for performing fast-Fourier transforms (FFTs), convolution and correlation operations, by taking advantage of the reconfigurability and parallel processing of crossbar interconnect. Also, the optical crossbar switch can be employed to implement a programmable logic array (PLA). The electronic PLA contains a two-level, AND–OR circuit on a single chip. The number of AND and OR gates and their inputs is fixed for a given PLA. A PLA can be used as a read-only memory (ROM) for the implementation of combinational logic.

The matrix–matrix multiplier is a mathematical extension of the matrix–vector multiplier. In contrast, the implementation of the matrix–matrix multiplier requires a more complex optical arrangement. Matrix–matrix multipliers may be needed to change or process matrices that will eventually be used in a matrix–vector multiplier. Matrix–matrix multiplication can be computed by successive outer-product operations as follows:

$$
\begin{bmatrix} a_{11} & a_{12} & a_{13} \\ a_{21} & a_{22} & a_{23} \\ a_{31} & a_{32} & a_{33} \end{bmatrix}
\begin{bmatrix} b_{11} & b_{12} & b_{13} \\ b_{21} & b_{22} & b_{23} \\ b_{31} & b_{32} & b_{33} \end{bmatrix} =
\begin{bmatrix} a_{11} \\ a_{21} \\ a_{31} \end{bmatrix} [b_{11} b_{12} b_{13}] +
\begin{bmatrix} a_{12} \\ a_{22} \\ a_{32} \end{bmatrix} [b_{21} b_{22} b_{23}]
$$

$$
+ \begin{bmatrix} a_{13} \\ a_{23} \\ a_{33} \end{bmatrix} [b_{31} b_{32} b_{33}].
\tag{16.100}
$$

Since the outer product can be expressed as

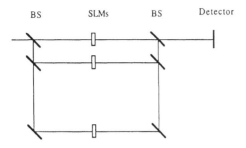

Figure 16.40 Matrix–matrix multiplier based on outer product. BS, beamsplitter; SLMs, spatial light modulators.

$$
\begin{bmatrix} a_{11} \\ a_{21} \\ a_{31} \end{bmatrix} [b_{11} b_{12} b_{13}] = \begin{bmatrix} a_{11}b_{11} & a_{11}b_{12} & a_{11}b_{13} \\ a_{21}b_{11} & a_{21}b_{12} & a_{21}b_{13} \\ a_{31}b_{11} & a_{31}b_{12} & a_{31}b_{13} \end{bmatrix},
\tag{16.101}
$$

it can be obtained by simply butting two SLMs against each other. Equation (16.101) can be realized optically, as shown in Fig. 16.40. Each pair of the SLMs performs the multiplication of the outer products, while beamsplitters perform the addition of the outer products.

Thus we see that the basic operations of matrix–matrix multiplication can be performed by using a pair of SLMs, while the addition is performed by using a beamsplitter. The whole combinational operation can be completed in one cycle. If the addition is performed electronically in a sequential operation, then only a pair of SLMs is required; if the multiplication is performed sequentially then more than one cycle of operation is needed, and the approach is known as *systolic processing*, and is discussed in Section 16.5.3. The obvious advantage of systolic processing is that fewer SLMs are needed. The trade-off is the increase in processing time.

Figure 16.41 illustrates an example of a systolic array matrix operation. Two transmission-type SLMs are placed close together and in registration at the input plane. By successively shifting the A and B systolic array matrix formats into two SLMs, one can obtain the product of matrices A and B with a time-integrating CCD at the output plane. Although only two SLMs are required, the computation time needed for performing an $n \times n$ matrix–matrix multiplication with b-bit numbers would be $(2nb - 1) + (n - 1)$ times that needed by an outer-product processor.

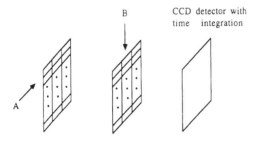

Figure 16.41 Matrix–matrix multiplier based on systolic array.

Figure 16.42 shows an example of a systolic outer-product processing. The optical processor consists of $n \times n = 3 \times 3 = 9$ pieces of SLM, with $b \times b = 5 \times 5$ pixels each. Note that the systolic array representation of matrices A and B differs from the systolic formats described previously. By sequentially shifting the row and column elements of A and B into the SLMs, we can implement the $a_{ij}b_{kl}$ multiplication at each step, with an outer-product operation that has been performed in the 5×5 SLM. The result can be integrated in relation to time with a CCD detector at the output plane. Since more SLMs are employed in parallel, the matrix–matrix multiplication can be completed in fewer steps.

16.5.3 Systolic Processor

The engagement matrix–vector multiplier, in fact a variation of a systolic processor, is illustrated in Fig. 16.43. The components of vector B are shifted into multiplier-added modules starting at time t_0. Subsequent vector components are clocked in contiguously at t_1 for b_2, t_2 for b_3, and so on. At time t_0, b_1 is multiplied with a_{11} in module 1. The resultant b_1a_{11} is retained within the module to be added to the next product. At time t_1, b_1 is shifted to module 2 to multiply with a_{21}. At the same time, b_2 enters module 1 to multiply with a_{12}, which forms the second product of the output vector component. Consequently, module 1 now contains the sum $b_1a_{11} + b_2a_{12}$. This process continues until all the output vector components have been formed. In all, $(2N - 1)$ clock cycles that employ N multiplier-adder modules are required. The main advantage of a systolic processor is that optical matrix–vector multiplication can be performed in the high-accuracy digital mode.

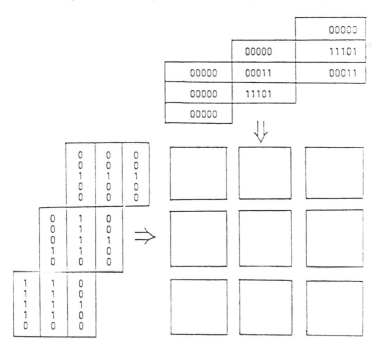

Figure 16.42 Matrix–matrix multiplication based on the systolic outer-product method.

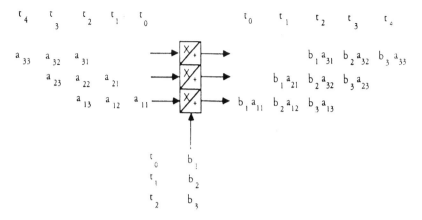

Figure 16.43 Conceptual diagram of an engagement systolic array processor.

A discrete linear transformation (DLT) system can be characterized by an impulse response h_{mn}. The input–output relationship of such a system can be summarized by the following equation:

$$g_m = \sum f_n h_{mn} \tag{16.102}$$

Since the output g_m and the input h_n can be considered vectors, the preceding equation can be represented by a matrix–vector multiplication, where h_{mn} is known as a transform matrix. Thus, the different DLTs would have different matrices. The discrete Fourier transform (DFT) is one of the typical examples for DLT, as given by

$$F_m = (1/N) \sum f_n \exp[-i2\pi mn/N] \tag{16.103}$$

where $n = 0, 1, \ldots, N$, and

$$h_{mn} = \exp[-i2\pi mn/N], \tag{16.104}$$

is also known as the *transform kernel*. To implement the DFT transformation in an optical processor, we present the complex transform matrix with real values, for which the real transform matrices can be written

$$\text{Re}[h_{mn}] = \cos\frac{2\pi mn}{N}, \quad \text{Im}[h_{mn}] = \sin\frac{2\pi mn}{N} \tag{16.105}$$

which are the well-known *discrete cosine transform* (DCT) and the *discrete sine transform* (DST).

The relationship between the real and imaginary parts of an analytic signal can be described by the *Hilbert transform*. The discrete Hilbert transform (DHT) matrix can be written as

$$h_{mn} = \begin{cases} \dfrac{2\sin^2[\pi(m-n)/2]}{\pi(m-n)} & m - n \neq 0, \\ 0, & m - n = 0. \end{cases} \tag{16.106}$$

Another frequently used linear transform is the chirp-Z transform, which can be used to compute the DFT coefficients. The discrete chirp-Z transform (DCZT) matrix can be written as

$$h_{mn} = \exp \frac{i\pi(m-n)^2}{N}. \tag{16.107}$$

Since the DLT can be viewed as the result of a digital matrix–vector multiplication, systolic processing can be used to implement it. By combining the systolic array processing technique and the *two's complement representation*, a DLT can be performed with a digital optical processor. As compared with the analog optical processor, the technique has high accuracy and a low error rate. Also, it is compatible with other digital processors.

Two's complement representation can be applied to improving the accuracy of matrix multiplication. Two's complement numbers provide a binary representation of both positive and negative values, and facilitate subtraction by the same logic

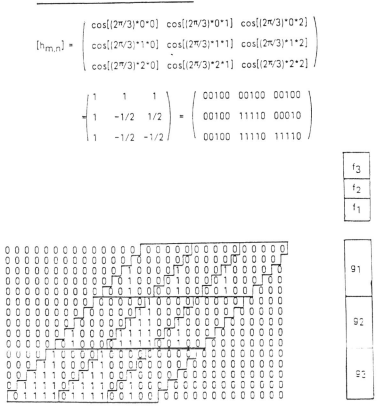

Figure 16.44 Transformation matrix for a discrete cosine transform using two-complement representation.

hardware that is used for addition. In a *b*-bit two's complement number, the most significant bit is the sign. The remaining $b - 1$ bits represent its magnitude. An example for DCT is given in Fig. 16.44 in which we see that the digital transform matrix is encoded in two's complement representation and arranged in the systolic engagement format.

To conclude this chapter we remark that it was not our intention to cover the vast domain of Fourier optics. Instead, we have provided some basic principles and concepts. For further reading, we refer the readers to the list of references as cited at the end of this chapter. Nevertheless, several basic applications of Fourier optics are discussed, including coherent and incoherent light image processing, optical neural networks, wavelet transform, and computing with optics.

REFERENCES

1. Yu, F. T. S. and S. Jutamulia, *Optical Signal Processing, Computing, and Neural Networks*, Wiley-Interscience, New York, 1992.
2. Yu, F. T. S. and X. Yang, *Introduction to Optical Engineering*, Cambridge University Press, Cambridge, 1997.
3. Reynolds, G. O., J. B. Develis, G. B. Parrent, Jr., and B. J. Thompson, *Physical Optics Notebook: Tutorials in Fourier Optics*, SPIE Optical Engineering Press, Bellingham, WA, 1989.
4. Salch, B. E. A. and M. C. Teich, *Fundamentals of Photonics*, Wiley-Interscience, New York, 1991.
5. Yu, F. T. S. and S. Jutamulia (eds), *Optical Pattern Recognition*, Cambridge University Press, Cambridge, 1998.
6. Yu, F. T. S., *Entropy and Information Optics*, Marcel Dekker, New York, 2000.

17

Electro-Optical and Acousto-Optical Devices

MOHAMMAD A. KARIM

The City College of the City University of New York, New York, New York

17.1 INTRODUCTION

In this chapter, we introduce and discuss the basic role of photoconduction, photo-detection, and electro-optic and acousto-optic modulation. These concepts are vital in the design and understanding of various detection, imaging, amplification, modulation, and signal processing systems, many of which are either electronic or hybrid in nature.

Section 17.2 introduces the concepts of photoconduction. It is followed by Section 17.3, where we discuss design, characteristics, and applications of *p–n* and *p–i–n* photodiodes, avalanche photodiodes, vacuum photodiodes, and photomultipliers. The concept of metal oxide semiconductor (MOS) capacitor and its application in the design of charge-coupled device (CCD) structure, MOS read-out scanner, and CCD imager are introduced in Section 17.4. Next, in Section 17.5, we describe cathode-ray tube (CRT) technology and various imaging tube technologies, such as vidicon, plumbicon, and image intensifier. Section 17.6 introduces the physics of electro-optic (EO) modulation. Section 17.7 discusses the working of EO amplitude modulator, EO phase modulator, Pockels read-out optical modulator, Kerr modulator, liquid-crystal light valve, spatial light modulator, and liquid-crystal display devices. Finally, in Section 17.8, the concept of acousto-optical modulation and its application to a few hybrid systems are elaborated.

17.2 PHOTOCONDUCTORS

Almost all semiconductors exhibit a certain degree of photoconductivity. A photoconductor is a simple photodetection device built exclusively of only one type of semiconductor that has a large surface area and two ohmic contact points. In the presence of an energized incident photon, the excited valence band electron of the photoconductor leaves behind a hole in the valence band. Often an extrinsic semiconductor is better suited for the purpose of photoconduction. For example, a far-infrared (IR) sensitive photoconductor can be designed by introducing an acceptor level very close to the valence band or by introducing a donor level very close to the conduction band. Consequently, photoconduction may have two causes. It is caused either by the absorption of photons at the impurity levels in an extrinsic semiconductor or due to band-gap transition in an intrinsic semiconductor. Typically, photoconductors are cooled in order to avoid excessive thermal excitation of carriers.

Figure 17.1 shows a typical photoconductor circuit where R_s is the series resistance. Assume further that the resistance of the photoconductor is larger than R_s, so that most of the bias voltage appears across the photoconductor surface. To guarantee that in the absence of incoming light the number of carriers is a minimum, the operating temperature is maintained sufficiently low. Incident light continues to affect both generation and recombination of carriers until the photoconductor has reached a new equilibrium at higher carrier concentration. A change in carrier density causes a reduction in the photoconductor's resistance. In fact, there are a great many commercial applications of photoconductors where the fractional change in resistance value is significant. In the presence of an electrical field, the generated excess majority carriers drift away from the appropriate terminals.

The absorbed portion of the incoming monochromatic light that falls normally onto the photoconductor is determined in terms of the absorption coefficient α. In the case of extrinsic semiconductors, α is typically very small (1/cm to 10/cm) since the number of available impurity levels is rather small. But in the case of an intrinsic photoconductor, α is large ($\cong 10^4$/cm) in comparison, as the number of available electron states is very large. The absorbed optical power $P_{abs}(y)$ is given by

$$P_{abs}(y) = P_{in}(1 - R)e^{-\alpha y}, \tag{17.1}$$

where P_{in} represents the incoming optical power and R is the surface reflectance of the photoconductor. At steady state, the generation and recombination rates are equal to each other. Consequently,

$$\alpha P_{abs}(y)/h\nu\lambda\omega = \{\alpha P_{in}(1 - R)e^{-\alpha y}\}/h\nu\lambda\omega = p(y)/\tau_p, \tag{17.2}$$
$$= n(y)/\tau_n$$

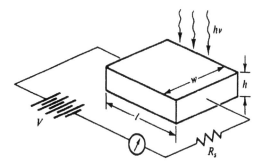

Figure 17.1 A photoconductor circuit.

where τ_n and τ_p are the mean lifetimes, respectively, of electrons and holes; $n(y)$ and $p(y)$ are the carrier densities, respectively, of electrons and holes; and the product lw represents the surface area of the photoconductor.

The total drift current passing through the intrinsic photoconductor is determined using Eq. (17.2):

$$i_s = (\eta_{pc} e P_{in}/h\nu l) E(\mu_n \tau_n + \mu_p \tau_p), \tag{17.3}$$

where the quantum efficiency η_{pc} is defined as

$$\eta_{pc} = \alpha(1 - R)(1 - e^{-\alpha h})/\alpha \tag{17.4}$$

and E is the electric field. In the case of an extrinsic photoconductor, the signal current of Eq. (17.3) reduces to

$$i_s = \begin{cases} (\eta_{pc} e P_{in}/h\nu)[(E\mu_n \tau_n/l)], & n\text{-type} \\ (\eta_{pc} e P_{in}/h\nu)[(E\mu_p \tau_p/l)], & p\text{-type.} \end{cases} \tag{17.5}$$

In either case, the quantity within the square bracket is generally referred to as the photoconductive gain G, as given by

$$G = \begin{cases} \tau_n/\tau_d, & n\text{-type} \\ \tau_p/\tau_d, & p\text{-type,} \end{cases} \tag{17.6}$$

where τ_d is the average carrier drift time or transit time between the two metal contacts, since drift velocity is given by the product of electrical field and carrier mobility. The photoconductive gain generally measures the effective charge transferred through the external circuit due to each of the photoinduced carriers. A high gain is attained by reducing τ_d. It is accomplished by increasing the volume of the photoconductor and decreasing the separation between the metal contacts. Accordingly, photoconductive ribbons are often prepared in the shape of a long ribbon with metal contacts along its edges. However, it should be noted that the carrier lifetime will affect the device response. The current diminishes at a faster rate if light is withdrawn at any instant. Consequently, the device is not sufficiently effective unless the duration of exposure exceeds the carrier lifetime.

Photoconductors are relatively easy to construct, but they are relatively slow in their operation. They require external voltage sources and in most cases are cryogenically cooled to minimize the effect of thermally generated charge carriers. Thus what appears to be a less-expensive detector in the beginning becomes quite expensive when all the peripherals are taken into account. Some of the common photoconductor materials are PbS, CdS, CdSe, InSb, and $Hg_x Cd_{1-x} Te$. While InSb has a good response ($\cong 50\,ns$), CdS and CdSe have poor responses ($\cong 50\,ms$). CdS and CdSe are used for detecting visible light, and both have very high photoconductive gain ($\cong 10^4$).

17.3 PHOTODIODES

In general, a photovoltaic detector consists of a semiconductor junction so that the equilibrium energy bands on the two sides of the junction are shifted relative to one another. If a sufficiently energized photon strikes a junction, it will result in the generation of an electron–hole pair, which in turn results in a current flowing through the wire that connects the two components of the junction. Such a mode of operation, which requires no external bias, is said to be photovoltaic. Photovoltaic detectors have large surface areas so that they can generate a large photocurrent in the absence of a bias. However, they are nonlinear in their responses. Light-powered meters and solar cells are common examples of this type of detector. Interestingly, we may use photovoltaic detectors in the so-called photoconductive mode by applying a reverse bias. When used in this mode, the detector has a rather linear response.

Photodiodes are examples of bipolar semiconductor junctions that are operated in reverse bias. These photodetectors are generally sensitive, easily biased, small in area, and compatible with integrated optics components. Consequently, they are suitable for use in systems like those of fiber communication links. Beyond a certain bias voltage, the detector response is generally improved at higher bias values. The frequency response is often limited by two factors: carrier diffusion time across the depletion layer and junction capacitance of the diode. The carrier diffusion time is generally reduced by increasing the bias voltage but without exceeding the value of the breakdown voltage, whereas the junction capacitance is improved by incorporating an intrinsic layer between the p and n regions as in the p–i–n photodiode. Our attention later in this section is geared toward the details of such semiconductor devices.

As soon as a semiconductor junction is established, electrons start flowing from the n-region to the p-region, leaving behind donor ions, and holes start flowing from the p-region to the n-region, leaving behind acceptor ions. This flow of electrons and holes builds up a depletion layer at the junction. In the absence of any bias, however, the drift and diffusion components of the total current balance each other out. A reverse bias, on the other hand, greatly reduces the diffusion current across the junction but leaves the drift component relatively unaltered.

The photodiode is reverse-biased such that a current (generated by incoming photons) proportional to the number of absorbed photons can be generated. With an optical energy in excess of the band-gap energy, electron and hole pairs are generated in photodiodes. Those pairs that are generated in the depletion region are driven by the electric field through the junction, thus contributing to reverse current. In addition, those pairs that are generated in the bulk regions, but within the diffusion length of the depletion region, diffuse into the depletion region and also contribute to the reverse current. If we neglect the amount of recombination loss in the depletion region, we can estimate photocurrent by

$$I_\lambda = e\eta_{pn}P_{abs}/h\nu, \tag{17.7}$$

where P_{abs} is the absorbed optical power and η_{pn} is the conversion efficiency. The effective conversion efficiency is reduced by the fact that some of the electron–hole pairs of the bulk areas diffuse into the depletion region.

The number of minority holes generated in the n-side but within the diffusion length of the depletion region is $AL_p g$, where g is the generation rate and A is the cross-sectional area of the junction. Similarly, the number of minority electrons generated in the p-side but within the diffusion length of the depletion region is $AL_n g$. The net photocurrent in the reverse-biased photodiode is thus given by

$$I = eA[(L_p p_{n0}/\tau_p) + (L_n p_{p0}/\tau_n)][\exp(eV_A/kT) - 1] - eAg(L_p + L_n), \tag{17.8}$$

where the first term refers to photodiode dark current i_d and the second term accounts for the oppositely directed diffusion photocurrent.

When a photodiode is short-circuited (i.e., when $V_A = 0$), the photocurrent is not any more absent since the current caused solely by the collection of optically generated carriers in the depletion region is nonzero. The equivalent circuit of the photodiode and the corresponding V–I characteristics are shown in Fig. 17.2. If the photodiode is open-circuited (i.e., when $I = 0$) in the presence of illumination, an open-circuit photovoltage $V_A = V_{oc}$ appears across the photodiode terminals.

The magnitude of the open-circuit photovoltage, V_{oc} is found from Eq. (17.8):

$$V_{oc} = (kT/e)\ln[g(L_p + L_n)/\{(L_p p_{n0}/\tau_p) + (L_n p_{p0}/\tau_n)\} + 1]$$
$$= (kT/e)\ln[(I_\lambda/I_0) + 1], \tag{17.9}$$

where $-I_0$ is the peak reverse dark current. The open-circuit voltage is thus a logarithmic function of the incidental optical power P_{abs}. In a symmetrical p–n photodiode, I_λ/I_0 approaches the value g/g_{th}, where $g_{th} = p_{no}/\tau_p$ is the equilibrium thermal generation-recombination rate. Thus, as the minority carrier concentration is increased, g_{th} increases due to the decrease in the carrier lifetime. Consequently, the increase of the minority carrier concentration does not allow V_{oc} to grow indefinitely. In fact V_{oc} is limited by the value of the equilibrium junction potential.

The power delivered to the load is given by

$$P_L = IV_A = I_0 V_A[\exp(eV_A/kT) - 1] - I_\lambda V_A. \tag{17.10}$$

Thus the particular voltage V_{Am}, corresponding to the maximum power transfer, is found by setting the derivative of P_L to zero. Consequently, we obtain

$$[1 + (eV_{Am}/kT)\exp(eV_{Am}/kT) = 1 + (I_\lambda/I_0). \tag{17.11}$$

Since a p–n photodiode is also used as a solar cell for converting the sunlight to electrical energy, we may increase the value of V_{Am} as well as the corresponding photocurrent I_m. Note that the photodiode can achieve a maximum current of I_λ and a maximum voltage of V_{oc}. Often, therefore, the efficiency of the photodiode is measured in terms of the ratio, $(V_{Am}I_m/V_{oc}I_\lambda)$, also known as the fill factor. The present thrust of solar-cell research is thus directed toward increasing this ratio. By cascading thousands of individual solar cells, we can generate an enormous amount of power that is sufficient for energizing orbiting satellites.

The mode of operation where the photodiode circuit of Fig. 17.2(a) is applied across a simple load is photovoltaic. The voltage across the load R_L can be used to evaluate the current flowing through it. However, if the photodiode in conjunction

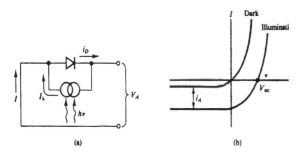

Figure 17.2 (a) Equivalent photodiode circuit and (b) its V–I characteristics.

with a load is subjected to a relatively large external bias, the operation will be referred to as photoconductive. The latter mode is preferred over the photovoltaic because the current flowing through the load is generally large enough and, therefore, approaches I_λ. Thus, while the current-to-optical power relationship in the photovoltaic mode is logarithmic, it is linear in the photoconductive mode. Since the depletion-layer junction-capacitance C_j in an abrupt junction is proportional to $A[(N_d N_a)/\{V_a(N_d + N_a)\}]^{1/2}$, the photovoltaic mode contributes to a larger capacitance and, therefore, to a slower operation. In comparison, the photoconductive photodiode has a faster response.

A cut-off frequency, f_c, is generally defined as the frequency when the capacitive impedance of the photodiode equals the value of the load resistance. Therefore,

$$f_c = \frac{1}{2\pi R_L C_j}.$$
(17.12)

Thus, the junction capacitance has to be decreased to increase the frequency response. This is achieved by decreasing the junction area, by reducing the doping, or by increasing the bias voltage. While there is a physical limit to the smallest junction area, the other two requirements in effect tend to increase depletion width, drift transit time, bulk resistance, none of which is truly desirable.

The p–n photodiode discussed so far has one weakness: the incident optical power is not fully utilized in the optical-to-electrical conversion process because the depletion width of a p–n junction is extremely small. Because of this physical limitation, the p–n photodiodes do not have a desirable response time. This obstacle is overcome by introducing a semi-insulating thick intrinsic (lightly doped) semiconductor layer between its p-layer and its n-layer, as shown in Fig. 17.3. Such especially organized photodiodes are referred to as p–i–n photodiodes.

In p–i–n photodiodes, the separating electric field occupies a large fraction of the device. The wider the thickness of the intrinsic layers the higher the quantum efficiency. High field strength in the intrinsic layer allows the electron–hole pairs to be driven rapidly towards the respective extrinsic regions. However, the carrier transit time is generally proportional to the thickness of the intrinsic layer. Accordingly, there is a design compromise between the expected quantum efficiency and the desirable response time. For a typical p–n photodiode, the response time is in the order of 10^{-11} s, whereas that for a p–i–n photodiode is about 10^{-9} s. The quantum efficiency of a p–i–n photodiode can be anywhere in the range of 50%

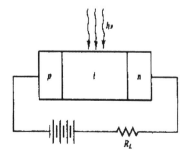

Figure 17.3 Reverse-biased p–i–n photodiode.

through 90%. Usually, indirect band-gap semiconductors are preferred over direct band-gap semiconductors as photodiode materials because otherwise there is a significant conversion loss due to surface recombination. Indirect band-gap materials engage photons to conserve momentum during the transfer. A p–i–n configuration eliminates part of this problem because of its longer absorption length.

Si photodiodes (having a maximum quantum efficiency of 90% at 0.9 μm) are used mostly in the wavelength region below 1 μm, whereas Ge photodiodes (having a maximum quantum efficiency of 50% at 1.35 μm) are preferred in the ranges above 1 μm. In addition to the single-element semiconductor photodiodes, there are many ternary (e.g., InGaAs, HgCdTe, and AlGaSb) photodiodes that are commercially produced.

Photodiodes have proven to be very successful in their applications in the background-limited photodetection. However, photodiodes lack internal gain and in many cases require an amplifier to provide noticeable signal currents. The avalanche photodiode (APD) is a specific photodiode that makes use of the avalanche phenomenon. By adjusting the bias voltage to a level where it is on the verge of breakdown, we can accelerate the photogenerated carriers. The accelerated carriers, in turn, produce additional carriers by collision ionization.

Avalanche gain is generally dependent on impact ionization encountered in the regions having sufficiently high electric field. This gain is achieved by subjecting the reverse-biased semiconductor junction to voltage below its breakdown field ($= 10^5$ V/cm). Electrons and holes thereby acquire sufficient kinetic energy to collide inelastically with a bound electron and ionize it generating an extra electron–hole pair. These extra carriers, in turn, may have sufficient energy to cause further ionization until an avalanche of carriers has resulted. Such a cumulative avalanche process is normally represented by a multiplication factor M that turns out to be an exponential function of the bias. Gains of up to 1000 can be realized in this way. This makes an APD that competes strongly with another high-gain photodetector device, known as a photomultiplier tube, in the red and near-infrared.

The probability that carrier ionization will occur depends primarily on the electric field in the depletion layer. Again, since the electric field in the depletion layers is a function of position, the ionization coefficients, α and β, respectively, for the electrons and the holes, turn out also to be functions of position. The ionization coefficients are particularly low at lower values of electric field, as shown in Fig. 17.4, for the case of silicon.

Consider the reverse-biased p–n junction of depletion width shown in Fig. 17.5. The entering hole current $I_p(0)$ increases as it travels toward the p-side, and the entering electron current $I_n(W)$ increases as it travels toward the n-side. In addition, hole and electron currents due to generation in the depletion layer also move in their respective directions. Thus, for the total hole and electron currents we can write

$$\frac{dI_p(x)}{dx} = \alpha(x)I_n(x) + \beta(x)I_p(x) + g(x),$$
$$-\frac{dI_n(x)}{dx} = \alpha(x)I_n(x) + \beta(x)I_p(x) + g(x),$$

(17.13)

where $g(x)$ is the rate per unit length with which the pairs are generated thermally and/or optically. With first of Eq. (17.13) integrated from $x = 0$ to $x = x$ and similarly, the second of Eq. (17.13) integrated from $x = x$ to $x = W$ we obtain

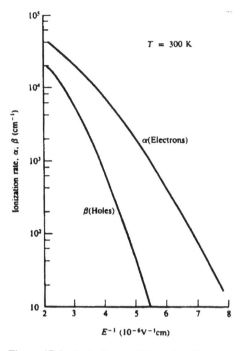

Figure 17.4 Ionization coefficients for silicon at 300°K.

$$I_p(x) - I_n(0) = \int_0^x [\alpha(x) - \beta(x)]I_n(x)\mathrm{d}x + I \int_0^x \beta(x)\mathrm{d}x + \int_0^x g(x)\mathrm{d}x \qquad (17.14a)$$

and

$$-I_n(W) + I_n(x) = \int_x^w [\alpha(x) - \beta(x)]I_n(x)\mathrm{d}x + I \int_x^w \beta(x)\mathrm{d}x + \int_x^w g(x)\mathrm{d}x \qquad (17.14b)$$

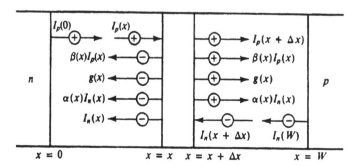

Figure 17.5 Avalanche in a reverse-biased *p–n* junction.

where the sum $I \cong I_n(x) + I_p(x)$ is independent of position. Note, however, that I is equivalent to the saturation current I_0. Adding Eqs (17.14a) and (17.14b), we obtain

$$I = [(I_0 + I_g) + \text{Int}(\{\alpha(x) - \beta(x)\}I_n(x))]/[1 - \text{Int}(\beta(x))], \tag{17.15}$$

where I_g represents the total generation current and $\text{Int}(\xi)$ is the integral of ξ with respect to x when evaluated from $x = 0$ to $x = W$.

For a very special case when $\alpha(x) = \beta(x)$, the total current is given by

$$I = M(I_0 + I_g), \tag{17.16}$$

where M is the avalanche multiplication factor as defined by

$$M = 1/[1 - \text{Int}(\alpha(x))] = 1/(1 - \delta). \tag{17.17}$$

Ideally speaking, the avalanche condition is thus given by

$$\text{Int}(\alpha(x)) = 1, \tag{17.18}$$

when M becomes infinite. In most practical cases, electron and hole coefficients are not equal and these coefficients vary with the electric field. A practical avalanche photodiode is thus referred to by its ionization rate ratio, k ($\equiv \beta/\alpha$). We can then arrive at an expression for M after going through some extremely cumbersome mathematics and multiple assumptions:

$$M = \frac{k - 1}{k - e^{(k-1)\delta}}. \tag{17.19}$$

We may note from Fig. 17.6 that for most electrical fields of interest, k is negligible. Thus, the avalanche multiplication factor reduces to

$$M \approx e^\delta. \tag{17.20}$$

When $k = 0$, the gain increases exponentially with δ, but it does not necessarily become infinite. As shown in Fig. 17.6, with k approaching unity, the gain approaches infinity at a still smaller value of the field.

It is appropriate to note that changing the level of doping can easily alter the electric field. The ionization coefficient α is often given by

$$\alpha = Ae^{-B/|\varepsilon|}, \tag{17.21}$$

where A and B are material constants and ε is the electric field in terms of doping level. For silicon, A and B are, respectively, 9×10^5/cm and 1.8×10^6 V/cm. At a gain of 100 when $k = 0.01$, for example, a 0.5% alteration in the doping changes the gain would have changed by almost 320% if $\alpha = \beta$. The choices of k and doping are, therefore, critical in the design of an APD. APDs are meant for use with small signals, and they require special power supplies to maintain them in their avalanche mode.

The phototransistor, like APD, is a detector that exhibits current gain. It can be regarded as a combination of a simple photodiode and a transistor. Phototransistors are photoconductive bipolar transistors that may or may not have base lead. Light is generally absorbed in the base region. A p–n–p phototransistor is shown in Fig. 17.7. When there is no light, no current flows because there is no base control current. Upon illumination, holes that are excited in the base diffuse out leaving behind an overall negative charge of the excess base electrons that are

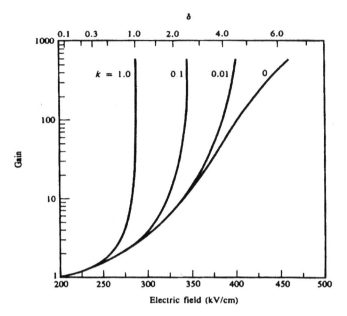

Figure 17.6   Gain versus electric field in an APD.

neutralized by recombination. In the photodiode, a much larger current flows through the device. Thus, while the phototransistor works very much like the photodiode, it amplifies the photogenerated current. Also, a longer recombination time for the excess base electrons contributes to higher gain.

For the phototransistors,

$$I_E = I_C + I_B, \tag{17.22}$$

where I_E, I_C, and I_B are emitter, collector, and base currents, respectively. In the presence of illumination, the base current is given by $\eta I_{abs} A e\lambda/hc$, where A is the junction area, η is the internal quantum efficiency, and I_{abs} is the intensity of the absorbed light. The collector current I_C has two components: (a) the standard diodes reverse saturation current, I_{CBO} and (b) the portion of the emitter current αI_E that crosses into the collector where $\alpha < 1$. The leakage current I_{CBO} corresponds to the collector current at the edge of the cutoff when $I_E = 0$. Thus,

$$I_E = (I_B + I_{CBO})\left[1 + \frac{\alpha}{1 - \alpha}\right]. \tag{17.23}$$

The ratio $\alpha/1 - \alpha$ is an active region performance parameter of a phototransistor. This ratio is usually in the order of $\sim 10^2$. In the absence of light, the current flowing in a phototransistor is $I_{CBO}[1 + \{\alpha(1 - \alpha)\}]$, which is much larger than that in a photodiode under similar (dark) conditions. When illuminated, phototransistor current approaches $I_B[1 + \{\alpha(1 - \alpha)\}]$, thus contributing to significant gain like that of an APD. The only limitation of the phototransistor happens to be its response time, which is about $5\,\mu s$, whereas that in a photodiode is on the order of $0.01\,\mu s$.

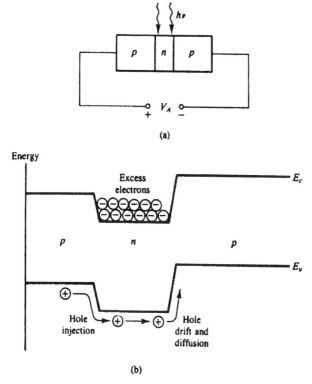

Figure 17.7 (a) A biased phototransistor and (b) its energy diagram.

Electrons may be emitted when light of an appropriate frequency v strikes the surface of solids. Such light-emitted solids are called photocathodes. The minimum energy necessary for the emission of an electron is referred to as the work function ϕ of the solid. In the specific case of semiconductors, electron affinity (energy difference between the vacuum level and E_c) plays the role of work function. The behavior of an electron in solids is like that of an electron in a finite potential well, where the difference between the highest occupied (bound) level and vacuum (free) level is ϕ, as shown in Fig. 17.8(a). The Fermi level is equivalent to the highest possible bound energy. Thus, the kinetic energy E of an emitted electron is given by

$$E = hv - \phi. \tag{17.24}$$

Since electrons reside at or below the Fermi level, E corresponds to the maximum possible kinetic energy. The emission of an electron thus requires a minimum of $\phi \, (= hv)$ photoenergy. However, in case of a semiconductor, this minimum energy is equivalent to $E_g + E_a$, where E_a is the electron affinity energy. Often it may become necessary to reduce the value of E_a, which is accomplished by making the semiconductor surface p-type. The band bending at the surface results in a downward shift of the conduction band by an amount E_b, as shown in Fig. 17.8(b). Consequently, the effective electron affinity becomes

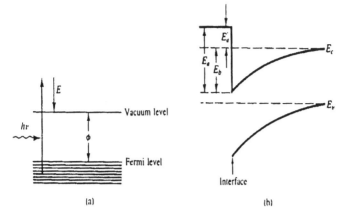

Figure 17.8 Energy level diagram in (a) a solid–vacuum interface and (b) a band-bended semiconductor–vaccuum interface.

$$E'_a = E_a - E_b. \tag{17.25}$$

In certain semiconductors – those referred to as having negative electron affinity – E_b exceeds E_a. Semiconductors such as these are used for infrared photocathodes.

The vacuum photodiode shown in Fig. 17.9 is a quantum detector designed by placing a photocathode along with another electrode (referred to as an anode) within a vacuum tube. In prctice, the photocathode consists of a semicylindrical surface while the anode is placed along the cylindrical axis. When an optical energy in excess of the work function illuminates the photocathode, current begins to flow in the circuit. When the bias voltage V_A is large enough ($\cong 100$ V), the emitted electrons are collected at the anode. When optical energy falls below the work function level, current ceases to exist, irrespective of the bias voltage. For efficient collection of elcctrons, the distance between the anode and the photodiode is kept to a minimum by making sure that the associated capacitance value remains reasonable. Often the anode is made of highly grid-like wires so as not to impede the incoming optical energy. In comparison, solid-state photodetectors are not only smaller, faster, and less power consuming but also more sensitive. Consequently, vacuum photodiodes are used only when the incoming optical energy is more than a certain maximum that may otherwise damage the solid-state photodetectors.

The characteristic curve of a vacuum phototube shows that the photocurrent for a given illumination is invariant above the saturation voltage. The saturation voltage is mildly wavelength- and illumination-sensitive. Since the operating voltage of a phototube is usually larger than the saturation voltage, minor fluctuations in the supply voltage do not cause any discrepancy in the phototube's performance. An important feature of a phototube is that the photocurrent varies linearly with light flux. A slight departure from linearity occurs at high enough flux values and is caused by the space-charge effects. This nonlinearity is avoided by using a large anode-to-photocathode voltage. In practice, the flux level sets a lower limit on the value of the load resistance. The load, used to produce a usable signal voltage, in turn, sets a lower limit on the time constant.

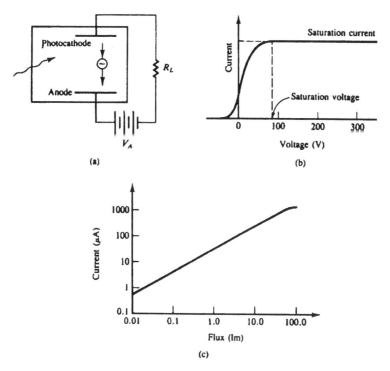

Figure 17.9 A vacuum photodiode: (a) circuit; (b) characteristic curve; and (c) current versus light flux.

Gas-filled phototubes are identical to vacuum phototubes except that they contain approximately 0.1 mm of an inert gas. The inert gas provides a noise-free amplification ($\cong$ 5–10) by means of ionization of the gas molecules. However, inert gases have poor frequency responses. Again the response, which is basically non-linear, is a function of the applied voltage. The phototubes are, therefore, used in applications where the frequency response is not critical.

Photoemissive tube technology is used to develop an alternative but quite popular high-gain device known as a photomultiplier. In a photomultiplier tube (PMT), the photoelectrons are accelerated through a series of anodes (referred to as dynodes) housed in the same envelope; these dynodes are maintained at successively higher potentials. A photoelectron emitted from the photocathode is attracted to the first dynode because of the potential difference. Depending on the energy of the incident electron and the nature of the dynode surface, secondary electrons are emitted upon impact at the first dynode. Each of these secondary electrons produces more elecrons at the next dynode, and so on, until the elecrons from the last dynode are collected at the anode. The dynodes are constructed from materials that, on average, emit δ (> 1) electrons for each of the incident electrons. One such photomultiplier is shown in Fig. 17.10(a), where δ is a function of the interdynode voltage. When the PMT is provided with N such dynodes, the total current amplification factor is given by

$$G = (i_{\text{out}}/i_{\text{in}})\delta^{n}. \tag{17.26}$$

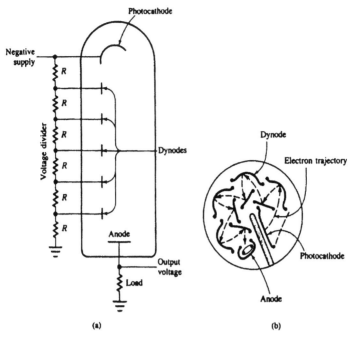

Figure 17.10 A photomultiplier tube: (a) schematic of a five-stage PMT and (b) focusing dynode structure.

Thus, with fewer than 10 dynodes and $\delta < 5$, the gain can easily approach 10^6.

The problems of a PMT are quite the same as those of a vacuum photodiode. However, a PMT is undoubtedly more sensitive. The response of a PMT is comparatively slower since electrons have to move through a longer distance. In addition, there is a finite spread in the transit time because all of the electrons may not have identical velocities and trajectories. This transit-time spread is often reduced, not by reducing the number of dynodes but by increasing the value of δ. However, it must be noted that for the most photocathode materials, the maximum wavelength of incoming light is permitted to be about 1200 nm. Thus, for the detection of longer wavelength radiation, a solid-state detector is preferred. PMTs are commonly operated with $\approx 10^2$ V between the dynodes, which is advantageous because the overall gain of the tube may be varied over a wide range by means of a relatively small voltage adjustment. But, at the same time, it is also disadvantageous because the voltage supply for the PMT must be extremely stable for the calibration to be reliable. We can show that in an N-stage PMT operating at an overall voltage V, a fluctuation ΔV in the voltage produces a change ΔG in the gain G such that

$$\Delta g = GN \frac{\Delta V}{V}. \tag{17.27}$$

Consequently, a 1% fluctuation in a 10-stage PMT will cause a 10% change in the gain.

Different types of PMTs are distinguishable by their geometrical arrangement of dynodes. In particular, the focusing-type PMT, as shown in Fig. 17.10(b), employs electrostatic focusing between the adjacent dynodes and thereby reduces the spread in the transit time. These PMTs are, however, somewhat more noisy and unstable than the unfocused types. Like phototubes, PMTs have an exceptionally linear response.

It is appropriate to introduce a solid-state equivalent of PMT, known as the staircase avalanche photodiode (SAPD) that has been added lately to the list of photodetectors. The noise is an APD increases with the increase in the ratio of the ionization coefficient $k\ (\equiv \beta/\alpha)$. On the other hand, a high k is required for a higher gain. An SAPD provides a suitable solution to this apparent anomaly by incorporating PMT-like stages in the solid-state APDs. An unbiased SAPD consists of a graded-gap multilayer material (almost intrinsic) as shown in Fig. 17.11(a). Each dynode-like stage is linearly graded in composition from a low band-gap value E_{gl} to a high band-gap value E_{gh}. The materials are chosen so that the conduction band drops ΔE_c at the end of each stage equals or just exceeds the electron ionization energy. Note, however, that ΔE_c is much larger than the valence-band rise ΔE_v. Consequently, only electrons contribute to the impact ionization provided the SAPD, is biased as shown in Fig. 17.11(b).

A photoelectron generated next to p^+-contact drifts toward the first conduction band under the influence of the bias field and the grading field (given by $\Delta E_c/l$, where l is the width of each step). But this field value is not large enough for the electrons to impact ionize. In this device, only the bias field is responsible for the hole-initiated ionization. The actual impact ionization process in each stage occurs at the very end of the step when the conduction-band discontinuity undergoes a ΔE_c change. The total SAPD gain becomes $(2 - f)^N$, where N is the number of stages and f is the average fraction of electrons that do not impact ionize in each of the stages. The critical bias SAPD field just exceeds $\Delta E_c/l$ so as to provide the electrons with necessary drift through l but not impact ionize.

17.4 CHARGE-COUPLED IMAGERS

An important solid-state photodetecting device is the charge-coupled imager, which is composed of a closely spaced array of charge-coupled devices (CCDs) arranged

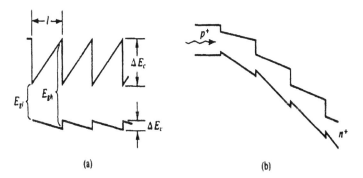

Figure 17.11 Staircase APD: (a) unbiased and (b) biased.

in the form of a register. Each of the CCD units is provided with a storage potential well and is, therefore, able to collect photogenerated minority carriers. The collected charges are shifted down the array and converted into equivalent current or voltage at the output terminal. To understand the overall function of such repetitive storage and transfer of charge packets, consider the metal oxide semiconductor (MOS) structure of Fig. 17.12, where the metal electrode and the *p*-type semiconductor are separated by a thin SiO_2 layer of width x_0 and dielectric constant K_0. Silicon nitride, Si_3N_4, is also used for the insulating layer. The capacitance of such a structure depends on the voltage between the metal plate and the semiconductor.

In thermal equilibrium, the Fermi level is constant all throughout the device. For simplicity, we may assume (a) that the work function difference between the metal and the semiconductor is zero and (b) that there is no charge accumulated in the insulator or at the junction between the insulator and the semiconductor. Consequently, the device may be considered to have no built-in potential.

A biased MOS capacitor results in two space-charge regions by the displacement of mobile carriers, as shown in Fig. 17.13. The total bias voltage V_G applied at the gate input G is shared between the oxide layer and the semiconductor surface, whereas there is only a neglible voltage across the metal plate. Under reverse bias the surface potential gives rise to an upward bend in the energy diagram, as shown in Fig. 17.13(a). At the edges E_i–E_F becomes comparatively larger, thus resulting in a higher hole density at the surface than that within the bulk region. This condition generally increases the surface conductivity. Figure 17.13(b) shows the forward-biased case, where a decrease of $E_i - E_F$ at the edges causes a depletion of holes at the semiconductor surface. The total charge per unit area in the bulk semiconductor is given by

$$Q_B = -eN_A x_d, \tag{17.28}$$

where x_d is the width of the depletion layer. Using the depletion approximation in Poisson's equation, we can arrive at the potential within the semiconductor as

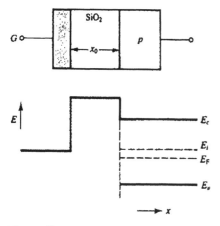

Figure 17.12 MOS capacitor and its unbiased energy-band diagram.

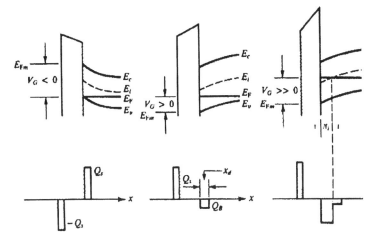

Figure 17.13 Energy-band diagram and charge distribution of an unbiased MOS capacitor: (a) $V_G < 0$, (b) $V_G > 0$, and (c) $V_G \gg 0$.

$$V_s(x) = V_s(0)\left[1 - \frac{x}{x_d}\right]^2,$$

(17.29)

where

$$V_s(0) = \frac{eN_A}{2K_s\varepsilon_0}x_d^2$$

(17.30)

is often referred to as the surface potential. The voltage characteristic is similar to that of a step junction having a highly doped p-side. Note, however, that as the bias voltage V_G is increased further, the band bending could result in a crossover of E_i and E_F within the semiconductor, as shown in Fig. 17.13(c). Consequently, the carrier depletion gives rise to an extreme case of carrier inversion whereby electrons are generated at the junction and holes are generated inside the semiconductor with two regions being separated by the crossover point. Therefore, a p–n junction is induced under the metal electrode. The effect of the gate voltage is to remove the majority carriers from the semiconductor region that is closest to the gate and introduce a potential well. Absorbed photons contribute to the freeing of minority carriers that are collected in the well. The resulting output signal corresponds to the photoinduced charge.

If the semiconductor were approximated as a borderline conductor, the metal–semiconductor structure could be envisioned as a parallel-plate capacitor with the oxide layer working as its dielectric material. However, in forward bias, the MOS structure is modeled by incorporating an additional capacitor in series with the oxide capacitor to accommodate the presence of a surface space-charge layer in the semiconductor. The overall MOS capacitance c is thus given by

$$\frac{1}{c} = \frac{1}{c_0} + \frac{1}{c_s},$$

(17.31)

where

$$c_0 = \frac{K_0 \varepsilon_0}{x_0} \qquad (17.32a)$$

and

$$c_s = \frac{K_s \varepsilon_0}{x_d}. \qquad (17.32b)$$

Neglecting the voltage drop in the metal plate, the forward bias can be expressed as

$$V_G = V_s(0) - \frac{Q_s}{c_0} \qquad (17.33)$$

where Q_s is the density of induced charge in the semiconductor surface region and $V_s(0)$ is the surface potential. The gradient of the surface potential generally determines the minority carrier movements. The depth of the potential well is often decreased either by decreasing the oxide capacitance – i.e., by increasing the oxide thickness – or by increasing the doping level of the p-type material.

Sufficient forward bias may eventually induce an inversion layer. With the passing of time, electrons accumulate at the oxide–semiconductor junction, and a saturation condition is reached when the electron drift current arriving at the junction counterbalances the electron diffusion current leaving the junction. The time required to reach this saturation condition is referred to as the thermal relaxation time. The net flow of electrons is directed toward the junction prior to from the thermal-relaxation time, whereas the net flow of electrons is directed away from the junction after the thermal-relaxation time has elapsed. Since there was no inversion layer prior to the saturation, the induced charge Q_s is obtained by summing A_B and the externally introduced charge Q_e. Equations (17.28) and (17.29) can be incorporated into Eq. (17.33) to give surface potential as

$$V_s(0) = V_G - \frac{Q_e}{c_0} + \frac{eK_s\varepsilon_0}{c_0^2}\left[1 - \left\{1 + \frac{2c_0^2\left(V_G - \frac{Q_e}{c_0}\right)}{eK_s\varepsilon_0 N_A}\right\}^{1/2}\right]. \qquad (17.34)$$

The depth of the potential well x_d is often evaluated using Eqs (17.24) and (17.29). The value of x_d is used in turn to evaluate c_s using Eq. (17.32b) and, consequently, we can determine the overall MOS capacitance as

$$c = \frac{c_0}{\left[1 + \frac{2c_0^2}{eN_A K_s\varepsilon_0}V_G\right]^{1/2}}. \qquad (17.35)$$

The MOS capacitor in effect serves as a storage element for some period of time prior to reaching the saturation point.

A CCD structure formed by cascading an array of MOS capacitors, as shown in Fig. 17.14(a), is often referred to as a surface channel charge-coupled device (SCCD). Basically, the voltage pulses are supplied in three lines, each connected to every third gate input (and consequently this CCD is called a three-phase CCD). In the beginning, G_1 gates are turned on, resulting in an accumulation and storage of charge under the gates. This step is followed by a step whereby G_2 is turned on, thus resulting in a charge equalization step across two-thirds of each cell.

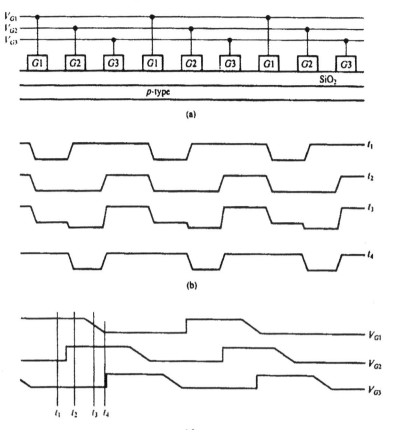

Figure 17.14　(a) CCD structure; (b) potential wells at different times; and (c) timing diagram.

Subsequently G_1 is turned off, resulting in a complete transfer of all charges to the middle one-third of each cell. This process is repeated to transfer charge to the last one-third of the CCD cell. Consequently, after a full cycle of clock voltages has been completed, the charge packets shift to the right by one cell, as illuminated in Fig. 17.14(b) and (c). When the CCD structure is formed using an array of photosensors, charge packets proportional to light intensity are formed and these packets are shifted to a detector for the final readout.

The CCD signal readout is also accomplished by using either two-phase or four-phase clocking schemes. In each of the cases, however, transfer of charges is accomplished by means of the sequentially applied clock pulses. There are three phenomena that enhance the transfer of charges in the SCCD: (a) self-induced drift; (b) thermal diffusion; and (c) fringe-field drift. The self-induced drift, responsible for most of the transfer, is essentially a repulsion effect between the like charges. The thermal diffusion component makes up for most of the remaining signal charge. It can be shown that for most materials the thermal time constant is longer than self-induced time constant. The upper frequency limit for switching operations is thus determined by the thermal time constant. For the SCCDs, this upper limit can be in the order of 10 MHz. The fringe-field drift is determined by the spacing of the electrodes and results in a smoothing out of the transitional potential fields. This third effect is responsible for the transfer of the final few signal electrons.

Figure 17.15(a) shows a system of MOS transistors along with a photodiode array, both of which can be embedded under the same monolithic structure. The system is able to perform sequential readout. A voltage pattern can be generated from the shift register so as to turn on only one transistor at a time. The switching voltage is shifted serially around all diodes. This scheme can also be extended to two

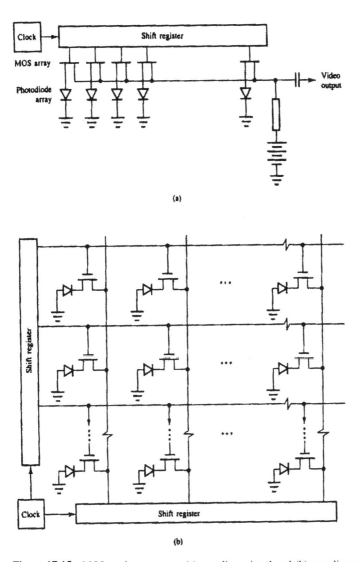

Figure 17.15 MOS readout scanner: (a) one-dimensional and (b) two-dimensional.

dimensions, as shown in Fig. 17.15(b), where one row is switched on and all columns are then scanned serially. The process is repeated for the remaining rows until all the photodiodes have experienced the scanning.

The primary item that hinders transfer of charge is surface-state trapping that occurs along the semiconductor–oxide interface. These trapping energy levels are introduced by nonuniformities in the interface. These energy levels tend to trap and re-emit electrons at a rate that is a function of clocking frequency and their positions relative to the Fermi level. Transferring a constant amount of charge in each of the CCD wells reduces this hindrance of the transfer of charge. This charge fills most of the trapping states, as a result of which interaction of trapping levels and charge is minimized.

A popular method used to control the problem of surface trapping is accomplished by having what is known as a buried channel CCD (BCCD). It involves implementing a thin secondary layer that is opposite polarity to that of the substrate material along the oxide surface. The fringe fields are much smoother in a BCCD than in an SCCD, but the well depth is smaller. Hence, a BCCD can switch information at a faster rate, but it cannot hold as much signal. Switching speeds of up to 350 MHz are not uncommon for BCCDs.

In applications where a semiconductor depletion region is formed, electron–hole pairs are generated due to the thermal vibration of the crystal lattice at any temperature above 0 K. This generation of carriers constitutes a dark current level and determines the minimum frequency with which the transfer mechanism can occur. The time taken by a poential well to fill up with dark electrons can be quite long in some of CCDs. There are two basic types of CCD imagers: the line imager and the area imager. In the line imager, charge packets are accumulated and shifted in one direction via one or two parallel CCD shift registers, as shown in Fig. 17.16(a), where the CCD register is indicated by the shaded regions. The two basic types of CCD area imagers are shown in Figs 17.16(b) and (c).

In the interline transfer CCD (ITCCD), photocells are introduced between the vertical CCD shift registers. Polysilicon strips are placed vertically over each line of the photocells to provide shielding. During an integration period (referred to as one frame), all of the cells are switched with a positive voltage. The ITCCD is obtained by extending the line imager to a two-dimensional matrix of line imagers in parallel. The outputs of the line imagers in parallel are fed into a single-output CCD register. For the case of the frame transfer (CCD (FTCCD), the sensor is divided into two halves: a photosensitive section and a temporary storage section. The charge packets of the photosensing array are transferred over to the temporary storage array as frame of picture. Subsequently, the information is shifted down one-by-one to the output register and is then shifted horizontally.

Besides CCDs, a different family of MOS charge transfer devices, referred to as charge-injection devices (CIDs), can also be an integral part of the focal plane array. CIDs involve exactly the same mechanism for detecting photons as CCDs. They differ only in the methods used for reading out the photoinduced charges. The mechanism of charge injection is illustrated in Fig. 17.17, where the CID consists of an *n*-substrate, for example. Application of a negative gate voltage causes the collection of photoinduced minority carriers in the potential well adjacent to the semiconductor–oxide interface. This accumulation of charge is directly proportional to the incident optical irradiance. Once the gate voltage is withdrawn, the potential

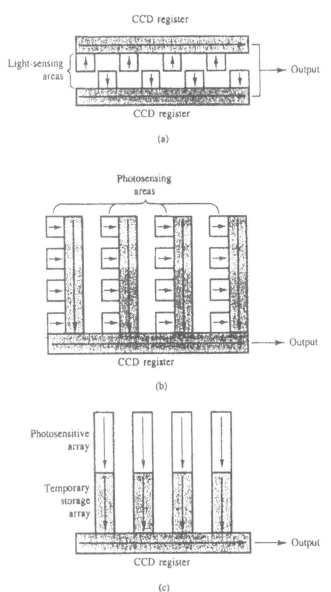

Figure 17.16 CCD imager: (a) line, (b) interline transfer, and (c) frame transfer.

well dissipates and the minority carriers are injected promptly into the substrate, resulting in current flow in the circuit.

Because of the serial nature of the CCDs, optical input signals cause the resulting charge to spill over into adjacent cells. This effect, referred to as blooming, causes the image to appear larger than its actual magnitude. In comparison, CIDs are basically *x–y* addressable such that any one of their pixels can be randomly

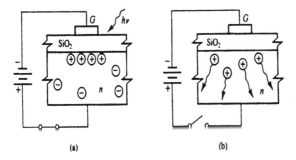

Figure 17.17 CID charge: (a) accumulation and (b) injection.

accessed with little or no blooming. However, the CID capacitance (sum of the capacitance of a row and a column) is much larger than the CCD capacitance and, therefore, the CID images tend to be noisier.

17.5 CRT AND IMAGING TUBES

The cathode-ray tube (CRT) is a non-solid-state display that is deeply entrenched in our day-to-day video world. Even though other competing technologies such as flat-panel display or making serious inroads, it is unlikely that CRTs will be totally replaced. In spite of its large power consumption and bulky size, it is by far the most common display device found in both general- and special-purpose usage, aside from displaying small alphanumeric. The cost factor and the trend toward using more and more high-resolution color displays are the key factors that guarantee the CRT's longevity. CRTs have satisfactory response speed, resolution, design, and life. In addition, there are very few electrical connections, and CRTs can present more information per unit time at a lower cost than any other display technology. A CRT display is generally subdivided into categories such as having electrostatic or magnetic deflection, monochromatic or color video, and single or multiple beams.

Figure 17.18(a) shows the schematic of a CRT display where the cathod luminescent phosphors are used at the output screen. Cathodoluminescent refers to the emission of radiation from a solid when it is bombarded by a beam of electrons. The electrons in this case are generated by thermionic emission of a cathode and are directed onto the screen by means of series of deflection plates held at varying potentials. The electron beam is sequentially scanned across the screen in a series of lines by means of electrostatic or electromagnetic fields (introduced by deflection plates) orthogonal to the direction of electron trajectory. The bulb enclosing the electron gun, the deflectors, and the screen are made air-free for the purpose of having an electron beam and a display area. The video signals to be displayed are applied to both the electron gun and deflectors in synchronization with the scanning signals. The display is usually refreshed 60 times a second to avoid having a flickering image. While in the United States CRT displays consist of 525-scan line, the number is about 625 overseas. The phosphor screen is often treated as being split into two interlaced halves. Thus, if a complete refreshing cycle takes t_r time, only odd-numbered lines are scanned during the first $t_r/2$ period, and the even-numbered

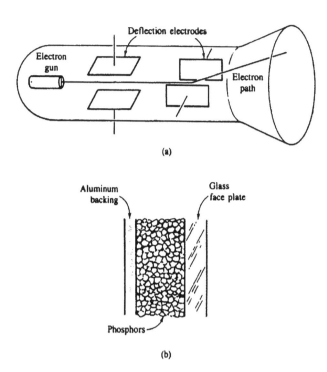

Figure 17.18   (a) A CRT schematic and (b) a CRT screen.

lines are scanned during the remaining half. Consequently, our eyes treat the refreshing rate as if it were $2/t_r$ Hz instead of only $1/t_r$ Hz.

The deflected electron strikes the CRT phosphor screen causing the phosphors of that CRT location to emit light. It is interesting to note that both absorption and emission distributions of phosphors are bell shaped, but the distribution peaks are relatively displaced in wavelength. When compared with the absorption distribution, the emission distribution peaks at a higher wavelength. This shift toward the red end of the spectrum is referred to as the Stokes' shift. It is utilized to convert ultraviolet radiation to useful visible radiation. It is used in fluorescent lamps to increase their luminous efficiencies. In particular, the CRT illumination, caused by the cathodoluminescent phosphors, is a strong function of both current and accelerating voltage and is given by

$$L_e = Kf(i)V^n, \qquad (17.36)$$

where K is a constant, $f(i)$ is a function of current and n ranges between 1.5 and 2. With larger accelerating voltages, electrons penetrate further into the phosphor layer, causing more phosphor cells to irradiate.

The factors that are taken into consideration in selecting a particular phosphor are, namely, decay time, color, and luminous efficiency. Note that even though a phosphor may have lower radiant efficiency in the green, the latter may have a more desirable luminous efficiency curve. Usually, the phosphor screen consists of a thin layer ($\sim 5\,\mu m$) of phosphor powder placed between the external glass faceplate and a very thin ($\sim 0.1\,\mu m$) aluminium backing as shown in Fig. 17.18(b). The aluminium backing prevents charge build-up and helps in redirecting light back toward the glass plate. The aluminum backing is thin enough so that most of the electron beam energy can get through it. A substantial amount of light that reaches the glass at normal (beyond the critical angle of incidence) may be totally internally reflected at the glass–air interface, some of which may again get totally internally reflected at the phosphor–glass interface. Such physical circumstances produce a series of concentric circles of reduced brightness instead of one bright display spot. The combination of diffused display spots results in a display spot that has a Gaussian-distribution profile.

Of the many available methods, the most common one for introducing color in a CRT display involves the use of a metal mask and three electron guns, each corresponding to a primary phosphor granule (red, blue, and green), as shown in Fig. 17.19. The three electron guns are positioned at different angles, so that while each of the electron beams is passing through a particular mask-hole strikes a particular primary phosphor dot. All three beams are deflected simultaneously. In addition, the focus elements for the three guns are connected in parallel so that a single focus control is sufficient to manipulate all beams. The three primary dots are closely packed in the screen so that proper color can be generated for each signal. Misalignment of the three beams causes a loss of purity for the colors. In any event, when compared with the monochrome display, the CRT color reproduction process involves a loss of resolution to a certain degree because the primary phosphor cells are physically disjointed.

Imaging tubes convert a visual image into equivalent electrical signals that are used thereafter for viewing the image on display devices such as CRTs. They are used as in a television camera tube in which a single multilayer structure serves both as an image sensor and as a charge storage device. A beam of low-velocity electrons to produce a video signal scans the single multilayer structure. In particular, when the characteristics of the photosensor layer during the optical-to-electrical conversion depend on photosensor's photoconductive property, the imaging tube is referred as a vidicon. Figure 17.20 shows a typical vidicon whose thin target material consists of a

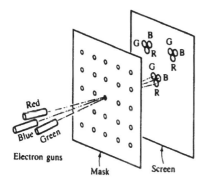

Figure 17.19 Shadow masking in a color CRT.

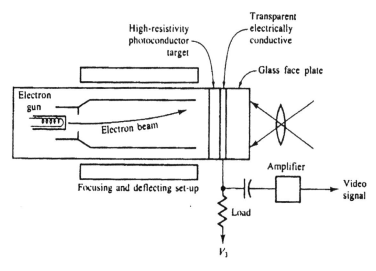

Figure 17.20 A vidicon structure.

photoconductive layer, such as selenium or antimony trisulphide, placed immediately behind a transparent conductive film of tin oxide. Its charge-retention quality is very good since the resistivity of the photoconductor material is very high ($\sim 10^{12}\,\Omega$-cm). The conducting layer is connected to a positive potential V_B via a load resistor. The other side of the target is scanned with an electron beam almost in the same way as CRT scanning. The output video signal is generally capacitively coupled, and the majority of vidicons employ magnetic focusing and deflection schemes.

The vidicon target is often modeled as a set of leaking capacitors, each of which corresponds to a minute area of the target. One side of these capacitors is tied together by means of a transparent conductive layer and is connected to a bias voltage V_B. The low-velocity scanning electron beam makes momentary contact with each of the miniature areas, charging them negatively. The target has high resistance in the dark, but when it is photoirradiated its resistance drops significantly. In the absence of illumination, the scanning beam drives the target to a potential value close to that of the cathode, which allows for a small amount of dark current to flow when the beam is removed. The decrease in resistivity due to photoirradiance causes the capacitor-like target to discharge itself in the absence of the electron beam. However, when the electron beam scans this discharged area, it will recharge the target. More current is being taken away from the illuminated area than from unilluminated area, thus generating a video signal at the output. The video signal is found to be proportional to nth power of illumination, where n is a positive number less than unity.

The dark current is rather large in vidicons. Again, the spectral response of antimony trisulphide is very poor at wavelengths greater than 0.6 μm. An imaging tube referred to as a plumbicon is often used to overcome the aforementioned shortcomings of a vidicon. The plumbicon is essentially identical to the vidicon, except that the photoconductive target is replaced by a layer of lead oxide that behaves like *p–i–n* diode and not like a photoconductor.

The band-gap energy of lead oxide is 2 eV and, therefore, it is not too sensitive to red. However, the introduction of a thin layer of lead sulfide (with a band gap of about 0.4 eV), along with lead oxide, eliminates the problem of red insensitivity in the plumbicon. The transparent conductive film acts like an *n*-type region, and the PbO layer acts like a *p*-type region, whereas the region between the two behaves like an intrinsic semiconductor. Photoirradiance of this *p–i–n* structure generates carriers in the plumbicon. But the flow of carriers in the opposite direction generally reduces the amount of stored charge. However, in the absence of photoirradiance, the reverse bias gives rise to a dark current that is negligible in comparison to that encountered in the vidicon. There are two serious disadvantages using plumbicons because their resolution (fewer than 100 lines) is octane limited by the thick lead oxide layer and because the change in target voltage cannot be used to control their sensitivity. In spite of these demerits, they are used widely in color TV studios. However, the lead oxide layer can be replaced by an array formed by thousands of silicon diodes to increase the sensitivity of plumbicons.

Another important imaging tube, referred to as an image intensifier, is of significant importance in the transmittal of images. In principle, it is vacuum photodiode equipped with a photocathode on the input window and a phosphor layer on the output window. Image intensifiers are devices in which the primary optical image is formed on a photocathode surface (with an S20 phosphor layer backing), and the resulting photocurrent from each of the image points is intensified by increasing the energy of the electrons, as shown in Fig. 17.21. The windows are made of the fiber-optic plates, so that the plane image surface of the input can be transformed to the curve object and image surfaces of a simple electrostatic lens. The electrons strike a luminescent screen and the intensified image is produced by cathodoluminescent. It is possible to cascade more than one such intensifier with fiber-optic coupling between them, making sure that an accelerating potential is applied between the photocathode and the screen. Such a cascade device, along with an objective lens and an eyepiece, is used in the direct-view image intensifier. In any event, it is possible to achieve a luminance gain of up to 1000 with each image intensifier. An image intensifier such as this can also be designed by increasing the number of electrons (as in a photomultiplier).

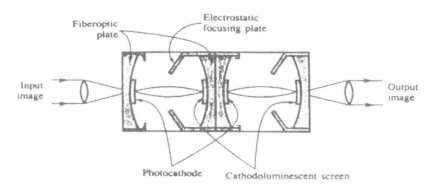

Figure 17.21 Two cascaded image intensifiers.

17.6 ELECTRO-OPTIC MODULATION

Electro-optic (EO) modulators consist of a dielectric medium and a means of applying an electric field to it. Application of an electric field causes the refractive index of the dielectric medium to be modified. The mechanism involved in this phenomenon is referred to as the electro-optic effect of the dielectric medium. The mechanism is often used for realizing both amplitude as well as the phase modulation of optical signals. The application of such a modulation scheme exists in optical communications, Q-switching of lasers, and beam deflections. In principle, electro-optic modulators cause a change in phase shift, which in turn is either a linear or a quadratic function of the applied electric field. This change in phase shift implies a change in optical length or index of refraction.

In an electro-optic crystal, such as those mentioned in the last section, the change in the index of refraction n along a crystal axis may be expressed in series form as

$$\Delta\left(\frac{1}{n^2}\right) = pE + kE^2 + \cdots, \tag{17.37}$$

where E is the electric field, p is the linear electro-optic coefficient, and k is the quadratic electro-optic coefficient. In useful crystal, either the linear electro-optic effect (referred to as the Pockels effect) or the quadratic electro-optic effect (referred to as the Kerr effect) is predominant. In either case, the index of refraction will change at the modulation rate of the electric field. The effect allows a means of controlling the intensity or phase of the propagating beam of light. A Pockets cell uses the linear effect in crystals, whereas a Kerr cell uses the second-order electro-optic effect in various liquids and ferroelectrics; however, the former requires far less power than the latter to achieve the same amount of rotation and thus is used more widely. The Pockets effect, in particular, depends on the polarity of the applied electric field.

In the absence of an external electric field, the indices of refraction along the rectangular coordinate axes of a crystal are related by the index ellipsoid:

$$(x/n_x)^2 + (y/n_y)^2 + (z/n_z)^2 = 1. \tag{17.38}$$

In the presence of an arbitrary electric field, however, the linear change in the coefficients of the index ellipsoid can be represented by

$$\Delta\left(\frac{1}{n^2}\right)_i = \sum_j p_{ij} E_j, \tag{17.39}$$

where p_{ij}'s are referred to as Pockels constants, $i = 1, 2, 3, \ldots, 6$; and $j = x, y, z$. The 6×3 electro-optic matrix having p_{ij} as its elements is often called the electro-optic tensor. In centro-symmetric crystals, all 18 elements of the tensor are zero, whereas in the triclinic crystals, all elements are nonzero. But in the great majority of crystals, while some of the elements are zero many of the elements have identical values. Table 17.1 lists the characteristics of the electro-optic tensors for some of the more important noncentrosymmetric crystals.

Determining the electro-optic effect in a particular modulator thus involves using the characteristics of the crystal in question and finding the allowed polarization directions for a given direction of propagation. Knowledge of refractive indices

Table 17.1 Some of the Important Electro-optic Constants

Crystal	Nonzero elements (in 10^{12} m/V)	Refractive index
BaTiO$_3$	$p_{13} = p_{23} = 8.0$ $p_{33} = 23.0$ $p_{42} = p_{51} = 820.0$	$N_x = n_y = n_0 = 2.437$ $N_E = n_z = 2.365$
KDP	$p_{41} = p_{52} = 8.6$ $p_{63} = 10.6$	$N_x = n_y = n_0 = 1.51$ $N_F = n_r = 1.47$
ADP	$p_{41} = p_{52} = 28.0$ $p_{63} = 8.5$	$N_x = n_y \equiv n_0 = 1.52$ $N_E = n_z = 1.48$
GaAs	$p_{41} = p_{52} = p_{63} = 1.6$	$N_x = n + y \equiv n_0 = 3.34$
CdTe	$p_{41} = p_{52} = p_{63} = 6.8$	$N_x = n_y = n_0 = 2.6$
Quartz	$p_{11} = 0.29; p_{21} = p_{62} = -0.29$ $p_{41} = 0.2; p_{52} = -0.52$	$N_x = n_y \equiv n_0 = 1.546$ $N_z = n_E = 1.555$
LiNbO$_3$	$p_{13} = p_{23} = 8.6; p_{33} = 30.8$ $p_{22} = 3.4; p_{12} = p_{61} = -3.4$ $p_{42} = p_{51} = 28$	$N_x = n_y = n_0 = 2.286$ $N_z = n_E = 2.200$
CdS	$p_{13} = p_{23} = 1.1$ $p_{33} = 2.4; p_{42} = p_{51} = 3.7$	$N_x = n_y = n_0 = 2.46$ $N_z = n_E = 2.48$

along the allowed directions can be used to decompose the incident optical wave along those allowable polarization directions. Thereafter, we can determine the characteristics of the emergent wave. Consider, for example, the case of KDP, whose nonzero electro-optic tensor components are p_{41}, p_{52}, and p_{63}. Using Eqs (17.38) and (17.39), we can write an equation of the effective index ellipsoid for KDP as

$$\frac{x^2 + y^2}{n_0^2} + \left(\frac{z}{n_E}\right)^2 + 2p_{41}yzE_x + 2p_{63}xyE_z = 1, \tag{17.40}$$

with $n_x = n_y = n_0$ and $n_z = n_E$ for this uniaxial crystal where $p_{41} = p_{52}$. To be specific, let us restrict ourselves to the case in which the external field is directed along only the z-direction. Equation (17.40), for such a case, reduces to

$$\frac{x^2 + y^2}{n_0^2} + \left(\frac{z}{n_E}\right)^2 + 2p_{63}xyE_z = 1. \tag{17.41}$$

Equation (17.41) can be transformed to have a form of the type

$$\left(\frac{x'}{n_{x'}}\right)^2 + \left(\frac{y'}{n_{y'}}\right)^2 + \left(\frac{z'}{n_{z'}}\right)^2 = 1, \tag{17.42}$$

which has no mixed terms. The parameters x', y', and z' of Eq. (17.42) denote the directions of the major axes of the index ellipsoid in the presence of the external field; $2n_{x'}'$, $2n_{y'}'$, and $2n_{z'}'$ give the lengths of the major axes of the index ellipsoid, respectively.

By comparing Eqs (17.41) and (17.42), it is obvious that z and z' are parallel to each other. Again, the symmetry of x and y in Eq. (17.41) suggests that x' and y' are related to x and y by a rotation of $45°$. The transformation between the coordinates (x, y) and (x', y') is given by

$$\begin{bmatrix} x \\ y \end{bmatrix} = \begin{bmatrix} \cos\left(\frac{1}{4}\pi\right) - \sin\left(\frac{1}{4}\pi\right) \\ \sin\left(\frac{1}{4}\pi\right) \cos\left(\frac{1}{4}\pi\right) \end{bmatrix} \begin{bmatrix} x' \\ y' \end{bmatrix}, \tag{17.43}$$

which, when substituted in Eq. (17.41), results in

$$\left[\frac{1}{n^2} + p_{63}E_z\right]x'^2 + \left[\frac{1}{n_0^2} - p_{63}E_z\right]y'^2 + \left(\frac{z'}{n_E}\right)^2 = 1. \tag{17.44}$$

Comparing Eqs (17.42) and (17.44) and using the differential relation $dn = \frac{1}{2}n^3 d(1/n^2)$, we find that

$$n_{x'} = n_0 - \frac{1}{2}n_0^3 p_{63}E_z \tag{17.45a}$$

$$n_{y'} = n_0 + \frac{1}{2}n_0^3 p_{63}E_z \tag{17.45b}$$

$$n_{z'} = n_E \tag{17.45c}$$

when $(1/n_0^2) \gg p_{63}E_z$. The velocity of propagation of an emerging wave polarized along the x' axis differs from that of an emerging wave polarized along y' axis. The corresponding phase shift difference between the two waves (referred to as electro-optic retardation) after having traversed a thickness W of the crystal is given by

$$\Delta\phi = \frac{2\pi W}{\lambda}|n_{x'} - n_{y'}| - \frac{2\pi W}{\lambda}n_0^3 p_{63}E_z. \tag{17.46}$$

Provided that V is the voltage applied across the crystal, retardation is then given by

$$\Delta\phi = \frac{2\pi}{\lambda}n_0^3 p_{63}V. \tag{17.47}$$

The emergent light is in general elliptically polarized. It becomes circularly polarized only when $\Delta\phi = 1/2\pi$ and linearly polarized when $\Delta\phi = \pi$. Often the retardation is also given by $\pi V/V_\pi$, where $V_\pi (\equiv \lambda/2n_0^3 p_{63})$ is the voltage necessary to produce a retardation of π.

17.7 ELECTRO-OPTIC DEVICES

In the last section, we showed that it is possible to control optical retardation and, thus, flow of optical energy in non-centrosymmetric cyrstals by means of voltage. This capability of retardation serves as the basis of modulation of light. Figure 17.22 shows a schematic of a system showing how a KDP modulator can be used to achieve amplitude modulation. This particular setup is also known as Pockels electro-optic (EO) amplitude modulator. The crossed polarizer-analyzer combination of the setup is necessary for the conversion of phase-modulated light into amplitude-modulated light. Further, in this setup, the induced electro-optic axes of the crystal make an angle of $45°$ with the analyzer–polarizer axes.

Elliptically polarized light emerges from the EO crystal since, upon modulation, the two mutually orthogonal components of the polarized beams travel inside the crystal with different velocities. To feed an electric field into the system, the end faces of the EO crystal are coated with a thin conducting layer that is transparent to optical radiation. As the modulating voltage is changed, the eccentricity of the ellipse

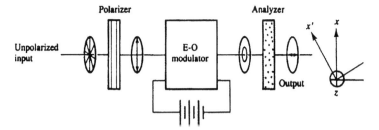

Figure 17.22 Pockels EO amplitude modulator: (a) system and (b) orientation of crystal axes.

changes. The analyzer allows a varying amount of outgoing light in accordance with the modulating voltage applied across the EO crystal.

The electric field components associated with the optical wave, immediately upon emerging from the EO crystal, are respectively

$$E'_x = A, \tag{17.48a}$$

$$E'_y = Ae^{-j\Delta\phi}. \tag{17.48b}$$

Thus the total field that emerges out of the analyzer is evaluated by summing the E'_x and E'_y components (along the analyzer axis), which gives us

$$
\begin{aligned}
E_o &= [Ae^{-j\Delta\phi} - A]\cos(1/4\pi) \\
&= \frac{A}{\sqrt{2}}e^{-j\Delta\phi} - 1.
\end{aligned}
\tag{17.49}
$$

The resulting irradiance of the transmitted beam is therefore given by

$$I_0 = \text{const}(E_0)(E_0^*) = \text{const } 2A^2 \sin^2\left(\frac{\Delta\phi}{2}\right) \equiv I_i \sin^2\left(\frac{\Delta\phi}{2}\right), \tag{17.50}$$

where I_i is the irradiance of the light incident on the input side of the EO crystal. One can rewrite Eq. (17.50) as

$$\frac{I_0}{I_i} = \sin^2\left[\frac{1}{2}\pi\frac{V}{V_\pi}\right], \tag{17.51}$$

where $V_\pi = |\lambda/2p_{63}n_0^3|$ is the voltage required for having the maximum transmission. Often V_π is also referred to as the half-wave voltage because it corresponds to a relative spatial displacement of $\lambda/2$ or to an equivalent phase difference of π.

Figure 17.23 shows the transmission characteristics of the cross-polarized EO modulator as a function of the applied voltage. It can be seen that the modulation is nonlinear. In fact, for small voltages, the transmission is proportional to V^2. The effectiveness of an EO modulator is often enhanced by biasing it with a fixed retardation of $\pi/2$. A small sinusoidal voltage will then result in a nearly sinusoidal modulation of the transmitted intensity. This is achieved by introducing a quarter-wave plate between the polarizer and the EO crystal. The quarter-wave plate shifts the EO characteristics to the 50% transmission point. With the quarter-wave plate in

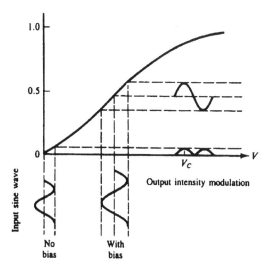

Figure 17.23 Transmission versus voltage in a Pockels EO amplitude modulator.

place, the net phase difference between the two emerging waves becomes $\Delta\phi = \Delta\phi + (\pi/2)$. Thus the output transmission is then given by

$$\frac{I_0}{I_i} = \sin^2\left(\frac{\pi}{4} + \frac{1}{2}\pi\frac{V}{V_\pi}\right) \approx \frac{1}{2}\left(1 + \sin\left(\pi\frac{V}{V_\pi}\right)\right). \tag{17.52}$$

Note that, with no modulating voltage, the modulator output intensity transmission reduces to 0.5 gain. Again, for small V, the transmission factor varies linearly as the crystal voltage.

For an input sine wave modulating voltage, the transmission can be expressed as

$$\frac{I_0}{I_i} = \frac{1}{2}[1 + \sin\{mV_p \sin(\omega_p t)\}], \tag{17.53}$$

where m is a constant of proportionality, V_p is the peak modulating voltage, and ω_m is the modulation angular frequency. When $mV_p \ll 1$, the intensity modulation becomes a replica of the modulating voltage. The irradiance of the transmitted beam begin to vary with the same frequency as the sinusoid voltage. If, however, the condition $mV_p \ll 1$ is not satisfied, the intensity variation becomes distorted. Note that Eq. (17.53) can be expanded in terms of Bessel functions of the first kind to give

$$\frac{I_0}{I_i} = \frac{1}{2} + J_1(mV_p)\sin(\omega_m t) + J_3(mV_p)\sin(3\omega_m t) + J_5(mV_p)\sin(5\omega_m t) + \cdots \tag{17.54}$$

since

$$\sin(x\sin y) = 2[J_1(x)\sin(y) + J_3(x)\sin(3y) + J_5\sin(5y) + \cdots] \tag{17.55}$$

when $J_0(0) = 1$ and, for nonzero n, $J_n(x) = 0$. The ratio between the square root of the sum of harmonic amplitude squares and the fundamental term amplitude often characterizes the amount of distortion in the modulation process. Therefore,

$$\text{Distortion (\%)} = \frac{\left\{[J_3(mV_p)]^2 + [J_5(mV_p)]^2 + \cdots\right\}^{1/2}}{J_1(mV_p)} \times 100. \tag{17.56}$$

Consider the setup of Fig. 17.24 where the EO crystal is oriented in such a way that the incident beam is polarized along one of the birefringence axes, say x'. In this specific case, the state of polarization is not changed by the applied electric field. However, the applied field changes the output phase by an amount

$$\Delta\phi_{x'} = \frac{\omega W}{c}|\Delta n_{x'}| \tag{17.57}$$

where W is the length of the EO crystal. For a sinusoidal bias field $E_z = E_{z,p}\sin(\omega_p t)$ and an incident $E_{in} = E_{in,p}\cos(\omega t)$, the transmitted field is given by

$$\begin{aligned}
E_{out} &= E_{in,p}\cos\left[\omega t - \frac{\omega W}{c}\{n_0 - \Delta\phi_{x'}\}\right] \\
&= E_{in,p}\cos\left[\omega t - \frac{\omega W}{c}\left\{n_0 - \frac{1}{2}n_0^3 p_{63}E_{z,p}\sin(\omega_m t)\right\}\right].
\end{aligned} \tag{17.58}$$

If the constant phase factor $\omega W n_0/c$ is neglected, the transmitted electric field can be rewritten as

$$E_{out} = E_{in,p}\cos[\omega t + \delta\sin(\omega_m t)], \tag{17.59}$$

where $\delta \equiv 1/2(\omega W n_0^3 p_{63}E_{z,p}/c)$ is the phase-modulation index. Note that this phase-modulation index is one-half of the retardation $\Delta\phi$ (as given by Eq. (17.47)). Using Eq. (17.55) and the relationship

$$\cos(x\sin y) = J_0(x) + 2[J_2(x)\cos(2y) + J_4(x)\cos(4y) + \cdots], \tag{17.60}$$

we can rewrite Eq. (17.59) as

$$\begin{aligned}
E_{out} = E_{in,p}[&J_0(\delta)\cos(\omega t) + J_1(\delta)[\cos\{(\omega + \omega_m)t\} - \cos\{(\omega - \omega_m)t\}] \\
&+ J_2(\delta)[\cos\{(\omega + 2\omega_m)t\} - \cos\{(\omega - 2\omega_m)t\}] \\
&+ J_3(\delta)[\cos\{(\omega + 3\omega_m)t\} - \cos\{(\omega - 3\omega_m)t\}] + \cdots].
\end{aligned} \tag{17.61}$$

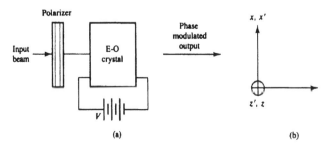

Figure 17.24 A phase modulator using an EO crystal: (a) the system and (b) the orientation of crystal axes.

Accordingly, in this case, the optical field is seen to be phase modulated with energy distribution in the side bands varying as a function of the modulation index δ. We observe that while the EO crystal orientation of Fig. 17.23 provides an amplitude modulation of light, the setup of Fig. 17.24 can provide a phase modulation of light. Both of these modulators are called longitudinal effect devices because in both cases the electric field is applied in the same direction as that of propagation. In both of these cases, the electric field is applied either by electrodes with small apertures in them or by making use of semitransparent conducting films on either side of the crystal. This arrangement, however, is not too reliable because the field electrodes tend to interfere with the optical beam.

Alternatively, transverse electro-optic modulators can be used, as shown by the system shown in Fig. 17.25. The polarization of the light lies in the $x'-z$ plane at a $45°$ angle to the x'-axis while light propagates along the y'-axis and the field is applied along z. With such an arrangement, the electrodes do not block the incident optical beam and, moreover, the retardation can be increased by introducing longer crystals. In this longitudinal case, the amount of retardation is proportional to V and is independent of the crystal length W according to Eq. (17.47). Using Eqs (17.45a) and (17.45c), the retardation caused by the transverse EO amplitude modulator is given by

$$\Delta\phi_t = \phi_{x'} - \phi_{z'} = \frac{2\pi W}{\lambda}\left[(n_0 - n_E) - n_0^3 p_{63}\frac{V}{d}\right], \tag{17.62}$$

where n_0 and n_E are refractive indices, respectively, along ordinary and extraordinary axes. Note that $\Delta\phi_t$ has a voltage-independent term that can be used to bias the irradiance transmission curve. Using a long and thin EO crystal can reduce the half-wave voltage. Such an EO crystal allows the transverse EO modulators to have a better frequency response but at the cost of having small apertures.

There are occasions when we might be interested in driving the modulating signals to have large bandwidths at high frequencies. This can happen when we decide to use the wide-frequency spectrum of a laser. To meet the demand of such a scenario, the modulator capacitance that is caused by the parallel-plate electrodes and the finite optical transit time of the modulator limits both bandwidth and the maximum modulation frequency. Consider the equivalent circuit of a highfrequency, electro-optic modulator as shown in Fig. 17.26. Let R_{in} be the total internal resis-

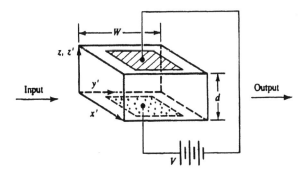

Figure 17.25   A transverse EO amplitude modulator.

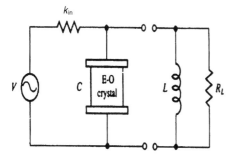

Figure 17.26   A circuit equivalent to an EO modulator.

tance of the modulating source V, while C represents the parallel-plate capacitance of the EO crystal. When R_{in} is greater than the capacitance impedance, a significant portion of the modulating voltage is engaged across the internal resistance, thus making the generation of electro-optic retardation relatively insignificant. In order to increase the proportion of the modulating voltage that is engaged across the EO crystal, it is necessary to connect a parallel resistance-inductance circuit in parallel with the modulator. The load R_L is chosen to be very large when compared with R_{in}. The choice guarantees that most of the modulating voltage is employed across the EO crystal. At the resonant frequency $v_0 = [1/\{2\pi(LC)\}^{1/2}]$, the circuit impedance is equivalent to load resistance. However, this system imposes a restriction on the bandwidth and makes it finite. The bandwidth is given by $[1/\{2\pi R_L C\}]$ and is centered at the resonant frequency v_0. Beyond this bandwidth, the modulating voltage is generally wasted across R_{in}. Consequently, for the modulated signal to be an exact replica of the modulating signal, the maximum modulation bandwidth must not be allowed to exceed Δv, where Δv is usually dictated by the specific application.

The power P needed to drive the EO crystal is given by $1/2(V_{max}^2/R_L)$ where V_{max} $(= E_{z,max} W)$ is the peak modulating voltage that produces the peak retardation $\Delta\phi_{max}$. Using Eq. (17.47) we can relate the driving power and the modulation bandwidth by

$$P = \frac{(\Delta\phi_{max})^2 \lambda^2 C \Delta v}{4\pi p_{63}^2 n_0^6},$$

$$= \frac{(\Delta\phi_{max})^2 \lambda^2 A K_s \varepsilon_0 \Delta v}{4\pi p_{63}^2 n_0^6 W}, \tag{17.63}$$

since at the modulation frequency v_0, the parallel-plate capacitance C is given by $A K_s \varepsilon_0 / W$, where A is the cross-sectional area of the crystal, K_s is the relative dielectric constant for the material, and W is the plate separation between the two electrodes.

As long as the modulating frequency is relatively low, the modulating voltage remains appreciably constant across the crystal. If the above condition is not fulfilled, however, the maximum allowable modulation frequency is restricted substantially by the transit time of the crystal. To overcome this restriction, the modulating signal is applied transversely to the crystal in the form of a traveling wave with a velocity equal to the phase velocity to the optical signal propagating through the crystal. The transmission line electrodes are provided with matched impedance at the termination point, as shown in Fig. 17.27(a). The optical field is then subjected to a constant refractive index as it passes through the modulator, thus making it possible to have higher modulation frequencies. The traveling modulation field at a time t will have the form

$$
\begin{aligned}
E(t, z(t)) &= E_p \exp\left[j\omega_m \left\{ t - \frac{z}{v_p} \right\} \right] \\
&= E_p \exp\left[j\omega_m \left\{ t - \frac{c}{nv_p}(t - t_0) \right\} \right],
\end{aligned}
\tag{17.64}
$$

where v_p is the phase velocity of the modulation field, ω_m is the modulating angular frequency, and t_0 is defined as a reference to account for the time when the optical wavefront enters the EO modulator. The electro-optic retardation due to the field can be written in accordance with Eq. (17.47) as

$$
\Delta\phi = \Phi \frac{c}{n} \int_{t_0}^{t_0+t_t} E(t, z(t)) dt,
\tag{17.65}
$$

where $\Phi = 2\pi n_0^3 p_{63}/\lambda$ and $t_t = nW/c$ is the total transit time (i.e., time taken by light to travel through the crystal). Equation (17.65) can be evaluated to give the traveling wave retardation as

$$
\Delta\phi_{travel} = (\Delta\phi)_0 \frac{e^{j\omega_m t_0}[e^{j\omega_m t_t[1-(c/nv_p)]} - 1]}{j\omega_m t_t \left(1 - \dfrac{c}{nv_p}\right)},
\tag{17.66}
$$

where $(\Delta\phi)_0 = (\Phi c t_t E_p/n)$ is the peak retardation. The reduction factor

$$
F_{travel} = \frac{e^{j\omega_m t_t[1-(c/nv_p)]} - 1}{j\omega_m t_t \left(1 - \dfrac{c}{nv_p}\right)}
\tag{17.67}
$$

provides the amount of reduction in the maximum retardation owing to transit time limitation. If instead we had begun to calculate the retardation for a sinusoidal

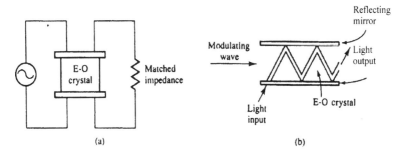

Figure 17.27 (a) Ideal traveling wave EO modulator and (b) zig-zag traveling wave modulator.

modulation field that has the same value throughout the modulator, the retardation would have been

$$\Delta\phi_{\text{travel}} = (\Delta\phi)_0 \exp(j\omega_m t_0)\left[\frac{e^{j\omega_m t} - 1}{j\omega_m t_t}\right]. \tag{17.68}$$

By comparing Eqs. (17.66) and (17.68), we find that the two expressions are identical except that in the traveling wave, the EO modulator t_t is replaced by $t_t\{1 - (c/nv_n)\}$. The reduction factor in this latter case is given by

$$F_{\text{nontravel}} = \frac{e^{j\omega_m t} - 1}{j\omega_m t_t}. \tag{17.69}$$

In the case of a nontraveling system, the reduction factor is unity only when $\omega_m t_t \ll 1$, i.e., when the transit time is smaller than the smallest modulation period. But in the case of the traveling system, F approaches unity whenever the two-phase velocities are equal – that is, when $c = nv_p$. Thus, in spite of the limitation of transit time, the maximum retardation is realized using the traveling wave modulator.

In practice, it might become very difficult to synchronize both electrical and optical waves. For a perfect synchronization, we expect to use an EO crystal for which $n = K_s^{1/2}$. But for most naturally occurring materials, n is less than $K_s^{1/2}$. Thus, synchronization is achieved either by including air gaps in the electrical waveguide cross section or by slowing down the optical wave by means of a zigzag modulator, as shown in Fig. 17.27(b).

A useful amplitude modulator involving waveguides, known as the Mach–Zehnder modulator, consists of neither polarizer nor analyzer but only EO material. The system, as shown in Fig. 17.28, splits the incoming optical beam into a "waveguide Y" and then recombines them. If the phase shift present in both of the arms is identical, all of the input power minus the waveguide loss reappears at the output. One arm of the Mach–Zehnder system is provided with an electric field placed across it such that the amplitude of the field can be varied. Changing the voltage across the waveguides modulates the output power. When a sufficiently high voltage is applied, the net phase shift difference between the arms can become 180°, thus canceling the power output altogether. Because of the small dimensions involved in electrode separations, relatively small switching voltages are used in such a modulation scheme.

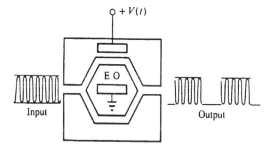

Figure 17.28 Mach–Zehnder waveguide EO modulator.

Figure 17.29(a) shows a prism deflector that uses a bulk electro-optic effect so that we can deflect an optical beam by means of an externally applied electric voltage. The electric voltage induces regions having different refractive indices, as a result of which the prism steers the refracted beam of light. Stacks of prisms can be used to provide a larger angle of deflection. Figure 17.29(b) shows an interesting application where birefringent crystals are combined with a Pockels cell modulator to form a digital EO deflector. By applying an electric field, we can rotate the direction of polarization by $\pi/2$. Accordingly, by manipulating voltage (V_1 and V_0) we can shift the input optical beam to any one of the four spatial locations at the output. Similarly, by using n Pockels cell modulators and n birefringent crystals, we can shift light to a total of 2^n spatial locations. A system such as that of Fig. 17.29(b) can be considered for various optical computing applications.

A device by the name of Pockels read-out optical modulator (PROM) can be fabricated by having an EO crystal, such as Bi_2SiO_{20} or ZnSe, sandwiched between two transparent electrodes. There is also an insulating layer between the EO crystal and the electrode. A dc voltage can create mobile carriers, which in turn causes the voltage in the active crystal to decay. The device is normally exposed with the illumination pattern of a blue light. The voltage in the active area that corresponds to the brightest zones of the input pattern decays, while that corresponding to the comparatively darker area either does not change or changes very little. On the other

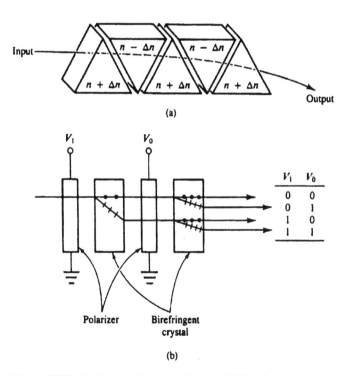

Figure 17.29 (a) Prism deflector and (b) digital EO deflector.

hand, the read-out beam usually uses a red light. Note that the sensitivity of an EO crystal is much higher in the blue region than in the red region. Such a choice of read-out beam ensures that the read-out beam may not cause a change in the stored voltage pattern. In the read-out mode, the regions having the least amount of voltage act like a half-wave retarder. The reflected light, whose polarization is thus a function of the voltage pattern, is then passed through a polarizer to reproduce the output. For an eraser light E, the amplitude of the output is found to be

$$A = A_0 \sin\left(\frac{\pi V_0}{V_{1/2}}\right) e^{-KE}, \tag{17.70}$$

where A_0 is the amplitude of the input read-out beam, V_0 is the voltage applied across the EO crystal, $V_{1/2}$ is the half-wave voltage, and K is a positive constant. In the reflection read-out mode, $V_{1/2} = 2V_0$. The amplitude of the reflectance of the PROM when plotted against the exposure is surprisingly found to be similar to that of a photographic film with a nearly linear region between $E = 0$ and $E = 2/K$, as shown in Fig. 17.30.

There are many isotropic media available that behave like uniaxial crystals when subjected to an electric field E. The change in refractive index Δn of those isotropic media varies as the square of the electric field. Placing one of these media between crossed polarizers produces a Kerr modulator. Modulation at frequencies up to 10^{10} Hz has been realized using a Kerr modulator. The difference between the two refractive indices that corresponds respectively to light polarized perpendicular to the induced optic axis and light polarized perpendicular to the induced optic axis is provided by

$$\Delta n = |n_{\parallel} - n_{\perp}| = k\lambda E^2, \tag{17.71}$$

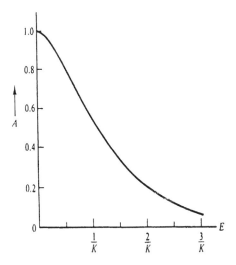

Figure 17.30 The characteristics of a PROM device.

where k is the Kerr constant of the material. At room temperature, the Kerr constant has a typical value of 0.7×10^{-12}, 3.5×10^{-10}, and 4.5×10^{-10} cm/V^2, respectively, for benzene, carbon disulphide, and nitrobenzene.

The applied electric field induces an electric moment, which in turn reorients the molecules in a manner so that the medium becomes anisotropic. The delay between the application of the field and the appearance of the effect is, though not negligible, on the order of 10^{-12} s. A liquid Kerr cell containing nitrobenzene, as shown in Fig. 17.31, has been used for many years, but it has the disadvantage of requiring a large driving power. This problem is often overcome by using, instead, mixed ferroelectric crystals at a temperature near the Curie point, where ferroelectric materials start exhibiting optoelectric properties. Potassium tantalate niobate (KTN) is an example of such a mixture of two crystals, where one has a high Curie point and the other has a low Curie point, but the Curie point of the compound lies very close to room temperature. The transmittance characteristics of a Pockels cell and a Kerr cell are shown, for comparison, in Fig. 17.32.

One of the ways one can realize a Q-switched laser involves subjecting either an EO crystal or a liquid Kerr cell to an electric field. Such a nonmechanical system is shown in Fig. 17.33. When there is no electric field, there is no rotation. But in the presence of the field, an EO device can introduce a rotation $\pi/2$. Thus, when the EO device is assembled along with a polarizer, the combination works as a shutter. Because of the vertical polarizer, the light coming out of the lasing medium is plane-polarized. Because the polarized light has to traverse through the EO device

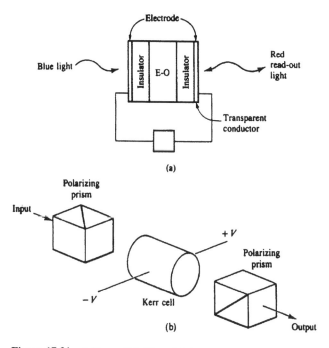

Figure 17.31 A Kerr cell light modulator.

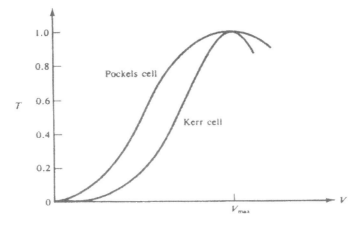

Figure 17.32 Transmittance versus voltage curve for the EO modulators.

twice before coming back to the polarizer, only half of the voltage required to produce a rotation of $\pi/2$ is applied to the system. Accordingly, the polarizer of the EO system blocks the light from coming to the main chamber of the laser. This is equivalent to causing a loss in the laser resonator. Such a loss, when suitably timed, can be made to produce a pulsed laser output with each pulse having an extremely high intensity. The voltages required to introduce appropriate fields are usually in the order of kilovolts, and thus it becomes possible to Q-switch at a rate of only nanoseconds or less.

The liquid-crystal light valve (LCLV) is a specific spatial light modulator (SLM) with which one can imprint a pattern on a beam of light in nearly real time. The two aspects of this device that are particularly important are the modulation of light and the mechanism for addressing the device. Besides LCLVs, there are many SLMs that are currently being considered for electro-optic applications. However, most of these devices follow only variations of the same physical principle.

The term liquid crystal (LC) refers to a particular class of materials whose rheological behavior is similar to that of liquids, but whose optical behavior is

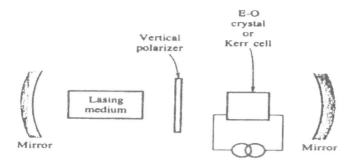

Figure 17.33 A Q-switched laser system.

similar to that of crystalline solids over a given range of temperature. In particular, a type of LC, referred to as the nematic LC, is commonly used in LCLV as well as in most other LC devices. Lately, ferroelectric LCs are also being used in real-time display devices. In comparison, ferroelectric LC-based devices respond at a faster rate than nematic LC-based devices. Nematic LCs generally consist of thin, long molecules, all of which tend to be aligned in the same direction, as shown in Fig. 17.34. When an electric field is applied to an LC layer, the molecules are reoriented in space due to both field and ionic conduction effects. The fact that LCs have a rod-like cylindrical molecular shape and that they tend to be aligned are prime reasons for yielding two EO effects: electric field effect and birefringence. In the "off" state, an LCLV utilizes the properties of a nematic cell, while in the "on" state, it utilizes the birefringence properties.

The shape of the LC molecule introduces a polarization-dependent variation of the index of refraction that contributes to its birefringence characteristics. The difference between the two indices of refraction given by $\Delta n = |n_\parallel - n_\perp|$ is a measure of the anisotropy of the material, where $n_\parallel$ represents the index of refraction for the component of light parallel to the molecular axis, and $n_\perp$ represents the index of refraction for the light component having an orthogonal polarization. Since LC molecules tend to be aligned, a bulk LC sample exhibits the same anisotropy as that exhibited by an individual LC molecule. In fact, a birefringent LC cell is normally formed by stacking LC layers parallel to the cell wall. Accordingly, all of the LC molecules are aligned along only one direction. When linearly polarized (at 45° to the alignment axis) light enters an LC cell of thickness D, the parallel polarization component lags behind the orthogonal component (due to positive anisotropy) by $\Delta\phi = [2\pi D(\Delta n)]/\lambda$. In general, the transmitted light turns out to be elliptically polarized. With a suitable choice of D, we can force the LC cell to behave like a half-wave or a quarter-wave plate. When compared with other traditional materials, LCs are preferable for making such retardation plates. In a typical LC material, Δn is non-negligible quantity. For example, for a typical LC, Δn is about 0.2 in the infrared, whereas Δn is only about 0.0135 in CdS in the infrared; in the visible, say, for quartz, Δn can be as low as 0.009. Thinner LC cells produce a comparatively large value of $\Delta\phi$. Note that a thinner LC cell allows for a larger acceptance angle for the incoming light, whereas a thicker, solid crystal like quartz forms a cell that is extremely sensitive to the angle of incidence. Further, LC cells can be grown to have reasonably large aperture sizes, suitable for handling higher laser power; however, the size will be limited by the tolerance for optical flatness.

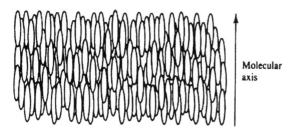

Molecular
axis

Figure 17.34 An aggregate of nematic liquid crystals.

For positive anisotropy, there is a region of applied voltage over which the LC molecules may gradually rotate, introducing a variable phase delay in the output light. This feature, as illustrated in Fig. 17.35, can be used to create a voltage-control phase shifter. As stated earlier, a typical LC device combines the characteristics of both birefringence and field effects. A typical SLM structure is shown in Fig. 17.36, where the cell is organized in the form of a quarter-wave plate. Consider an incoming light that is linearly polarized at an angle of 45° to the direction of alignment. The transmitted light is then found to be circularly polarized, but the reflected light that passes back through the quarter-wave cell becomes polarized in a direction perpendicular to that of the incident light. The first polarizer acts as an analyzer and thus blocks the light. In the presence of an external voltage, however, the birefringence can be reduced. The polarization characteristics of the resultant reflected light are changed so that the analyzer cannot block all of the reflected light. The external voltage can thus be used to control the transmission of light. In particular, transmission is zero when voltage is zero, but transmission reaches a maximum at high enough voltage. It is possible for the transmission to have a nonbinary value when voltage is set between the two extremes. Devices made using this SLM configuration are generally very sensitive to variations in cell thickness and light wavelength.

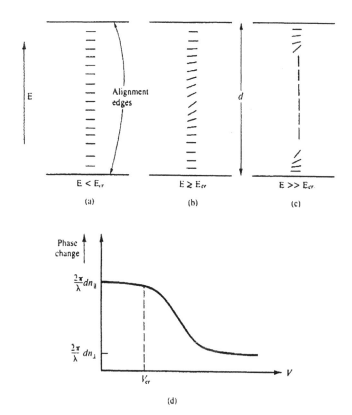

Figure 17.35 Field effects in liquid crystals.

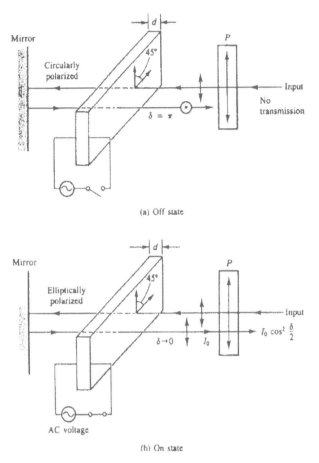

(a) Off state

(b) On state

Figure 17.36   A spatial light modulator configuration.

Interestingly, the twisted nematic LCs are also used in the common liquid-crystal display (LCD), as shown in Fig. 17.37. Instead of having parallel layers of LC stacks, the alignment layers at the opposite faces of the cell are maintained at 90° to each other. The remaining layers, depending on their positions, are oriented in a manner such that there is a gradual change in orientation from one end of the cell to the other. The molecules of a stack tend to line up in the same common direction, but they tend to align themselves with those of the neighboring stacks. The tilt of the molecular axes changes gradually between the two edges. An externally applied voltage can generally overcome the effects of alignment force and, with sufficient voltage, the molecules can fall back to their isotropic states. When voltages are withdrawn, the light going through the cell undergoes a rotation of 90° and thus passes through the analyzer. But in the presence of voltage, light falls short of 90° and, as a result, the analyzer can block most of the light.

When the voltage is low or completely withdrawn, LC molecules remain gradually twisted across the cell from the alignment direction of one wall to that of the

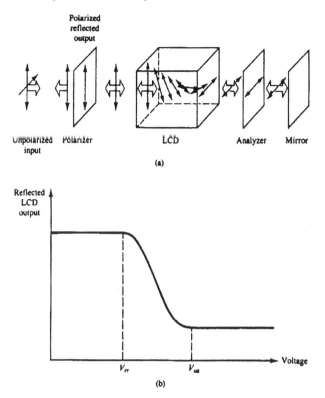

Figure 17.37 The LCD display: (a) the system and (b) its characteristics.

other wall. Such smooth transition is referred to as adiabatic. The light polarization is able to follow this slowly varying twist primarily because the cell width is larger than the light wavelength. As voltage is increased, LC molecules tend to be reorganized along the direction of the applied field. But in order for this to happen, the LC molecules have to overcome the alignment forces of the cell walls. The molecules located at the very center are farthest away from the walls and are, therefore, more likely to be reorganized.

As the tilt angle approaches $\pi/2$, molecules fail to align themselves with their immediate neighbors. In the extreme case of a $\pi/2$ tilt, the cell splits into two distinct halves. While the molecules in one half are aligned with one wall, the molecules in the other half are aligned with the other wall. Such a nonadiabatic system is modeled simply as a birefringent cell with an elliptically polarized transmission. Consequently, this state corresponds to maximum transmission. Other than two extremes, slowly varying and abrupt, there exists transmitted light with an intermediate degree of elliptical polarization. The amount of polarization in the transmission is thus controllable by an externally applied voltage that allows the LCLV to operate with gray levels. Note that the maximum birefringence occurs when the incident light forms an angle of $\pi/4$ with the molecular axis. Thus, nematic LCs having a 45° twist provide the maximum LCLV transmission.

The LCLV functions quite similarly to an LCD with only slight differences. The transmission characteristics of the analyzer in the case of an LCLV are the inverse of those in an LCD. This is indicated in Fig. 17.38. The LCLV requires only a little energy to produce an output, but it requries a large energy to yield its maximum output. This allows the device to produce a positive image when it is addressed optically. Again, while an LCD deals with binary transmission, an LCLV utilizes gray levels (corresponding to intermediate levels of transmission) to accurately represent an image. Thus, an LCLV is often operated in the transition region that exists between the minimum and the maximum transmission. To produce such an operating characteristic, therefore, an LCLV is organized differently from an LCD.

The light incident on a cell normally strikes its surface along the normal. The incident beam of light becomes linearly polarized in the direction of the molecular alignment of the first layer. In the absence of an applied voltage, the polarization direction rotates through an angle of $\pi/4$ along the helical twist of the LCs. The returning reflected light from the mirror undergoes a further twist of $\pi/4$, amounting to a total of $\pi/2$ rotation, and is thus blocked by the analyzer. With external voltage, however, light transmission increases because light can no longer follow the twist. Thus, by a combination of both birefringence and field effects, the twisted nematic cell can produce EO modulation.

Typical LCDs are addressed via electrode leads; each is connected to only one display segment. Such an addressing technique poses a serious problem when the number of leads begins to increase. LCLVs are, however, addressed differently – by means of optics. Optical addressing allows the image information to be fed into an LCLV in parallel. Thus, in the case of an LCLV, the frame time is the same as the response time of only one pixel. By comparison, in a scanning display, a frame time may equal the response time of one pixel multiplied by the total number of pixels. For example, a typical 20 inch × 20 inch flat panel display may consist of up to 1000×1000 pixels, with 2000–3000 pixels per dimension possible. The simplicity of optical addressing is thus obvious. Optical addressing is also preferable because it provides better resolution.

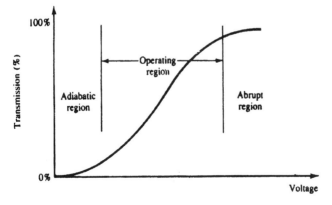

Figure 17.38   The transmission characteristics of an LCLV.

A typical optically addressed LCLV is shown in Fig. 17.39. A photoconductive material is used to transfer the optical input to an equivalent voltage on the LC layer. The photoconductor is highly resistive, and it thus utilizes most of the voltage when there is no incident light. Very little or no voltage can be applied across the LC layer to limit the transmission of the read-out light. In the presence of incident light, the resistivity of the photoconductor decreases, as a result of which more voltage appears across the LC layer. The intensity of the incident light thus engages a proportional amount of voltage across the LC layer. Accordingly, an input intensity variation will manifest itself as a voltage variation across the LC layer. Again, the coherent read-out beam illuminating the back of the LCLV is reflected back but modulated by the birefringent LC layer. Thus the input optical image is transferred as a spatial modulation of the read-out beam. The dielectric mirror present in the device provides optical isolation between the input in coherent beam and the coherent read-out beam. In practice, however, the LCLV is driven by the ac voltage. For frequencies with periods less than the molecular response time, the LCLV responds to the rms value of the voltage. An ac-driven LCLV allows flexibility in choosing both the type of photoconductor as well as the arrangement of the intermediate layers.

For the ac-driven LCLV, CdS is generally chosen as the photoconductor, while CdTe is used as the light-blocking material. The CdTe layer isolates the photoconductor from any read-out light that gets through the mirror. It is possible to feed the optical input data by means of a CRT, fixed masks, or even an actual real-time imagery. A typical LCLV has a 25-mm diameter aperture, a 15 ms response time, 60 lines/mm resolution at 50% modulation transfer function, 100:1 average contrast ratio, and a lifetime of several thousand operating hours.

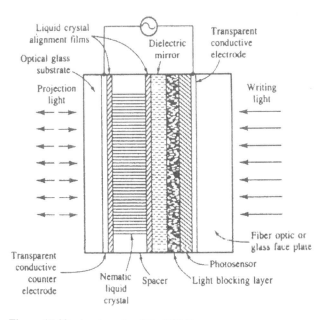

Figure 17.39 A schematic of the LCLV.

A more recent innovation is the CCD-addressed LCLV, as shown in Fig. 17.40. The CCD structure is introduced at the input surface of the semiconductor wafer in the LCLV. The first CCD register is fed with charge information until it is full. The content of this CCD register is then clocked into the next CCD register. The serial input register is again filled and then emptied in parallel. This process is repeated until the total CCD array is loaded with a complete frame of charge information. An applied voltage can then cause the charge to migrate across the silicon to mirror surfaces. The CCD-addressed LCLV generally requires a positive pulse to transfer the charge. This particular display device is very attractive because of its high speed, high sensitivity, high resolution, and low noise distortion.

The LCLV is able to provide image intensification because the read-out beam may be as much as five orders of magnitude brighter than the write beam. The efficient isolation provided by both the mirror and the light-blocking layer is responsible for such intensification as well as for the wavelength conversion between two beams. An LCLV is thus ideally suited to process infrared imagery. Infrared images typically consist of weak signals that require amplification before being processed further. In addition, it is easier to perform optical processing in the visible domain than in the infrared. LCLVs have already been applied to radar-signal processing, digital-optical computation, optical correlation, and optical image processing; in fact, they have many more applications. However, the response time of an LCLV (determined by the finite time it takes to rotate the LC molecules) is questionable for many of the operations. New LC materials that may improve the response time are currently being developed. The LCLV can be operated with all-coherent, all-incoherent, or mixed read-and-write beams. But the real limitation of this device happens to be its inflated cost.

Often in an adaptive system it becomes necessary to measure the difference between a signal and an estimate of the signal. This measured difference is often used

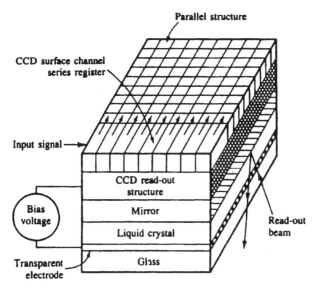

Figure 17.40   CCD-addressed LCLV.

to improve future estimates of the signal. Figure 17.41 shows such a feedback system, which uses a special spatial light modulator, known as a microchannel spatial light modulator, MSLM. The MSLM consists of an EO crystal, a dielectric mirror, a microchannel plate, and a photocathode. The signal wavefront passes through the beamsplitter (BS) and is then reflected by an optically controlled MSLM that serves as a phase shifter. After further reflections, a part of the beam passes through a second beamsplitter to generate a compensated signal beam. The remaining portion of the beam is mixed with a local oscillator to produce an error signal which in turn is used to control further phase shifting as well as to maintain phase compensation.

17.8 ACOUSTO-OPTIC MODULATION AND DEVICES

The terms acousto-optic (AO) or elasto-optic effects are used interchangeably to indicate that the refractive index of a medium is being changed either by a mechanically applied strain or by ultrasonic waves. Accordingly, a cousto-optic modulator consists of a medium whose refractive index undergoes a sinusoidal variation in the presence of an externally applied ultrasonic signal, as shown in Fig. 17.42. The solid lines indicate the regions of maximum stress, and the dashed lines indicate the regions of minimum stress. There are many materials, such as water, lithium niobate, lucite, cadmium sulphide, quartz, and rutile, that exhibit changes in the refractive index once they are subjected to strain.

As the light enters an AO medium, it experiences a higher value of refractive index at the region of maximum stress, and thus advances with a relatively lower velocity than those wavefronts that encounter the regions of minimum stress. The resultant light wavefront thus inherits a sinusoidal form. The variation in the acoustic wave velocity is generally negligible, and so we may safely assume that the variation of refractive index in the medium is stationary as far as the optical wavefront is concerned. A narrow collimated beam of light incident upon such a medium is thus scattered into primary diffraction orders. In most practical cases, higher-diffraction orders have negligible intensities associated with them. The zero-order beam generally has the same frequency as that of the incident beam, while the frequencies of $+1$ and -1 orders undergo a frequency modulation.

In order to appreciate the basics of acousto-optic effects, we can consider the collisions of photons and phonons. Light consists of photons that are characterized

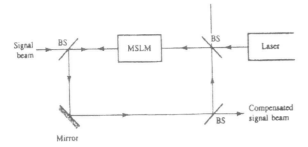

Figure 17.41 An adaptive compensating system.

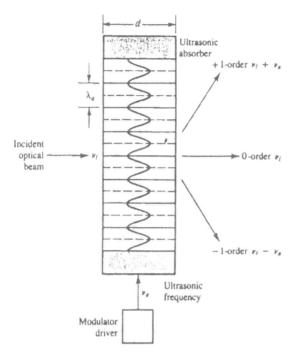

Figure 17.42 An acousto-optic modulator.

by their respective momentum, $hk_l/2\pi$ and $hk_a/2\pi$, where k_a and k_l are the respective wave vectors. Likewise, photon and phonon energies are given respectively by hv_l and hv_a, where v_l and v_a are the respective frequencies. Consider the scenario where a wave vector is given by k_l', as illustrated in Fig. 17.43. The condition for the conversation of momentum, when applied to this collision, yields

$$k_l \cos \theta = k_l' \cos \theta' \tag{17.72a}$$

and

$$k_a = k_l \sin \theta + k_l' \sin \theta', \tag{17.72b}$$

where θ and θ' are the angles formed by the incident and the scattered photons, respectively. Consequently, the angle of scattered photon is evaluated to give

$$\theta' = \arctan\left[\frac{k_a}{k_l} \sec \theta - \tan \theta\right]. \tag{17.73}$$

It is reasonable to assume that $k_a \ll k_l$ and, thus, for small values of θ, Eq. (17.73) reduces to

$$\theta' = \left[\frac{c}{v_a v_l}\right] v_a - \theta \tag{17.74}$$

where v_a is the acoustic velocity. Equation (17.74) explicitly shows that the angle formed by the scattered photon is proportional to the acoustic frequency. By mea-

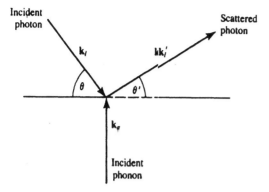

Figure 17.43 Photon–phonon collison resulting in the annihilation of a phonon.

suring the deflection angle, we can estimate the acoustic frequency. Further, Eq. (17.73) reveals that for an incident angle $\theta = \theta_B = \sin^{-1}(k_a/2k_1)$ in an isotropic medium, $\theta = \theta'$ and $k_1 = k_1'$. At this particular angle of incidence, referred to as the Bragg angle, photon momentum is conserved and the diffraction efficiency reaches a maximum. Note that the power in the scattered beam varies with θ and reaches a maximum when θ is equal to the Bragg angle.

It is important to realize that the acousto-optical effect is produced by multiple collisions of photons and phonons. In any event, the scope of Eq. (17.74) is somewhat valid in most practical devices. The condition for conservation of energy is only approximately valid in photon–phonon collision. However, in practice, the frequency of the scattered photon $\nu_1 = \nu_1'$, since $\nu_a \ll \nu_1$. In anisotropic materials, k_1' approaches rk_1, where r is the ratio of the refractive indices corresponding to the diffracted and incident waves, respectively. Equations (17.72a) and (17.72b) can be modified to give

$$\theta = \sin^{-1}\left[\frac{k_a}{2k_1}\left\{1 + \left(\frac{k_1}{k_a}\right)^2 (1 - r^2)\right\}\right] \tag{17.75a}$$

and

$$\theta' = \sin^{-1}\left[\frac{k_a}{2rk_1}\left\{1 - \left(\frac{k_1}{k_a}\right)^2 (1 - r^2)\right\}\right]. \tag{17.75b}$$

But to have valid solutions, the condition

$$1 - \frac{k_a}{k_1} \ll r \ll 1 + \frac{k_a}{k_1} \tag{17.76}$$

must be satisfied. It is obvious that θ and θ' are equal only when $r = 1$ because it is not possible to have $r = (k_a/k_1) - 1$ when $k_a \ll k_1$. Thus, the phenomenon $\theta = \theta'$ is associated only with the Bragg angle of incidence and the condition $r = 1$. In general, for an incident wave vector, there are two values of k_a (and thus k_1') that satisfy the condition of conservation of momentum. Note that in anisotropic media, the conservation of momentum is satisfied over a wider range of acoustic frequencies or incident light beam directions than is normally realizable in isotropic materials. Consequently, in acousto-optical devices, birefringent diffraction plays a dominant role in determining modulation.

The diffraction of the light beam in AO modulators is justifiably associated with a diffraction grating set up by the acoustic waves. The exact characteristics of this diffraction are indicated by the parameter $Q = k_a^2 d/k_1$, where d is the width of the acoustic-optic device. When $Q < 1$, the diffraction is said to operate in the Raman–Nath regime, and when $Q \gg 1$, the diffraction is said to operate in the Bragg regime. In the region where $1 \ll Q \ll 10$, the diffraction has a characteristic that is a mixture of the two extremes. Since Q is directly proportional to d, a higher Q requires lesser drive power for any given interaction efficiency. In the Raman–Nath regime, the acoustic-optic grating can be treated as a simple grating, such that

$$m\lambda = \lambda_a \sin\theta_m, \tag{17.77}$$

where λ_a is the acoustic wavelength, m is an integer, and θ_m is the corresponding angle of diffraction. By comparison, in the Bragg regime, the acoustic field acts very much like a "thick" diffraction grating, requiring that

$$\theta = \theta' = \sin^{-1}\left(\frac{m\lambda}{2\lambda_a}\right). \tag{17.78}$$

Bragg diffraction is identical to that of a plane grating when the angle of incidence equals the diffracting angle. Reflected waves, except those for which $\theta = \theta'$, interfere constructively, producing a very strong first-order component.

The fraction of the light diffracted is often characterized by the diffraction efficiency η, defined as $(I_0 - I)/I_0$, where I is the output irradiance in the diffraction orders and I_0 is the output irradiance in the absence of the acoustic waves. While the diffraction efficiency of the Raman–Nath grating is only about 0.35, it approaches 1.00 for the Bragg case. At the Bragg angle, the diffraction efficiency is given by $\sin^2[(\pi\Delta nd)/(\lambda\cos\theta_B)]$, where Δn is the amplitude of the refractive index fluctuation.

A Bragg cell can be used to switch light beam directions by turning on and off the acoustic source. The intensity of the diffracted light, however, depends on the amplitude of the acoustic wave. An amplitude modulation of the acoustic wave will, therefore, produce amplitude-modulated light beams. But again the movement of the acoustic waves produces a moving diffraction grating, as a result of which the frequencies of the diffracted beams are Doppler-shifted by an amount $+/- mv_a$. This frequency shifting can be effectively manipulated to design frequency modulators. AO modulator transfer function is sinusoidally dependent on the input voltage; however, this presents no difficulty in on-and-off modulation. For analog modulation, it is necessary to bias only the modulator at a carrier frequency such that the operating point is in an approximately linear region of operation. When compared with an EO modulator that consumes voltage on the order of 10^3 V, an AO modulator requires only a couple of volts. But since the acoustic wave propagation is slow, the AO devices are often limited by the frequency response of the acoustic source, figure of merit, and acoustic attenuation. Most of the AO materials are lossy. Materials with high figures of merit normally have a high attenuation. The most commonly used AO materials are quartz, tellurium dioxide, lithium niobate, and gallium phosphide. The materials with lower figures of merit are also used, but they operate with a higher drive power. A practical limit for small devices is a drive power density of 100–500 W/mm^2, provided there is a proper heat sink. Bandwidths of up to 800 MHz are common in most commercial AO modulators.

AO modulators are used widely in a large number of applications, such as laser ranging, signal-processing systems, optical computing, medium-resolution high-speed optical deflectors, acoustic traveling-wave lens devices, and mode-locking. Figure 17.44 shows a system where AO modulators are used to support beam scanning of a laser printer. In the laser printers, a rotating drum with an electrostatically charged photosensitive surface (a film of cadmium sulfide or selenium on an aluminum substrate) is used so that a modulated laser beam can repeatedly scan across the rotating surface to produce an image. The most commonly used beam-scanning system utilizes an He–Ne laser, a modulator, and a rotating polygonal prism. The He–Ne laser is preferred over other lasers because the photosensitive layer is sensitive to its output. But since it is difficult to modulate an He–Ne laser internally, the mdoulation is done externally. We can also use an EO cell for the modulator, but an AO modulator is preferred because of its ability to operate with unpolarized light, and also because it requires a low-voltage power supply.

AO modulators are also used in systems involving optical disks. Quite like their audio counterparts, optical disks store information in optical tracks. But in the case of optical disks, there is neither a groove nor a continuous line present, but rather "pits" that are small areas providing an optical contrast with respect to their surroundings. The varied reflectance along the track represents the information stored. These disks are versatile in the sense that they can be used for both direct read-out and recording. Figure 17.45 shows one such direct-read-after-write optical disk system. The write laser usually has more power than the read laser. The more sophisticated systems are arranged to have angular and polarization separation of beam to ensure that the read beam does not interfere with the reflections of the write beam.

17.9 SUMMARY

This chapter provides only a brief introduction and discussion of the basic role of photoconduction, photodetection, and electro-optic and acousto-optic modulation.

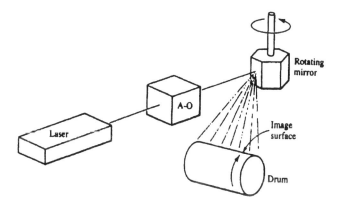

Figure 17.44 A beam-scanning laser printer.

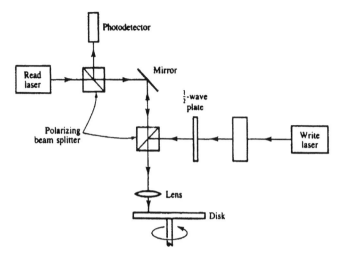

Figure 17.45   Direct-read-after-write optical disk system.

Some of the more significant applications of these concepts have contributed to various detection, imaging, amplification, modulation, and signal-processing systems which have also been discussed in this chapter. For details on many of the systems considered in this chapter, as well as for elaboration on other variations, one will need to use other reference sources and public-domain publications.

REFERENCES

Banerjee, P. P. and T. C. Poon, *Principles of Applied Optics*, Irwin, Boston, Mass, 1991.

Berg, N. J. and J. N. Lee, *Acousto-Optic Signal Processing*, Marcel Dekker, New York, 1983.

Karim, M. A., *Electro-Optical Devices and Systems*, PWS-Kent Pub, Boston, Mass, 1990.

Karim, M. A. and A. A. S. Awwal, *Optical Computing: An Introduction*, John Wiley, New York, 1992.

Lizuka, K., *Engineering Optics*, Springer-Verlag, New York, 1983.

Pollock, C. R., *Fundamentals of Optoelectronics*, Irwin, Boston, Mass, 1995.

Tannas, L. E., Jr., ed., *Flat-Panel Displays and CRTs*, Van Nostrand Reinhold, New York, 1985.

Wilson, J. and J. F. B. Hawkes, *Optoelectronics: An Introduction*, Prentice-Hall International, Englewood Cliffs, New Jersey, 1985.

Yu, F. T. S., *Optical Information Processing*, John Wiley, New York, 1983.

Yu, F. T. S. and I. C. Khoo, *Principles of Optical Engineering*, John Wiley, New York, 1990.

18

Radiometry

MARIJA STROJNIK and GONZALO PAEZ
Centro de Investigaciones en Optica, León, Mexico

18.1 INTRODUCTORY CONCEPTS

Radiometry refers to the measurement of radiation. The word radiation is closely associated with radio waves, because Hertz generated radio waves in 1887 when electromagnetic waves were first generated in a laboratory using electronics. Based on the earlier theoretical work by Maxwell, we know today that the radiation of different wavelengths spans the whole electromagnetic spectrum; see Fig. 18.1 and Table 18.1. Within the broad area of optical engineering, we are interested in the ultraviolet (UV), visible, infrared (IR), and often, millimeter waves. Many of the concepts apply equally well to the X-ray region, which has recently been gaining in importance. Historically, humans have more narrowly focused their interest on the visible wavelength region, the portion of the electromagnetic spectrum where the human eyes are sensitive. This has resulted in the development of a special branch of radiometry, tailored to the human eyes as detectors, referred to as illumination engineering, together with the development of specific units related to human vision.

As we learned that radiation existed outside the visible region, we found it advantageous to apply a uniform terminology for the quantities related to the optical radiation and to the units applicable to the radiation of all wavelengths. Thus, we prefer to use the terminology for the optical radiation which is based on power (the standard terminology in physics) for the shorter wavelengths, and that based on frequency (or wavelength), used traditionally in electrical engineering and infrared, for the longer wavelengths. We refer to those as radiometric quantities and units. The MKS units have been widely accepted in the international communities, based on the proposition that we can measure the radiative power at specific wavelengths or frequencies in watts. When the visible radiation is considered just a portion of the electromagnetic spectrum, the radiometric terminology and units may also be

649

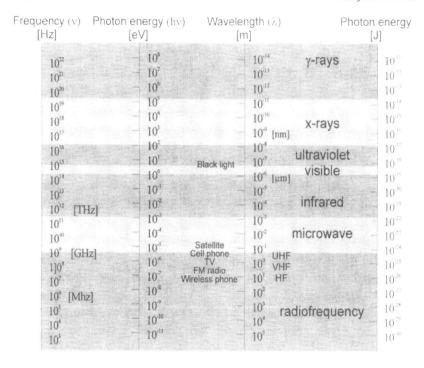

Figure 18.1 The electromagnetic spectrum.

applied. However, when referring to the effect of the visible radiation on the particular detector of the *human* visual system, the *photometric* terminology and units are preferred. They are discussed in Section 18.4.

The advantages of using the concept of power are obvious when we think of the radiation as the carrier of information from one point in space to a different point in space. The information may be coded in terms of the wavelength, or its spectral content, or, even better, in terms of the amount of the radiation at the specific wavelength that is being transferred. The amount of radiation of a specific wavelength that is being transferred is the spectral radiative power.

While the amount of radiation may be characterized by its power, its spectral characteristics may be defined by the wavelength region: we are familiar with the rainbow produced by water drops or dispersion generated with a glass prism. The concept of wavelength has been used advantageously in the infrared, or the long-wavelength region of the electromagnetic spectrum because this radiation has very low energy. The simplest way of visualizing the wavelength is by considering the generation of the radiation with an oscillating dipole, as in radio transmission. A wavelength is that distance in space for which the electromagnetic radiation has the same phase and the same algebraic sign of its derivative. For the radio waves, we actually prefer to use the concept of frequency, measuring the number of waves that pass a given position in space per unit time:

$$\nu = c/\lambda, \qquad [\text{Hz}] \tag{18.1}$$

where c is speed of the radiation in [m/s], in vacuum $= 2.9972 \times 10^8$ m/s, λ is the wavelength in [m], or usually in [μm] for convenience, and ν is frequency in [s^{-1}] or [Hz]. We use a square bracket to denote the units. In radiometry, a consistent use of units helps avoid confusion due to varied, and oftentimes, confusing, terminology.

In the shorter wavelength regions, we are familiar with the effects of UV radiation, which is powerful enough to produce changes in material physical characteristics, ranging from burning the human skin to causing glass to darken. In the short-wavelength region, we characterize the radiation by energy carried by the smallest packet of such radiation, called a photon. Its energy is

$$Q_q = h\nu = hc/\lambda_q, \qquad [\text{J}] \tag{18.2}$$

where Q is energy [J] and h is Planck's constant ($= 6.6266176 \times 10^{-34}$ Js).

There are two ways of looking at the matter. In the first one, referred to as the microscopic point of view, we consider the individual atoms, even if there are many of them. The second one, called the macroscopic view, deals with the collective behavior of very many atoms together, in an assembly.

Radiation may be viewed similarly. When we think of *the power of the radiation*, or the radiative power, we can define it macroscopically as the power of the radiation of a certain wavelength, or of a certain frequency. When we assume this view, we do not have to accept the concept of the minimum energy carried by a single photon. This point of view is most widely accepted in optical *engineering* applications. It deals with the type of power that is defined in mechanical and electrical engineering. It is the power that may be used to generate work; it is the power that generates heat when not used wisely; it is the power that is the energy expanded per unit time. This concept of power is used in Section 18.2 to define the radiometric concepts,

$$P = dQ/dt. \qquad [\text{W}] \tag{18.3}$$

where P is the power in [W], Q is energy in [J], t is the time in [s], and d(quantity A)/d(quantity B) = derivative of quantity A with respect to quantity B in units of [A/B].

Another concept of power may be introduced when we consider the quantum nature of light, with the number of quanta carrying energy per unit time to produce the "photon" power P_q. This power, and the quantities associated with it, are presented in Section 18.3. To distinguish them from the macroscopic, the more commonly used quantities, we denote them with the subscript q; i.e.,

$$P_q = dn/dt, \qquad [\#/\text{s}] \tag{18.4}$$

where P_q is the photon power, the power transferred by N photons per unit time, and $n =$ the number of photons in [#].

18.2 STANDARD TERMINOLOGY FOR RADIOMETRIC QUANTITIES

Unfortunately, a standard terminology for radiation quantities does not exist. Every field and just about every application has found its particular nomenclature and

Table 18.1 The Electromagnetic Spectrum, Its Natural Sources, Detection, and Manmade Generators

Natural sources	Detection	Artificial sources	Wavelength (λ) [m]		Photon energy [J]
			10^{14}	γ-rays	10
Atomic nuclei	Scintillation and Geiger counters	Accelerators	10^{13}		10^{16}
			10^{12}		10^{17}
	Ionation Chamber	X-ray tubes	10^{11}		10^{14}
Inner			10^{10}	x-rays	10^{15}
electrons	Photografic film		10^{9}		10^{16}
	Photomultiplier	Synchrotons	10^{8}		10^{17}
Outer electrons	GaN photodetector Eye	Hg lamps LEDs	10^{7}	ultraviolet	10^{18}
Molecular vibration	Bolometer		10^{6}	visible	10^{19}
	HgCdTe photovoltaic	Hot bodies	10^{5}		10^{20}
Molecular rotaion	Thermopile		10^{4}	infrared	10^{21}
		Magnetron	10^{3}		10^{22}
Electron spin			10^{2}	microwave	10^{23}
Nuclear spin	Resonant		10^{1}		10^{24}
	electronic circuits	Electronic resonator	10^{0}		10^{25}
			10^{1}		10^{26}
			10^{2}		10^{27}
			10^{1}	radiofrequency	10^{28}
			10^{4}		10^{29}
		AC power lines	10^{5}		10^{31}

symbols. Here, we adopt the terminology recommended by Nicodemus [1] and made popular by William L. Wolfe and George J. Zissis' *The Infrared Handbook*. [2] Table 18.2 includes a list of the commonly used radiometric terms.

18.2.1 Power

We consider that the primary characteristics of the radiation is to propagate in free space: it cannot stand still. Due to its "motion" or "propagation," it is often referred to as the radiant flux, emphasizing the transfer of energy past some imaginary surface in space. When it encounters a material surface it is either reflected or absorbed by it. We can say that the electromagnetic radiation carries the optical energy across space in time – *it moves with the speed of light*. The amount of energy transferred per unit time is defined as power, given in Eq. (18.3). Additional terms that may be found in the literature, describing the fleeting nature of radiation, include the radiative power and the radiative flux.

The terms derived from the concept of the power are related to the geometrical consideration and the spectral content of the radiative power. We refer to the spectral power as the power of wavelength λ found in the narrow wavelength interval δλ:

$$\delta P(\lambda) = P_{\lambda}(\lambda)\,\delta\lambda, \qquad [W] \tag{18.5}$$

Table 18.2 List of Commonly Used Radiometric Terms

Term	Symbol	Defining equation	Units
Radiant energy			J
Radiant power	P		$J/s = W$
Radiant exitance	M	$M = \dfrac{dP}{dA}$	W/m^2
Radiant incidance	E	$E = \dfrac{dP}{dA}$	W/m^2
Radiant intensity	I	$I = \dfrac{dP}{d\omega}$	W/sr
Radiance	L	$L = \dfrac{\partial^2 P}{\partial A_p \partial \omega}$	$W/(m^2\ sr)$
Solid angle	ω	$\omega = \dfrac{A}{r^2}$	sr
Frequency	v		1/s
Wavelength	λ	$\lambda = \dfrac{c}{v}$	m
Wavenumber	m	$m = \lambda^{-1}$	1/m

where $\delta P(\lambda)$ is the infinitesimal element of power of the wavelength λ in the wavelength interval $\delta\lambda$, $\delta\lambda$ is the width of the infinitesimal wavelength interval in [μm], and $P_\lambda(\lambda)$ is the spectral power of wavelength λ in [W/μm].

We denote spectral quantities with the subscript λ, explicitly indicating a derivative with respect to the wavelength λ. The laser is one popular source of coherent, narrow wavelength-band radiation. Equation (18.5) describes well the quasi-monochromatic power output of a laser source, due to its power output in the narrow wavelength range. The majority of sources, though, emit the radiation in a wide wavelength interval. For those, the spectral power is defined as follows:

$$P_\lambda(\lambda) = dP(\lambda)/d\lambda. \quad [W/\mu m] \tag{18.6}$$

Most of the natural sources, and, therefore, most of the naturally occurring radiation is broadband. Additionally, the electro-optical systems used to transfer the radiation do so effectively only between two wavelengths λ_1 and λ_2, or within a wavelength interval $[\lambda_1, \lambda_2]$.

$$\Delta\lambda = \lambda_2 - \lambda_1. \quad [\mu m] \tag{18.7}$$

The radiative power within a wavelength interval depends not only on the width of the wavelength interval $\Delta\lambda$, but also on specific choice of the wavelength limits λ_1 and λ_2. The radiative power within a wavelength interval is obtained by integrating Eq. (18.6) over the wavelength, from wavelength λ_1 to wavelength λ_2,

$$W_{[\lambda_1, \lambda_2]} = \int W_\lambda(\lambda)\, d\lambda. \quad [W] \tag{18.8}$$

We find it advantageous to record the wavelength subscripts explicitly for the wavelength interval to remind us that the integrated power (or any other radiometric quantity), is considered, evaluated, or integrated over a given wavelength interval. We refer to such power as the power in the wavelength interval from λ_1 to λ_2.

18.2.2 Exitance and incidance

One of the important concepts to keep in mind is that the radiation is generated by matter, whose spatial extent is defined by its surfaces. The radiation may similarly be absorbed by matter. The creation and destruction of radiation requires the presence of matter. The radiative power may be generated within the boundaries defined and limited by surfaces. Similarly, it may be incident on a surface.

The surface characteristics most likely vary from point to point. In both of these cases, the properties of the surface vary as a function of position – generating or absorbing a different amount of radiation depending on the surface spatial coordinates. We can thus define the power (area) density as the infinitesimal amount of power incident on the infinitesimal amount of area δA:

$$\delta P = p \, \delta A, \qquad [\text{W}] \tag{18.9}$$

where δP is the infinitesimal radiative power incident on the infinitesimal area δA in [W], p is the radiative power (area) density in [W/m^2], and δA is the infinitesimal area in [m^2].

It is important to distinguish radiative power (area) density from radiative power (volume) density. The volume power density is the relevant quantity when discussing the radiation in a cavity or the radiation generated by the volume, or bulk, radiators. From Eq. (18.9) we define the radiative power (area) density as follows:

$$p = dP/dA. \qquad [\text{W/m}^2] \tag{18.10}$$

This quantity is often called irradiance, and denoted by E with units [W/m^2]. In radiometry, we do not use this term, due to the possibility of ambiguities.

When we consider a specific spectral component of the radiative power, Eq. (18.10) may be modified so that it refers to the spectral power area density

$$p_\lambda = dP_\lambda/dA. \qquad [\text{W/(m}^2\,\mu\text{m})] \tag{18.11}$$

In optical engineering applications, the radiation leaving the surface (or, *exiting* the surface) is usually different from the radiation incident on the surface. For example, the spatial characteristics of radiation may be different when we use optical components to collect or redistribute the radiation. The radiation leaves the natural, nonpolished surface dispersed in all directions. The radiation may be incident on the surface within only a narrow cone specified by the F-number of the optical system, when the surface is the detector element in a radiometer.

Let us consider a beam of radiation incident on a specific surface. A portion of it is reflected from the surface, a portion is transmitted, and a portion is absorbed, depending on the material characteristics. The surface spectral reflectivity modifies the amount of reflected radiation, and, thus, its spectral distribution. The surface shape and its finish modify the directional characteristics of the reflected and transmitted radiation. For these reasons, the term irradiance is considered ambiguous within the radiometric community.

The nature of the surface shape, finish, and composition changes the spectral and directional characteristics of the reflected or transmitted radiation. As the surface modifies the radiation, we find it preferable to differentiate between the radiation before it is incident on the surface and after it is incident on and reflected off the surface. The power (area) density for the radiation incident on the surface is called the incidance, and is denoted by E in [W/m^2]:

$$E(x, y, z) = dP/dA. \quad [\text{W/m}^2] \tag{18.12}$$

The spectral incidance is similarly defined for each spectral component:

$$E_\lambda(x, y, z) = dP_\lambda/dA. \quad [\text{W/(m}^2\,\mu\text{m)}] \tag{18.13}$$

The radiation leaving the surface, either generated by the matter or reflected from it, is described by the term exitance, and is denoted by $M(x, y, z)$ with units [W/m^2]:

$$M(x, y, z) = dP/dA. \quad [\text{W/m}^2] \tag{18.14}$$

The spectral exitance $M_\lambda(x, y, z)$ is defined for each spectral component

$$M_\lambda(x, y, z) = dP_\lambda/dA. \quad [\text{W/(m}^2\,\mu\text{m)}] \tag{18.15}$$

18.2.3 Radiance

The propagation of the radiation may be guided or manipulated using one or more optical components, resulting in a change in power density from one point to the next along the propagation path. The imaging components, such as lenses and mirrors, change the angular distribution of the radiation in order to modify the spatial extent and direction of propagation of radiation. Thus, the incidance is not a particularly suitable parameter to characterize the radiation in an optical system.

The objects that generate radiation are called sources. They typically emit radiation with an angular dependence that is a consequence of the source physical characteristics, shape, form, layout, and construction. Thus, the directional characteristics of emitted radiation determine the source capacity to produce useful radiative power. Additionally, the sources typically do not generate the power uniformly over its surface area, due to the surface nonuniformity.

Thus, a quantity that is suitable to characterize spatial and angular radiative properties of an extended (surface) area source is required. It is called the radiance, L in [W/(m^2 sr)]:

$$L(x, y, z; \theta, \varphi) = \partial^2 P/(\partial A_\text{p}\,\partial\omega), \quad [\text{W/(m}^2\text{ sr)}] \tag{18.16}$$

where ω is the solid angle in steradians, [sr], and A_p is projected area in [m^2]; $\partial A/\partial B$ denotes a partial derivative of quantity A with respect to quantity B. The use of partial derivatives is required when a function depends on several variables.

The source radiance may also be given as a spectral quantity when the spectral power is considered:

$$L_\lambda(x, y, z; \theta, \varphi) = \partial^2 P_\lambda/(\partial A_\text{p}\,\partial\omega). \quad [\text{W/(m}^2\text{ sr }\mu\text{m)}] \tag{18.17}$$

Because of the directional nature of the emitted radiation from the extended area sources, the spectral radiance is the quantity used to describe the radiative characteristics of such a source.

In an optical system the beam cross section often decreases at the same time as the solid angle increases, resulting in simultaneous changes in area and solid angle in an optical beam. Thus, the radiance is also the most appropriate radiometric quantity to characterize the transfer of radiative power in an optical system.

18.2.4 Intensity

The radiation may propagate through the free space, but it cannot be generated or annihilated (aborbed) in free space. For these two phenomena to take place, the presence of matter is required. In particular, when talking about a radiative source, matter is required, so that its energy level is changed in order to generate the radiation. A material body is always of finite dimensions. As already mentioned in the previous section, the radiance is used to describe the spatial and directional characteristics of the radiation emitted from an extended-area source.

Often, though, we find it convenient to talk about point sources. While they do not exist in the physical world, we define the point source as the source whose spatial extent is not resolved by the resolution of the optical system used to image it. For this reason, the concept of the point source depends solely on the sensor resolution, and not on the source spatial extent. We may consider stars as point sources, even though their size is quite large. The human visual system resolves only the nearest star, the sun. Therefore, it is usually not considered a point source, while the other stars are.

Even though the characteristics of the area distribution of the radiation emitted by a point source cannot be resolved with the sensor, the directional characteristics of its radiation are of uttermost importance. There may be another reason why we may want to talk about point sources – we are not interested in their spatial characteristics. This may be because we are far away from the source, because its spatial characteristics are within the uniformity tolerances of our application, or because the system analysis is significantly simplified if the source is treated as a point. This is often the simplifying assumption when analyzing laser beam propagation: we define an imaginary point source, such that the laser beam appears to originate from it.

We do not consider that the spatial distribution of the power emitted by the point source is important; however, its angular distribution is significant. Thus, we define the concept of intensity with the symbol I [W/sr], the power per unit solid angle:

$$I(\theta, \varphi) = dP/d\omega. \qquad [\text{W/sr}] \qquad (18.18)$$

Often, the quantity of interest is the spectral intensity $I_\lambda(\theta, \varphi)$

$$I_\lambda(\theta, \varphi) = dP_\lambda/d\omega. \qquad [\text{W/(sr \mu m)}] \qquad (18.19)$$

The angular dependence of the power radiated by the point source into a solid angle is given by the spectral intensity. A radiator is said to be isotropic if its intensity does not depend on direction.

The intensity is the most appropriate quantity to characterize a point source.

18.3 PHOTON-BASED RADIOMETRIC QUANTITIES

There are two ways of looking at radiation. They were considered mutually exclusive prior to the acceptance of the quantum theory at the beginning of the 20th century; now they are considered complementary. Radiation may be considered as a continuous flow of waves that carry energy (see Eq. (18.1)). The alternative, equally useful, way of looking at radiation is that this same energy is carried in quanta, each containing the minimum packet of energy equal to the change between specific (electronic) energy states in an atom, molecule, or a solid. These changes between the energy levels, E_1 and E_2, are also referred to as energy transitions.

$$\Delta E_q = E_1 - E_2 = h\nu = hc/\lambda_q, \qquad [\text{J}] \tag{18.20}$$

where E_1 and E_2 are the two energy levels involved in the transition. Radiation may be considered to consist of packets and waves, both carrying the energy, which is referred to as the dual nature of the radiation. The smallest packet of radiation with energy ΔE_q is called a photon.

Instead of describing the radiation in terms of the energy transferred across an imaginary surface per unit time, or power, we can take advantage of the quantum representation to emphasize its other aspect. Sometimes, it is more appropriate or convenient to count the number of photons that pass an imaginary surface per unit time. The photon-based quantities are defined similarly to the power-based terms. We use a subscript q to distinguish the photon-based quantities from the power-based quantities. When it is necessary to emphasize the power-based quantities the subscript e is used to indicate the energy or power-based units. Usually, though, the subscript e is omitted for the power-based quantities.

The photon flux P_q is the number of photons crossing an imaginary surface per unit time:

$$P_q = dn_q/dt. \qquad [\text{\#/s}] \tag{18.21}$$

Here, dn_q/dt is the number of photons per unit time, which may be evaluated by considering that the power transferred is equal in both representations:

$$P = P_q(h\nu) = (dn_q/dt)(h\nu) = (dn_q/dt)(hc/\lambda). \qquad [\text{W}] \tag{18.22}$$

To obtain the last two equalities we used Eq. (18.4). From here we obtain for the photon flux or photon power

$$P_q = dn_q/dt = P/(h\nu) = P(\lambda/hc). \qquad [\text{\#/s}] \tag{18.23}$$

The spectral photon flux is the power that includes only the photons of the same energy, or with the same wavelength:

$$P_{q,\lambda} = dn_{q,\lambda}/dt = P_\lambda/(h\nu) = P_\lambda(\lambda/hc). \qquad [\text{\#/(s\,µm)}] \tag{18.24}$$

The photon incidance E_q is defined similarly:

$$E_q = (\partial/\partial A)(dn_q/dt) = E/(h\nu) = E(\lambda/hc). \qquad [\text{\#/(s\,m}^2)] \tag{18.25}$$

The photon spectral incidance is, correspondingly,

$$E_{q,\lambda} = (\partial/\partial A)(dn_{q,\lambda}/dt) = E_\lambda/(h\nu) = E_\lambda(\lambda/hc). \qquad [\text{\#/(s\,m}^2\,\text{µm)}] \tag{18.26}$$

Photon exitance M_q is defined as

$$M_q = (\partial/\partial A)(dn_q/dt) = M/(h\nu) = M(\lambda/hc). \qquad [\text{\#/(s\,m}^2)] \tag{18.27}$$

The photon spectral exitance $M_{q,\lambda}$ is given by

$$M_{q,\lambda} = (\partial/\partial A)(dn_{q,\lambda}/dt) = M_\lambda/(h\nu) = M_\lambda(\lambda/hc). \qquad [\text{\#/(s\,m}^2\,\text{µm)}] \tag{18.28}$$

Additionally, the photon intensity $I_q(\theta, \varphi)$ may be defined as

$$I_q(\theta, \varphi) = dP_q/d\omega. \qquad [\#/\text{sr}] \qquad (18.29)$$

Likewise, photon spectral exitance $I_{q,\lambda}(\theta, \varphi)$ is correspondingly,

$$I_{q,\lambda}(\theta, \varphi) = dP_{q,\lambda}/d\omega. \qquad [\#/(\text{s sr} \, \mu\text{m})] \qquad (18.30)$$

In the photon-based formalism, the radiance is also the most appropriate quantity to characterize an extended-area source

$$L_q(x, y, z; \theta, \varphi) = \partial^2 P_q/(\partial A \, \partial \omega). \qquad [\#/(\text{s m}^2 \, \text{sr})] \qquad (18.31)$$

Similarly, the photon spectral radiance $L_{q,\lambda}(x, y, z; \theta, \varphi)$ may be defined in the same way:

$$L_{q,\lambda}(x, y, z; \theta, \varphi) = \partial^2 P_{q,\lambda}/(\partial A \, \partial \omega). \qquad [\#/(\text{s m}^2 \, \text{sr} \, \mu\text{m})] \qquad (18.32)$$

18.4 PHOTOMETRIC QUANTITIES AND UNITS

Photometric quantities, with their corresponding units, have been developed to deal with the use of a specific detector, i.e., the human eye. [3] They incorporate the response of the human eye to the incident radiation. Thus, they are defined to deal with the radiation only within the wavelength interval where the human eye is sensitive. For a typical observer, this wavelength region is from $0.380 \, \mu\text{m}$ to $0.760 \, \mu\text{m}$.

Photometric quantities and units have been developed to describe the sensation that reports the standard observer in response to the radiation of 1 W at a specific wavelength. Table 18.3 lists the basic photometric quantities, their defining equations, and units. In this system of units, therefore, there is no radiation if it cannot be detected by the average human observer.

Figure 18.2 shows the nominal eye response of a standard observer to a constant amount of radiation of 1 W, as a function of wavelength, for both photopic (the $K(\lambda)$ curve) and the scotopic (the $K'(\lambda)$ curve) vision. Photopic vision refers to vision under conditions of high illumination and includes color vision. Scotopic vision does not permit color recognition and is effective under conditions of low illumination.

The $K(\lambda)$ curve is also called the luminous efficacy, as it describes the ability or the efficiency of a human eye to sense the incident radiation of different wavelengths in the visible range. Specifically, the meaning of this curve is as follows: the response of a human observer to the constant monochromatic radiation is solicited as the wavelength of the incident radiation is varied.

The standard human observer reports the sensation of much more radiation when the wavelength of the incident light is $0.55 \, \mu\text{m}$ than when the wavelength is only $0.4 \, \mu\text{m}$, even though the incident (physical) radiative power is the same. Specifically, the observers were able to quantify their perceptions that the radiation of one given wavelength is sensed as twice as intense as radiation of another wavelength, even though the incident radiation had the same (physical) power in both cases. Thus, the response curve changes as a function of the wavelength. There is no response to the radiation outside the interval [0.380–0.760 μm]. Thus, photometric units may not be used to characterize the radiation outside this

Table 18.3 Basic Photometric Quantities, Their Defining Equations, and Units

Term	Symbol	Defining equation	Units
Radiative power (luminous flux)	P_v	$P_v = K_m \int_{0.380\,\mu m}^{0.760\,\mu m} P_\lambda V(\lambda)d\lambda \frac{dP}{dA}$	lm
Luminous intensity	I_v	$I_v = \dfrac{dP_v}{d\omega}$	lm/sr, cd
Luminance (luminous radiance)	L_v	$L_v = \dfrac{\partial^2 P_v}{\partial A_p \partial \omega}$	lm/(m² sr), cd/m²
Illuminance	E_v	$E_v = \dfrac{dP_v}{dA}$	lm/m²
Luminous exitance	M_v	$M_v = \dfrac{dP_v}{dA}$	lm/m²
Luminous efficacy function	$K(\lambda)$	$K(\lambda) = K_m V(\lambda)$	

interval; i.e., the radiation outside this interval expressed in photometric units has a value of zero.

The efficacy curves $K(\lambda)$ normalized to 1 at their peak responsivity are called the relative spectral luminous efficiency for the CIE-standard photometric observer.

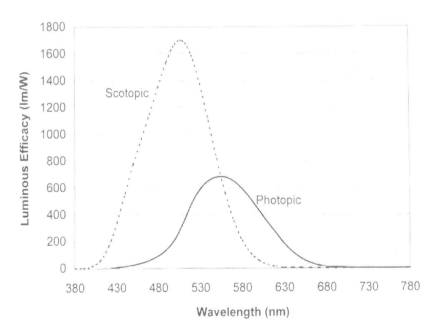

Figure 18.2 Spectral luminous efficacy functions $K(\lambda)$ and $K'(\lambda)$ for photopic and scotopic vision.

They are designated as $V(\lambda)$ and $V'(\lambda)$ curves, respectively, and are shown in Fig. 18.3. At the wavelength of 0.555μm, where the $V'(\lambda)$ curve attains a peak, 1 W of (physical) radiative power equals 683 lumens. At other wavelengths, the curve provides the corrective factor by which the source spectral power in watts needs to be multiplied to obtain the source luminous power in lumens.

More precisely, the luminous power P_v in lumens [lm] is the radiative power detected by a standard human observer:

$$P_v = K_m \int_{\lambda_1=0.380\,\mu m}^{\lambda_2=0.760\,\mu m} P_\lambda V(\lambda) d\lambda, \qquad [lm] \tag{18.33}$$

where P_v is the luminous power in lumens [lm] or talbots per second [talbot/s], K_m is the maximum luminous efficacy, equal to 683 lm/W, $V(\lambda)$ is the relative spectral luminous efficacy for the CIE-standard photometric observer, with the tabulated values given in Table 18.4, and $\lambda_1 = 0.380\,\mu m$, $\lambda_2 = 0.760\,\mu m$ is the wavelength interval within which $V(\lambda)$ and $V'(\lambda)$ have values different from zero. The subscript v is used to designate the quantities related to the human visual system and for dealing with the illumination (the radiation as sensed by the human eye). The source luminous power depends on its spectral distribution, weighted with respect to that of the spectral luminous efficiency.

The luminous energy ε_v in lumen-second [lm s] (also called talbot) is the radiative energy detected by the standard observer:

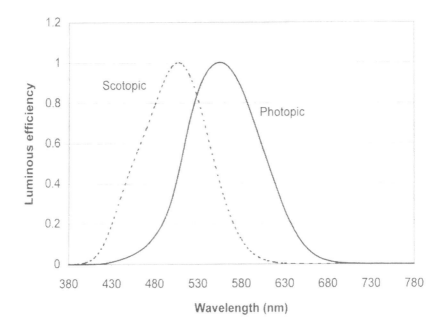

Figure 18.3 Spectral luminous efficiency $V(\lambda)$ and $VK'(\lambda)$ for the standard photometric observer, for scotopic and photopic vision.

Table 18.4 Luminous Efficiency Functions $V(\lambda)$ and $V'(\lambda)$ for Photopic and Scotopic Vision

Wavelength λ (nm)	Photopic $V(\lambda)$	Scotopic $V'(\lambda)$
380	0.00004	0.000589
390	0.00012	0.002209
400	0.0004	0.00929
410	0.0012	0.03484
420	0.0040	.0.0966
430	0.0116	0.1998
440	0.023	0.3281
450	0.038	0.455
460	0.060	0.567
470	0.091	0.676
480	0.139	0.793
490	0.208	0.904
500	0.323	0.982
510	0.503	0.997
520	0.710	0.935
530	0.862	0.811
540	0.954	0.650
550	0.995	0.481
560	0.995	0.3288
570	0.952	0.2076
580	0.870	0.1212
590	0.757	0.0655
600	0.631	0.03315
610	0.503	0.01593
620	0.381	0.00737
630	0.265	0.003335
640	0.175	0.001497
650	0.107	0.000677
660	0.061	0.0003129
670	0.032	0.0001480
680	0.017	0.0000715
690	0.0082	0.00003533
700	0.0041	0.00001780
710	0.0021	0.00000914
720	0.00105	0.00000478
730	0.00052	0.000002546
740	0.00025	0.000001379
750	0.00012	0.000000760
760	0.00006	0.000000425
770	0.00003	0.0000002413
780	0.000015	0.0000001390

$$\varepsilon_v = K_m \int_{\lambda_1=0.380\,\mu m}^{\lambda_2=0.760\,\mu m} Q_\lambda V(\lambda)\mathrm{d}\lambda. \qquad \text{[lm s]} \tag{18.34}$$

The luminous energy is related to the luminous power in the same way as energy is related to power. Luminous energy in lumen-second is luminous power integrated over time.

$$\varepsilon_v = \int P_v \,\mathrm{d}t. \qquad \text{[lm s]} \tag{18.35}$$

Luminous power (volume) density $p_{v,V}$ in lumens per cubic meter $[\mathrm{lm/m^3}]$ is luminous power per unit volume. It is also called luminous flux (volume) density.

$$p_{v,V} = \mathrm{d}P_v/\mathrm{d}V. \qquad [\mathrm{lm/m^3}] \tag{18.36}$$

Here V is the volume. The luminous power (area) density or luminous irradiance I_v is the luminous power per unit area:

$$I_v = \mathrm{d}P_v/\mathrm{d}A. \qquad [\mathrm{lm/m^2}], [\mathrm{lux}] \tag{18.37}$$

There are two units for the luminous power density: lumen per square meter $[\mathrm{lm/m^2}]$ is also called lux [lux]; and for smaller areas, we use lumen per square centimeter $[\mathrm{lm/cm^2}]$, also called phot [phot].

Luminous exitance M_v is luminous power (area) density for the luminous radiation per unit area leaving the surface:

$$M_v = \mathrm{d}P_v/\mathrm{d}A. \qquad [\mathrm{lm/cm^2}], [\mathrm{phot}] \tag{18.38}$$

Luminous exitance is measured in lumens per square meter $[\mathrm{lm/m^2}]$. A preferred unit is 1 phot [phot], equal to 1 lumen per square centimeter, $[\mathrm{lm/cm^2}],[\mathrm{phot}]$.

Illuminance E_v is the luminous power (area) density or luminous incidance for the luminous radiation incident per unit surface area:

$$E_v = \mathrm{d}P_v/\mathrm{d}A. \qquad [\mathrm{lm/ft^2}], [\mathrm{ft\text{-}c}] \tag{18.39}$$

The established unit for the illuminance is lumen per square foot, also known as foot-candle [ft-c]. Taking into consideration that the definition assumes the human eye as a detecting element, the illuminance measures the same physical quantity as luminous irradiance (Eq. (18.37)).

Luminous intensity I_v is the luminous power per unit solid angle:

$$I_v = \mathrm{d}P_v/\mathrm{d}\omega. \qquad [\mathrm{lm/sr}] \tag{18.40}$$

Its units are lumens per steradian $[\mathrm{lm/sr}]$, commonly known as candela [cdla].

Luminance, luminous radiance, or photometric brightness L_v is the luminous radiation, either incident or exiting, per unit area per unit solid angle:

$$L_v = \partial^2 P_v/(\partial A_p \,\partial \omega). \qquad [\mathrm{lm/(m^2\,sr)}] \tag{18.41}$$

The most common unit for the luminance is lumen per square meter per steradian $[\mathrm{lm/(m^2\,sr)}]$, also called nit. This unit is called candela per square meter $[\mathrm{cdla/m^2}]$, taking into account that lumen per steradian is a candela. A smaller unit is candela per square centimeter $[\mathrm{cdla/cm^2}]$, also known as stilb [stlb].

An additional photometric unit, important in particular for characterizing the luminance, is a foot-lambert [ft-lb], generally applicable to describing diffuse materials (those that behave like Lambertian reflectors and radiators). One foot-lambert is equal to $(1/\pi)$ candela per square foot. A somewhat smaller unit is a lambert, which is equal to $(1/\pi)$ candela per square centimeter.

Here we note once again that a great number of units have developed over time in various branches of photometric applications. Table 18.5 lists the conversion factors between illuminance and luminance.

18.5 RADIATIVE POWER TRANSFER IN A THREE-DIMENSIONAL SPACE

18.5.1 Geometrical Factors

Proper functioning of the radiation source, as well as its detectors, requires the presence of matter. A material object is limited by its surfaces: therefore, it is desirable to define the geometrical characteristics of the radiation emitting and absorbing surfaces. Toward this goal, we first review some basic, but important geometrical concepts.

We start with the distinction between a surface and an area. A geometrical surface is an assembly of points in a three-dimensional (3-D) space in contact with one another, that separate two regions with different physical properties. In optics, materials are differentiated by their complex indices of refraction which, in general, are wavelength-dependent. The real part of the index of refraction is the index of refraction as used in optical design applications; its complex part is related to the absorption coefficient. Figure 18.4(a) shows a surface separating two regions of space.

Sometimes, we find it convenient to extend further this rather theoretical description of a surface by defining it as a continuous assembly of points separating two regions of space. Therefore, there exists no requirement that the surface separate two regions of space having different physical characteristics. An area, on the other hand, is a purely geometrical property of a surface, similar to its orientation or position.

Let us now consider a special case as shown in Fig. 18.4(a): region 1 of space is filled with (made of) opaque material, raised to some temperature above that of its background. Consequently, the surface emits radiation into the region of space according to Planck's radiation law; thus, it is a radiative source. Region 2 is a vacuum, devoid of any material presence. The light emitted from the source propagates in a straight line through the vacuum, until it encounters an obstruction in the form of a physical object.

18.5.1.1 Area and Projected Area

Area

A surface of an arbitrary shape may be thought of as being covered by small, infinitesimal elements of area inclined with respect to one another (see Fig. 18.4(b)). One of the intrinsic characteristics of a surface is its total area. It is obtained by adding up all the infinitesimal elements of the area that encompass the surface, i.e., by performing the integration over the surface. The total, actual, area is obviously larger than the *apparent* surface area, seen from a particular direction.

Table 18.5 Conversion Factors for Different Units Used for Illuminance and Luminance

1. Illuminance conversion factors

	Lux	Phot	Foot-candle	Lumen
1 lux	1	10^{-4}	9.290×10^{-2}	$= 1\,\text{lm/m}^2$
1 phot	10^4	1	9.290×10^2	$= 1\,\text{lm/cm}^2$
1 foot-candle	10.76	1.076×10^{-3}	1	$= 1\,\text{lm/ft}^2$

2. Luminance conversion factors

	Nit	Stilb	Apostilb	Lambert	Foot-lambert	Candle/ft²	Candle/in²	Candle
1 nit	1	10^{-4}	3.142	3.142×10^{-4}	2.919×10^{-1}	9.290×10^{-2}	6.452×10^{-4}	cd/m^2
1 stilb	10^4	1	3.142×10^4	4.142	2.919×10^3	9.290×10^2	6.452	cd/cm^2
1 apostilb	0.3183	3.183×10^{-5}	1	10^{-4}	9.290×10^{-2}	2.957×10^{-2}	2.054×10^{-4}	$(1/\pi)\ \text{cd/m}^2$
1 lambert	3183	3.183×10^{-1}	10^4	1	9.290×10^2	2.957×10^2	2.054	$(1/\pi)\ \text{cd/cm}^2$
1 foot-lambert	3.426	3.426×10^{-4}	10.76	1.076×10^{-3}	1	0.3183	2.210×10^{-3}	$(1/\pi)\ \text{cd/ft}^2$
1 candle/ft²	1.076	1.076×10^{-3}	33.82	0.3382	3.142	1	6.944×10^{-3}	cd/ft^2
1 candle/in²	1550	1.550×10^{-1}	4869	0.4868	4.524×10^2	1.44×10^2	1	cd/in^2

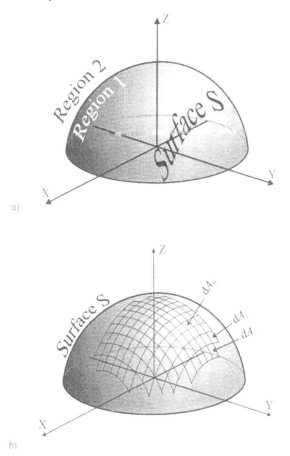

Figure 18.4 (a) Surface S separates two regions of space. (b) Surface S, separating two regions of space, is composed of small area elements.

In radiometry, we are more interested in how an area appears from a certain direction, or from a particular view point, rather than in its exact, absolute size. When an element of area serves as a radiation collector, the intrinsic characteristics of its surface area are less significant than how large this surface "appears" to the incident radiation. Similarly, the actual area of a radiative source is not as important as the "projected" area, seen from the direction of the collector.

We consider the geometrical configuration of Fig. 18.5, which shows a differential element of area ΔA_1 in space and an arbitrary point P. The line connecting the center of the area and the point P is referred to as the line of sight between the area and the point P. We characterize the area by its unit normal **N**. (Vectors are denoted with a bold letter.)

The line directed from the center of the area to the point of observation P is referred to as the direction of observation, characterized by a unit vector **S**. It may be considered as defining the z-axis of a hypothetical spherical coordinate system,

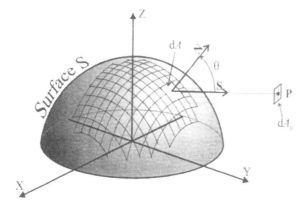

Figure 18.5   Area and projected area, as seen from the point of observation P.

erected at the center of the area ΔA_1. Thus, θ is the angle between the direction of observation and the normal to the area ΔA_1, $\mathbf{N}$. The cosine of the angle θ then is the dot or the scalar product of the unit normal to the area $\mathbf{N}$ and the direction of the line of sight $\mathbf{S}$:

$$\cos\theta = \mathbf{N} \cdot \mathbf{S}. \quad \text{[unitless]} \tag{18.42}$$

Projected Area

We define the projected area ΔA_p as the projection of the area ΔA on the plane normal to vector $\mathbf{S}$.

$$\Delta A_p = \Delta A \cos\theta. \quad [\text{m}^2] \tag{18.43}$$

In this plane, an area of size ΔA_p will appear to the observer at point P as having the same size as the area ΔA. The plane normal to the unit vector $\mathbf{S}$ through the point B on the area ΔA is called a projection plane; and the area in this plane ΔA_p is called the projected area. Its magnitude depends only on the orientation of the area ΔA with respect to the direction of observation. When we substitute Eq. (18.42) into Eq. (18.43) we obtain

$$\Delta A_p = \Delta A \mathbf{N} \cdot \mathbf{S}. \quad [\text{m}^2] \tag{18.44}$$

This equation applies equally well to infinitesimal areas. The infinitesimal area is that assembly of points that all lie in the same plane, i.e., they may all be characterized by the same unit normal:

$$dA_p = dA \cos\theta_A = dA \mathbf{N} \cdot \mathbf{S}. \quad [\text{m}^2] \tag{18.45}$$

In this equation, the subscript A on θ has been included explicitly to emphasize that each differential area has its own normal. If a surface consists of many elements of area with different orientations, then the projected area is an integral over the individual projected surfaces. Each infinitesimal element of the area contributes an infinitesimal projected area to the total projected area. The total projected area is obtained by integrating over all infinitesimal surface areas:

$$A_p = \int_{\text{Surface}} dA_p. \quad [\text{m}^2] \tag{18.46}$$

Substituting the expression for the infinitesimal projected area from Eq. (18.44) we obtain

$$A_p = \int_{\text{Surface}} dA \cos\theta_A = \int_{\text{Surface}} \mathbf{N}_A \cdot \mathbf{S} \, dA. \quad [\text{m}^2] \tag{18.47}$$

Here we show explicitly that the angle θ depends on the orientation of each individual element of area dA and that, in general, this angle will vary. Next, some familiar examples of projected areas are presented for a few well-known surfaces: the projected area of a cone is a triangle or a circle, depending on the direction of observation; that of a sphere is a circle; that of an ellipsoid of revolution is an ellipse or a circle; that of a cube is a square. There exist more complicated projections of these geometrical entities when the direction of observation does not coincide with one of the principal axes.

The difference between the area and the projected area is important only for angles θ larger than $10°$. In many radiometric applications, the angle θ is quite small. The approximation of replacing the projected area with the area is acceptable for the vast majority of cases. The error in using the area instead of the projected area is less than 5% for values of angle θ equal to or less than $10°$.

18.5.1.2 Solid Angle and the Projected Solid Angle

The solid angle is an even more important concept in radiometry than the projected area, because it describes real objects in 3-D space. It is an extension into three dimensions of a two-dimensional angle. For this reason, it will help us to appreciate the idea of a solid angle if we first review the formal definition of an angle.

Angle
Let us refer to Fig. 18.6. We define the angle with the help of a point P located at the center of curvature of a unit circle and an arc $\underline{AB}$ on this circle. Angle α, measured in

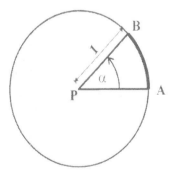

Figure 18.6 Definition of an angle in radians as a length on a unitary circle.

radians, is equal to the length of the arc on the unit circle with the center of the curvature at the point P; i.e.,

$$\alpha = \underline{AB} =\sim s_u, \qquad [\text{rad}] \tag{18.48}$$

where $\underline{AB}$ is the length of the arc. The subscript u refers to the unit radius of the sphere: s_u is the length of a curve, along the arc on the unit circle. In a more general case, when the distance PA is not unitary, but rather has a general value r, the angle α may be interpreted as the following ratio:

$$\alpha = \underline{AB}/r. \qquad [\text{rad}] \tag{18.49}$$

The infinitesimal angle $d\alpha$ is even more easily interpreted with the help of Fig. 18.7. When the angle $d\alpha$ is infinitesimally small, the corresponding arc ds approaches a straight line dd, which is parallel to the tangent at the midpoint of the arc. Then the length of the curve at the point C, ds, may be replaced by the tangent at this point:

$$d\alpha = dd/r. \qquad [\text{rad}] \tag{18.50}$$

For the differential quantities, we have a similar relationship:

$$\Delta\alpha = \Delta d/r. \qquad [\text{rad}] \tag{18.51}$$

The unit for the angle is radian [rad]. It is defined as follows: a circle of unit radius subtends an angle of 2π radians. Equally, we know that a circle of unit radius subtends an angle of 360 degrees [°] or [deg]. Thus, we obtain

$$1 \, \text{rad} = (180/\pi) \, \text{deg}, \tag{18.52}$$

and

$$1 \, \text{deg} = (\pi/180) \, \text{rad}. \tag{18.53}$$

Referring back to Fig. 18.7, we reiterate that the arc $\underline{AB}$ subtends an angle α at the point P. The point of observation P may be thought of as coinciding with the center of curvature of the circle. Also, the arc or the element of path along the arc Δs is the distance that the angle subtends at the point of observation P. Figure 18.8 shows a number of arcs or (incomplete) circles with different radii of curvature, r, all centered at P. The radius r_1 is smaller than the unitary radius r_2, while the radii r_3 and r_4 are larger than it. We note that the arcs $\underline{A_1B_1}$, $\underline{A_2B_2}$, $\underline{A_3B_3}$, $\underline{A_4B_4}$ all subtend the same

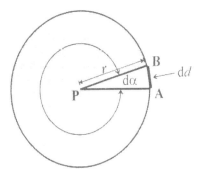

Figure 18.7 Definition of a differential angle in radians as a length on a unitary circle.

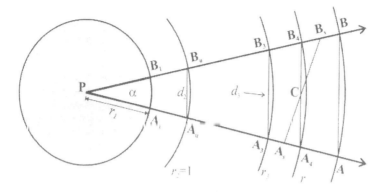

Figure 18.8 Generalized definition of a solid angle as an area on a sphere of arbitrary radius.

angle α. Thus, we conclude correctly that the two definitions presented earlier are completely equivalent.

We further see that the tangent d_2, the straight line segment connecting point A_2 with B_2, subtends the same angle as the arc $\underline{A_2 B_2}$. Thus, for all practical purposes we can replace the arc length with the arc segment in the functional definition of the angle. So, we obtain a new definition for an angle:

$$\alpha = \underline{A_2 B_2} = \underline{A_u B_u}, \qquad \text{[rad]} \tag{18.54}$$

where $\underline{A_u B_u}$ is the distance along the arc on a circle with a unitary radius. For the arc $\underline{AB}$ on a circle with an arbitrary radius r, we obtain

$$\alpha = \underline{AB}/r. \qquad \text{[rad]} \tag{18.55}$$

Line segments may be described according to their construction as tangents at the midpoint of the arc $\underline{AB}$. The tangent to an arc is normal to the radius at that point. Similarly, the line segment C is normal to the line connecting its midpoint with the point of observation P. The point of observation P defines the location of the vertex of the angle.

We now examine in more detail the general line segment $A_{4s}B_{4s}$, skewed or inclined with respect to the segment $A_4 B_4$, but intersecting it at its midpoint C_4. We note that the line segment $A_{4s}B_{4s}$ subtends the same angle α at the point P as the line segment $\underline{A_4 B_4}$, parallel to the tangent at the midpoint C_4 of the arc $\underline{A_4 B_4}$.

We may generalize this observation as follows: any arbitrary curve segment whose endpoints lie on lines PA and PB, respectively, and whose midpoint C defines the distance to the origin CP equal to radius r, may be projected on the tangent at C with the same projected length AB. All such curve segments subtend the same angle α, given as

$$\alpha = \underline{AB}/r. \qquad \text{[rad]} \tag{18.56}$$

A segment whose endpoints lie on lines PA and PB, respectively, and whose midpoint C defines a distance to the origin CP equal to radius r, may be projected on the tangent at C, having the same length, according to

$$d = \underline{AB} = A_s B_s \cos \theta_s. \quad [\text{rad}] \tag{18.57}$$

We now see that an angle may be defined using only two geometrical quantities – a point and a line segment oriented in an arbitrary, but specified direction, as illustrated in Fig. 18.9. These two quantities in space determine a single plane in which the angle is defined. The angle α is given as follows:

$$\alpha = \underline{A_s B_s} \cos \theta_s / PC. \quad [\text{rad}] \tag{18.58}$$

This, indeed, is the definition of an angle that we will find most useful in the interpretation of a solid angle in three dimensions, a basic concept in radiometric analysis. By examining Eq. (18.57) and Fig. 18.9, we conclude that the general definition of an angle does not call for the circle or its radius of curvature.

Often, the geometrical definitions in the 3-D space are simplified in the vector notation. A line segment d_s defines an angle at the point C at a distance r, equal to

$$\alpha = (\mathbf{N}_d \cdot \mathbf{S} \, d_s)/r. \quad [\text{rad}] \tag{18.59}$$

We employed two unitary vectors in this definition: $\mathbf{N}_d$ is a unitary vector, normal to the line segment d_s, that defines a line segment in a plane; $\mathbf{S}$ is a unitary vector directed from the central point on the line segment to the point of observation.

Field of View

The field of view of an optical instrument is usually given as an angle. Its numerical value may be given as half of the apex angle of a cone whose base is the focal plane array and height is the focal distance. The cone may be inscribed inside a square field of view subtended by a (square) focal plane array. It is sometimes given as a half-angle of the cone circumscribed outside the square field of view subtended by a (square) focal plane array. One half of the length of the array diagonal then defines the half field of view. These two options are illustrated in Fig. 18.10. In both cases, the assumption is generally made that the optical system is rotationally symmetric, and that the field of view may be defined in a single (meridional) plane.

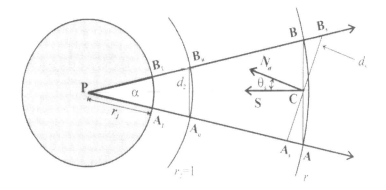

Figure 18.9 Definition of an angle using a line, its normal, and the point of observation.

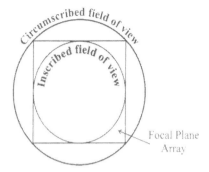

Figure 18.10 Inscribed and circumscribed field of view of an optical instrument with respect to a square focal plane array.

The casual use of the term "field of view" often leaves out "half" when referring to this concept.

Solid Angle

The solid angle ω is an angle in a 3-D space. It is subtended by an area at a distance r from a point of observation P, in the same manner as a line at distance r subtends an angle α in a plane.

The unit of a solid angle is a steradian [sr]. A (full) sphere subtends 4π steradians. An infinitesimal area dA at a distance r from the origin subtends an infinitesimal solid angle $d\omega$ equal to the size of the projection of this small area dA on a unit sphere. This is illustrated in Fig. 18.11:

$$d\omega = dA/r^2. \quad [\text{sr}] \tag{18.60}$$

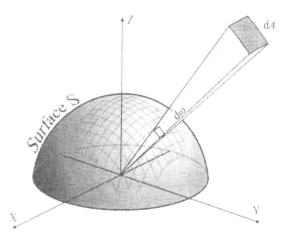

Figure 18.11 Differential solid angle subtended by an element of area dA at a distance r from the point of observation P.

The concept of a solid angle in 3-D space is similar to that of an angle in two dimensions. The definition of a solid angle is given correctly only in its differential form. The infinitesimal element of a solid angle may also be given in spherical coordinates:

$$d\omega = \sin\theta d\theta d\phi. \quad [\text{sr}] \tag{18.61}$$

The factor r^2 in the numerator and the denominator has been canceled in the derivation leading to Eq. (18.61).

We are most often interested in the solid angle subtended by an optical component, such as a lens, or a mirror, at a particular point along the optical axis as, for example, at the image (or object) location. The solid angle is completely characterized by the area that an optical element subtends at a particular point of observation. It may be found by integrating Eq. (18.61) over the appropriate limits. In an optical system with a cylindrical symmetry, an optical element with the area A subtends a circular cone with the apex half-angle Θ at the image. We orient the z-axis along the cone axis of symmetry, as illustrated in Fig. 18.12:

$$\omega = \int_0^{2\pi} \left[\int_0^{\Theta} \sin\theta d\theta \right] d\phi = 2\pi(1 - \cos\Theta). \quad [\text{sr}] \tag{18.62}$$

In an optical system without an axis of symmetry, the solid angle must be found by numerical integration.

Projected Solid Angle

The projected solid angle Ω is neither a physical nor a geometrical concept. Rather, it is a convenient mathematical abstraction: it appears in many radiometric expressions, even though it has no physical significance. It may best be understood by extending the analogy between the area and the projected area to the concept of the solid angle: the projected solid angle is related to the solid angle in the same manner

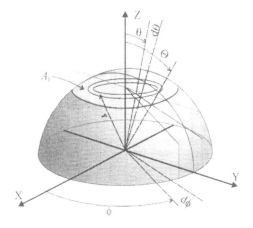

Figure 18.12 Optical surface of A_o area subtends a circular cone with a half-angle (apex angle) Θ at the image.

as the projected area is related to the area. A factor of $\cos\theta$, or obliquity factor, has to multiply an area to convert it to the "projected" area. The same obliquity factor, $\cos\theta$, has to multiply the solid angle to obtain the projected solid angle. It is visualized in the following manner, as illustrated in Fig. 18.13.

First we consider the solid angle as being equal to the area on a unit sphere, in units of steradians. Then we recall that a projected area depends on the direction of observation from the point P, defining the angle θ.

The projected solid angle is smaller in comparison with the solid angle by the factor of $\cos\theta$, the angle between the normal to the surface and the direction of observation.

The element of the projected solid angle $d\Omega$ is given in spherical coordinates:

$$d\Omega = \cos\theta d\omega = \sin\theta\cos\theta d\theta d\phi. \quad [\text{sr}] \tag{18.63}$$

The projected solid angle Ω for a circular disc subtending the apex semi-angle of Θ, shown in Fig. 18.12, is obtained by integrating Eq. (18.63) over angles:

$$\Omega = \int_0^{2\pi}\left[\int_0^{\Theta}\sin\theta\cos\theta d\theta\right]d\phi = \pi\sin^2\Theta. \quad [\text{sr}] \tag{18.64}$$

The numerical value of the projected solid angle is equal to the value of the solid angle with less than 5% error for the values of angle Θ equal to or less than $10°$. The unit of the projected solid angle is obviously the same as that of the solid angle, steradians, [sr].

A projected solid angle may be visualized as the projection of the area on a unit hemisphere onto the base of the hemisphere. With this interpretation we can easily appreciate that the solid angle ω, subtended by the hemisphere, is equal to 2π, the area of a half-sphere. This solid angle corresponds to the apex semi-angle Θ of $\pi/2$.

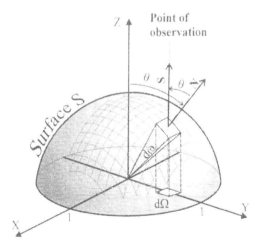

Figure 18.13 A projected solid angle as an area on a unit hemisphere projected on the base of the hemisphere.

However, the projected solid angle Ω, subtended by the hemisphere, is equal to only π, the area of the base of the hemisphere (the area of the circle of unit radius).

18.5.2 Radiative Power

Radiation carries energy from one point in space to another: being nothing other than a portion of the electromagnetic spectrum, it travels rapidly across space, transferring the energy with "the speed of light." The only time when the radiation does not move, it is either being generated or absorbed, which is done instantaneously for all practical considerations. The amount of radiation emitted by a specific surface element and the direction of the radiation propagation depends on the source's geometrical characteristics.

18.5.3 Sources and Collectors

Radiation is generated in a specific region of space. After having been created, it propagates somewhere else. A radiative source is a physical body of finite extent, occupying a specific region of space that generates the radiative energy. From each source point, the radiation propagates in all directions not obstructed by (another) physical body, in accordance with the source's radiance – the amount of radiative power emitted per unit source area per unit solid angle. In a homogeneous, transparent medium, the radiation propagates in a straight line. (In this context, we consider the vacuum, the absence of material, a special case of homogeneous transparent medium.) The radiation may be absorbed, reflected (scattered) or, most commonly both, after it has encountered another physical body or particle. At this instant, the radiation no longer propagates in a straight line.

The physical parameters that determine the source spectral distribution and the amount of the emitted radiation are the source material composition, its finish, and its temperature. The source surface geometry additionally determines the directional characteristics of the emitted radiation.

The area that intercepts and collects the radiation is referred to as a collector. It is not to be confused with a detector. A detector is a transducer that changes the absorbed radiation into some other physical parameter, amenable to further processing. This includes, for example, the electrical current, voltage, polarization, displacement, etc.

We consider two small area elements ΔA_s and ΔA_c, each with an arbitrary orientation in space, whose centers are separated by a distance d, shown in Fig. 18.14. We may think of the element of area ΔA_s to be a part of the source with the area A_s, and the element of area ΔA_c to be a part of the collector with the area A_c, illustrated in Fig. 18.15. We place the source and the collector into their respective coordinate systems. The element of area of the source is located at coordinates $S(x_s, y_s, z_s)$ and the element of area of the collector at the point $C(x_c, y_c, z_c)$.

The element of source area ΔA_s is considered so small that it is emitting the radiation whose radiance does not change as a function of position over the element of surface area ΔA_s or the direction of emission. This means that the radiance is constant over the surface element ΔA_s. Similarly, the elements of the area ΔA_s and ΔA_c are so small that all the points on the element of the area of the collector ΔA_c may be considered equidistant from all the points on the element of the area of the

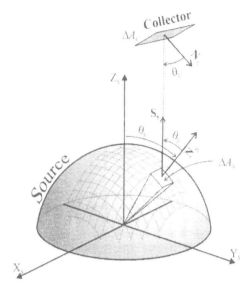

Figure 18.14 Geometry for the power transfer from the incremental source to the incremental image area.

source ΔA_s. The orientation of the planar surface elements ΔA_s and ΔA_c, respectively, is defined by their surface normals $\mathbf{N}_s$ and $\mathbf{N}_c$.

We connect the centers of the incremental area elements with a straight line of length d, which indeed depends on the specific choice of the area elements. The line connecting them is also called *the line of sight* from the source to the collector, or from the collector to the source, depending on the "point of view" that we wish to assume. We consider the direction of the line connecting these two points as being from S to P at the source, and as being from P to S at the collector: θ_s is the angle that the normal at the source subtends with the line of sight at the source; similarly, θ_c is defined as the angle that the normal at the collector subtends with the line of sight at the collector.

In many, but not all, power transfer applications, the distance d is sufficiently large with respect to the size of the element of the area ΔA_s that the radiation-emitting area ΔA_s may be considered a point source from the point on the collector $C(x_s, y_s, z_s)$.

18.5.4 Inverse Power Square Law

The power density (the power per unit area, also called the incidance) falloff for a radiating point source is described by the inverse square law. This law may be easily understood if we consider the radiation emitted from a point source into a full sphere, as illustrated in Fig. 18.16. By conservation of power, we see that the same amount of power passes through the sphere of radius r_1 as through the sphere of radius r_2:

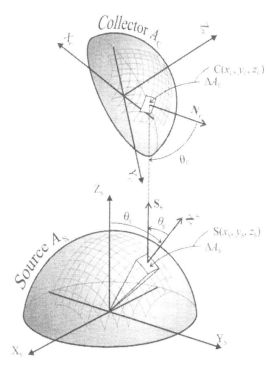

Figure 18.15 The (spectral) power transfer equation gives the amount of (spectral) power generated and emitted by a source of area A_s oriented in the direction N_s and intercepted by a collector of area A_c, oriented in direction N_c.

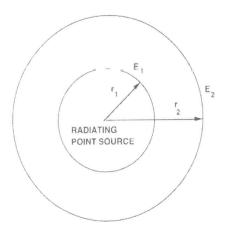

Figure 18.16 The radiative power passing through the sphere of radius r_1 is equal to that through the sphere of radius r_2. They both are equal to the power radiated by the point source, due to the absence of energy sources and sinks.

$$P = (4\pi r_1^2)E_1 = (4\pi r_2^2)E_2. \qquad [\text{W}] \tag{18.65}$$

Equation (18.65) is a conservation of energy, expressed in terms of power (the energy per unit time). We can solve it for the power density across the sphere of radius r_2, denoted as E_2 in Eq. (18.65):

$$E_2 = E_1(r_1^2/r_2^2) = (E_1 r^2)/r_2^2 = 4\pi I_s/r_2^2 = P/r_2^2. \qquad [\text{W/m}^2] \tag{18.66}$$

For a point source the incidence E decreases with the distance from the source according to the inverse power square law; I_s is the intensity of the presumed point source at the center of the sphere. It was defined previously as the power per unit solid angle, the radiative quantity most appropriate for the characterization of point sources. Upon regrouping the terms in Eq. (18.66), we obtained several different ways of expressing the inverse square law. The *power* in this case refers to the exponent of the sphere radius, as opposed to its normal use in radiometry.

While Eq. (18.66) has been derived for simplicity for a full sphere, the same relationship may be shown to be true also for cones bounded by areas and subtending solid angles.

18.5.5 Power Transfer Equation for the Incremental Surface Areas

The radiance, i.e., the incremental power that the source area ΔA_s emits into the incremental solid angle $\Delta\omega$, may be found by considering the definition of radiance. In this section only, we show explicitly the dependence of the radiometric quantities, radiance in particular, on the independent variables.

Referring to Fig. 18.17, we define two coordinate systems to describe the source radiance. The source is shown as an extended body in a plane (x_s, y_s) with the irregularly shaped boundary. The center of the incremental area ΔA_s is located at the point S with coordinates (x_s, y_s). This point forms one end of the line of sight in the direction toward the collector (not shown). The radiation is emitted into an incremental projected solid angle $\Delta\Omega_s(\theta_s, \varphi_s)$, subtended by an imaginary collector at the source. The subscript s on the (projected) solid angle indicates that the apex of the cone defining the solid angle is at the source. The direction of the line-of-sight forms two angles with respect to the normal to the incremental area ΔA_s, N_s. These two angles are the angle coordinates in a spherical coordinate system erected at each point $S(x_s, y_s)$ whose z-axis is along the normal to the incremental area ΔA_s.

We replace the differential quantities with the increments in the definition of radiance to obtain

$$L_\lambda(x_s, y_s; \theta_s, \varphi_s) = \Delta^2 P_\lambda(x_s, y_s; \theta_s, \varphi_s)/[\Delta A_s(x_s, y_s)\Delta\Omega_s(\theta_s, \varphi_s)]. \qquad [\text{W}/(\text{m}^2\,\text{sr}\,\mu\text{m})]$$
$$\tag{18.67}$$

We observe that the increments in the numerator are balanced with those in the denominator. It is important to keep in mind that the radiance is defined for every point on the source $S(x_s, y_s)$ and for every direction of emission (θ_s, φ_s). Here we note again that the subscript on the solid angle $\Omega_s(\theta_s, \varphi_s)$ refers to the point of observation, i.e., the apex of the cone that the solid angle may be seen to subtend. The spherical coordinate system (θ_s, φ_s) is erected at any source point $S(x_s, y_s)$. From Eq. (18.67) we solve for the double increment of spectral power $\Delta^2 P_\lambda$:

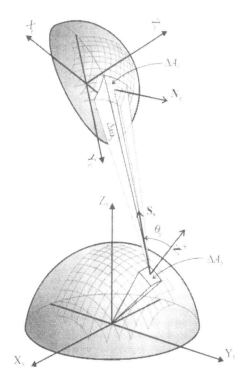

Figure 18.17   Incremental power transferred into the incremental solid angle $\Delta\omega_s(\theta_s, \varphi_s)$.

$$\Delta^2 P_\lambda(x_s, y_s; \theta_s, \varphi_s) = L_\lambda(x_s, y_s; \theta_s, \varphi_s)\Delta A_s(x_s, y_s)\Delta\Omega_s(\theta_s, \varphi_s). \qquad [\text{W}/\mu\text{m}]$$
$$(18.68)$$

We refer to Fig. 18.17 to obtain the full appreciation of the significance of Eq. (18.68). It gives us the (doubly) incremental amount of power that an incremental area ΔA_s of the source with spectral radiance L_λ emits into the incremental, projected solid angle $\Delta\Omega_s$ in the direction of (θ_s, φ_s). The assumption is made again that the incremental projected solid angle $\Delta\Omega_s$ is so small that the radiance may be considered constant within it.

18.5.6   Power Transfer Equation in Integral Form

If the incremental source area ΔA_s and the incremental projected solid angle $\Delta\Omega_s$ are simultaneously allowed to decrease to infinitesimally small quantities, we obtain the expression for the infinitesimal amount of power emitted by the infinitesimal surface area dA_s into the infinitesimal element of a projected solid angle $d\Omega_s$. Formally, this means that all the incremental (physical) quantities are changed into the infinitesimal (mathematical) quantities:

$$d^2 P_\lambda(x_s, y_s; \theta_s, \varphi_s) = L_\lambda(x_s, y_s; \theta_s, \varphi_s)\, dA_s(x_s, y_s)\, d\Omega_s(\theta_s, \varphi_s). \quad [\text{W}/\mu\text{m}]$$

$$(18.69)$$

The infinitesimal amount of radiative power dP_λ, emitted by the total source area A_s into an infinitesimal projected solid angle, may then be obtained by integrating both sides of Eq. (18.69) over the source area:

$$d\left[\int_{A_s} dP_\lambda(x_s, y_s; \theta_s, \varphi_s) \right] = \int_{A_s} L_\lambda(x_s, y_s; \theta_s, \varphi_s)\, dA_s(x_s, y_s)\, d\Omega_s(\theta_s, \varphi_s). \quad [\text{W}/\mu\text{m}]$$

$$(18.70)$$

The infinitesimal projected solid angle that the collector subtends at the source, $d\Omega_s(\theta_s, \varphi_s)$, is independent of the source Cartesian coordinates. Therefore, it is placed outside the integral:

$$d[(P_\lambda(\theta_s, \varphi_s)] = \left[\int_{A_s} L_\lambda(x_s, y_s; \theta_s, \varphi_s)\, dA_s(x_s, y_s) \right] d\Omega_s(\theta_s, \varphi_s). \quad [\text{W}/\mu\text{m}]$$

$$(18.71)$$

We may perform the integrals over the area on both sides of Eq. (18.71):

$$dP_\lambda(\theta_s, \varphi_s) = d\Omega_s(\theta_s, \varphi_s) \left[\int_{A_s} L_\lambda(x_s, y_s; \theta_s, \varphi_s)\, dA_s(x_s, y_s) \right]. \quad [\text{W}/\mu\text{m}] \quad (18.72)$$

In Eq. (18.72) one power of the infinitesimal symbol "d" has been integrated (out) over the source area:

$$P_\lambda(\theta_s, \varphi_s) = \int_{A_s} dP_\lambda(x_s, y_s; \theta_s, \varphi_s). \quad [\text{W}/\mu\text{m}]$$

$$(18.73)$$

For this reason the spectral power $P_\lambda(\theta_s, \varphi_s)$ no longer shows the explicit dependence on the source Cartesian coordinates x_s, y_s. The infinitesimal amount of radiative power, emitted by the area A_s of the source with radiance L_λ into an infinitesimal solid angle $d\Omega_s$ is obtained by evaluating the integral in the parenthesis over the source area. This may only be found if the functional form of the source radiance is known.

Special Case 1:   Radiance Independent of the Source Coordinates

We assume that the radiance is independent of the source position in order to simplify further the expression for the power transfer in Eq. (18.72):

$$L_\lambda(x_s, y_s; \theta_s, \varphi_s) = L_\lambda(\theta_s, \varphi_s). \quad [\text{W}/(\text{m}^2\,\text{sr}\,\mu\text{m})]$$

$$(18.74)$$

For the case that the source is characterized by the radiance that is independent of the source position, Eq. (18.72) is appreciably simplified. The radiance may be placed outside the integral:

$$dP_\lambda(\theta_s, \varphi_s) = d\Omega_s(\theta_s, \varphi_s) \left[L_\lambda(\theta_s, \varphi_s) \int_{A_s} dA_s(x_s, y_s) \right]. \quad [\text{W}/\mu\text{m}] \qquad (18.75)$$

The integral over the differential of the source area is just the source area

$$A_s = \int_{A_s} dA_s(x_s, y_s). \quad [\text{m}^2] \qquad (18.76)$$

The infinitesimal power dP_λ emitted into the infinitesimal projected solid angle $d\Omega_s$ by the source of area A_s, whose radiance is independent of the source coordinates, is obtained upon substitution of Eq. (18.76) into Eq. (18.75):

$$dP_\lambda(\theta_s, \varphi_s) = A_s L_\lambda(\theta_s, \varphi_s) \, d\Omega_s(\theta_s, \varphi_s). \quad [\text{W}/\mu\text{m}] \qquad (18.77)$$

This brings the special case to an end.

We now return to the general radiation transfer presented in Eq. (18.72). In order to find the power emitted by the source of area A_s into the projected solid angle Ω_s, we integrate both sides of Eq. (18.72) over the solid angle Ω_s:

$$\int_{\Omega_s} dP_\lambda(\theta_s, \varphi_s) = \int_{\Omega_s} \left[\int_{A_s} L_\lambda(x_s, y_s; \theta_s, \varphi_s) \, dA_s(x_s, y_s) \right] d\Omega_s(\theta_s, \varphi_s). \quad [\text{W}/\mu\text{m}]$$

$$(18.78)$$

First, we comment on the left side of Eq. (18.78). The integral over the infinitesimal power emitted into the projected solid angle Ω_s is just the total power emitted by the source:

$$P_\lambda = \int_{\Omega_s} d[P_\lambda(\theta_s, \varphi_s)]. \quad [\text{W}/\mu\text{m}] \qquad (18.79)$$

The right side of Eq. (18.78) is more difficult to evaluate. The functional form of the dependence of the radiance on the set of four coordinates is, in general, not known. Therefore, we leave it in the form of the integral over the solid angle:

$$P_\lambda = \int_{\Omega_s} \left[\int_{A_s} L_\lambda(x_s, y_s; \theta_s, \varphi_s) \, dA_s(x_s, y_s) \right] d\Omega_s(\theta_s, \varphi_s). \quad [\text{W}/\mu\text{m}] \qquad (18.80)$$

We have chosen a particular order of integration, indicated by the square bracket in Eq. (18.80), although we could have chosen a different order by first integrating over the projected solid angle and then over the surface area. By examining Eq. (18.80), we see that the order of the integration may be interchanged. Thus, the general equation for the power transfer is usually written without prescribing the order of integration:

$$P_\lambda = \int_{\Omega_s} \int_{A_s} L_\lambda(x_s, y_s; \theta_s, \varphi_s) \, dA_s(x_s, y_s) \, d\Omega_s(\theta_s, \varphi_s). \quad [\text{W}/\mu\text{m}] \qquad (18.81)$$

This equation usually appears in a much simplified form; it does not show explicitly that the source area is a function of the source coordinates and that the solid angle depends on the spherical coordinates erected at a source point, with the z-axis along the line of sight to the collector. Its familiar, but less explicit form is given next:

$$P_\lambda = \int\limits_{\Omega_s} \int\limits_{A_s} L_\lambda(x_s, y_s; \theta_s, \varphi_s)\, dA_s\, d\Omega_s. \qquad [\text{W}/\mu\text{m}] \qquad (18.82)$$

The radiance is usually given as a function of position and angle. It depends on the source Cartesian coordinates and the spherical coordinate system subtended at the specific source point. As this is implicitly understood, the subscripts are not shown explicitly.

$$P_\lambda = \int\limits_{\Omega_s} \int\limits_{A_s} L_\lambda(x, y; \theta, \varphi)\, dA\, d\Omega. \qquad [\text{W}/\mu\text{m}] \qquad (18.83)$$

Equation (18.81) is an exact, informative, but somewhat busy equation. As before, we try to simplify the power transfer equation, Eq. (18.81).

Special Case 1: Radiance Independent of the Source Coordinates
Once again we make the assumption that the radiance is independent of the source position, given previously in Eq. (18.74). We evaluate Eq. (18.81) for the radiance that is independent of the source coordinates. The integral over the area of the source is just the source area, as in Eq. (18.76):

$$P_\lambda = A_s \int\limits_{\Omega_s} L_\lambda(\theta_s, \varphi_s)\, d\Omega_s(\theta_s, \varphi_s). \qquad [\text{W}/\mu\text{m}] \qquad (18.84)$$

If we wish to further simplify the power transfer equation, we must make an additional assumption about the radiance.

Special Case 2: Constant Radiance
We next assume that the radiance is also independent of the direction of observation, in addition to the source coordinates:

$$L_\lambda(x_s, y_s; \theta_s, \varphi_s) = L_\lambda. \qquad [\text{W}/(\text{m}^2\,\text{sr}\,\mu\text{m})] \qquad (18.85)$$

When the radiance is additionally independent of the angle coordinates, it may be placed outside the integrals:

$$P_\lambda = L_\lambda A_s \int\limits_{\Omega_s} d\Omega_s(\theta_s, \varphi_s). \qquad [\text{W}/\mu\text{m}] \qquad (18.86)$$

The integral over the projected solid angle in Eq. (18.86) is the projected solid angle:

$$\Omega_s = \int\limits_{\Omega_s} d\Omega_s(\theta_s, \varphi_s). \qquad [\text{sr}] \qquad (18.87)$$

Upon the substitution of Eq. (18.87) into Eq. (18.86), we obtain the simplest form of the power transfer equation, applicable to the case when the radiance is independent of the source coordinates and the direction of observation:

$$P_\lambda = L_\lambda A_s \Omega_s. \quad [\text{W}/\mu\text{m}] \tag{18.88}$$

For a surprisingly large number of radiometric problems this equation produces adequate results. For other applications, this equation represents the first-order approximation, which may be used to estimate the order-of-magnitude results. This form is identical to the power transfer equation in the incremental form, given in Eq. (18.68), but with the symbols for increments missing.

In an even greater number of radiometric problems, the distance between the source and the collector is so large that Eq. (18.88) (or Eq. (18.68)) is given correctly for the incremental quantities. This implies that the transverse dimensions of the source and collector are appreciably smaller than their separation. Similarly, for such large separation distances, the radiance variation across the source and its angular dependence are justifiably assumed negligible.

In the special case when the source and collector are relatively small with respect to their separation, the power transfer equation (Eq. (18.88)) is applicable in its incremental form:

$$\Delta^2 P_\lambda = L_\lambda \, \Delta A_s \Delta \Omega_s. \quad [\text{W}/\mu\text{m}] \tag{18.89}$$

We may substitute different expressions for the solid angle to present Eq. (18.89) in different forms.

18.6 LAMBERT'S LAW

18.6.1 Directional Radiator

In Section 18.2.3, we defined the radiance as the second derivative of the power with respect to both the projected solid angle and the area. The radiance depends on the Cartesian coordinates of the source and the spherical coordinates whose *z*-axis is along the normal to the source area element. These two coordinate systems are illustrated in Fig. 18.17:

$$L(x_s, y_s; \theta, \phi) = \partial^2 P/(\partial \Omega \, \partial A_s). \quad [\text{W}/(\text{m}^2 \, \text{sr})] \tag{18.90}$$

We show the subscripts explicitly only for the Cartesian coordinates, indicating that we are in the plane of the source. This *spectral* radiance is indeed the most general radiometric quantity to characterize a radiative source:

$$L_\lambda(x_s, y_s; \theta, \phi) = \partial^2 P_\lambda/(\partial \Omega \, \partial A_s). \quad [\text{W}/(\text{m}^2 \, \text{sr} \, \mu\text{m})] \tag{18.91}$$

While it is easy to understand that the radiance depends on the coordinates of the point on the source $S(x_s, y_s)$, its directional properties are a bit more difficult to visualize: the implication is that the source radiance has a different angular dependence for each source point. The direction of observation has been defined as the angle that the local normal to the surface makes with the line of sight (the line connecting the source with the collector). The angle θ, belonging to a spherical coordinate system erected at the source point $S(x_s, y_s)$, is then the angle of observation.

The angular dependence of the source radiance may be measured only with a great deal of difficulty even just for a few representative incremental source areas. In principle, we would like to know its value for all source *points*, but the point is an abstract, geometrical entity. The measurement of the radiation emission from an

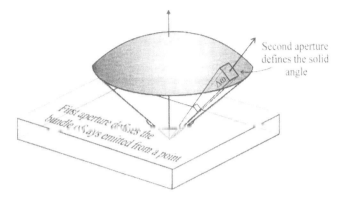

Second aperture
defines the solid
angle

First aperture defines the
bundle (S, α)'s emitted from a point

Figure 18.18 The experimental setup to measure the emission of radiation from an incremental area of the source.

incremental area of the source into a specific direction toward the collector may be accomplished using two small apertures, as illustrated in Fig. 18.18: the first aperture defines the size of the radiation-emitting surface area; the second aperture specifies the size of the incremental solid angle into which the radiation is being emitted.

The aperture that defines the incremental source area may actually interfere with the normal operation of the source if it comes in contact with it. If it does not come in contact, the aperture may limit the solid angle into which the elemental area of the source is radiating.

The important point to keep in mind here is that the very process of measurement is bound to introduce errors because of the finite size of the apertures and their unavoidable interference with the normal operation of the source.

Fortunately, most sources behave in a somewhat predictable manner that may be described sufficiently well with a few parameters for approximate radiometric analysis. Only a few sources fall into one of the two extreme cases of directional and nondirectional or isotropic radiators. The isotropic radiator emits the same amount of power in all directions. A point source in a homogeneous medium or a vacuum is an example of an isotropic radiator. A laser is an example of a directional source, emitting the radiation only within a very narrow angle.

The majority of natural sources tend to be adequately well described as Lambertian radiators.

18.6.2 Lambertian Radiator

Natural sources most often have directional characteristics; they emit strongly in the forward direction, i.e., in the direction along the normal to the radiative surface. The amount of radiation usually decreases with increasing angle until, at the angle of 90 degrees with respect to the surface normal, its emission reduces to zero. This angular dependence is characteristic of the so-called Lambertian radiator. Most natural sources are similar to Lambertian radiators, even though there are very few perfect ones to be found.

A Lambertian radiator is a good example of why it is so difficult to measure radiative characteristics of a source. The finite apertures used in the measurement setup on the one hand limit the solid angle being measured, and, on the other hand, average the results due to their size.

A Lambertian radiator is an extended area source characterized by a cosinusoidal dependence of the (spectral) radiance on the angle θ, according to the equation

$$L_\lambda(x_s, y_s; \theta, \phi) = L_\lambda(\theta) = L_{\lambda 0} \cos\theta. \qquad [\text{W}/(\text{m}^2 \,\text{sr}\,\mu\text{m})] \qquad (18.92)$$

Here $L_{\lambda 0}$ is the constant spectral radiance, which does not depend on the source coordinates. This dependence is shown in Fig. 18.19. So, the radiance of a Lambertian radiator is independent of the source coordinates. Its radiance has azimuthal symmetry. A Lambertian radiator may also be specified with the integrated radiance:

$$L(x_s, y_s; \theta, \phi) = L(\theta) = L_0 \cos\theta. \qquad [\text{W}/(\text{m}^2 \,\text{sr})] \qquad (18.93)$$

Here, the integrated radiance L_0 is a constant. Thus, a Lambertian radiator is a source whose emission depends only on the cosine of the azimuthal angle. We may also say that a Lambertian source emits radiation according to the (Lambert's) cosine law.

When the source radiance obeys Lambert's cosine law, the source is referred to as a Lambertian radiator, or a perfectly diffuse source. A diffuse source radiates in all directions within a hemisphere. The amount of radiation emitted by an incremental area ΔA_s in the direction θ is decreased by the obliquity factor, the very same factor that decreases the apparent size of the area viewed from this direction.

A Blackbody Radiator

The radiation that is established inside an evacuated cavity with completely absorbing walls at temperature T is referred to as a blackbody radiator (Fig. 18.20(a)). It is completely isotropic, i.e., the same in all directions. If an infinitesimally small hole of area δA_s is punched in a wall, as shown in Fig. 18.20(b), then the radiation escaping from the blackbody cavity is referred to as a blackbody radiation.

The radiation that escapes from such a cavity, even if somewhat idealized, is Lambertian. The radiation that is incident on any area of the wall, including the area

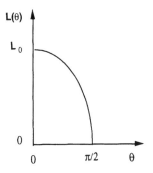

Figure 18.19 The radiance of a Lambertian radiator.

PERFECTLY ABSORBING
WALLS AT TEMPERATURE T

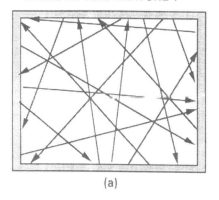

(a)

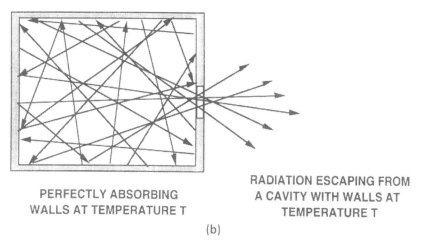

PERFECTLY ABSORBING
WALLS AT TEMPERATURE T

RADIATION ESCAPING FROM
A CAVITY WITH WALLS AT
TEMPERATURE T

(b)

Figure 18.20 (a) Blackbody radiation is established inside an evacuated cavity with completely absorbing walls at temperature T. (b) The radiation escaping from the cavity at equilibrium temperature T is referred to as a blackbody radiation.

of the hole δA_s, is isotropic. However, the apparent size of the opening as seen by the radiation incident from direction θ changes, according to the $\cos\theta$ obliquity factor. The reduced area has been called the projected area in the previous section. The small opening in the cavity where the isotropic blackbody radiation leaves the cavity with walls at temperature T is a Lambertian source – the projected area has an obliquity factor $\cos\theta$ when viewed from the direction θ.

Nonplanar Sources

In general, portions of a source may also lie outside the x–y plane. Nevertheless, if the source emission is given by Eq. (18.93), it is considered a diffuse or a Lambertian radiator. Our sun is an example of such a Lambertian radiator: its emission surface is

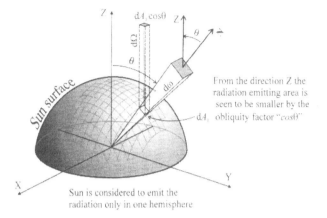

Figure 18.21 Our sun is an example of a Lambertian radiator: its emission surface is a sphere, with each surface element ΔA_s at the angle θ offering a projected area of $\Delta A_s \cos \theta$.

a sphere, with each surface element ΔA_s at the angle θ offering a projected area of $\Delta A_s \cos \theta$ (Fig. 18.21). Thus, the sun appears as a disk of uniform radiance or brightness. (Brightness is the term used for the radiance when dealing with the visible radiation.)

18.6.3 Relationship Between the Radiance and Exitance for a Lambertian Radiator

The problem that we are tying to solve is the following: consider a small area $\Delta A(x_s, y_s)$ on the x_s–y_s plane, radiating into a hemisphere as a Lambertian radiator. How much power does this area $\Delta A(x_s, y_s)$ emit into the space? Figure 18.22 illustrates this problem.

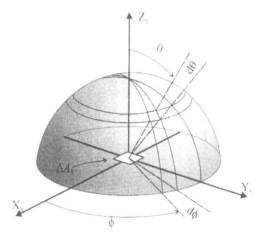

Figure 18.22 An increment of solid angle intercepts the radiation that a small area $\Delta A(x_s, y_s)$ on the x_s–y_s plane emits as a Lambertian radiator into the hemisphere.

We first recall that an extended source emits the radiation only into a hemisphere. The increment of area ΔA_s is located in the x–y plane; z is equal to zero, and, therefore, is omitted in the analytical development. We present the development for a general radiator. Finally, we evaluate the power emitted into the hemisphere for a Lambertian radiator, given in Eq. (18.92):

$$L_\lambda(x_s, y_s, z_s\theta; \varphi) = L_\lambda(x_s, y_s, 0; \theta, \varphi) = L_\lambda(x_s, y_s; \theta, \varphi). \quad [\text{W}/(\text{m}^2 \, \text{sr} \, \mu\text{m})]$$

$$(18.94)$$

The mathematical relationship between the radiance and exitance for an infinitesimal source area $\Delta A(x_s, y_s)$ has been found in Section 18.2.3. The (spectral) radiance of a source is its (spectral) exitance per unit solid angle:

$$L_\lambda(x_s, y_s; \theta, \varphi) = dM_\lambda((x_s, y_s; \theta, \varphi)/d\omega. \quad [\text{W}/(\text{m}^2 \, \text{sr} \, \mu\text{m})] \quad (18.95)$$

Thus, the source (spectral) exitance is the (spectral) radiance integrated over the solid angle Ω. From Eq. (18.95) we solve for the spectral exitance dM_λ:

$$dM_\lambda(x_s, y_s; \theta, \varphi) = L_\lambda(x_s, y_s; \theta, \varphi) \, d\omega. \quad [\text{W}/(\text{m}^2 \, \mu\text{m})] \quad (18.96)$$

Another way of interpreting this problem is to emphasize its physical significance. How much power does an incremental surface area at $O(x_0, y_0)$ emit, having Lambertian radiance, into an incremental solid angle subtended by the area dA on the unit sphere (equal to $d\omega$, by the definition of a solid angle)?

We consider the specific point (x_0, y_0) on the source x_s–y_s plane:

$$M_\lambda(x_0, y_0) = M_\lambda(x_s, y_s). \quad [\text{W}/(\text{m}^2 \, \text{sr} \, \mu\text{m})] \quad (18.97)$$

We substitute Eqs. (18.97) and (18.92) into Eq. (18.96):

$$dM_\lambda(x_0, y_0; \theta, \varphi) = L_\lambda(\theta, \phi) \, d\omega. \quad [\text{W}/(\text{m}^2 \, \mu\text{m})] \quad (18.98)$$

Then we integrate over the solid angle:

$$M_\lambda(x_0, y_0) = \int_{\text{Full hemisphere}} L_\lambda(\theta, \phi) \, d\omega. \quad [\text{W}/(\text{m}^2 \, \mu\text{m})] \quad (18.99)$$

Next we substitute the differential angles for the differential solid angle and specify the limits of integration. First, we evaluate this integral for general angles:

$$M_\lambda(x_0, y_0) = \int_{\phi_{min}}^{\phi_{max}} \int_{\theta_{min}}^{\theta_{max}} L_\lambda(\theta, \phi) \sin\theta d\theta d\phi. \quad [\text{W}/(\text{m}^2 \, \mu\text{m})] \quad (18.100)$$

To determine the relationship between the radiance and exitance of a radiating source we consider the geometry shown in Fig. 18.22. We assume that the infinitesimally small radiation-emitting area δA is located in the plane of the source. A planar Lambertian radiator emits the radiation in all directions defined above the plane, into the whole hemisphere. (A source of infinitesimal area cannot emit in more than half of the hemisphere, because the infinitesimal area is assumed to be planar.)

Thus, the limits of integration in Eq. (18.100) for the Lambertian radiative emitters are 0 to 2π for the coordinate ϕ and 0 to $\pi/2$ for the coordinate θ:

$$M_\lambda(x_o, y_o) = \int_0^{2\pi} \int_0^{\pi/2} L_\lambda(\theta, \phi) \sin\theta d\theta d\phi. \qquad [\text{W}/(\text{m}^2\,\mu\text{m})] \qquad (18.101)$$

This integral is generally difficult to evaluate, unless the (spectral) radiance is a particularly simple function of θ and ϕ, such as the Lambertian radiator. Usually, we have to resort to the methods of numerical integration. There are a number of approximations to the real radiation emitters of the form

$$L_\lambda(\theta, \phi) = L_{\lambda o} \cos^n\theta. \qquad [\text{W}/(\text{m}^2\,\text{sr}\,\mu\text{m})] \qquad (18.102)$$

Here n is a rational number and $L_{\lambda o}$ is a constant that depends only on wavelength. Fortunately, there are a large number of natural emitters that are well represented by this expression. The source characterized by the angular dependence given in Eq. (18.97) reduces to a Lambertian radiator when $n = 1$. We substitute Eq. (18.102) into Eq. (18.101):

$$M_\lambda(x_o, y_o) = \int_0^{2\pi} \int_0^{\pi/2} L_{\lambda o} \cos^n\theta \sin\theta d\theta d\phi. \qquad [\text{W}/(\text{m}^2\,\mu\text{m})] \qquad (18.103)$$

This integral is easy to integrate once we change the variable

$$\sin\theta d\theta = -d[\cos\theta]. \qquad (18.104)$$

The limits are changed correspondingly: when $\theta = 0$, $\cos\theta = 1$; when $\theta = \pi/2$, $\cos\theta = 0$. Using Eq. (18.103) we get for (spectral) exitance, given in Eq. (18.99),

$$M_\lambda(x_o, y_o) = L_{\lambda o} \int_0^{2\pi} \int_0^1 \cos^n\theta d[\cos\theta] d\phi. \qquad [\text{W}/(\text{m}^2\,\mu\text{m})] \qquad (18.105)$$

We changed the limits of integration to eliminate the negative sign. With this change of variables, Eq. (18.105) is easily evaluated:

$$M_\lambda(x_o, y_o) = 2\pi L_{\lambda o}/(n+1). \qquad [\text{W}/(\text{m}^2\,\mu\text{m})] \qquad (18.106)$$

We may assume that this relationship holds for any point x_o, y_o. For a Lambertian radiator, $n = 1$ (see Eq. (18.92)). Thus, Eq. (18.106) reduces to

$$M_\lambda(x_o, y_o) = \pi L_{\lambda o}. \qquad [\text{W}/(\text{m}^2\,\mu\text{m})] \qquad (18.107)$$

The source coordinates are omitted in the familiar form of the relationship between the radiance and exitance for a Lambertian radiator. So, Eq. (18.106) becomes

$$M_\lambda = 2\pi L_{\lambda o}/(n+1). \qquad [\text{W}/(\text{m}^2\,\mu\text{m})] \qquad (18.108)$$

Similarly, for a Lambertian radiator, we obtain

$$M_\lambda = \pi L_{\lambda o}. \qquad [\mathrm{W}/(\mathrm{m}^2\,\mu\mathrm{m})] \qquad\qquad (18.109)$$

This is a very significant relationship, stating that the spectral exitance of a Lambertian radiator into the full hemisphere is the spectral radiance multiplied by a factor π.

At first glance, this result may appear somewhat surprising, because the volume of the hemisphere equals 2π, rather than π. This apparent discrepancy can be understood quite easily when we remember functional dependence of a Lambertian radiator. We note that the amount of radiation is maximum for normal emission, decreasing with increasing angle θ to the point of being zero for the tangential emission of radiation. Thus, the exitance may be interpreted as the angle-average radiance over the whole hemisphere.

We have tried to indicate throughout this section that the relationships presented here are valid for the spectral and integrated quantities. If the relationship is valid for any wavelength, it is also valid for the sum of wavelengths, or their integrals. So, for the integrated exitance for a general radiator, we obtain

$$M = 2\pi L_o/(n+1). \qquad [\mathrm{W}/\mathrm{m}^2] \qquad\qquad (18.110)$$

Similarly, the integrated exitance is obtained for a Lambertian radiator:

$$M = \pi L_o. \qquad [\mathrm{W}/\mathrm{m}^2] \qquad\qquad (18.111)$$

Thus, for a Lambertian radiator the exitance is equal to $\pi\times$ the radiance. This is to be contrasted with the amount of radiation emitted into a hemisphere by a point source of constant intensity I_o equal to $4\pi I_o$. The factor of 4 in the expression for the point source may be attributed to the different geometries of these two sources. A point source of constant intensity omits radiation uniformly in all directions with the solid angle 4π. An extended area source emits the radiation from its planar surface and experiences a diminished projected area. The projected area $\Delta A_s\cos\theta$ averaged over a hemisphere is $\Delta A_s/2$. The radiance of the point source differs from that of an extended source by a factor 2. The other factor of 2 may be explained away by the fact that 2 is the ratio of the solid angle of a sphere over that of a hemisphere.

18.7 POWER TRANSFER ACROSS A BOUNDARY

18.7.1 Polished and Diffuse Surface

Radiometry concerns itself with the transfer of information in the form of the electromagnetic radiation. The radiative power is transferred from the point on the source to the point on the detector using an optical system consisting of beam-shaping elements such as mirrors, prisms, and gratings, whose function is to reshape and redistribute the radiation. These components are made of different materials with distinct indices of refraction. We refer to the boundary as that surface which separates two regions of space with different indices of refraction and absorption.

In terms of its response to the incident radiation, a surface may be reflective, transmissive, or both. A reflective surface may be specularly reflective, diffusive, or both. In fact, the majority of surfaces are specularly reflective and diffusive at the same time. Surfaces are said to be specularly reflective when the specular component of the reflected light is the more prominent one, while for the diffusive surfaces the majority of the radiation is reflected in all directions.

An optical surface is said to be specularly reflective when it reflects the beam of incident radiation in accordance with Snell's law of reflection. In this case, the angle that the reflected beam forms with the surface normal is equal to the angle that the incident beam makes with the surface normal. Also, the reflected beam lies in the plane defined by the incident beam and the surface normal, called the plane of incidence. An example of a reflective surface is a polished optical component such as a plane or a curved mirror. A magnified image of a polished surface shows the actual surface lying within two bounding surfaces that are separated by a very small distance and with approximately the same slope. An unpolished surface has large deviations from the reference surface, with high peaks and deep valleys, and a wide distribution of slopes. The two types of surfaces are illustrated in Fig. 18.23.

A polished surface has a prevailing amount of surface area at approximately the same height above the reference surface and has about the same slope; thus, it reflects the incident pencil of light as a collimated light beam, acting as a specular reflector. On the other hand, an unpolished surface has a wide range of slope values and heights, resulting in the random redistribution of the parallel beam of light into all directions.

A diffuse surface reflects the parallel beam of radiation in all directions. When the reflected radiation examined at a short distance from a surface shows no angular preference, then the surface is said to be a diffuse reflector. A perfectly diffuse reflector is also referred to as a Lambertian reflector, and exhibits only the $\cos \theta$ dependence for a planar surface. This is due purely to the geometrical effects of the projected area. A nonspecularly reflecting surface is often referred to as a scattering surface.

Light scattering is an optical phenomenon that takes place at a rough (planar) surface that reflects the incident collimated beam into all directions. Most often, though, it happens within solids, liquids, and gases that offer scattering centers to the incident radiation. The angle that each randomly reflected (scattered) ray makes with the surface normal is referred to as a scattering angle. Within the realm of radiometry, the scattered light nonetheless follows the laws of geometrical optics: due to the surface texture, Snell's law applies on a microscopic scale.

18.7.2 BRDF

The surface reflection characteristic is formally described by a functional dependence that includes two angles of incidence (θ_i, ϕ_i), and two reflecting or scattering angles (θ_r, ϕ_r), which specify the direction into which the light reflects or scatters (Fig. 18.24). This function of four angles, the bidirectional reflectivity distribution function, $\text{BRDF}(\theta_i, \phi_i; \theta_r, \phi_r)$, is the surface reflection coefficient that relates the radiance reflected into an element of solid angle $\Delta\omega_r$ along a particular angular direction (θ_r, ϕ_r) with the beam incidence $M(\theta_i, \phi_i)$ on a small surface area ΔA being characterized:

$$\text{BRDF}(\theta_i, \phi_i; \theta_r, \phi_r) = [L(\theta_r, \phi_r)\, \Delta\omega_r]/M(\theta_i, \phi_i), \qquad \text{[unitless]} \qquad (18.112)$$

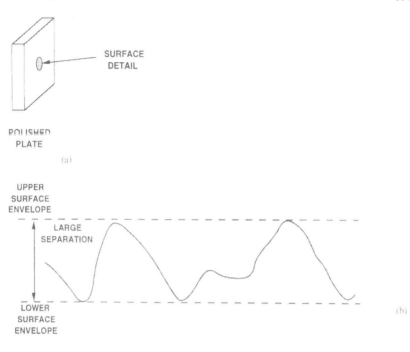

Figure 18.23 (a) A small surface detail on a polished plate. (b) A surface much magnified shows large peaks and valleys. (c) A polished surface displays small separation between the upper surface envelope and the lower surface envelope.

where $\Delta\omega_r$ is the incremental solid angle subtended by the detecting surface. For a specularly reflecting surface, the BRDF is a delta function:

$$\mathrm{BRDF}(\theta_i, \phi_i; \theta_r, \phi_r) = \delta(\theta_r - \theta_i, \phi_r - \phi_i) = 1 \text{ when } \theta_r = \theta_i \text{ and } \phi_r = \phi_i$$
$$= 0 \text{ otherwise.} \quad \text{[unitless]}$$

$$(18.113)$$

For a perfectly diffuse or a Lambertian surface, the BRDF has a $\cos\theta$ dependence.

$$\mathrm{BRDF}(\theta_i, \phi_i; \theta_r, \phi_r) = \cos\theta_r \Delta\omega_r. \quad \text{[unitless]} \quad (18.114)$$

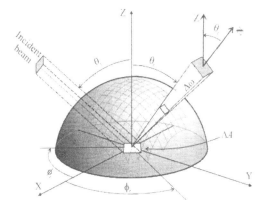

Figure 18.24 The reflectivity of a surface is determined by measuring the amount of the radiation incident as a collimated beam and reflected from an increment of area into an element of solid angle.

The subscript r is usually left out for diffuse surfaces. The component of the radiation that is not reflected according to Snell's law in a reflective (polished) component is referred to as the scattered light. This fraction of light may be decreased by careful polishing, handling, and storage in a dust-free environment, but it cannot ever be completely eliminated. It is one of the major contributors to the stray light noise in an optical system.

The angular dependence of the light transmitted into the second medium may behave in one of two ways: it may obey Snell's law for refraction if the surface is polished, as in the case of a lens or a prism; it may be scattered, as in the case of highly absorbing coating materials designed to increase the absorption of the incident radiation by the detector.

The material is said to be reflecting, when only an negligibly small amount of radiation is transmitted into it. A material is transparent if the surface is transmissive and the material is not absorptive. With the exception of a few highly reflecting interfaces [4], a significant fraction of the radiation is transmitted into the second medium, especially in the case of transparent materials.

18.7.3 Surface or Fresnel Losses

When the incident light passes from one transparent region with a given index of refraction n_1 to another transparent region with a different index of refraction n_2, only a small fraction of light is reflected, while the major part of the light is transmitted, as illustrated in Fig. 18.25.

The portion of the radiative energy lost to crossing the boundary is referred to as the Fresnel loss. It depends only on the indices of refraction on each side of the interface and the ray angle of incidence. The fraction of energy reflected at a single surface separating two different regions of space, characterized by two indices of refraction n_1 and n_2, respectively, is simplified for the normally incident beam of light:

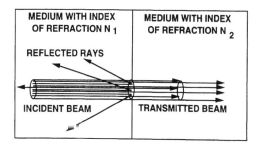

Figure 18.25 The surface or Fresnel loss is that fraction of incident power that is reflected back into the medium of exitance when the incident light passes from one transparent region with a given index of refraction n_1 to another one, also transparent, but with a different index of refraction n_2.

$$R_{12} = r_{12}^2 = [(n_1 - n_2)/(n_1 + n_2)]^2, \quad \text{[unitless]} \tag{18.115}$$

where R is the energy (or intensity) reflection coefficient, while r is the field reflection coefficient. The fraction of the radiative energy that remains in the main beam and is transmitted into the second medium, τ, is called the transmission coefficient. Using the conservation of energy and assuming that the surface does not absorb any energy, we get the single-surface transmission factor:

$$\tau_{12} = 1 - R_{12}. \quad \text{[unitless]} \tag{18.116}$$

Upon the substitution of Eq. (18.115) into Eq. (18.116), we obtain the desired result:

$$\tau_{12} = 1 - [(n_1 - n_2)/(n_1 + n_2)]^2 = (4n_1 n_2)/(n_1 + n_2)^2. \quad \text{[unitless]} \tag{18.117}$$

Here we note the symmetry of indices of two media, indicating that the transmission (losses) from medium 1 to medium 2 are equal to the transmission (losses) from medium 2 to medium 1:

$$\tau_{12} = \tau_{21} = (4n_1 n_2)/(n_1 + n_2)^2. \quad \text{[unitless]} \tag{18.118}$$

The transmission losses of a complete optical system are a product of the losses at each individual surface. The reflection losses for an optical system consisting of a number of surfaces made of the same material with the same index of refraction are equal to that of a single surface raised to the exponent of the number of surfaces.

18.7.4 Radiation Propagating in a Medium

In the previous section, we saw that the radiation incident on the boundary between two media may be reflected into the medium from where it came, or it may be transmitted.

We consider a beam of light with incidance M_I incident on a medium. The amount of light transmitted into the medium is given by

$$M_0 = \tau_{12} M_I = (1 - R_{12}) M_I = (1 - r_{12}^2) M_I. \quad \text{[W/m}^2\text{]} \tag{18.119}$$

Here R denotes the energy (or power) reflection coefficient. The parallel beam of light incident normally on a polished interface will continue as such even inside the

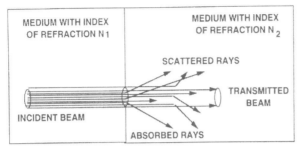

Figure 18.26   In a material, some rays are scattered and others are absorbed out of the main beam, resulting in a decrease in the exitance as a function of propagation distance.

medium, unless the medium is scattering, i.e., it includes particles that deviate the beam of light in a random fashion.

Generally, the beam transmitted into the second medium M_0 will consist of a component in the direction of the incident beam M_T and a component scattered out of the beam in all directions. The scattered component M_S represents losses for the information-carrying main beam. The amount of radiation absorbed in the medium M_A represents additional losses. These beams are illustrated in Fig. 18.26:

$$M_0 = M_T + M_S + M_A. \quad [\text{W/m}^2] \tag{18.120}$$

This relation expresses the conservation of radiative energy inside the medium.

18.7.5   Radiation Scattered in a Medium

The amount of radiation that is scattered within a medium may be described by an exponential function. We consider a beam of radiation incident on an imaginary surface at a distance z from the surface delineating the absorbing material, as illustrated in Fig. 18.27. In the transverse section of material of width Δz, a fraction of the radiation $\Delta M(z)$ is scattered outside the beam. This exitance is equal to the difference between the exitance at z, $M(z)$, and that at $z + \Delta z$, $M(z + \Delta z)$; i.e.,

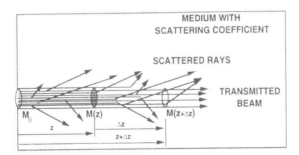

Figure 18.27   A beam of radiation is incident on an imaginary surface at a distance z from the surface. In the section of material of width Δz, a fraction of the radiation $\Delta M(z)$ is scattered outside the beam. This exitance is equal to the difference between the exitance at z, $M(z)$, and that at $z + \Delta z$, $M(z + \Delta z)$.

$$\Delta M(z) = M(z) - M(z + \Delta z). \quad [\text{W/m}^2] \tag{18.121}$$

The amount of radiation scattered out of the beam along the path from z to $z + \Delta z$ is proportional to the interval Δz and the exitance at z, $M(z)$. The proportionality constant is called the (linear) scattering coefficient, k_s, in $[\text{m}^{-1}]$. The negative sign indicates that the exitance remaining in the beam decreases with the propagation distance z:

$$\Delta M(z) \quad k_s \Delta z M(z). \quad [\text{W/m}^2] \tag{18.122}$$

The volume scattering coefficient is important when considering the integrated amount of radiation scattered out of or into a 3-D optical beam. In such a case, the radiation incident from all directions is evaluated. For this problem, the differential equation has to be formulated in three dimensions.

Equation (18.22) is integrated after the differentials have been replaced with the infinitesimals:

$$\int_{M_0}^{M(z)} dM(z)/M(z) = -k_s \left\{ \int_0^z dz \right\}. \quad [\text{unitless}] \tag{18.123}$$

The lower limits on the definite integrals are set as follows: at $z = 0$ the exitance is intact, $M(0) = M_0$: at zero propagating distance, no light has been scattered out of the beam as yet. So, the definite integrals in Eq. (18.123) are evaluated:

$$M(z) = M_0 e^{-k_s z}. \quad [\text{W/m}^2] \tag{18.124}$$

The scattered radiation is "lost" to the main b eam as to its capacity to transfer information efficiently: within the first-order analysis, it gets neither reflected nor transmitted outside the medium. However, the scattered radiation is not associated with the image-carrying optical beam. Its contribution is just the opposite: first of all, it reduces the signal; secondly, if it is finally incident on the image plane, most likely its location does not correspond to the conjugate object point. Thus, it increases the optical noise. Both of these effects diminish the signal-to-noise ratio. [5]

In fact, there are two phenomena characteristic of the interaction of the radiation with matter that decrease the signal-to-noise ratio of the information-carrying optical beam: in addition to scattering, there is also the radiation absorption within the matter.

18.7.6 Radiation Absorbed in a Medium

The absorption of radiation is the other phenomenon characteristic of light interaction with matter. Similarly to scattering, the amount of radiation absorbed upon propagation is also described by an exponential function. We consider a beam of radiation incident on an imaginary surface at a distance z from the surface delincating the absorbing material, as illustrated in Fig. 18.28. In material of width Δz, a fraction of the radiation incident on this imaginary surface $\Delta M(z)$ is absorbed from the beam. This exitance is equal to the difference between the exitance at z, $M(z)$, and that at $z + \Delta z$, $M(z + \Delta z)$:

$$\Delta M(z) = M(z) - M(z + \Delta z). \quad [\text{W/m}^2] \tag{18.125}$$

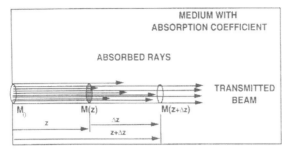

Figure 18.28 A beam of radiation is incident on an imaginary surface at a distance z from the surface. In the section of material of width Δz, a fraction of the radiation $\Delta M(z)$ is absorbed from the beam. This exitance is equal to the difference between the exitance at z, $M(z)$, and that at $z + \Delta z$, $M(z + \Delta z)$.

The amount of radiation absorbed from the beam along the path from z to $z + \Delta z$ is proportional to the width of the interval of propagation Δz and the exitance at z, $M(z)$. The proportionality constant is called the (linear) absorption coefficient, α, in $[\mathrm{m}^{-1}]$. The negative sign indicates that the exitance remaining in the beam decreases with the propagation distance z:

$$\Delta M(z) = -\alpha \Delta z M(z). \quad [\mathrm{W/m^2}] \tag{18.126}$$

Equation (18.126) is easily integrated, as for the case of scattering,

$$M(z) = M_0 e^{-\alpha z}. \quad [\mathrm{W/m^2}] \tag{18.127}$$

The absorbed radiation is also lost to the information-transmitting beam. It increases the internal energy of the material, raising ever so slightly its temperature. This may be significant for those materials whose index of refraction, expansion, or the absorption coefficient are temperature-dependent. Germanium is an example of such a material.

Some materials exhibit both absorption and scattering. It is easy to show that the exitance at the distance z in the medium $M(z)$ becomes

$$M(z) = M_0 e^{-(\alpha + k_s)z}. \quad [\mathrm{W/m^2}] \tag{18.128}$$

When the absorption or scattering coefficients become appreciable, the material no longer functions as a transmissive medium.

18.7.7 External and Internal Transmittance

First, we consider the case of internal transmittance. We refer to Fig. 18.29, which shows a parallel beam of light with exitance M_I incident on a plane of thickness d. We evaluate Eq. (18.128) for the beam-propagation distance $z = d$ to find the internal transmittance in the case of the absorbing and scattering plate. Thus, we obtain

$$M(d) = M_0 e^{-(\alpha + k_s)d}. \quad [\mathrm{W/m^2}] \tag{18.129}$$

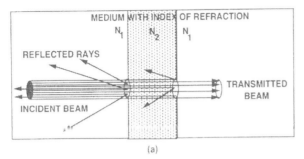

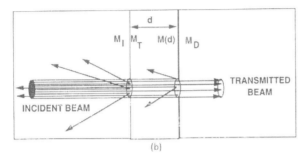

Figure 18.29 (a) Internal transmittance includes only the losses inside the medium. (b) External transmittance includes, in addition to the internal transmittance, the Fresnel losses at both interfaces.

The internal transmittance τ_i is defined as the ratio of the exitance at the end of the propagation distance d within the medium $M(d)$ to the exitance after the beam has entered the medium M_0. It is given as follows:

$$\tau_i = M(d)/M_0. \quad \text{[unitless]} \tag{18.130}$$

Thus, the internal transmittance is equal to

$$\tau_i = e^{-(\alpha+k_s)d}. \quad \text{[unitless]} \tag{18.131}$$

The internal transmittance depends on the thickness over which the beam is propagating, the material properties, and the absorption and scattering coefficients. If we know the plate thickness and the absorption and scattering coefficients, we can calculate the internal transmittance. The internal transmission of a plate is increased when the material has low absorption and scattering coefficients, for any plate thickness. When a plate of a thickness d_1 is replaced by a plate of the same material but with a different thickness d_2, the internal transmittance is changed.

The external transmittance is the ratio of the exitance leaving the plate M_D to the exitance incident on the plate M_I:

$$\tau_e = M_D/M_I. \quad \text{[unitless]} \tag{18.132}$$

To find the external transmittance, we need to determine the amount of light transmitted through the plate to the other side, M_D. The external transmittance includes, in addition to the internal transmittance, the Fresnel reflection losses at both boundaries. Using Eq. (18.119) again, we obtain at each surface:

$$M_0 = \tau_{12} M_{\mathrm{I}}, \qquad [\mathrm{W/m^2}] \tag{18.133}$$

and

$$M_{\mathrm{D}} = \tau_{21} M(d). \qquad [\mathrm{W/m^2}] \tag{18.134}$$

The exitances on the inside of the slab, M_0 and $M(d)$, are related through Eq. (18.129). We substitute Eq. (18.129) into Eq. (18.134):

$$M_{\mathrm{D}} = \tau_{21} M_0 \mathrm{e}^{-(\alpha+k_{\mathrm{s}})d}. \qquad [\mathrm{W/m^2}] \tag{18.135}$$

We now substitute Eq. (18.133) into Eq. (18.135) to obtain an explicit result:

$$M_{\mathrm{D}} = \tau_{21} \tau_{12} M_{\mathrm{I}} \mathrm{e}^{-(\alpha+k_{\mathrm{s}})d}. \qquad [\mathrm{W/m^2}] \tag{18.136}$$

The expression for the external transmission is simplified when the exitance leaving the plate, given in Eq. (18.136), is substituted into Eq. (18.132). The common factor M_{I} cancels in the numerator and denominator; i.e.,

$$\tau_{\mathrm{e}} = \tau_{21} \tau_{12} \mathrm{e}^{-(\alpha+k_{\mathrm{s}})d}. \qquad [\mathrm{unitless}] \tag{18.137}$$

Equation (18.137) is the most physically intuitive expression for the external transmittance. It may also be given in terms of internal transmittance, when Eq. (18.131) is substituted in Eq. (18.137):

$$\tau_{\mathrm{e}} = \tau_{21} \tau_{12} \tau_{\mathrm{i}}. \qquad [\mathrm{unitless}] \tag{18.138}$$

This expression shows quite clearly that the external transmittance is the product of the internal transmittance multiplied by the Fresnel losses at each surface. It is important to keep in mind that we may only measure the external transmittance.

For the normal angle of incidence, the Fresnel losses are given in terms of the indices of refraction, Eq. (18.116). Then, Eqs (18.137) and (18.138) become

$$\tau_{\mathrm{e}} = \tau_{12}^2 \mathrm{e}^{-(\alpha+k_{\mathrm{s}})d} = (1 - R_{12})^2 \mathrm{e}^{-(\alpha+k_{\mathrm{s}})d}, \qquad [\mathrm{unitless}] \tag{18.139}$$

and

$$\tau_{\mathrm{e}} = \tau_{12}^2 \tau_{\mathrm{i}} = (1 - R_{12})^2 \tau_{\mathrm{i}}. \qquad [\mathrm{unitless}] \tag{18.140}$$

Using the second equality in Eq. (18.117), we obtain another set of expressions:

$$\tau_{\mathrm{e}} = [(4n_1 n_2)/(n_1 + n_2)^2]^2 \mathrm{e}^{-(\alpha+k_{\mathrm{s}})d}, \qquad [\mathrm{unitless}] \tag{18.141}$$

and

$$\tau_{\mathrm{e}} = [(4n_1 n_2)/(n_1 + n_2)^2]^2 \tau_{\mathrm{i}}. \qquad [\mathrm{unitless}] \tag{18.142}$$

Only in the case when the plate is made of material that is neither scattering nor absorbing is the internal transmittance equal to 1, and the transmission losses of a plate are due solely to the inevitable Fresnel losses at the two surfaces.

REFERENCES

1. Nicodemus, F., *Self Study Manual on Optical Radiation Measurements, Part 1, Concepts*, Superintendent of Documents, US Government Printing Office, Washingtion, DC, 20402 (1976).

2. Wolfe, W. L. and G. J. Zissis, *The Infrared Handbook*, Office of Naval Research, Washington, DC, 1978.

3. Wyszecki, G. and W. S. Stiles, *Color Science*, John Wiley & Sons, New York, 1982.

4. Scholl, M. S. "Figure Error Produced by the Coating Thickness Error," *Infr. Phys. & Tech.*, **37**, 427–437 (1996).

5. Scholl, M. S. and G. Paez Padilla, "Using the y, y-bar Diagram to Control Stray Light Noise in IR Systems," *Infr. Phys. & Tech.*, **38**, 25–30 (1997).

19

Incoherent Light Sources

H. ZACARIAS MALACARA
Centro de Investigaciones en Optica, León, Mexico

19.1 INTRODUCTION

A light source is a necessary component in most optical systems. Except for those systems that use natural light, all others must include an artificial light source. In more than 100 years, a very large variety of light sources have been developed, and still new ones are currently being designed. Five main types of artificial sources are available:

(a) Light sources that emit from a thermally excited metal. Most of these sources are made from a tungsten filament. The spectrum of light corresponds to a quasi-blackbody emitter at the emitter temperature.

(b) Light emitted by an electrically produced arc in a gap between two electrodes. The arc can be produced in open air, although most modern arc lamps are enclosed within a transparent bulb in a controlled atmosphere. The spectrum is composed of individual lines from the gas, superimposed to a continuous spectrum emitted by the hot electrode.

(c) Light produced by the excitation of a material by ultraviolet energy in a long discharge tube, generically known as fluorescent lamps.

(d) Light emitted due to a recombination of charge carriers in a semiconductor gap. A semiconductor pair is needed for the light to be emitted. These light sources receive the generic name of *light-emitting diodes* (LEDs).

(e) Light emitted as a result of the stimulated radiation of an excited ensemble of atoms. This emission results in laser radiation with light having both spatial coherence (collimated light) and temporal coherence (monochromaticity). Due to its importance in optical instruments, these devices are described in another chapter in this book and will not be described in this chapter.

Some Basic Concepts

(a) *Luminous efficacy*. All light sources emit only a small amount of visible power from their input power. *Luminous efficacy* $\mathscr{E}$ is defined as the ratio of the total luminous flux F to the total power input P, measured in lumen per watt (lm/W); i.e.,

$$\mathscr{E} = \frac{F}{P} \tag{19.1}$$

Assuming an ideal white source, which is one with constant output power over all the visible portions only, the luminous efficacy will be about 220 lm/W. [20]

(b) *Color temperature*. For a blackbody emitter, the color temperature corresponds to the spectral energy distribution for a blackbody at that temperature. The Kelvin temperature scale is used to describe color temperature for a source.

(c) *Correlated color temperature*. When the emitter is not a perfect blackbody, but the color appearance resembles that of a blackbody, the correlated color temperature is the closest blackbody temperature found in the CIExy color diagram.

(d) *Color rendering index (CRI)*. This is a property of a light source to reproduce colors as compared with a reference source. [16] This figure reflects the capability of a light source to faithfully reproduce colors. The color rendering index is 100 for daylight.

19.2 TUNGSTEN FILAMENT SOURCES

The tungsten filament source, which is now more than 100 years old, is also the first reliable light source for optical devices. The basic lamp has evolved since its invention by Edison. An historical account is described by Anderson and Saby [1], Josephson [10] and Elenbaas [5]. The basic components for a tungsten lamp are (Fig. 19.1): an incandescent electrically heated filament; supporting metal stems for the filament, two of them used to conduct electrical current to the filament; a glass envelope; filling gas; and a base to support and make the electrical contact. These basic components, with variations according to their applications, are now considered.

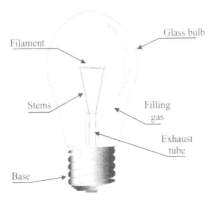

Figure 19.1 Basic components for a tungsten lamp.

(a) *Filament.* Electrical power heats the filament with a spectral distribution that follows a gray body. Since high temperatures are attained, filaments must support the highest possible temperatures. The higher the temperature, the higher the light efficacy. Carbon is capable of sustaining the highest temperature; unfortunately, it also evaporates too fast. After testing several materials, such as osmium and tantalum, tungsten is the metal most used for filaments. High temperature has the effect of vaporizing the filament material, until after some time, the thinning filament breaks. The evaporation rate is not constant and depends on the impurity contents. [13] Most optical applications require a point-like source. Ideally, filaments should have low extension. To reduce the emitter extension, and at the same time increase the emissivity, filaments are coiled and, in some cases, double coiled. The reason is that by coiling the filament, the surface exposed to cool gas is reduced, decreasing the convection cooling. [18] Some typical filaments are shown in Fig. 19.2. Additional filament shapes can be seen in the book by Rea. [20] Where light must be concentrated, filaments have low extension, as in projection, searchlights, or spotlights. If light is needed to cover a large area, large filaments are used instead. Some optical instruments require a line source. Straight filaments are used: for example, in hand-held ophthalmoscopes. Filaments operate at a temperature of about 3200 K. Tungsten emissivity ranges typically from 0.42 to 0.48. Spectral emissivity is reported in the literature. [6, 13]

(b) *Wire stems.* Besides carrying electrical current to the filament, stems are used to hold the filament in position. Several different kinds of stems are used, depending on the filament and bulb shape. Mechanical vibration from the outside can create vibration modes in the filament and stems, reducing its life by metal fatigue. Some lamps have a design to reduce vibration but they must be mounted according to manufacturer specifications. Lead-in wires are chosen as to have a similar thermal coefficient to pinch the glass at the electric seal to avoid breaking the glass.

Figure 19.2 Some typical filaments.

(c) *Filling gas.* The purpose of filling with gas is twofold: (i) to provide an inert atmosphere and avoid filament burning and (ii) to exert a pressure over the filament and delay the evaporation process. A negative effect of the gas is convective cooling, reducing the lamp efficacy. Low current lamps are made with a vacuum instead of a filling gas because the gas may give a negative effect for small filaments. Nitrogen and argon are the most commonly used gases, but some small lamps use the more expensive krypton gas. A common lamp is the quartz-halogen lamp. [28] A mixture of a halogen and a gas are enclosed in a low-volume quartz envelope. As the lamp burns, the tungsten filament is evaporated over the inner bulb surface, but due to the high temperatures attained by the quartz envelope, a reaction occurs between the halogen and the evapaorated tungsten, removing it from the bulb. At the arrival of this mixture to the filament, the tungsten is captured again by the filament. Due to this effect, called the halogen cycle, the filament has a much longer life than any standard filament lamp for a filament at a higher temperature. Actually, the hot quartz envelope reacts easily with the stems, eroding the wire, and the breakage is usually produced in the stem. Quartz halogen lamps do not usually show bulb blackening. Due to the high chemical reactivity of the quartz bulb, care must be taken to avoid any grease deposition over the bulb surface; otherwise, a hot center is developed, resulting in the bulb breaking. It has been found that halogen vapor absorbs some radiation in the yellow-green part of the spectrum and emits in the blue and ultraviolet. [26]

(d) *Glass envelope.* The glass bulb prevents the oxygen from burning the filament and allows a wide light spectrum to leave from the lamp. Most lamps are made of soft lime-soda glass. Outdoor lamps have impact or heat-resisting glass. Halogen and tubular heat lamps are made of quartz. As mentioned, glass bulbs are chosen to transmit most of the visible spectrum. The transmission spectrum for a glass bulb is reported by Kreidl and Rood [12] and Wolfe [27]. Optical transmittance is affected by temperature. For lamps for use at near-infrared wavelengths, a window is placed at some distance from the source, like Osram WI 17/g (Osram 2000, internet site). Tungsten lamps are manufactured with diverse bulb shapes, designated by their shape: A for arbitrary shape, R for reflector, T for tubular, and PAR for parabolic reflector (Fig. 19.3). New shapes are brought to the market constantly, and others are discontinued. The most recent catalog from the manufacterer is recommended for current bulb shapes. Some lamps are made transparent while others are frosted by acid etching. Some are covered with white silica powder on the inner surface for a better diffusing light. Acid-etched lamps do not absorb significant amounts of light; silica-covered lamps absorb about 2% of the light. [18]

(e) *Supporting bases.* Besides the electrical contact, the base must support the lamp in place. The traditional all-purpose screw base is used for most general lighting applications. For most optical instruments, where the filament position is critical, the so-called prefocus base is used. Other bases are also used, such as bayonet-type bases. Quartz-halogen lamps are subjected to high temperature, and ceramic bases are used in those cases.

Some optical applications where tungsten lamps are used are described as follows:

(i) *Spectral irradiance standard lamps.* Since quartz halogen lamps have a high stability for a relative long period of time (about 3%), it has been proposed for use as

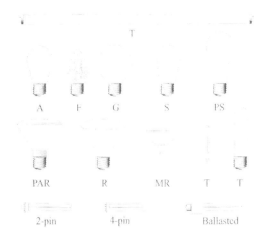

Figure 19.3 Bulb shapes for lamps.

a secondary standard for spectral irradiance. [Osram, 2000] Later, an FEL lamps has been proposed as a secondary standard. [8] This type of lamp is placed at a fixed distance and at a specified orientation, where the spectral irradiance is known. This lamp provides a handy reference to check for the calibration of some light sensors. Some laboratories make a traceable calibration for every lamp they sell.

(ii) *Standard type A source.* The International Committee for Illumination (CIE) has defined a standard source called a Type A source. This source is a reference for a color specimen to be observed. The original definition describes a tungsten lamp with quartz window operating at a correlated color temperature of 2856 K. Originally, this description was for a 75 W lamp under a fixed supply voltage. Any tungsten lamp with a carefully controlled filament temperature could be used; also, it can be purchased with a certification to be used for this purpose, like the Osram WI 9 (Osram, 2000).

(iii) *Photographic flash lamps and photolamps.* Daylight color films are made for a color temperature of 6500 K. The maximum attainable temperatures for photolamps used to be 3200 and 3400 K. For this reason, a film with a color balance of 3200 or 3400 K were selected for indoor use of color light. The mean life for a 3200 K lamp was about 6 hours, while a 3400 K lamp lasted for about 100–150 hours. [2] Now, quartz halogen lamps have a life about 10 times longer, with a higher light output. For movies, TV studio, and photographic use, tungsten halogen lamps are always used.

Old photoflash lamps were built to produce the highest possible light output. To make it possible, a glass bulb was made with a long filament and filled with oxygen instead of a vacuum. The filament was quickly burned, producing a high light output. Flash lamps were usually covered with a lacquer to avoid an explosion and to support a color-balancing dye. Since the light power curve increased at a fixed rate, the shutter had to be synchronized for the highest power output. For filament flash lamps, the synchronization is called M synchronization. Other disposable lamps had a piezoelectric element to fire the lamp and yet others, a mechanical firing device started the chemical filament reaction for filament burning. [2–4]

(iv) *Projection lamps.* Old projection lamps were made with a screen-shaped filament, to form on the film gate a blurred image of the filament, producing an almost uniform illumination at the gate. This resulted in a large filament with a complicated support and a reduced color temperature. Recent projection lamps rely on lamps with a single coiled filament and an elliptic reflector (Fig. 19.4). The advantages for the new system results in the following: (a) the integrated elliptical reflector has a thin film cold mirror to reduce heat at film gate; (b) short single-coiled filaments increase color temperature; (c) quartz halogen lamps increase life and reduce bulb blackening; and (d) nonuse of condenser lenses. The condenser system brings about 55–60% of total flux to the objective, while an elliptical mirror without a condenser lens can bring up to 80% of total flux. [2]

19.3 ARC SOURCES

Among other mechanisms that can produce light, is the electron de-excitation in gases. An electron is excited by an electron or ion collision, which is achieved by the following means:

1. *Thermal electron emission by a hot cathode.* A heating incandescent filament emits some electrons that statistically overcome the metal's work function. In the presence of an electrical potential, the electron is accelerated and, by collision, new ions are produced. This method is used to start a discharge.

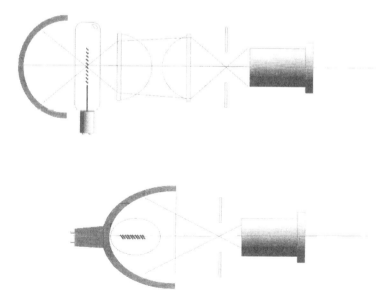

Figure 19.4 Old and new projection optics.

2. *Secondary emission.* Several electrons are emitted from a metal that is struck by a colliding electron. Once a discharge begins, it is maintained by continuous ion bombardment. This mechanism is responsible for the arc maintaining in arc sources.

3. *Field emission.* A strong electrical field applied in a relatively cold cathode can be high enough for electrons to be emitted from the metal. This mechanism is used to start the discharge in an initially cold lamp.

The oldest form of an arc can be found in the now-obsolete carbon arc. Two carbon electrodes were brought into contact and in series to a limiting resistor. There are three types of carbon arcs: low intensity, high intensity, and flame. *Low-intensity arcs* operate by circulating a current high enough to reach the sublimation point for the carbon (3700 K). The emission characteristics of a carbon arc is that of a blackbody. Although simple in operation, carbon electrodes are consumed fast and new electrodes should be replaced. In old movie theater projectors, a device was made to maintain a constant current (constant luminance) on the electrodes. For some special applications, where a carbon arc was to be operated for long time, an automatic electrode supply was devised. [14] In *flame arcs*, an additional compound was included with the carbon electrode to modify the emission characteristics of the light, increasing at the same time the efficiency of the source. *High-intensity arcs* are a special kind of flame arc with an increased current. Because of its low reliability and short light cycle, arc lamps are not currently used; they have been replaced by short arc lamps. Emissivity for carbon arcs is about 0.98–0.99. [17]

Short arc lamps: modern arc sources are made from tungsten electrodes enclosed in a large spherical or ellipsoidal fused silica envelope. In this case, light is produced by electron de-excitation in gases. Gases are at about atmospheric pressure, but when hot, pressure may increase up to 50 atmospheres. Thoriated tungsten electrodes have a typical gap between 0.3 and 12 mm. Since short arc lamps have a negative resistance region, once the arc is started, a very low impedance appears at the electrodes. Short arc lamps' lives are rated at more than 1000 hours, when the bulb blackens or the electrode gap increases and the lamp cannot start. Commercially, lamps are available from 75 W up to 30,000 W. Short arc or compact arc lamps are used in motion picture and television illumination, movie theater projection, solar simulators, and TV projection.

The starting voltage may raise up to 40 kV. To avoid a rapid destruction of the lamp due to a high current, a ballast must be used. For ac operation, an inductive ballast is used, but for many applications where line current modulation is not allowed, an electronic current limiter must be provided after the start. Three short arc lamps are available: mercury and mercury–xenon; xenon lamps; and metal halide lamps.

1. *Mercury and mercury–xenon lamps.* Short arc mercury lamps have at low temperature and a pressure of about 20–60 Torr [2, 20] of argon with traces of mercury. After the initial pulse starts the arc, mercury is vaporized, the pressure increases and the emission spectrum is of mercury, but with broad lines because of high gas pressure (Fig. 19.5(a)). It takes several minutes to reach full power, but if the lamp is turned off, it may take up to 15 min to cool down for restart. By adding xenon to the lamp, the warm-up time is

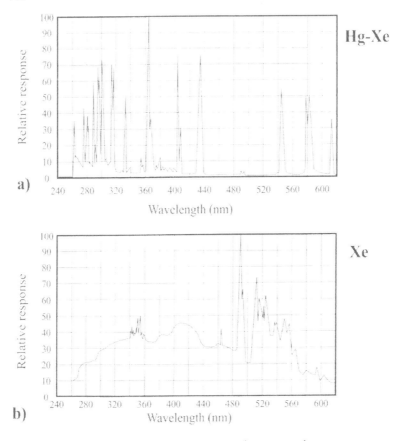

Figure 19.5   Emission spectra for (a) mercury and (b) xenon arc lamps.

reduced by about 50%. The spectral light distribution is the same as the mercury lamp. The luminous efficacy is about 50 lm/W for a 1000 W lamp.

2. *Xenon lamps.* These lamps have a continuous spectral distribution with some lines from the xenon (Fig. 19.5(b)); maximum power is obtained a few seconds after the start. Correlated color temperature for a xenon lamp is 5000 K. The luminous efficacy ranges from 30 to 50 lm/W.

3. *Metal halide lamps.* The inclusion of rare earth iodides and bromides to a mercury short arc lamp results in a lamp with a full-spectrum emission, a high color-rendering index, and a high luminous efficacy. These lamps are used mainly in TV and movie lighting, and some fiber optics illuminating devices.

19.4   DISCHARGE LAMPS

High-intensity discharge lamps produce light by the electrical current flowing through a gas. It is necessary to reach the gas ionization for the gas to glow. As already mentioned, gas discharges have negative resistance characteristics; then, discharge lamps need a ballast to limit current once it is started. Most high-intensity discharge lamps operate from an ac supply. Three main types of lamps are produced for illumination purposes: mercury, metal halide, and high pressure.

1. *Mercury lamps.* Mercury lamps are made of two glass bulbs, one within the other. The inner bulb contains two tungsten electrodes for electrical contact. Argon gas is used to initiate the discharge, but small amounts of mercury are quickly vaporized to produce a broad line spectrum with the lines at 404.7, 432.8, 546.1, 577, and 579 nm. This results in a blue-green light. In contrast to a short arc lamp, the distance between the electrodes is several centimeters. The outer glass bulb serves as a filter for the ultraviolet light and contains some nitrogen to reduce pressure differences with the atmosphere. Sometimes the inner surface for this bulb is covered with a phosphor to convert ultraviolet radiation to visible light. The color is selected in the orange-red portion to improve the color-rendering properties of the lamp. A version of this lamp includes a phosphor to convert the 253.7 nm UV radiation to near-UV light (black light). High-intensity mercury discharge lamps have a light efficacy of 30–65 lm/W. Most mercury lamps operate with a 220 V supply voltage, but can also function with 127 V with an auto transformer-ballast.

2. *Metal halide.* The inclusion of some metal halide in a mercury lamp adds several spectral lines to an almost continuous spectrum. [21] The effect is a better color-rendering index and an improvement in the luminous efficacy (75–125 lm/W). Scandium iodide and sodium iodide are two of the added materials.

3. *High-pressure sodium lamps.* Sodium vapor at high pressure is used for discharge lamps. Spectral width is strongly dependent on gas pressure, so that a high-pressure sodium lamp gives a broad spectrum dominant at the yellow line. [25]

Most high-intensity discharge lamps are operated with an inductive ballast; this drops the power factor up to 50%, but with a power factor correcting capacitor, it increases up to 90%. [18] Dimming of discharge lamps imposes several design restrictions as for power factor correction and harmonics control. Electronic systems for this purpose are described in RCA. [19]

Discharge lamps, which are used for general lighting and are not frequently used for optical instruments, include the following types.

1. *Low-pressure discharge lamps.* Small low-pressure discharge lamps are made for spectroscopic application. Due to its low pressure, spectral lines are sharp and can be used for spectral line calibrations. Table 19.1 shows some spectral lamps. [Osram, 2000]

2. *Long-discharge lamps.* Low-pressure long-arc lamps are also made. They are used for solar simulators. Since they have low luminous efficacy (30 lm/W), they are not used for general lighting. Long-arc xenon lamps produce a color similar to daylight, rich in UV content. These lamps are used for ageing chambers, as recommended by ASTM G-181 and G155-98 and ISO 4582 standards.

Table 19.1 Characteristics for Spectral Low Pressure Lamps [Osram, 2000]

Designation	Elements	Lamp voltage (V)	Lamp current (A)	Type of current	Lamp wattage (W)	Luminous area $H \times B$ (mm)
Cd/10	Cadmium	10	1.0	ac	15	15×6
Cs/10	Caesium	10	1.0	ac	10	15×6
He/10	Helium	60	1.0	ac	55	15×8
Hg/100	Mercury	45	0.6...1	ac/dc	22...44	20×3
HgCd/10	Mercury + cadmium	30	1.0	ac	25	20×8
K/10	Potassium	10	1.0	ac	10	15×6.5
Na/10	Sodium	15	1.0	ac	15	15×6.5
Na/10FL	Sodium	16	0.57	ac	9	
Ne/10	Neon	30	1.0	ac	30	15×8
Rb/10	Rubidium	10	1.0	ac	10	15×6
Tl/10	Thallium	15	1.0	ac	15	8×3
Zn/10	Zinc	15	1.0	ac	15	15×6

3. *Flash lamps.* Flash lamps are discharge lamps with a long arc that emits a fast flash of light for a short time. Flash lamps are used in photography, stroboscopic lamps, warning lights in aviation and marine, and laser pumping. Flash lamps uses low-pressure xenon as the active gas, although some have traces of hydrogen to change its spectral content. Tubes for photoflash are usually a long arc in a long straight tube; sometimes the tube is coiled or U-shaped. A ring electrode is wrapped to the tube to trigger the flash.

Lamp electrodes are at high impedance, but the trigger electrode increases the gas conductivity to almost zero impedance. A typical circuit for a flash is shown in Fig. 19.6. A capacitor is charged in a relatively long time; after charging the capacitor C2, another capacitor, also previously charged is discharged through a transformer to produce a high voltage at the trigger electrode. Capacitor C1 is discharged through the lamp. The luminous efficacy is about 50 lm/W for a typical flash lamp. Loading in joules for a flash tube is: [20]

$$\text{Loading} = \frac{CV^2}{2} \tag{19.2}$$

Figure 19.6 A typical circuit for a photographic flash lamp.

The flash tube impedance is

$$Z = \frac{\rho L}{A},\tag{19.3}$$

where ρ is the plasma impedance in ohm-cm, L is the tube length, and A is the cross-sectional area in cm.

A version of a xenon arc lamp is used for photography. Xenon flash lamps have spectral distribution that closely resembles CIE D65 illuminant or daylight. Photographic flash lamps can stand more than 10,000 flashes. Flash tubes cannot be connected in parallel, since each lamp must have its own capacitor and trigger circuit. For multiple lamp operation, a slave flash lamp is designed to trigger with the light from another lamp. Flash lamps are synchronized to the camera in such a way that lamps are triggered when the shutter is fully opened. This is called the X-synchronization.

19.5 FLUORESCENT LAMPS

Fluorescent lamps are a general kind of lamp that produce a strong mercury line at 253.7 nm from low–pressure mercury gas. This excites a fluorescent material to emit a continuum of visible radiation.

(i) *Physical construction.* Fluorescent lamps are made mainly in tubular form, the diameter is specified in eighths of an inch, starting with 0.5 inch designated to T-4 to 2.125 inches or T-17. The length, including lamp holders, ranges from 4 inches (100 mm) to 96 inches (2440 mm). Lamp tubes are usually made from soft lime soda glass, and at each end, a small tungsten filament is used as electrode and as preheater to start the lamp. Alternatively, fluorescent lamps are also made in circular form, U-shaped and quad or double parallel lamps for compact fluorescent lamps. At each end of the tube, a base for electrical contact is provided. For circular lamps, at some point on the cicle, a connector for both ends is located.

As a filling gas, low-pressure mercury with some argon or argon and krypton is added to initiate the discharge. At this low pressure (200 Pa) most of the mercury is vaporized, but this depends on the ambient temperature.

(ii) *Electrical characteristics.* When cold, the electrical impedance of the gas is very high, but as soon as the lamp ignites, the conductivity decreases suddenly to a very low value. A means must be provided for current limitation. The principle of the electrical operation is shown in Fig. 19.7. Both filaments, an inductive ballast, and the starter are all in a single series circuit. Initially, the starter is closed, and both filaments in series are heated to vaporize the mercury gas and to ionize the gas. A rapid break in the starter circuit produces a high voltage from the ballast, enough to initiate the discharge. Multiple lamps can operate from a single ballast assembly designed for such a case. Electronic ballasts work on high line frequency (~ 20–60 kHz), since the efficacy increases about 10% for frequencies above 10 kHz. Electronic ballasts have better ballast efficiency, less weight, less noise, and some other advantages. There are two main modes of operation in fluorescent lamps, glow and arc.

In the glow mode, electrodes are made of single-ended cylinders; the inside is covered with an emissive material. In this mode of operation, the current in the lamp is less than 1 A, and the voltage through the lamp is about 50 V.

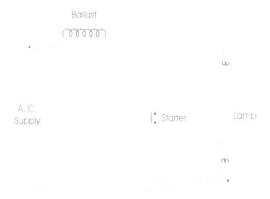

Figure 19.7   Electrical operation of a fluorescent lamp.

In the arc mode, the electrodes are tungsten, at a temperature of 1000°C. Electrons are emitted thermically. The current increases up to more than 1.5 A and the lamp voltage across the lamp is around 12 V. [20] This mode of operation is more efficient than the glow mode, and most of the fluorescent lamps work on this principle.

(iii) *Optical characteristics.* The optical spectrum for a fluorescent lamp is composed mostly of the continuous emission from the fluorescent material and a small amount of the line spectra from mercury. Fluorescent materials for lamps determine the light color. Table 19.2 shows some common phosphors and its resulting color.

Table 19.3 shows some of the most common lamp color designations and characteristics. CIExy color coordinates for each phosphor are referenced in Philips [18] as well as the spectral distribution for several phosphors.

(iv) *Applications.* One of the most common applications for a fluorescent lamp is found in color evaluation boots. Some industries have standardized a light that is commonly found in stores. This lamp is the *cool white* lamp. For graphic arts industries, a correlated color temperature of 5000 K, is selected; then some lamps are specifically designed for this purpose, like the Ultralume 85® or equivalent. For a good color consistency, it is recommended to replace lamps well before they cease to emit light. Another use for fluorescent lamps is found in photography, but since the color is not matched for any commercial film, a color compensating filter must be used for most fluorescent lamps. [23] Since phosphors have a relatively low time

Table 19.2   Phosphor Properties for Fluorescent Lighting

Phosphor	Color of fluorescence
Zinc silicate	Green
Calcium tungstate	Blue
Calcium borate	Pink
Calcium halo phosphate	Whites of various color temp.
Magnesium tungstate	Bluish white

Table 19.3 Fluorescent Lamp Designation Characteristics

Lamp description	Designation	Light output (%)	Color-rendering index	Color temperature (K)
Cool white	CW	100	67	4100
Cool white deluxe	CWX	70	89	4200
White	W	102	58	3500
Warm white	WW	102	53	3000
Warm white deluxe	WWX	68	79	3000
Daylight	D	83	79	6500
Colortone 50	C50	70	92	5000
Cool green	CG	83	70	6100
Sign white	SGN	75	82	5300
Natural	N	66	81	3400
Supermarket white	SMW	74	85	4100
Modern white deluxe	MWX	77	80	3450
Soft white	SW	68	79	3000
Lite white	LW	105	51	4100
Ultralume 83	83U	105	85	3000
Ultralume 84	84U	105	85	4100
Ultralume 85	85U	105	85	5000
Red	R	6	—	—
Pink	PK	35	—	—
Gold	GO	60	—	—
Green	G	140	—	—
Blue	B	35	—	—

constant, some flickering from line frequency can be observed. To avoid flickering in some fast optical detectors, like photodetectors and some video cameras, a high line frequency must be provided, such as the ones provided by electronic ballasts. Fluorescent lamps emit light in a cylindrically symmetric pattern. For some purposes, this can be adequate, especially for diffuse illumination. In some other cases, lamps with an internal reflector have been designed to send most of the light in some preferred direction. [4, 20] These lamps are used in desktop scanners and photocopiers.

Germicidal lamps are lamps with the same construction as any fluorescent lamp except that they have no fluorescent phosphor. This kind of lamp does not have high luminance since most (~ 95%) of the radiated energy is at the UV line of 253.7 nm. This radiation is harmful, since it produces burning to the eyes and skin. These lamps are used for air and liquid disinfection, lithography, and EPROM erasure.

The so-called black light lamp has a phosphor that emits UV radiation at the UV-A band (350–400 nm) and peaks at 370 nm. Two versions are available for these lamps: unfiltered lamps with a white phosphor that emits a strong blue component; a filtered one, with a filter to block most of the visible part. Uses for these lamps are found in theatrical scenery lighting, counterfeit money detection, stamp examination, insect traps, fluorescence observing in color inspection boots, and mineralogy. These lamps are manufactured in tubular form, from T5 to T12, compact fluorescent lamps, and a high-intensity discharge lamp.

19.6   LIGHT-EMITTING DIODES

Light-emitting diodes (LEDs) are light sources in which light is produced by the phenomenon of luminescence. An LED is made from a semiconductor device with two doped zones: one positive, or p-region, and the other negative, or n-region, as in any semiconductor diode. Electrons are injected in the n-region, while holes are injected in the p-region. At the junction, both holes and electrons are annihilated, producing light in the process. This particular case of luminescence is called electroluminescence and is explained in several books. [9, 15, 22, 24] Recombined electrons and holes release some energy in either a radiative or nonradiative process. For the latter, a phonon is produced and no visible energy is produced, whereas for the first case, the energy is released in some form of electromagnetic radiation or photon. The emitted photon has an energy that is equal to the energy difference between the conduction and the valence band minus the binding energy for the isoelectronic centers for the crystal impurities in the semiconductor. [9] Hence, the photon has a wavelength

$$\lambda = \frac{1240}{\Delta E} \text{ nm} \tag{19.4}$$

where ΔE is the energy transition in electron volts.

Historically, the first commercial LED was made in the late 1960s by combining three primary elements; gallium, arsenic, and phosphor (GaAsP), giving a red light at 655 nm. Galium phosphide LEDs were developed with a near-IR emission at 700 nm. Both found applications for indicators and displays, although the latter has poor luminance, since its spectral emission is in a region where the eye has a poor sensibility.

Since light is produced in an LED at the junction, this device has to show this junction to the detecting area. Figure 19.8(a) shows a cross section of an LED. These devices are found also in fiber optics. To make an efficient coupling to the fiber, most of the emitting area must be within the acceptance cone for the fiber. The Burrus LED was developed for optical fibers and its cross section is shown in Fig. 19.8(b).

(a) *Optical characteristics.* As mentioned, peak wavelength is a function of the band gap in the semiconductor. Several materials are used for making LEDS. Table 19.4 shows some materials used for making LEDS and their corresponding wavelengths.

Spectral bandwidth for most LEDs lie between 30 and 50 nm. Figure 19.9 shows the relative intensity for some LEDs. The first LEDs were red, then infrared and amber. Later green was obtained, but although blue LEDs were made on the laboratory scale, [11] life for these devices was very short due to the high photon energy.

At the emission point, light is emitted in all directions; hence at this point, an LED is a Lambertian emitter. Some LEDs have reflective electrical contacts, and the light is confined to emit along the junction. This device is called an edge-emitting LED, and is appropriate for fiber optics and integrated optics.

Some devices include two LEDs with different color in a single package. Current flowing in one direction produces light from one emitter; the other emitter lights by reversing the current. Mixed light color is obtained with bipolar square waves, the exact hue depending on the duty cycle for the square wave.

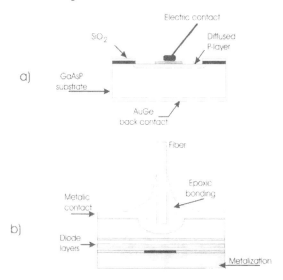

Figure 19.8 Cross section of (a) a typical LED and (b) a Burrus LED.

(b) *Electrical characteristics.* An LED electrically is a diode operated in direct polarization; by reversing polarization it does not emit light. To limit current, a limiting resistor is placed in series with the source and the diode. Heat dissipated by the device, may eventually end the diode's life. To avoid this, instead of operating the diode from direct current, a pulsed current operation dissipates less power for a given retinal perception. This is due to the eye retentivity to a rapid light pulse or light enhancement factor. [7] Different LED materials have different threshold vol-

Table 19.4 Materials for Making LEDs

Material	Wavelength (nm)	Color
SiC	470	Deep blue
GaN	490	Blue
GaP	560	Green
GaP:N	565	Green
SiC	590	Yellow
$GaAs_{0.15}P_{0.85}$:N	590	Yellow
$GaAs_{0.35}P_{0.65}$:N	630	Orange
$GaAsP_{0.5}$	610	Amber
$GaAs_{0.6}P_{0.4}$	660	Red
GaAlAs	660	Red super bright
GaP	690	Red
GaP:Zn,N	700	Deep red
GaAlAs	880	Infrared super bright
GaAs:Zn	900	Infrared
GaAs:Si	940	Infrared

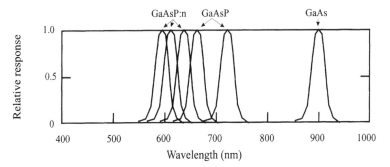

Figure 19.9   Relative intensity for some LEDs.

tage: consult manufacturers' data for proper operating levels and maximum power dissipation.

(c) *Applications.* The main application of LEDs are found in economical fiber-optic links. This is an economic alternative to lasers, where bit rate is lower than high-performance laser systems. LEDs for optical fiber communications can be purchased both connectorized or without the connector included. Wavelength can be chosen to fit the optimum transmission for the fiber. Some LED combinations can give red, green, and blue light. Some colorimeters, as an alternative to tristimulus colorimetry, can use these LEDs instead of filters. Since spectral distribution is not similar to tristimulus values, some error exists in the color measurement. Also, LED triads are used in some color document scanners for image capture.

High efficiency–low-cost LEDs are now produced. They can be used for illumination as well as indicators. Currently, a green LED can have luminous efficiency of about 100 lm/w, while a blue LED has only about 0.01 lm/w.

REFERENCES

1. Anderson, J. M. and J. S. Saby, "The Electric Lamp: 100 Years of Applied Physics," *Physics Today*, **32**(10), 33–40 (1979).
2. Anstee, P. T., "Light Sources for Photography," in *Photography for the Scientist*, 2nd edn, Morton, R. A., ed., Academic Press, London, 1984.
3. Carlson, F. E. and C. N. Clark, "Light Sources for Optical Devices," in *Applied Optics and Optical Engineering*, Vol 1, Academic Press, New York, 1975, Chapter 2.
4. Eby, J. E. and R. E. Levin, "Incoherent Light Sources," in *Applied Optics and Optical Engineering*, Vol 7, Academic Press, New York, 1979, Chapter 1.
5. Elenbaas, W., *Light Sources*, Crane Russak and Co., New York, 1972.
6. Forsythe, W. E. and E. Q. Adams, "Radiating Characteristics of Tungsten and Tungsten Lamps," *J. Opt. Soc. Am.*, **35**, 108–113 (1945).
7. General Instruments Co., *Catalog of Optoelectronic Products 1983*, Optoelectronic Division, General Instruments, Palo Alto, California, 1983.
8. Grum, F. and R. J. Becherer, "Optical Radiation Measurements," in *Radiometry*, Vol 1, Academic Press, New York, Chapter 5.
9. Hewlett-Packard Co., *Optoelectronics/Fiber Optics Application Manual*, 2nd edn, McGraw-Hill, New York, 1981.

10. Josephson, M., "The Invention of the Electric Light," *Scientific American*, **201**, 98–114 (1959).

11. Kawabata, T., T. Matsuda, and S. Koike, "GaN Blue Light Emitting Diodes Prepared by Metalorganic Chemical Vapor Deposition," *J. Appl. Phys.*, **56**, 2367–2368 (1984).

12. Kreidl, N. J. and J. L. Rood, "Optical Materials," in *Applied Optics and Optical Engineering*, Vol. 1, Kingslake, R., ed., Academic Press, New York, 1965, Chapter 5.

13. Larrabee, R. D., "Spectral Emissivity of Tungsten," *J. Opt. Soc. Am.*, **49**, 619–625 (1959).

14. Latil, J. P., "A Fully Automatic Continually Operating, Very High Intensity, Carbon Arc Lamp," *J. Opt. Soc. Am.*, **44**, 1–4 (1954).

15. Motorola Semiconductor Prods., *Optoelectronics Device Data*, 2nd printing, Phoenix, Arizona, 1981.

16. Nickerson, D., "*Light Sources* and Color Rendering," *J. Opt. Soc. Am.*, **50**, 57–69 (1960).

17. Null, M. R. and W. W. Lozier, "Carbon Arc as a Radiation Standard," *J. Opt. Soc. Am.*, **52**, 1156–1162 (1962).

18. Philips Lighting Co., *Lighting Handbook*, Philips Lighting Co., Eindhoven, 1984.

19. RCA, Co., *RCA Solid-State Power Circuits, Designer's Handbook SP-52*, RCA, Co., Sommerville, New Jersey, 1971.

20. Rea, M. S., ed., *Lighting Handbook, Reference and Application*, 8th edn, Illuminating Engineering Society of America, New York, 1993.

21. Reiling, G. H., "Characteristics of Mercury Vapor–Metallic Iodide Arc Lamps," *J. Opt. Soc. Am.*, **54**, 532–540 (1964).

22. Seipel, R. G., *Optoelectronics*, Reston Pub. Co., Reston, Virginia, 1981.

23. *SPSE Handbook of Photographic Science and Engineering*, Thomas, W., ed., Wiley, New York, 1973.

24. Sze, S. M., *Physics of Semiconductor Devices*, Wiley, New York, 1969.

25. van Vliet, J. A. J. M. and J. J. de Groot, "High Pressure Sodium Discharge Lamps," *IEE Proc.*, **128A**, 415–441 (1981).

26. Studer, F. J. and R. F. VanBeers, "Modification of Spectrum of Tungsten Filament Quartz–Iodine Lamps due to Iodine Vapor," *J. Opt. Soc. Am.*, **54**, 945–947 (1964).

27. Wolfe, W. L., "Properties of Optical Materials," in *Handbook of Optics*, 1st edn, Driscoll, W. G., ed., McGraw-Hill, New York, 1978.

28. Zubler, E. G. and F. A. Mosby, "An Iodine Incandescent Lamp with Virtually 100 Per Cent Lumen Maintenance," *Illuminating Engineering*, **54**, 734–740 (1959).

INTERNET SITES

http://www.astm.org   site for the American Society for Testing and Materials, where some testing methods are established in illumination and color evaluation.

http://www.hike.te.chiba-u.ac.jp/ikeda/CIE/ International commission for illumination. Tables for illuminants, sources, and tristimulus values.

http://iso.ch International standards organization, with standards for color and lighting.

http://www.osram.de Commercial lamps catalog.

http://www.lighting.philips.com Commercial lamps catalog.

http://www.sylvania.com Commercial lamps catalog.

20

Lasers

VICENTE ABOITES
Centro de Investigaciones en Optica, León, Mexico

20.1 TYPES OF LASERS, MAIN OPERATION REGIMES, AND EXAMPLES

Based on the quantum idea used by Max Planck [49] to explain blackbody radiation emission, in 1917 Albert Einstein proposed the processes of stimulated emission and absorption of radiation [14]. Light amplification by stimulated emission of radiation (laser) [59] was first demonstrated by Maiman [42] in 1960 using a ruby crystal pumped with a xenon flash lamp. Since then, laser coherent emission has been generated in thousands of substances using a wide variety of pumping mechanisms. Lasers are normally classified according to their active medium: solid, liquid, and gas. Table 20.1 shows examples of some of the most important used lasers according to the nature of the active media; their typical operation wavelengths and temporal operation regimes are also shown.

20.1.1 Solid-State Lasers

There are essentially two types of solid lasers media: impurity-doped crystals and glasses. They are almost exclusively optically pumped with flash lamps, continuous wave arc lamps or with other laser sources such as semiconductor lasers. Well-known examples are the Nd^{3+}:YAG and the Nd^{3+}:glass lasers.

20.1.2 Semiconductor Lasers

Even though these are also "solid-state lasers," for classification purposes they are generally considered different due to the difference in the inversion mechanism. These lasers are characterized in terms of the way by which the hole–electron pair population inversion is produced. They can be optically pumped by other laser

Table 20.1 Representative Examples of Laser Sources

Class	Laser medium	Nominal operating wavelength (nm)	Typical output power or energy	Typical temporal regime
Solid	Nd^{3+}:YAG	1064	10–100 W	CW
	Nd^{3+}:glass	1064	50 J	Pulsed
	Ti^{3+}:Al_2O_3	660–1180	10 W	CW
Semiconductor	InGaAsP	1300	10 mW	CW
Gas	He–Ne	633	5 mW	CW
	Ar^+	488	10 W	CW
	KrF	248	0.5 J	Pulsed
	CO_2	10600	100–1000 W	CW
	Kr^+	647	0.5 W	CW
	HCN	336.8×10^3	1 mW	CW
Liquid	Rhodamine-5G	560–640	100 mW	CW
Plasma	C^{6+}	18.2	2 mJ	Pulsed
FEL	Free electrons	300–4000	1 mJ	Pulsed

sources, by electron beams, or more frequently by injection of electrons in a *p–n* junction. A common example is the GaAs laser.

20.1.3 Gas Lasers

There are essentially six different types of gas lasers, involving

1. electronic transitions in neutral atomic active media
2. electronic transitions in ionized atomic active media
3. electronic transitions in neutral molecular active media
4. vibrational transitions in neutral molecular active media
5. rotational transitions in neutral molecular active media
6. electronic transitions in ionic molecular active media.

These lasers are pumped by several methods including continuous wave (CW), pulsed, and rf electrical discharges, optical pumping, chemical reaction, and gasdynamic expansion. Common examples for each of the above lasers are Ne–He, Ar^+, KrF, CO_2, CH_3F, and N_2^+.

20.1.4 Liquid Lasers

The active medium is a solution of a dye compound in alcohol or water. There are three main types: organic dyes, which are well known because of their tunability; rare-earth chelate lasers using organic molecules; and lasers using inorganic solvents and trivalent rare earth ion active centers. Typically, they are optically pumped by flash lamps or using other lasers. Common examples are Rh6G, TTF, and $POCl_4$.

20.1.5 Plasma Lasers

These lasers use as active medium a plasma typically generated by a high-power, high-intensity laser (or a nuclear detonation). They radiate in the UV or X-ray region of the spectrum. A typical example is the C^{6+} laser.

20.1.6 Free Electron Lasers

These lasers make use of a magnetic "wiggler" field produced by a periodic arrangement of magnets of alternating polarity. The active medium is a relativistic electron beam moving in the wiggler field. The most important difference in relation to other lasers is that the electrons are not bound to any active center such as an atom or a molecule. The amplification of an electromagnetic field is due to the energy that the electromagnetic laser beam takes from the electron beam.

20.1.7 Temporal Laser Operation

Any laser can be induced to produce output radiation with specific temporal characteristics. This can be achieved by proper design of the excitation source and/or by controlling the Q of the laser resonator. Table 20.2 describes the most common temporal operation regimes. Tables 20.3 and 20.4 show examples and the performance of some important CW and pulsed lasers, respectively.

Table 20.2 Main Temporal Operation Regimes

Temporal operation	Technique	Pulse width (s)
Continuous wave (CW)	Continuous pumping; resonator Q is held constant	∞
Pulsed	Pulsed pumping, resonator Q is held constant	10^{-8}–10^{-3}
Q-switched	Pumping is continuous or pulsed; resonator Q varies between a low and a high value	10^{-8}–10^{-6}
Mode-locked	Excitation is continuous or pulsed; a modulation rate related to the transit time in the resonator is introduced	10^{-12}–10^{-9}

20.2 LASER MEDIUM

The amplification of electromagnetic radiation takes place in a laser medium which is pumped in order to obtain a population inversion. Next are described the basic terms used to characterize a laser medium.

20.2.1 Unsaturated Gain Coefficient

The *unsaturated gain coefficient* or *unsaturated gain per unit length* α is given by

$$\alpha = \frac{c^2}{8\pi n^2 f^2 \tau_R}[N_2 - (g_2/g_1)N_1]g(f), \tag{20.1}$$

where $g(f)$ is the *lineshape function*; N_2, N_1, g_2, and g_1 are the population inversion densities and the degeneracies of levels 22.2 and 22.1, respectively; and n, f, c, and τ_R are the refractive index, the frequency of the laser radiation, the speed of light, and the *radiative lifetime* of the upper laser level, which is given as

$$\tau_R = \left(\frac{\varepsilon_0}{2\pi}\right)\frac{m_e c^3}{f_{12}e^2 f_0^2}, \tag{20.2}$$

where ε_0, e, m_e, f_0, and f_{12} are the permitivity in vacuum, the electronic charge, the elctronic mass, the frequency at the line center, and the oscillator strength of the transition between levels 22.2 and 22.1.

The *stimulated transition cross section* σ is

$$\sigma = \frac{c^2}{8\pi n^2 f^2 \tau_R}\, g(f). \tag{20.3}$$

Therefore, the unsaturated gain per unit length can also be written as

$$\alpha = [N_2 - (g_2/g_1)N_1]\sigma. \tag{20.4}$$

The increase in the intensity I per unit length dI/dz is described by the equation

Table 20.3 Properties and Performance of Some Continuous Wave (CW) Lasers

Parameter	Unit	Gas		Liquid (Rhodamine-6G)	Solid (Ti:sapphire)
		Neon–helium	CO_2		
Excitation method		Dc discharge	RF Excited		Ar^+ Laser pump
Gain medium composition		Neon–helium	$CO_2(1):N_2(1):He(3):Xe(0.5)$	R6G:sol–gel	$Ti:Al_2O_3$
Wavelength	nm	632.8	10,600	560	790
Laser cross-section	$\times 10^{-19}$ cm^2	3×10^6	1.5×10^{-16}	1.8×10^3	2.8
Radiative lifetime (upper level)	μs	≈ 0.1	4×10^3	6.7×10^{-3}	3.2
Decay lifetime (lower level)	μs	≈ 0.1	$\approx 4 \times 10^3$	6×10^{-3}	
Gain bandwidth	nm	2×10^{-3}	1.6×10^{-2}	3.4×10^{-3}	180
Type, gain saturation		Inhomogeneous	Homogeneous	Homogeneous	Homogeneous
Homogeneous saturation flux	W cm^{-2}	—	≈ 20	—	0.9
Inversion density	cm^{-3}	$\approx 1 \times 10^9$	2×10^{10}	$\approx 2 \times 10^{10}$	—
Small signal gain coefficient	cm^{-1}	$\approx 1 \times 10^{-3}$	$\approx 3 \times 10^{-2}$	1×10^{-2}	—
Pump power	W	—	900	4×10^{-3}	5.5
Output power	W	0.03	46	2.0×10^{-3}	0.27
Efficiency	%	0.1	12	60	12
Reference		Xianshu et al. [72]	Chernikov et al. [10]	Lo et al. [40]	Shieh et al. [61]

Table 20.4 Properties and Performance of Some Pulsed Lasers

		Gas		Liquid	Solid	
Parameter	Unit	XeCl	CO_2	(Rh6G:methanol)	Nd:YVO	Nd:YLF
Excitation method		E-beam	Traverse dc with RF preionization	Frequency-doubled Nd:YAG	Diode pump laser	Diode pump laser
Gain medium composition		XeCl	$CO_2(1):N_2(1):H_3(3)$	Rh6G:methanol	$Nd(3\%):YVO_4$	$Nd(1.5\%):YLF$
Wavelength	nm	308	10,600	563–604	1064	1047
Laser cross section	cm^{-2}	4.5×10^{-16}	2×10^{-18}	1.8×10^{-16}	25×10^{-19}	0.4×10^{-19}
Radiative lifetime (upper level)	μs	11×10^3	4×10^3	6.5×10^{-3}	50	480
Decay lifetime (lower level)	μs	—	5×10^{-2}	6×10^{-3}	92	—
Gain bandwidth	nm	2	1	80	1	1.3
Homogeneous saturation flux	$W\,cm^{-2}$	—	0.2	2×10^{-3}	0.037	—
Inversion density	cm^{-3}	—	3×10^{17}	2×10^{16}	—	—
Small signal gain coefficient	cm^{-1}	—	2×10^{-2}	4×10^{-2}	—	25×10^{-2}
Pump power	W	—	—	2	2	4
Excitation current/ voltage	A/V	$210 \times 10^3/800 \times 10^3$	$3.6 \times 10^4/15 \times 10^3$	—	—	—
Excitation current density	$A\,cm^{-2}$	5.25×10^2	180	—	—	—
Pump power density	$W\,cm^{-3}$	2×10^6	5.75×10^5	—	—	—
Output pulse energy	J	136	87.8×10^{-3}	100×10^{-3}	53×10^{-9}	30×10^{-9}
Output pulse length	Ns	100	36.5	3×10^{-3}	0.0037	8×10^{-3}
Output pulse power	W	1.36×10^9	0.9×10^9	3.3×10^{10}	0.46	3.75×10^3
Reference		Jingru et al. [31b]	Jiang et al. [31a]	Christophe et al. [11b]	Spühler et al. [62]	Hönninger et al. [29]

$$\frac{dI}{dz} = \alpha I. \tag{20.5}$$

For an initial intensity $I(0)$ at $z = 0$, the beam intensity varies along the propagation distance z according to

$$I(z) = I(0)e^{\alpha z}. \tag{20.6}$$

For a laser medium of length l the *total unsaturated gain* G_{db} (in decibels) is

$$G_{db} = 4.34\alpha l \tag{20.7}$$

From Eq. (20.6) it is clear that when the total gain per pass αl is sufficiently small the following approximation is valid:

$$\frac{I(l)}{I(0)} \approx 1 + \alpha l \tag{20.8}$$

and the gain is equal to the fractional increase in intensity:

$$[I(l) - I(0)]/I(0) \tag{20.9}$$

Therefore, the *percentage gain* is

$$G \text{ (in percent)} = 100 \, \alpha l. \tag{20.10}$$

The *threshold population inversion* ΔN_{th} needed to sustain laser oscillation in a resonator with output mirror reflectivity R is

$$\Delta N_{th} = [N_2 - (g_2/g_1)N_1]_{th} = \left[\gamma - \left(\frac{1}{2L}\right)\ln R\right]\Big/ \sigma, \tag{20.11}$$

where γ is the absorption coefficient of the host medium at frequency f and L is the distance between the mirror resonator (assumed equal to the laser medium length l). The *threshold gain coefficient* to start laser oscillation is given as

$$\alpha_{th} = \gamma - \frac{\ln R}{2L}. \tag{20.12}$$

20.2.2 Lineshape Function

To describe the gain distribution as a function of the frequency of the *lineshape function* $g(f)$ is used. The function $g(f)$ is normalized:

$$\int_{-\infty}^{\infty} g(f)df = 1. \tag{20.13}$$

The normalized *Lorentzian lineshape function* $g_L(f)$ is

$$g_L(f) = \frac{\Delta f}{2\pi\left[(f - f_0)^2 + \left(\frac{\Delta f}{2}\right)^2\right]}. \tag{20.14}$$

For natural broadened transitions the Lorentzian lineshape has a linewidth $\Delta f = \Delta f_N$, where

$$\Delta f_N = \frac{1}{\pi\tau_F} = \frac{1}{\pi}\left(\frac{1}{\tau_R} + \frac{1}{\tau_{NR}}\right). \tag{20.15}$$

In this expression τ_F is the fluorescent lifetime and τ_{NF} is the nonradiative decay time constant given by

$$\frac{1}{\tau_{NR}} = N_b Q_{ab} \sqrt{\frac{8kT}{\pi}} (M_a^{-1} + M_b^{-1}) \tag{20.16}$$

where M_a, M_b, and Q_{ab} are the masses of the atoms (or molecules) of species a and b, respectively, and its collision cross section, N is the number density of atoms (or molecules), k is the Boltzmann constant, and T the temperature. For collision-broadened transitions, the Lorentzian lineshape has a linewidth $\Delta f = \Delta f_{coll}$, where

$$\Delta F_{coll} = \frac{NQ}{\pi} \sqrt{\frac{16kT}{\pi M}}. \tag{20.17}$$

This bandwidth arises from elastic collisions between like atoms (or molecules) of atomic mass M and collision cross section Q.

The normalized *Gaussian lineshape function* $g_G(f)$ is given as

$$g_G(f) = \frac{2(\ln 2)^{1/2}}{\pi^{1/2} \Delta f} e^{-[4(\ln 2)(f - f_0)^2 / (\Delta f)^2]}. \tag{20.18}$$

For a Doppler-broadened transition the linewidth $\Delta f = \Delta f_D$ is

$$\Delta f_D = 2f_0 \sqrt{\frac{2kT \ln 2}{Mc^2}}. \tag{20.19}$$

The Gaussian and Lorentzian lineshape are shown in Fig. 20.1. The linewidths at half-maximum are shown as Δf_G (for Gaussian profile) and as Δf_L (for the Lorentzian profile).

20.2.3 Saturated Gain Coefficient

The *saturated gain coefficient for a homogeneously broadened* line is

$$\alpha_s = \frac{\alpha}{1 + (I/I_s)}, \tag{20.20}$$

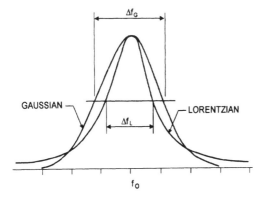

Figure 20.1 Gaussian and Lorentzian lineshapes.

where I_s is the saturation intensity:

$$I_s = \frac{4\pi n^2 hf}{\lambda^2 g(f)} \left(\frac{\tau_R}{\tau_F}\right) \tag{20.21}$$

where τ_F is the fluorescence lifetime of the upper laser level, h is the Planck constant, and λ is the wavelength of the laser radiation.

The *saturated gain coefficient for an inhomogeneously broadened line* is

$$\alpha_s = \frac{\alpha}{[1 + (I/I_s)]^{1/2}} \tag{20.22}$$

where the saturation intensity is

$$I_s = \frac{2\pi^2 n^2 hf \Delta f}{\lambda^2} \left(\frac{\tau_R}{\tau_F}\right) \tag{20.23}$$

The bandwidth Δf is the homogeneous linewidth of the inhomogeneously broadened transition.

20.3 RESONATORS AND LASER BEAMS

Most lasers have an optical resonator consisting of a pair of mirrors facing each other. In this way it is possible to maintain laser oscillation due to the feedback provided to the active (amplifying) medium and it is also possible to sustain well-defined longitudinal and transversal oscillating modes. An *optical resonator* is shown in Fig. 20.2. One of the mirrors has an ideal optical reflectivity of 100% and the other less than 100% (the laser beam is emitted through this second mirror). The mirrors are separated by a distance L and the radii of curvature of the mirrors are R_1 and R_2.

20.3.1 Stability, Resonator Types, and Diffraction Losses

A resonator is stable if the *stability condition*:

$$0 < g_1 g_2 < 1 \tag{20.24}$$

is satisfied, where g_1 and g_2 are the *resonator parameters*:

$$g_1 = 1 - \frac{L}{R_1} \quad \text{and} \quad g_2 = 1 - \frac{L}{R_2}. \tag{20.25}$$

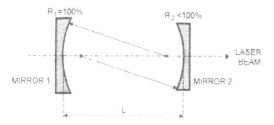

Figure 20.2 Optical resonator.

The radius of curvature R is defined as positive if the mirror is concave with respect to the resonator interior and the radius of curvature is negative if the mirror is convex. In Fig. 20.2, both R_1 and R_2 are positive.

Figure 20.3 draws the hyperbola defined by the stability condition (Eq. (20.24)) and is called the *stability diagram*. Since an optical resonator can be represented by its coordinates g_1 and g_2 in the stability diagram, the resonator is *stable* if the point (g_1, g_2) falls within the shaded region. Figure 20.3 also shows the location of several resonator types. Clearly, if

$$g_1 g_2 < 0 \quad \text{or} \quad g_1 g_2 > 1, \tag{20.26}$$

the resonator is *unstable* and upon multiple reflections a ray will diverge from the cavity axis.

The diffraction losses in a laser are characterized by the *resonator Fresnel number N*. This is a dimensionless parameter given as

$$N = \frac{a^2}{L\lambda}, \tag{20.27}$$

where a is the radius of the mirror resonator. A large Fresnel number implies small diffraction losses.

Some important *resonator types* (shown in Fig. 20.4) are the following:

- *Plane parallel resonator*, shown in Fig. 20.4(a). This resonator has the largest mode volume, but is difficult to keep in alignment and is used only with high gain medium. Its diffraction losses are larger than those of stable resonators with spherical mirrors.
- *Symmetrical confocal resonator*, shown in Fig. 20.4(b). The spot sizes at the mirrors are the smallest of any stable symmetric resonator. With a Fresnel number larger than unity, the diffraction losses of this resonator are essentially negligible.

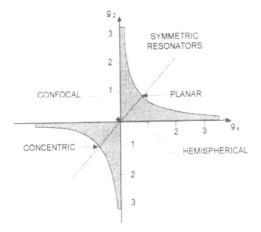

Figure 20.3 Stability diagram.

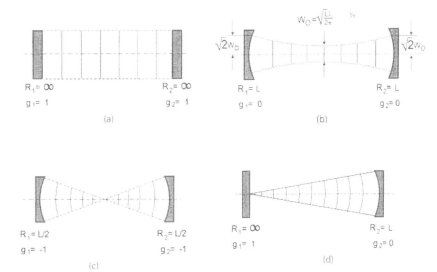

Figure 20.4 Resonator types: (a) plane parallel resonator, (b) symmetrical confocal resonator, (c) symmetrical concentric resonator, and (d) confocal-planar resonator.

- *Symmetrical concentric resonator*, shown in Fig. 20.4(c). As with the plane resonator, the exactly concentric resonator is relatively difficult to keep in alignment; however, when L is slightly less than $2R$, the alignment is no longer critical. The TEM_{00} mode in this resonator has the smallest beam waist.
- *Confocal-planar resonator*, shown in Fig. 20.4(d). This resonator is simple to keep in alignment, especially when L is slightly less than R. Also, small variations in the mirror spacing allows the adjustment of the spot size w_2, so that only the TEM_{00} mode fills the laser medium or mirror. It is widely used in low-power gas lasers.

There are some useful empirical formulas to find the *diffraction losses* in a resonator. The one-way power loss per pass δ in a real finite-diameter resonator for two common cases are:

For a confocal resonator:

$$\delta \approx \pi^2 2^4 N \exp(-4\pi N) \qquad \text{for } N \geq 1 \tag{20.28}$$

$$\delta \approx 1 - (N\pi^2) \qquad \text{for } N \to 0$$

For a plane parallel resonator:

$$\delta \approx 0.33 N^{-3/2} \qquad \text{for } N \geq 1 \tag{20.29}$$

20.3.2 Axial Modes

The *axial* or *longitudinal modes* of the resonator are the resonant frequencies of the cavity f_q, where q is the number of half-waves along the resonator axis.

$$f_q = q(c/2L). \tag{20.30}$$

The *frequency spacing* $f_{\Delta q}$ between two axial resonances of the laser cavity is

$$f_{\Delta q} = f_{(q+1)} - f_q = \frac{c}{2nL}. \tag{20.31}$$

For a TEM$_{mnq}$ mode the *resonance frequency* of the qth axial mode with the *mn*th transverse mode is

$$f_{mnq} = \left[q + (m+n+1)\frac{\cos^{-1}\sqrt{g_1 g_2}}{\pi} \right] \frac{c}{2nL}. \tag{20.32}$$

The frequency spacing $f_{\Delta mnq}$ between transverse modes is

$$f_{\Delta mnq} = \left(\cos^{-1}\sqrt{g_1 g_2} \right) \frac{c}{2\pi nL}. \tag{20.33}$$

The above expression is represented in Fig. 20.5. The bandwidth of a resonant mode of frequency f_{mnq} is

$$\Delta f_{mnq} = \frac{1}{2\pi\tau_c} = \frac{c(\gamma L - \ln\sqrt{R})}{2\pi nL} \tag{20.34}$$

Substituting in Eq. (20.32) the parameters g_1 and g_2 for particular resonators we obtain, for a plane parallel resonator,

$$f_{mnq} = \frac{qc}{2nL}; \tag{20.35}$$

for a symmetrical concentric resonator,

$$f_{mnq} = [q - (m+n+1)]\frac{c}{2nL}; \tag{20.36}$$

for a symmetrical confocal and a confocal-planar resonator,

$$f_{mnq} = \left[q + \frac{(m+n+1)}{2} \right] \frac{c}{2nL}. \tag{20.37}$$

20.3.3 Transverse Modes

For rectangular coordinates (x, y), the *transverse field distribution* $E(x, y)$ of a TEM$_{mnq}$ mode is given as

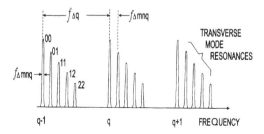

Figure 20.5 Frequency spacing between transverse modes.

$$E(x, y) = E_o H_m\left(\frac{\sqrt{2}x}{w(z)}\right) H_n\left(\frac{\sqrt{2}y}{w(z)}\right) \exp\left(-\frac{x^2 + y^2}{w^2(z)}\right), \tag{20.38}$$

where $H_n(x)$ are the Hermite polynomials defined by

$$H_n(x) = (-1)^n e^{x^2} \frac{d^n}{dx^n} e^{-x^2}. \tag{20.39}$$

The first four Hermite polynomials are

$$\begin{aligned}
&H_0(x) = 1, \\
&H_1(x) = 2x, \\
&H_2(x) = 4x^2 - 2, \\
&H_3(x) = 8x^3 - 12x;
\end{aligned} \tag{20.40}$$

these polynomials obey the recursion relation

$$H_{n+1}(x) = 2xH_n(x) - 2nH_{n-1}(x), \tag{20.41}$$

which also provides a useful way of calculating the higher-order polynomials. Figure 20.6 shows some transverse mode patterns.

For polar coordinates (r, ϕ), the transverse field distribution $E(r, \phi)$ of a TEM_{mnq} mode is given as

$$E(r, \phi) = E_0\left(\frac{\sqrt{2}r}{w(z)}\right)^l L_p^l\left(\frac{2r^2}{w^2(z)}\right) \exp\left(-\frac{r^2}{w^2(z)}\right)\left(\left\{\begin{matrix}\sin \\ \cos\end{matrix}\right\}l\phi\right) \tag{20.42}$$

where $L_p^l(x)$ are the Laguerre polynomials defined by

$$L_p^l(x) = e^x \frac{x^{-1}}{p!} \frac{d^p}{dx^p}(e^{-x}x^{p+1}) \tag{20.43}$$

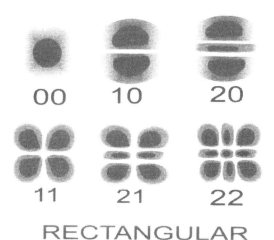

RECTANGULAR

Figure 20.6   Transverse mode patterns TEM_{mn}.

The first three Laguerre polynomials are

$$L_0^l(x) = 1,$$
$$L_1^l(x) = l + 1 - x,$$
$$L_2^l(x) = \tfrac{1}{2}(l+1)(l+2) - (l+2)x + \tfrac{1}{2}(x^2);$$

(20.44)

these polynomials obey the recursion relation:

$$(p+1)L_{p+1}^l(x) = (2p+l+1-x)L_p^l(x) - (p+l)L_{p-1}^l(x),$$

(20.45)

which also provides a useful way of calculating the higher-order polynomials. Figure 20.7 shows some transverse mode patterns. A mode may be described as a superposition of two like modes. For example, the TEM_{01}^* is made up of a rectangular TEM_{01} and TEM_{10} modes.

20.3.4 Resonator Quality Parameter

The *quality factor* Q of a laser resonator is defined as

$$Q = 2\pi \frac{\text{energy stored in the resonator}}{\text{energy lost in one cycle}},$$

(20.46)

$$Q = 2\pi f_{mnq} \frac{E}{|dE/dt|}.$$

(20.47)

The loss of energy dE/dt is related to the energy decay time or photon lifetime τ_c by

$$\frac{dE}{dt} = -\frac{E}{\tau_c}$$

(20.48)

where

$$\tau_c = \frac{nL}{c(\gamma L - \ln \sqrt{R})}.$$

(20.49)

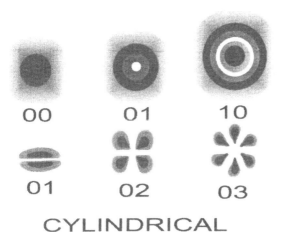

Figure 20.7 Some transverse mode patterns TEM_{mn}.

This expression can also be written as

$$Q = \frac{f_{mnq}}{\Delta f_{mnq}}. \tag{20.50}$$

20.4 GAUSSIAN BEAMS

As can be seen from Eq. (20.38) with $n = m = 0$, the transverse field distribution TEM_{00} has a bell-shaped Gaussian amplitude:

$$E(x, y) \propto E_o \exp\left(-\frac{x^2 + y^2}{w^2(z)}\right). \tag{20.51}$$

Taking as w_0 the $1/e$ transversal spot size at $z = 0$, the description of a *Gaussian beam* at any position z is given by the parameters $w(z)$ describing the spot size and $R(z)$ describing the wavefront radius of curvature. This is schematically shown in Fig. 20.8:

$$w(z) = w_0\sqrt{1 + \left(\frac{\lambda z}{\pi w_0^2}\right)^2} = w_0\sqrt{1 + \left(\frac{z}{z_R}\right)^2} \tag{20.52}$$

$$R(z) = z\left[1 + \left(\frac{\pi w_0^2}{\lambda z}\right)^2\right] = z\left[1 + \left(\frac{z_R}{z}\right)^2\right] \tag{20.53}$$

In the above equations z_R is the *Rayleigh distance*, defined as

$$z_R = \frac{\pi w_0^2}{\lambda}. \tag{20.54}$$

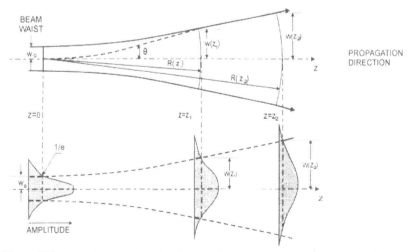

Figure 20.8 Gaussian beam showing the spot size $w(z)$ and the wavefront radius of curvature $R(2)$.

For a laser resonator (Fig. 20.9), with parameters g_1 and g_2, the spot size at the beam waist w_0 is given by

$$w_0 = \left(\frac{\lambda L}{\pi}\right)^{1/2} \frac{[g_1 g_2 (1 - g_1 g_2)]^{1/4}}{(g_1 + g_2 - 2g_1 g_2)^{1/2}}. \tag{20.55}$$

The position of the beam waist w_0 relative to the resonator mirrors (Fig. 20.9) is

$$z_1 = \frac{-g_2(1 - g_1)L}{g_1 + g_2 - 2g_1 g_2}, \tag{20.56}$$

$$z_2 = \frac{g_1(1 - g_2)L}{g_1 + g_2 - 2g_1 g_2} = z_1 + L. \tag{20.57}$$

The spot sizes w_1 and w_2 at each of the resonator mirrors are

$$w_1 = \left[\frac{L\lambda}{\pi}\right]^{1/2} \left[\frac{g_2}{g_1(1 - g_1 g_2)}\right]^{1/4}, \tag{20.58}$$

$$w_2 = \left[\frac{L\lambda}{\pi}\right]^{1/2} \left[\frac{g_1}{g_2(1 - g_1 g_2)}\right]^{1/4}. \tag{20.59}$$

The half-angle *beam divergence* in the far field (for $z \gg z_R$), shown in Fig. 20.8, is given by

$$\theta = \frac{\lambda}{\pi w_0}. \tag{20.60}$$

The focusing of a Gaussian laser beam by a thin lens is shown in Fig. 20.10. The position of the focused beam waist is given by

$$z_2 = f + \frac{(z_1 - f)f^2}{(z_1 - f)^2 + \left(\dfrac{\pi w_{01}^2}{\lambda}\right)^2}. \tag{20.61}$$

The spot size of the focused laser beam is obtained from

$$\frac{1}{w_2^2} = \frac{1}{w_1^2}\left(1 - \frac{z_1}{f}\right)^2 + \frac{1}{f^2}\left(\frac{\pi w_1}{\lambda}\right)^2. \tag{20.62}$$

In most practical applications, this expression can be approximated to

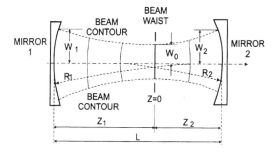

Figure 20.9   Optical resonator with parameters g_1 and g_2.

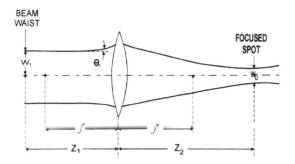

Figure 20.10 Focusing of a Gaussian beam by a thin lens.

$$w_2 \cong f\theta_1, \tag{20.63}$$

where (from Eq. (20.60)), $\theta_1 \equiv \lambda/\pi w_1$.

20.5 LASER APPLICATIONS AND INSTRUMENTS

There are at least as many laser instruments as there are laser applications. This only mirrors the fact that all laser instruments (the laser itself) started as research tools. What follows is a general listing of some important laser applications and instruments along with some significant published results in the following areas: laser Doppler velocimetry and its medical applications; laser radar (LIDAR) and tunable laser radars for applications to pollution detection, laser thermometry, laser applications to the electronics and solid-state industry, and laser applications to art. For these and other areas not mentioned here the reader is also referred to general references. [13, 24, 25, 27, 30, 32, 35, 37, 45, 46, 48, 50, 52, 54, 63, 68, 75]

20.5.1 Laser Doppler Velocimetry

Laser Doppler velocimetry is a well-established technique widely applied in many engineering areas. Examples of recent research work applied to fluid dynamics can be found in references. [4, 31, 33, 39, 41, 53, 57, 58, 67] The use of this technique in medical applications is a wide and promising area of research where many new instruments are currently developed for specific problems. [2, 55, 69, 73]

20.5.2 LIDAR

The short and intense laser pulses produced by a Q-switched laser are ideal for optical ranging. These instruments are also called optical radar or LIDAR. Small solid-state lasers with the associated detection electronics are available in small and rugged units for military field applications. Laser ranging systems making use of tunable lasers whose wavelengths can be tuned to specific molecular or atomic transitions can be used for pollution detection, ranging of clouds, and aerosol measurements, etc. These instruments placed in orbiting satellites may also be used for weather forecasting applications. [3, 5, 16, 17, 21, 23, 28, 30, 69, 71]

20.5.3 Laser Thermometry

The fact that the laser can be used as a nonintrusive measurement instrument is widely used in many applications where a measurement instrument would be damaged or plainly destroyed by the studied system. Flames and air jets are good examples of systems that can be studied using laser-based techniques such as spectroscopy and holography. [8, 9, 20, 26, 36, 44, 64]

20.5.4 Laser Applications to the Electronic Industry

The electronics industry has widely benefited from the use of lasers. Many new laser instruments and applications are currently being designed. Nowadays, the laser is used in automatic microsoldering, in laser recrystalization of semiconductor substrates, and in laser ablation of thin-films deposition among other applications. [11, 12, 15, 18, 19, 22, 34, 38, 43, 47, 56, 57, 60, 65, 66, 74]

20.5.5 Laser Applications to Art

Lasers are being used for diagnostic, conservation, and restoration of great masterworks. Until recently the cleaning of painted surfaces required the use of solvents to remove old varnish or encrustations from the painted surfaces. This is a difficult process because the solvents may also damage the paint layer itself, causing soluble materials in the paint to diffuse out in a process known as leaching. This restoration process is now carried out using lasers without any effect on the original paint layers. [1, 6, 7]

REFERENCES

1. Anglos, D., M. Solomidou, and C. Fotakis, "Laser-Induced Fluorescence in Artwork Diagnostics: An Application in Pigment Analysis," *Appl. Spect.*, **50**, 1331–1335 (1996).
2. Arbit, E. and R. G. DiResta, "Application of Laser Doppler Flowmetry in Neurosurgery," *Neurosurgery Clin. N. Am.*, 7, 741–744 (1996).
3. Arev, N. N., B. F. Gorbunov, and G. V. Pugachev, "Application of a Laser Ranging System to the Metrologic Certification of Satellite Radar Measurement Systems," *Measurement Techniques*, **36**, 524–527 (1993).
4. Belorgey, M. B., J. Le, and A. Grandjean, "Application of Laser Doppler Velocimetry to the Study of Turbulence Generated by Swell in the Vicinity of Walls or Obstacles," *Coas. Eng.*, **13**, 183–187 (1989).
5. Bulatov, V., V. V. Gridin, and L. Schechter, "Application of Pulsed Laser Methods to in situ Probing of Highway Originated Pollutants," *An. Chem. Acta*, **343**, 93–96 (1997).
6. Castellini, P., N. Paone, and E. P. Tomasini, "The Laser Doppler Vibrometer as an Instrument for Non-intrusive Diagnostic of Works of Art: Application to Fresco Paintings," *Opt. Las. Eng.*, **25**, 227–231 (1996).
7. Cruz, A., S. A. Hauger, and M. L. Wolbarsht, "The Role of Lasers in Fine Arts Conservation and Restoration," *Opt. & Photo. News*, **10**(7), 36–40 (1999).
8. Chang, K.-C., J.-M. Huang, and S.-M. Tieng, "Application of Laser Holographic Interferometry to Temperature Measurements in Buoyant Air Jets," *J. Heat Transfer*, **6**, 377–381 (1992).
9. Chen, C.-C. and K.-C. Chang, "Application of Laser Holographic Interferometry to Temperature Measurements in Premixed Flames with Different Flame Structures," *J. Chinese Inst. Eng.*, **14**, 633–638 (1991).

10. Chernikov, S. B., A. L. Karapuzikov, and S. A. Stojanov, "RF Exited CO_2 Slab Waveguide Laser," *Proc. SPIE*, **2773**, 52–56 (1995).

11a. Choi, D.-H., E. Sadayuki, and O. Sugiura, "Lateral Growth of Poly-Si Film by Excimer Laser and its Thin Film Transistor Application," *Jap. J. Appl. Phys. Part 1*, **33**, 70–75 (1994).

11b. Christophe, J. B. et al., "Stabilization of terahertz beats from a pair of picosecond dye lasers by coherent photon seeding," *Optic Comm.*, **161**, 31–36, (1999).

12. Deneffe, K., V. Mieghem, and Piet, Brijs, Bert, "As-Deposited Superconducting Thin Films by Electron Cyclotron Resonance-Assisted Laser Ablation for Application in Micro-Electronics," *Jap. J. Appl. Phys. Part 1*, **30**, 1959–1963 (1991).

13. Dong Hou, J., Z. Tianjin, L. Chonghong et al., "Study on Mini High Repetition Frequency Sealed-Off TEA CO_2 Laser," *Proc. SPIE*, **2889**, 392–397 (1996).

14. Einstein, A., "Zur Quantentheorie des Strahlung," *Physikalische Zeitschrift*, **18**, 121–129 (1917).

15. Engel, C., P. Baumgartner, and G. Abstreiter, "Fabrication of Lateral NPN- and PNP-Structures on Si/SiGe by Focused Laser Beam Writing and Their Application as Photodetectors," *J. Appl. Phys.*, **81**, 6455–6459 (1997).

16. Frejafon, E., J. Kasparian, and J. P. Wolf, "Laser Application for Atmospheric Pollution Monitoring," *Euro. Phys. J.*, **4**, 231–234 (1998).

17. Fried, A., J. R. Drummond, and B. Henry, "Versatile Integrated Tunable Diode Laser System for High Precision: Application for Ambient Measurements of OCS," *Appl. Opt.*, **30**, 1916–1919 (1991).

18. Fujii, E., K. Senda, and F. Emoto, "A Laser Recrystallization Technique for Silicon-TFT Integrated Circuits on Quartz Substrates and Its Application to Small-Size Monolithic Active-Matrix LCD's," *IEEE Transactions on Electron Devices*, **37**, 121–124 (1990).

19. Fujii, E., K. Senda, and F. Emoto, "Planar Integration of Laser-Recrystallized SOI-Tr's Fabricated by Lateral Seeding Process and Bulk-Tr's and Its Application to Fabrication of a Solid-State Image Sensor," *Elect. & Comm. Jap. Part 2*, **72**, 77–81 (1989).

20. Robert, G. D., L. McNesby, and W. A. Miziolek, "Application of Tunable Diode Laser Diagnostics for Temperature and Species Concentration Profiles of Inhibited Low-Pressure Flames," *Appl. Opt.*, **35**, 4018–4020 (1996).

21. Gaft, M., "Application of Laser-Induced Luminescence in Ecology," *Water Sci. Tech.*, **27**, 547–551 (1993).

22. Garrett, L., J. Argenal, A. Rink, and L. Dan, "A Fully Automatic Application for Bonding Surface Assemblies: Laser Microsoldering is an Alternative Method of Integrated Circuit Manufacture," *Assembly Automation*, **8**, 195–199 (1988).

23A. Gavan, J., "Optimization for Improvement of Laser Radar Propagation Range in Detecting and Tracking Cooperative Targets," *J. Elec. Waves Appl.*, **5**, 1055–1059 (1991).

23B. Gordon, J. P., H. J. Zeiger, and C. H. Townes, "Molecular Microwave Spectrum of HN_3," *Phys. Rev.*, **95**, 282–287 (1954).

23C. Gherezghiher, T., "Choroidal and Ciliary Body Blood Flow Analysis: Application of Laser Doppler Flowmetry in Experimental Animals," *Experimental Eye Research*, **53**, 151–154 (1991).

24. Greulich, K. O. and D. L. Farkas, eds, *Micromanipulation by Light in Biology and Medicine*, Birkhauser, Boston, 1998.

25. Grundwald, E., D. F. Dever, and P. M. Keehn, *Megawatt Infrared Laser Chemistry*, John Wiley, New York, 1978.

26. Hencke, H., "The Design and Application of Honeywell's Laser-Trimmed Temperature Sensors," *Measurement and Control*, **22**, 233–236 (1989).

27. Hinkley, E. D., ed., *Laser Monitoring of the Atmosphere*, Springer-Verlag, Berlin, 1976.

28. Hoffmann, E., C. Ludke, and H. Stephanowitz, "Application of Laser-ICP-MS in Environmental Analysis," *Fresenius' J. An. Chem.*, **355**, 900–904 (1996).

29. Hönninger, C., R. Paschotta, F. Morier-Genound, M. Moser, and U. Keller, "Q-Switching Stability Limits of Continuous-Wave Passive Mode Locking," *J. Opt. Soc. Am.*, **16**, 46–56 (1999).

30. Jelalian, A. V., *Laser Radar Systems*, Artech House, Cambridge, 1992.

31a. Jiang, S., Y. Gorai, and X. Shun-Zhao, "Development of a New Laser-Doppler Microvelocimetering and its Application in Patients with Coronary Artery Stenosis," *Angiology*, **45**, 225–228 (1994).

31b. Jingru et al. "High power eximer laser and application," *SPIE-High-Power Lasers*, **2889**, pp. 98–103, (1996).

32. Kamerman, X. X., *Applied Laser Radar Technology*, Society for Photo-Optical Instrumentation Engineers, 1993.

33. Karwe, M. V. and V. Sernas, "Application of Laser Doppler Anemometry to Measure Velocity Distribution Inside the Screw Channel of a Twin-Screw Extruder," *J. Food Proc. Eng.*, **19**, 135–141 (1996).

34. Katz, A. and S. J. Pearton, "Single Wafer Integrated Processes by RT-LPMOCVD Modules – Application in the Manufacturing of InP-Based Laser Devices," *Mat. Chem. and Phys.*, **32**, 315–321 (1992).

35. Koebner, H. K., ed., *Lasers in Medicine*, John Wiley, New York, 1980.

36. Kroll, S., "Influence of Laser-Mode Statistics on Noise in Nonlinear-Optical Processes – Application to Single-Shot Broadband Coherent Anti-Stokes Raman Scattering Thermometry," *J. Opt. Soc. Am.*, **5**, 1910–1913 (1988).

37. Kuhn, K. J., *Laser Engineering*, Prentice Hall, New York, 1998.

38. Lamond, C., E. A. Avrutin, and J. H. Marsh, "Two-Section Self-Pulsating Laser Diodes and Their Application to Microwave and Millimetre-Wave Optoelectronics," *Int. J. Opt.*, **10**, 439–444 (1995).

39. Lemoine, E., M. Wolff, and M. Lebouche, "Simultaneous Concentration and Velocity Measurements Using Combined Laser-Induced Fluorescence and Laser Doppler Velocimetry: Application to Turbulent Transport," *Exp. Fluids*, **20**, 319–324 (1996).

40. Lo, D., S. K. Lam, C. Ye, and K. S. Lam, "Narrow Linewidth Operation of Solid State Dye Laser Based on Sol-Gel Silica," *Optic Comm.*, **156**, 316–320 (1998).

41. Louranco, L. M. and A. Krothapalli, "Application of On-Line Particle Image Velocimetry to High-Speed Flows," *Laser Anemometry*, **5**, 683–689 (1993).

42. Maiman, T., H. R. H. Hoskins, I. J. D'haenens, C. K. Asawa, and V. Evtuhov, "Stimulated Optical Emission in Fluorescent Solids. II. Spectroscopy and Stimulated Emission in Ruby," *Phys. Rev.*, **123**, 1151–1159 (1961).

43. Matsumura, M., "Application of Excimer-Laser Annealing to Amorphous, Poly-Crystal and Single-Crystal Silicon Thin-Film Transistors," *Phys. S. Sol. A: Appl. Res.*, **166**, 715–721 (1998).

44. Meier, W., I. Plath, and W. Stricker, "The Application of Single-Pulse CARS for Temperature Measurements in a Turbulent Stagnation Flame," *Appl. Phys.*, **53**, 339–343 (1991).

45. Metcalf, H. J. and H. E. Stanley, eds, *Laser Cooling*, Springer Verlag, Berlin, 1999.

46. Motz, H., *The Physics of Laser Fusion*, Academic Press, San Diego, 1979.

47. Niino, H., Y. Kawabata, and A. Yabe, "Application of Excimer Laser Polymer Ablation to Alignment of Liquid Crystals: Periodic Microstructure on Polyethersulfone," *Jap. J. Appl. Phys., Part 2*, **28**, 2225–2229 (1989).

48. Papannareddy, R., *Introduction to Lightwave Communication Systems*, Artech House, Cambridge, 1997.

49. Planck, M., "Uber das Gesetz der Enerpreverteilung in Normal Spectrum," *Ann. Physik*, **4**, 553–563 (1901).

50. Powell, J., *CO₂ Laser Cutting*, Springer Verlag, Berlin, 1998.

51. Prudhomme, S., A. Seraudie, and A. Mignosi, "A Recent Three Dimensional Laser Doppler Application at the T2 Transonic Wind Tunnel: Optimization, Experimental Results, Measurement Accuracy," *Laser Anemometry*, **3**, 57–63 (1991).

52. Ready, J. F., *Industrial Applications of Lasers*, Academic Press, San Diego, 1997.

53. Reisinger, D., W. Heiser, and D. Olejak, "The Application of Laser Doppler Anemometry in a Trisonic Windtunnel," *Laser Anemometry*, **1**, 217–224 (1991).

54. Riviere, C. J. B., R. Baribault, D. Gay, N. McCarthy, and M. Piché, "Stabilization of Terahertz Beats From a Pair of Picosecond Dye Lasers by Coherent Photon Seeding, *Optic Comm.*, **161**, 31–36 (1999).

55. Ruggero, M. A. and N. C. Rich, "Application of a Commercially-Manufactered Doppler-Shift Laser Velocimeter to the Measurement of Basilar-Membrane Vibration," *Hearing Res.*, **51**, 215–221 (1991).

56. Sameshima, T., "Laser Beam Application to Thin Film Transistors," *Appl. Surf. Sci.*, **96**, 352–355 (1996).

57. Sanada, M., "New Application of Laser Beam to Failure Analysis of LSI with Multi-Metal Layers," *Microelectronics and Reliability*, **33**, 993–996 (1993).

58. Scharf, R., "A Two-Component He–Ne Laser-Doppler Anemometer for Detection of Turbulent Reynold's Stresses and its Application to Water and Drag-Reducing Polymer Solutions," *Means. Sci. & Tech.*, **5**, 1546–1551 (1994).

59. Schawlow, A. L. and C. H. Towes, "Infrared and Optical Masers," *Phys. Rev.*, **112**, 1940–1949 (1958).

60. Seddon, B. J., Y. Shao, and J. Fost, "The Application of Excimer Laser Micromachining for the Fabrication of Disc Microelectrodes," *Electrochim. Acta*, **39**, 783–787 (1994).

61. Shieh, J.-M., T. C. Huang, C.-L. Wang, and C.-L. Pan, "Broadly Tunable Self-Starting Mode-Locked Ti:sapphire Laser with Triple-Strained Quantum-Well Saturable Bragg Reflector," *Optic Comm.*, **156**, 53–57 (1998).

62. Spühler, G. J., R. Paschotta, R. Fluck, *et al.*, "Experimentally Confirmed Design Guidelines for Passively Q-Switched Microchip Lasers Using Semiconductor Saturable Absorbers," *J. Opt. Soc. Am.*, **12**, 376–388 (1999).

63. Stenholm, S., *Laser in Applied and Fundamental Research*, Hilger, London, 1985.

64. Su, K.-D., C.-Y. Chen, and K.-C. Lin, "Application of Laser-Enhanced Ionization to Flame Temperature Determination," *Appl. Spec.*, **45**, 1340–1345 (1991).

65. Takai, M., S. Natgatomo, and H. Kohda, "Laser Chemical Processing of Magnetic Materials for Recording-Head Application," *Appl. Phys.*, **58**, 359–362 (1994).

66. Takasago, H., E. Gofuku, and M. Takada, "An Advanced Laser Application for Controlling of Electric Resistivity of Thick Film Resistors," *J. Elec. Mat.*, **18**, 651–655 (1989).

67. Thiele, B. and H. Eckelmann, "Application of a Partly Submerged Two-Component Laser-Doppler Anemometer in a Turbulent Flow," *Exp. Fluids*, **17**, 390–395 (1994).

68. Ven Hecke, G. R. and K. K. Karukstis, *A Guide to Lasers in Chemistry*, Jones and Bartlett, New York, 1997.

69. Wagener, J. T., N. Demma, and T. Kubo, "2 µm LIDAR for Laser-Based Remote Sensing: Flight Demonstration and Application Survey," *IEEE Aerospace and Electronic Systems Magazine*, **10**, 23–26 (1995).

70. Wey, V. D., P. L. Polder, W. Theo, J. Gabrcels, and M. Fons, "Peripheral Nerve Elongation by Laser Doppler Flowmetry–Monitored Expansion: An Experimental Basis for Future Application in the Management of Peripheral Nerve Defects," *Plastic and Reconstructive Surgery*, **97**, 568–571 (1996).

71. Wulfmeyer, V. and J. Bosenberg, "Single-Mode Operation of an Injection-Seeded Alexandrite Ring Laser for Application in Water-Vapor and Temperature Differential Absorption Lidar," *Opt. Lett.*, **21**, 1150–1153 (1996).

72. Xianshu, L., C. Ychuan, L. Taouyu, *et al.*, "A Development of Red Internal He–Ne Lasers with Near Critical Concave–Convex Stable Resonator," *Proc. SPIE*, **2889**, 358–366 (1996).

73. Yokomise, H., H. Wada, and K. Inui, "Application of Laser Doppler Velocimetry to Lung Transplantation," *Transplantation*, **48**, 550–559 (1989).

74. Yong, J., "Laser Application in Packaging of Very Large Scale Integrated Chips," *J. Vacuum Sci. & Tech.*, **8**, 1789–1793 (1990).

75. Zuev, V. E., *Laser Beams in the Atmosphere*, Plenum, New York, 1982.

FURTHER READING

Kneubuhl, F. K. and M. W. Sigrist, *Laser*, B. G., Teubner, Stuttgart, 1989.

Koechner, W., *Solid-State Laser Engineering*, Springer-Verlag, Berlin, 1976.

Marschall, T. C., *Free Electron Lasers*, McMillan, London, 1985.

Pressley, R. J., ed., *CRC Handbook of Lasers*, The Chemical Rubber Co., Boca Raton, Florida, 1971.

Saleh, B. E. A. and M. C. Teich, *Fundamentals of Photonics*, John Wiley, New York, 1991.

Sargent, M., III, M. O. Scully, and W. E. Lamb, Jr., *Laser Physics*, Addison-Wesley, New York, 1974.

Schafer, F. P., ed., *Dye Lasers*, Springer-Verlag, Berlin, 1977.

Schultz, D. A., ed., *Laser Handbook*, North-Holland, Amsterdam, 1972.

Siegman, A. E., *Lasers*, University Science Books, Mill Valley, 1986.

Silfvast, W. I., *Laser Fundamentals*, Cambridge University Press, London, 1996.

Smith, W. V. and P. P. Sorokin, *The Laser*, McGraw-Hill, New York, 1966.

Svelto, O., *Principles of Lasers*, Plenum, New York, 1976.

Tarasov, L., *Physique des Processus dans les Generateurs de Rayonnement Optique Coherent*, Mir, Moscow, 1985.

Weber, M. J., *Handbook of Laser Science and Technology*, Vol. 1, CRC Press, Boca Raton, Florida, 1982.

Weber, M. J., *Handbook of Laser Science and Technology*, Supplement 1, CRC Press, Boca Raton, Florida, 1991.

Witteman, W. J., *The CO_2 Laser*, Springer-Verlag, Berlin, 1987.

Yariv, A., *Quantum Electronics*, 2nd edn, John Wiley, New York, 1975.

APPENDIX: LIST OF SYMBOLS

c	Speed of light $c = 2.998 \times 10^8$ m/s
e	Electronic charge $e = 1.602 \times 10^{-19}$ coulomb
f	Focal length of thin lens
f	Frequency of laser radiation
f_0	Frequency at line center
f_{12}	Oscillator strength of the transition between levels 2 and 1
Δf	Frequency bandwidth at half-maximum
f_{mnq}	Frequency of TEM_{mnq} mode
g_1, g_2	Degeneracies of lower and upper laser levels, respectively
g_1, g_2	Resonator g parameters for mirrors 1 and 2, respectively
$g(f)$	Lineshape function

I	Beam intensity
I_s	Saturation beam intensity
l	Transverse mode number (radial geometry)
l	Length of gain medium
L	Distance between resonator mirrors
m	Transverse mode number (rectangular coordinates)
m_e	Electronic rest mass $m_e = 9.109 \times 10^{-31}$ kg
M	Mass of atom (or molecule)
M_a, M_b	Mass of atom (or molecule) of species a and b, respectively
n	Transverse mode number in rectangular coordinates
n	Index of refraction
N	Number density of atoms (or molecules)
N_i	Number density of atoms (or molecules) in level i
ΔN_{th}	Threshold population inversion
N_0	Number density of laser atoms (or molecules)
p	Transverse mode number (radial geometry)
Q_{ab}	Collision cross section
r	Radial coordinate
R_1, R_2	Radii of curvature of resonator mirrors 1 and 2, respectively
$R(z)$	Radius of curvature of wavefront
R	Reflectivity of output mirror
t	Time
T	Temperature
w_0	Spot size at beam waist
w_1, w_2	Spot sizes at mirrors 1 and 2, respectively
$w(z)$	Spot size at a distance z from beam waist
x, y, z	Rectangular coordinates
z_R	Rayleigh range

Greek Symbols

α	Unsaturated gain coefficient per unit length
α_s	Saturated gain coefficient per unit length
α_{th}	Threshold gain coefficient per unit length
γ	Absorption coefficient of laser medium
ε_0	Permittivity in vacuum
θ	Far-field beam divergence half-angle
λ	Wavelength of laser radiation
σ	Stimulated transition cross section
τ_F	Fluorescence lifetime of the upper laser level
τ_{NR}	Nonradiative decay time constant
τ_R	Radiative lifetime of the upper laser level

21

Spatial and Spectral Filters

ANGUS MACLEOD

Thin Film Center, Inc., Tucson, Arizona

21.1 INTRODUCTION

A filter is a device that modifies a distribution of entities by varying its acceptance or rejection rate according to the value of a prescribed attribute. A true filter simply selects but does not change the accepted entities, although those rejected may be changed. Power may be supplied to aid in the filtering process. A simple form of filter is a screen that separates particles according to their size. The filters that are to be described here select according to the spectral characteristics or the spatial characteristics of a beam of light. The selected light is unmodified as far as wavelength or frequency is concerned, although the direction may be changed. In most cases the objective is to render the light more suitable for carrying out a prescribed task, with the penalty of a loss in available energy. The filters that we are concerned with in this chapter are all linear in their operation; that is, their response is independent of the actual magnitude of the input and for any value of the prescribed attribute the ratio of the magnitude of the output to that of the input is constant. We exclude tuned amplifiers, fluorescence filters, spatial modulators, and other active components as outside the scope of this chapter.

21.2 SOME FUNDAMENTAL IDEAS

A propagating electromagnetic disturbance that is essentially in the optical region and that can be described as a single entity is usually loosely referred to as a beam of light. In the general case, a complete description of the attributes of an arbitrary light beam would be impossibly complicated. More often, however, the beams of light that concern us in optical systems have a regularity in character that allows us to make good use of them and, at the same time, allows us to describe them in reasonably uncomplicated terms. There are, of course, many properties that can be involved in the description of a regular beam of light. Here we are interested in only two of these: the spatial distribution and the spectral distribution of the energy carried by the beam. These two characteristics are not strictly completely separable in any given beam but in the simple cases that we use most often they can profitably be considered separately.

We imagine a beam of light that has the simplest spatial distribution possible. The beam can be considered to be propagating in a well-defined direction in an isotropic medium at levels of energy that are well below any that might produce nonlinear effects. We choose a set of coordinate axes such that the z-axis is along the direction of propagation of the beam of light, and then the x- and y-axes are normal to it. We can make measurements of the attributes of the beam in any plane parallel to the x- and y-axes, that is normal to the direction of propagation. If we find that the attributes of the beam at any point are completely independent of the values of the x- and y-coordinates of that point and depend only on the values of the z-coordinate and the time, then we describe the beam of light as a *plane wave* propagating in the z-direction. See Heavens and Ditchburn [1] for a detailed description of fundamental optical ideas.

At any value of z we can measure the temporal variation of the energy, or of the fields of the plane wave. This yields the temporal profile of the wave. When we compare the temporal profile of the wave at one value of z with that measured at another, we should expect to see the same general shape but with a separation in time equal to the separation in z divided by the velocity of the wave. However, we see this simple relationship only when the wave is propagating through free space. For all other media there is a distortion of the temporal profile that increases with the distance between the two measurement points. Furthermore, because of this effect it becomes impossible to assign a definite velocity to the wave. All media, except free space, exhibit an attribute known as *dispersion* and it is dispersion that is responsible for the profile shape change and the uncertainty in the velocity measurement.

Fortunately, we find that there is one particular temporal profile that propagates without change of shape even in dispersive media. This is a profile that can be described as a sine or cosine function. A wave that possesses such a profile is known as a *harmonic* or *monochromatic* wave. In the simplest form of the plane, monochromatic wave there is a consistency in the directions of the electric and magnetic fields of the wave which is known by the term *polarization*. Polarization is described elsewhere in the handbook. Here we adopt the simplest form known as *linear* or, sometimes, *plane*, where the directions of the electric and magnetic vectors are constant. Our simple wave is now a linearly polarized, plane, monochromatic wave. The electric field, magnetic field, and direction of propagation are all mutually perpendicular and form a right-handed set.

Such a monochromatic wave travels at a well-defined velocity in a medium. The ratio of the velocity of the wave in free space to that in the medium is known as the refractive index, written n. The ratio of the magnetic field magnitude to the electric field magnitude of the wave is another constant of the medium known as the *characteristic admittance*, usually written as y. A great simplification is possible at optical frequencies. There the interaction of a wave with a propagation medium is entirely through the electric field, which can exert a force on even a stationary electron. The magnetic field can interact only with moving electrons, and at optical frequencies any direct magnetic interaction with the electrons is negligible. Then

$$y = n\mathscr{Y}, \tag{21.1}$$

$\mathscr{Y}$ being the admittance of free space, $1/377$ siemens. This relationship is usually expressed as

$$y = n \text{ free space units}, \tag{21.2}$$

allowing the same number to be used for both quantities.

In the linear regime the combination of two separate waves is quite simple. Whatever the characteristics and directions of the individual waves, the resultant electric and magnetic fields are simply the vector sum of the individual fields. This is the *Principle of Linear Superposition*. Irradiances, in general, do not combine in such a simple way.

Since the interaction with the medium is through the electric field, and since the magnetic field of a harmonic wave can be found readily from the direction of propagation and the characteristic admittance, y, it is normal to write the analytical expression for a harmonic wave in terms of the electric field only. Moreover, since the combinations of waves are linear, we can profitably use a complex form for the harmonic wave expression:

$$\begin{aligned} E &= |\mathscr{E}| \exp\{i(\omega t - \kappa z + \varphi)\} = [|\mathscr{E}| \exp(i\varphi)] \cdot [\exp\{i(\omega t - \kappa z)\}] \\ &= \mathscr{E} \exp\{i(\omega t - \kappa z)\}, \end{aligned} \tag{21.3}$$

where the relative phase is usually incorporated into what is known as the complex amplitude, ε, and the remaining exponential is known as the phase factor; ω and κ are the temporal and spatial angular frequencies, κ usually being known as the wavenumber. We can write

$$\omega = 2\pi/\tau \text{ and } \kappa = 2\pi n/\lambda, \tag{21.4}$$

where λ is the wavelength that the light has in free space.

The expression for the wave is usually written

$$E = \mathscr{E} \exp\left\{i\left(\omega t - \frac{2\pi n z}{\lambda}\right)\right\}. \tag{21.5}$$

If the medium is absorbing, then there will be a fall in amplitude of the wave as it propagates. This can be accommodated in the expression by introducing a complex form for n:

$$n \to (n - ik) \text{ and } y \to (n - ik)\mathscr{Y}, \tag{21.6}$$

$$E = \mathscr{E} \exp\left\{i\left(\omega t - \frac{2\pi(n - ik)z}{\lambda}\right)\right\} = \mathscr{E} \exp\left(-\frac{2\pi k z}{\lambda}\right) \exp\left\{i\left(\omega t - \frac{2\pi n z}{\lambda}\right)\right\}. \tag{21.7}$$

The irradiance of the harmonic wave in its complex form is given by

$$\text{Irradiance} = I = \tfrac{1}{2} \operatorname{Re}\{EH^*\} = \tfrac{1}{2} \operatorname{Re}\{\mathscr{E}\mathscr{H}\} = \tfrac{1}{2} n\mu|\mathscr{E}|^2. \tag{21.8}$$

Now let us return to the plane wave of arbitrary profile and let us insist that it be linearly polarized. In the first instance we consider its propagation through free space. The electric field, the magnetic field, and the direction of propagation will form a right-handed set. Let us freeze the electric field and magnetic field at a particular instant. By a Fourier decomposition process we can construct a continuous set of harmonic profiles that, when added together at each point, will yield the instantaneous profile. In free space there is no dispersion and we can show that the magnetic and electric field profiles are always in the ratio of the admittance of free space, $\mathcal{Y}$. The magnetic field decomposition therefore mimics the electric and when we pair frequencies we construct a complete set of monochromatic component waves that make up the primary wave. The set of attributes of the component waves as a function of wavelength or frequency is known as the spectrum of the primary wave and the process of finding the spectrum is sometimes called spectral decomposition. The spectrum is most often expressed in terms of the varying irradiance of the component waves per frequency or wavelength unit. Real optical sources exhibit fluctuations in the nature of their output. These translate into fluctuations, either of amplitude or phase or both, in the spectral components. Such fluctuations are best handled in a statistical context. The theory of coherence that deals with such ideas is discussed elsewhere in this handbook. Here we use only a very elementary treatment of waves.

Now let us remove our restriction of linear polarization. We can represent our wave now as made up of two major components linearly polarized in orthogonal directions. If there is no consistent relationship between these components – in other words they are unpolarized – then although the irradiance spectrum of each component may be similar, the individual components will have no consistent phase relationship that will allow their combination into a consistent polarization state and neither the primary wave nor the component waves will exhibit polarization. If, however, there is a consistent polarization in the primary beam, then this will be reflected also in the components.

Now let the wave enter a dispersive material. Since the primary wave and the spectral component waves are entirely equivalent we can look most closely at the latter and follow their progress through the medium individually. To find the net disturbance we combine them. Each experiences a refractive index and a characteristic admittance that are functions of the properties of the medium. In a dispersive medium these attributes show a dependence on frequency or wavelength. Thus the components are not treated equally. As they propagate through the medium, therefore, their resultant changes, and this is the reason for the changes in the profile in dispersive media mentioned earlier.

A plane wave strictly has no center or axis and really does not correspond at all to our ideas of a light ray. It has a direction but there is nothing that corresponds to a lateral position or lateral displacement. Yet beams of light emitted from lasers have all the attributes of spectral purity and potential for interference that we would expect from a plane wave, and yet they have a position as well as direction and act in many ways as if they were light rays rather than plane waves. These are better described as Gaussian beams. A Gaussian beam has a direction and phase factor not unlike the plane wave but it is limited laterally and the irradiance expressed in terms of lateral distance from the center of the beam is a Gaussian expression. If we define the beam radius, ρ as the diameter of the ring where the irradiance has fallen to $1/e^2$ of the irradiance in the center of the beam, I_0, then we can write for the irradiance

$$I = I_0 e^{-2r^2/\rho^2} \tag{21.9}$$

where r is the radius of the point in question. Strictly speaking the Gaussian beams described by Eq. (21.9) are in the fundamental mode, the TEM_{00} mode. There can be higher-order modes that have more complicated variations of irradiance as a function of displacement from the axis.

Equation (21.9) is insufficient for a complete description of the beam. We need also the wavelength, λ, and the beam divergence (or convergence) ϑ, which may be defined as the angle between the axis and the $1/e^2$ irradiance ring (Fig. 21.1). (The divergence is also defined sometimes as the total angle.) The Gaussian beam appears to emanate from or converge to a point. However, in the neighborhood of the point, the beam – instead of contracting completely – reaches a minimum size and then expands to take up the cone again. The minimum is called the beam waist and it has a radius that is a function both of the wavelength and the divergence of the beam:

$$\rho_0 = \frac{\lambda}{\pi\vartheta}. \tag{21.10}$$

The diameter of the beam at distance d from the beam waist is given by

$$\rho^2 = \rho_0^2 + \vartheta^2 d^2. \tag{21.11}$$

Normal thin lens theory where the positions of the beam waists are taken as the image points applies to Gaussian beams. An excellent account of Gaussian beams and their manipulation is given by O'Shea [2].

21.3 SPATIAL AND SPECTRAL FILTERING

Filtering is performed to render light more suitable for a given purpose. Spatial filters accept light on the basis of positional information in the filter plane, while spectral filters operate according to the wavelength or frequency. Although they are separate attributes of a light beam they may be required to operate together to achieve the desired result. Frequently, for example, spatial filters are a necessary part of spectral filters. Rather less often, spectral filters may be involved in spatial filtering. We shall consider the two operations separately but it will become clear in the section on spectral filtering that they are often connected.

A simple example of a common spatial filter is a field stop. This is usually an aperture that is placed in an image plane to limit the extent of the image that is selected. Normally the boundaries between that part of the image that is selected and that rejected should be sharp but this may not always be the case. Especially in the early part of the 20th century, vignettes were a popular form of photographic portrait. There, a defocused field stop, usually elliptical in shape, graded the boundary

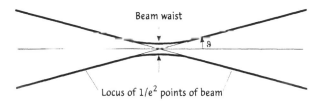

Figure 21.1 Schematic of the parameters of a Gaussian beam.

of a head and shoulders image so that in the final print it was detached from the background and appeared to be surrounded by a gradually thickening white mist.

An improved aesthetic effect may demand a change in the color balance of illumination. An interferometric application may require a greater degree of spectral purity. A luminous light beam may contain also appreciable infrared energy that will damage the object under illumination. An interaction with a material may demand a particular frequency. Operation of a spectrum analyzer may demand elimination of all light of frequency higher than a given limit. The signal-to-noise ratio of a measuring apparatus may need elimination of white background light and acceptance of a narrow emission line. A bright line creating problems in the measurement of a dim background may require removal. The list of possibilities is enormous. All these involve spectral filters that use wavelength or frequency as the operating criterion.

Spatial filters will usually either accept light, which is then used by the system, or reject it completely so that it takes no part in what follows. Spectral filters may have a similar role but they may also be required to separate the light into different spectral bands, each of which is to be used separately. A bandpass filter is typical of the first role and a dichroic beamsplitter of the second.

We consider linear filters only, i.e., filters with properties independent of irradiance but dependent on wavelength or frequency or position. We exclude tuned receivers, wavelength shifting filters, tuned amplifiers, and the like. The output of the filters is derived by the removal of light from the input beam and can be described by a response function that is the ratio of output to input as a function of wavelength or frequency or position and is a number between 0 and 1 or between 0 and 100%. Usually the parameter characterizing input and output will be the irradiance per element of wavelength or frequency or area. The response will normally be a transmittance T or reflectance R. Typical ideal response functions for spectral filters are sketched in Fig. 21.2. Spatial filters would be similar with wavelength replaced by position.

A beamsplitter will have two such responses. Usually one will be in reflection and the other in transmission.

There are many classifications of spectral filter: those that transmit narrow bands of wavelength are usually called narrow band pass filters; those that reflect narrow bands are notch filters; those that transmit broad regions at wavelengths longer or shorter than a rapid transition between acceptance and rejection are edge filters and are either long-wave pass or short-wave pass.

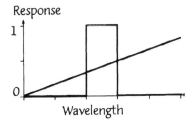

Figure 21.2 Typical ideal responses: one shows a response that varies gradually with wavelength; the other shows a response that accepts light with wavelength within a given band and rejects all others.

The light may be removed by redirection or by conversion or by both.

A process of absorption is really one of conversion. The light energy is converted into kinetic and potential energy of molecules, atoms, and electrons. Some of this energy may then be re-emitted in a changed form or dissipated as heat.

An important question that should always be asked about a filter is: "Where does the unwanted energy go?"

Most filters exhibit the various modifications of the input shown in Fig. 21.3. For example an absorption filter to eliminate short wavelengths will reflect residual light (redirection), will convert some light (fluorescence), will scatter some light, and will emit thermal radiation – all unwanted. If this light is accepted back into the system, performance will suffer.

Figure 21.4 shows a sketch of a very simple instrument showing the stops. The image of the aperture stop in source space is the entrance pupil and in receiver space the exit pupil. Note that the performance of a filter inserted in the instrument cannot be separated from the details of stops and pupils. These are part of the overall system design.

In the design of an instrument that includes spatial or spectral filters the behavior of the unwanted light, and especially any redirected light, is very important. For example, can rejected light be scattered back into the acceptance zone of the system?

Baffles, a special form of spatial filter, can help to prevent return of unwanted energy and their correct design is as important as the design of the primary components such as lenses.

Since performance is system-dependent, an optical filter is usually specified with regard to standard (and that usually means ideal) conditions. For spectral filters, for example, entrance and exit pupils are usually considered to be at infinity (light ideally collimated), and scattered, redirected, converted, and emitted light are assumed lost to the system. A real system will rarely have this ideal arrangement and, therefore, the performance may not correspond to the standard specified performance of the filter.

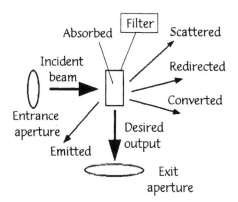

Figure 21.3 The character of the desired output light is unchanged, except perhaps in direction, from its character as part of the input beam. The rejected light may simply be redirected or may suffer a conversion; some possible mechanisms are illustrated here.

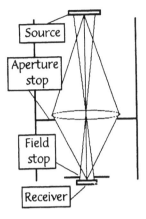

Figure 21.4   A very simple optical instrument showing an aperture and a field stop.

21.4   SPATIAL FILTERS

Any optical instrument that has a limited field of view is effectively a spatial filter. In the vast majority of instruments a deliberate limitation of the field of view is desirable. If it is not well defined, then uncontrolled and unwanted input may be received that will interfere with the correct operation.

In a very simple case a stop is placed in a field plane somewhere in the instrument. Microscopes are frequently fitted with a variable field stop in the form of an adjustable diaphragm that can be used to select a particular part of an image and eliminate the remainder, which is useful when precise measurements on a selected object have to be made.

Spatial filtering is often a vital instrumental feature. An instrument that is to be used to examine images near the sun, for example, may even be destroyed if light from the sun can reach the image plane. Such demanding spatial filtering tasks usually involve not just the elimination of the direct light by a suitably shaped aperture but also the elimination of scattered light as well, using complicated assemblies of spatial filters known as baffles. A solar telescope acquires an actual image of the sun, but when details of the limb must be investigated the solar image must be removed and this is achieved by a special spatial filter called an occluding disk that fits exactly the image of the disk of the sun but allows the surrounding regions to be examined. Elimination of stray light in this configuration is of critical importance.

Apart from its instrumental uses, probably the commonest form of spatial filtering is in the collimation of light: i.e., the attempt to construct a beam of light that is as near as possible to a plane wave. Plane waves are necessary inputs to all kinds of optical instruments. To construct a plane wave we first of all create a point source. The point source emits a spherical wave and the spherical wave can be converted into a plane wave by passing it through a suitable lens. Unfortunately, a point source is an unattainable ideal. The best we can do is to make the source very small. We do this by first producing an image of a real source and in the image plane we introduce a diaphragm with a very small aperture. This aperture is then used as our point source. A lens is then placed such that the point source is at the focus. This is illustrated in Fig. 21.5.

The radiance, L, of this source is the important quantity. This is the power in unit solid angle leaving the unit projected area of a source. It is measured in watt/steradian meter^2 (W sr^{-1} m^{-2}). The total power accepted by the collimating lens will be given by

$$\text{Power} = L \cdot A_S \cdot A_L / R^2, \tag{21.12}$$

where A_S is the area of the effective source, A_L is the area of the collimating lens and R is the distance from effective source to the collimating lens. If Ω_L and Ω_S are the solid angles subtended by the lens at the effective source and by the effective source at the lens, respectively, then we can write

$$\text{Power} = L \cdot A_S \cdot \Omega_L = L \cdot A_L \cdot \Omega_S; \tag{21.13}$$

here $A_L \Omega_S$ and $A_S \Omega_L$ are equal and are known as the $A\Omega$ product. For a well-designed system this should be a constant. Unfortunately it is very easy to make a mistake in pairing the correct A with the correct Ω and so Eq. (21.12), which is completely equivalent, is a safer expression to start with.

The degree to which the light departs from perfect collimation may be defined as the semiangle of the vertex of the cone – i.e., the solid angle subtended by the effective source at the center of the collimating lens. It is easy to see that a factor of 2 improvement in this angle is accompanied by a factor of 4 reduction in power. Thermal sources, in particular, give disappointing performance when the collimated light must also be of narrow spectral bandwidth. A common use of collimators is in dispersive filters such as grating or prism monochromators, where the dispersion can be arranged to be in one well-defined direction and the high degree of collimation is demanded only in this direction. This permits the use of a long slit as aperture in the spatial filter of the collimator. The degree of collimation can then be varied by changing the width of the slit but not its length. This makes the power proportional to the collimation angle rather than its square, and permits much more satisfactory use of thermal sources.

A Gaussian beam from a laser is already collimated with a degree of collimation defined by the divergence, ϑ. This divergence may not be acceptable. Since ϑ and ρ_0 are interrelated, a reduction in ϑ implies an increase in ρ_0, and vice versa. A large degree of spatial magnification implies an object at or near the focal length of a

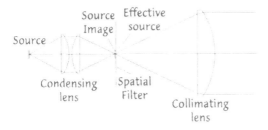

Figure 21.5 A simple collimator showing the light source, condenser, collimating lens, and spatial filter to produce the effective point source.

positive lens and the image at near infinity. Since the Gaussian beam at the input to the collimator will almost certainly be some distance from its beam waist it is necessary to insert a leading lens to form a waist at a suitable distance from the following collimating lens. Any beam that propagates through an optical system will tend to accumulate spurious stray light. It is usual, then, to insert a spatial filter with the purpose of cleaning up the beam. Any light representing a departure from regularity will tend to appear off axis at the beam waist. A small-diameter filter, usually called a pinhole, selects the ideal image and rejects the rest. Usually the pinhole is quite large compared with the Gaussian beam waist, since the usual objective is the elimination of spurious light rather than the creation of the diffraction-limited spot itself. The assembly of input and collimator lenses is also sometimes called a beam expander, because this is effectively what it does. O'Shea [2] gives more information on designing systems using Gaussian beams.

A rather different type of spatial filter is sometimes used in wide-angle aircraft cameras. Figure 21.6 shows a typical arrangement. If we assume that the ground is flat and that its emission characteristic is independent of direction, that is a Lambertian surface, then the energy in an element of area on the photographic plate in the image plane will be proportional to the $A\Omega$ product for the areal element on the ground and the pupil of the camera. If we assume the ground is completely flat, then this will be easily shown to be given by

$$A\Omega = \frac{(A_{\text{pupil}}/\cos\vartheta)(A_{\text{element}}/\cos\vartheta)}{(h\cos\vartheta)^2} \propto \frac{1}{\cos^4\vartheta}. \tag{21.14}$$

However, since the surface is Lambertian and the focus of the camera is fixed, the energy on unit element area on the photographic plate will follow the rule in Eq. (21.14) even if the terrain height varies across the image. The fall-off in energy with $\cos^4\vartheta$ applies to all cameras with a flat focal plane, but it is only in wide-angle cameras that the problem can be severe. These may have acceptance angles higher than 45°. At 45° the ratio of energies is 1:4 and such a large ratio cannot be accommodated in the optical design of the camera lens. A simple solution, again involving rejection of otherwise useful light, is a spatial filter fixed in front of the lens, usually on the hood, where there is an appreciable spatial separation of marginal from axial light. An evaporated film of Inconel or Nichrome with a radial distribution so that it

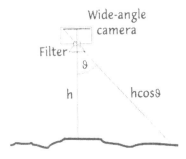

Figure 21.6 A wide-angle camera. The irradiance at the image plane is proportional to $1/\cos^4\vartheta$, assuming constant radiance and a flat horizontal Lambertian surface for the terrain.

has a higher density in the center than at the periphery is usually sufficient. This is often combined with the light yellow antihaze filter that is normal in high-altitude photography. The correction is rarely exact.

A circular pupil, with sharp boundaries is illuminated in collimated light that is imaged to a single spot, as shown in Fig. 21.7. If the rays obeyed the laws of geometry, the spot would be a single point, but diffraction causes a spreading of the point. The resulting distribution of energy in the focal plane is known as the point-spread function. Note that the point-spread function may be the result of aberrations in the system rather than diffraction. When an image is produced in the focal plane of any system the result is a convolution of the image and the point-spread function. The point-spread function for a sharply defined circular pupil illuminated by a perfectly uniform collimated beam, with no other aberrations, is the Airy distribution. This has a pronounced outer ring that can be an undesirable feature in diffraction-limited systems. A special type of spatial filter, known as an apodizing filter, inserted in the plane of the pupil can modify this distribution. It operates by varying the distribution of irradiance across the aperture. As with all the other filters described, it is rejecting energy. A simple form of apodizing filter yields a linear fall in irradiance from a maximum in the center to zero at the outer edge. This gives a point-spread function that is the square of the Airy distribution and the first diffraction ring is now much reduced. An apodizing filter that has a Gaussian distribution will produce a spot with a Gaussian distribution of energy. Although in all cases the central part of the spot is broader (we are rejecting some of the otherwise useful light) the resulting image can be much clearer.

The ideas of the previous paragraph are carried much further in a class of spatial filters that uses a property of a lens and its front and back focal planes. It can be shown that the light distribution in the rear focal plane is the two-dimensional spatial Fourier transform of the distribution in the front focal plane. If the front focal plane contains an image, then the rear focal plane will contain a distribution that is the two-dimensional spectrum of the image in spatial frequency terms. Higher frequency terms in the image are translated into light in the back focal plane that appears at points further from the axis of the system. If, now, a second similar lens is added so that its front focal plane coincides with the current back plane, then the image will be recovered in the back focal plane of the second lens: Fig. 21.8 shows the arrangement.

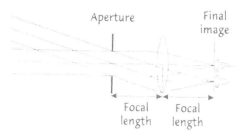

Figure 21.7 A simple telecentric imaging system with object at infinity. The distribution of energy in the images is indicated to the right of the focal plane.

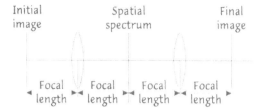

Figure 21.8 The two lens system showing the Fourier plane where the spatial spectrum is located and the final image plane.

Since this system has, at its final output, the original image, but has also at an intermediate plane, the spatial spectrum of the image, it becomes possible to make modifications to the spectrum so that desirable changes may be made in the final image (Fig. 21.9).

The spatial filters may be used in many different ways but a common application is in removing high spatial frequency structure from images. For instance, the raster lines in a television frame could readily be removed using this technique. The apodizing filters mentioned previously are a special case of such spatial filters.

More advanced filters of this kind also operate to change the phase of selected frequencies but are completely beyond the scope of this chapter. Much more information is given by VanderLugt [3].

Phase contrast microscopy uses an interesting kind of spatial filter. Objects that are completely transparent and are immersed in a fluid of only slightly different refractive index are difficult to see. If the light that is transmitted by the phase objects is arranged to interfere with that transmitted by the medium the result is a very slight difference in contrast, because the phase difference is sufficient for just a slight change at the peak of a cosine fringe. Small changes in relative phase are much more visible if they are occurring on the side of a fringe peak where there is much larger variation with small phase differences. In the phase contrast microscope the sample is illuminated by light derived from a ring source created by a spatial filter admitting a narrow circle only. Beyond the illuminated specimen, the optical system is arranged to produce an image of the circular light source. However, the light that passes through the objects that are the subject of examination is scattered and although it passes through the optical

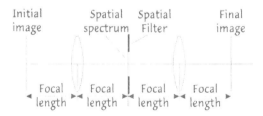

Figure 21.9 The two lens system of Fig. 21.8 showing a spatial spectrum filter inserted in the Fourier plane. The filter is simply known as a spatial filter.

system it does not form any image of the circular ring light source. Light that passes through the supporting medium of the specimen is scattered to a much smaller extent and forms a reasonable ring image. It is arranged that the ring is formed exactly over a narrow spatial filter that alters the phase of the light by one-quarter of a wavelength with respect to the scattered unaffected light. All the light then goes on to form the final image where there is interference between the light that propagated through the medium and through the phase objects but now, because of the quarter-wave phase difference, the variation is on the side of the cosine fringes and the variation in radiance large. This brilliant idea conceived by Fritz Zernike in the 1930s has inspired other optical instruments where a quarter-wave phase difference reveals small phase excursions as much larger differences in the radiance of an image: see, for example, Hecht and Zajac [4].

A class of lasers uses what are known as unstable resonators. In such lasers the gain is extremely high and so there is no need for the confining effect of the regular modes of stable resonators. It is possible with such constructions to make exceedingly efficient use of the gain medium. The penalty is that the output of the laser is rather far from ideal and certainly not at all like the regular Gaussian beam profile. In particular, much energy propagates rapidly away from the resonator axis making it difficult to use the beams efficiently. It has been found that a rather special type of spatial filter renders the output beams much nearer the ideal Gaussian profile. The special type of spatial filter involved is the graded coating filter.

Even the unstable resonators must be in the form of a cavity. The light must be reflected back and forth between opposite mirrors, which need reflecting coatings for their correct operation. It has been shown that if the profile of the reflectance of the output mirror is Gaussian then the output beam will also have a Gaussian, or near-Gaussian, profile. The Gaussian profile is achieved by a process of grading the thicknesses of the reflecting coating. Laser systems using such graded mirrors are already commercially available.

How do we make a reflector with a radial Gaussian profile? The reflectance of a quarter-wave stack as a function of the thicknesses of all the layers is shown in Fig. 21.10. This means that a suitable radial grading of the thicknesses of the layers can result in a radial variation of reflectance at the reference wavelength that is Gaussian. Similarly, Fig. 21.11 shows another arrangement where just one of the layers is varied in thickness radially, the other layers remaining of constant thickness.

A Gaussian profile is ideal but the real mirrors are more or less successful in achieving it and, provided the agreement is reasonable, exact correspondence is not necessary nor is it really practical. Usually masks are used during the appropriate part of the deposition process that assures the correct thickness distribution either of the chosen layer or even the entire multilayer.

Note that the variation represented by Fig. 21.11, where one layer only is varied, is rather easier than that of Fig. 21.10, where the percentage variation to change from almost zero to almost 100% reflectance is quite small. On the other hand, the process that uses Fig. 21.11 must mask only one of the layers, a difficult mechanical problem, whereas the other process masks all of them. Both types of graded coatings are actually used in practice. For a much more detailed description see Piegari [5].

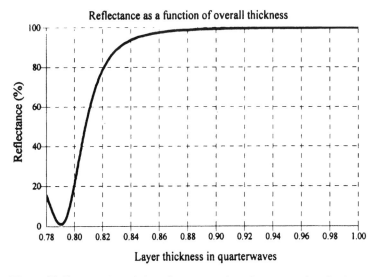

Figure 21.10   Variation of the reflectance at the reference wavelength of a quarter-wave stack as a function of the thicknesses of the layers. The stack is made up of 13 alternating quarterwave layers of titanium and silica.

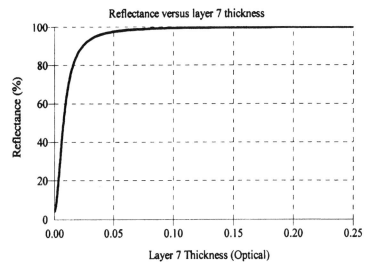

Figure 21.11   Variation of the reflectance at the reference wavelength of a quarterwave stack in which one of the layers is varied in thickness. The quarterwave stack is identical to that of Fig. 21.10 but in this case only the thickness of the third layer from the incident medium is varied. The thickness scale is in units of the reference wavelength.

21.5   SPECTRAL FILTERS

The operation of different filter types is often a mixture of several mechanisms, some of which include absorption, refraction, reflection, scattering, polarization, interference, and diffraction. We exclude neutral density and similar filters because they cannot be described as spectral filters since their attempted purpose is to treat all spectral elements equally.

21.5.1   Absorption Filters

Semiconductors are intrinsically long-wave pass absorption filters. Photons of energy greater than the gap between valence and conduction bands of the electrons are absorbed by transferring energy to electrons in the valence band to move them into the conduction band. High transmittance implies intrinsic semiconductors of high resistivity. Gallium arsenide, silicon, germanium, indium arsenide, and indium antimonide are all useful (Fig. 21.12). Note that these semiconductors have a high refractive index so must be antireflected in the pass region. They are rarely antireflected in the absorbing region and so a large amount of the rejected light is actually reflected (Fig. 21.13).

Some colored filter glasses of the long-wave pass type contain colloidal semiconductors.

Other colored glasses have metallic ions dispersed in them. In general, colored glass filters make excellent long-wave pass filters but it is not possible to find short-wave pass filters with the same excellent edge steepness. In fact, short-wave pass filters present an almost universal problem. For more information see Dobrowolski [6, 7].

Strong absorbers are also strong reflectors. In the far infrared, the reststrahlen bands associated with very strong lattice resonances show strong reflectance and are sometimes used as filters. Beryllium oxide reflects strongly in the 8–12 μm atmospheric window. Because its thermal emittance is therefore very low in the atmospheric window, it is sometimes used as a glaze on electrical insulators to keep them warmer and therefore frost-free during cold nights.

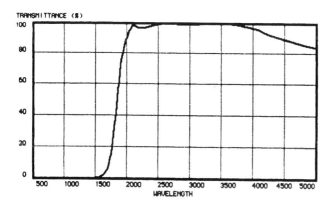

Figure 21.12   Transmittance of a germanium filter with antireflection coatings.

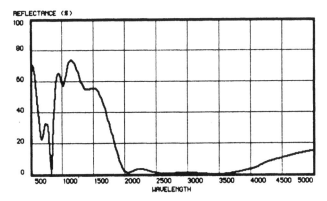

Figure 21.13   Reflectance of the same filter as in Fig. 21.12.

21.5.2   Refraction and/or Diffraction

A prism monochromator is a variable spectral filter of a very inexpensive nature. The prisms operate due to refraction that varies because of the dispersion of the index of the prism material. The optical arrangement usually assures that the spatial distribution of the component waves will vary according to wavelength or frequency. A spatial filter of a simple kind is then used to select the appropriate wavelength. Usually, although not always, the light source is in the form of a narrow slit that is then intentionally imaged to enhance chromatic effects in the image. A spatial filter, also in the shape of a slit, is then used to select the light in the image having the correct wavelength.

Diffraction gratings are used in a similar way to refractive prisms. They are essentially multiple-beam interference devices that use diffraction to broaden the beams to given reasonable efficiency over slightly more than an octave. Disadvantages are the low throughput because of the narrow entrance and exit slits, although they are superior in this respect to prisms, and the need for mechanical stability of a high degree. A further problem with diffraction gratings is that they admit multiple interference orders so that harmonics of the desired wavelength may also be selected. Order-sorting filters may be necessary to remove the unwanted orders. Figure 21.14 shows a sketch of a grating monochromator where the wavelength can be changed by rotating the diffraction grating, all other elements remaining unchanged. High-performance monochromators invariably use mirrors rather than refracting elements for collimating and collecting since they have no chromatic aberrations.

21.5.3   Scattering

Christiansen filters consist of dispersed fragments or powder in a matrix. The dispersion curves of the two materials differ but cross at one wavelength at which the scattering disappears. The undesired scattered light is then removed from the desired specular light by a simple spatial filter (Fig. 21.15)

Christiansen filters have been constructed for virtually the entire optical spectrum. An excellent and compact account is given by Dobrowolski [6].

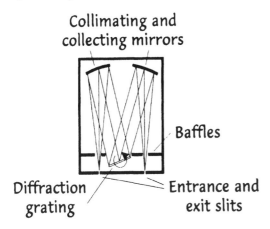

Figure 21.14 Sketch of a grating monochromator. Additional optics focus a light source on the entrance slit and collect the light from the exit slit.

21.5.4 Polarization

Retardation produced by thickness d of a birefringent material is given by

$$\varphi = \frac{2\pi(n_1 - n_2)d}{\lambda}, \tag{21.15}$$

so that a half-wave retarder made of birefringent material is correct for only one wavelength. Let there be a polarizer, a retarder at 45°, and an orthogonal polarizer in series. All the light transmitted by the first polarizer will be transmitted without loss through the remainder of the system provided the retardation is equivalent to an odd number of half-wavelengths half-wave plate. For any other value of retardation the light will be stopped by the second polarizer either totally, if the retardation is an integral number of wavelengths, or partially, if not. The irradiance is given by an expression of the style:

$$\text{Irradiance} \propto \sin^2 \frac{\varphi}{2}. \tag{21.16}$$

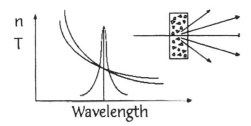

Figure 21.15 Schematic diagram of the principle of the Christiansen filter. Where the dispersion curves of the matrix and scattering particles cross, there is no scattering loss and the transmittance of the device is specular; elsewhere, the light is scattered. A spatial filter is then used to select the specular light and reject the scattered.

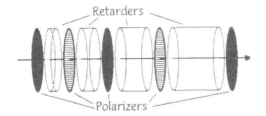

Figure 21.16 Schematic of a Lyot filter. Real filters have rather greater numbers of elements and must be tightly controlled in temperature.

This effect can be used in filters either in a single element or in a series of such elements. The Lyot filter consists of a series, each member having an increased half-wave plate order and so a narrower, more rapidly changing response (Fig. 21.16). The responses of the various elements combine to give a very narrow band width.

The Solc filter has only two polarizers and a set of identical retarder plates in between with the axes arranged in a fan. For a detailed theoretical analysis of such filters see Yariv and Yeh [8].

21.5.5 Acousto-optic Filters

In acousto-optic filters the periodic strain caused by an acoustic wave alters the refractive index in step with the wave. This impresses a thick Bragg phase grating on the material. The light interacts with this grating, which effectively becomes a narrow band filter. In the collinear case, the light is polarized and a narrow band is scattered into the orthogonal plane of polarization where it is selected by a suitably oriented polarizer (Fig. 21.17). Variation of the frequency of the RF drive for the acoustic transducer varies the wavelength of the filter.

In the noncollinear type of filter, the Bragg grating deflects the light with high efficiency. The light need not be polarized. Unwanted light is obscured from the receiver by a spatial filter (not usually as crude as in the diagram, Fig. 21.18).

Bandwidths of a few nanometers with an aperture of a few degrees with an area of almost $1\,cm^2$ and with tuning over the visible region are possible. The collinear

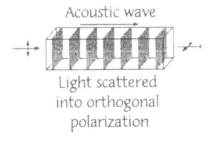

Figure 21.17 Schematic of a collinear acousto-optic filter.

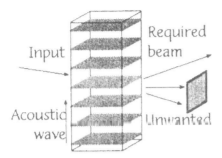

Figure 21.18 Schematic of a noncollinear acousto-optic filter.

type has yielded bandwidths of 0.15 nm with tuning range over the visible and near ultraviolet. For more information see Chang [9].

21.5.6 Interference

The bulk of filters that depend on interference for their operation are of the thin-film class. Because they form a rather special and very important category they are considered separately. Here we examine filters that operate by processes of interference, usually multiple beam, but are not made up of assemblies of thin films, nor can be included in the other classes we have considered.

Any interferometer is potentially a narrow-band filter. Since the fringe positions normally vary with wavelength, a spatial filter can be used to select the light of a particular wavelength. The commonest interference filter is probably based on a Fabry–Perot etalon. The fringes are localized at infinity. Thus, if a spatial filter is placed in the focal plane of an objective lens, then it can be used to select the appropriate wavelength. The etalon is usually arranged so that the central fringe is of the correct wavelength and then the spatial filter can be a simple circular aperture of the correct diameter. The filter can be tuned by varying the optical thickness of the spacer layer in the etalon and this can be achieved either by varying the refractive index, as for example by altering the gas pressure, or by physically moving the interferometer plates themselves. Over rather small tuning ranges, tilting the etalon is another arrangement that can be, but is less frequently, used.

Most filters that employ interference are thin-film filters, which are discussed separately. However, there is a variant of the thin-filter that is usually known as a holographic filter. The holographic filter is produced in a completely different way but, in principle, it can be thought of as a thin-film reflecting filter that has been sliced across its width at an angle and then placed on a substrate. This is illustrated in Fig. 21.19. We take a thin-film reflecting filter that consists of alternate high and low index quarter-waves or, in the case of the hologram a sinusoidal variation of index in the manner of a rugate filter (described later). The thin-film reflecting filter is sliced by two parallel cuts at angle ϑ to the normal and the slice is laid over its substrate. The hologram now reflects strongly at the angle ϑ to the hologram surface. As in the rugate, the sinusoidal variation assures that there are no higher orders that are

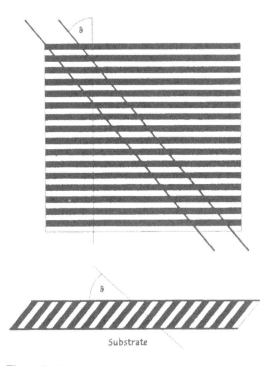

Substrate

Figure 21.19 Diagram of the way in which a holographic filter may be imagined to be produced from a thin-film filter. The thin-film reflecting filter, usually a rugate filter, is sliced by two parallel cuts at angle ϑ to the normal. The slice is then placed on a substrate. Both filters reflect strongly, the thin-film filter at normal incidence but the holographic filter at the angle ϑ to the hologram surface.

reflected at the same angle. A simple spatial filter that selects light propagating at the correct angle completes the filter. The variation in refractive index in the hologram is small and this means that the reflecting band is very narrow in terms of wavelength.

Another variant of the rugate filter is in the form of an optical fiber with a propagation constant that varies in a sinusoidal fashion along the core. This reflects a very narrow band of wavelengths and transmits all others. Such fiber filters are, therefore, very narrow notch filters.

21.5.7 Thin Films

A thin film is a sheet of material defined by surfaces that are sufficiently parallel for interference to exist between light reflected at the surfaces. Usually this means that the films will be at most a few wavelengths in thickness. They are usually deposited from the vapor or liquid phase directly on a substrate. The optical thickness of the individual films and the wavelength determine the phase differences between the beams that interfere. Assemblies of many films exhibiting complicated but engineered interference effects are common. Frequently, the interference effects are supplemented by the intrinsic properties of the materials, especially their absorption behavior. Thin-film optical coatings, therefore, operate by a mixture of interference and material properties.

We can classify the materials that are used in optical coatings into three principal classes: dielectrics, metals, and semiconductors. The main characteristics of these materials are summarized in Table 21.1. Note that these are somewhat idealized. For example there is slight variation of n with λ in the case of the dielectrics but it is negligible compared with the enormous changes in k with λ in metals. Similarly, there is a residual small finite n even in high-performance metals

With increasing wavelength:

Dielectrics	Metals
Become weaker	Become stronger
T increases	R increases
Dielectrics have lower losses	Metals have higher losses

Note that for interference coatings we require the presence of dielectric (or transparent semiconductor) layers. Metals by themselves are not enough, although they may act as broadband absorbers, reflectors, or beamsplitters.

In a device that operates by optical interference, the path differences vary with angle of incidence. Although at first sight it may seem anomalous, path differences actually are smaller when the angle of incidence moves away from normal. Thus, all filters of the thin-film interference type exhibit a shift of their characteristics towards shorter wavelengths with increasing angle of incidence. Polarization effects also gradually become important. For small tilts of a few degrees in air the wavelength shift is proportional to the square of the angle of incidence:

$$\frac{\Delta\lambda}{\lambda} = \frac{1.5 \times 10^{-4}}{n_{\mathrm{e}}^2}\vartheta^2, \tag{21.17}$$

where n_{e} is the effective index of the coating (a value in between the highest and lowest indices in the coating) and where ϑ is the angle of incidence in degrees.

Table 21.1 Principal Characteristics of Dielectrics, Metals, and Semiconductors

Dielectrics	Metals	Semiconductors
n real	$y = -ik\mu$	Usually classified as either
$y = n\mu$	$k \propto \lambda$	metal or dielectric, depending
n independent of λ		on the spectral region
$\delta = \dfrac{2\pi nd}{\lambda} \propto \dfrac{1}{\lambda}$	$\beta = \dfrac{2\pi kd}{\lambda} = \text{constant}$	
$y - \text{constant}$	$y \propto \lambda$	
Progressive wave	Evanescent wave	
$\varepsilon e^{i(\omega t - \kappa z)}$	$\varepsilon e^{-\alpha z} e^{i\omega t}$	
	In all cases $R = \left\|\dfrac{y_0 - y}{y_0 + y}\right\|^2$ and $T = \dfrac{4y_0\,\text{Re}\,y}{\|y_0 + y\|^2}$	

21.5.7.1 Multilayer Coatings

Combinations of purely metal layers show no interference effects and so multilayer filters are constructed either from purely dielectric layers or from combinations of dielectrics and metals. Figures 21.20 and 21.21 show some idealized characteristics. Because of their nature the natural type of filter for dielectric layers is one that reflects over a limited region at shorter wavelengths and transmits well to longer and longer wavelengths: in other words, a long-wave pass filter. Metal–dielectric systems, on the other hand, lend themselves to short-wave transmittance and stronger reflection toward longer and longer wavelengths: i.e., a short-wave pass filter. Any requirement that demands the opposite of these simple characteristics presents formidable difficulties.

Transparent conductors such as indium tin oxide (ITO) or ZnO:Al appear dielectric at shorter wavelengths and metallic at longer wavelengths (Fig. 21.22). They are much used as heat-reflecting short-wave pass filters.

A metal film can be antireflected on either side by a dielectric system consisting of a phase-matching layer and a reflector (Fig. 21.23). The reflector generates a reflected beam that is capable of destructively interfering with the beam reflected by the metal surface, provided the phases are opposite. The parameters of the outermost reflector are adjusted until the amplitudes of the beams are correct and then correct choice of the thickness of the phase-matching layer assures the necessary relative phase. The greater the thickness of the primary metal film, the greater must be the reflectance of the outer reflector. Provided the primary metal layer is not too thick, good transmittance within the antireflected region can result.

For a very thin metal the mismatch between high-admittance phase-matching layers and the surrounding media can give a sufficiently high reflectance for the antireflection condition. This leads to a very simple design consisting of a thin silver

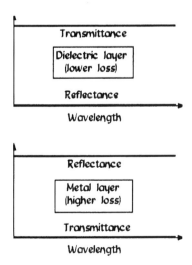

Figure 21.20 Ideal characteristics of a dielectric layer (top) and a metal layer (bottom). Dielectrics transmit well but reflect poorly. Metals reflect more and more strongly with increasing wavelength but have low transmittance.

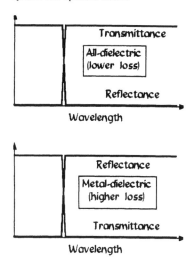

Figure 21.21 Idealized characteristics of an interference filter made up of dielectric layers (top) and of a filter made of metal and dielectric layers. The dielectric system is able to reflect by an interference process effective over a limited region only. Similarly, the metal is induced to transmit by an interference process that is likewise limited.

layer surrounded by two thin titania layers (Fig. 21.24). This is a structure that has been known for some time. Early filters of this kind used gold and bismuth oxide but the performance of silver is better. The metal assures the high reflectance at long wavelengths and the antireflection coating the transmittance at shorter wavelengths, as in the idealized Fig. 21.21. The silver layer is rather thin and there are some difficulties in achieving metallic performance from it.

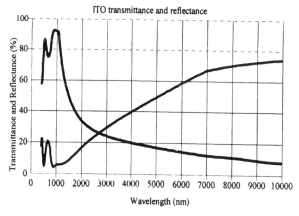

Figure 21.22 Calculated transmittance and reflectance of a layer of indium tin oxide (ITO), showing high transmittance in the visible region and gradually increasing reflectance and reducing transmittance in the infrared. The refractive index in the visible region is around 2.0 and, consequently, interference fringes are pronounced.

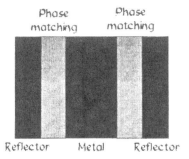

Figure 21.23 A central metal layer is antireflected by a system of phase matching layer and reflector on each side.

To improve the properties of this filter more silver is needed in the central layer and this implies a more powerful outer matching reflector. For this we make use of the properties of quarter-wave dielectric layers.

A quarter-wave dielectric layer acts as an admittance transformer. If the admittance of the emergent medium is y_{sub} and that of the film y_f, then the addition of a quarter-wave film to the substrate will transform the admittance according to the quarter-wave rule:

$$Y = \frac{y_f^2}{y_{sub}}. \tag{21.18}$$

An additional quarter-wave of magnesium fluoride on either side, making a five-layer coating, effectively reduces the admittance of the substrate and incident medium (Fig. 21.25). This increases the contrast between the titania and the outer media and increases the reflectance, permitting the use of a thicker silver layer. This thicker silver layer then reflects more strongly in the infrared and gives a steeper transition between transmitting and reflecting.

Increasing the outer reflectance still further, and therefore permitting still greater thickness of silver narrows the transmittance zone further and steepens the

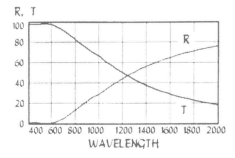

Figure 21.24 A three-layer heat-reflecting filter consisting of a silver layer surrounded by two titania layers and a glass cover cemented over it:
Glass|TiO$_2$(20 nm)Ag(10 nm)TiO$_2$(20 nm)|Glass

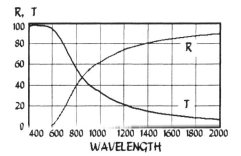

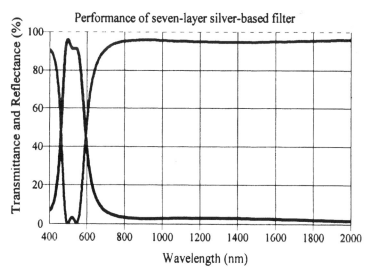

Figure 21.25 The design of Fig. 21.24 with additional quarter-waves of magnesium fluoride added between the titania layers and the glass media permitting the use of thicker silver. The additional silver makes the reflectance rise much more rapidly into the infrared, and improves the performance of the coating.

edge between transmission and reflection. This increased reflectance is achieved by adding a further titania quarter-wave layer to each side of the design of Fig. 21.25. This gives the seven-layer coating shown in Fig. 21.26, where the performance is now much more than that of a bandpass filter.

With a reflecting system consisting of four quarter-waves of titania and magnesium fluoride and the silver thickness increased to 60 nm the coating is now a narrowband filter (Fig. 21.27).

Such filters are used by themselves as useful bandpass filters in their own right and, also, because of their low infrared transmittance that is difficult to achieve in

Figure 21.26 A bandpass filter consisting of seven layers. A reflector of titania and magnesium fluoride bounds the silver and phase-matching layer system on either side:

Glass|HLH$'$(34 nm)Ag(34 nm)H$''$(34 nm)LH|Glass.

H and L represent quarter-waves at 51 nm.

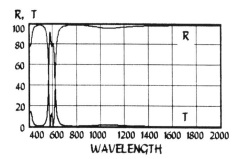

Figure 21.27 The performance of the 11-layer silver-based narrowband filter.

any other way, as blocking filters for use with other types of filters with less acceptable infrared performance such as all-dielectric narrowband filters.

A coating to transmit the infrared and reflect a band in the visible would best be constructed from dielectric materials. In the quarter-wave stack, repeated use of the quarter-wave transformer achieves high admittance mismatch and hence high reflectance (Fig. 21.28).

Ripple in the long-wave pass band is clearly a severe problem. This can be reduced by changing the thicknesses of the outermost layers to eighth waves (Fig. 21.29).

Losses in the central metal layers are the major limitations in the narrowband filters discussed so far. We can use the quarter-wave stack structure to replace the central metal layer and reduce losses. This allows us to make still narrower filters (Fig. 21.30).

Figure 21.30 has the design:

Air | HLHLHHLHLHLHLHLHHLHLH | Glass

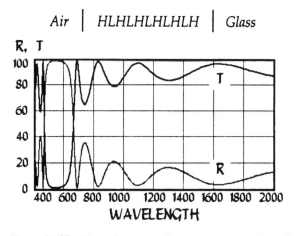

Figure 21.28 The performance of a quarter-wave stack consisting of 11 alternate quarter-wave layers of titania and silica with the titanium outermost. The characteristic is a long-wave pass filter, but the pronounced oscillatory ripple is a problem.

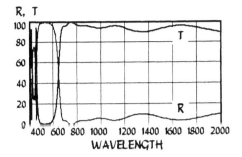

Figure 21.29 The greatly improved performance produced by the change in the thicknesses in the outermost layers of Fig. 21.28 to eighth waves.

Because the losses are so much lower, it is possible to achieve exceedingly narrow filters. The half-wave layers, those designated as HH, and derived from the original phase-matching layers, are called cavity layers because their function is really that of a tuned cavity. The filter of Fig. 21.30 is a two-cavity filter. The greater the number of cavities, the steeper become the passband sides.

A narrowband filter design involving 59 layers and three cavities is shown in Fig. 21.31.

Antireflection coatings are not strictly spectral filters in the normal sense of the word because their purpose is not to filter but to reduce losses due to residual reflection. But they are essential components of filters and so a few words about them is in order. An antireflection coating with low loss implies the use of dielectric layers.

The simplest antireflection coating is a single quarter-wave layer. The quarter-wave rule permits us to write an expression for the reflectance of a quarter-wave of material on a substrate:

$$R = \left\{ \frac{y_0 - y_f^2/y_{\text{sub}}}{y_0 + y_f^2/y_{\text{sub}}} \right\}^2, \tag{21.19}$$

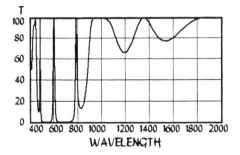

Figure 21.30 An all-dielectric narrowband filter based on the metal–dielectric arrangement of Fig. 21.27. The central metal layer is now replaced by an all-dielectric quarter-wave stack.

Glass | *HLHLHLHLH LL HLHLHLHLHLHLHLHLHLHLH LL*
HLHLHLHLHLHLHLHLHLH LL HLHLHLHLH | *Glass*

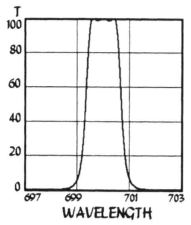

Figure 21.31   The performance of the 59-layer narrowband three-cavity filter.

and if $y_0 = y_f^2/y_{sub}$ then the reflectance will be zero. Unfortunately, suitable materials with such a low refractive index as would perfectly antireflect glass, of admittance 1.52, in air, 1.00, are lacking. Magnesium fluoride, with admittance 1.38, is the best that is available. The performance, at 1.25% reflectance, is considerably better than the 4.25% reflectance of an uncoated glass surface (Fig. 21.32).

This performance is good enough for many applications and the single-layer antireflection coating is much used. It has a characteristic magenta color in reflection, because of the rising reflectance in the red and blue regions of the spectrum.

There are applications where improved performance is required. If zero reflectance at just one wavelength, or over a narrow range, then the two-layer V-coat is the preferred solution. This consists of a thin high-index layer, say roughly one-sixteenth of a wave in optical thickness of titania, followed by roughly five-sixteenths of a

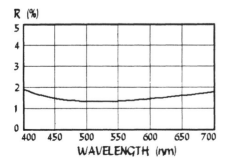

Figure 21.32   Computed performance of a single-layer antireflection coating consisting of a quarter-wave of MgF_2 on glass with air as the incident medium.

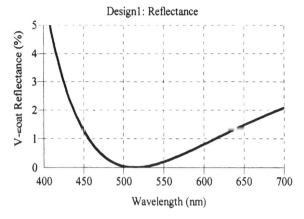

Figure 21.33 The performance of a typical V-coat, a two-layer antireflection coating to give (near) zero reflectance for just one wavelength.

wave of low-index material, magnesium fluoride or, sometimes, silica. The thicknesses are adjusted to accommodate the particular materials that are to be used in the coating. A typical performance is illustrated in Fig. 21.33.

Then there will be other applications demanding low reflectance over the visible region, 400–700 nm. These requirements are usually met with a four-layer design based on the V-coat with a half-wave flattening layer inserted just one quarter-wave behind the outer surface. The design is of the form

Air | LHHL′H | Glass

A slight adjustment by refinement then yields the performance shown in Fig. 21.34.

The ultimate antireflection coating is a layer that represents a gradual, smooth transition from the admittance of the substrate to the admittance of the incident medium. Provided this layer is thicker than a half-wave, the reflectance

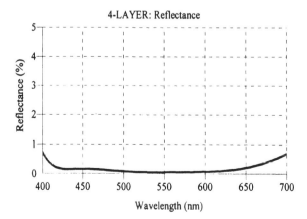

Figure 21.34 The performance of the four-layer two-material design over the visible region.

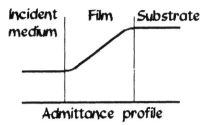

Figure 21.35 Sketch of the variation of optical admittance as a function of optical thickness through an inhomogeneous matching layer.

will be very low. This implies that for all wavelengths shorter than a maximum, the layer will be virtually a perfect antireflection coating. The profile of such a layer is shown in Fig. 21.35. Figure 21.36 shows the performance as a function of g, i.e., λ_0/λ, where λ_0 is the wavelength for which the layer is a half-wave. Best results are obtained if the derivative of admittance as a function of distance can be smoothed along with the admittance. This implies a law of variation that is a fifth-order polynomial.

For an air incident medium the inhomogeneous layer can be achieved only by microstructural variations (Fig. 21.37). Processes involving etching, leaching, sol–gel deposition, and photolithography have all been used. This does give the expected reduction in reflectance, but as the wavelength gets shorter, the microstructural features become comparatively larger and so there is a short-wave limit determined by scattering. For a low reflectance range of reasonable width, the features must be very long and thin. Because of the inherent weakness of the film, this solution is limited to very special applications.

The quarter-wave stack, already mentioned, is a useful rejection filter for limited spectral regions. Unfortunately, for some applications, the interference conditions that assures high reflectance repeats itself for discrete bands of shorter wavelengths. Thus, the first-order reflection peak at λ_0 is accompanied by peaks at $\lambda_0/3, \lambda_0/5$, and so on. There are many applications where such behavior is unde-

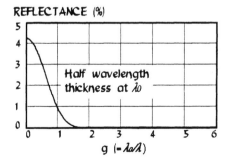

Figure 21.36 The reflectance of an inhomogeneous layer with fifth-order polynomial variation of admittance throughout. The substrate has admittance 5.1 and the incident medium, 1.00.

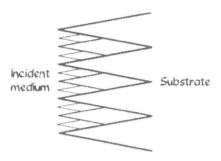

Figure 21.37 Microstructural approach to an inhomogeneous layer.

sirable. The inhomogeneous layer can be used to advantage in the suppression of these higher-order peaks because it can act as an antireflection coating that suppresses all reflection at wavelengths shorter than a long-wavelength limit. A rugate filter, the word *filter* often being omitted, consists of essentially a quarter-wave stack that has been antireflected by an inhomogeneous layer at each of the interfaces between the original quarter-wave layers. The result is a cyclic variation of admittance that can be considered close to sinusosidal throughout the structure. All higher orders of reflectance are suppressed. The width of the zone of high reflectance, now only the fundamental, can be adjusted by varying the amplitude of the cycle of admittance. The smaller this amplitude, the narrower the reflectance peak. Of course when the amplitude is small many cycles are needed to achieve reasonable reflectance (Fig. 21.38). Rugate filters are used in applications where a narrow line must be removed from a background. They can simply be suppressing a bright laser line, as in Raman spectroscopic applications, or they can be actually

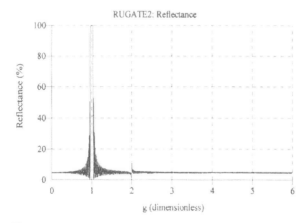

Figure 21.38 Performance of a rugate filter comprising 40 cycles of sinusoidal admittance variation from 1.45 to 1.65. The horizontal axis is in terms of g, which is a dimensionless variable given by λ_0/λ, where λ_0 is the reference wavelength. Note the absence of higher-order peaks. The slightly more pronounced oscillation at $g = 2$ is a real feature of the design.

reflecting the line and transmitting all others. Head-up displays can make use of narrowband beamsplitters of this type.

For more information on thin film filters, see Dobrowolski [10] or Macleod [11].

REFERENCES

1. Heavens, O. S. and R. W. Ditchburn, *Insight into Optics*, John Wiley and Sons, Chichester, 1991.
2. O'Shea, D. C., *Elements of Modern Optical Design*, 1st edn, Wiley Series in Pure and Applied Optics, Goodman, J. W., ed., John Wiley and Sons, New York, 1985, p. 402.
3. VanderLugt, A., *Optical Signal Processing*, 1st edn, Wiley Series in Pure and Applied Optics, Goodman, J. W., ed., John Wiley and Sons, New York, 1992, p. 604.
4. Hecht, E. and A. Zajac, *Optics*, 1st edn, Addison-Wesley, Reading, Massachussetts, 1974, p. 565.
5. Piegari, A., "Graded Coatings," in *Thin Films for Optical Systems*, Flory, F., ed., Marcel Dekker, New York, 1995, pp. 475–519.
6. Dobrowolski, J. A., "Coatings and Filters," in *Handbook of Optics*, Driscoll, W. G. and W. Vaughan, eds, McGraw-Hill, New York, 1978, pp. 8.1–8.124.
7. Dobrowoski, J. A., G. E. Marsh, D. G. Charbonneau, J. Eng, and P. D. Josephy, "Colored Filter Glasses: An Intercomparison of Glasses Made by Different Manufacturers," *Applied Optics*, **16**(6), 1491–1512 (1977).
8. Yariv, A. and P. Yeh, "Optical Waves in Crystals," 1st edn, in *Pure and Applied Optics*, Goodman, J. W., ed., John Wiley and Sons, New York, 1984, p. 589.
9. Chang, I. C., "Acousto-Optic Devices and Applications," in *Handbook of Optics*, Bass, M., *et al.*, eds, McGraw-Hill, New York, 1995, pp. 12.1–12.54.
10. Dobrowolski, J. A., "Optical Properties of Films and Coatings," in *Handbook of Optics*, Bass, M., *et al.*, eds, McGraw-Hill, New York, 1995, pp. 42.1–42.130.
11. Macleod, H. A., *Thin-Film Optical Filters*, 2nd edn, Adam Hilger, Bristol, 1986, p. 519.

22

Optical Fibers and Accessories

A. N. STARODUMOV
Centro de Investigaciones en Optica, León, Mexico

22.1 INTRODUCTION

Historically, light-guiding effects were demonstrated in the mid-19th century when Swiss physicist D. Collodon and French physicist J. Babinet showed that light can be guided in jets of water for fountain displays. In 1854 the British physicist J. Tyndall demonstrated this effect in his popular lectures on science, guiding light in a jet of water flowing from a tank. In 1880, in Massachusetts, an engineer W. Wheeler patented a scheme for piping light through buildings. He designed a net of pipes with reflective linings and diffusing optics to carry light through a building. Wheeler planned to use light from a bright electric arc to illuminate distant rooms. However, this project was not successful. Nevertheless, the idea of light piping reappeared again and again until it finally converted into the optical fiber.

During the 1920s, J. L. Baird in England and C. W. Hansell in the United States patented the idea of image transmission through arrays of hollow pipes or transparent rods. In 1930, H. Lamm, a medical student in Munich, had demonstrated image transmission through a bundle of unclad optical fibers. However, the major disadvantage of this device was the poor quality of the transmitted images. Light "leaked" from fiber to fiber, resulting in degradation of the quality of the image.

Modern optical fibers technology began in 1954 when A. van Heel of the Technical University of Delft in Holland covered a bare fiber of glass with a transparent cladding of lower refractive index. This protected the reflection surface from contamination, and greatly reduced cross-talk between fibers.

The invention of the laser in 1957 provided a promising light source for an optical communication system and stimulated research in optical fibers. In 1966 Kao and Hockman pointed out that purifying glass could dramatically improve its transmission properties. In 1970 the scientists from Corning Glass reported fibers made from extremely pure fused silica with losses below 20 dB/km at 633 nm. Over the next few years fiber losses dropped dramatically. In 1976, Bell Laboratories combined in a laboratory experiment all the components needed for an optical communication system, including lasers, detectors, fibers, cables, splices, and connectors. Since 1980, the growth in the number of installed communications systems has been extremely rapid. The improvement of the quality of optical fibers and fast development of the fiber analogs of bulk optical elements stimulated nontelecommunication applications of fiber optics in such areas as aircraft and shipboard control, sensors, optical signal processors, displays, delivery of high-power radiation, and medicine.

Starting in the 1970s a number of good books have been published that discuss the theoretical aspects of fiber optics. In the last ten years practical guides of fiber optics for telecommunication applications have appeared. The objective of this chapter is to describe the principles of fiber optics, with an emphasis on basic fiber optical elements such as fibers, spectrally selective and polarization-sensitive fiber elements, and couplers.

22.2 OPTICAL FIBERS

22.2.1 Single-Mode and Multimode Fibers

Optical fiber is the medium in which radiation is transmitted from one location to another in the form of guided waves through glass or plastic fibers. A fiber waveguide is usually cylindrical in form. It includes three layers: the center core that carries the light, the cladding layer covering the core, and the protection coating. The core and cladding are commonly made from glass, while the coating is plastic or acrylate.

Figure 22.1 shows a fiber structure. The core of radius a has a refractive index n_1. The core is surrounded by a dielectric cladding with a refractive index n_2 that is less than n_1. The silica core and cladding layers differ slightly in their composition due to small quantities of dopants such as boron, germanium, or fluorine. Although light can propagate in the core without cladding, the latter serves several purposes. The cladding reduces scattering loss on a glass–air surface, adds mechanical strength to the fiber, and protects the core from surface contaminants. An elastic plastic material that encapsulates the fiber (buffer coating) adds further strength to the fiber and mechanically isolates the fiber from adjacent surfaces, protecting from physical damage and moisture.

Glass optical fibers can be *single-mode fibers* (SMFs) or *multimode fibers* (MMFs). A single-mode fiber sustains only one mode of propagation and has a small core of 3–10 μm. In an MMF, the core diameter is usually much larger than that of an SMF. In an MMF, light can travel many different paths (called modes)

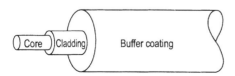

Figure 22.1 Fiber structure.

through the core of the fiber. The boundary between the core and the cladding may be sharp (*step-index profile*) or graduated (*graduated-index profile*). The core index profile of an SMF is usually a step-index type (Fig. 22.2(a), while in an MMF it can be either step-index or a graded-index type (Fig. 22.2(b) and (c)). Step-index fiber has a core composed completely of one type of glass. Straight lines, reflecting off the core–cladding interface can describe mode propagation in such a fiber. Since each mode travels at different angles to the axis of the fiber and has a different optical path, a pulse of light is dispersed while traveling along the fiber. This effect called *intermodal dispersion* limits the bandwidth of a step-index fiber.

In graded-index fibers, the core is composed of many different layers of glass. The index of refraction of each layer is chosen to produce an index profile approximating a parabola with maximum in the center of the core (Fig. 22.2(c)). In such a fiber, low-order modes propagate close to the central part of the core with higher refractive index. Higher-order modes, although traversing a much longer distance than the central ray, does so in a region with less refractive index and hence the velocities of these modes are greater. Thus, the effects of these two factors can be made to cancel each other out, resulting in very similar propagation velocities. A properly constructed index profile will compensate for different path lengths of each mode, increasing the bandwidth capacity of the fiber by as much as 100 times over that of step-index fiber.

Multimode fibers offer several advantages over single-mode fibers. The larger core radius of MMFs makes it easier to launch optical power into the fiber and facilitate the connecting together of similar fibers. Another advantage is that radia-

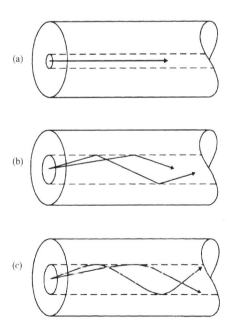

Figure 22.2 Single-mode and multimode fibers: (a) single-mode step-index fiber, (b) multimode-step-index fiber, and (c) multimode graded-index fiber.

tion from light-emitting diodes can be efficiently coupled to an MMF, whereas SMFs must generally be excited with laser diodes. However, SMFs offer higher bandwidths in communication applications.

22.2.2 Fiber Modes

When light travels through a medium with a high refractive index n_1 to a medium with a lower refractive index n_2, the optical ray is refracted at the boundary into the second medium. According to the Snell's law $n_1 \sin \alpha_1 = n_2 \sin \alpha_2$, the angle of refraction α_2 is greater than the angle of incidence α_1 in this case. As the incident angle increases, a point is reached at which the optical ray is no longer refracted into the second medium ($\alpha_2 = \pi/2$). The optical radiation is completely reflected back into the first medium. This effect is called *total internal reflection.*

When total internal reflection occurs, a phase shift is introduced between the incident and reflected beams. The phase shift depends on polarization of the incident beam and on the difference of refractive indices. Two different cases should be considered: when electric field vector **E** is in the plane of incidence ($E_{\parallel}$), and when **E** is perpendicular to this plane ($E_{\perp}$). These two situations involve different phase shifts when total internal reflection takes place, and hence they give rise to two independent sets of modes. Because of the directions of **E** and **H** with respect to the direction of propagation down the fiber, the two sets of modes are called transverse magnetic (TM) and transverse electric (TE), which correspond to the $E_{\parallel}$ and $E_{\perp}$, respectively. Two integers, l and m, are required to completely specify the modes. This is because the waveguide is bounded in two dimensions. Thus we refer to TM_{lm} and TE_{lm} modes.

In fiber waveguides the core–cladding index difference is rather small and does not exceed a few percent. The phase shift acquired under total internal reflection is practically equal for both sets of modes. Thus, the full set of modes can be approximated by a single set called *linearly polarized* (LP_{lm}) modes. Such a mode in general has m field maxima along a radius vector and $2l$ field maxima round a circumference. Figure 22.3 shows an electric field distribution for LP_{01} and LP_{21} modes.

22.2.3 Fiber Parameters

The light can enter and leave a fiber at various angles. The maximum angle (acceptance angle θ_{max}) that supports total internal reflection inside the fiber defines the

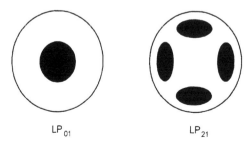

LP_{01} LP_{21}

Figure 22.3   Electric field distribution for LP_{01} mode (a) and LP_{21} mode (b).

numerical aperture (NA). Numerical aperture can be expressed in terms of core–cladding refractive index difference as

$$\text{NA} = n_0 \sin \theta_{\max} = (n_1^2 - n_2^2)^{1/2} \approx n_1 (2\Delta)^{1/2}, \tag{22.1}$$

where $\Delta = (n_1 - n_2)/n_1$ is the *relative (or normalized) index difference* and n_0 is the refractive index. This representation is valid when $\Delta \ll 1$. Since the numerical aperture is related to the maximum acceptance angle, it is commonly used to describe the fiber characteristics and to calculate source-to-fiber coupling efficiencies.

A second important fiber parameter is the normalized frequency V, which defines the number of modes supported by a fiber. This parameter depends on optical wavelength λ, a core radius a, and a core–cladding refractive index difference, and can be written as

$$V = \frac{2\pi a}{\lambda} \text{NA}. \tag{22.2}$$

With the parameters V, Δ, and index profile optical fibers can be classified more precisely. For $V < 2.405$, the fiber sustains only one mode and is a single mode. Multimode fibers have values $V > 2.405$ and can sustain many modes simultaneously. The number 2.405 corresponds to the first zero of the Bessel function, which appears in the solution of the wave equation for the fundamental mode in a cylindrical waveguide. A fiber can be multimode for short wavelengths and, simultaneously, can be single mode for longer wavelengths. The wavelength corresponding to $V = 2.405$ is known as the cutoff wavelength of the fiber and is given by

$$\lambda_c = \frac{2\pi a}{2.405} \text{NA}. \tag{22.3}$$

The fiber is single mode for all wavelengths longer than λ_c and is multimode for shorter wavelengths. By decreasing the core diameter or the relative index difference such that $V < 2.405$, the cutoff wavelength can be shifted to shorter wavelengths, and single-mode operation can be realized. Usually the fiber diameters are greater for longer wavelengths.

The small core diameter of an SMF makes it difficult to couple light into the core and to connect two fibers. To increase the effective core diameter, fibers with multiple cladding layers were designed. In such a fiber, two layers of cladding, an inner and an outer cladding surround its core with a barrier in between. The index profiles of single and multiple cladding fibers are shown in Fig. 22.4(a) and (b). The multiple cladding schemes permit the design of an SMF with a relatively larger core diameter, facilitating fiber-handling and splicing. The additional barrier provides an

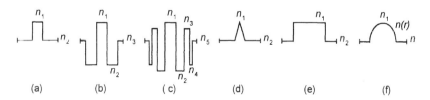

Figure 22.4 Typical refractive index profiles of fibers.

efficient control of dispersion properties of a fiber and permits us to vary the total dispersion of an SMF.

A more complicated index profile called a quadrupole clad (Fig. 24.4(c)) gives even more flexibility in handling of fiber dispersion. The quadrupole-clad fiber has low dispersion (< 1 ps/km-nm) over a wide wavelength range extending from $1.3\,\mu$m to $1.6\,\mu$m.

Figure 22.4(d) shows a triangle index profile. A fiber with triangle profile called a T-fiber provides a much higher second-order mode cutoff wavelength (for LP_{11} mode) and lower attenuation than does a step-index fiber.

If the parameter V increases much above 2.405, the step-index fiber can support a large number of modes. The maximum number of modes propagating in the fiber can be calculated as

$$N_{\text{mode}} \approx \frac{V^2}{2}. \tag{22.4}$$

When many modes propagate through a fiber, carrying the same signal but along different paths, the output signal is the result of interference of different modes. If a short pulse enters into the fiber, at the output this pulse has longer length because of the temporal delay between carrying modes. As has been mentioned above, this effect is called intermodal dispersion. The quality of the signal deteriorates as the fiber length increases, limiting the information capacity of the step-index fiber.

To reduce the effect of intermodal dispersion, a graded-index fiber has been proposed. A typical core profile is shown in Fig. 22.4(f). The most often used index profile is the power law profile designed according to the following expression

$$n(r) = n_1\left[1 - 2\Delta\left(\frac{r}{a}\right)^g\right]^{1/2}, \tag{22.5}$$

where g is usually chosen to reduce the effect of intermodal dispersion. For optimum effect

$$g = 2 - \left(\tfrac{12}{5}\right)\Delta. \tag{22.6}$$

The refractive index of the cladding is maintained at a constant value.

22.2.4 Optical Losses (Attenuation)

Light traveling in a fiber loses power over distance. If P_0 is the power launched at the input of a fiber, the transmitted power P_t at a distance L is given by

$$P_t = P_0 \exp(-\alpha L), \tag{22.7}$$

where α is the attenuation constant, known also as the fiber loss. Attenuation is commonly measured in decibels (dB). In fibers, the loss is expressed as attenuation per 1 km length, or dB/km, using the following relationship:

$$\alpha_{dB} = -10\frac{10}{L}\log\left[\frac{P_t}{P_0}\right] = 4.343\alpha. \tag{22.8}$$

Fiber loss is caused by a number of factors that can be divided into two categories: *induced* and *inherent* losses. The induced losses may be introduced during manufacturing processes. These losses are caused by inclusions of contaminating atoms or ions, by geometrical irregularities, by bending and microbending, by splicing, by connectors, and by radiation. The fabrication process is aimed at reducing these losses as much as possible.

Bend losses occur due to the change of the angle of incidence at the core–cladding boundary and depend strongly on the radius of curvature. The loss will be greater (a) for bends with smaller radii of curvature, and (b) for those modes that extend most into the cladding. The bending loss can generally be represented by a loss coefficient α_B, which depends on fiber parameters and radius of curvature R, and is given as

$$\alpha_B = C \exp\left(-\frac{R}{R_c}\right) \tag{22.9}$$

where C is a constant, R_c is given by $R_c = r/(NA)^2$, and r is the fiber radius. As is seen from Eq. (22.9), the attenuation coefficient depends exponentially on the bend radius. Thus, decreasing the radius of curvature drastically increases the bending losses. The bending loss decreases as the core–cladding index difference increases. The optical radiation at wavelengths close to cutoff will be affected more than that at wavelengths far from cutoff.

Bending loss can also occur on a smaller scale due to fluctuations of core diameter and perturbations in the size of the fiber, caused by buffer, jacket, fabrication, and installation practice. This loss is called microbending and can contribute significantly over a distance.

Splice and connector loss can also add to the total induced loss. The mechanism of these losses will be discussed below.

Light loss that cannot be eliminated during fabrication process is called inherent losses. The inherent losses have two main sources: (a) Rayleigh scattering and (b) ultraviolet and infrared absorption losses.

Scattering

Glass fibers have a disordered structure. Such a disorder results in variations in optical density, composition, and molecular structure. These types of disorder, in turn, can be described as fluctuations in refractive index of the material. If the scale of these fluctuations is of the order of $\lambda/10$ or less, then each irregularity can be considered as a point scattering center. The light is scattered in all directions by each point center. The intensity of scattered light and the power loss coefficient vary with wavelength as λ^{-4}. The term λ^{-4} is the characteristic wavelength-dependence factor of Rayleigh scattering. The addition of dopants into the silica glass increases the scattering loss because the microscopic inhomogeneities become more important. Rayleigh scattering is a fundamental process limiting the minimum loss that can be obtained in a fiber.

Absorption

The absorption of light in the visible and near-infrared regions at the molecular level arises mainly from the presence of impurities such as transition metal ions (Fe^{3+}, Cu^{2+}) or hydroxyl (OH^-) ions. The OH radical of the H_2O molecule vibrates at a fundamental frequency corresponding to the infrared wavelength of 2.8 μm. Because the OH radical is an anharmonic oscillator, the overtones can occur, producing absorption peaks at 0.725 μm, 0.95 μm, 1.24 μm, and 1.39 μm. Special precautions must be taken during the fiber-fabrication process to diminish the impurity concentrations. The OH concentration should be kept at levels below 0.1 ppm if ultralow losses are desired in the 1.20–1.60 µm range.

Ultraviolet absorption produces an absorption tail in the wavelength region below 1 µm. This absorption decreases exponentially with increasing wavelength and is often negligible in comparison with Rayleigh scattering within the visible wavelength range.

At wavelengths greater than about 1.6 µm the main contribution to the loss is due to transitions between vibrational states of the lattice. Although the fundamental absorption peaks occur at 9 µm, overtones and combinations of these fundamental vibrations lead to various absorption peaks at shorter wavelengths. The tails of these peaks result in typical values of 0.02 dB/km at 1.55 µm, and 1 dB/km at 1.77 µm. Figure 22.5 shows the total loss coefficient (solid curve) as a function of wavelength for a silica fiber. The absorption peak at 1.39 µm corresponds to OH radicals. A properly chosen fiber-fabrication process permits suppression of this peak. The inherent loss level is shown by a dashed curve. The minimum optical losses of 0.2 dB/km have been obtained with GeO_2-doped silica fiber at 1.55 µm.

22.2.5 Dispersion, Fiber Bandwidth

When an optical pulse propagates along a fiber, the shape of the pulse changes. Specifically, a pulse of light gets broader. There are three main sources of such changes: *intermodal* dispersion, *material* dispersion, and *waveguide* dispersion. Intermodal dispersion can be avoided by using single-mode fibers or can be diminished by using graded-index multimode fibers.

Material (or chromatic) dispersion is due to the wavelength dependence of the refractive index. On a fundamental level, the origin of material dispersion is related to the electronic absorption peaks. Far from the medium resonances, the refractive index is well approximated by the Sellmeier equation:

$$n^2(\omega) = 1 + \sum_{i=1}^{l} \frac{B_i \omega_i^2}{\omega_i^2 - \omega^2}, \tag{22.10}$$

where ω_i is the resonance frequency and B_i is the strength of ith resonance. For bulk fused silica these parameters are found to be $B_1 = 0.6961663$, $B_2 = 0.4079426$, $B_3 = 0.8974794$, $\lambda_1 = 0.0684043$ µm, $\lambda_2 = 0.1162414$ µm, $\lambda_3 = 9.896161$ µm, where

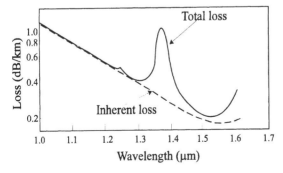

Figure 22.5 The total loss coefficient as a function of wavelength for a silica fiber.

$\lambda_i = 2\pi c/\omega_i$. Fiber dispersion plays an important role in propagation of short optical pulses since the different spectral components travel at different velocities, resulting in pulse broadening. Dispersion-induced pulse broadening can be detrimental for optical communication systems.

Mathematically, the effects of fiber dispersion are accounted for by expanding the mode-propagation constant $\beta(\omega)$ in a Taylor series near the central frequency ω_0:

$$\beta(\omega) = n(\omega)\frac{\omega}{c} = \beta_0 + \beta_1(\omega - \omega_0) + \frac{1}{2}\beta_2(\omega - \omega_0)^2 + \cdots, \tag{22.11}$$

where

$$\beta_m = \left(\frac{d^m \beta}{d\omega^m}\right)_{\omega_0} \qquad (m = 0, 1, 2, \ldots) \tag{22.12}$$

The parameter β_1 characterizes a group velocity $v_g = d\omega/d\beta = (\beta_1)^{-1}$. Although individual plane waves travel with a phase velocity $v_p = \omega/\beta$, a signal envelope propagates with a group velocity. The parameter β_2 is responsible for pulse broadening.

The transit time required for a pulse to travel a distance L is $\tau = L/v_g$. Since the refractive index depends on wavelength, the group velocity is also a function of λ. The travel time per unit length τ/L may be written as

$$\frac{\tau}{L} = \frac{1}{c}\left(n - \lambda\frac{dn}{d\lambda}\right). \tag{22.13}$$

The difference in travel time $\Delta\tau$ for two pulses at wavelengths λ_1 and λ_2, respectively, is a measure of dispersion. In a dispersive medium, the optical pulse of a spectral width $\Delta\lambda$, after traveling a distance L, will spread out over a time interval

$$\Delta\tau = \frac{d\tau}{d\lambda}\Delta\lambda. \tag{22.14}$$

The derivative $d\tau/d\lambda$ describes the pulse broadening and can be expressed through parameter β_2 as

$$\frac{1}{L}\frac{d\tau}{d\lambda} = -\frac{\lambda}{c}\frac{d^2 n}{d\lambda^2} = -\frac{2\pi c}{\lambda^2}\beta_2 = D. \tag{22.15}$$

The parameter D (in units ps/nm-km), called also dispersion parameter or dispersion rate, is commonly used in fiber-optics literature instead of β_2. The parameter β_2 is generally referred to as the group velocity dispersion (GVD) coefficient. For a bulk silica, β_2 vanishes with wavelength and is equal to zero at 1.27 μm. This wavelength is often called zero-dispersion wavelength λ_D. An interesting fact is that the sign of the dispersion term changes in passing through zero-dispersion point.

Waveguide dispersion is due to the dependence of propagation constant on fiber parameters when the index of refraction is assumed to be constant. The reason for this is that the fiber parameters, such as the core radius and the core–cladding index difference, cause the propagation constant of each mode to change for different wavelengths. The sign of waveguide dispersion is opposite to the sign of material dispersion at wavelengths above 1.3 μm. This feature is used to shift the zero-dispersion wavelength λ_D in the vicinity of 1.55 μm, where the fiber loss has a minimum value. Such *dispersion-shifted* fibers have found numerous applications in optical communication systems. It is possible to design *dispersion-flattened* optical fibers having low dispersion ($|D| \leq 1$ ps/nm-km) over a relatively large wavelength range. This is achieved by the use of multiple cladding layers. Figure 22.6 shows the dependence of the dispersion parameter D on wavelength for a single-clad fiber (dashed curve) and a quadrupole-clad fiber (solid curve) with flattened dispersion in the wavelength range extending from 1.25 to 1.65 µm.

Fiber Bandwidith

Optical fiber bandwidth is a measure of the information-carrying capacity of an optical fiber. The fiber's total dispersion limits the bandwidth of the fiber. This occurs because pulses distort and broaden, overlapping one another and become indistinguishable at a receiver. To avoid an overlapping, pulses should be transmitted at less repetition rate (thereby reducing bit rate). Use of these terms (bandwidth and bit rate) is technically difficult because of two factors: link length and dispersion. To calculate a desired bandwidth or bit rate, the fiber provider must know the length of the link. In addition, the provider does not know the spectral bandwidth of the optical source to be used in the system. The spectral bandwidth of the light source determines the amount of chromatic dispersion in the link. Because of these two difficulties, instead of terms "bandwidth" and "bit rate" two other terms are used: *bandwidth-distance product*, in MHz-km, for multimode fibers; *dispersion rate*, in ps/nm/km, for single-mode fibers.

The bandwidth-distance product is the product of the fiber length and the maximum bandwidth that the fiber can transmit. For example, a fiber with a bandwidth-distance product of 100 MHz-km, can transmit a signal of 50 MHz over a distance of 2 km or a signal of 100 MHz over a distance of 1 km. It should be noted that graded-index MMF have an information-carrying capacity 30 to 50 times greater than step-index MMF because of diminished intermodal dispersion. In single-mode fibers, the information-carrying capacity is approximately two orders of magnitude greater than that of graded-index MMF.

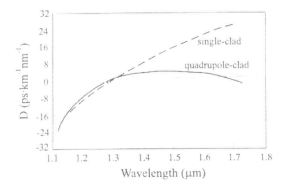

Figure 22.6 Dispersion parameter D as a function of wavelength for a single-clad fiber (dashed curve) and a quadrupole-clad fiber (solid curve).

22.2.6 Typical Fibers and Fiber Parameters

Optical fibers used for telecommunications and other applications are manufactured with different core and cladding diameters. Fiber size is specified in the format "core/cladding." A 100/140 fiber means the fiber has a core diameter of 100 μm and a cladding diameter of 140 μm. A polymer coating covers the cladding and can be either 250 or 500μm. For a tight-buffered cable construction, a 900-μm-diameter plastic buffer covers the coating. Table 22.1 shows typical fiber core, cladding and coating diameters.

Most fibers have a glass core and glass or plastic cladding. These fibers can be classified in four types: all glass, plastic clad silica, hard clad silica, and plastic optical fibers.

Table 22.1   Fiber Parameters

Core/cladding/ coating diameter (μm)	Wavelength (nm)	Optical loss (dB/km)	Bandwidth–distance product (MHz-km)	Numerical aperture
2.4/65/190	400	60	—	0.13
3.3/80/200	500	22	—	0.13
4.0/125/250	630	10	—	0.12
5.5/125/250	820	3.5	—	0.12
6.6/125/250	1060	2	—	0.13
6.6/80/200	1300	1	—	0.16
7.8/125/250	1550	1	—	0.16
9/125/250 or 500	780	—	< 800	0.11
	850	—	2000	
	1300	0.5–0.8	20000	
	1550	0.2–0.3	4000–20000	
50/125/250 or 500	780	4.0–8.0	150–700	0.20
	850	3.0–7.0	200–800	
	1300	1.0–3.0	400–1500	
	1550	1.0–3.0	300–1500	
62.5/125/250 or 500	780	4.0–8.0	100–400	0.275
	850	3.0–7.0	100–400	
	1300	1.0–4.0	200–1000	
	1550	1.0–4.0	150–500	
100/140/250 or 500	780	4.5–8.0	100–400	0.29
	850	3.5–7.0	100–400	
	1300	1.5–5.0	100–400	
	1550	1.5–5.0	10–300	
110/125/250 or 500	780	—	—	0.37
	850	15.0	17	
	1300	—	—	
	1550	—	—	
200/230/500	780	—	—	0.37
	850	12.0	17	
	1300	—	—	
	1550	—	—	

Glass Fibers

The most popular fibers are all-glass fibers, especially single-mode fibers. These fibers are widely used because of low attentuation rates and high information-carrying capacity. Single-mode fibers are less expensive than multimode fibers, but optoelectronic elements and connectors for single-mode systems are more expensive than those for multimode systems. The majority of single-mode fibers have a core diameter of 5–10 μm and a cladding diameter of 125 μm.

Most multimode telecommunications fibers are graded-index fibers with a cladding diameter of 125 μm. The typical core diameters with the 125 μm cladding are 50 μm, 62.5 μm, 85 μm, and 110 μm.

The 50/125-μm diameter fiber has a low numerical aperture of 0.2 but the highest bandwidth-distance product between MMF. The 62.5/125-μm fiber is the most popular for multimode transmission; its higher numerical aperture means that, this fiber provides better light-coupling efficiency and is less sensitive to microbending losses. The large core diameter fibers, such as the 85/12-, 110/125-, and 100/140-μm fibers, have a good light-coupling ability, but have less bandwidth-distance product than fibers with small core diameter. Table 22.1 shows typical fiber optical losses, numerical apertures, and bandwidth-distance products. It should be noted that there are other fibers with larger core diameters, which find applications in fiber sensors and medicine.

Plastic-Clad Silica (PCS)

PCS consists of a step-index silica core surrounded by a soft plastic cladding of silicone rubber. This fiber combines the low attenuation of a glass core with a soft plastic cladding; however, this fiber needs a buffer coating to protect the soft cladding.

Hard Clad Silica (HCS)

Hard clad silica fiber includes a step-index silica core surrounded by a hard plastic cladding. This plastic cladding has important advantages compared to a glass cladding, providing a high strength of the fiber, small bending radius, and a high resistance to surface damage.

Plastic Optical Fibers (POF)

Plastic fibers have a step-index or graded-index core surrounded by a plastic clad. The core material of POF is normally made of acrylic resin, while the cladding is made of fluorinated polymer. The POF diameter is usually of around 1 mm, which is many times larger than a glass fiber and the light-transmission core section accounts for the 96% of the cross-sectional area. Compared with all glass fibers POFs do not suffer from the problem of breakage. In Section 22.3.6 POFs are presented in more detail.

22.3 SPECIAL FIBERS

22.3.1 Erbium-Doped Fibers

Incorporating rare-earth elements into glass gives the resulting material new optical properties that allow the material to perform amplification and generation of optical light. Doping can be done both for silica and for halide glasses.

Three commonly used rare-earth materials for silica fiber lasers are erbium, neodymium, and ytterbium. Erbium-doped fibers have been a key element in the transformation of modern optical communication systems. Erbium-doped fiber amplifiers and lasers operating at a wavelength of 1.55 μm have attracted most attention because their amplification band coincides with the least-loss region of silica fibers used for telecommunication systems. In particular, erbium-doped fiber amplifiers (EDFAs) are used for the amplification of lightwave signals purely in the optical domain. They can be used as power amplifiers to boost transmitted power, as repeaters or in-line amplifiers to increase the transmission distance, or as preamplifiers to enhance receiver sensitivity. Figure 22.7 shows the basic configuration of an EDFA. The wavelength-division multiplexer (WDM) combines the light from the high-pump power laser diode (with wavelength of 980 nm or 1480 nm) and the signal to be amplified (in the wavelength region of 1530–1570 nm) into an Er-doped silica fiber. The optical isolators prevent any back reflections. To adjust the gain of EDFA, a part of the output signal is compared with the reference level. The produced control signal goes back to the pump diode to adjust the current.

The key element in an EDFA is a short-length (5–200 m) silica fiber doped with about 200 mole ppm or erbium, which corresponds to an erbium concentration of about 10^{19} ions/cm^3. The gain characteristics of EDFAs depend on the pumping scheme as well as on the various codopants (GeO_2, Al_2O_3, and P_2O_5) that are used to make the fiber core. Table 22.2 shows typical parameters of an erbium-doped fiber (available from 3M). Efficient pumping may be obtained by using semiconductor lasers operating at 980 nm and 1480 nm, where the excited state absorption (ESA) – the excitation to higher levels than pump wavelength – is not present. A broadband gain of EDFAs permits amplification of multiple optical channels in the bandwidth that ranges from 1 to 5 THz (~ 40 nm).

One of the advantages of EDFA amplifiers over electronic amplifiers is that EDFAs can amplify many signals at different wavelengths, which is used to expand the capacity of fiber-optic communication systems. Since the optical signals are directly amplified without conversion to electrical signals, the amplifier will work efficiently even at higher bit rates. This is in contrast to electronic repeaters, which work only at the fixed bit rate. A small signal gain of 40 dB can be achieved in EDFAs, while the noise, added by the amplifier, is close to the lowest level (3–4 dB). It should be noted also that the gain is polarization-insensitive, providing equal amplification for all polarization states of a signal. In long transmission sys-

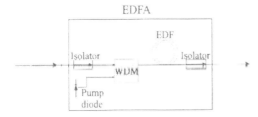

Figure 22.7 Basic configuration of an erbium-doped fiber amplifier: EDF = erbium-doped fiber, WDM = wavelength-division multiplexer).

Table 22.2 Parameters of an Erbium-Doped Fiber (3M)

Pumping wavelength (nm)	Operating wavelength (nm)	Core/cladding/ jacket diameter (μm)	Attenuation maximum (dB/km)	Numerical aperture
980–1480	1530–1560	5/125/245	15 (1200 nm)	0.28

tems, EDFAs are used to periodically restore the power level, after it has decreased due to attenuation in the fiber.

22.3.2 Powerful Double-Clad Fibers

The power levels generated by conventional fiber lasers, pumped with diode sources that couple light directly into the single-mode core, are relatively low, and currently limited to fractions of a watt. Double-clad fibers offer a solution to increasing the amount of pump power in a fiber laser. Double-clad fibers comprise a rare-earth-doped single-mode core within a multimode waveguide, which enables light pumping from a low-brightness multimode pump source such as a diode array to be efficiently absorbed by a single mode core. The geometry of a high-power Yb^{3+} cladding-pumped fiber laser is shown in Fig. 22.8. The inner rectangular silica cladding with refractive index $n = 1.46$ acts as a waveguide for the pump light. In Fig. 22.8 the silica rectangular waveguiding region has dimensions 360×120 μm, and is referred to as the pump cladding. The noncircular shape of the pump cladding eliminates helical rays, which have poor overlap with the core. The pump-cladding region is typically surrounded by a low-index polymer ($n = 1.39$) which acts as a cladding for the inner cladding, providing a high numerical aperture (NA = 0.48) for the rectangular waveguide. This permits an efficient light coupling from a diode laser into the inner cladding. A second protective polymer surrounds the low-index polymer.

The pumping light is absorbed when optical rays cross the rare-earth-doped single-mode core. Then the excited ions emit the light at lower frequency, which is amplified through stimulated emission in the single-mode core with much lower numerical aperture. Thus, pumping double-clad fiber lasers with low-brightness

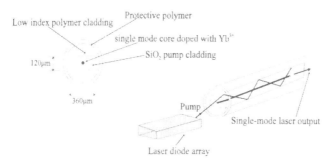

Figure 22.8 Double-clad fiber structure.

beams from a pump laser diode array may result in enhancing of the brightness by a factor in excess of 1000. This increased brightness is the significant advantage of double-clad fibers over both diodes and other types of fiber lasers. The necessary feedback elements for laser operation of the doped core may be formed by Bragg gratings directly written into the doped single-mode core, or by using mirrors deposited on or attached to the fiber ends. Slope efficiency approaches 70% and, output power is limited only by the pump power.

Different geometry has been developed for double-clad fiber lasers, some of which are shown in Fig. 22.9. The main purpose of the proposed design is to break a circular symmetry and to provide an efficient pump absorption in the core. Although the core can be doped with various rare-earth ions such as Er^{3+}, Nd^{3+}, Tm^{3+}, Ho^{3+}, the highest output power has been obtained from Yb^{3+}-doped single-mode core. Using tens of meters of Yb-doped double-clad fiber, continuous wave (CW) power of 200 W has been achieved in a single-mode beam at 1064 nm (IRE-Polus Group).

Potential applications for high-power lasers are in medicine, laser cutting, pumping other lasers, and in satellite-to-satellite communications links. Polaroid developed a high-power double-clad Yb^{3+}-doped fiber laser for a printing system. High-power lasers are also of great interest in telecommunication networks, since they can provide the necessary pump power, for example, for the practical implementation of cascaded Raman lasers for optical amplification. Laboratory experiments have demonstrated that the use of Raman gain devices can quadruple communication system capacity. Furthermore, in medical applications high-power 2-μm fiber sources may be useful in microsurgical applications. Also, medical spectroscopic applications in areas such as dermatology and diagnostic imaging, should benefit from double-clad fiber lasers.

22.3.3 Infrared Optical Fibers

In 1979 optical fibers made from silica and silica-based glasses reached their limit of transparency. Transmission losses as low as 0.2 dB/km have been obtained. This value almost corresponds to the ultimate inherent (intrinsic) loss value for a silica-based fiber. However, the demand for further improvements in trasmission capacity require the realization of ultra-low loss optical fibers with losses far below those of the silica-based optical fibers. Moreover, fibers with low loss in mid-infrared are required in medicine and industrial applications. The solution is in nonsilica infrared fiber materials, which offer the possibility of an ultra-low loss of less than 10^{-2} dB/km.

In principle, infrared optical fibers can be classified into two groups: dielectric optical fibers, based on the total internal reflection in solid cores; hollow waveguides,

Figure 22.9 Cross-sectional view of various double-clad fibers.

whose core regions are hollow. Optical materials studied to date for infrared fibers are halides, chalcogenides, and heavy-metal oxides.

Oxide Glass Fibers

Infrared oxide glass fibers are mainly based on heavy-metal oxides such as GeO_2, $GeO_2-Sb_2O_3$, and TeO_2. The minimum losses typically occur at wavelengths of around 2–3 μm. In metal glasses, infrared absorption due to lattice vibration (Ge–O) can be shifted toward a longer wavelength, since their constituent metals (such as Ge) are heavier than Si in SiO_2 glass. As a result, an ultra-low loss in the infrared region is expected. Theoretical predictions give a value of 0.1 dB/km for minimum intrinsic loss of these fibers. However, the experimentally obtained values are more than one order of magnitude larger: in particular, losses of 4 dB/km have been reported in $GeO_2-Sb_2O_3$.

Fluoride Glass Fibers

Fluoride glasses are the most promising candidates for the ultra-low loss optical fibers in long-distance optical communication. The initial system discovered by Poulain and coworkers in France in 1974 were fluorozirconates, where ZrF_4 was the primary constituent (> 50 mol%), BaF_2 the principal modifier (~ 30 mol%) and various metal fluorides, such as ThF_4 and LaF_3, were tertiary constituents. Depending on the composition, fluoride glasses have various desirable optical characteristics, such as a broad transparency range extending from the mid-infrared (~ 7 μm) to near ultraviolet (0.3 μm), low refractive index and dispersion, low Rayleigh scattering, and the potential of ultra-low absorption and ultra-low thermal distortion. Recent progress in reducing transmission losses of fluoride fibers to less than 1 dB/km strongly encourages the realization of ultra-low loss fibers of 0.01 dB/km, or less. Intrinsic losses in fluoride glasses are estimated to be 0.01 dB/km at 2–4 μm. In particular, fibers based on ZrF_4-BaF_2 have intrinsic losses in the vicinity of 0.01 dB/km at around 2.5 μm.

The visible refractive index of most fluoride glasses lies in the range 1.47–1.53, which is comparable to silicates but much lower than chalcogenides and can be tailored by varying composition. The zero of the material dispersion for fluorozirconates occurs in the region of 1.6–1.7 μm, as opposed to the loss minimum in the 3–4 μm regime. Nevertheless, the magnitude of the material dispersion is small through an extended range of wavelengths, including the minimum loss region, so that respectable values of pulse broadening, on the order of several picoseconds per angstrom-km are calculated for typical fluoride glasses near their loss minimum.

Chalcogenide Glass Fibers

Chalcogenides are compounds composed of chalcogen elements, i.e., S, Se, and Te, and elements such as Zn, Cd, and Pb. They are available in a stable vitreous state and have a wide optical transmission range. Chalcogenide glasses are advantageous because they exhibit no increase in scattering loss due to the plastic deformation that usually occurs in crystalline fibers. Chalcogenide glasses are divided into sulfides, selenides, and tellurides.

Sulfide glasses are divided into arsenic-sulfur and germanium-sulfur glasses. The optical transmission range of As-S and Ge-S glasses is almost the same, giving a broad transmission in the mid-infrared region. Transmission loss of 0.035 dB/m has been achieved for an As-S glass without a cladding, while in an As-S glass fiber with teflon FEP cladding transmission loss of 0.15 dB/m has been obtained.

Various selenide glasses have been studied mainly in order to achieve lower loss at the wavelengths of 5.4 μm (CO laser) and 10.6 μm (CO_2 laser). Selenide glasses are divided into As-Se As-Ge-Se (As rich) glasses, Ge-Se, Ge-As-Se (Ge rich) glasses, La-Ga-Ge-Se glasses and Ge-Sb-Se glasses. Selenide glasses have a wide transparency region compared with sulfide glasses. They exhibit a stable vitreous state, resulting in flexibility of the fiber and, thus, these fibers are the candidates for infrared laser power transmission and wide-bandwidth infrared light transmission, such as in radiometric thermometers.

Although the Se-based chalcogenide glasses have a wide transparency range, their losses at the wavelength of 10.6 μm for CO_2 laser power transmission are still higher than the 1 dB/m required for practical use. To lower transmission loss caused by lattice vibration, the atoms must be introduced to shift the infrared absorption edge toward a longer wavelength. Transmission losses of 1.5 dB/m at 10.6 μm in $Ge_{22}Se_{20}Te_{58}$ telluride glass have been reported.

Polycrystalline Fibers

The technology of fabrication of polycrystalline fibers by extrusion has allowed the preparation of TlBr-TlI (KRS-5) fibers with total optical loss of 120–350 dB/km. This crystalline material is known to transmit from 0.6 to 40 μm, and has a theoretical transmittance loss of $\sim 10^{-3}$ dB/km at 10.6 μm due to low intrinsic scattering losses from Rayleigh and Brillouin mechanisms and a multiphonon edge that is shifted to longer wavelengths. Fiber-optic waveguides made from KRS-5, however, have optical losses at 10.6 μm that are orders of magnitude higher than those predicted by theory. In addition, the optical loss increases with time (~ 6 dB/m per year) via the mechanism of water incorporation. Transmission of a 138 W (70 kW/cm^2) beam of a CO_2 laser through a 0.5-mm diameter and 1.5-m length fiber made from KRS-5 has been achieved with transmitted power of 93%.

On the other hand, polycrystalline silver halide fibers, formed by extrusion of mixed crystals of $AgCl_xBr_{1-x}$ ($0 \leq x \leq 1$), are transparent to 10.6 μm CO_2 laser radiation. The total loss coefficient at this wavelength is in the range of 0.1 to 1 dB/m, and the fiber transmission is independent of power levels up to at least 50 W total power input, which is an input power density of about 10^4 W/cm^2. In addition, the fibers are nontoxic, flexible and insoluble in water. They have been successfully used in many applications such as directing high-power CO_2 laser radiation for surgical and industrial applications, IR spectroscopy, and radiometry.

Single-Crystalline Fibers

Single crystal fibers have the potential of eliminating all the deleterious effects of extruded fiber, and, therefore, of having much lower loss than polycrystalline fibers. This material possesses a wide transparent wavelength region from visible to far-infrared, so that it is possible to transmit both visible and infrared light. Materials for single-crystalline fibers are almost the same as those for polycrystalline fibers: TlBr-TlI, AgBr, KCl, CsBr, and CsI have been mainly studied. The lowest losses at the CO_2 laser wavelength of 10.6 μm have been attained in thallium halide fibers (0.3–0.4 dB/m). Also, a transmission loss of 0.3 dB/m and maximum transmitted power of 47 W were obtained at a 10.6 μm wavelength by using a 1-mm diameter CsBr fiber.

22.3.4 Metallic and Dielectric Hollow Waveguides

In the mid-infrared region (10.6 μm and 5 μm), hollow-core waveguides have advantages over solid-core fibers in high power transmission of CO_2 and CO lasers. Lower loss and higher laser-induced-damage threshold are attainable because of the highly transparent airy core. Various types of hollow infrared waveguides have been fabricated, including cylindrical and rectangular metallic, cylindrical dielectric, and rectangular metallic with inner dielectric coatings (Fig. 22.10).

In cylindrical hollow waveguides, transverse electric modes TE_{0n} propagate with the least attenuation, while hollow rectangular waveguides propagate TE_{m0} and TM_{m0} modes most efficiently because these modes interact less with the waveguide inner walls. Metal material in hollow waveguides provides high infrared reflectivity. Transmission of more than 95%/m in straight rectangular hollow metal waveguides has been reported. Transmission losses of dielectric-coated metallic waveguides are expected to be low for the HE_{11} mode. Theory predicts that the power loss of the HE_{11} mode is 3.2×10^{-2} dB/m. The total transmission loss of fabricated waveguides are always below 0.5 dB/m, in contrast to 2.5 dB/m for a metal hollow waveguide with no dielectric layer. Metallic hollow fibers are not as flexible as conventional optical fibers, and so they are rather restricted.

In dielectric hollow-core fibers the glass (e.g., oxide glasses such as SiO_2 and GeO_2) acts as a cladding to the fiber core, which is air. To provide a condition of total internal reflection, and thus high transmission, the glass cladding must retain a refractive index of less than unity. This means selecting a dielectric with the anomalous dispersion of refractive index at an operating wavelength. Many inorganic materials, either in vitreous (i.e., amorphous) or crystalline form exhibit strong anomalous dispersion in the mid-infrared, providing their refractive indices are less than unity at certain frequencies. However, it should be noted that a refractive

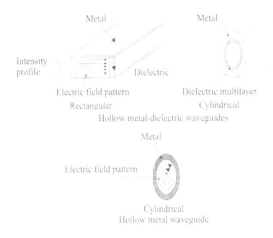

Figure 22.10 Hollow metal waveguides.

index of below unity at the middle region originates from the strong absorption due to lattice vibration and results in increase of transmission losses and limits the length of the waveguide. A transmission loss of 0.1 dB/m at 10.6 μm is expected in a fiber with a cladding composition of 80 mol% GeO_2–10 mol% ZnO–10 mol% K_2O and a bore size of 1 mm. The experimental loss is about 20 times larger than the theoretical one. It is expected that lower transmission losses can be obtained by smoothing the inner surface.

For a given bore size, for example 1 mm, metal circular waveguides will have the highest attenuation, and this attenuation decreases significantly for those materials exhibiting anomalous dispersion. Hollow dielectric fibers with polycrystalline hexagonal GeO_2 cladding can have loss below the 0.5 dB/m generally regarded as an acceptable loss for the CO_2 laser power delivery. High-power transmission hollow fibers are suitable for applications in laser surgery, cutting, welding, and heat treatment.

Liquid-Core Fibers

Liquid-filled hollow fibers contain liquid materials in the hollow cores. These fibers are advantageous because there are no stress effects leading to birefringence, and wall imperfections and scattering effects are negligible. The transmission losses in the transparent wavelength regions depend on the liquids used. Liquid bromine has been inserted into a Teflon tube with attenuation of 0.5 dB/m at 9.6 μm. For silica glass tubes with diameter of 125 μm filled with bromobenzene, the transmission loss of 0.14 dB/km at 0.63 μm has been reported. Although these fibers may be useful for the intrared region, fabrication and toxicity are still serious difficulties.

22.3.5 ZBLAN Fiber Based Up-Conversion Lasers

Frequency up-conversion is a term that is associated with a variety of processes whereby the gain medium, a trivalent rare-earth ion, in a crystal or glass host, absorbs two or more infrared pump photons to populate high-lying electronic states. Visible light is then produced by one-photon transitions to low-lying electronic levels. Up-conversion lasers appear to be attractive candidates for compact, efficient visible laser sources for applications in optical data storage, full-color displays, color printing, and biomedical instrumentation because of the relative simplicity (the gain and frequency conversion material are one and the same) of these devices.

In 1990 several CW, room temperature up-conversion lasers in the ZBLAN single-mode fibers were demonstrated. Many up-conversion lasers based on doped optical fibers have produced output wavelengths ranging from the near infrared to the ultraviolet, but the greatest advantage offered by the optical fiber geometry is that room temperature operation is much easier to obtain than in bulk media. In addition to their simplicity and compactness, up-conversion fiber lasers are efficient and tunable. Slope efficiencies (pump power to output power conversion) of up to 50% have been obtained in the two-photon 550 nm Er^{3+} ZBLAN fiber laser and 32% in the three-photon pumped Tm^{3+} ZBLAN fluoride laser. Table 22.3 shows a summary of up-conversion fiber lasers that demonstrated generation at room temperature in rare-earth-doped fluorozirconate glass. The key to up-conversion laser operation in single-mode optical fibers has been the use of low phonon energy ($\hbar\omega < 660 \, cm^{-1}$) fluorozirconate glasses as hosts for the rare-earth ions.

Table 22.3   Rare-Earth-Doped ZBLAN Up-Conversion Fiber Lasers

Rare-earth ions	Pump wavelength (nm)	Laser wavelength (nm)	Slope efficiency (%)
Er	801	544, 546	15
	970		> 40
Tm	1064, 645	455	1.5
	1112, 1116, 1123	480, 650	32 (480 nm)
	1114–1137	480	13
Ho	643–652	547.6–594.5	36
Nd	582–596	381, 412	0.5 (412 nm)
Pr	1010, 835	491, 520, 605, 635	12 (491nm)
Pr/Yb	780–885	491, 520, 605, 635	3 (491 nm)
			52 (635 nm)

Because the fluorozirconate host is a disordered medium, rare-earth-doped fibers fabricated from these glasses exhibit absorption and emission profiles that are broad compared with those characteristic of a crystalline host. The broad emission profile has a negative impact on the stimulated emission cross section but is more than compensated for by the advantages of the fiber laser geometry, the high pump intensities, and the maintenance of this intensity over the entire device length. The broad emission linewidths permit tuning continuously of the wavelength of generation over 10 nm.

22.3.6   Plastic Optical Fibers

Copper and glass have been the traditional solutions to data communication and they are well suited to specific applications. However, for high-speed data transmission copper is unsuitable because of its susceptibility to interference. On the other hand, the small diameter and fragility of glass elevates a cost of installation in fast-growing local-area networks. Plastic optical fibers (POFs) provide an alternative and fill some of the void between copper and glass.

POF shares many of the advantages and characteristics of glass fibers. The core material of POF is normally made of acrylic resin, while the cladding is made of fluorinated polymer. POF diameter is usually of around 1 mm, which is many times larger than a glass fiber and the light-transmission core section accounts for 96% of the cross-sectional area. The first POF available had a step-index profile, in which high-speed transmission was difficult to achieve. In 1995, Mitsubishi Rayon announced the first graded-index plastic optical fiber (GIPOF) with transmission speeds in excess of 1 Gb/s. GIPOF in conjunction with high-speed 650 nm light-emitting diode (LED) provides an ultimate solution for high-speed, short- or moderate distance, low-cost, electromagnetic-interference-free data links demanded by desktop local-area network specifications such as asynchronous transfer-mode-local area network (ATM-LAN) and fast ethernet.

Because of its bandwidth capability, POF is a much faster medium than copper, allowing for multitasking and multimedia applications that today's copper supports at slower, less productive speed. The advantage of plastic optical fiber over glass fibers is in its low total-system cost. In multimode glass fiber, a precision technology is needed to couple the light effectively into a 62.5 μm core. The larger diameter of plastic fiber allows relaxation of connector tolerances without sacrificing optical coupling efficiency, which simplifies the connector design. In addition, the plastic fiber and large core diameter permit termination procedures other than polishing, which requires an expensive tool. For quick and easy termination of plastic fiber, a handheld hot-plate terminator is available, so that even a worker with no installation training can terminate and assemble links within a minute. Because POF transmission losses are higher than for silica fiber, it is not suitable for long distance, but is, however, suitable for home and office applications. Typical losses are of the order of 140–160 dB/km.

Applications of POF

The largest application of POF has been in digital audio interfaces for short-distance (5 m), low-speed communication between amplifiers and built-in digital-to-analog converter and digital audio appliances such as CD/MD/DAT players and BS tuners. The noise-immune nature of POF contributes to creating sound of high quality and low jitter.

Because of its flexibility and immunity to factory floor noise interference, rugged and robust POF communication links have been successfully demonstrated in tough industrial manufacturing environment.

Lightweight and durable POF networks could link the sophisticated systems and sensors used in automobiles, which would increase performance and overall efficiency. Also, POF could be used to incorporate video, minicomputers, navigational equipment, and fax machines into a vehicle. Another short-haul application for POF is home networking, where appliances, entertainment and security systems, and computers are linked to create a smart home.

22.3.7 Fiber Bundles

A fiber bundle is made up of many fibers that are assembled together. In a flexible bundle the separate fibers have their ends fixed and the rest of their length unattached to each other. On the other hand, in a rigid or fused bundle, the fibers are melted together into a single rod. Fused bundles have lower cost than flexible bundles, but because of their rigidity they are unsuitable for some applications.

The fiber cores of the bundle must occupy as much of the surface area as possible, in order to minimize the losses in their claddings. Thus, bundled fibers must have thin claddings to maximize the packing fraction, i.e., the portion of the surface occupied by fiber cores. The optics of the fiber bundles is the same as that of a single fiber. In a good approximation, it may be assumed that a single light ray may represent the light entering the input end of the bundle. If a light ray enters the fiber at an angle θ within the acceptance angle of the fiber, it will emerge in a ring of angles centered on θ, as shown in Fig. 22.11. It should be pointed out that in fiber bundles formed by step-index fibers with constant-diameter cores, the light entering the fiber emerges at roughly the same angle it entered.

If light focusing (and magnification and demagnification of objects) is needed, a tapered fiber may be used. Figure 22.12 shows a schematic representation of a tapered fiber. If a ray entering a fiber at an angle θ_1 meets criteria for total internal

Figure 22.11 The light rays emerge from the fiber in a diverging ring.

reflection, it is confined in the core. However, it meets the core–cladding boundary at different angles on each bounce so it emerges at a different angle θ_2,

$$d_1 \sin \theta_1 = d_2 \sin \theta_2,\tag{22.16}$$

where d_1 is the input core diameter and d_2 is the output core diameter. The same relationship holds for the fiber's outer diameter as long as core and outer diameter change by the same factor d_2/d_1.

Bundles of step-index fibers can be used for imaging. In this case, each fiber core of the bundle will carry some segment of the image, so that the complete image is formed by the different segments of the fibers of the bundle. As long as the fibers are aligned in the same way on both ends, the bundle will form an image of an object. Typical losses of fiber bundles are around 1 dB/m. Since fiber bundles are required in short-distance applications, such a large loss is not a limitation.

The majority of bundles are made from step-index multimode fibers, which are easy to make and have large numerical apertures. The higher NA of these fibers ($\gtrsim 0.4$) gives large acceptance angles, which in turn decreases coupling losses.

The simplest application of optical fibers of any type is light piping, i.e., transmission of light from one place to another. A flexible bundle of optical fibers, for example, can efficiently concentrate light in a small area or deliver light around corners to places it could not otherwise reach. Application of fiber bundles includes illumination in various medical instruments, including endoscopes, in which they illuminate areas inside the body.

Fiber bundles also may be used for beam-shaping by changing the cross section of the light beam. It is possible to array the fibers in one of the ends of the bundle to form a circle or a line as shown in Fig. 22.13. This may be important if the fibers are being used for illumination in medical instruments, where special arrangements of the output fibers helps in the design and may result in more uniform illumination of the field of view.

Figure 22.12 Light propagation through a tapered fiber.

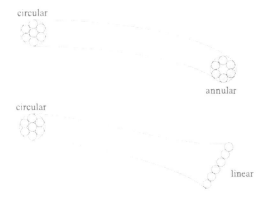

Figure 22.13 Beam-shaping by nonordered bundles of fibers.

The other application of optical fiber bundles is for beamsplitting or beam combining. For example, a Y-guide is formed by two fiber bundles combined into one bundle near a sample, while they are still separated at the other ends (Fig. 22.14). If the NA of the individual fibers is large, the light-collection efficiency (the ratio between the input and output energy) of the Y-guide is fairly high.

In image transmission, the fibers must maintain identical relative positions on input and output faces. These are called ordered or coherent bundles. Images can be viewed through coherent fiber bundles by placing the bundle's input end close or directly on the object, or by projecting an image onto the input end. Light from the object or the image is transmitted along the bundle, and the input image is reproduced on the output face. Coherent fiber bundles are very valuable in probing otherwise inaccessible areas such as inside machinery or inside the human body.

The easiest way to make coherent fiber bundles is to fuse fibers together throughout the length of the bundle. However, such fiber bundles are not usable in many situations because of the lack of flexibility. In flexible coherent bundles, the fibers are bonding together at the two ends, so they maintain their relative alignment, but they are free to move in the middle. Individual fibers, unlike fused bundles, are more flexible. The imaging transmission bundles are the basic building blocks of fiberscopes and endoscopes. The purpose of a coherent fiber bundle is to transmit the

Figure 22.14 Y-fiber bundles.

full range of an illuminated object. Normally both the coherent and the nonordered bundles are incorporated into an endoscope. A nonordered bundle (Fig. 22.15) illuminates the object inside the body. The imaging bundle must then transmit the color image of the object with adequate resolution.

22.4 FIBER OPTIC COMPONENTS

22.4.1 Optical Fiber Couplers

The optical directional fiber coupler is a waveguide equivalent of a bulk beamsplitter and is one of the basic in-line fiber components. When two (or more) fiber cores are placed sufficiently close to each other, the evanescent tail of an optical field in one fiber extends to a neighboring core and induces an electric polarization in the second core. In its turn, the polarization generates an optical field in the second core, which also couples back to the core of the first fiber. Thus, the modes of different fibers become coupled trough their evanescent fields, resulting in a periodical power transfer from one fiber to the other. If the propagation constants of the modes of the individual fibers are equal, then this power exchange is complete. If their propagation constants are different, then exchange of power between the fibers is still periodic, but incomplete.

The basic mechanism of a directional coupler can be understood on an example of a coupler formed by a pair of identical symmetric single-mode waveguides. The system of two coupled waveguides can be considered as a single waveguide with two cores. Such a system supports two modes, the fundamental being the symmetric mode and the first excited being the antisymmetric mode. These two modes have different propagation constants. Light launched in one waveguide excites a linear combination of the symmetric and antisymmetric modes (Fig. 22.16) in both cores. The interference of two modes at the input is constructive in the first waveguide and is destructive in the second waveguide, resulting in the absence of the field in the last one. In the coupling region the two modes propagate at different velocities, acquiring a phase difference. When the phase difference becomes π, then the superposition of these two modal fields will result in a destructive interference in the first waveguide and constructive in the second. Further propagation over an equal length will result in the phase difference of 2π, leading to a power transfer back to the first waveguide. Thus, the optical power exchanges periodically between the two waveguides. By an

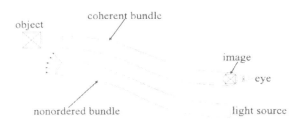

Figure 22.15 Illumination and image transmission through a nonordered and coherent bundle of fibers.

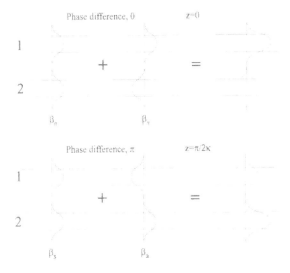

Figure 22.16 Symmetric and antisymmetric modes.

appropriate choice of the coupler length, one can fabricate couplers with an arbitrary splitting ratio.

A power transfer ratio depends on the core spacing and interaction length. If $P_1(0)$ is the power launched into fiber 1 at $z = 0$, then the transmitted power $P_1(z)$ and the coupled power $P_2(z)$ for two nonidentical single-mode fibers are given by

$$\frac{P_1(z)}{P_1(0)} = 1 - \frac{k^2}{\gamma^2} \sin^2 \gamma z,$$
$$\frac{P_2(z)}{p_1(0)} = \frac{k^2}{\gamma^2} \sin^2 \gamma z,$$

$$(22.17)$$

where

$$\gamma^2 = k^2 + \tfrac{1}{4}(\Delta\beta) \tag{22.18}$$

and $\Delta\beta = \beta_1 - \beta_2$ is the difference of propagation constants of the first and the second fiber, respectively, called also as a phase mismatch; k is the coupling coefficient, which depends on the fiber parameters, the core separation, and the wavelength of operation. If the two fibers are separated by a distance much greater than the mode size, then there would be no interaction between the two fibers.

In the case of two identical fibers, the phase mismatch is equal to zero and the power oscillates between two fibers. The coupling coefficient is given by

$$k(d) = \frac{\lambda_0}{2\pi n_1} \frac{U^2}{a^2 V^2} \frac{K_0(W d/a)}{K_1^2(W)}, \tag{22.19}$$

where λ_0 is the free space wavelength, n_1 and n_2 are the core and cladding refractive indices, respectively, a is the fiber core radius, d is the separation between the fiber axes, K_v is the modified Bessel function of order v, $k_0 = 2\pi/\lambda_0$, $n_e = \beta/k_0$, n_e is the mode effective

$$U = k_0 a(n_1^2 - n_e^2)^{1/2}$$
$$W = k_0 a(n_e^2 - n_2^2)^{1/2} \tag{22.20}$$
$$V = k_0 a(n_1^2 - n_2^2)^{1/2}$$

index. Knowing a coupling coefficient, one can easily calculate the corresponding coupling length.

Parameters of a Coupler

The 2×2 coupler is shown schematically in Fig. 22.17. For an input power P_i, transmitted power P_t, coupled power P_c, and back-coupled power P_r we can determine the main characteristics of the coupler as follows:

$$\textit{Power-splitting ratio } R(\%) = \frac{P_t}{P_c} \times 100;$$
$$\textit{Excess loss } L_e \text{ (dB)} = 10 \log\left[\frac{P_i}{P_c + P_t}\right];$$
$$\textit{Insertion loss } L_i \text{ (dB)} = 10 \log\left[\frac{P_i}{P_t}\right]; \tag{22.21}$$
$$\textit{Directivity } D \text{ (dB)} = 10 \log\left[\frac{P_r}{P_i}\right].$$

The popular power-splitting ratios between the output ports are 50% : 50%, 90% : 10%, 95 : 5%, and 99% : 1%; however, almost any value can be achieved on a custom basis. Excess loss in a fiber coupler is the intrinsic loss of the coupler when not all input power emerges from the operation ports of the device. Insertion loss is the loss of power that results from inserting a component into a previously continuous path. Couplers should have low excess loss and high directivity. Commercially available couplers exhibit excess loss ≤ 0.1 dB, and directivity of better than -55 dB.

A coupler is identified by the number of input and output ports. In the $N \times M$ coupler, N is the number of input fibers and M is the number of output fibers. The simplest couplers are fiber-optic splitters. These devices have at least three ports but may have more than 32 for more complex devices. In a three-port device (tee coupler), one fiber is called the common fiber, while the other two fibers are called input or output ports. A common application is to inject light into the common port and to split it into two independent output ports.

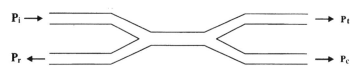

Figure 22.17 A 2×2 coupler.

Fabrication of Fiber Couplers

Practical single-mode fibers have a thick cladding to isolate the light propagating in the core. Hence, to place two cores close to each other, it is necessary to remove a major portion of the cladding. Two methods have been developed for the fabrication process. The first one consists of polishing the cladding on one side of the core of both fibers, and then bringing the cores in close proximity. A technique for polishing the cladding away consists of fixing the fibers inside grooves in glass blocks, and polishing the whole block to remove the cladding on one side of the core (Fig. 22.18(a)). The two blocks are then brought into contact (Fig. 22.18(b)). Usually the space between the blocks is filled with an index-matching liquid. By laterally moving one block with respect to the other, one can change the core separation, resulting in changes of the coupling constant k. This, in turn, changes the power-splitting ratio. Such couplers, called tunable, permit smooth tuning of the characteristics of a coupler.

Polished couplers exhibit excellent directivity, better than -60 dB. Their splitting ratio can be continuously tuned. The insertion losses of such couplers are very low (≈ 0.01 dB). One of the important characteristics of couplers is sensitivity to the polarization state of the input light. The polished couplers have the advantage of being low polarization sensitive. The variation in splitting ratio for arbitrary input polarization states can be less than 0.5%. The performance of such couplers can be affected by temperature variations, because of the temperature dependence of the refractive index of the index-matching liquid.

Fabrication of polished fiber couplers is a time-consuming operation. Hence, more popular couplers today are fused directional couplers. Such couplers are easier to fabricate, and the fabrication process can be automated. In this type of coupler, two or more fibers with removed protecting coatings are twisted together and melted in a flame. After the pulling, the fiber cores approach each other, resulting in over-lapping of the evanescent fields of the fiber modes. The coupling ratio can be monitored online as the fibers are drawn. Fused couplers exhibit low excess loss (typically less than 0.1 dB) and directivity better than -55 dB.

22.4.2 Wavelength-Division Multiplexers

A more complex coupler is the wavelength-division multiplexer. A WDM is a passive device that allows two or more different wavelengths of light to be split into multiple fibers or combined onto one fiber. Let us consider a directional coupler made of two

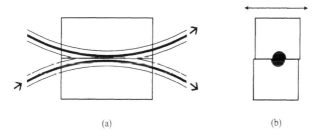

(a)   (b)

Figure 22.18   Polished fiber coupler.

identical fibers with coupling coefficients k_1 and k_2 at wavelengths λ_1 and λ_2, respectively. As was shown above, the optical power at each wavelength exchanges periodically between the two waveguides. Since a coupling coefficient depends on wavelength, the period of oscillations at each wavelength is different. By an appropriate choice of the coupler's length L the two conditions

$$K_1 L = m\pi, \qquad k_2 L = (m - \tfrac{1}{2})\pi \qquad\qquad (22.22)$$

can be satisfied simultaneously. In this case, if the total optical power at wavelength λ_1 and λ_2 is launched in the port 1 (Fig. 22.19), then the WDM will sort radiation at λ_1 and λ_2 between ports 3 and 4, respectively.

Two important characteristics of WDM device are cross-talk and channel separation. Cross-talk refers to how well the demultiplexed channels are separated. Optical radiation at each wavelength should appear only at its intended port and not at any other output port. Channel separation describes how well a coupler can distinguish wavelengths. Fused WDMs are more appropriate in applications where the wavelengths must be widely separated, for example in commercially available 980/1550-nm single-mode WDMs. Such a WDM exhibits an excess loss of 0.3 dB, an insertion loss of 0.5 dB, and a cross-talk better than 20 dB.

The communication WDM often needs a channel separation of about 1 nm. A sharp cutoff slope of the channel transmission characteristics is also required to achieve interchannel isolation (cross-talk) of 30 dB. The interference-filter-based WDMs are more suitable for this application. Figure 22.20 shows a multiplexer and demultiplexer using graded-index rod lenses and interference filters. The filter is designed so that it passes radiation at wavelength λ_2 and reflects at λ_1. The GRIN-rod lens collects the transmitted and reflected radiation into fibers.

22.4.3 Switches

Fiber-optical switches selectively reroute optical signals between different fiber paths. The performance of a switch is characterized by an insertion loss, cross-talk (back reflection), and speed. Figure 22.21 shows a typical 1×2 switch which can have an output with two positions, "on" (port 2) and "off" (port 3). For such a switch, the insertion loss can be determined as

$$\text{Insertion loss} = -10 \log \frac{P_2}{P_1}; \qquad\qquad (22.23)$$

the cross-talk determines the isolation between the input and the "off" port:

$$\text{Cross-talk} = -10 \log \frac{P_3}{P_1}, \qquad\qquad (22.24)$$

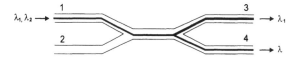

Figure 22.19   Wavelength-division multiplexer (WDM).

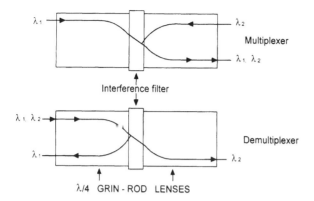

Figure 22.20 Interference-filter-based wavelength-division multiplexer (WDM).

where P_1, P_2, and P_3 are the input power, power at the output "on," and power at the output "off," respectively. Typical switches have low insertion loss (0.5 dB), and low cross-talk (55 dB). The speed of the switch depends on mechanisms involved in switching. Switches can be classified as optomechanical, electronic, and photonic (or optical) switches. Optomechanical switches include a moving optical element such as a fiber, a prism, or a lens assembly in the path of a beam to deflect the light. The speed of operation in this case is limited by millisecond speeds, so the optomechanical switches are not suitable for fast switching.

Electronic switches use an electronic-to-optical conversion technique, which can be fast enough for communication systems. However, the complexity of electronic-to-optical conversion limits the applications of electronic switching.

Photonic switching uses an integrated optic technology to operate in high-speed regimes. Electro-optic and acousto-optic phenomena are usually used to

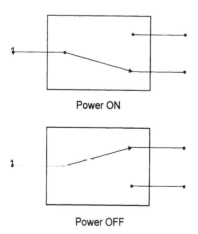

Figure 22.21 Typical 1 × 2 switch.

actuate the switching. The most advanced electro-optic waveguide technology utilizes LiNbO$_3$ crystals. This material provides necessary low insertion loss, high switching speed, and large bandwidth.

The waveguide configurations based on LiNbO$_3$ can be classified into directional couplers (Fig. 22.22(a)), Mach–Zehnder interferometer (Fig. 22.22(b)), and intersecting waveguides (Fig. 22.22(c)). The directional coupler (Fig. 22.22(a)) consists of a pair of optical waveguides placed in close proximity. Light propagating in one waveguide gets coupled to the second waveguide through an evanescent field. The coupling coefficient depends on refractive indices of the waveguides. By placing electrodes over the waveguides and varying applied voltage, the coupling coefficient can be efficiently tuned over a relatively large range. The length of the coupler is chosen in such a way that when no voltage is applied the switch is in the cross state (the input and output ports are from different waveguides). With applied voltage the switch changes to the bar state when the input and output ports are on the same waveguide. Splitting the electrode into two sections, one can tune both the cross and the bar states.

The Mach–Zehnder interferometer (Fig. 22.22(b)) consists of a pair of 3 dB couplers connected by two waveguides. With no voltage applied to the electrodes, the optical paths of the two arms are equal, so there is no phase shift between light in the waveguides. In this case the light entered in the port 1 goes out through the port 3. By applying a voltage, a π phase difference is introduced between light in two arms, resulting in switching of optical power from the port 3 to the port 2. These

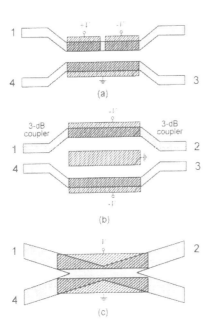

Figure 22.22 Photonic switching configurations: (a) directional couplers; (b) Mach–Zehnder interferometer; and (c) intersecting waveguides.

switches are typically 15–20 mm long and require a phase shift voltage of 3–5 V. High cross-talk levels of 8–20 dB is a disadvantage of these switches.

The switch based on intersecting waveguides (Fig. 22.22(c)) can be considered either as a modal interferometer or as a zero-gap directional coupler. Such a switch offers a topological flexibility, moderate voltage requirements, and simple electrode configurations.

In truly optical switching a control optical pulse switches a signal pulse from one channel to the other. A control pulse changes the conditions of propagation for a signal pulse due to nonlinear optical effects. The response time of nonlinear optical effects (for example, the Kerr and Raman effects) in fibers and waveguides is in the femtosecond range, providing the highest switching speeds without the need for electronics. The insertion loss can be very low.

22.4.4 Attenuator

An optical attenuator is a passive device placed into a transmission path to control an optical loss. Both fixed and variable attenuators are available.

In fixed attenuators an absorbing layer is inserted between two fibers. GRIN-rod or ball lenses are used to collimate the light between fibers. In variable attenuators a wedged-shaped absorber element whose position can be adjusted accurately is placed between the fibers. Figure 22.23 shows a variable attenuator with a wedge-shape absorber (shadowed) and a wedge-shape transparent element. The latter is needed to avoid beam displacement.

22.4.5 Polarization Fiber Components

Polarization is a property that arises because of the vector nature of the electro-magnetic waves. Electromagnetic light waves are represented by two vectors: the electric field strength **E** and the magnetic field strength **H**. Since the interaction of light with material media is mainly through the electric field strength **E**, the state of polarization of light is described by this field. Polarization refers to the behavior with time of the electric field vector **E**, observed at a fixed point in space. Time-harmonic transverse plane optical waves can be represented by

$$
\begin{aligned}
\mathbf{E}(\mathbf{r}; t) &= \mathbf{x}E_x(x, y, z; t) + \mathbf{y}E_y(x, y, z; t), \\
E_x(x, y, z; t) &= E_x(x, y)\cos(\omega t - kz + \delta_x), \\
E_y(x, y, z; t) &= E_y(x, y)\cos(\omega t - kz + \delta_y),
\end{aligned}
\tag{22.25}
$$

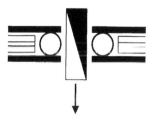

Figure 22.23 Attenuator.

where **x** and **y** are unit vectors along the transverse direction; E_x and E_y are the amplitudes of the waves along the transverse directions; k ($= 2\pi/\lambda$) is the propagation constant of the wave; ω ($= 2\pi\nu$) is the angular frequency of the wave; and λ and ν are the wavelength and the frequency of the light, respectively. Phase shift $\delta_x - \delta_y$ between the orthogonal components defines the polarization state of light.

Whenever a light wave of arbitrary polarization propagates through optical media, the optical properties of such media induce changes in its polarization state. In isotropic media, the induced polarization in the medium is parallel to the electric field of the optical wave. In many media, however, the induced polarization depends on the magnitude and direction of the applied field. One of the most important consequences of this anisotropy is the phenomenon of birefringence in which the phase velocity of an optical wave propagating in the medium depends on the direction of polarization of its vector **E** (i.e., $k_x \neq k_y$). This anisotropy changes the state of polarization of the wave after propagating in the medium. For example, an input linear polarization state becomes elliptically polarized after propagating by some distance through the medium.

The ideal optical fiber is perfectly cylindrical and invariant by translation along the propagation axis, and hence isotropic. In particular, single-mode optical fibers have been designed for supporting only one mode. Because no real fiber has a perfect circular symmetry (due to fabrication defects or to environmental conditions), the fundamental mode splits into two submodes, orthogonally polarized and propagating with different velocities. This is of little consequence in applications where the fiber transmits signals in the form of optical power with pulse-code or intensity modulation, as the devices detecting the transmitted light are not sensitive to its polarization state. However, in modern applications such as the fiber-optic interferometric sensor and coherent communication systems, polarization of the light is important. In these latter cases it is necessary to be able to control polarization of light, and compensate the changes of the polarization along the fiber. There are various fiber-optics components that are used to control the polarization. Here, we focus on three types of fiber polarization components: namely, polarizers, polarization splitters, and polarization controllers.

22.4.6   Polarizer

A polarizer is an optical device that transmits (or reflects) a state of polarization and that suppresses any transmission (or reflection) of the orthogonal state of polarization. An ideal linear polarizer is a device that transforms any input state of polarization of light to a linear output state. A linear polarizer can also be defined as a device whose eigenpolarizations (for example, the two orthogonal polarization modes in a single-mode fiber) are linear with one eigenvalue (one of the orthogonal modes) equal to zero. In optical fibers, the main method used to eliminate one of the two orthogonal modes is a loss process, in which one of the modes is coupled towards the outer medium or providing larger radiation loss for one mode than the other.

There are two general classes of fiber polarizers: invasive and noninvasive. Invasive polarizers require polishing away of a small section of fiber to expose the core. Such polarizers can be prepared directly in the system fiber and require no splices. These polarizers utilize the differential attenuation of transversal electric (TE) and transversal magnetic (TM) modes. (In optical fibers the fundamental mode is hybrid, i.e., both modes TE and TM exist and are orthogonally polarized.) Figure 22.24 shows a scheme of such a polarizer where a dielectric layer is deposited on the polished half-fiber, followed by the deposition of a metallic layer. The dielectric layer matches the propagation of one polarization to the TM plasmon wave (spatial oscillations of electric charge) at the metal interface, thus providing effective coupling to the lossy mode. An alternative design is the so-called cutoff polarizer, in which the fiber is polished all the way down to the core and a very thin film metal is deposited on the polished fiber.

Another polarizer replaces the fiber cladding by a birefringent cladding with refractive indices such that polarization-selective coupling from the fiber core occurs. This method therefore uses the evanescent field of the guided waves that exists in the cladding. If the refractive index of the birefringent cladding is greater than the effective index of the guided wave for a given polarization, then this polarization is coupled out of the fiber core. Also, if the refractive index of the birefringent cladding as seen by the other guided polarization is lower than the effective index of the light wave in the core, then this polarization remains guided. When both these conditions are simultaneously achieved, one polarization radiates while the other remains guided in the fiber. Figure 22.25 shows a device of this type.

Noninvasive fiber polarizers are made directly in the fiber. Such a polarizer can be cut out of one fiber and spliced into devices or systems. Most noninvasive polarizers work by differential tunneling loss after high stress or geometrical birefringence splits the propagation constants. Figure 22.26 shows a polarizer that uses a W-structure index profile. In such a profile, the unfavored polarization is attenuated when its effective index falls below the refractive index of the second cladding. A noninvasive polarizer can also be made by bending a highly birefringent fiber. The principle of operation is similar to that of the W-structure polarizer because bending can be viewed as modifying the index profile. The two polarizations then suffer differential attenuation.

22.4.7 Polarization Splitter

Polarization-sensitive couplers or polarization state splitters are usually realized using two face-to-face half-couplers (Fig. 22.27). The coupling coefficient is adjusted in such a way that light with one polarization state from the first half-coupler totally propagates in the fiber of the second coupler, whereas the coupling of the light of the other polarization state is not possible. In other words, there is a phase matching for

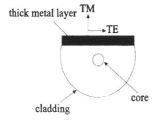

Figure 22.24 Transverse section of a metal film polarizer: transversal electric (TE) and transversal magnetic (TM) modes.

birefringent crystal

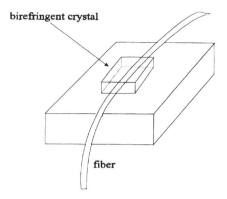

fiber

Figure 22.25 Polarizer with oriented crystal.

one state of polarization between the guided modes of the two fibers and there is not for the other.

Polished directional couplers have been made with polarization-preserving fiber to achieve both polarization-preserving and polarization-selective coupling. In a polarization-preserving coupler, the propagation constants of the two polarizations are matched across the coupler; in a polarization-selective coupler, only one of the propagation constants is matched while the mismatch is as large as possible for the other polarization.

22.4.8 Polarization Controller

A common problem is to transform the state of polarization in a fiber from an arbitrary polarization state to linear polarization with a proper orientation for a polarization-sensitive optical component. Whenever an initially isotropic fiber is bent, with or without axial tension, it becomes linearly birefringent. This peculiarity permits the production of phase shifters of an angle $\pi/2$ or π, i.e., the equivalent of a quarter- and half-wave plates of crystalline optics. Polarization controllers are made by bending ordinary nonbirefringent fiber in coils (Fig. 22.28). Rotating the coil is equivalent to turning the bulk waveplate.

A section of birefringent fiber can also act as a high-order waveplate if both polarization axes are equally excited. This property is commonly used to convert linear polarization to elliptical polarization. The characteristics of this polarization element are strongly dependent on temperature, pressure, and applied stresses. The

$n_\parallel(\text{eff})$

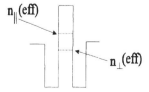

$n_\perp(\text{eff})$

Figure 22.26 Fiber polarizer with W-structure index profile.

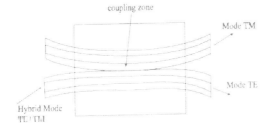

Figure 22.27 Linear polarization states splitter using two half-couplers.

addition of a polarizer converts such a variable waveplate into a variable in-line attenuator. Variable waveplates separated by polarizers can also form a tunable Lyot filter.

In an all-fiber system, the fiber itself defines the optical path and is subject to perturbations that in conjunction with the intrinsic birefringence can affect the state of polarization in complicated and unpredictable ways. The birefringence perturbations can be originated by twisting (circular birefringence) and lateral stress (linear birefringence). In applications such as Mach–Zehnder interferometers and fiber-rotation sensors it is important that this birefringence is compensated. Polarization controllers involving electromagnetically induced stress twisting have been proposed for an accurate control of birefringence effects.

22.4.9 Fiber Bragg Gratings

Photosensitivity is a phenomenon specific to germanium-doped silica fibers in which the exposition of the fiber core to blue-green or ultraviolet (UV) radiation induces changes in the refractive index of the core. If the changes in the refractive index are periodic, a grating is formed. The stronger changes occur when the fiber is exposed to UV radiation close to the absorption peak at 240 nm of a germania-related defect.

Two widely used methods for grating fabrication are the holographic side exposure and the phase-mask imprinting. In the holographic side exposure, two beams from a UV laser interfere in the plane of the fiber (Fig. 22.29). The periodicity of the interference pattern created in the plane of the fiber is determined by the angle between the two beams and the wavelength of the UV laser. The regions of constructive interference cause an increase in the local refractive index of the photo-

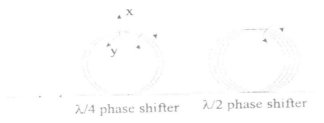

Figure 22.28 Half- and quarter-wave phase shifter realized with fiber loops.

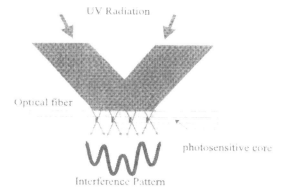

Figure 22.29 Holographic side exposure method.

sensitive core, while dark regions remain unaffected, resulting in the formation of a refractive Bragg grating.

In the phase-mask fabrication process, light from a UV source passes through a diffractive phase mask that is in contact with the fiber (Fig. 22.30). The phase mask has the periodicity of the desired grating. Light diffracted in orders $(+1, -1)$ of the mask interferes in the plane of the fiber, providing periodical modulation of the refractive index. The phase-mask technique allows fabrication of fiber gratings with variable spacing (chirped gratings). Fiber Bragg gratings can be routinely fabricated to operate over a wide range of wavelengths, extending from the ultraviolet to the infrared region.

22.4.10 Fiber Bragg Grating as a Reflection Filter

If a broadband radiation is coupled into the fiber, only an appropriate wavelength matching the Bragg condition is reflected, permitting a reflection filter to be made. The Bragg condition determines the wavelength of the reflected light λ_{Bragg}, referred as the Bragg wavelength, as

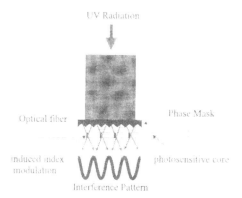

Figure 22.30 Fiber Bragg grating fabrication by the phase-mask method.

$$\lambda_{\text{Bragg}} = 2\Lambda n_{\text{eff}}, \tag{22.26}$$

where n_{eff} is the refractive index of the mode and Λ is the period of the Bragg grating. A strong reflection can be understood from the fact that at each change in the refractive index, some light is reflected. If the reflections from points that are a spatial period apart are in phase, then the various multiple reflections add in phase, leading to a strong reflection. The peak reflectivity R of a grating may be calculated as

$$R = \tanh^2(\kappa L), \tag{22.27}$$

where L is the length of the fiber grating, and the coupling coefficient κ is defined by

$$\kappa = \frac{\pi n \delta n \eta}{\lambda_{\text{Bragg}} n_{\text{eff}}}, \tag{22.28}$$

where n is the cladding index, η is the overlap integral of the forward and backward propagating modes over the perturbed index within the core, and δn is the magnitude of the refractive index modulation.

The bandwidth of the reflection spectrum can be calculated as

$$\Delta\lambda = \frac{\lambda_{\text{Bragg}}^2}{\pi n_{\text{eff}} L} \left(\pi^2 + \kappa^2 L^2\right)^{1/2}. \tag{22.29}$$

The simplest application of fiber Bragg gratings is as reflection filters with bandwidths of approximately 0.05–20 nm. Multiple reflection gratings may be written into a piece of fiber to generate a number of reflections, each at different wavelength. Fiber Bragg gratings may be used as narrow-band filters and reflectors to stabilize semiconductor lasers or DBR lasers, narrow-band reflectors for fiber lasers, simple broad- and narrow-band-stop reflection filters, radiation mode taps, bandpass filters, fiber grating Fabry–Perot etalons, in dispersion compensation schemes, narrow-linewidth dual-frequency laser sources, nonlinear pulsed sources, optical soliton sources, and applications in sensor networks.

22.4.11 Fiber Bragg Grating-Based Multiplexer

Fiber Bragg gratings may also be used as multiplexers, demultiplexers, or as add/drop elements. Figure 22.31 shows a simple scheme used for demultiplexing that includes a fiber Bragg grating in conjunction with an optic circulator. The optical

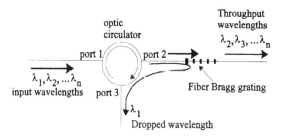

Figure 22.31 Demultiplexing of optical signals by using a fiber Bragg grating and an optic circulator.

radiation at different wavelengths enters into the port 1 of the circulator, comes out through the port 2, and passes through the fiber grating with the reflection peak corresponding to the wavelength λ_1. As a result, the optical signal at λ_1 goes back to the port 2 of the circulator and emerges at the port 3.

Figure 22.32 shows a scheme used for add and drop several wavelengths. It includes three-port optic circulators with a series of electrically tunable fiber gratings for each wavelength. The demultiplexer separates the dropped wavelengths into individual channels and the multiplexer combines wavelengths into transmission fiber line. In the normal state the gratings are transparent to all wavelengths. If a grating is tuned to a specific wavelength, this signal is reflected back to port 2 of the first circulator and exits from port 3. The signals at transmitted wavelengths enter into port 1 of the second optic circulator and exit from port 2. To add or reinsert wavelengths that were dropped, one injects these into port 3 of the second circulator. They first come out of port 1 and travel toward the tuned gratings. Each grating of the array reflects a specific wavelength back to port 1. The reflected signals exit from port 2, where all channels are recovered.

22.4.12   Chirped Fiber Bragg Gratings

In a chirped grating the period of the modulation of the refractive index varies along the grating length. This type of grating can be used for the compensation of the dispersion, which occurs when an optical signal propagates through a fiber. The dispersion associated with transmission through a Bragg grating may be used to compensate for dispersion of the opposite sign. Figure 22.33 shows a schematic of one of the methods used for dispersion compensation. In this method, the propagation delay through a grating is used to provide large dispersion compensation. The chirped grating reflects each wavelength from a different point along its length. Thus, the dispersion imparted by the grating depends on the grating length and is given by

$$\tau = \frac{2L}{v_\mathrm{g}}, \tag{22.30}$$

where v_g is the group velocity of the pulse incident on the grating. Therefore, a grating with a linear wavelength chirp of $\Delta\lambda$ nm will have a dispersion of

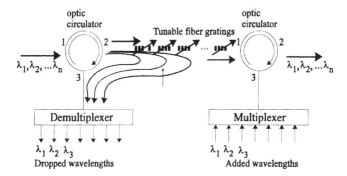

Figure 22.32   Tunable grating filters to add and drop different wavelengths.

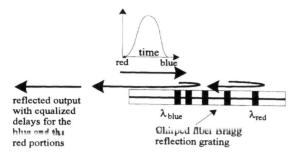

Figure 22.33 Chirped fiber Bragg grating for dispersion compensation.

$$D = \frac{\tau}{\Delta\lambda} \text{ ps/nm.} \qquad (22.31)$$

22.4.13 Fiber Connectors

A fiber connection is defined as the point where two optical fibers are joined together to allow a light signal to propagate from one fiber to the next continuing fiber. Fiber connection has three basic functions: precise alignment, fiber retention, and end protection. To achieve low losses at the connection a precise alignment is required because of extremely small size of the fiber cores. Fiber retention prevents any misalignment or separation that may be caused by tension or manipulation of a cable. Ends protection ensures the optical quality of fiber ends against environmental factors such as humidity and dust.

Fiber connections generally fall into two categories: the permanent, which uses a fiber splice; temporary (nonfixed), which uses a fiber-optic connector. Splices and connectors are used in different places. Typical uses of splices include pigtail vault splices, distribution breakouts, and reel ends. On the other hand, connectors are used as interfaces between terminal on LANs, patch panels, and terminations into transmitters and receivers. The quality of fiber connections whether splices or connectors is estimated from the point of view of signal loss or attenuation.

Interconnection Losses

Interconnection loss is the loss of signal or light intensity as it travels through the fiber joints. These losses are caused by several factors and can be classified into two categories: extrinsic and intrinsic losses. Intrinsic losses are related to the mismatches between fiber parameters and do not depend on applied technique as extrinsic losses do. The differences include variations in core and/or outer diameter, NA mismatch, and differences in the fiber's index profile, core ellipticity, and core eccentricity (Fig. 22.34).

If the core diameter of the receiving fiber is smaller than that of the emitting fiber, some light is lost in the cladding of the receiving fiber. The same is true for NA mismatch. When connecting fibers having different NAs, testing will show a significant loss when the receiving fiber's NA is smaller than that of the emitting fiber, because a fiber with a lower NA cannot capture all light coming out from a fiber with a greater NA. If the core of either transmitting or receiving fiber is elliptic rather than

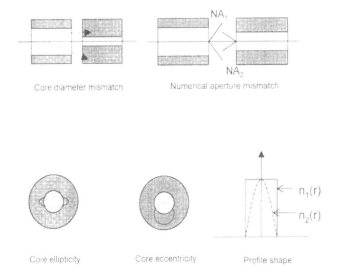

Figure 22.34 Intrinsic losses mechanism.

circular, the collecting area of the receiving core is reduced, which results in less captured light in the receiving fiber. Any core eccentricity variations in both fibers also reduce the collecting area of the receiving fiber.

Extrinsic losses are due to alignment errors and poor fiber-end quality, and depend on connection technique. Frequent causes of extrinsic losses include fiber-end misalignment, poor quality of the fiber-end face, longitudinal separation of fiber ends, contamination of fiber ends, and angular misalignment of bonded fibers. Figure 22.35 shows typical extrinsic loss mechanisms. In addition, back reflection caused by the abrupt change of refractive index at the end of each fiber also contributes to connection loss.

Connectors

A connector is a demountable device used to connect and disconnect fibers. The connector permits two fibers to be connected and disconnected hundred of times easily. Connectors can be divided into two basic categories: expanded beam-coupled (lens-coupled) and butt-coupled. Expanded beam coupling requires a lens system in the connector to increase the size of the beam of light at one connector and to reduce it at the other. In butt coupling, the connectors are mechanically positioned close enough for the light to pass from one fiber to another.

Expanded beam connectors are frequently not used because of cost and performance disadvantages: they never achieve the low losses of butt connectors. In addition, such connectors use index-matching oil, which attracts dust and can affect the characteristics of the connector. Thus, fabrication of the majority of expanded beam connectors has been discontinued.

Although there are different types of butt connectors, the rigid ferrule designs are the most popular butt connectors used today. Regardless of type and manufac-

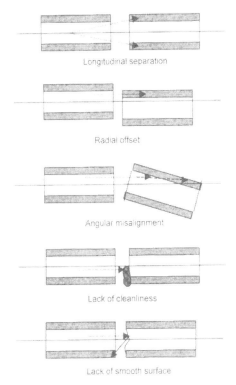

Figure 22.35   Extrinsic losses mechanism.

turer, a rigid ferrule connector comprises five basic structural elements: ferrule, retaining nut, backshell, boot, and cap (Fig. 22.36).

A ferrule is a rigid tube within a connector with a central hole that contains and aligns a fiber. It can be made of different materials, such as ceramic, steel and plastic. Ceramic ferrules offer the lowest insertion loss and the best repeatability, while steel ferrules permit re-polishing and plastic ferrules have the advantages of low cost. In addition, ferrules can have different shapes – straight thin cylinder, conical, or stepped. The retaining nut provides the mechanism by which the connector is secured to an adapter. It can be made of either steel or plastic. Retaining nuts are threaded, bayonet, or push–pull to make a connection. The backshell is the portion

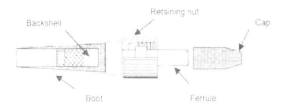

Figure 22.36   Connector.

of the connector in the back of the retaining nut for attaching the cable to the connector. The attachment is made of metal or plastic and provides the principal source of strength between the cable and connector. The boot is a plastic component that slides over the cable and the backshell of the connector. It limits the radius of curvature of the cable and relieves strains on the optical fiber. The cap is a plastic cover that protects the end of a connector ferrule from dust contamination and damage. In addition to these structural elements, some connectors have a crimp ring or other mechanism for attaching the cable to the backshell.

It should be noted that most fiber-optical connectors do not use the male–female configuration common to electronics connectors. Instead, a coupling device called a mating (or alignment) sleeve is used to mate the connectors (Fig. 22.37). There are two kinds of mating sleeves: the flouting style adapter to connect two single-mode or multimode cables and the panel mount style adapter to mate fiber-optical transmitters and receivers to the optical cable via a connector. Other coupling devices are hybrid adapters that permit mating of different connector types.

One important criterion is connector performance. When selecting a connector, the following characteristics should be analyzed: insertion loss (usually 0.1–0.6 dB per connection); back reflection (-20 to -60 dB); repeatability of connection. In the short life of fiber-optic systems, fiber-optic connectors have gone through four generations. Fiber-optic connectors developed in the first generation were mostly screw-type (e.g., SMA 905 and SMA 906). Because of the lack of rotational alignment, these connectors have a large amount of variation in the insertion loss as the connector is unmated and remated.

Connectors of the second generation, such as bayonet type [e.g., straight tip (ST)], biconic, and FC types, solved many problems associated with earlier connectors, providing rotational alignment and greatly improving repeatability. ST is a keyed and contact connector with low average losses (0.5 dB). ST design includes a spring-loading twist-and-lock bayonet coupling that keeps the fiber and the ferrule from rotating during multiple connections. However, this connector is not pull-proof.

The biconic connector is non-keyed, non-contact with rotational sensibility and susceptible to vibrations. Originally it was developed as a multimode fiber connector; later, it was the first successful single-mode connector used by the telecommunications industry. Actually, biconic connectors are suitable for single-mode and multimode fibers.

In the FC connector, a new method called face contact (FC) has been introduced to reduce backreflections. FC provides a flat-end face contact between joining connectors with low average losses (0.4 dB). FC/PC is a new version of FC that

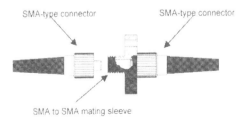

Figure 22.37 Mating (or alignment) sleeve.

includes a new polishing method called PC (physical contact). The PC uses a curved polishing that dramatically reduces back reflections. PC style offers very good performance for single-mode and multimode fiber connectors. They are commonly used in analog systems (CATV) and high bit-rate systems. The majority of the connectors of the second generation are threaded connectors. This is inconvenient and decreases packing density.

Connectors of the third and fourth generation tend to be push–pull types. This type of connector has shorter connection time and allows a significant increase in packing density. Since packing density is becoming more important, fourth-generation connectors are focused on ease of use and reduced size. One of the new designs is the LC connector. This connector offers an easy installation and high pack density because of a reduced connector's diameter.

Table 22.4 describes some of the widespread connector types.

Table 22.4 Connector Types

CONNECTOR	FIBER TYPE	TYPICAL LOSS (dB)	APPLICATIONS	POPULARITY
SMA	MM	0.25 - 0.60	Military	Fading
Biconic	SM, MM	0.20 - 0.60	Telecommunications	Fading
FC	SM, MM	0.20 - 0.50	Telecommunications and Datacom	Widely used
D4	SM, MM	0.3-0.50	Telecommunications	Fading
ST	SM, MM	0.15 - 0.50	Inter/intra-building	Widely used
SC	SM, MM	0.20 - 0.40	Telecommunications	Growing
FDDI	SM, MM	0.20 - 0.50	Fiber Optic Networks	Widely used
ESCON	SM, MM	0.15 - 0.30	Fiber Optic Networks	Widely used
LC	SM, MM	0.10 - 0.15	Telecommunications and Fiber Optics Networks	Newer

22.4.14   Fiber Splicers

Splices are permanent connections between fibers. Splices are used in two situations: mid-span splices, which connect two lengths of cable; and pigtails, at the ends of a main cable, when rerouting of optical paths is not required or expected. Splices offer lower attenuation, easier installation, lower backreflection, and greater physical strength than connectors, and they are generally less expensive. In addition, splices can fit inside cable, offer a better hermetic seal, and allow either individual or mass splicing. There are two basic categories of splices: fusion splices and mechanical splices.

Fusion Splicing

The most common type of splice is a fusion splice, formed by aligning and welding the ends of two optical fibers together. Usually a fusion splicer includes an electric arc welder to fuse the fibers, alignment mechanisms, and a camera or binocular microscope to magnify the alignment by 50 times or more. The fusion parameters can usually be changed to suit particular types of fibers, especially if it is necessary to fuse two different fibers. After the splicing procedure, the previously removed plastic coating is replaced with a protective plastic sleeve.

Fusion splicing provides the lowest connection loss, keeping losses as low as 0.05 dB. Also, fusion splices have lower consumable cost per connection than mechanical splices. However, the capital investment in equipment to make fusion splices is significantly higher than that for mechanical splices. Fusion splices must be performed in a controlled environment, and should not be done in open spaces because of dust and other contamination. In addition, fusion splices cannot be made in an atmosphere that contains explosive gasses because of the electric arc generated during this process.

Mechanical Splices

A mechanical splice is a small fiber connector that precisely aligns two fibers together and then secures them by clamping them within a structure or by epoxying the fibers together. Because tolerances are looser than in fusion splicing, this approach is used more often with multimode fibers than single-mode fibers. Although losses tend to be slightly higher than those of fusion splices and back reflections can be a concern, mechanical splices are easier to perform and the requirements for the environment are looser for mechanical splicing than those for fusion splicing. Generally, the consumables for a mechanical splice results in a higher cost than consumables for fusion splices; however, the equipment needed to produce a mechanical splice is much less expensive than the equipment for fusion splices.

To prepare mechanical splices, the fibers are first stripped of all buffer material, cleaned, and cleaved. Cleaving a fiber provides a uniform surface, which is perpendicular to a fiber axis, needed for maximum light transmission to the other fiber. Then the two ends of the fibers are inserted into a piece of hardware to obtain a good alignment and to permanently hold the fibers' ends. Many hardware devices have been developed to serve these goals. The most popular devices can be divided in two broad categories: capillary splices and V-groove splices.

Capillary splice is the simplest form of mechanical splicing. Two fiber ends are inserted into a thin capillary tube made of glass or ceramic, as illustrated in

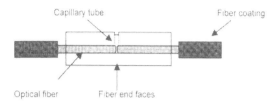

Figure 22.38 Capillary splice.

Fig. 22.38. To decrease backreflections from the fiber ends, an index-matching gel is typically used in this splice. The fibers are held together by compression or friction, although epoxy may be used to permanently secure the fibers.

The V-groove splice is probably the oldest and still most popular method, especially for multifiber splicing of ribbon cable. This type of splice is either crimped or snapped to hold the fibers in place. Many types of V-groove splices have been developed using different techniques. The simplest technique confines the two fibers between two plates, each one containing a groove into which the fiber fits. This approach centers the fiber core, regardless of variation in the outer diameter of the fiber (Fig. 22.39(a)).

The popular V-groove technique uses three precision rods tightened by means of an elastic band or shrinkable tube (Fig. 22.39(b)). The splice loss in this method depends strongly on the fiber size (core and cladding diameter variations) and eccentricity (the position of the core relative to the center of the fiber). V-groove techni-

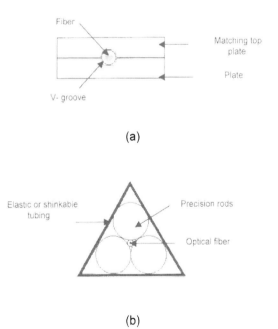

(a)

(b)

Figure 22.39 V-groove splice cross sections: (a) V-groove using two plates and (b) V-groove using three rods.

ques properly carried out with multimode fibers result in splice losses of the order of 0.1 dB or less. Some of them can be applied to single-mode fibers as well.

The borderline connectors and splices is rather indefinite. One example is the rotary mechanical splice (RMS), which is a disconnectable splice made by attaching ferrules to the two fiber ends, joining the ferrules in a housing, and holding the assembly together with a spring clip. The assembly can be disconnected by removing the slip and is rated to suvive 250 mating cycles. Rotary mechanical splices provide a simple and quick method of joining single-mode and multimode fibers with mean losses less than 0.2 dB without the need for optical or electronic monitoring equipment.

Splices, once completed, whether fusion or mechanical, are then placed into splicing trays designed to accommodate the particular type of splice in use. On the other hand, fiber-optic splices require protection from the environment, so they are stored in a splice enclosure. These special boxes are available for indoors as well as outdoor mounting. The outdoor type should be weatherproof, with a watertight seal. Additionally, splices enclosures protect stripped fiber-optic cable and splices from strain and help organize spliced fibers in multifiber cables.

REFERENCES

Optical Fibers

1. Agrawal, G. P., *Nonlinear Fiber Optics*, 2nd edn, Academic Press, San Diego, 1995.
2. Ghatak, A. and K. Thyagarajan, *Introduction to Fiber Optics*, Cambridge University Press, Cambridge, 1998.
3. Goff, D. R., *Fiber Optic Reference Guide*, Focal Press, Boston, 1996.
4. Hecht, J., *Understanding Fiber Optics*, SAMS, Indiana, 1987.
5. Keiser, G., *Optical Fiber Communications*, McGraw-Hill, Singapore, 1991.
6. Midwinter, J., *Optical Fibers for Transmission*, John Wiley and Sons, New York, 1979.
7. Wilson, J. and J. F. B. Hawkes, *Optoelectronics: An Introduction*, Prentice Hall, London, 1983.
8. Yeh, C., *Handbook of Fiber Optics: Theory and Applications*, Academic Press, San Diego, 1990.

Special Fibers

9. Agrawal, G. P., *Nonlinear Fiber Optics*, 2nd edn, Academic Press, San Diego, 1995.
10. Armitage, J. R., R. Wyatt, B. J. Ainslie, and S. P. Craig-Ryan, "Highly Efficient 980 nm Operation of an Yb^{3+}-Doped Silica Fiber Laser," *Electron. Lett.*, **25**, 298–299 (1989).
11. Artjushenko, V. G., L. N. Butvina, V. V. Vojtsekhovsky, E. M. Dianov, and J. G. Kolesnikov, "Mechanisms of Optical Losses in Polycrystalline Fibers," *J. Lightwave Technol.*, **LT-4**, 461–464 (1986).
12. Berman, I. E., "Plastic Optical Fiber: A Short-Haul Solution," *Optics & Photonics News*, **9**, 29 (1998).
13. Bornstein, A. and N. Croitoru, "Experimental Evaluation of a Hollow Glass fiber," *Appl. Opt.*, **25**, 355–358 (1986).
14. Desurviere, E., *Erbium-Doped Fiber Amplifiers*, John Wiley and Sons, New York, 1994.
15. DiGiovanni, D. J. and M. H. Muendel, "High Power Fiber Lasers," *Optics & Photonics News*, 26–30 (January 1999).
16. Drexhage, M. G. and C. T. Moynihan, "Infrared Optical Fibers," *Scientific American*, **259**, 76–81 (November 1988).

17. France, P. W., S. F. Carter, M. W. Moore, and C. R. Day, "Process in Fluoride Fibers for Optical Communications," *British Telecom Technology Journal*, **5**(2), 28–44 (1987) in Selected Papers on Infrared Fiber Optics, *SPIE*, **MS 9**, 59–75 (1990).

18. Garmire, E., T. McMahon, and M. Bass, "Flexible Infrared Waveguides for High-Power Transmission, " *IEEE J. Quantum Electron.*, **QE-16**(1), 23–32 (1980).

19. Ghatak, A. and Thyagarajan, K., *Introduction to Fiber Optics*, Cambridge University Press, New York, 1998.

20. Harrington, J. A., A. G. Standlee, A. C. Pastor, and L. G. DeShazer, "Single-Crystal Infrared Fibers Fabricated by Traveling-Zone Melting," *Infrared Optical Materials and Fibers III, Proc. SPIE*, **484**, 124–127 (1984).

21. Hecht, J., *Understanding Fiber Optics*, 1st edn, SAMS, Carmel, Indiana, 1992.

22. Hecht, J., "Perspectives: Fiber Lasers Prove Versatile," *Laser Focus World*, 73–77 (1998).

23. Ikedo, M., M. Watari, F. Ttateishi, and H. Ishiwatari, "Preparation and Characteristics of the TlBr-TlI Fiber for a High Power CO_2 Laser Beam," *J. Appl. Phys.*, **60**(9), 3035–3039 (1986).

24. Kaminov, I. P. and T. L. Koch, *Optical Fiber Telecommunications IIIB*, Academic Press, San Diego, 1997.

25. Katsuyama, T. and H. Matsumura, "Low-Loss Te-Based Chalcogenide Glass Optical Fibers," *Appl. Phys. Lett.*, **49**(1), 22–23 (1986).

26. Katsuyama, T. and H. Matsumura, *Infrared Optical Fibers*, Adam Hilger, Bristol, 1989.

27. Katzir, A., *Lasers and Optical Fibers in Medicine*, Academic Press, San Diego, 1993.

28. Keiser, G. E., "A Review of WDM Technology and Applications," *Optical Fiber Technology*, **5**, 3–39 (1999).

29. Klocek, P. and G. H. Sigel, Jr., *Infrared Fiber Optics*, SPIE Optical Engineering Press, Bellingham, 1989.

30. Lüthy, W. and H. P. Weber, "High-Power Monomode Fiber Lasers," *Opt. Eng.*, **34**(8), 2361–2364 (1995).

31. Matsuura, Y., M. Saito, M. Miyagi, and A. Hongo, "Loss Characteristics of Circular Hollow Waveguides for Incoherent Infrared Light," *JOSA A*, **6**(3), 423–427 (1989).

32. Minelly, J. D. and P. R. Morkel, "320-mW Nd^{3+}-Doped Single-Mode Fiber Superfluorescent Source," *CLEO'93*, **CFJ6**, 624–626 (1993).

33. Miyashita, T. and T. Manabe, "Infrared Optical Fibers, *IEEE J. Quantum Electron.*, **QE-18**(10), 1432–1450 (1982).

34. Ono, T., "Plastic Optical Fiber: The Missing Link For Factory, Office Equipment," *Photonics Spectra*, **29**, Issue 11, 88–91 (1995).

35. *Optical Fiber Lasers and Amplifiers*, France, P. W., ed., Blackie and Son, Glasgow, 1991.

36. Pask, H. M., R. J. Carman, D. C. Hanna, *et al.*, "Ytterbium-Doped Silica Fiber Lasers: Versatile Sources for the 1–1.2 μm Region," *IEEE J. Selected Topics in Quantum Electron.*, **1**(1), 2–13 (1995).

37. Sa'ar, A. and A. Katzir, "Intrinsic Losses in Mixed Silver Halide Fibers," *Infrared Fiber Optics, Proc. SPIE*, **1048**, 24–32 (1989).

38. Sa'ar, A., N. Barkay, F. Moser, I. Schnitzer, A. Levite, and A. Katzir, "Optical and Mechanical Properties of Silver Halides Fibers," *Infrared Optical Materials and Fibers V, Proc. SPIE*, **843**, 98–104 (1987).

39. Saito, M. and M. Takizawa, "Teflon-Clad As-S Glass Infrared Fiber with Low-Absorption Loss," *J. Appl. Phys.*, **59**(5), 1450–1452 (1986).

40. Sakaguchi, S. and S. Takahashi, "Low-Loss Fluoride Optical Fibers for Midinfrared Optical Communication," *J. Lightwave Technol.*, **LT-5**(9), 1219–1228 (1987).

41. Tran, D. C., G. H. Sigel, Jr., and B. Bendow, "Heavy Metal Fluoride Glasses and Fibers: A Review," *J. Lightwave Technol.*, **LT-2**(5), 566–586 (1984).

42. Vasil'ev, A. V., E. M. Dianov, L. N. Dmitruk, V. G. Plotnichenko, and V. K. Sysoev, "Single-Crystal Waveguides for the Middle Infrared Range," *Soviet Journal of Quantum Electronics*, **11**(6), 834–835 (1981).

43. Worrell, C. A., "Infra-Red Optical Properties of Glasses and Ceramics for Hollow Waveguides Operating at 10.6 μm Wavelength," *Infrared Optical Materials and Fibers V, Proc. SPIE*, **843**, 80–87 (1987).

44. Wysocki, J. A., R. G. Wilson, A. G. Standlee, *et al.*, "Aging Effects in Bulk and Fiber TlBr-TlI," *J. Appl. Phys.*, **63**(9), 4365–4371 (1988).

45. Zyskind, J. L., Nagel, J. A., and H. D. Kidorf, "Erbium-Doped Fiber Amplifiers for Optical Communications," in *Optical Fiber Telecommunications IIIB*, Kaminov, I. P. and T. L. Koch, eds, Academic Press, San Diego, 1997.

Fiber Optic Components

46. Agrawal, G. P., *Nonlinear Fiber Optics*, 2nd edn, Academic Press, San Diego, 1995.

47. Allard, F. C., *Fiber Optics Handbook: for Engineers and Scientists*, McGraw-Hill, New York, 1990.

48. Azzam, R. M. A. and N. M. Bashara, *Ellipsometry and Polarized Light*, North-Holland, Amsterdam, 1989.

49. Bergh, R. A., M. J. F. Digonnet, H. C. Lefevre, S. A. Newton, and H. J. Shaw, "Single Mode Fiber Optic Components," *Fiber Optics-Technology, Proc. SPIE*, **326**, 137–142 (1982).

50. Born, M. and E. Wolf, *Principles of Optics*, 6th edn, Cambridge University Press, Cambridge, 1997.

51. Calvani, R., R. Caponi, and F. Cisternino, "Polarization Measurements on Single-Mode Fibers," *J. Lightwave Technol.*, **7**(8), 1187–1196 (1989).

52. Chomycz, B., *Fiber Optic Installations*, McGraw-Hill, New York, 1996.

53. Culshaw, B., C. Michie, P. Gardiner, and A. McGown, "Smart Structures and Applications in Civil Engineering," *Proc. IEEE*, **84**(1), 78–86 (1996).

54. Ghatak, A. and K. Thyagarajan, *Introduction to Fiber Optics*, Cambridge University Press, Cambridge, 1998.

55. Goff, D. R., *Fiber Optic Reference Guide*, Focal Press, Boston, 1996.

56. Hayes, J., *Fiber Optics Technician's Manual*, International Thomson Publishing Company, New York, 1996.

57. Huard, S., *Polarization of Light*, John Wiley and Sons, Masson, Belgium, 1997.

58. Kaminov, I. P., "Polarization in Optical Fibers," *IEEE J. Quantum Electron.*, **QE-17**(1), 15–22 (1981).

59. Kashyap, R., "Photosensitive Optical Fibers: Devices and Applications," *Optical Fiber Technology*, **1**, 17–34 (1994).

60. Keiser, G., *Optical Fiber Communications*, McGraw-Hill, Singapore, 1991.

61. Keiser, G. E., "A Review of WDM Technology and Applications," *Optical Fiber Technology*, **5**, 3–39 (1999).

62. Kersey, A. D., "A Review of Recent Developments in Fiber Optic Sensor Technology," *Optical Fiber Technology*, **2**, 291–317 (1996).

63. Miller, C., *Optical Fiber Splicers and Connectors*, Marcel Dekker, New York, 1986.

64. Murata, H., *Handbook of Optical Fibers and Cables*, Marcel Dekker, New York, 1996.

65. Noda, J., K. Okamoto, and I. Yokohama, "Fiber Devices Using Polarization-Maintaining Fibers," Reprinted from *Fiber and Integrated Optics*, **6**(4), 309–330 (1987) in *On Selected Papers on Single-Mode Optical Fibers*, Brozeit, A., K. D. Hinsch, and R. S. Sirohi, eds, *SPIE*, **MS 101**, 23–33 (1994).

66. Pearson, E. R., *The Complete Guide to Fiber Optic Cable System Installation*, Delmar Publishers, Albany, New York, 1997.

67. Rashleigh, S. C., "Origins and Control of Polarization Effects in Single-Mode Fibers," *J. Lightwave Technol.*, **LT-1**(2), 312–331 (1983).
68. Sirkis, J. S., "Unified Approach to Phase-Strain-Temperature Models for Smart Structure Interferometric Optical Fiber Sensors: part 1, Development," *Optical Engineering*, **32**(4), 752–761 (1993).
69. Stolen, R. H. and R. P. De Paula, *Proceeding of the IEEE*, **75**(11), 1498–1511 (1987).
70. Todd, D. A., G. R. J. Robertson, and M. Failes, "Polarization-Splitting Polished Fiber Optic Couplers," *Opt. Eng.*, **32**(9), 2077–2082 (1993).
71. Yariv, A., *Optical Electronics in Modern Communications*, 5th edn, Oxford University Press, New York, 1997.

Erbium-Doped Fibers

72. Agrawal, G. P., *Nonlinear Fiber Optics*, 2nd edn, Academic Press, San Diego, 1995.
73. Desurviere, E., "Erbium-Doped Fiber Amplifiers," John Wiley and Sons, New York, 1994.
74. Ghatak, A. and K. Thyagarajan, *Introduction to Fiber Optics*, Cambridge University Press, New York, 1998.
75. Keiser, G. E., "A Review of WDM Technology and Applications," *Optical Fiber Technology*, **5**, 3–39 (1999).
76. Zyskind, J. L., J. A. Nagel, and H. D. Kidorf, "Erbium-Doped Fiber Amplifiers for Optical Communications," in *Optical Fiber Telecommunications IIIB*, Kaminov, I. P. and T. L. Koch, eds, Academic Press, San Diego, 1997.

Double-Clad Fiber Lasers

77. Armitage, J. R., R. Wyatt, B. J. Ainslie, and S. P. Craig-Ryan, "Highly Efficient 980 nm Operation of an Yb^{3+}-Doped Silica Fiber Laser," *Electron. Lett.*, **25**(5), 298–299 (1989).
78. Bell Labs Ultra-High-Power, "Single-Mode Fiber Lasers," Technical Information.
79. DiGiovanni, D. J. and M. H. Muendel, "High Power Fiber Lasers," *Optics & Photonics News*, 26–30 (1999).
80. Hecht, J., "Perspectives: Fiber Lasers Prove Versatile," *Laser Focus world*, 73–77 (1998).
81. Kaminov, L. P. and T. L. Koch, *Optical Fiber Telecommunications IIIB*, Academic Press, San Diego, 1997.
82. Lüthy, W. and H. P. Weber, "High-Power Monomode Fiber Lasers," *Opt. Eng.*, **34**(8), 2361–2364 (1995).
83. Minelly, J. D. and P. R. Morkel, "320-mW Nd^{3+}-Doped Single-Mode Fiber Superfluorescent Source," *CLEO'93*, **CFJ6**, 624–626 (1993).
84. Pask, H. M., R. J. Carman, D. C. Hanna, *et al.*, "Ytterbium-Doped Silica Fiber Lasers: Versatile Sources for the 1–1.2 μm Region," *IEEE J. Selected Topics in Quantum Electron.*, **1**(1), 2–13 (1995).

Infrared Fibers

85. Artjushenko, V. G., L. N. Butvina, V. V. Vojtsekhovsky, E. M. Dianov, and J. G. Kolesnikov, "Mechanisms of Optical Losses in Polycrystalline Fibers," *J. Lightwave Technol.*, **LT-4**(4), 461–464 (1986).
86. Drexhage, M. G. and C. T. Moynihan, "Infrared Optical Fibers," *Scientific American*, 110–115 (1988).
87. France, P. W., S. F. Carter, M. W. Moore, and C. R. Day, "Progress in Fluoride Fibers for Optical Communications," *British Telecom Technology Journal*, **5**(2), 28–44 (1987) in *Selected Papers on Infrared Fiber Optics, SPIE*, **MS 9**, 59–75 (1990).

88.   Harrington, J. A., A. G. Standlee, A. C. Pastor, and L. G. DeShazer, "Single-Crystal Infrared Fibers Fabricated by Traveling-Zone Melting," *Infrared Optical Materials and Fibers III, Proc. SPIE*, **484**, 124–127 (1984).

89.   Kedo, M., M. Watari, F. Ttateishi, and H. Ishiwatari, "Preparation and Characteristics of the TlBr-TlI Fiber for a High Power CO_2 Laser Beam," *J. Appl. Phys.*, **60**(9), 3035–3039 (1986).

90.   Kaminov, I. P. and T. L. Koch, *Optical Fiber Telecommunications IIIB*, Academic Press, San Diego, 1997.

91.   Kanamori, T., Y. Terunuma, S. Takahashi, and T. Miyashita, *J. Lightwave Technol.*, **LT-2**, 607 (1984).

92.   Katsuyama, T. and H. Matsumura, "Low-Loss Te-Based Chalcogenide Glass Optical Fibers," *Appl. Phys. Lett.*, **49**(1), 22–23 (1986).

93.   Katsuyama, T. and H. Matsumura, *Infrared Optical Fibers*, Adam Hilger, Bristol, 1989.

94.   Miyashita, T. and T. Manabe, "Infrared Optical Fibers," *IEEE J. Quantum Electron.*, **QE-18**(10), 1432–1450 (1982).

95.   Sa'ar, A. and A. Katzir, "Intrinsic Losses in Mixed Silver Halide Fibers," *Infrared Fiber Optics, Proc. SPIE*, **1048**, 24–32 (1989).

96.   Sa'ar, A., N. Barkay, F. Moser, I. Schnitzer, A. Levite, and A. Katzir, "Optical and Mechanical Properties of Silver Halides Fibers," *Infrared Optical Materials and Fibers V, Proc. SPIE*, **843**, 98–104 (1987).

97.   Saito, M. and M. Takizawa, "Teflon-Clad As-S Glass Infrared Fiber with Low-Absorption Loss," *J. Appl. Phys.*, **59**(5), 1450–1452 (1986).

98.   Sakaguchi, S. and S. Takahashi, "Low-Loss Fluoride Optical Fibers for Midinfrared Optical Communication," *J. Lightwave Technol.*, **LT-5**(9), 1219–1228 (1987).

99.   Tran, D. C., G. H. Sigel, Jr., and B. Bendow, "Heavy Metal Fluoride Glasses and Fibers: A Review," *J. Lightwave Technol.*, **LT-2**(5), 566–586 (1984).

100. Vasil'ev, A. V., E. M. Dianov, L. N. Dmitruk, V. G. Plotnichenko, and V. K. Sysoev, "Single-Crystal Waveguides for the Middle Infrared Range," *Soviet Journal of Quantum Electronics*, **11**(6), 834–835 (1981).

101. Wysocki, J. A., R. G. Wilson, *et al.*, "Aging Effects in Bulk and Fiber TlBr-TlI," *J. Appl. Phys.*, **63**(9), 4365–4371 (1988).

Hollow Waveguides

102. Bornstein, A. and N. Croitoru, "Experimental Evaluation of a Hollow Glass Fiber," *Appl. Opt.*, **25**(3), 355–358 (1986).

103. Garmire, E., T. McMahon, and M. Bass, "Flexible Infrared Waveguides for High-Power Transmission," *IEEE J. Quantum Electron.*, **QE-16**(1), 23–32 (1980).

104. Katsuyama, T. and H. Matsumura, *Infrared Optical Fibers*, Adam Hilger, Bristol, 1989.

105. Klocek, P. and G. H. Sigel, Jr., *Infrared Fiber Optics*, SPIE Optical Engineering Press, Bellingham, 1989.

106. Matsuura, Y., M. Saito, M. Miyagi, and A. Hongo, "Loss Characteristics of Circular Hollow Waveguides for Incoherent Infrared Light," *JOSA A*, **6**(3), 423–427 (1989).

107. Worrell, C. A., "Infra-Red Optical Properties of Glasses and Ceramics for Hollow Waveguides Operating at 10.6 μm Wavelength," *Infrared Optical Materials and Fibers V, Proc. SPIE*, **843**, 80–87 (1987).

Plastic Fibers

108. Berman, L. E., "Plastic Optical Fiber: A Short-Haul Solution," *Optics & Photonics News*, **9**(2), 29 (1998).

109. Mitsubishi Rayon Homepage.

110. Ono, T., "Plastic Optical Fiber: The Missing Link For Factory, Office Equipment," *Photonics Spectra*, **29**(11), 88–91 (1995).

Fiber Bundles

111. Hecht, J., *Understanding Fiber Optics*, 1st edn, SAMS, Carmel, Indiana, 1992.
112. Katzir, A., *Lasers and Optical Fibers in Medicine*, Academic Press, San Diego, 1993.

Polarization Fiber Components

113. Azzam, R. M. A. and N. M. Bashara, *Ellipsometry and Polarized Light*, North-Holland, Amsterdam, 1989.
114. Bergh, R. A., M. J. F. Digonnet, H. C. Lefevre, S. A. Newton, and H. J. Shaw, "Single Mode Fiber Optic Components," *Fiber Optics-Technology, Proc. SPIE*, **326**, 137–142 (1982).
115. Born, M. and E. Wolf, *Principles of Optics*, 6th edn, Cambridge University Press, Cambridge, 1997.
116. Calvani, R., R. Caponi, and F. Cisternino, "Polarization Measurements on Single-Mode Fibers," *J. Lightwave Technol.*, **7**(8), 1187–1196 (1989).
117. Huard, S., *Polarization of Light*, John Wiley and Sons, Masson, Belgium, 1997.
118. Kaminov, I. P., "Polarization in Optical Fibers," *IEEE J. Quantum Electron.*, **QE-17**(1), 15–22 (1981).
119. Noda, J., K. Okamoto, and I. Yokohama, "Fiber Devices Using Polarization-Maintaining Fibers," Reprinted from *Fiber and Integrated Optics*, **6**(4), 309–330 (1987) in *Selected Papers on Single-Mode Optical Fibers*, Brozeit, A., K. D. Hinsch, and R. S. Sirohi, eds, *SPIE*, **MS 101**, 23–33 (1994).
120. Rashleigh, S. C., "Origins and Control of Polarization Effects in Single-Mode Fibers," *J. Lightwave Technol.*, **LT-1**(2), 312–331 (1983).
121. Stolen, R. H. and R. P. De Paula, *Proceeding of the IEEE*, **75**(11), 1498–1511 (1987).
122. Todd, D. A., G. R. J. Robertson, and M. Failes, "Polarization-Splitting Polished Fiber Optic Couplers," *Opt. Eng.*, **32**(9), 2077–2082 (1993).
123. Yariv, A., *Optical Electronics in Modern Communications*, 5th edn, Oxford University Press, New York, 1997.

Bragg Gratings

124. Agrawal, G., *Nonlinear Fiber Optics*, 2nd edn, Academic Press, San Diego, USA, 1995.
125. Culshaw, B., C. Michie, P. Gardiner, and A. McGown, "Smart Structures and Applications in Civil Engineering," *Proc. IEEE*, **84**(1), 78–86 (1996).
126. Ghatak, A. and K. Thyagarajan, *Introduction to Fiber Optics*, Cambridge University Press, New York, 1998.
127. Kashyap, R., "Photosensitive Optical Fibers: Devices and Applications," *Optical Fiber Technology*, **1**, 17–34 (1994).
128. Keiser, G. E., "A Review of WDM Technology and Applications," *Optical Fiber Technology*, **5**, 3–39 (1999).
129. Kersey, A. D., "A Review of Recent Developments in Fiber Optic Sensor Technology," *Optical Fiber Technology*, **2**, 291–317 (1996).
130. Sirkis, J. S., "Unified Approach to Phase-Strain-Temperature Models for Smart Structure Interferometric Optical Fiber Sensors: Part 1, Development," *Opt. Eng.*, **32**(4), 752–761 (1993).

23

Isotropic Amorphous Optical Materials

LUIS EFRAIN REGALADO and DANIEL MALACARA
Centro de Investigaciones en Optica, León, Mexico

23.1 INTRODUCTION

The materials used in optical instruments and research can be characterized by many physical and chemical properties [1, 6, 9, 18, 27, 31, 32]. The ideal material is determined by the specific intended application. Optical materials can be crystalline or amorphous. Crystalline materials can be isotropic or anisotropic, but amorphous materials can only be isotropic. In this chapter some of the main isotropic amorphous materials used to manufacture optical elements are described.

The optical materials used to make optical elements such as lenses or prisms have several important properties to be considered, most important of which are described below.

23.1.1 Refractive Index and Chromatic Dispersion

The refractive index of a transparent isotropic material is defined as the ratio of the speed of light in vacuum to the speed of light in the material. With this definition, Snell's law of refraction gives

$$n_1 \sin \theta_1 = n_2 \sin \theta_2, \tag{23.1}$$

where n_1 and n_2 are the refractive indices in two transparent isotropic media separated by an interface. The angles θ_1 and θ_2 are the angles between the light rays and the normal to the interface at the point where the ray passes from one medium to the other.

The refractive index for most materials can vary from values close to 1 to values greater than 2, as shown in Table 23.1.

Table 23.1 Refractive Indices of Some
Materials

Material	Refractive index
Air	1.0003
Water	1.33
Acrylic	1.49
Crown glass	1.48–1.70
Flint glass	1.53–1.95
Diamond	2.42

The refractive index n of a given optical material is not the same for all colors: the value is greater for smaller wavelengths. The refractive indices of optical materials are measured at some specific wavelengths, as shown in Table 23.2.

The chromatic dispersion is determined by the principal dispersion $(n_F - n_C)$. Another quantity that determines the chromatic dispersion is the Abbe value for the line d, given by

$$V_d = \frac{n_d - 1}{n_F - n_C}. \tag{23.2}$$

The Abbe value expresses the way in which the refractive index changes with wavelength. Optical materials are mainly determined by the value of these two constants.

Two materials with different Abbe numbers can be combined to form an achromatic lens with the same focal length for red (C) and for blue (F) light. However, the focal length for yellow (d) can be different. This is called the secondary spectrum. The secondary spectrum produced by an optical material or glass is

Table 23.2 Spectral Lines Used to Measure Refractive
Indices

Wavelength (nm)	Spectral line	Color	Element
1013.98	t	Infrared	Hg
852.11	s	Infrared	Cs
706.52	r	Red	He
656.27	C	Red	II
643.85	C'	Red	Cd
589.29	D	Yellow	Na
587.56	d	Yellow	He
546.07	e	Green	Hg
486.13	F	Blue	H
479.99	F'	Blue	Cd
435.83	g	Blue	Hg
404.66	h	Violet	Hg
365.01	i	Ultraviolet	Hg

determined by its partial dispersion. The partial dispersion $P_{x,y}$ for the lines x and y is defined as

$$P_{x,y} = \frac{n_x - n_y}{n_F - n_C}. \tag{23.3}$$

An achromatic lens for the lines C and F without secondary spectrum for yellow light d can be made only if the two transparent materials being used have different Abbe numbers V_d but the same partial dispersion number $P_{d\,F}$.

23.1.2 Other Optical Characteristics

Spectral Transmission

The light transmittance through an optical material is affected by two factors, i.e., the Fresnel reflections at the interfaces and the transparency of the material. The Fresnel reflections in a dielectric material like glass are a function of the angle of incidence, the polarization state of the incident light beam, and the refractive index. At normal incidence in air the irradiance reflectance ρ_λ is a function of the wavelength, thus, the irradiance transmittance T_R due to the reflections at the two surfaces of the glass plate is

$$T_R = [1 - \rho_\lambda^2]^2. \tag{23.4}$$

The spectral transparency has large fluctuations among different materials and is also a function of the wavelength. Any small impurities in a piece of glass with concentrations as small as ten parts per billion can introduce noticeable absorptions at some wavelengths. For example, the well-known green color of window glass is due to ion oxides. The effect of impurities can be so high that the critical angle for prism-shaped materials can vanish for nonoptical grade materials. High-index glasses have a yellowish color due to absorptions in the violet and ultraviolet regions, from the materials used to obtain the high index of refraction. At the spectral regions where a material has absorption the refractive index is not a real number but a complex number n^* that can be written as

$$n^* = n(1 - i\kappa_\lambda), \tag{23.5}$$

where κ_λ is the absorption index, which is related to an extinction coefficient α_λ by

$$\alpha_\lambda = \frac{4\pi n}{\lambda} \kappa_\lambda. \tag{23.6}$$

If α_λ is the extinction coefficient for the material, the irradiance transmittance T_A due to absorption is

$$T_A = e^{-\alpha_\lambda t}, \tag{23.7}$$

where t is the thickness of the sample. Figure 23.1 shows the typical variations of the absorption index κ with the wavelength for metals, semiconductors, and dielectrics. [7] Dielectrics have two characteristic absorption bands. The absorption in the ultraviolet band is due to the lattice structure vibrations. Metals are highly absorptive at long wavelengths, due to their electrical conductivity, but they become transparent at short wavelengths. The absorption band in semiconductors is around the visible region.

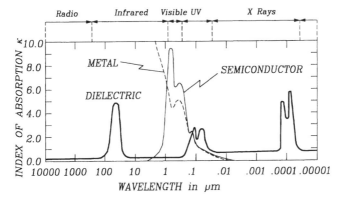

Figure 23.1 Absorption index as a function of the wavelength for optical materials. (Adapted from Kingery [7].)

The total transmittance T_λ of a sample with thickness t talking into consideration surface reflections as well as internal absorption is

$$T_\lambda = T_R T_A = [1 - \rho_\lambda^2]^2 e^{-\alpha_\lambda t}. \tag{23.8}$$

For dielectric materials far from an absorption region the refractive index n_λ is real and the irradiance reflectance ρ is given by

$$\rho_\lambda = \left[\frac{n_\lambda - 1}{n_\lambda + 1}\right]^2. \tag{23.9}$$

Optical Homogeneity

The degree to which the refractive index varies from point to point within a piece of glass or a melt is a measure of its homogeneity. A typical maximum variation of the refractive index in a melt is $\pm 1 \times 10^{-4}$, but more homogeneous pieces can be obtained. For the case of optical glasses the homogeneity is specified in four different groups, as shown in Table 23.3.

Table 23.3 Homogeneity Groups for Optical Glasses

Homogeneity group	Maximum n_d variation
H1	$\pm 2 \times 10^{-5}$
H2	$\pm 5 \times 10^{-6}$
H2	$\pm 2 \times 10^{-6}$
H3	$\pm 1 \times 10^{-6}$

Table 23.4 Bubble Classes for Optical Glasses

Bubble class	Total area of bubbles (mm^2)
B0	0–0.029
B1	0.03–0.10
B2	0.11–0.25
B3	0.26–0.50

Stresses and Birefringence

The magnitude of the residual stresses within a piece of glass depends mainly on the annealing of the glass. Internal stresses produce birefringence. The quality of the optical instrument may depend very much on the residual birefringence.

Bubbles and Inclusions

Bubbles are not frequent in good-quality optical glass, but they are always present in small quantities. When specifying bubbles and inclusions, the total percentage covered by them is estimated, counting only those $\geq$ 0.05 mm. Bubble classes are defined as in Table 23.4.

23.1.3 Physical and Chemical Characteristics

Thermal Expansion

The dimensions of most optical materials, increase with temperature. The thermal expansion coefficient is also a function of temperature. If we plot the natural algorithm of L/L_0 vs. the temperature T ($^{\circ}$C), we obtain a graph (Fig. 23.2) for glass

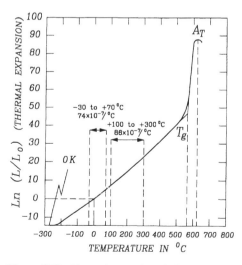

Figure 23.2 Thermal expansion of glass BK7 as a function of the temperature.

BK7. Then, we can easily show that the slope of this curve at the temperature T is equal to the thermal expansion coefficient $\alpha(T)$ defined by

$$\alpha(T) = \frac{1}{L}\frac{dL}{dT}, \tag{23.10}$$

where L is the measured length at the temperature T and L_0 is the length at $0°C$. For the case of glass, typically the expansion coefficient is zero at $0\,K$, increasing steadily until near room temperature; then, the coefficient continues to increase almost linearly until about $700\,K$, after which it increases very sharply.

It is customary to specify two values of the thermal expansion coefficient $\alpha(T)$ for two linear ranges of the plot of $\ln(L/L_0)$ vs. T, where this coefficient remains constant. One interval is from -30 to $+70°C$ and the other from $+100$ to $+300°C$. The transformation temperature T_g is that where the glass begins to transform from a solid to a plastic state. The yield point A_T is the temperature at which the thermal expansion coefficient becomes zero.

Thermal expansion is nearly always very important, but for mirrors, especially astronomical mirrors, it becomes very critical, because any thermal expansion deforms the optical surface.

Hardness

The hardness of materials is important: hard materials are more difficult to scratch but also more difficult to polish. The Moh scale, introduced more than 100 years ago by Friedrich Mohs (1773–1839) is based on which materials scratch others and ranges from 1 to 10, in unequal steps. Table 23.5 shows the materials used to define this scale.

There are many other ways to define the hardness of a material, but another common method used mainly for optical glasses is the Knoop scale, which is defined by the dimensions of a small indentation produced with a diamond under a known pressure. A rhomboidal diamond with vertex angles $172°30'$ and $130°00'$ with respect to the vertical direction is used to produce a mark on the polished surface. Then, the Knoop hardness HK is defined as

$$HK = 1.45\frac{F_N}{d^2} = 14.23\frac{F_K}{d^2}, \tag{23.11}$$

Table 23.5 Moh Hardness Scale

Hardness	Material name	Chemical compound
1	Talc	$Mg_3Si_4O_{10}(OH)_2$
2	Gypsum	$CaSO_4 \cdot 2H_2O$
3	Calcite	$CaCO_3$
4	Fluorite	CaF_2
5	Apatite	$Ca_5(PO_4)_3(F, Cl, OH)$
6	Felspar	$KAlSi_3O_8$
7	Quartz	SiO_2
8	Topaz	$Al_2SiO_4(OH, F)_2$
9	Ruby	Al_2O_3
10	Diamond	C

where F_N is the force in newtons, F_K is the weight (force) of 1 kg, and d is the length of the longest diagonal of the indentation in millimeters. In glasses, a pressure of 1.9613 newtons (the weight of 200 g) is applied to the surface for 10 s. This number is important because it indicates the sensitivity to scratches and is also directly related to the grinding and lapping speeds.

The Rockwell hardness is defined by the depth of penetration under the load of a steel ball or a diamond cone. This method is used for materials softer than glass, such as metals or plastics. There are separate scales for ferrous metals, nonferrous metals, and plastics. The most common Rockwell hardness scales are B and C for metals: the B scale uses a ball, while the C scale uses a cone. The M and R scales are employed for polymers: the M scale is for harder materials and the R scale is for the softer ones.

Elasticity

Elasticity and rigidity are specified by Young's modulus E, which is related to the torsional rigidity modulus G and to the Poisson's ratio μ. Young's modulus E in glasses is measured from induced transverse vibrations of a glass rod at audio frequencies. The torsional rigidity is calculated from torsional vibrations of the rods. The relation between the rigidity modulus, the Poisson's ratio, and Young's modulus is

$$\mu = \frac{E}{2G} - 1. \tag{23.12}$$

The Young's modulus is directly related to the hardness of the material, as can be seen in most glass manufacturer's specifications. Young's modulus is also important when considering thermal and mechanical stresses.

Density

The density is the mass per unit volume. This quantity is quite important in many applications: for example, in space instruments and in ophthalmic lenses. Crown glasses are in general less dense than Flint glasses.

Chemical Properties

Resistance to stain and corrosion by acids or humidity is also important, mainly if the optical glass is going to be used in adverse atmospheric conditions. The water vapor present in the air can produce stains on the glass surface that cannot be wiped out. An accelerated procedure is used to test these properties of optical materials by exposing them to a water vapor saturated atmosphere. The temperature are alternated between 45°C and 55°C in a cycle of about 1 hour. Then, a condensate forms on the material during the heating phase and dries during the cooling phase. With this test the climatic resistance of optical materials is classified in four different classes:

- Class CR1: after 180 hours (1 week) there is no sign of deterioration of the surface of the material.
- Class CR4: after less than 30 hours there are signs of scattering produced by the atmospheric conditions.
- Classes CR2 and CR3 are intermediate to Classes CR1 and CR4.

In a similar manner the resistances to acids and alkalis is tested and classified.

23.1.4 Cost

The cost of the optical material or glass to be used should also be considered when selecting a material for an optical design. The range of prices is extremely large. Glasses and plastics produced in large quantities are cheaper than specially produced materials.

23.2 OPTICAL GLASSES

Glass is a material in a so-called glassy state, structurally similar to a liquid, that at ambient temperature reacts to the impact of force with elastic deformations. Thus, it can be considered as an extremely viscous liquid. It is always an inorganic compound made from sand and sodium and calcium compounds. Plastics, on the other hand, are organic. Glass can also be in natural forms, such as obsidian, found in volcanic places. Obsidian was fashioned into knives, arrowheads, spearheads, and other weapons in ancient times.

Glass is made by mixing silica sand (SiO_2) with small quantities of some inorganic materials such as soda and lime and some pieces of previously fabricated glass, called glass cullet. This glass cullet acts as a fluxing agent that accelerates the melting of the sand. With the silica and the soda, sodium silicate is formed, which is soluble in water. The presence of the lime reduces the solubility of the sodium silicate. This mixture is heated to about 1500°C and then cooled at a well-controlled rate to prevent crystallization. To release any internal stresses, the glass is cooled very slowly at the proper speed in a process called annealing. [5] The quartz sands used to make glass have to be quite free from iron (less than 0.001%), otherwise a greenish-colored glass is obtained. The optical glass used for refracting optical elements such as lenses or prisms has several important properties to be considered, the most important of which is described here. The refractive indices vs. the wavelength for some optical glasses are shown in Fig. 23.3. A diagram of the Abbe number V_d vs. the refractive index n_d, for Schott glasses is shown in Fig. 23.4. The glasses with a letter K at the end of the glass type name are crown glasses and those with a letter F are flint glasses.

The chromatic variation in the refractive index of glasses for the visual spectral range can be represented by several approximate expressions. The simplest and probably oldest formula was proposed by Cauchy [3]:

$$n = A_0 + \frac{A_1}{\lambda^2} + \frac{A_2}{\lambda^4}. \tag{23.13}$$

This formula is accurate to the third or fourth decimal place in some cases. An empirically improved formula was proposed by Conrady [2] as

$$n = A_0 + \frac{A_1}{\lambda^2} + \frac{A_2}{\lambda^{3.5}} \tag{23.14}$$

with an accuracy of one unit in the fifth decimal place. Better formulas have been proposed by several authors: for example, by Herzberger.

From a series expansion of a theoretical dispersion formula, a more accurate expression was used by Schott for many years. Recently, Schott has adopted for glasses a more accurate expression called the Sellmeier formula [25], which is derived

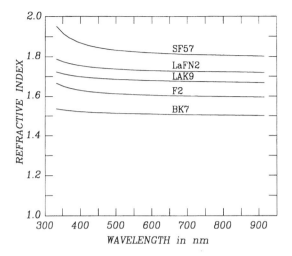

Figure 23.3 Refractive indices as a function of the wavelength for some optical glasses.

from the classical dispersion theory. This formula permits the interpolation of refractive indices in the entire range, from infrared to ultraviolet, with a precision better than 1×10^{-5}, and it is written as

$$n^2 = \frac{B_1 \lambda^2}{\lambda^2 - C_1} + \frac{B_2 \lambda^2}{\lambda^2 - C_2} + \frac{B_3 \lambda^2}{\lambda^2 - C_3}. \tag{23.15}$$

The coefficients are provided by the glass manufacturers, using the refractive indices' values from several melt samples. The values of these coefficients for each type of glass are supplied by the glass manufacturers.

The refractive index and the chromatic dispersion of optical materials are not the only important factors to be considered when designing an optical instrument.

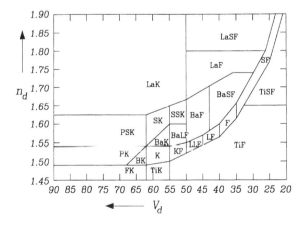

Figure 23.4 Abbe number vs. refractive index chart for optical glasses.

There is such a large variety of optical glasses that to have a complete stock of all types in any optical shop is impossible. Many lens designers have attempted to reduce the list to the most important glasses, taking into consideration important factors such as optical characteristics, availability, and price. Table 23.6 lists some of the most commonly used optical glasses.

The location of these glasses in a diagram of the Abbe number V_d vs. the refractive index n_d, is given in Fig. 23.5. Figure 23.6 shows a plot of the partial dispersion $P_{g,F}$ vs. the Abbe number V_d for these glasses. Table 23.7 shows some physical properties for the glasses in Table 23.6. The visible transmittance for some optical glasses is shown in Fig. 23.7, which can be considered as an expanaion in the visible region of Fig. 23.1.

Ophthalmic glasses are cheaper than optical glass, since their quality requirements are in general lower. They are produced to make spectacles. Table 23.8 lists some of these glasses.

Finally, Table 23.9 lists some other amorphous isotropic materials used in optical elements.

23.3 VITREOUS SILICA

Vitreous silica is a natural material formed by silicon dioxide (SiO_2): it is noncrystalline and isotropic. It is also known as fused quartz or fused silica. This material can be fabricated by fusion of natural crystalline quartz or synthesized by thermal

Table 23.6 Refractive Indices for Some Optical Glasses

Schott name	V_d	n_C	n_d	n_F	n_g
BaF4	43.93	1.60153	1.60562	1.61532	1.62318
BaFN10	47.11	1.66579	1.67003	1.68001	1.68804
BaK4	56.13	1.56576	1.56883	1.57590	1.58146
BaLF5	53.63	1.54432	1.54739	1.55452	1.56017
BK7	64.17	1.51432	1.51680	1.52238	1.52668
F2	36.37	1.61503	1.62004	1.63208	1.64202
K4	57.40	1.51620	1.51895	1.52524	1.53017
K5	59.48	1.51982	1.52249	1.52860	1.53338
KzFSN4	44.29	1.60924	1.61340	1.62309	1.63085
LaF2	44.72	1.73905	1.74400	1.75568	1.76510
LF5	40.85	1.57723	1.58144	1.59146	1.59964
LAK9	54.71	1.68716	1.69100	1.69979	1.70667
LLF1	45.75	1.54457	1.54814	1.55655	1.56333
PK51A	76.98	1.52646	1.52855	1.53333	1.53704
SF2	33.85	1.64210	1.64769	1.66123	1.67249
SF8	31.18	1.68250	1.68893	1.70460	1.71773
SF10	28.41	1.72085	1.72825	1.74648	1.76198
SF56A	26.08	1.77605	1.78470	1.80615	1.82449
SK6	56.40	1.61046	1.61375	1.62134	1.62731
SK16	60.32	1.61727	1.62041	1.62756	1.63312
SK18A	55.42	1.63505	1.63854	1.64657	1.65290
SSKN5	50.88	1.65455	1.65844	1.66749	1.67471

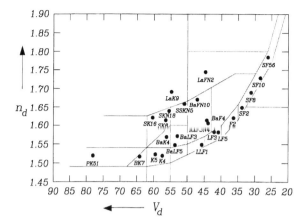

Figure 23.5 Abbe number vs. refractive index for some common optical glasses shown in Table 23.6.

decomposition and oxidation of SiCl$_4$. The optical properties of these types of vitreous silica are similar but not identical.

The fusion of crystalline quartz can be made by direct fusion of large pieces. This fused quartz has a low ultraviolet transmittance due to metallic impurities. On the other hand, it has a high infrared transmittance due to its low content of hydroxyl. For this reason this material is frequently used for infrared windows.

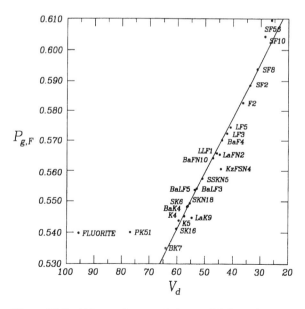

Figure 23.6 Abbe number vs. relative partial dispersion of some common optical glasses shown in Table 23.6.

Table 23.7 Physical Properties for Glasses in Table 23.6

Schott name	Thermal expansion coefficient α (1/°C)	Density ρ (g/cm^3)	Young's modulus E (N/mm^2)	Knoop hardness HK
BaF4	7.9×10^{-6}	3.50	66×10^3	400
BaFN10	6.8×10^{-6}	3.76	89×10^3	480
BaK4	7.0×10^{-6}	3.10	77×10^3	470
BaLF5	8.1×10^{-6}	2.95	65×10^3	440
BK7	7.1×10^{-6}	2.51	81×10^3	520
F2	8.2×10^{-6}	3.61	58×10^3	370
K4	7.3×10^{-6}	2.63	71×10^3	460
K5	8.2×10^{-6}	2.59	71×10^3	450
KzFSN4	4.5×10^{-6}	3.20	60×10^3	380
LaF2	8.1×10^{-6}	4.34	87×10^3	480
LF5	9.1×10^{-6}	3.22	59×10^3	410
LaK9	6.3×10^{-6}	3.51	110×10^3	580
LLF1	8.1×10^{-6}	2.94	60×10^3	390
PK51A	12.7×10^{-6}	3.96	73×10^3	340
SF2	8.4×10^{-6}	3.86	55×10^3	350
SF8	8.2×10^{-6}	4.22	56×10^3	340
SF10	7.5×10^{-6}	4.28	64×10^3	370
SF56A	7.9×10^{-6}	4.92	58×10^3	330
SK6	6.2×10^{-6}	3.60	79×10^3	450
SK16	6.3×10^{-6}	3.58	89×10^3	490
SK18A	6.4×10^{-6}	3.64	88×10^3	470
SSKN5	6.8×10^{-6}	3.71	88×10^3	470

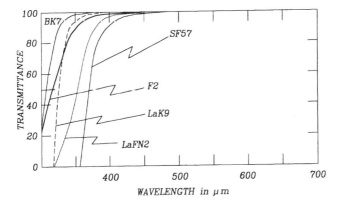

Figure 23.7 Internal transmittance as a function of the wavelength for some optical glasses (without the surface reflections), with a thickness of 5 mm.

Table 23.8 Refractive Indices for Some Ophthalmic Glasses

Glass type	V_d	n_C	n_d	n_F	Density (g/cm^3)
Crown	58.6	1.5204	1.5231	1.5293	2.54
Compatible flint	42.3	1.5634	1.5674	1.5768	3.12
Compatible flint	38.4	1.5926	1.5972	1.6082	3.40
Compatible flint	36.2	1.6218	1.6269	1.6391	3.63
Compatible flint	32.4	1.6656	1.6716	1.6803	3.86
Soft barium flint	33.2	1.6476	1.6533	1.6672	3.91
Soft barium flint	29.6	1.6944	1.7013	1.7182	4.07
Low-density flint	31.0	1.6944	1.7010	1.7171	2.99

If the fusion is made with powdered quartz in the presence of chlorine gases, the transmission in the ultraviolet region is improved, but it is worsened in the infrared. The reason is that the water content increases.

If the hydrolysis of an organosilicon compound in the vapor phase is performed, a synthetic high-purity fused silica free of metals is obtained. It has a high transmission in the ultraviolet but the infrared transmission is low due to its high water content. Some manufacturers have improved their processes to reduce the water, and obtained good transparency in the ultraviolet as well as in the infrared. These good transmission properties make this fused quartz ideal for infrared and ultraviolet windows and for the manufacturing of optical fibers.

The transmittance of three different types of fused quartz are shown in Fig. 23.8. Table 23.10 shows the refractive indices for this material. The thermal expansion of fused silica is $0.55 \times 10^{-6}/°C$ in the range from 0 to 300°C. Below 0°C this coefficient decreases, until it reaches a minimum near $-120°C$.

23.4 MIRROR MATERIALS

Glasses or other materials used to make metal [26] or dielectric coated mirrors do not need to be transparent. Instead, the prime useful characteristic is thermal stability. The thermal expansion coefficient for mirrors has to be lower than for lenses for two reasons: first, a change in the figure of a surface affects the refracted or reflected wavefront four times more in reflecting surfaces than in refracting surfaces; secondly, mirrors are frequently larger than lenses, especially in astronomical instrumentation. Some of the materials used for mirrors are shown in Table 23.11 with the thermal expansion coefficients plotted in Fig. 23.9. They are described below.

Table 23.9 Refractive Indices for Some Optical Isotropic Materials

Material	V_d	n_C	n_d	n_F	n_g
Fused rock crystal	67.6	1.45646	1.45857	1.46324	1.46679
Synthetic fused silica	67.7	1.45637	1.45847	1.46314	1.46669
Fluorite	95.3	1.43249	1.43384	1.43704	1.43950

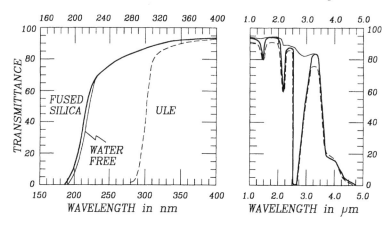

Figure 23.8 Ultraviolet and infrared transmittances for three types of vitreous silica, including surface reflections, with a thickness of 10 mm.

23.4.1 Low-Expansion Glasses

The most well known of these glasses are Pyrex (produced by Corning Glass Works), Duran 50 (Schott), and E-6 (Ohara). These borosilicate glasses in which B_2O_3 replaces the CaO and MgO of soda-lime window glass. The softening temperature is higher in these glasses than in normal window glass.

23.4.2 Very Low Expansion Glasses

The thermal expansion of fused silica can be lowered even more by shifting the zero expansion point to 300 K by adding 7% of titania-silica glass. This glass is a synthetic amorphous silica known as Corning ULE^{TM} titanium silicate.

23.4.3 Glass Ceramics

Glass ceramic materials contain both crystalline and glass phases. They are produced as lithia-alumina glasses where a microcrystalline structure is produced with a special thermal procedure. As opposed to ceramic materials, glass ceramics are nonporous. In general they are not transparent, although sometimes they could be, depending on the size of the crystals, which is about 0.05 µm. Frequently they are turbid with an amber color. The Pyroceram technology developed by Corning led to the production of a glass ceramic with almost zero thermal expansion coefficient, which is manufactured by several companies, such as CER-VIT (produced by Owens-Illinois), Zerodur (Schott), and Cryston-Zero (Hoya).

Table 23.10 Refractive Indices for Fused Quartz

Material	V_d	n_C	n_d	n_F	n_g
Fused silica Corning 7940	67.8	1.4564	1.4584	1.4631	1.4667

Table 23.11 Physical Properties for Some Mirror Materials

Material	Thermal expansion coefficient α $(1/^\circ C)$	Density ρ (g/cm^3)	Young's modulus E (N/cm^2)
Pyrex	2.5×10^{-6}	2.16	5.4×10^6
ULE™	$\pm 0.03 \times 10^{-6}$	2.21	6.7×10^6
Fused quartz	0.55×10^{-6}	2.19	7.2×10^6
Zerodur	$\pm 0.05 \times 10^{-6}$	2.53	9.1×10^6
Beryllium	12.0×10^{-6}	1.85	28.7×10^6
Aluminum	24.0×10^{-6}	2.70	6.8×10^6

23.4.4 Beryllium

Beryllium is a very light and elastic metal. Aluminum, copper, or most other metal mirrors are fabricated by casting the metal. However, for beryllium this is not possible because cast grains are very large and brittle with little intergranular strength. The method of producing beryllium mirrors is by direct compaction of powder.

Hot isostatic pressing has been used in one single step for many years since the late 1960s [21], using pressures up to 50,000 psi to obtain a green compact blank. A multistep cold isostatic pressing in a rubber container followed by vacuum sintering or hot isostatic pressing has also been recently used [22].

Beryllium is highly toxic in powder form. It can be machined only in specially equipped shops using extreme precautions when grinding and polishing. In many respects – mainly its specific weight, elasticity, and low thermal distortion – this is the ideal material for space mirrors.

23.4.5 Aluminum

Aluminum is a very light metal than can also be easily polished and machined. Many mirrors are being made with aluminum for many applications. Even large astronom-

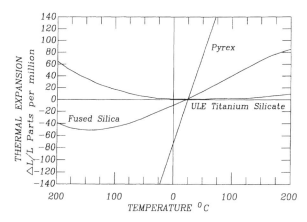

Figure 23.9 Thermal expansion for fused silica, ULE™ titanium silicate, and Pyrex.

Table 23.12 Refractive Indices for Some Optical Plastics

Material	V_d	n_C	n_d	n_F
Acrylic	57.2	1.488	1.491	1.497
Polystyrene	30.8	1.584	1.590	1.604
Polycarbonate	30.1	1.577	1.583	1.597
CR-39	60.0	1.495	1.498	1.504

ical mirrors for telescopes that do not need a high image resolution have been made with this material.

23.5 OPTICAL PLASTICS

There are a large variety of plastics, with many different properties, used to make optical components [10, 19, 30]. Plastics have been greatly improved recently, mainly because of the high-volume production of plastics for CD-roms [23] and spectacle lenses [8, 28]. The most common plastics used in optics, whose properties are given in Table 23.12 are:

(a) Methyl methacrylate, also called acrylic, is the most desirable of all plastics. It is the common equivalent to crown glass. It is relatively inexpensive and can be easily molded and polished. It also has many other advantages such as transmission, resistance to scratch, and moisture absorption.

(b) Polystyrene or styrene is the most common equivalent to flint glass. It is the cheapest plastic and the easiest to mold, but it is difficult to polish. It is frequently used for color-correction lenses.

(c) Polycarbonate has a very high impact strength as well as high temperature stability and very low moisture absorption. However, it is expensive, difficult to polish, and scratches quite easily.

(d) Methyl methacrylate or allyldiglycol carbonate commonly known as CR-39 is a thermosetting material that should be casted and cured with highly controlled conditions. It can be easily machined, molded, and polished. It is as transparent as acrylic. Its most frequent use is in ophthalmic lenses.

Some of the most important physical characteristics of these plastics are given in Table 23.13 and their spectral transmission is shown in Fig. 23.10. (See also Figs 23.7 and 23.8.)

Table 23.13 Physical Characteristics for Some Optical Plastics

Material	Thermal expansion coefficient α (1/°C)	Density ρ (g/cm^3)	Rockwell hardness
Acrylic	65×10^{-6}	1.19	M97
Polystyrene	80×10^{-6}	1.06	M90
Polycarbonate	70×10^{-6}	1.20	M70
CR-39	100×10^{-6}	1.09	M75

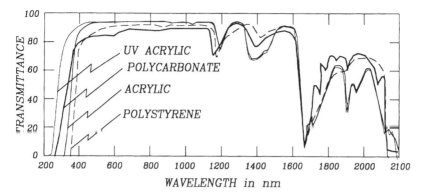

Figure 23.10 Ultraviolet and infrared transmittances of some plastics, including surface reflections, with a thickness of 3.22 mm.

Plastic lenses [4] are very cheap compared with glass lenses if made in large quantities. However, they cannot be used in hostile environments, where the lens is exposed to chemicals, high temperatures, or abrasion. Their use in high-precision optics is limited because of their physical characteristics. There is a lack of plastics with high refractive indices; their homogeneity is not as high as in optical glasses and this is a field needing intense investigation. Plastics have a low specific gravity and their thermal expansion is high.

23.6 INFRARED AND ULTRAVIOLET MATERIALS

Most glasses are opaque to infrared and ultraviolet radiation. If a lens or optical element has to be transparent at these wavelengths special materials have to be selected.

The infrared spectral regions from red light to about 4 μm are quite important. The research on materials that are transparent at these regions is making them more available and better every day [11–17, 20]. Infrared materials are in the form of glasses, semiconductors, crystals, metals, and many others. In the case of optical glasses, the absorption bands due to absorbed water vapor need to be avoided. Common optical glasses transmit until about 2.0 μm and their absorption becomes very high in the water band region 2.7–3.0 μm. Special manufacturing techniques are used to reduce the amount of water in glasses. Unfortunately, these processes also introduce some scattering. An example is the fused quartz manufactured with the name of Vycor®. Table 23.14 shows the properties of some of the many available infrared materials in the form of glasses or hot-pressed polycrystalline compacts (crystals and semiconductors are excluded).

The ultraviolet region also has many interesting applications. Pellicori [24] has described the transmittances of some optical materials for use in the ultraviolet region, between 190 and 340 nm. In general, materials for the ultraviolet region are more rare and expensive than those for the infrared. Table 23.15 shows some of these materials: the first three are highly soluble in water.

Table 23.14 Characteristics of Some Infrared
Materials

Material	n_d	Wavelength cutoff (μm)
Vycor®	1.457	3.5
Germanate	1.660	5.7
CaAl$_2$O$_4$	1.669	5.8
As$_2$S$_3$	2.650	12.5
Irtran 1 MgF$_2$	1.389	7.5
Irtran 2 ZnS	2.370	14.0
Irtran 3 CaF$_2$	1.434	10.0
Irtran 5 MgO	1.737	8.0

Figure 23.11 shows the infrared transmittances for some other infrared materials.

23.7 OPTICAL COATINGS AND OTHER AMORPHOUS MATERIALS

In thin-film work, the properties of materials are also very important. Most of the materials used in the production of optical coatings, single layer or multilayers, i.e., mirrors, antireflective, broadband and narrow filters, edge filters, polarizers, etc., must be performed with a knowledge of the refractive index and the region of transparency, hardness or resistance to abrasion, magnitude of any built-in stresses, solubility, resistance to attack by the atmosphere, compatibility with other materials, toxicity, price, and availability. Also, sometimes, the electrical conductivity or dielectric constant or emissivity for selective materials used in solar absorption (only from elements of groups IV, V, and VI of the periodic system) have to be known.

Most of the techniques used to prepare thin films yield amorphous or poly-crystalline materials in a two-dimensional region, because the thickness is always of the order of a wavelength. The most common materials used in optical coatings are metals, oxides, fluorides, and sulfides. Their optical properties may vary because the method of evaporation produces different density packing in thin films than in bulk

Table 23.15 Characteristics of Some Ultraviolet Materials

Material	n_d	V_d	Wavelength cutoff (nm)
Sodium chloride	1.544	42.8	250
Potassium bromide	1.560	33.4	210
Potassium iodide	1.667	23.2	250
Lithium fluoride	1.392	99.3	110
Calcium fluoride	1.434	95.1	120
Fused quartz	1.458	67.8	220
Barium fluoride	1.474	81.8	150

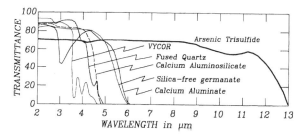

Figure 23.11 Infrared transmittance as a function of wavelength for several materials.

materials. The dielectric function may be studied *in situ* or *ex situ* by analyzing the spectral transmittance and reflectance and by weighing the deposited materials.

Sol–gel glasses are of high purity, but at the present stage of development there are only applications in large thin-film coatings as heat or IR reflecting coatings on windows or coatings on rear-view to reduce glare and reflections.

Inhomogeneous films or rugates [29] are a kind of multilayer film with sinusoidal or other functional refractive index profiles avoiding the presence of very abrupt interfaces. They act as a single layer with the same properties of some multilayers and show no harmonic peaks in the case of optical filters. These materials are mainly isotropic and amorphous.

Only deposits made on crystal substrates, at high substrate temperatures, with special techniques such as Knudsen cells or molecular or liquid beam epitaxy (MBE, LBE), produce crystalline films epitaxially, but these are extreme conditions searched for very specific applications such as semiconductor devices, and even the lattice parameters of the materials available are very restrictive.

23.8 CONCLUSIONS

The field of optical materials is a very active one and every day we have new materials that permit us to make much better optical instruments.

REFERENCES

1. Barnes, W. P., Jr., "Optical Materials – Reflective," in *Applied Optics and Optical Engineering*, Vol. VII, Shannon, R. R. and J. C. Wyant, eds, Academic Press, San Diego, 1979.
2. Conrady, A. E., *Applied Optics and Optical Design Part II*, Dover, New York, 1960.
3. Calve, J. G., *Mémoire sur la Dispersion de la Lumiere*, Prague, 1936.
4. Greis, U. and G. Kirchhof, "Injection Molding of Plastic Optics," *Proc. SPIE*, **381**, 69–71 (1983).
5. Horne, D. F., *Optical Production Technology*, Adam Hilger, Bristol, 1972.
6. Kavanagh, A. J., "Optical Material," in *Military Standardization Handbook: Optical Design, MIL-HDBK 141*, US Defense Supply Agency, Washington, DC, 1962.
7. Kingery, W. D., H. K. Bowen, and D. R. Uhlmann, *Introduction to Ceramics*, John Wiley and Sons, New York, 1976.
8. Koppen, W., "Optical Surfaces Out of Plastic for Spectacle Lenses," *Proc. SPIE*, **381**, 78–80 (1983).

9. Kreidl, N. J. and J. L. Rood, "Optical Materials," in *Applied Optics and Optical Engineering*, Vol. I, Kingslake, R., ed., Academic Press, San Diego, 1965.

10. Lytle, J. D., "Aspheric Surfaces in Polymer Optics," *Proc. SPIE*, **381**, 63–65 (1983).

11. Malitson, I. H., "A Redetermination of Some Optical Properties of Calcium Fluoride," *Appl. Opt.*, **2**, 1103–1107 (1963).

12. McCarthy, D. E., "The Reflection and Transmission of Infrared Materials, Part 1, Spectra From 2 μm to 50 μm," *Appl. Opt.*, **2**, 591–595 (1963a).

13. McCarthy, D. E., "The Reflection and Transmission of Infrared Materials, Part 2. Bibliography," *Appl. Opt.*, **2**, 596–603 (1963b).

14. McCarthy, D. E., "The Reflection and Transmission of Infrared Materials, Part 3, Spectra From 2 μm to 50 μm," *Appl. Opt.*, **4**, 317–320 (1965a).

15. McCarthy, D. E., "The Reflection and Transmission of Infrared Materials, Part 4, Bibliography," *Appl. Opt.*, **4**, 507–511 (1965b).

16. McCarthy, D. E., "The Reflection and Transmission of Infrared Materials, Part 5, Spectra From 2 μm to 50 μm," *Appl. Opt.*, **7**, 1997–2000 (1965c).

17. McCarthy, D. E., "The Reflection and Transmission of Infrared Materials, Part 6, Bibliography," *Appl. Opt.*, **7**, 2221–2225 (1965d).

18. Musikant, S., *Optical Materials. An Introduction to Selection and Application*, Marcel Dekker, New York, 1985.

19. Palmer, A. L., "Practical Design Considerations for Polymer Optical Systems," *Proc. SPIE*, **306**, 18–20 (1981).

20. Parker, C. J., "Optical Materials – Refractive," in *Applied Optics and Optical Engineering*, Vol. VII, Shannon, R. R. and J. C. Wyant, eds, Academic Press, San Diego, 1979.

21. Paquin, R. A., "Hot Isostatic Pressed Beryllium for Large Optics," *Opt. Eng.*, **25**, 1003–1005 (1986).

22. Paquin, R. A., "Hot Isostatic Pressed Beryllium for Large Optics," *Proc. SPIE*, **1168**, 83–87 (1989).

23. Peck, W. G. and C. Tribastone, "Issues in Large Scale Production of Plastic Lenses," *Proc. SPIE*, **2622**, 144–146 (1995).

24. Pellicori, S. F., "Transmittances of Some Optical Materials for Use Between 1900 and 3400 Å," *Appl. Opt.*, **3**, 361–366 (1964).

25. Tatian, B., "Fitting Refractive-Index Data with the Sellmeier Dispersion Formula," *Appl. Opt.*, **23**, 4477–4485 (1984).

26. Taylor, H. D., "Metal Mirrors in the Large," *Opt. Eng.*, **14**, 559–561 (1975).

27. Tropf, W. J., M. E. Thomas, and T. J. Harris, "Properties of Crystals and Glasses," in *Handbook of Optics*, 2nd edn, Part II, McGraw-Hill, New York, 1995.

28. Ventures, G. M., "Improved Plastic Molding Technology for Ophthalmic Lens & Contact Lens," *Proc. SPIE*, **1529**, 13–21 (1991).

29. Villa, F., A. Martinez, and L. E. Regalado, "Correction Masks for Thickness Uniformity in Large Area Thin Films," *Appl. Opt.*, **39**, 1602–1610 (2000).

30. Welham, B., "Plastic Optical Components," in *Applied Optics and Optical Engineering*, Vol. VII, Shannon, R. R. and J. C. Wyant, eds, Academic press, San Diego, 1979.

31. Weber, M. J., *CRC Handbook of Laser Science and Technology Supplement 2: Optical Materials*, CRC Press, LLC, Boca Raton, Florida, 1995.

32. Wolfe, W. L., "Optical Materials," in *The Infrared and Electro Optics Systems Handbook*, Rogatto, W. D., ed., Infrared Information Analysis Center and SPIE, Ann Arbor and Bellingham, 1993.

24

Anisotropic Materials

DENNIS H. GOLDSTEIN
Air Force Research Laboratory, Eglin AFB, Florida

24.1 INTRODUCTION

In this chapter, we discuss the interaction of light with anisotropic materials. An anisotropic material has properties (thermal, mechanical, electrical, optical, etc.) that are different in different directions. Most materials are anisotropic. This anisotropy results from the structure of the material, and our knowledge of the nature of that structure can help us to understand the optical properties.

The interaction of light with matter is a process that is dependent upon the geometrical relationships of the light and matter. By its very nature, light is asymmetrical. Considering light as a wave, it is a transverse oscillation in which the oscillating quantity, the electric field vector, is oriented in a particular direction in space perpendicular to the propagation direction. Light which crosses the boundary between two materials, isotropic or not, at any angle other than normal to the boundary, will produce an anisotropic result. The Fresnel equations illustrate this. Once light has crossed a boundary separating materials, it experiences the bulk properties of the material through which it is currently traversing, and we are concerned with the effects of those bulk properties on the light.

The study of anisotropy in materials is important to understanding the results of the interaction of light with matter. For example, the principle of operation of many solid state and liquid crystal spatial light modulators is based on polarization modulation [5]. Modulation is accomplished by altering the refractive index of the modulator material, usually with an electric or magnetic field. Crystalline materials are an especially important class of modulator materials because of their use in electro-optics and in ruggedized or space-worthy systems, and also because of the potential for putting optical systems on integrated circuit chips.

We briefly review the electromagnetics necessary to the understanding of anisotropic materials, and show the source and form of the electro-optic tensor. We discuss crystalline materials and their properties, and introduce the concept of the index ellipsoid. We show how the application of electric and magnetic fields alters the properties of materials, and give examples. Liquid crystals are also discussed.

A brief summary of electro-optic modulation modes using anisotropic materials concludes the chapter.

24.2 REVIEW OF CONCEPTS FROM ELECTROMAGNETISM

Recall from electromagnetics (see, for example, [10, 19, 26]) that the electric displacement vector $\vec{D}$ is given by (MKS units):

$$\vec{D} = \varepsilon \vec{E}, \tag{24.1}$$

where ε is the permittivity and $\varepsilon = \varepsilon_0(1 + \chi)$, where ε_0 is the permittivity of free space, χ is the electric susceptibility, $(1 + \chi)$ is the dielectric constant, and $\mathbf{n} = (1 + \chi)^{1/2}$ is the index of refraction. The electric displacement is also given by

$$\vec{D} = \varepsilon_0 \vec{E} + \vec{P}, \tag{24.2}$$

but

$$\vec{D} = \varepsilon_0(1 + \chi)\vec{E} = \varepsilon_0\vec{E} + \varepsilon_0\chi\vec{E}, \tag{24.3}$$

so $\vec{P}$, the polarization (also called the electric polarization or polarization density) is $\vec{P} = \varepsilon_0\chi\vec{E}$.

The polarization arises because of the interaction of the electric field with bound charges. The electric field can produce a polarization by inducing a dipole moment, i.e., separating charges in a material, or by orienting molecules that possess a permanent dipole moment.

For an isotropic, linear medium,

$$\vec{P} = \varepsilon_0\chi\vec{E} \tag{24.4}$$

and χ is a scalar; but note that in

$$\vec{D} - \varepsilon_0\vec{E} + \vec{P} \tag{24.5}$$

the vectors do not have to be in the same direction and, in fact, in anisotropic media, $\vec{E}$ and $\vec{P}$ are not in the same direction (and so $\vec{D}$ and $\vec{E}$ are not in the same direction). Note that χ does not have to be a scalar nor is $\vec{P}$ necessarily related linearly to $\vec{E}$. If the medium is linear but anisotropic,

$$\mathbf{P}_i = \sum_j \varepsilon_0\chi_{ij}\mathbf{E}_j, \tag{24.6}$$

where χ_{ij} is the susceptibility tensor, i.e.,

$$\begin{pmatrix} \mathbf{P}_1 \\ \mathbf{P}_2 \\ \mathbf{P}_3 \end{pmatrix} = \varepsilon_0 \begin{pmatrix} \chi_{11} & \chi_{12} & \chi_{13} \\ \chi_{21} & \chi_{22} & \chi_{23} \\ \chi_{31} & \chi_{32} & \chi_{33} \end{pmatrix} \begin{pmatrix} \mathbf{E}_1 \\ \mathbf{E}_2 \\ \mathbf{E}_3 \end{pmatrix} \tag{24.7}$$

and

$$\begin{pmatrix} \mathbf{D}_1 \\ \mathbf{D}_2 \\ \mathbf{D}_3 \end{pmatrix} = \varepsilon_0 \begin{pmatrix} 1 & 0 & 0 \\ 0 & 1 & 0 \\ 0 & 0 & 1 \end{pmatrix} \begin{pmatrix} \mathbf{E}_1 \\ \mathbf{E}_2 \\ \mathbf{E}_3 \end{pmatrix} + \varepsilon_0 \begin{pmatrix} \chi_{11} & \chi_{12} & \chi_{13} \\ \chi_{21} & \chi_{22} & \chi_{23} \\ \chi_{31} & \chi_{32} & \chi_{33} \end{pmatrix} \begin{pmatrix} \mathbf{E}_1 \\ \mathbf{E}_2 \\ \mathbf{E}_3 \end{pmatrix}$$

$$= \varepsilon_0 \begin{pmatrix} 1+\chi_{11} & \chi_{12} & \chi_{13} \\ \chi_{21} & 1+\chi_{22} & \chi_{23} \\ \chi_{31} & \chi_{32} & 1+\chi_{33} \end{pmatrix} \begin{pmatrix} \mathbf{E}_1 \\ \mathbf{E}_2 \\ \mathbf{E}_3 \end{pmatrix}, \tag{24.8}$$

where the vector indices 1, 2, 3 represent the three Cartesian directions.

This can be written

$$\mathbf{D}_i = \varepsilon_{ij}\mathbf{E}_j, \tag{24.9}$$

where

$$\varepsilon_{ij} = \varepsilon_0(1 + \chi_{ij}) \tag{24.10}$$

is variously called the dielectric tensor, or permittivity tensor, or dielectric permittivity tensor. Equations (24.9) and (24.10) use the Einstein summation convention, i.e., whenever repeated indices occur, it is understood that the expression is to be summed over the repeated indices. This notation is used throughout this chapter.

The dielectric tensor is symmetric and real (assuming the medium is homogeneous and nonabsorbing), so that

$$\varepsilon_{ij} = \varepsilon_{ji}, \tag{24.11}$$

and there are at most six independent elements.

Note that for an isotropic medium with nonlinearity (which occurs with higher field strengths)

$$\mathbf{P} = \varepsilon_0(\chi\mathbf{E} + \chi_2\mathbf{E}^2 + \chi_3\mathbf{E}^3 + \cdots) \tag{24.12}$$

where χ_2, χ_3, etc., are the nonlinear terms.

Returning to the discussion of a linear, homogeneous, anisotropic medium, the susceptibility tensor

$$\begin{pmatrix} \chi_{11} & \chi_{12} & \chi_{13} \\ \chi_{21} & \chi_{22} & \chi_{23} \\ \chi_{31} & \chi_{32} & \chi_{33} \end{pmatrix} = \begin{pmatrix} \chi_{11} & \chi_{12} & \chi_{13} \\ \chi_{12} & \chi_{22} & \chi_{23} \\ \chi_{13} & \chi_{23} & \chi_{33} \end{pmatrix} \tag{24.13}$$

is symmetric so that we can always find a set of coordinate axes (that is, we can always rotate to an orientation) such that the off-diagonal terms are zero and the tensor is diagonalized thus

$$\begin{pmatrix} \chi'_{11} & 0 & 0 \\ 0 & \chi'_{22} & 0 \\ 0 & 0 & \chi'_{33} \end{pmatrix}. \tag{24.14}$$

The coordinate axes for which this is true are called the principal axes, and these χ' are the principal susceptibilities. The principal dielectric constants are given by

$$\begin{pmatrix} 1 & 0 & 0 \\ 0 & 1 & 0 \\ 0 & 0 & 1 \end{pmatrix} + \begin{pmatrix} \chi_{11} & 0 & 0 \\ 0 & \chi_{22} & 0 \\ 0 & 0 & \chi_{33} \end{pmatrix} = \begin{pmatrix} 1+\chi_{11} & 0 & 0 \\ 0 & 1+\chi_{22} & 0 \\ 0 & 0 & 1+\chi_{33} \end{pmatrix}$$

$$= \begin{pmatrix} \mathbf{n}_1^2 & 0 & 0 \\ 0 & \mathbf{n}_2^2 & 0 \\ 0 & 0 & \mathbf{n}_3^2 \end{pmatrix}$$
(24.15)

where $\mathbf{n}_1$, $\mathbf{n}_2$, and $\mathbf{n}_3$ are the principal indices of refraction.

24.3 CRYSTALLINE MATERIALS AND THEIR PROPERTIES

As we have seen above, the relationship between the displacement and the field is

$$\mathbf{D}_i = \varepsilon_{ij}\mathbf{E}_j,$$
(24.16)

where ε_{ij} is the dielectric tensor. The impermeability tensor η_{ij} is defined as

$$\eta_{ij} = \varepsilon_0(\varepsilon^{-1})_{ij},$$
(24.17)

where ε^{-1} is the inverse of the dielectric tensor. The principal indices of refraction, $\mathbf{n}_1$, $\mathbf{n}_2$, and $\mathbf{n}_3$ are related to the principal values of the impermeability tensor and the principal values of the permittivity tensor by

$$\frac{1}{\mathbf{n}_1^2} = \eta_{ii} = \frac{\varepsilon_0}{\varepsilon_{ii}}; \qquad \frac{1}{\mathbf{n}_2^2} = \eta_{jj} = \frac{\varepsilon_0}{\varepsilon_{jj}}; \qquad \frac{1}{\mathbf{n}_3^2} = \eta_{kk} = \frac{\varepsilon_0}{\varepsilon_{kk}}.$$
(24.18)

The properties of the crystal change in response to the force from an externally applied electric field. In particular, the impermeability tensor is a function of the field. The electro-optic coefficients are defined by the expression for the expansion, in terms of the field, of the change of the impermeability tensor from zero field value, i.e.,

$$\eta_{ij}(\mathbf{E}) - \eta_{ij}(0) \equiv \Delta\eta_{ij} = r_{ijk}\mathbf{E}_k + s_{ijkl}\mathbf{E}_k\mathbf{E}_l + O(\mathbf{E}^n), n = 3, 4, \ldots,$$
(24.19)

where η_{ij} is a function of the applied field E, r_{ijk} are the linear, or Pockels, electro-optic tensor coefficients, and the s_{ijkl} are the quadratic, or Kerr, electro-optic tensor coefficients. Terms higher than quadratic are typically small and are neglected.

Note that the values of the indices and the electro-optic tensor coefficients are dependent on the frequency of light passing through the material. Any given indices are specified at a particular frequency (or wavelength). Also note that the external applied fields may be static or alternating fields, and the values of the tensor coefficients are weakly dependent on the frequency of the applied fields. Generally, low- and/or high-frequency values of the tensor coefficients are given in tables. Low frequencies are those below the fundamental frequencies of the acoustic resonances of the sample, and high frequencies are those above. Operation of an electro-optic modulator subject to low (high) frequencies is sometimes described as being unclamped (clamped).

The linear electro-optic tensor is of third rank with 3^3 elements and the quadratic electro-optic tensor is of fourth rank with 3^4 elements; however, symmetry reduces the number of independent elements. If the medium is lossless and optically inactive:

ε_{ij} is a symmetric tensor, i.e., $\varepsilon_{ij} = \varepsilon_{ji}$,

η_{ij} is a symmetric tensor, i.e., $\eta_{ij} = \eta_{ji}$,

$\mathbf{r}_{ijk}$ has symmetry where coefficients with permuted first and second indices are equal, i.e., $\mathbf{r}_{ijk} = \mathbf{r}_{jik}$, and

$\mathbf{s}_{ijkl}$ has symmetry where coefficients with permuted first and second indices are equal and coefficients with permuted third and fourth coefficients are equal, i.e., $\mathbf{s}_{ijkl} = \mathbf{s}_{jikl}$ and $\mathbf{s}_{ijkl} = \mathbf{s}_{ijlk}$.

Symmetry reduces the number of linear coefficients from 27 to 18, and reduces the number of quadratic coefficients from 81 to 36. The linear electro-optic coefficients are assigned two indices so that they are $\mathbf{r}_{lk}$ where l runs from 1 to 6 and k runs from 1 to 3. The quadratic coefficients are assigned two indices so that they become $\mathbf{s}_{ij}$ where i runs from 1 to 6 and j runs from 1 to 6. For a given crystal symmetry class, the form of the electro-optic tensor is known.

24.4 CRYSTALS

Crystals are characterized by their lattice type and symmetry. There are 14 lattice types. As an example of three of these, a crystal which has a cubic structure can be simple cubic, face-centered cubic, or body-centered cubic.

There are 32 point groups corresponding to 32 different symmetries. For example, a cubic lattice has five types of symmetry. The symmetry is labeled with point group notation, and crystals are classified in this way. A complete discussion of crystals, lattice types, and point groups is outside the scope of the present work, and will not be given here; there are many excellent references [13, 14, 15, 20, 24, 31, 34]. Table 24.1 gives a summary of the lattice types and point groups, and shows how these relate to optical symmetry and the form of the dielectric tensor.

In order to understand the notation and terminology of Table 24.1, some additional information is required which we now introduce. As we have seen in the previous sections, there are three principal indices of refraction. There are three types of materials: those for which the three principal indices are equal; those where two principal indices are equal; and those where all three principal indices are different. We will discuss these three cases in more detail in the next section. The indices for the case where there are only two distinct values are named the ordinary index (n_o) and the extraordinary index (n_e). These labels are applied for historical reasons [9]. Erasmus Bartholinus, a Danish mathematician, in 1669 discovered double refraction in calcite. If the calcite crystal, split along its natural cleavage planes, is placed on a typewritten sheet of paper, two images of the letters will be observed. If the crystal is then rotated about an axis perpendicular to the page, one of the two images of the letters will rotate about the other. Bartholinus named the light rays from the letters that do not rotate the ordinary rays, and the rays from the rotating letters he named the extraordinary rays; hence, the indices that produce these rays are named likewise. This explains the notation in the dielectric tensor for tetragonal, hexagonal, and trigonal crystals.

Let us consider such crystals in more detail. There is a plane in the material in which a single index would be measured in any direction. Light that is propagating in the direction normal to this plane with equal indices experiences the same refractive index for any polarization (orientation of the **E** vector). The direction for which this

Table 24.1 Crystal Types, Point Groups, and the Dielectric Tensors [37]

Symmetry	Crystal system	Point group	Dielectric tensor
Isotropic	Cubic	$\overline{4}3m$ 432 $m3$ 23 $m3m$	$\varepsilon = \varepsilon_0 \begin{pmatrix} n^2 & 0 & 0 \\ 0 & n^2 & 0 \\ 0 & 0 & n^2 \end{pmatrix}$
Uniaxial	Tetragonal	4 $\overline{4}$ $4/m$ 422 $4mm$ $\overline{4}2m$ $4/mmm$	
	Hexagonal	6 $\overline{6}$ $6/m$ 622 $6mm$ $\overline{6}m2$ $6/mmm$	$\varepsilon = \varepsilon_0 \begin{pmatrix} n_o^2 & 0 & 0 \\ 0 & n_o^2 & 0 \\ 0 & 0 & n_e^2 \end{pmatrix}$
	Trigonal	3 $\overline{3}$ 32 $3m$ $\overline{3}m$	
Biaxial	Triclinic	1 $\overline{1}$	
	Monoclinic	2 m $2/m$	$\varepsilon = \varepsilon_0 \begin{pmatrix} n_1^2 & 0 & 0 \\ 0 & n_2^2 & 0 \\ 0 & 0 & n_3^2 \end{pmatrix}$
	Orthorhombic	222 $2mm$ mmm	

occurs is called the optic axis. Crystals which have one optic axis are called uniaxial crystals. Materials with three principal indices have two directions in which the **E** vector experiences a single refractive index. These materials have two optic axes and are called biaxial crystals. This is more fully explained in Section 24.4.1. Materials that have more than one principal index of refraction are called birefringent materials and are said to exhibit double refraction.

Crystals are composed of periodic arrays of atoms. The lattice of a crystal is a set of points in space. Sets of atoms which are identical in composition, arrangement, and orientation are attached to each lattice point. By translating the basic structure attached to the lattice point, we can fill space with the crystal. Define vectors **a**, **b**, and **c** which form three adjacent edges of a parallelepiped which spans the basic atomic structure. This parallelepiped is called a unit cell. We call the axes that lie along these vectors the crystal axes.

We would like to be able to describe a particular plane in a crystal, since crystals may be cut at any angle. The Miller indices are quantities which describe the orientation of planes in a crystal. The Miller indices are defined as follows: (1) Locate the intercepts of the plane on the crystal axes: these will be multiples of lattice point spacing. (2) Take the reciprocals of the intercepts and form the three smallest integers having the same ratio. For example, suppose we have a cubic crystal so that the crystal axes are the orthogonal Cartesian axes. Suppose further that the plane we want to describe intercepts the axes at the points 4, 3, and 2. The reciprocals of these intercepts are $\frac{1}{4}$, $\frac{1}{3}$, and $\frac{1}{2}$. The Miller indices are then (3, 4, 6). This example serves to illustrate how the Miller indices are found, but it is more usual to encounter simpler crystal cuts. The same cubic crystal, if cut so that the intercepts are $1, \infty, \infty$ (defining a plane parallel to the *yz*-plane in the usual Cartesian coordinates) has Miller indices (1, 0, 0). Likewise, if the intercepts are $1, 1, \infty$ (diagonal to two of the axes), the Miller indices are (1, 1, 0), and if the intercepts are 1, 1, 1 (diagonal to all three axes), the Miller indices are (1, 1, 1).

Two important electro-optic crystal types have the point group symbols $\overline{4}3m$ (this is a cubic crystal, e.g., CdTe and GaAs) and $\overline{4}2m$ (this is a tetragonal crystal, e.g., AgGaS$_2$). The linear and quadratic electro-optic tensors for these two crystal types, as well as all the other linear and quadratic electro-optic coefficient tensors for all crystal symmetry classes, are given in Tables 24.2 and 24.3. Note from these tables that the linear electro-optic effect vanishes for crystals that retain symmetry under inversion, i.e., centrosymmetric crystals, whereas the quadratic electro-optic effect never vanishes. For further discussion of this point, see Yariv and Yeh [37].

24.4.1 The Index Ellipsoid

Light propagating in anisotropic materials experiences a refractive index and a phase velocity that depends on the propagation direction, polarization state, and wavelength. The refractive index for propagation (for monochromatic light of some specified frequency) in an arbitrary direction (in Cartesian coordinates)

$$\vec{a} = x\hat{i} + y\hat{j} + z\hat{k} \tag{24.20}$$

can be obtained from the index ellipsoid, a useful and lucid construct for visualization and determination of the index. (Note that we now shift from indexing the Cartesian directions with numbers to using x, y, and z.) In the principal coordinate system the index ellipsoid is given by

$$\frac{x^2}{n_x^2} + \frac{y^2}{n_y^2} + \frac{z^2}{n_z^2} = 1 \tag{24.21}$$

in the absence of an applied electric field. The lengths of the semimajor and semiminor axes of the ellipse formed by the intersection of this index ellipsoid and a plane normal to the propagation direction and passing through the center of the ellipsoid are the two principal indices of refraction for that propagation direction. Where there are three distinct principal indices, the crystal is defined as biaxial, and the above equation holds. If two of the three indices of the index ellipsoid are equal, the crystal is defined to be uniaxial and the equation for the index ellipsoid is

Table 24.2   Linear Electro-optic Tensors [37]

Crystal symmetry class	Symmetry group	Tensor
Centrosymmetric	$\overline{1}$ 2/m mmm 4/m 4/mmm $\overline{3}$ $\overline{3}$m 6/m 6/mmm m3 m3m	$\begin{pmatrix} 0 & 0 & 0 \\ 0 & 0 & 0 \\ 0 & 0 & 0 \\ 0 & 0 & 0 \\ 0 & 0 & 0 \\ 0 & 0 & 0 \end{pmatrix}$
Triclinic	1	$\begin{pmatrix} r_{11} & r_{12} & r_{13} \\ r_{21} & r_{22} & r_{23} \\ r_{31} & r_{32} & r_{33} \\ r_{41} & r_{42} & r_{43} \\ r_{51} & r_{52} & r_{53} \\ r_{61} & r_{62} & r_{63} \end{pmatrix}$
Monoclinic	$2\ (2\|x_2)$	$\begin{pmatrix} 0 & r_{12} & 0 \\ 0 & r_{22} & 0 \\ 0 & r_{32} & 0 \\ r_{41} & 0 & r_{43} \\ 0 & r_{52} & 0 \\ r_{61} & 0 & r_{63} \end{pmatrix}$
	$2\ (2\|x_3)$	$\begin{pmatrix} 0 & 0 & r_{13} \\ 0 & 0 & r_{23} \\ 0 & 0 & r_{33} \\ r_{41} & r_{42} & 0 \\ r_{51} & r_{52} & 0 \\ 0 & 0 & r_{63} \end{pmatrix}$
	$m\ (m \perp x_2)$	$\begin{pmatrix} r_{11} & 0 & r_{13} \\ r_{21} & 0 & r_{23} \\ r_{31} & 0 & r_{33} \\ 0 & r_{42} & 0 \\ r_{51} & 0 & r_{53} \\ 0 & r_{62} & 0 \end{pmatrix}$
	$m\ (m \perp x_3)$	$\begin{pmatrix} r_{11} & r_{12} & 0 \\ r_{21} & r_{22} & 0 \\ r_{31} & r_{32} & 0 \\ 0 & 0 & r_{43} \\ 0 & 0 & r_{53} \\ r_{61} & r_{62} & 0 \end{pmatrix}$

Table 24.2 Linear Electro-optic Tensors [37] (*contd.*)

Crystal symmetry class	Symmetry group	Tensor
Orthorhombic	222	$\begin{pmatrix} 0 & 0 & 0 \\ 0 & 0 & 0 \\ 0 & 0 & 0 \\ r_{41} & 0 & 0 \\ 0 & r_{52} & 0 \\ 0 & 0 & r_{63} \end{pmatrix}$
	2mm	$\begin{pmatrix} 0 & 0 & r_{13} \\ 0 & 0 & r_{23} \\ 0 & 0 & r_{33} \\ 0 & r_{42} & 0 \\ r_{51} & 0 & 0 \\ 0 & 0 & 0 \end{pmatrix}$
Tetragonal	4	$\begin{pmatrix} 0 & 0 & r_{13} \\ 0 & 0 & r_{13} \\ 0 & 0 & r_{33} \\ r_{41} & r_{51} & 0 \\ r_{51} & r_{41} & 0 \\ 0 & 0 & 0 \end{pmatrix}$
	$\bar{4}$	$\begin{pmatrix} 0 & 0 & r_{13} \\ 0 & 0 & -r_{13} \\ 0 & 0 & 0 \\ r_{41} & -r_{51} & 0 \\ r_{51} & r_{41} & 0 \\ 0 & 0 & r_{63} \end{pmatrix}$
	422	$\begin{pmatrix} 0 & 0 & 0 \\ 0 & 0 & 0 \\ 0 & 0 & 0 \\ r_{41} & 0 & 0 \\ 0 & -r_{41} & 0 \\ 0 & 0 & 0 \end{pmatrix}$
	4mm	$\begin{pmatrix} 0 & 0 & r_{13} \\ 0 & 0 & r_{13} \\ 0 & 0 & r_{33} \\ 0 & r_{51} & 0 \\ r_{51} & 0 & 0 \\ 0 & 0 & 0 \end{pmatrix}$
	$\bar{4}2m$ $(2\|x_1)$	$\begin{pmatrix} 0 & 0 & 0 \\ 0 & 0 & 0 \\ 0 & 0 & 0 \\ r_{41} & 0 & 0 \\ 0 & r_{41} & 0 \\ 0 & 0 & r_{63} \end{pmatrix}$

Table 24.2 Linear Electro-optic Tensors [37] (*contd.*)

Crystal symmetry class	Symmetry group	Tensor
Trigonal	3	$\begin{pmatrix} r_{11} & -r_{22} & r_{13} \\ -r_{11} & r_{22} & r_{13} \\ 0 & 0 & r_{33} \\ r_{41} & r_{51} & 0 \\ r_{51} & -r_{41} & 0 \\ -r_{22} & -r_{11} & 0 \end{pmatrix}$
	32	$\begin{pmatrix} r_{11} & 0 & 0 \\ -r_{11} & 0 & 0 \\ 0 & 0 & 0 \\ r_{41} & 0 & 0 \\ 0 & -r_{41} & 0 \\ 0 & -r_{11} & 0 \end{pmatrix}$
	3m $(m \perp x_1)$	$\begin{pmatrix} 0 & -r_{22} & r_{13} \\ 0 & r_{22} & r_{13} \\ 0 & 0 & r_{33} \\ 0 & r_{51} & 0 \\ r_{51} & 0 & 0 \\ -r_{22} & 0 & 0 \end{pmatrix}$
	3m $(m \perp x_2)$	$\begin{pmatrix} r_{11} & 0 & r_{13} \\ -r_{11} & 0 & r_{13} \\ 0 & 0 & r_{33} \\ 0 & r_{51} & 0 \\ r_{51} & 0 & 0 \\ 0 & -r_{11} & 0 \end{pmatrix}$
Hexagonal	6	$\begin{pmatrix} 0 & 0 & r_{13} \\ 0 & 0 & r_{13} \\ 0 & 0 & r_{33} \\ r_{41} & r_{51} & 0 \\ r_{51} & -r_{41} & 0 \\ 0 & 0 & 0 \end{pmatrix}$
	6mm	$\begin{pmatrix} 0 & 0 & r_{13} \\ 0 & 0 & r_{13} \\ 0 & 0 & r_{33} \\ 0 & r_{51} & 0 \\ r_{51} & 0 & 0 \\ 0 & 0 & 0 \end{pmatrix}$
	622	$\begin{pmatrix} 0 & 0 & 0 \\ 0 & 0 & 0 \\ 0 & 0 & 0 \\ r_{41} & 0 & 0 \\ 0 & -r_{41} & 0 \\ 0 & 0 & 0 \end{pmatrix}$

Table 24.2 Linear Electro-optic Tensors [37] (*contd.*)

Crystal symmetry class	Symmetry group	Tensor
Hexagonal	$\bar{6}$	$\begin{pmatrix} r_{11} & -r_{22} & 0 \\ -r_{11} & r_{22} & 0 \\ 0 & 0 & 0 \\ 0 & 0 & 0 \\ 0 & 0 & 0 \\ -r_{22} & -r_{11} & 0 \end{pmatrix}$
	$\bar{6}m2 \ (\mathbf{m} \perp \mathbf{x}_1)$	$\begin{pmatrix} 0 & -r_{22} & 0 \\ 0 & r_{22} & 0 \\ 0 & 0 & 0 \\ 0 & 0 & 0 \\ 0 & 0 & 0 \\ -r_{22} & 0 & 0 \end{pmatrix}$
	$\bar{6}m2 \ (\mathbf{m} \perp \mathbf{x}_2)$	$\begin{pmatrix} r_{11} & 0 & 0 \\ -r_{11} & 0 & 0 \\ 0 & 0 & 0 \\ 0 & 0 & 0 \\ 0 & 0 & 0 \\ 0 & -r_{11} & 0 \end{pmatrix}$
Cubic	$\bar{4}3m$ 23	$\begin{pmatrix} 0 & 0 & 0 \\ 0 & 0 & 0 \\ 0 & 0 & 0 \\ r_{41} & 0 & 0 \\ 0 & r_{41} & 0 \\ 0 & 0 & r_{41} \end{pmatrix}$
	432	$\begin{pmatrix} 0 & 0 & 0 \\ 0 & 0 & 0 \\ 0 & 0 & 0 \\ 0 & 0 & 0 \\ 0 & 0 & 0 \\ 0 & 0 & 0 \end{pmatrix}$

$$\frac{x^2}{n_o^2} + \frac{y^2}{n_o^2} + \frac{z^2}{n_e^2} = 1. \tag{24.22}$$

Uniaxial materials are said to be uniaxial positive when $n_o < n_e$ and uniaxial negative when $n_o > n_e$. When there is a single index for any direction in space, the crystal is isotropic and the equation for the ellipsoid becomes that for a sphere,

$$\frac{x^2}{n^2} + \frac{y^2}{n^2} + \frac{z^2}{n^2} = 1. \tag{24.23}$$

The index ellipsoids for isotropic, uniaxial, and biaxial crystals are illustrated in Fig. 24.1.

Table 24.3 Quadratic Electro-optic Tensors [37]

Crystal symmetry class	Symmetry group	Tensor
Triclinic	1 $\bar{1}$	$\begin{pmatrix} s_{11} & s_{12} & s_{13} & s_{14} & s_{15} & s_{16} \\ s_{21} & s_{22} & s_{23} & s_{24} & s_{25} & s_{26} \\ s_{31} & s_{32} & s_{33} & s_{34} & s_{35} & s_{36} \\ s_{41} & s_{42} & s_{43} & s_{44} & s_{45} & s_{46} \\ s_{51} & s_{52} & s_{53} & s_{54} & s_{55} & s_{56} \\ s_{61} & s_{62} & s_{63} & s_{64} & s_{65} & s_{66} \end{pmatrix}$
Monoclinic	2 $\mathbf{m}$ $2/\mathbf{m}$	$\begin{pmatrix} s_{11} & s_{12} & s_{13} & 0 & s_{15} & 0 \\ s_{21} & s_{22} & s_{23} & 0 & s_{25} & 0 \\ s_{31} & s_{32} & s_{33} & 0 & s_{35} & 0 \\ 0 & 0 & 0 & s_{44} & 0 & s_{46} \\ s_{51} & s_{52} & s_{53} & 0 & s_{55} & 0 \\ 0 & 0 & 0 & s_{64} & 0 & s_{66} \end{pmatrix}$
Orthorhombic	$\mathbf{2mm}$ 222 $\mathbf{mmm}$	$\begin{pmatrix} s_{11} & s_{12} & s_{13} & 0 & 0 & 0 \\ s_{21} & s_{22} & s_{23} & 0 & 0 & 0 \\ s_{31} & s_{32} & s_{33} & 0 & 0 & 0 \\ 0 & 0 & 0 & s_{44} & 0 & 0 \\ 0 & 0 & 0 & 0 & s_{55} & 0 \\ 0 & 0 & 0 & 0 & 0 & s_{66} \end{pmatrix}$
Tetragonal	4 $\bar{4}$ $4/\mathbf{m}$	$\begin{pmatrix} s_{11} & s_{12} & s_{13} & 0 & 0 & s_{16} \\ s_{12} & s_{11} & s_{13} & 0 & 0 & -s_{16} \\ s_{31} & s_{31} & s_{33} & 0 & 0 & 0 \\ 0 & 0 & 0 & s_{44} & s_{45} & 0 \\ 0 & 0 & 0 & -s_{45} & s_{44} & 0 \\ s_{61} & -s_{61} & 0 & 0 & 0 & s_{66} \end{pmatrix}$
	422 $\mathbf{4mm}$ $\bar{4}2\mathbf{m}$ $4/\mathbf{mm}$	$\begin{pmatrix} s_{11} & s_{12} & s_{13} & 0 & 0 & 0 \\ s_{12} & s_{11} & s_{13} & 0 & 0 & 0 \\ s_{31} & s_{31} & s_{33} & 0 & 0 & 0 \\ 0 & 0 & 0 & s_{44} & 0 & 0 \\ 0 & 0 & 0 & 0 & s_{44} & 0 \\ 0 & 0 & 0 & 0 & 0 & s_{66} \end{pmatrix}$
Trigonal	3 $\bar{3}$	$\begin{pmatrix} s_{11} & s_{12} & s_{13} & s_{14} & s_{15} & -s_{61} \\ s_{12} & s_{11} & s_{13} & -s_{14} & -s_{15} & s_{61} \\ s_{31} & s_{31} & s_{33} & 0 & 0 & 0 \\ s_{41} & -s_{41} & 0 & s_{44} & s_{45} & -s_{51} \\ s_{51} & -s_{51} & 0 & -s_{45} & s_{44} & s_{41} \\ s_{61} & -s_{61} & 0 & -s_{15} & s_{14} & \frac{1}{2}(s_{11}-s_{12}) \end{pmatrix}$
	32 $\mathbf{3m}$ $\bar{3}\mathbf{m}$	$\begin{pmatrix} s_{11} & s_{12} & s_{13} & s_{14} & 0 & 0 \\ s_{12} & s_{11} & s_{13} & -s_{14} & 0 & 0 \\ s_{13} & s_{13} & s_{33} & 0 & 0 & 0 \\ s_{41} & -s_{41} & 0 & s_{44} & 0 & 0 \\ 0 & 0 & 0 & 0 & s_{44} & s_{41} \\ 0 & 0 & 0 & 0 & s_{14} & \frac{1}{2}(s_{11}-s_{12}) \end{pmatrix}$

Table 24.3 Quadratic Electro-optic Tensors [37] (*contd.*)

Crystal symmetry class	Symmetry group	Tensor
Hexagonal	6 $\bar{6}$ $6/m$	$$\begin{pmatrix} s_{11} & s_{12} & s_{13} & 0 & 0 & -s_{61} \\ s_{12} & s_{11} & s_{13} & 0 & 0 & s_{61} \\ s_{31} & s_{31} & s_{33} & 0 & 0 & 0 \\ 0 & 0 & 0 & s_{44} & s_{45} & 0 \\ 0 & 0 & 0 & -s_{45} & s_{44} & 0 \\ s_{61} & -s_{61} & 0 & 0 & 0 & \frac{1}{2}(s_{11}-s_{12}) \end{pmatrix}$$
	622 $6mm$ $\bar{6}m2$ $6/mmm$	$$\begin{pmatrix} s_{11} & s_{12} & s_{13} & 0 & 0 & 0 \\ s_{12} & s_{11} & s_{13} & 0 & 0 & 0 \\ s_{31} & s_{31} & s_{33} & 0 & 0 & 0 \\ 0 & 0 & 0 & s_{44} & 0 & 0 \\ 0 & 0 & 0 & 0 & s_{44} & 0 \\ 0 & 0 & 0 & 0 & 0 & \frac{1}{2}(s_{11}-s_{12}) \end{pmatrix}$$
Cubic	23 $m3$	$$\begin{pmatrix} s_{11} & s_{12} & s_{13} & 0 & 0 & 0 \\ s_{13} & s_{11} & s_{12} & 0 & 0 & 0 \\ s_{12} & s_{13} & s_{11} & 0 & 0 & 0 \\ 0 & 0 & 0 & s_{44} & 0 & 0 \\ 0 & 0 & 0 & 0 & s_{44} & 0 \\ 0 & 0 & 0 & 0 & 0 & s_{44} \end{pmatrix}$$
	432 $m3m$ $\bar{4}3m$	$$\begin{pmatrix} s_{11} & s_{12} & s_{12} & 0 & 0 & 0 \\ s_{12} & s_{11} & s_{12} & 0 & 0 & 0 \\ s_{12} & s_{12} & s_{11} & 0 & 0 & 0 \\ 0 & 0 & 0 & s_{44} & 0 & 0 \\ 0 & 0 & 0 & 0 & s_{44} & 0 \\ 0 & 0 & 0 & 0 & 0 & s_{44} \end{pmatrix}$$
Isotropic		$$\begin{pmatrix} s_{11} & s_{12} & s_{12} & 0 & 0 & 0 \\ s_{12} & s_{11} & s_{12} & 0 & 0 & 0 \\ s_{12} & s_{12} & s_{11} & 0 & 0 & 0 \\ 0 & 0 & 0 & \frac{1}{2}(s_{11}-s_{12}) & 0 & 0 \\ 0 & 0 & 0 & 0 & \frac{1}{2}(s_{11}-s_{12}) & 0 \\ 0 & 0 & 0 & 0 & 0 & \frac{1}{2}(s_{11}-s_{12}) \end{pmatrix}$$

Examples of isotropic materials are CdTe, NaCl, diamond, and GaAs. Examples of uniaxial positive materials are quartz and ZnS. Materials that are uniaxial negative include calcite, $LiNbO_3$, $BaTiO_3$, and KDP (KH_2PO_4). Examples of biaxial materials are gypsum and mica.

24.4.2 Natural Birefringence

Many materials have natural birefringence, i.e., they are uniaxial or biaxial in their natural (absence of applied fields) state. These materials are often used in passive devices such as polarizers and retarders. Calcite is one of the most important naturally birefringent materials for optics, and is used in a variety of well-known polar-

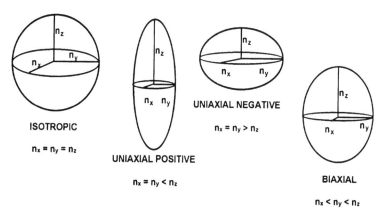

Figure 24.1 Index ellipsoids.

izers, e.g., the Nichol, Wollaston, or Glan–Thompson prisms. As we shall see later, naturally isotropic materials can be made birefringent, and materials that have natural birefringence can be made to change that birefringence with the application of electromagnetic fields.

24.4.3 The Wave Surface

There are two additional methods of depicting the effect of crystal anisotropy on light. Neither is as satisfying or useful to this author as the index ellipsoid; however, both are mentioned for the sake of completeness and in order to facilitate understanding of those references that use these models. They are most often used to explain birefringence, e.g., in the operation of calcite-based devices [2, 11, 16].

The first of these is called the wave surface. As a light wave from a point source expands through space, it forms a surface that represents the wavefront. This surface is composed of points having equal phase. At a particular instant in time, the wave surface is a representation of the velocity surface of a wave expanding in the medium; it is a measure of the distance through which the wave has expanded from some point over some time period. Because the wave will have expanded further (faster) when experiencing a low refractive index and expanded less (slower) when experiencing a high index, the size of the wave surface is inversely proportional to the index.

To illustrate the use of the wave surface, consider a uniaxial crystal. Recall that we have defined the optic axis of a uniaxial crystal as the direction in which the speed of propagation is independent of polarization. The optic axes for positive and negative uniaxial crystals are shown on the index ellipsoids in Fig. 24.2, and the optic axes for a biaxial crystal are shown on the index ellipsoid in Fig. 24.3.

The wave surfaces are now shown in Fig. 24.4 for both positive and negative uniaxial materials. The upper diagram for each pair shows the wave surface for polarization perpendicular to the optic axes (also perpendicular to the principal section through the ellipsoid), and the lower diagram shows the wave surface for polarization in the plane of the principal section. The index ellipsoid surfaces are shown for reference. Similarly, cross sections of the wave surfaces for biaxial materi-

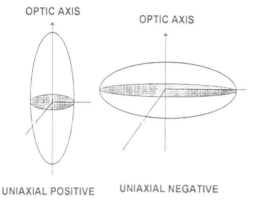

Figure 24.2 Optic axis on index ellipsoid for uniaxial positive and uniaxial negative crystals.

als are shown in Fig. 24.5. In all cases, polarization perpendicular to the plane of the page is indicated with solid circles along the rays, whereas polarization parallel to the plane of the page is shown with short double-headed arrows along the rays.

24.4.4 The Wavevector Surface

A second method of depicting the effect of crystal anisotropy on light is the wavevector surface. The wavevector surface is a measure of the variation of the value of **k**, the wavevector, for different propagation directions and different polarizations. Recall that

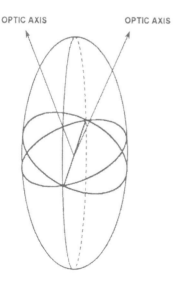

Figure 24.3 Optic axes on index ellipsoid for biaxial crystals.

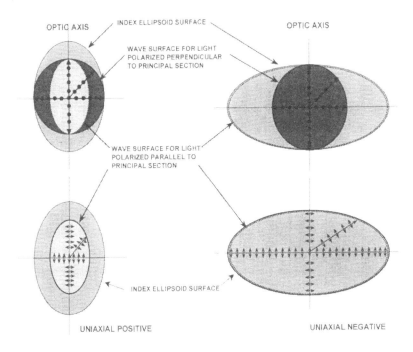

Figure 24.4   Wave surfaces for uniaxial positive and negative materials.

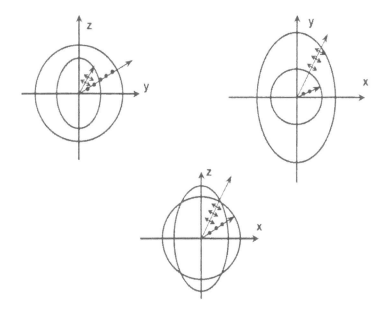

Figure 24.5   Wave surfaces for biaxial materials in principal planes.

$$\mathbf{k} = \frac{2\pi}{\lambda} = \frac{\omega \mathbf{n}}{\mathbf{c}}, \tag{24.24}$$

so $\mathbf{k} \propto \mathbf{n}$. Wavevector surfaces for uniaxial crystals will then appear as shown in Fig. 24.6. Compare these to the wave surfaces in Fig. 24.4.

Wavevector surfaces for biaxial crystals are more complicated. Cross sections of the wavevector surface for a biaxial crystal where $\mathbf{n_x} < \mathbf{n_y} < \mathbf{n_z}$ are shown in Fig. 24.7. Compare these to the wave surfaces in Fig. 24.5.

24.5 APPLICATION OF ELECTRIC FIELDS: INDUCED BIREFRINGENCE AND POLARIZATION MODULATION

When fields are applied to materials, whether isotropic or anisotropic, birefringence can be induced or modified. This is the principle of a modulator; it is one of the most important optical devices, since it gives control over the phase and/or amplitude of light.

The alteration of the index ellipsoid of a crystal on application of an electric and/or magnetic field can be used to modulate the polarization state. The equation for the index ellipsoid of a crystal in an electric field is

$$\eta_{ij}(\mathbf{E})\mathbf{x}_i\mathbf{x}_j = 1 \tag{24.25}$$

or

$$(\eta_{ij}(0) + \Delta\eta_{ij})\mathbf{x}_i\mathbf{x}_j = 1. \tag{24.26}$$

This equation can be written as

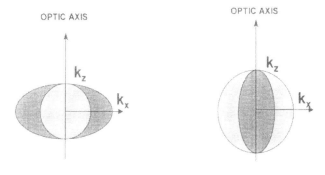

POSITIVE UNIAXIAL   NEGATIVE UNIAXIAL

Figure 24.6 Wavevector surfaces for positive and negative uniaxial crystals.

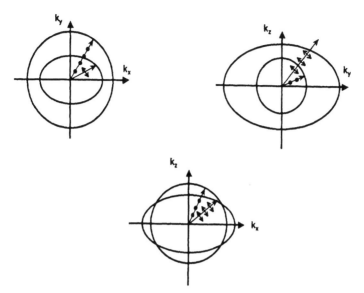

Figure 24.7 Wavevector surface cross sections for biaxial crystals.

$$
x^2\left(\frac{1}{n_x^2} + \Delta\left(\frac{1}{n}\right)_1^2\right) + y^2\left(\frac{1}{n_y^2} + \Delta\left(\frac{1}{n}\right)_2^2\right) + z^2\left(\frac{1}{n_z^2} + \Delta\left(\frac{1}{n}\right)_3^2\right)
$$
$$
+ 2yz\left(\Delta\left(\frac{1}{n}\right)_4^2\right) + 2xz\left(\Delta\left(\frac{1}{n}\right)_5^2\right) + 2xy\left(\Delta\left(\frac{1}{n}\right)_6^2\right) = 1
$$

$$(24.27)$$

or

$$
x^2\left(\frac{1}{n_x^2} + r_{1k}E_k + s_{1k}E_k^2 + 2s_{14}E_2E_3 + 2s_{15}E_3E_1 + 2s_{16}E_1E_2\right)
$$
$$
+ y^2\left(\frac{1}{n_y^2} + r_{2k}E_k + s_{2k}E_k^2 + 2s_{24}E_2E_3 + 2s_{25}E_3E_1 + 2s_{26}E_1E_2\right)
$$
$$
+ z^2\left(\frac{1}{n_z^2} + r_{3k}E_k + s_{3k}E_k^2 + 2s_{34}E_2E_3 + 2s_{35}E_3E_1 + 2s_{36}E_1E_2\right)
$$
$$
+ 2yz\left(r_{4k}E_k + s_{4k}E_k^2 + 2s_{44}E_2E_3 + 2s_{45}E_3E_1 + 2s_{46}E_1E_2\right)
$$
$$
+ 2zx\left(r_{5k}E_k + s_{5k}E_k^2 + 2s_{54}E_2E_3 + 2s_{55}E_3E_1 + 2s_{56}E_1E_2\right)
$$
$$
+ 2xy\left(r_{6k}E_k + s_{6k}E_k^2 + 2s_{64}E_2E_3 + 2s_{65}E_3E_1 + 2s_{66}E_1E_2\right) = 1,
$$

$$(24.28)$$

where the E_k are components of the electric field along the principal axes and repeated indices are summed.

   If the quadratic coefficients are assumed to be small and only the linear coefficients are retained, then

$$\Delta\left(\frac{1}{n}\right)_l^2 = \sum_{k=1}^{3} r_{lk} E_k \tag{24.29}$$

and $k = 1, 2, 3$ corresponds to the principal axes x, y, and z. The equation for the index ellipsoid becomes

$$x^2\left(\frac{1}{n_x^2} + r_{1k}E_k\right) + y^2\left(\frac{1}{n_y^2} + r_{2k}E_k\right) + z^2\left(\frac{1}{n_z^2} + r_{3k}E_k\right) + 2yz(r_{4k}E_k) \tag{24.30}$$
$$+ 2zx(r_{5k}E_k) + 2xy(r_{6k}E_k) = 1.$$

Suppose we have a cubic crystal of point group $\overline{4}3m$, the symmetry group of such common materials as GaAs. Suppose further that the field is in the z-direction. Then the index ellipsoid is

$$\frac{x^2}{n^2} + \frac{y^2}{n^2} + \frac{z^2}{n^2} + 2r_{41}E_z xy = 1. \tag{24.31}$$

The applied electric field couples the x-polarized and y-polarized waves. If we make the coordinate transformation

$$x = x'\cos 45° - y'\sin 45°,$$
$$y = x'\sin 45° - y'\cos 45°, \tag{24.32}$$

and substitute these equations into the equation for the ellipsoid, the new equation for the ellipsoid becomes

$$x'^2\left(\frac{1}{n^2} + r_{41}E_z\right) + y'^2\left(\frac{1}{n^2} - r_{41}E_z\right) + \frac{z^2}{n^2} = 1, \tag{24.33}$$

and we have eliminated the cross term. We want to obtain the new principal indices. The principal index will appear in Eq. (24.33) as $1/n_{x'}^2$ and must be equal to the quantity in the first parenthesis of the equation for the ellipsoid, i.e.,

$$\frac{1}{n_{x'}^2} = \frac{1}{n^2} + r_{41}E_z. \tag{24.34}$$

Consider the derivative of $1/n^2$ with respect to n: $d(n^{-2})/dn = -2n^{-3}$, or, rearranging, $dn = -\frac{1}{2}n^3 d(1/n^2)$. Assume $r_{41}E_z \ll n^{-2}$. Now $dn = n_{x'} - n$ and $d(1/n^2) = ((1/n_{x'}^2) - (1/n^2)) = r_{41}E_z$ and we can write $n_{x'} - n = -\frac{1}{2}n^3 r_{41}E_z$. The equations for the new principal indices are

$$n_{x'} = n - \frac{1}{2}n^3 r_{41}E_z$$
$$n_{y'} = n + \frac{1}{2}n^3 r_{41}E_z \tag{24.35}$$
$$n_{z'} = n.$$

As a similar example for another important material type, suppose we have a tetragonal (point group $\overline{4}2m$) uniaxial crystal in a field along z. The index ellipsoid becomes

$$\frac{x^2}{n_0^2} + \frac{y^2}{n_0^2} + \frac{z^2}{n_e^2} + 2r_{63}E_z xy = 1. \tag{24.36}$$

A coordinate rotation can be done to obtain the major axes of the new ellipsoid. In the present example, this yields the new ellipsoid

$$\left(\frac{1}{n_0^2} + r_{63}E_z\right)x'^2 + \left(\frac{1}{n_0^2} - r_{63}E_z\right)y'^2 + \left(\frac{z^2}{n_e^2}\right) = 1. \tag{24.37}$$

As in the first example, the new and old z-axes are the same, but the new x' and y' axes are 45° from the original x- and y-axes (see Fig. 24.8).

The refractive indices along the new x- and y-axes are

$$\begin{aligned}
n_x' &= n_0 - \tfrac{1}{2}n_0^3 r_{63}E_z \\
n_y' &= n_0 + \tfrac{1}{2}n_0^3 r_{63}E_z
\end{aligned} \tag{24.38}$$

Note that the quantity $n^3 rE$ in these examples determines the change in refractive index. Part of this product, $n^3 r$, depends solely on inherent material properties, and is a figure of merit for electro-optical materials. Values for the linear and quadratic electro-optic coefficients for selected materials are given in Tables 24.4 and 24.5, along with values for n and, for linear materials, $n^3 r$. While much of the information from these tables is from Yariv and Yeh [37], materials tables are also to be found in Kaminow [13, 14]. Original sources listed in these references should be consulted on materials of particular interest. Additional information on many of the materials listed here, including tables of refractive index versus wavelength and dispersion formulae, can be found in Tropf *et al.* [33].

For light linearly polarized at 45°, the x and y components experience different refractive indices n_x' and n_y'. The birefringence is defined as the index difference $n_y' - n_x'$. Since the phase velocities of the x and y components are different, there is a phase retardation Γ (in radians) between the x and y components of E, given by

$$\Gamma = \frac{\omega}{c}(n_y' - n_x')d = \frac{2\pi}{\lambda}n_0^3 r_{63}E_z d, \tag{24.39}$$

where d is the path length of light in the crystal. The electric field of the incident light beam is

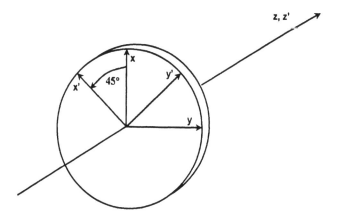

Figure 24.8   Rotated principal axes.

$$\vec{\mathbf{E}} = \frac{1}{\sqrt{2}}\mathbf{E}(\hat{\mathbf{x}} + \hat{\mathbf{y}}). \tag{24.40}$$

After transmission through the crystal, the electric field is

$$\frac{1}{\sqrt{2}}\mathbf{E}(e^{i\Gamma/2}\hat{\mathbf{x}}' + e^{-i\Gamma/2}\hat{\mathbf{y}}'). \tag{24.41}$$

If the path length and birefringence are selected such that $\Gamma = \pi$, the modulated crystal acts as a half-wave linear retarder and the transmitted light has field components

$$\begin{aligned}\frac{1}{\sqrt{2}}\mathbf{E}(e^{i\pi/2}\hat{\mathbf{x}}' + e^{-i\pi/2}\hat{\mathbf{y}}') &= \frac{1}{\sqrt{2}}\mathbf{E}(e^{i\pi/2}\hat{\mathbf{x}}' - e^{i\pi/2}\hat{\mathbf{y}}')\\ &= \mathbf{E}\frac{e^{i\pi/2}}{\sqrt{2}}(\hat{\mathbf{x}}' - \hat{\mathbf{y}}').\end{aligned} \tag{24.42}$$

The axis of linear polarization of the incident beam has been rotated by 90° by the phase retardation of π radians or one-half wavelength. The incident linear polarization state has been rotated into the orthogonal polarization state. An analyzer at the output end of the crystal aligned with the incident (or unmodulated) plane of polarization will block the modulated beam. For an arbitrary applied voltage producing a phase retardation of Γ, the analyzer transmits a fractional intensity $\cos^2 \Gamma$. This is the principle of the Pockels cell.

Note that the form of the equations for the indices resulting from the application of a field is highly dependent upon the direction of the field in the crystal. For example, Table 24.6 gives the electro-optical properties of cubic $\overline{4}3\mathbf{m}$ crystals when the field is perpendicular to three of the crystal planes. The new principal indices are obtained in general by solving an eigenvalue problem. For example, for a hexagonal material with a field perpendicular to the (111) plane, the index ellipsoid is

$$\left(\frac{1}{n_0^2} + \frac{r_{13}E}{\sqrt{3}}\right)x^2 + \left(\frac{1}{n_0^2} + \frac{r_{13}E}{\sqrt{3}}\right)y^2 + \left(\frac{1}{n_e^2} + \frac{r_{33}E}{\sqrt{3}}\right)z^2 + 2yzr_{51}\frac{E}{\sqrt{3}} + 2zxr_{51}\frac{E}{\sqrt{3}} = 1, \tag{24.43}$$

and the eigenvalue problem is

$$\begin{pmatrix} \dfrac{1}{n_0^2} + \dfrac{r_{13}E}{\sqrt{3}} & 0 & \dfrac{2r_{51}E}{\sqrt{3}} \\[2mm] 0 & \dfrac{1}{n_0^2} + \dfrac{r_{13}E}{\sqrt{3}} & \dfrac{2r_{51}E}{\sqrt{3}} \\[2mm] \dfrac{2r_{51}E}{\sqrt{3}} & \dfrac{2r_{51}E}{\sqrt{3}} & \dfrac{1}{n_e^2} + \dfrac{r_{33}E}{\sqrt{3}} \end{pmatrix} \mathbf{V} = \frac{1}{n'^2}\mathbf{V}. \tag{24.44}$$

The secular equation is then

Table 24.4 Linear Electro-optic Coefficients

Substance	Symmetry	Wavelength (μm)	Electrooptic coefficients r_{lk} (10^{-12} m/V)	Indices of refraction	$n^3 r$ (10^{-12} m/V)
CdTe	$\bar{4}3m$	1.0	$r_{41} = 4.5$	$n = 2.84$	103
		3.39	$r_{41} = 6.8$		
		10.6	$r_{41} = 6.8$	$n = 2.60$	120
		23.35	$r_{41} = 5.47$	$n = 2.58$	94
		27.95	$r_{41} = 5.04$	$n = 2.53$	82
GaAs	$\bar{4}3m$	0.9	$r_{41} = 1.1$	$n = 3.60$	51
		1.15	$r_{41} = 1.43$	$n = 3.43$	58
		3.39	$r_{41} = 1.24$	$n = 3.3$	45
		10.6	$r_{41} = 1.51$	$n = 3.3$	54
ZnSe	$\bar{4}3m$	0.548	$r_{41} = 2.0$	$n = 2.66$	
		0.633	$r_{41}{}^a = 2.0$	$n = 2.60$	35
		10.6	$r_{41} = 2.2$	$n = 2.39$	
ZnTe	$\bar{4}3m$	0.589	$r_{41} = 4.51$	$n = 3.06$	
		0.616	$r_{41} = 4.27$	$n = 3.01$	
		0.633	$r_{41} = 4.04$	$n = 2.99$	108
			$r_{41}{}^a = 4.3$		
		0.690	$r_{41} = 3.97$	$n = 2.93$	
		3.41	$r_{41} = 4.2$	$n = 2.70$	83
		10.6	$r_{41} = 3.9$	$n = 2.70$	77
$Bi_{12}SiO_{20}$	23	0.633	$r_{41} = 5.0$	$n = 2.54$	82
CdS	6mm	0.589	$r_{51} = 3.7$	$n_o = 2.501$	
				$n_e = 2.519$	
		0.633	$r_{51} = 1.6$	$n_o = 2.460$	
				$n_e = 2.477$	
		1.15	$r_{31} = 3.1$	$n_o = 2.320$	
			$r_{33} = 3.2$	$n_e = 2.336$	
			$r_{51} = 2.0$		
		3.39	$r_{13} = 3.5$	$n_o = 2.276$	
			$r_{33} = 2.9$	$n_e = 2.292$	
			$r_{51} = 2.0$		
		10.6	$r_{13} = 2.45$	$n_o = 2.226$	
			$r_{33} = 2.75$	$n_e = 2.239$	
			$r_{51} = 1.7$		
CdSe	6mm	3.39	$r_{13}{}^a = 1.8$	$n_o = 2.452$	
			$r_{33} = 4.3$	$n_e = 2.471$	
PLZT[b] ($Pb_{0.814}La_{0.124}$ $Zr_{0.4}Ti_{0.6}O_3$)	∞m	0.546	$n_e^3 r_{33} - n_o^3 r_{13} =$ 2320	$n_o = 2.55$	

Table 24.4 Linear Electro-optic Coefficients (*contd.*)

Substance	Symmetry	Wavelength (μm)	Electrooptic coefficients r_{lk} (10^{-12} m/V)	Indices of refraction	$n^3 r$ (10^{-12} m/V)
$LiNbO_3$	3**m**	0.633	$r_{13} = 9.6$	$n_o = 2.286$	
			$r_{22} = 6.8$	$n_e = 2.200$	
			$r_{33} = 30.9$		
			$r_{51} = 32.6$		
		1.15	$r_{22} = 5.4$	$n_o = 2.229$	
				$n_e = 2.150$	
		3.39	$r_{22} = 3.1$	$n_o = 2.136$	
				$n_e = 2.073$	
$LiTaO_3$	3**m**	0.633	$r_{13} = 8.4$	$n_o = 2.176$	
			$r_{33} = 30.5$	$n_e = 2.180$	
			$r_{22} = -0.2$		
		3.39	$r_{33} = 27$	$n_o = 2.060$	
			$r_{13} = 4.5$	$n_e = 2.065$	
			$r_{51} = 15$		
			$r_{22} = 0.3$		
KDP (KH_2PO_4)	$\bar{4}$2**m**	0.546	$r_{41} = 8.77$	$n_o = 1.5115$	
			$r_{63} = 10.3$	$n_e = 1.4698$	
		0.633	$r_{41} = 8$	$n_o = 1.5074$	
			$r_{63} = 11$	$n_e = 1.4669$	
		3.39	$r_{63} = 9.7$		
			$n_o^3 r_{63} = 33$		
ADP ($NH_4H_2PO_4$)	$\bar{4}$2**m**	0.546	$r_{41} = 23.76$	$n_o = 1.5079$	
			$r_{63} = 8.56$	$n_e = 1.4683$	
		0.633	$r_{63} = 24.1$		
$RbHSeO_4$ [c]		0.633			13,540
$BaTiO_3$	4**mm**	0.546	$r_{51} = 1640$	$n_o = 2.437$	
				$n_e = 2.365$	
KTN ($KTa_x Nb_{1-x}O_3$)	4**mm**	0.633	$r_{51} = 8000$	$n_o = 2.318$	
				$n_e = 2.277$	
$AgGaS_2$	$\bar{4}$2**m**	0.633	$r_{41} = 4.0$	$n_o = 2.553$	
			$r_{63} = 3.0$	$n_e = 2.507$	

[a] These values are for clamped (high-frequency field) operation.
[b] PLZT is a compound of Pb, La, Zr, Ti, and O [8, 18]. The concentration ratio of Zr to Ti is most important to its electro-optic properties. In this case, the ratio is 40:60.
[c] Salvestrini *et al.* [29].

Table 24.5 Quadratic Electro-optic Coefficients [37]

Substance	Symmetry	Wavelength (μm)	Electro-optic coefficients s_{ij} ($10^{-18}\,m^2/V^2$)	Index of refraction	Temperature (°C)
$BaTiO_3$	**m3m**	0.633	$s_{11} - s_{12} = 2290$	$n = 2.42$	$T > T_c$ ($T_c = 120°C$)
PLZT[a]	∞**m**	0.550	$s_{33} - s_{13} = 26000/n^3$	$n = 2.450$	Room temperature
KH_2PO_4 (KDP)	$\overline{4}$**2m**	0.540	$n_e^3(s_{33} - s_{13}) = 31$ $n_o^3(s_{31} - s_{11}) = 13.5$ $n_o^3(s_{12} - s_{11}) = 8.9$ $n_o^3 s_{66} = 3.0$	$n_o = 1.5115$[b] $n_e = 1.4698$[b]	Room temperature
$NH_4H_2PO_4$ (ADP)	$\overline{4}$**2m**	0.540	$n_e^3(s_{33} - s_{13}) = 24$ $n_o^3(s_{31} - s_{11}) = 16.5$ $n_o^3(s_{12} - s_{11}) = 5.8$ $n_o^3 s_{66} = 2$	$n_o = 1.5266$[b] $n_e = 1.4808$[b]	Room temperature

[a]PLZT is a compound of Pb, La, Zr, Ti, and O [8, 18]. The concentration ratio of Zr to Ti is most important to its electro-optic properties. In this case, the ratio is 65:35.
[b]At 0.546 μm.

Table 24.6 Electro-optic Properties of Cubic $\overline{4}$**3m** Crystals [6]

E field direction	Index ellipsoid	Principal indices
E perpendicular to (001) plane: $E_x = E_y = 0$ $E_z = E$	$\dfrac{x^2 + y^2 + z^2}{n_0^2} + 2r_{41}Exy = 1$	$n_x' = n_0 + \frac{1}{2}n_0^3 r_{41}E$ $n_y' = n_0 - \frac{1}{2}n_0^3 r_{41}E$ $n_z' = n_0$
E perpendicular to (110) plane: $E_x = E_y = E/\sqrt{2}$ $E_z = 0$	$\dfrac{x^2 + y^2 + z^2}{n_0^2} + \sqrt{2}r_{41}E(yz + zx) = 1$	$n_x' = n_0 + \frac{1}{2}n_0^3 r_{41}E$ $n_y' = n_0 - \frac{1}{2}n_0^3 r_{41}E$ $n_z' = n_0$
E perpendicular to (111) plane: $E_x = E_y = E_z = E/\sqrt{3}$	$\dfrac{x^2 + y^2 + z^2}{n_0^2} + \dfrac{2}{\sqrt{3}}r_{41}E(yz + zx + xy) = 1$	$n_x' = n_0 + \dfrac{1}{2\sqrt{3}}n_0^3 r_{41}E$ $n_y' = n_0 - \dfrac{1}{2\sqrt{3}}n_0^3 r_{41}E$ $n_z' = n_0 - \dfrac{1}{\sqrt{3}}n_0^3 r_{41}E$

$$\begin{pmatrix} \left(\dfrac{1}{n_0^2}+\dfrac{r_{13}E}{\sqrt{3}}\right)-\dfrac{1}{n^{'2}} & 0 & \dfrac{2r_{51}E}{\sqrt{3}} \\[3mm] 0 & \left(\dfrac{1}{n_0^2}+\dfrac{r_{13}E}{\sqrt{3}}\right)-\dfrac{1}{n^{'2}} & \dfrac{2r_{51}E}{\sqrt{3}} \\[3mm] \dfrac{2r_{51}E}{\sqrt{3}} & \dfrac{2r_{51}E}{\sqrt{3}} & \left(\dfrac{1}{n_0^2}+\dfrac{r_{33}E}{\sqrt{3}}\right)-\dfrac{1}{n^{'2}} \end{pmatrix}=0 \qquad (24.45)$$

and the roots of this equation are the new principal indices

24.6 MAGNETO-OPTICS

When a magnetic field is applied to certain materials, the plane of incident linearly polarized light may be rotated in passage through the material. The magneto-optic effect linear with field strength is called the Faraday effect, and was discovered by Michael Faraday in 1845. A magneto-optic cell is illustrated in Fig. 24.9. The field is set up so that the field lines are along the direction of the optical beam propagation. A linear polarizer allows light of one polarization into the cell. A second linear polarizer is used to analyze the result.

The Faraday effect is governed by the equation

$$\theta = \mathbf{V}\mathbf{B}d, \qquad (24.46)$$

where $\mathbf{V}$ is the Verdet constant, θ is the rotation angle of the electric field vector of the linearly polarized light, $\mathbf{B}$ is the applied field, and d is the path length in the medium. The rotary power ρ, defined in degrees per unit path length, is given by

$$\rho = \mathbf{V}\mathbf{B}. \qquad (24.47)$$

A table of Verdet constants for some common materials is given in Table 24.7. The material that is often used in commercial magneto-optic-based devices is some formulation of iron garnet. Data tabulations for metals, glasses, and crystals, including many iron garnet compositions, can be found in Chen [3]. The magneto-optic effect is the basis for magneto-optic memory devices, optical isolators, and spatial light modulators [27, 28].

Other magneto-optic effects in addition to the Faraday effect include the Cotton–Mouton effect, the Voigt effect, and the Kerr magneto-optic effect. The Cotton–Mouton effect is a quadratic magneto-optic effect observed in liquids. The Voigt effect is similar to the Cotton–Mouton effect but is observed in vapors. The

POLARIZER   POLARIZER

Figure 24.9 Illustration of a setup to observe the Faraday effect.

Table 24.7 Values of the Verdet Constant at $\lambda = 5893$ Å

Material	T (°C)	Verdet constant (deg/G·mm)
Water[a]	20	2.18×10^{-5}
Air ($\lambda = 5780$ Å and 760 mmHg)[b]	0	1.0×10^{-8}
NaCl[b]	16	6.0×10^{-5}
Quartz[b]	20	2.8×10^{-5}
CS$_2$[a]	20	7.05×10^{-5}
P[a]	33	2.21×10^{-4}
Glass, flint[a]	18	5.28×10^{-5}
Glass, crown[a]	18	2.68×10^{-5}
Diamond[a]	20	2.0×10^{-5}

[a] Yariv and Yeh [37].
[b] Hecht [9].

Kerr magneto-optic effect is observed when linearly polarized light is reflected from the face of either pole of a magnet. The reflected light becomes elliptically polarized.

24.7 LIQUID CRYSTALS

Liquid crystals are a class of substances which demonstrate that the premise that matter exists only in solid, liquid, and vapor (and plasma) phases is a simplification. Fluids, or liquids, generally are defined as the phase of matter which cannot maintain any degree of order in response to a mechanical stress. The molecules of a liquid have random orientations and the liquid is isotropic. In the period 1888–1890 Reinitzer, and separately Lehmann, observed that certain crystals of organic compounds exhibit behavior between the crystalline and liquid states [7]. As the temperature is raised, these crystals change to a fluid substance which retains the anisotropic behavior of a crystal. This type of liquid crystal is now classified as thermotropic because the transition is effected by a temperature change, and the intermediate state is referred to as a mesophase [25]. There are three types of mesophases: smectic, nematic, and cholesteric. Smectic and nematic mesophases are often associated and occur in sequence as the temperature is raised. The term smectic derives from the Greek word for soap and is characterized by a more viscous material than the other mesophases. Nematic is from the Greek for thread and was named because the material exhibits a striated appearance (between crossed polaroids). The cholesteric mesophase is a property of the cholesterol esters, hence the name.

Figure 24.10(a) illustrates the arrangement of molecules in the nematic mesophase. Although the centers of gravity of the molecules have no long-range order as crystals do, there is order in the orientations of the molecules [4]. They tend to be oriented parallel to a common axis indicated by the unit vector $\hat{n}$.

The direction of $\hat{n}$ is arbitrary and is determined by some minor force such as the guiding effect of the walls of the container. There is no distinction between a positive and negative sign of $\hat{n}$. If the molecules carry a dipole, there are equal numbers of dipoles pointing up as down. These molecules are not ferroelectric. The molecules are achiral, i.e., they have no handedness, and there is no positional order of the molecules within the fluid. Nematic liquid crystals are optically uniaxial.

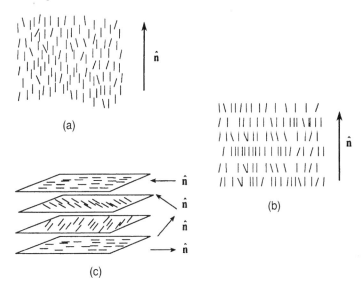

Figure 24.10 Schematic representation of liquid crystal order: (a) nematic, (b) smectic, and (c) cholesteric.

The temperature range over which the nematic mesophase exists varies with the chemical composition and mixture of the organic compounds. The range is quite wide: for example, in one study of ultraviolet imaging with a liquid crystal light valve, four different nematic liquid crystals were used [21]. Two of these were MBBA (*N*-(*p*-methoxybenzylidene)-*p-n* butylaniline) with a nematic range of 17–43°C, and a proprietary material with a range of −20 to 51°C.

There are many known electrooptical effects involving nematic liquid crystals [22, 25, 30]. Two of the more important are field-induced birefringence, also called deformation of aligned phases, and the twisted nematic effect, also called the Schadt–Helfrich effect. An example of a twisted nematic cell is shown in Fig. 24.11.

Figure 24.11(a) shows the molecule orientation in a liquid crystal cell, without and with an applied field. The liquid crystal material is placed between two electrodes. The liquid crystals at the cell wall align themselves in some direction parallel to the wall as a result of very minor influences. A cotton swab lightly stroked in one direction over the interior surface of the wall prior to cell assembly is enough to produce alignment of the liquid crystal [12]. The molecules align themselves with the direction of the rubbing. The electrodes are placed at 90° to each other with respect to the direction of rubbing. The liquid crystal molecules twist from one cell wall to the other to match the alignments at the boundaries as illustrated, and light entering at one cell wall with its polarization vector aligned to the crystal axis will follow the twist and be rotated 90° by the time it exits the opposite cell wall. If the polarization vector is restricted with a polarizer on entry and an analyzer on exit, only the light with the 90° polarization twist will be passed through the cell. With a field applied between the cell walls, the molecules tend to orient themselves perpendicular to the cell walls, i.e., along the field lines. Some molecules next to the cell walls remain

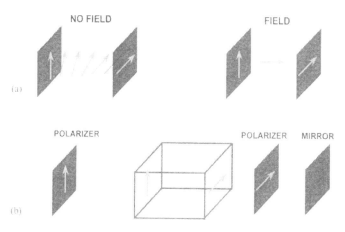

Figure 24.11 Liquid crystal cell operation: (a) molecule orientation in a liquid crystal cell, with no field and with field; (b) a typical nematic liquid crystal cell.

parallel to their original orientation, but most of the molecules in the center of the cell align themselves parallel to the electric field, destroying the twist. At the proper strength, the electric field will cause all the light to be blocked by the analyzer.

Figure 24.11(b) shows a twisted nematic cell as might be found in a digital watch display, gas pump, or calculator. Light enters from the left. A linear polarizer is the first element of this device and is aligned so that its axis is along the left-hand liquid crystal cell wall alignment direction. With no field, the polarization of the light twists with the liquid crystal twist, 90° to the original orientation, passes through a second polarizer with its axis aligned to the right-hand liquid crystal cell wall alignment direction, and is reflected from a mirror. The light polarization twists back the way it came and leaves the cell. Regions of this liquid crystal device that are not activated by the applied field are bright. If the field is now applied, the light does not change polarization as it passes through the liquid crystal and will be absorbed by the second polarizer. No light returns from the mirror, and the areas of the cell that have been activated by the applied field are dark.

A twisted nematic cell has a voltage threshold below which the polarization vector is not affected due to the internal elastic forces. A device 10 μm thick might have a threshold voltage of 3 V [25].

Another important nematic electro-optic effect is field-induced birefringence or deformation of aligned phases. As with the twisted nematic cell configuration, the liquid crystal cell is placed between crossed polarizers. However, now the molecular axes are made to align perpendicular to the cell walls and thus parallel to the direction of light propagation. By using annealed SnO_2 electrodes and materials of high purity, Schiekel and Fahrenschon [30] found that the molecules spontaneously align in this manner. Their cell worked well with 20 μm thick MBBA. The working material must be one having a negative dielectric anisotropy so that when an electric field is applied (normal to the cell electrodes) the molecules will tend to align themselves perpendicular to the electric field. The molecules at the cell walls tends to remain in their original orientation and the molecules within the central region will turn up to 90°; this is illustrated in Fig. 24.12.

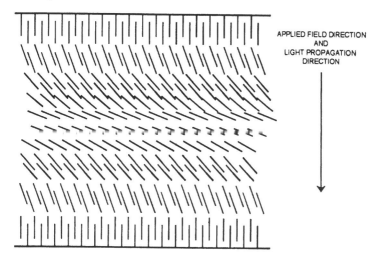

APPLIED FIELD DIRECTION
AND
LIGHT PROPAGATION
DIRECTION

Figure 24.12   Deformation of liquid crystal due to applied voltage [22].

There is a threshold voltage typically in the 4–6 V range [25]. Above the threshold, the molecules begin to distort and becomes birefringent due to the anisotropy of the medium. Thus with no field, no light exits the cell; at threshold voltage, light begins to be divided into ordinary and extraordinary beams, and some light will exit the analyzer. The birefringence can also be observed with positive dielectric anisotropy when the molecules are aligned parallel to the electrodes at no field and both electrodes have the same orientation for nematic alignment. As the applied voltage is increased, the light transmission increases for crossed polarizers [25]. The hybrid field-effect liquid crystal light valve relies on a combination of the twisted nematic effect (for the off state) and induced birefringence (for the on state) [1].

Smectic liquid crystals are more ordered than the nematics. The molecules are not only aligned, but they are also organized into layers, making a two-dimensional fluid. This is illustrated in Fig. 24.10(b). There are three types of smectics: A, B, and C. Smectic A is optically uniaxial; smectic C is optically biaxial; smectic B is most ordered, since there is order within layers. Smectic C, when chiral, is ferroelectric. Ferroelectric liquid crystals are known for their fast switching speed and bistability.

Cholesteric liquid crystal molecules are helical, and the fluid is chiral. There is no long-range order, as in nematics, but the preferred orientation axis changes in direction through the extent of the liquid. Cholesteric order is illustrated in Fig. 24.10(c).

For more information on liquid crystals and an extensive bibliography, see Wu [35, 36].

24.8   MODULATION OF LIGHT

We have seen that light modulators are composed of an electro- or magneto-optical material on which an electromagnetic field is imposed. Electro-optical modulators may be operated in a longitudinal mode or in a transverse mode. In a longitudinal mode modulator, the electric field is imposed parallel to the light propagating through the material, and in a transverse mode modulator, the electric field is imposed perpendicular to the direction of light propagation. Either mode may be used if the entire wavefront of the light is to be modulated equally. The longitudinal mode is more likely to be used if a spatial pattern is to be imposed on the modulation. The mode used will depend upon the material chosen for the modulator and the application.

Figure 24.13 shows the geometry of a longitudinal electro-optic modulator. The beam is normal to the face of the modulating material and parallel to the field imposed on the material. Electrodes of a material which is conductive yet transparent to the wavelength to be modulated are deposited on the faces through which the beam travels. This is the mode used for liquid crystal modulators.

Figure 24.14 shows the geometry of the transverse electro-optic modulator. The imposed field is perpendicular to the direction of light passage through the material. The electrodes do not need to be transparent to the beam. This is the mode used for modulators in laser beam cavities, e.g., a CdTe modulator in a CO_2 laser cavity.

24.9 CONCLUDING REMARKS

The origin of the electro-optic tensor, the form of that tensor for various crystal types, and the values of the tensor coefficients for specific materials have been discussed. The concepts of index ellipsoid, the wave surface, and the wavevector surface were introduced. These are quantitative and qualitative models that aid in the understanding of the interaction of light with crystals. We have shown how the equation for the index ellipsoid is found when an external field is applied, and how expressions for the new principal indices of refraction are derived. Magnetooptics and liquid

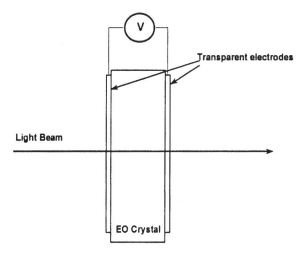

Figure 24.13 Longitudinal mode modulator.

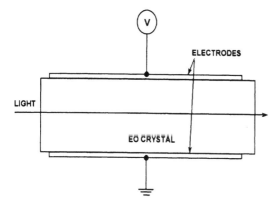

Figure 24.14   Transverse mode modulator.

crystals were described. The introductory concepts of constructing an electrooptic modulator were discussed.

While the basics of electro- and magnetooptics in bulk materials has been covered, there is a large body of knowledge dealing with related topics which cannot be included here. A more detailed description of electrooptic modulators is covered in Yariv and Yeh [37]. Information on spatial light modulators may be found in Efron [5]. Shen [32] describes the many aspects and applications of nonlinear optics, and current work in such areas as organic nonlinear materials can be found in SPIE Proceedings [17, 23].

REFERENCES

1. Bleha, W. P. *et al.*, "Application of the Liquid Crystal Light Valve to Real-Time Optical Data Processing," *Opt. Eng.*, **17**, 371 (1978).
2. Born, M. and E. Wolf, *Principles of Optics*, Pergamon Press, Oxford, 1975.
3. Chen, Di, "Data Tabulations," in *CRC Handbook of Laser Science and Technology, Volume IV, Optical Materials, Part 2: Properties*, Weber, M. J., ed., CRC Press, Boca Raton, Florida, 1986.
4. De Gennes, P. G., *The Physics of Liquid Crystals*, Oxford Univesity Press, Oxford, 1974.
5. Efron, U., ed., *Spatial Light Modulator Technology, Materials, Devices, and Applications*, Marcel Dekker, New York, 1995.
6. Goldstein, D., *Polarization Modulation in Infrared Electrooptic Materials*, PhD dissertation, 1990.
7. Gray, G. W., *Molecular Structure and the Properties of Liquid Crystals*, Academic Press, New York, 1962.
8. Haertling, G. H. and C. E. Land, "Hot-Pressed (Pb,La)(Zr,Ti)O₃ Ferroelectric Ceramics for Electrooptic Applications," *J. Am. Ceram. Soc.*, **54**, 1 (1971).
9. Hecht, E., *Optics*, Addison-Wesley, New York, 1987.
10. Jackson, J. D., *Classical Electrodynamics*, 2nd edn, Wiley, New York, 1975.
11. Jenkins, F. A. and H. E. White, *Fundamentals of Optics*, McGraw-Hill, New York, 1957.
12. Kahn, F. J., "Electric-Field-Induced Orientational Deformation of Nematic Liquid Crystals: Tunable Birefringence," *Appl. Phys. Lett.*, **20**, 199 (1972).

13. Kaminow, I. P., *An Introduction to Electrooptic Devices*, Academic Press, New York, 1974.

14. Kaminow, I. P., "Linear Electrooptic Materials," in *CRC Handbook of Laser Science and Technology, Volume IV, Optical Materials, Part 2: Properties*, Weber, M. J., ed., CRC Press, Boca Raton, Florida, 1986.

15. Kittel, C., *Introduction to Solid State Physics*, Wiley, New York, 1971.

16. Klein, M. V., *Optics*, Wiley, New York, 1970.

17. Kuzyk, M. G., ed., "Nonlinear Optical Properties of Organic Materials X," *Proc. SPIE*, **3147** (1997).

18. Land, C. E., "Optical Information Storage and Spatial Light Modulation in PLZT Ceramics," *Opt. Eng.*, **17**, 317 (1978).

19. Lorrain, P. and D. Corson, *Electromagnetic Fields and Waves*, 2nd edn, Freeman, New York, 1970.

20. Lovett, D. R., *Tensor Properties of Crystals*, Hilger, Bristol, 1989.

21. Margerum, J. D., J. Nimoy, and S. Y. Wong, *Appl. Phys. Lett.*, **17**, 51 (1970).

22. Meier, G., H. Sackman, and F. Grabmaier, *Applications of Liquid Crystals*, Springer-Verlag, Berlin, 1975.

23. Moehlmann, G. R., eds., "Nonlinear Optical Properties of Organic Materials IX," *Proc. SPIE*, **2852** (1996).

24. Nye, J. F., *Physical Properties of Crystals: Their Representation by Tensors and Matrices*, Oxford University Press, 1985.

25. Priestley, E. B., P. J. Wojtowicz, and P. Sheng, eds, *Introduction to Liquid Crystals*, Plenum Press, New York, 1974.

26. Reitz, J. R. and F. J. Milford, *Foundations of Electromagnetic Theory*, 2nd edn, Addison-Wesley, New York, 1967.

27. Ross, W. E., D. Psaltis, and R. H. Anderson, "Two-Dimensional Magneto-Optic Spatial Light Modulator for Signal Processing," *Opt. Eng.*, **22**, 485 (1983).

28. Ross, W. E., K. M. Snapp, and R. H. Anderson, "Fundamental Characteristics of the Litton Iron Garnet Magneto-Optic Spatial Light Modulator," *Proc. SPIE*, **388**, *Advances in Optical Information Processing* (1983).

29. Salvestrini, J. P., M. D. Fontana, M. Aillerie, and Z. Czapla, "New Material with Strong Electro-Optic Effect: Rubidium Hydrogen Selenate (RbHSeO₄)," *Appl. Phys. Lett.*, **64**, 1920 (1994).

30. Schiekel, M. F. and K. Fahrenschon, "Deformation of Nematic Liquid Crystals with Vertical Orientation in Electrical Fields," *Appl. Phys. Lett.*, **19**, 391 (1971).

31. Senechal, M., *Crystalline Symmetries: An Informal Mathematical Introduction*, Hilger, Bristol, 1990.

32. Shen, Y. R., *The Principles of Nonlinear Optics*, Wiley, New York, 1984.

33. Tropf, W. J., M. E. Thomas, and T. J. Harris, "Properties of Crystals and Glasses," in *Handbook of Optics, Volume II, Devices, Measurements, and Properties*, 2nd edn, McGraw-Hill, New York, 1995.

34. Wood, A., *Crystals and Light: An Introduction to Optical Crystallography*, Dover, London, 1977.

35. Wu, Shin-Tson, "Nematic Liquid Crystals for Active Optics," in *Optical Materials, A Series of Advances*, Vol. 1, S. Musikant, ed., Marcel Dekker, New York, 1990.

36. Wu, Shin-Tson, "Liquid Crystals," in *Handbook of Optics, Volume II, Devices, Measurements, and Properties*, 2nd edn, McGraw-Hill, New York, 1995.

37. Yariv, A. and P. Yeh, *Optical Waves in Crystals*, Wiley, New York, 1984.

25

Light-Sensitive Materials: Silver Halide Emulsions, Photoresist and Photopolymers

SERGIO CALIXTO
Centro de Investigaciones en Optica, León, Mexico

DANIEL J. LOUGNOT
UMR CNRS, Mulhouse, France

25.1 INTRODUCTION

The term "light sensitive" comprises a variety of materials, such as silver compounds [1], non-silver compounds [2], ferroelectric crystals, photochromic materials, and thermoplastics [3], and photopolymers and biopolymers [4]. Due to the scope of the handbook and limited space we review only a fundamental description of the three most widely used light-sensitive materials. Undoubtedly, the light-sensitive medium that is mostly used is photographic emulsion; this is because of its high sensitivity, resolution, availability, low cost, and familiarity with handling and processing. Two other popular emulsions are photoresist and photopolymers.

Photographic emulsion has come a long way since its introduction in the 19th century. Extensive research, basic and private, had led to its development. In contrast, photoresist began to be applied in the 1950s, when Kodak commercialized the KPR solution. Since then, due to its application in the microelectronics industry, a fair amount of research has been done. Finally, holographic photopolymers began to be developed more recently (1969) and, with a few exceptions, are still in the development stage.

Information presented in this chapter intends to provide a basic understanding to people who are not familiar with light-sensitive materials. With this foundation it will be possible to read and understand more detailed information. To be well acquainted with stock products and their characteristics, the reader should contact the manufacturers directly. We give a list of their addresses in Section 25.2.1.6.

25.2 COMMERCIAL SILVER HALIDE EMULSIONS

25.2.1 Black-and-White Films

The earliest observations of the influence of sunlight on matter were made on plants and on the coloring of the human skin [5]. In addition, in ancient times, people who lived on the Mediterraean coast dyed their clothes with a yellow substance secreted by the glands of snails. This substance (tyrian purple) develops under the influence of sunlight into a purple-red or violet dye. This purple dyeing process deserves much consideration in the history of photochemistry. The coloring matter of this dye was identified as being a 6-6″ dibromo indigo.

25.2.1.1 The Silver Halide Emulsion

The silver halide photographic emulsion [6] consists of several materials such as the silver halide crystals, a protective colloid, and a small amount of compounds such as sensitizers and stabilizers. Usually this emulsion is coated on some suitable support that can be a sheet of transparent plastic (e.g., acetate) or over glass plates, depending on the application. In the photographic emulsion, the silver halide crystals could consist of any one of silver chloride (AgCl), silver bromide (AgBr), or silver iodide (AgI). Silver halide emulsions can contain one class of crystals or mixed crystals such as $AgCl + AgBr$ or $AgBr + AgI$. Sensitivity of emulsions depends on the mixing ratio. The protective colloid is the second most important component in the emulsion. It supports and interacts with the silver halide grains. During the developing process it allows processing chemicals to penetrate, and eases the selective development of exposed grains. Different colloids have been used for the fabrication of photographic films; however, gelatin seems the most favorable material for emulsion making.

25.2.1.2 Photographic Sensitivity

Sensitometry is a branch of physical science that comprises methods of finding out how photographic emulsions respond to exposure and processing [6, 7]. Photographic sensitivity is the responsiveness of a sensitized material to electromagnetic radiation. Measurements of sensitivity are photographic speed, contrast, and spectral response [8].

Characteristic Curve, Gamma, and Contrast Index

After the silver halides in the emulsion have absorbed radiation, a latent image is formed. In the development step, certain agents react with the exposed silver halide in preference to the unexposed silver halide. In this form, the exposed silver is reduced to tiny particles of metallic silver. The unreduced silver halide is dissolved out in the fixing bath. The developed image can be evaluated in terms of its ability to block the passage of light; i.e., measuring its transmittance (T) [7]. As the amount of silver in the negative goes up, the transmittance goes down. Bearing this fact in mind, the reciprocal of transmittance (opacity) is more directly related to the amount of silver produced in the developing step. Although opacity increases as silver increases, it does not do so proportionally: the quantity that does, is called the density (density $= \log 1/T$), which is used as a measure of the responsiveness of the emulsion [8–14].

To quantify the amount of light that falls on the emulsion two quantities should be considered: the irradiance (E) of the light and the time (t) that the light beam illuminates the emulsion, which are related by the exposure (H), defined by $H = Et$.

The responsiveness of an emulsion is represented in a graphic form that illustrates the relationship between photographic densities and the logarithm of the exposures used to produce them. This graph is called the characteristic curve and varies for different emulsions, radiation sources, kinds of developers, processing times, and temperatures and agitation method. Published curves can be reliably useful only if the processing conditions agree in all essentials with those under which the data were obtained.

A typical characteristic curve for a negative photographic material [8] contains three main sections (Fig. 25.1): the toe, the straight-line portion, and the shoulder. Before the toe, the curve is parallel to the abscissa axis. This part of the curve represents the film where two densities are present even if illumination has not reached the film; they are the base density (density of the support) and the fog density (density in the unexposed but processed emulsion), i.e., the base-plus-fog density region. The toe represents the minimum exposure that will produce a density just greater than the base-plus-fog density. The straight-line portion of the curve density has a linear relationship with the logarithm of the exposure. The slope, gamma (γ), of this straight-line portion is defined as $\gamma = \Delta D / \Delta \log H$ and indicates the inherent contrast of the photographic material. Numerically, the value of γ is equal to the tangent of the angle α, the angle that makes the straight-line portion of the curve with the abscissa. It is usual that manufacturers present not only one characeristic curve for a given material, but a family of them. Each of these curves is obtained with a different developing time and presents a different value of γ. With

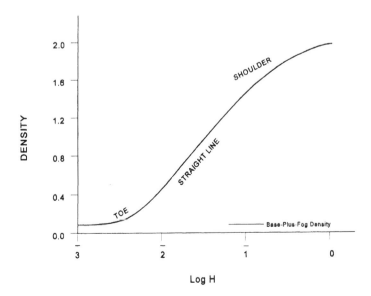

Figure 25.1 Characteristic curve.

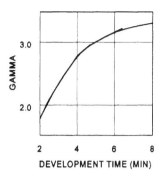

Figure 25.2 Slope γ as a function of the development time.

these data, a new graph can be made by plotting γ as a function of developing time (Fig. 25.2).

Another quantity that can be inferred from the characteristic curve is the contrast index [15] which is defined as the slope of the straight line drawn through three points on the characteristic curve. It is an average slope. The straight line makes an angle with the abscissa, and the tangent of this angle is the contrast index. This contrast also changes with development time (Fig. 25.3).

The slope of the characteristic curve changes with the wavelength of the light used to expose the emulsion. This behavior is evident if we plot the contrast index as a function of recording wavelength. This dependency of contrast varies considerably from one emulsion to another (Fig. 25.4).

Photographic Speed and Meter Setting

The speed of an emulsion [16–18] can be derived from the characteristic curve. If we denote the speed by S, the following relation with exposure will follow: $S = K/H$, where K is a constant and H is the exposure required to produce a certain density above base-plus-fog density. Because the characteristic curve depends on the spectral characteistics of the source used to illuminate the emulsion, the speed will depend also on this source. Therefore, indicating the exposure source when quoting speed values is essential. Speeds applied to pictorial photography (ISO speed) are obtained

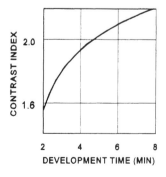

Figure 25.3 Behavior of contrast index with the developing time.

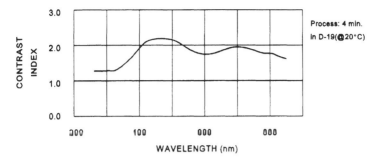

Figure 25.4 Behavior of contrast index with wavelength of the recording light.

by using the formula $S = 0.8/H$, where H is the exposure (to daylight illumination) that produces a density of 0.10 above the base-plus-fog density with a specified development. Several factors affect speed adversely, such as aging, high temperature, high humidity, and the spectral quality of the illumination.

For scientific photography, a special kind of speed value called meter setting is used [8]. The published meter setting value is calculated using the relation $M = k/H$, where k is a constant equal to 8 and H is the exposure that produces a reference density of 1.6 above base-plus-fog. This value is chosen because scientific materials are normally used in applications that require higher contrast levels than those used in pictorial photography. The reference density of 0.6 above base-plus-fog density is chosen for certain spectroscopic plates.

Exposure Determination

Parameters that are important when a camera, or a related instrument, is used, are time of exposure, lens opening, average scene luminance, and speed of the photographic material [19]. The four parameters are related by the equation $t = Kf/Ls$, where t is the exposure time (seconds), f is the f_{number}, L is the average scene luminance, S is the ISO speed, and K is related with the spectral response and transmission losses in the optical system [16, 17]. Other equations can be used to calculate the time of exposure [19].

Spectral Sensitivities of Silver Halide Emulsions

The response of silver halide emulsions to different wavelengths can vary. This response is shown in spectral sensitivity plots. These curves relate the logarithm of sensitivity as a function of the wavelength. Sensitivity is a form of radiometric speed and is the reciprocal of the exposure required to produce a fixed density above a base-plus-fog density [8] (Fig. 25.5).

25.2.1.3 Image Structure Characteristics

To select the best photographic emulsion for a specific application, in addition to the sensitometric data, image structure properties should also be considered.

Granularity

When a developed photographic image is examined with a microscope, it can be seen that the image is composed of discrete grains formed of filaments of silver [6, 8, 20–22].

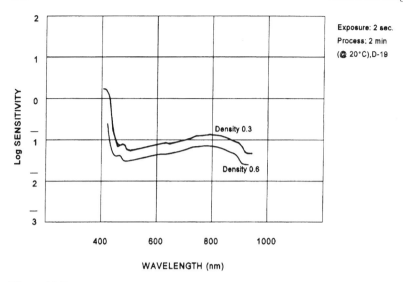

WAVELENGTH (nm)

Figure 25.5 Spectral sensitivity as a function of the recording wavelength. Density is a parameter.

The subjective sensation of the granular pattern is called the graininess. When a uniformly exposed and processed emulsion is scanned with a microdensitometer having a small aperture, a variation in density is found as a function of distance, resulting from the discrete granular structure of the developed image. The number of grains in a given area varies and causes density fluctuations that is called granularity. The microdensitometer shows directly the rms (root mean square) granularity [8], which has values ranging between 5 and 50; the lower numbers indicate finer grain. Granularity has been studied extensively. For a more in-depth knowledge we suggest consulting the references.

Photographic Turbidity and Resolving Power
The turbidity of a photographic emulsion results from light scattered by the silver halide grains and light absorption by the emulsion [23, 24]. This causes a gradual widening of the recorded image as the exposure is increased [8] (Fig. 25.6).

Resolving power is the ability of a photographic material to maintain in its developed image the separate identity of parallel bars when their relative displacement is small. Resolving power values specify the number of lines per millimeter that

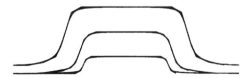

Figure 25.6 Profiles of a slit image given by a microdenistometer for different exposure time.

can be resolved in the photographic image of a test object, commonly named the test chart. The test object contrast (TOC) has a direct relation to the resolving power values that are often reported for test objcts with the low contrast of 1.6:1 and test objcts with the high contrast of 1000:1. Resolving power can be affected by factors such as turbidity, spectral quality of the exposed radiation, developer, processing conditions, exposure, and grain sizes.

Modulation Transfer Function, (MTF)

For light-sensitive emulsions, the MTF [0, 8, 25] is the function that relates the distribution of incident exposure in the image formed by the camera lens to the effective exposure of the silver halide grains within the body of the emulsion layer. To obtain these data, patterns having a sinusoidal variation in illuminance in one direction are exposed to the film. The "modulation" M_o for each pattern can be expressed by the formula $M_o = (H_{max} - H_{min})/(H_{max} + H_{min})$, where H is the exposure. After development the photographic image is scanned in a microdensitometer in terms of density. These densities of the trace are interpreted in terms of exposure, by means of the characteristic curve, and the effective modulation of the image M_i is calculated. The MTF (response) is the ratio M_i/M_o plotted (on a logarithmic scale) as a function of the spatial frequency of the patterns (cycles/mm). Parameters that should be mentioned when specifying an MTF are spatial frequency range, mean exposure level, color of exposing light, developer type, conditions of processing and, sometimes, the f_{number} of the lenses used in making the exposures, Fig. 25.7.

Sharpness

Sometimes the size of the recorded images is larger than the inverse of the highest spatial frequency that can be recorded in the film. However, the recorded image shows edges that are not sharp. The subjective impression of this phenomenon is called sharpness and the measurement of this property is the acutance. Several methods to measure acutance have been proposed [26].

25.2.1.4 Image Effects

Reciprocity

The law of reciprocity establishes that the product of a photochemical reaction is determined by the total exposure (H) despite the range of values assumed by either intensity or time. However, most photographic materials show some loss of sensi-

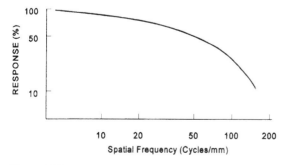

Figure 25.7 Modulation transfer function (MTF) of a film.

tivity (decreased image density) when exposed to very low or very high illuminance levels, even if the total exposure is held constant by adjusting the exposure time. This loss in sensitivity (known as reciprocity-law failure) means more exposure time (than normal calculations indicate) is needed at extreme levels of irradiance to produce a given density. Reciprocity effects can be shown graphically by plotting the log H vs. log intensity for a given density. Lines at 45° represent constant time (Fig. 25.8). Other image effects, present at exposure time, that affect the normal behaviors of emulsions are the following: intermittency effect, Clayden effect, Villard effect, solarization, Herschel effect, and Sabattier effect [8].

Adjacency Effects in Processing

These effects are due to the lateral diffusion of developer chemicals, reaction products, such as bromide and iodide ions, and exhausted developer within the emulsion layer in the border of lightly exposed and heavily exposed areas [6, 19, 21]. This phenomenon presents itself as a faint dark line just within the high-density (high-exposure) side of the border. This is the border effect. Related to this effect is the Eberhard effect, Kostsinky effect, and the MTF effect.

25.2.1.5 Development

The result of the interaction of light with the emulsion is the formation of a latent image. This image is not the cause of the reduction of the silver halide grains to silver by means of the developer but it causes the exposed grain to develop in less time than an underexposed grain. To form a latent image, only about 5–50 photons are needed. The development process [6, 28, 29] will amplify these phenomena by a factor of 10^7. This operation is done by the deposition of silver from the developer on the grain and is proportional, to some extent, to the amount of light action on the grain. The role of the developer on the emulsion has been studied widely and through many years and yet it seems it is not completely clear what happens in the process. A practical developing solution contains the developing agent, an alkali, a preservative, an antifoggant, and other substances such as wetting agents. For a more detailed information, the reader should consult the references.

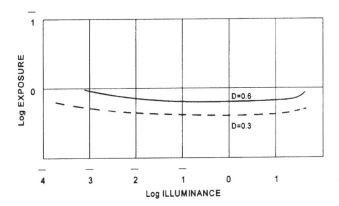

Figure 25.8 Plots showing the reciprocity-law failure.

25.2.1.6 Commercial Black-and-White, Color and Holographic Films

Kodak

Kodak manufactures a multitude of photographic products such as emulsions, books, guides, pamphlets, photographic chemicals, and processing equipment.

A partial list of Kodak photographic emulsions is given in the Kodak catalog L-9 [30]. There are listed about 63 black-and-white films and plates, 60 color films (color negative and reversal films). Also, are listed about 29 black-and-white and color papers.

For the scientific and technological community publication, Kodak P-315 [8] is highly recommended because it describes with detail the application of scientific films and plates. (Table IV on page 6d of this reference should be consulted because it exhibits different emulsion characteristics.) This publication is a good general reference; however, because it is old (1987), some films and plates mentioned in it are not fabricated anymore. The reason for these discontinuities is the wide use of CCDs. Another publication that is also useful for scientists and technicians is the Kodak Scientific Imaging products catalog. For more information write to Eastman Kodak Company, Information Center, 343 State Street, Rochester, NY 14650–0811.

Agfa

Agfa fabricate black-and-white, color negative, and reversal films; they do not fabricate holographic films and plates anymore [31]. To request information write to Agfa Corporation, 100 Challenge Road, Ridge Field Park, NJ 07660, USA. Tel. (201) 440-2500. 1-800-895-6806. In Europe, contact Agfa-Gevaert NV, Septstraat 27, B-2510 Mortsel, Antwerp, Belgium.

Fuji Film

Fuji Film fabricates black-and-white, color negative, and reversal films. Inquiries can be addressed to Fuji Film USA Inc., 555 Taxter Rd, Elsmford, NY 10523. (914) 789-8100, (800) 326-0800, ext 4223 (western USA).

Other manufacturers of black-and-white, color emulsion, and holographic emulsions are Ilford Limited, Mobberly, Knutsford, Cheshire WA16 7HA, UK and Polaroid Corp., 2 Osborn Street, Cambridge, MA 02139. *Note*: in 1992, Ilford stopped the production of holographic plates.

The following company fabricates the Omnidex photopolymer holographic film: du Pont de Nemours and Co., Imaging Systems Department, Experimental Station Laboratory, Wilmington, Delaware 19880-0352.

A list of commercial color and black-and-white films containing not only the main manufacturers listed above is presented in the magazine *Amateur Photographer*, February 15, 1992.

Copies of ANSI/ISO standards are available from the American National Standards Institute, 1430 Broadway, New York, NY 10018.

25.2.2 Color Films

Silver halide emulsions are primarily sensitive to ultraviolet and blue light. In 1873, H. W. Vogel [5] working with colloidon dry plates (fabricated by S. Wortley) noted that these plates presented a greatly increased sensitivity to the green part of the spectrum. These plates contained a yellowish-red dye (corallin) to prevent halation.

Vogel studied the role of several dyes in the emulsion and made the discovery of color sensitizers that extend the sensitivity of the silver halide emulsions into the red and near-infrared part of the spectrum [32].

Subtraction Process

Color reproduction processes can be divided into two: the direct process, in which each point of the recorded image will show the same spectral composition as the original image-forming light at that point; and the indirect process, in which the original forming light is matched with a suitable combination of colors. The direct process was invented by Gabriel Lippman in 1891 and is based on the interference of light. However, it has not been developed sufficiently to be practical. The indirect method to reproduce color comprises additive and subtractive processes [28]. The former process presents difficulties and is not used now. The subtractive process is the basis for most of the commercial products nowadays [21]. Next we describe briefly the structure of a simple color film, hoping this will clarify the color process recording [28].

The emulsion will be described in the sense that the light follows. Basically, color films consists of five layers, three of which are made sensitive by dyes to a primary color. Besides these dyes, layers contains color formers or color couplers. The first layer protects the emulsion from mechanical damage. The second layer is sensitive to blue light. A third layer is a yellow filter that lets pass red and green light that react with the following layers that are red and green sensitive. Finally, a base supports the integral tripack, as the structure containing the three light-sensitive emulsions, the yellow filter and the protecting layer is called. Green-sensitive and red-sensitive layers are also sensitive to blue light because no matter what additional spectral sensitization is used emulsions are still blue sensitive.

The most common procedure in the color development process is to form images by dye forming, a process called chromogenic development. This process can be summarized in two steps:

developing agent + Silver halide → Developer oxidation products

+ Silver metal + Halide ions

then a second reaction follows:

Developer oxidation products + Color couplers → Dye

The color of the dye depends mainly on the nature of the color coupler. At this step a silver image reinforced by a dye image coexist. Then, a further step removes the unwanted silver image and residual silver halide. In this process, there are dyes formed of just three colors: cyan, magenta, and yellow. With combinations of these colors the original colors in the recorded scene can be replicated.

A variety of other methods exist for the production of subtractive dye images such as Polacolor and Cibachrome.

25.2.2.1 Sensitometry [19, 33]

Characteristic Curves

Color films can be characterized in a similar way to black-and-white films. Sensitometric characterization of color films should consider that color films do not contain metallic silver, as black-and-white films do, but instead, they modulate the light by means of the dyes contained in each of the three layers. Because the reproduction of a given color is made by the addition of three wavelengths, $D \log H$ plots should show three curves, one for each primary color (Fig. 25.9).

Speed for color films is defined by the relation $S = K/H$ as for black-and-white films (see Section 25.2.1.2); however, this time, $K = 10$ and H is the exposure to reach a specified position on the D vs. $\log H$ curve.

Spectral Sensitivity Curves

As described above, color films comprise three emulsion layers. Each layer shows a response to certain wavelengths, or spectral sensitivity, expressed as the reciprocal of the exposure (in ergs/cm^2) required to produce a specified density (Fig. 25.10) [6].

25.2.2.2 Image Structure Characteristics [19]

Granularity, Resolving Power, and MTF

In the conventional color process the oxidized developing agent diffuses away from the developing silver grain until color coupling takes place at some distance. In fact each grain gives place to a roughly spherical dye cloud centered on the crystal and, after all unwanted silver has been removed, only the dye colors remain. Because these dye clouds are bigger than the developed grains, the process of dye development yields a more grainy image. The diffuse character of color images is also responsible for sharpness that is lower than that yielded by black-and-white development of the same emulsion.

In Section 25.2.1.3 we mentioned the terms granularity graininess, resolving power, and MTF for black-and-white films. These terms apply also to color films [19].

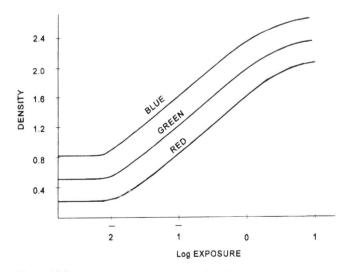

Figure 25.9 Characteristic curve for a color film.

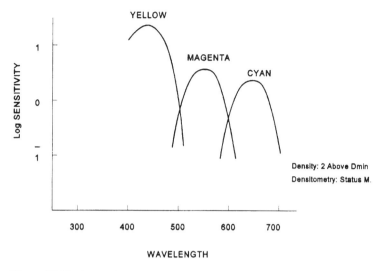

Figure 25.10 Spectral sensitivity as a function of recording wavelength.

25.3 SILVER HALIDE EMULSIONS FOR HOLOGRAPHY

25.3.1 Introduction

The objective of a photosensitive material to be used in holography is the recording of an interference pattern with the highest fidelity. The result of the action of light onto the light-sensitive material can be exposed in any of the three forms: local variations of refractive index, absorption, or thickness. At present, a variety of materials [3] are used to record the interference patterns; however, the most popular material is the holographic silver halide plate. This emulsion can record the high spatial frequencies found in the holographic setups. Interference patterns for transmission holograms can present spatial frequencies between a few lines/mm to about 4000 lines/mm. For reflection holograms, spatial frequencies presented by interference patterns will range between 4500 lines/mm and 6000 lines/mm [34]. Some silver halide holographic emulsions present maximum resolutions of 5000 lines/mm [35].

Full characterization of holographic emulsions cannot be carried out with the D vs. $\log H$ curves (Section 25.2). In holographic emulsions, to reconstruct linearly the object wave it is necessary that differences in amplitude transmittance should be proportional to the differences in exposure. Next, we describe briefly two characterizing methods based on the recoding of sinusoidal interference patterns.

25.3.2 Characterizing Methods for Holographic Emulsions

Several methods have been mentioned in the literature to characterize holographic emulsions. Some are based on the transmittance–exposure curves [36]. These methods emphasize experimental recording parameters such as exposure, intensity ratio of the reference and object beam, angle between recording beams, and recording wavelength. Parameters measured are transmittance and diffraction efficiency. Through them the optimum bias level, beam ratio, and spatial frequency of the interference pattern can be inferred. Also, the transfer function of the film can be inferred.

A more general method of characterizing holographic materials that is applicable to all types of holograms was proposed by Lin [37]. This method supposes that a series of sinusoidal interference patterns are recorded in the medium and later diffraction efficiencies (η) are measured. The material can be characterized by the relation $\sqrt{\eta} = SHV$, where S is a constant (holographic sensitivity), H is the average exposure, and V is the interference fringes visibility. For the ideal material, curves of the $\sqrt{\eta}$ as a function of H with constant V as parameter or curves of $\sqrt{\eta}$ as a function of V with H as parameter are families of straight lines. However, real materials only show curves with a limited linear section. Through them it is possible to know maximum attainable η, optimum exposure value, and holographic sensitivity. Also, these curves allow direct comparison of one material with another or with the ideal one. This cannot be done with the transmittance–exposure characteristic curves mentioned above. Two drawbacks of the $\sqrt{\eta}$ vs. H or $\sqrt{\eta}$ vs. V curves are that they cannot show the spatial frequency response and the noise in reconstruction.

25.3.3 Reciprocity Failure

Failure to follow the law of reciprocity (see Section 25.2.1.4) of holographic materials appears when lasers giving short pulses are employed, such as the Q-switched lasers. These lasers release energy during a very short time, producing a very high output peak power [38]. Contrary to this phenomenon, sometimes when continuous wave (CW) lasers are used and show an intensity-weak output, problems are created for interference pattern recording because of the very long exposure times needed.

25.3.4 Processing: Developing and Bleaching

Developing and bleaching processes recommended by the manufacturers have been optimized and should be followed strictly when a general work on holography is carried out. Much research in developing and bleaching has been done to obtain special effects such as high-diffraction efficiency, reduced developing time, increased emulsion speed and low noise in reconstruction, for example. It is advisable to consult the references and find the method that best suits the experimental needs [31]. In this last reference also a list or commercial holographic emulsions and some of their characteristics is exhibited (p. 114).

25.4 PHOTORESIST

Methods of coloring cloth in a pattern by pretreating designed areas to resist penetration by the dye are known as "resist printing" techniques. Batik is an example of this. Photoresists are organic light-sensitive resins suitable for the production of surface-relief patterns and selective stencils that protect the underlying surface against physical, electrical, or chemical attack.

25.4.1 Types of Photoresist

Many kinds of photoresists are commercially available today. Depending on their action, they may be divided into positive- and negative-working photoresists. In a negative-type photoresist the exposed areas become insoluble as they absorb light so that upon development only the unexposed are dissolved away. In a positive-type photoresist the exposed areas become soluble and dissolve away during the development process.

The first negative photoresists were produced in the 1950s, and were made possible by the evolution of the electronics industry. Kodak photoresist (KPR), produced by the Eastman Kodak Company (Rochester, NY), was the first of a range of high-resolution materials that polymerize under light and become insoluble in organic developer compounds. Polymerized photoresist is resistant to ammonium persulphate, which can be used for etching copper printed circuit boards. A higher-resolution thin-film resist (KTFR) was developed in the 1960s. These resists are members of two distinct families of negative photoresists: those based on an ester solvent (KPR) and those based on a hydrocarbon solvent (KTFR). Processing solutions such as developers and thinners are not interchangeable between the two families.

Positive photoresists do not polymerise upon exposure to light, but their structure releases nitrogen and forms carboxyl groups. The solubility acquired through the exposure enables the positive stencil to be washed out in aqueous alkali. Positive photoresists are basically composed of three components: a photoactive compound (inhibitor), a base resin, and a solvent. The base resin is soluble in aqueous alkaline developers and the presence of the photoactive compound strongly inhibits its dissolution. The light neutralizes the photosensitive compound and increases the solubility of the film. After development in such solutions, an intensity light pattern is converted into a relief structure. The Shipley Co. Inc. (Newton, MA 02162) produces good positive photoresists consisting of Novolak resins with naphthoquinone diazodes functioning as sensitizers (AZ formulations).

Most of the available photoresists are sensitive to light in the blue and ultraviolet regions (300–450 nm) of the spectrum, although deep UV (200–270 nm) photoresists are also obtainable.

The choice of photoresist for a particular application is based on a tradeoff among sensitivity, linearity, film thickness, resolution capability, and ease of application and processing. For instance, in holography and other optical applications, the resolution must be better than 1000 lines/mm and a linear response is usually desired. In microlithography one needs to fabricate line structures of micron (μm) and even submicron dimensions and it is preferable to have a highly nonlinear response of the material.

25.4.2 Application and Processing

Photoresist is provided in the form of a resin suitable for application on to a clean surface. The procedure of deposition, exposure, and development of a photoresist film can vary, depending on the particular kind of photoresist and the application. The preferred method of depositing a photoresist film on the substrate seems to be spin-coating. Fairly even films of photoresist, ranging in thickness from fractions of a micron, to more than 5 μm, can be produced with such a method. For the production of thin films, the substrate must be spun at speeds between 4000 and 7000 rpm. For thicker films, speeds as low as 1000 rpm can be used, at the expense of film uniformity. The quality and adhesion of the coated films to the substrate depends critically on its cleanliness. With positive photoresist the coatings should be allowed to dry, typically, for a few hours and be subsequently baked to allow for the evaporation of the remaining solvents. After exposure, the sample is developed in a solution of developer and water. The time and temperature of baking, the dilution ratio of the developer, the time and temperature of development, as well as the wavelength of the exposing light must be chosen for the particular kind of photoresist employed and determine its sensitivity and the linearity of its response.

25.4.3 Applications in Optics

Photoresists find their main application in the electronics industry, where they are used in the production of microcircuits [39]. However, they also have many important applications in optics. Among others, we can mention the fabrication of holographic gratings [40–42], holograms [43, 44], diffractive optical elements [45, 46], and randomly rough surfaces [47, 48]. Perhaps the most common and illustrative application of photoresist in optics is the production of the master plates for the fabrication of embossed holograms, which have become commonplace in modern life.

There is also a trend to create miniature opto-electro-mechanical instruments and subsystems. The existing technology to fabricate integrated circuits is being adapted to fabricate such devices and, since their manufacture is based on microlithography, the use of photoresist is a must. For example, refractive microlenses have been made by producing "islands" of photoresist on a substrate and baking them in a convection oven to reshape them. Other examples include microsensors, microactuators, and micromirrors. For a more complete picture of the development of this area of technological research, the interested reader is referred to the November 1994 issue of *Optical Engineering* (vol. 33, pp. 3504–3669).

25.5 PHOTOPOLYMER HOLOGRAPHIC RECORDING MATERIALS

25.5.1 Introduction

The principle of holography was described by Denis Gabor some 50 years ago [49]. However, because of the unavailability of monochromatic light sources exhibiting sufficient coherence length, this principle, which was expected to open new vistas in the field of information recording, remained dormant for some 15 years.

The laser effect obtained by Maiman in 1960 [50] from A. Kastler's pioneering theoretical work [51] triggered off a series of experiments that form the early references in practical holography. Among them, the off-axis recording technique described by Leith and Upatnieks [52] must be regarded as a key to the revival of interest in the field of holography. About the same time, Denisyuk's research work [53] on volume holography opened up another fascinating way that resuscitated Lippman's old proposals.

The actual possibility of recording and reconstructing three-dimensional images through a hologram brought about a recrudescence of activity in the scientific community all around the world. Finding out and optimizing new recording materials with a great variety of characteristics became the purpose of a great number of research programs. High-energy sensitivity and resolution, broad spectral response but also simple recording and processing procedure are considered to be indispensable. Furthermore, the demand for self-processing or erasability makes the design of an ideal recording material a challenge for the chemists specializing in photosensitive materials.

The most widely used recording material that has long been known before the emergence of holography is silver halide emulsion. It has been thoroughly investigated for the purpose of holographic recording. In spite of the multifarious improvements introduced during the last 30 years, the typical limitation of these media – the high grain noise resulting in poor signal-to-noise ratio – has prompted investigators to imagine new recording media. Their search has been focused on materials not suffering from this limitation and that would in essence exhibit a grainless structure.

The wide range of applications of holography is a powerful incentive for researchers concerned with this challenge. Indeed, holographic optical elements, holographic displays, videodiscs and scanners, integrated optic devices, optical computing functions, or large-capacity optical memories using volume holography techniques are, as many fields of optics, bursting with activity.

25.5.2 Fundamentals

Basically, photopolymer holographic recording systems are compositions consisting of, at least, a polymerizable substance and a photoiniator, the fate of which is to absorb the holographic photons and generate active species, radicals, or ions capable of triggering the polymerization itself. This composition can also contain a great variety of other structures that are not essential to the polymerization process [54–58]. They are generally incorporated in the formulation with the object of improving the recording properties. In this respect, it should be remembered that:

- the sensitizer is used to extend the spectral range of sensitivity of the sensitive material
- the binder introduces some flexibility in the viscosity of the formulation and may exercise some influence on the diffusive movements that take place during the recording process.

A large number of chemical structures containing either unsaturations, i.e. multiple bonds or cycles, are known to undergo polymerization when subjected to illumination [59]. This kind of process should, theoretically, achieve fairly high efficiency because a single initiating event can trigger a chain polymerization involving a large number of reactive monomer molecules. When the pocess is photochemically initiated, e.g. in holographic recording, the number of monomer units converted per initiating species is commonly referred to as the polymerization quantum yield. It stands to reason that the energy sensitivity of any photopolymer recording system depends to a great extent on this quantum yield.

It is, thus, important to be fully aware of the typical values of this parameter, that never exceeds 10^4 at the very best. Accordingly, as long as the process used for recording holograms with polymerizable material involves the growth of a polymer chain through a bimolecular-like process requiring the diffusion of a monomer to a living macroradical, the practical values of the energy sensitivity remain far lower than those of silver halide gelatins.

25.5.2.1 Photopolymers and Holographic Recording

Without getting to the roots of holography, it is nevertheless important to examine the adaptation of photopolymer materials to the different characteristics of the recording process. In this respect, a classification of holograms based on the geometry of recording, record thickness, and modulation type allows many specificities of polymer materials to be considered in the perspective of their application to holography.

Geometry

Basically, two different geometries are used to record holograms [60 61]. When a record is made by interfering two beams arriving at the same side of the sensitive layer from substantially different directions, the pattern is called an off-axis hologram. The planes of interference are parallel to the bisector of the reference-to-object beam angle. Since the volume shrinkage inherent in the polymerization appears predominantly as a contraction of the layer thickness, it does not substantially perturbate the periodic structure of the pattern of chemical composition of the polymer as the recording process proceeds to completion. The polymer materials are thus fully suitable for recording transmission holograms.

On the contrary, when a hologram is recorded by interfering two beams traveling in opposite direction, i.e., the Lippman or Denisyuk configuration, the interference planes are essentially parallel to the recording layer. Therefore, the shrinkage may cause their spacing to decrease, thus impairing the hologram quality. Consequently, the problem of volume shrinkage must be considered with great care in the formulation of photopolymerizable systems used to record reflection holograms.

Thickness

Holograms can be recorded either as thin or thick holograms, depending on the thickness of the recording medium and the recorded fringe width. In a simplified approach, these two types of hologram can be distinguished by the Q-factor defined as [62]:

$$Q = 2\pi\lambda d/n\Lambda^2,$$

where λ is the holographic wavelength, n is the refractive index, d is the thickness of the medium, and Λ is the fringe spacing. A hologram is generally considered thick when $Q \gg 10$, and considered thin otherwise.

The suitability of photopolymer materials for recording both thin and thick holograms must be examined with respect to the absorbancy of the corresponding layer at the holographic wavelength and the scattering that depends on the signal-to-noise ratio. As regards the absorbance parameter, it is related to the extinction coefficient of the initiator or the sensitizer. With a value of this coefficient in the order of $10^4 \, \text{M}^{-1}.\text{cm}^{-1}$, a concentration of 0.5 M, and an absorbance of at least 0.2 (a set of experimental parameters that corresponds to extreme conditions), the minimum attainable value of the film thickness is about 1 μm. Ultimately, the maximum thickness is limited by the inverse exponential character of the absorbed amplitude profile in the recording material: the thicker the sample is, the more the absorbancy has to be reduced. Although there is no limitation to this homographic interdependence between the thickness and the absorbance parameters, the maximum values are determined by the overall absorptivity that determines the photonic yield (absorbed to incident intensity ratio) and, correspondingly, by the energetic sensitivity. In addition, the impurities, the microheterogeneities as well as the submicroscopic grain structure of the glassy photopolymer material, ultimately introduce a practical limitation of a few millimeters.

Type of Modulation

Holograms are also classified according to the type of modulation they impose on a reconstructing wavefront depending on the change incurred by the recording medium during the holographic illumination. If the incident pattern is recorded by virtue of a density variation of the sensitive layer that modulates the amplitude of the reconstruction wave, the record is known as an amplitude hologram. If the information is stored as a density and/or thickness variation, the phase of the reading beam is modulated and the record is termed a phase hologram [60].

As a general rule, the photoconversion of a monomer to the corresponding polymer does not go along with any spectral change in the wavelength range used to play back holograms. Thus, polymerizable systems apart from some exotic formulations do not lend themselves to the recording of amplitude holograms. On the other hand, the differences in polarizability caused by different degrees of polymerization in the dark and the bright regions of recorded fringe patterns and differences in refractive index resulting therefrom, can be used for the storage of optical information.

25.5.2.2 The Different Models of Polymer Recording Materials

Without indulging in the details of macromolecular chemistry, it is important to clarify what is meant by a *polymerizable substance* in the composition of a recording medium. Depending on the mechanism of the polymerization process, three different meanings that support a classification in three broad categories have to be distinguished: (1) single monomer systems; (2) two-monomer systems; and (3) polymer systems. In addition, depending on the number of reactive functions carried by the various monomers, the different systems undergo either linear or crosslinking polymerization.

The Single-Monomer Linear Polymerization

Only one polymerizable monofunctional monomer is present in the sensitive formulation and gives rise to a linear polymerization. In this case, the recording process involves a spatially inhomogeneous degree of conversion of this monomer that images the amplitude distribution in the holographic field. A refractive index modulation (Δn) parallels, then, this distribution, the amplitude of which depends both on the exposure, the contrast of the fringes in the incident pattern, and the index change between the monomer and the corresponding polymer. If the recording layer has a free surface, a thickness modulation may also be created as the result of a spatially inhomogeneous degree of shrinkage. In spite of the large Δn that are attainable (~ 0.1), these systems are of little value, since they are essentially unstable and cannot lend themselves to a fixing procedure.

If, however, a neutral substance, i.e., non-chemically reactive, is added to such a monofunctional monomer, photopolymerization may interfere constructively with diffusion. Due to the gradient of the conversion rate resulting from the spatially inhomogeneous illumination, the photoconverted monomers accumulate in the bright fringes. Simultaneously, the neutral substance is driven off to the regions corresponding to the dark fringes. This coupled reaction–diffusion process can be maintained up to complete conversion of the monomer molecules. The segregation induced by the diffusion of the neutral substance results in a modulation of refractive index, the amplitude of which depends only on the difference in refractive index between this substance and the polymer. Values exceeding 0.2 are theoretically possible. In practice, the essential limitation is due to the solubility of the lost molecule in the monomer–polymer system.

Reactive formulations in which the neutral substance is a polymer binder have also been described. Such systems present the advantage of a highly viscous consistency that simplifies the coating procedure and facilitates the storage. In addition, the polymer structure of this binder often increases the compatibility with the recording monomer–polymer system. The price to pay for these improvements is a slowing down of the two-way diffusion process and, consequently, a decrease in energy sensitivity of the recording formulation.

Two Two-Monomer Linear Polymerization

In contrast with the previously described single-monomer polymerizable systems, it was surmised that using a mixture of monomers with different reactivity parameters and refractive indexes and, occasionally, diffusivity, shrinkage, and solubility parameters, could be advantageous. The mechanism of a hologram formation in the two-monomer systems (or by extension multicomponent systems) involves a linear photocopolymerization process leading to a different composition of the macromolecule, depending on the local rate of initiation that is, itself, controlled by the incident amplitude. The higher rate of polymerization in light-struck areas compared with dark regions associated with very different copolymerization ratios of the two monomers causes the more reactive one to be preferentially converted in the bright fringes, hence a gradual change in the average composition of the copolymer over the interfringe.

Compared with single-monomer/polymer binder systems, the multicomponent systems turn the coupling between diffusion and photochemical reactions to advantage, with an increased compatibility of the components that is related to the copolymer structure and without impairing the sensitivity.

The Single-Monomer Crosslinking Polymerization

The simplest system that works according to this model contains a single multifunctional monomer. When subjected to a normal exposure with a pattern of interference fringes, the monomer polymerizes, with the local rate of initiation and polymerization being a function of the illumination.

Consequently, a modulation of refractive index parallels the spatial distribution of the degree of conversion, which allows optical information to be recorded. This system suffers, however, from the same drawbacks as the single monomer system. When the conversion is carried out until the system vitrifies, both the light-struck and the non-light-struck regions of the record should finally reach the same maximum conversion. A careful examination of the polymer structure leads, however, to a less clear-cut conclusion. Indeed, the number of living macroradicals depends on the rate of initiation; hence, the indirect influence on the local value of the average length of the crosslinks and the architecture of the tridimensional net-work. Unfortunately, the amplitude of the refractive index resulting from such structural heterogeneities is too small ($< 10^{-4}$) to open up real vistas for practical use.

The Multicomponent Crosslinking Polymerization

Basically, the principle that governs the building up of the index modulation in these systems shows great similarity to that described in the two-monomer linear poly-merization. The only difference is the self-processing character that arises from the use of crosslinking monomers. In fact, substituting monofunctional monomers for a multicomponent mixture containing at least one polyfunctional monomer leads to the fast building up of a tridimensional network that converts the initial liquid system into a gel. This gel behaves, then, like a spongy structure in the holes and channels of which the less-reactive monomers are able to diffuse almost freely – hence a more pronounced effect of the transfer phenomena through diffusion pro-cesses. In addition to that very advantageous feature, the presence of crosslinking agents results in an entanglement of the polymer structure that finally annihilates the diffusion of unreacted species. The living macroradicals are, then, occluded in the structure; the record becomes insensitive to the holographic wavelength and can then be played back *in situ* without any degeneration of the hologram quality. Chemical development of a latent image, fixing, and repositioning are, therefore, not required. This can be most beneficial for real-time applications such as interferometry or information storage.

Photocrosslinking Polymers

These systems are another major class of recording materials that has been thor-oughly studied by many research groups. On exposure to an appropriate pattern of light distribution, active species are created that trigger, in the end, the geminate attachment of two linear polymer chains, thus generating a crosslinked material. Since they basically involve a starting material that is a linear polymer, most of the systems developed to date from this concept are used in the form of dry films.

Several variations of chemical structure and composition have been described. The chemical function used to crosslink the linear polymer chain can be part of this chain (residual unsaturtion or reactive pending group). In such a case, the coupling of the chain is more or less a static process that does not involve important changes of configuration or long-range migration, a feature that goes hand in hand with high-energy sensitivity. In return, since it is impossible to fix the record, the hologram has to be played back under inactinic conditions.

In a different approach, the crosslinking process may involve at least one component that is independent of the polymer chains and is, thus, able to diffuse freely in the matrix. In this case, the record can be made completely insensitive to the recording wavelength by chemical fixing. Systems can also be designed so as to obtain self-processing properties.

25.5.3 The Early History of Photopolymer Recording Materials

In 1969, Close *et al.* [63, 64] were the very first to report on the use of photopoly-merizable materials for hologram recording. Their pioneering experiment involved a mixture of metal acrylates, the polymerization of which was initiated by the methylene blue/sodium *p*-toluene sulfinate system. The holographic source was a ruby laser ($\lambda = 694.3\,\text{nm}$), and energy sensitivities of several hundreds of mJ/cm^2 with a diffraction efficiency of about 40% were reported. The holograms required a post-exposure to a mercury arc lamp to consume the unreacted sensitizer up to complete bleaching and, thus, prevent them from interacting with the reconstructing beam that could result in degeneration of the record.

Jenney *et al.* [65–68] carried out detailed experiments on the same material and investigated the influence of many optical photonic and chemical parameters. Energy sensitivities lower than $1\,mJ/cm^2$ were achieved on materials sensitized to both the red and green spectral range with various sensitizers. Many fixing methods were proposed, such as the flash lamp, thermal treatment or long-term storage in the dark.

The original materials devloped by E. I. Du Pont de Nemours & Co date also from the early 1970s [69–71]. They were based on acrylate monomers associated with a cellulose binder and a broadband sensitizing system (sensitive both to UV and blue-green light). These materials were reported to be able to record transmission gratings with 90% efficiency; compared with Jenney's formulation, they exhibited a relatively flat MTF up to 3000 lines/mm and 6–8 months shelf life when stored in a cool atmosphere.

Several authors investigated the mechanism of hologram formation [72, 73]. Although not completely understood, a complicated process involving both polymerization and monomer diffusion was assumed to account for the density changes and, thus, the refractive index variations that gradually convert the polymerizable layer into a hologram.

As possible substitutes for acrylate monomers, Van Renesse [74], and Sukagawara *et al.* [75, 76] and Martin *et al.* [77] used mixtures of acrylamides and bisacrylamides sensitized to red wavelengths by methylene blue. These systems, which exhibited fairly high energy sensitivity ($5\,mJ/cm^2$), were observed to suffer a pronounced reciprocity failure. From these new formulations, Sadlej and Smolinksa [78] introduced a series of variations intended to improve the shelf life of the recording layers and the dark stability of the hologram. Polymer binders such as polyvinylacetate and derivatives, polyvinyl alcohol, methyl cellulose, and gelatin were tested with various results. Besides an improved stability, the polymer binders were repsorted to induce transmission and sensitivity inhomogeneities. If one restricts the review of the early ages of photopolymer recording materials to the years previous to 1975, one cannot afford to ignore Tomlinson's research work on polymethylmethacrylate (PMMA) [79, 80]. Three-dimensional gratings were, thus, recorded in very thick PMMA samples (about $2\,mm$) by interfering two UV laser beams at $325\,nm$. Spatial frequencies up to 5000 lines/mm with diffraction efficiencies as large as 96% were reported.

The mechanism of hologram formation in PMMA was assigned to the cross-linking of homopolymer fragments by free radicals produced through photocleavage of peroxide or hydroperoxide groups present in the polymer structure as the result of its auto-oxidation during the polymerization of the starting material itself. It was also postulated that the local increase in density may be due to polymerization of small amounts of unreacted monomer trapped in the PMMA structure. No evidence, however, was offered in support of these statements. The most obvious disadvantage of this material was the very poor reproducibility of the holographic characteristics and the great difficulty of extenting its sensitivity to the visible range.

In conclusion, it is worthy of note that the foundations used nowadays to formulate polymerizable systems and to describe the mechanism of hologram formation have been laid from the mid-1970s. The improvements introduced since then are essentially concerned with the adaptability of the systems, their stability, the amplitude and the shape of the modulation of the refractive index and/or the thickness, and the signal-to-noise ratio.

25.5.4 The State of the Art

A great number of state-of-the-art reviews [81–91] on recording materials have been documented in detail over the past decades. Most of these reviews classify photopolymers in categories based either on the mechanism of the photochemical process taking place when holographic irradiation takes place or on the composition of the sensitive layer. Most frequently, they are divided into (i) photopolymerizable systems, (ii) photocrosslinking systems, and (iii) doped polymer systems. Another distinctive feature that is also taken into account to arrange them in categories is the state of the polymer formulation: i.e., (i) liquid composition and (ii) dry films. As will be discussed later, such classifications do not fully account for the specificities of all the physicochemical mechanisms used to record optical information in polymer materials.

Indeed, all the investigators concur in the belief that hologram formation is a complicated process involving both photopolymerization and mass transport by diffusion. Consequently, the distinction introduced between linear and crosslinking polymerization or copolymerization is hardly justifiable since it does not reflect basically different behaviors of the materials in terms of photoconversion or diffusion rates. Likewise, the subtle distinction between liquid and dry formulation insofar as this refers to the flowing or nonsticky character of a material is quite inappropriate, since it does not give any indication as regards the possibility of undergoing mass transport through diffusive movements. In fact, a dry system consisting of a crosslinked copolymer that contains a substantial percentage of residual monomers may well be the seat of faster liquid–liquid diffusive processes than a homogeneous liquid formulation with very high viscosity.

In the background of the statements developed in the second paragraph of the present review, the various recording systems that are in the process of development in the university research groups and those few ones that are commercially available can be arranged in three classes: (1) systems requiring development, (2) self-processing materials, and (3) dichromated gelatin mimetic systems. This classification does not take into account the physical state of the material before exposure. It puts forward the recording mechanism and its interrelation with the development and the processing step.

25.5.4.1 Systems Requiring Post-Treatment

As described earlier, the recording mechanism in photopolymers involves, with only a few exceptions, the coupling between a photoconversion process and a diffusion mechanism. On initiation of the polymerization in the region illuminated by a bright fringe, the monomer converts to polymer. As a result of several kinds of forcing functions (e.g., gradient of concentration or gradient of solubility), additional monomer diffuses to these regions from non-light-struck areas while the large size of the living polymer chains or elements of polymer network inhibits their diffusion. When the reactive mixture contains a binder, its solubility in the monomer may also decrease as the conversion proceeds. Consequently, there may be some net migration of binder away from the bright areas. This process continues until no monomer capable of migrating is left and/or until its mobility in the polymer network becomes negligible. At this stage, the information is recorded in the polymer layer as a modulation of the refractive index. The diffusive process may, yet, continue until any remaining monomer is present. In addition, the index modulation resulting from the partial segregation of the component of the mixture is often modest.

Several post-treatments aimed at increasing this modulation and/or at canceling any remaining photosensitivity were described. The following systems exemplify these different development, fixing, or enhancement treatments (Table 25.1).

du Pont's Omnidex System

This material, which is endowed with several outstanding features, is one of the most attractive photopolymers for holographic recording. Various formulations based on mixtures of aromatic aliphatic acrylates and polymer binders such as cellulose acetate–butyrate are available. These materials that can be sensitized either in the red or the blue-green wavelength range can record holograms of almost 100% efficiency with exposure energies down to 50–100 mJ/cm^2 [92–98].

When imaging such polymerizable formulations, the hologram builds up in real time, the film vitrifies, and finally the conversion stops. The resulting hologram is, then, stable on further illumination, and modulations of the refractive index in the order of 0.01 are typically attained.

A subsequent thermal processing by stoving at 80–120°C for 1–3 hours induces an important increase of Δn (a maximum of 0.073 has been achieved) while the tuning wavelength changes very little (a few nanometers). Polymer holograms can also be processed by immersing them in organic liquids that swell the coating or extract some elements of it. This treatment results in a shift of the playback wavelength and a bandwidth increase, so that reflection holograms reconstructed with white light appear brighter to the eye.

Table 25.1 Characteristics of the Recording Systems Requiring Development

Reference	Conditioning	Thickness (μm)	Recording wavelength (nm)	Sensitivity (mJ/cm^2)	Resolution (lines/mm)	Diffraction efficiency (%)
Dupont Omnidex	Coated on plastic film	5–100	488–633	10–100	6000	> 99
Polaroid DMP-128	Coated on plastic film	1–30	442–647	5–30	5000	80–95
PMMA/ titanocene	PMMA block	500–3000	514	4000	—	~ 100
Acrylamide	Coated on glass	100	633	100	3000	80

Polaroid's DMP-128 System

Another example of a polymer recording material with specially attractive features was developed at the Polaroid Corporation. This formulation was based, among others, on a mixture of acrylates, difunctional acrylamide, and a polyvinylpyrrolidone binder. The sequential processing steps include: an incubation in a humid environment prior to exposure; a holographic illumination with a 5–30 mJ/cm^2 exposure, depending on the type of hologram recorded; a uniform white light illumination for a few minutes; an incubation in a liquid developer/fixer; a rinse removing the processing chemicals; and, finally, a careful drying [99–107].

One of the outstanding characteristics of the holograms recorded with these materials is their insensitivity under high humidity environments (no significant alteration of the diffractive properties after 9 months incubation at 95% relative humidity at room temperature). In DMP-128, before chemical processing, the holograms formed shows a diffraction efficiency lower than 0.1%. The immersion in a developer/fixer bath removes soluble components and produces differential swelling and shrinkage that greatly amplify the modulation of the refractive index and cause very bright high diffraction efficiency holograms to appear.

Systems with Thermal or Photochemical Repolymerization

An original polymer material based on an advanced formulation that contains a variety of monomers (e.g., acrylate, methacrylate, vinylcarbazole), a binder, radical precursors, solvents, and a sensitizer was developed by Zhang *et al.* [108]. The fixing procedure of the holograms recorded with this material involves the overcoating of the record with a polymerizable liquid formulation, then its annealing and, finally, the curing of the layer by either a thermal or a photochemical treatment. Since the peak reflection wavelength depends on the condition of the post-treatment, these materials were used to fabricate trichromatic filters or pseudocolor reflection holograms.

PMMA/Titanocene Dichloride Systems

Thick PMMA samples containing various amounts of residual MMA and titanocene dichloride were also investigated for holographic recording in the blue-green wavelength range [109]. The recording mechanism was assumed to be due to a complex photoprocess in which photodegradation of the homopolymer chain, photopolymerization of residual monomers, as well as crosslinking of photogenerated fragments cannot be discriminated. This material presents the outstanding advantage of allowing the superposition of a large number of volume holograms without significant intertalk.

The Latent Imaging Polymers

A latent imaging polymer was described a few years ago [80]. It was based on a microporous glass substrate with a nanometric pore diameter on the surface of which a photosensitive molecule generating radicals was chemisorbed. The recording process involves: (1) illumination by a pattern of interference fringes, (2) filling the pores with a polymerizable formulation and, then, (3) uniform overall exposure.

25.5.4.2 Self-processing Materials

In spite of their very attractive features, the recording systems described above do not permit immediate and in-situ reconstruction. Since this property is a prerequisite to the use of a recording material for applications such as real-time holographic interferometry, a great deal of work has been devoted to the development of materials, a step that could dispense with any further processing after holographic illumination. The basic concept underlying the common approach developed by all the scientists involved in that field is the use of a formulation undergoing both crosslinking copolymerization and segregation of the polymerizable monomer units. The full self-processing and fixing character is achieved when the polymerization process terminates (due to the overall vitrification of the recording sample) just as the amplitude of the refractive index modulation resulting from microsyneresis and differences in the microscopic composition of the crosslinked polymer chains passes through its maximum.

A large number of materials developed according to this principle are described in the literature. They are generally liquid compositions that specially favor the diffusive motions in the early stages of the recording process. Since the initial viscosity of the formulation is a key factor, the major differences are in the choice of polymerizable structures that allows this parameter to be adjusted. The different systems are arranged in two categories: (1) the diluent + oligomer systems and (2) the prepolymerized multicomponent systems (Table 25.2).

Diluent and Oligomer Systems

These formulations, very popular in the former Soviet Union, were based mainly on oligourethane-, oligocarbonate-, and oligoether-acrylates and multiacrylates [110]. In some cases, inert components with high refractive indexes were also incorporated. These compositions are sensitive over the 300–500 nm spectral range, with a maximum resolution in the order of 2500 lines/mm. Similar materials sensitive to the He–Ne wavelength were also developed using acrylamide derivatives as the reactive monomers, several additives, and the methylene blue/amine system as a sensitizer. An efficiency of 60% was reported at 633 nm with an energy sensitivity of about 50 mJ/cm^2 [111–115].

Table 25.2 Characteristics for the Systems with Self-Processing Properties

Reference	Conditioning	Thickness (μm)	Recording wavelength (nm)	Sensitivity (mJ/cm^2)	Resolution (lines/mm)	Diffraction efficiency (%)
Diluent + oligomers (FPK-488)	Liquid between glass plates	20	300–500	20	1500–6000	80
Diluent + oligomers (FPK-488)	Liquid between glass plates	20	633	50	—	60
Pre polymerized multicomponents (PHG###)	Liquid between glass plates	20–100	450–800	100–500	> 3000	80

A dry acrylamide–polyvinylalcohol formulation based on the same approach was also used for optical information storage. This polymeric material exhibits a self-processing and fixing character, the completion of which requires a fairly long exposure. The records can, however, be played back with an attenuated reading beam power ($200 \, \mu W/cm^2$) before reaching complete inertness. It can also be fixed by illumination with the 365-nm line of a mercury arc, which terminates the polymerization [116, 117].

The Prepolymerized Multicomponent Systems

The mechanism of hologram formation in these multicomponent systems involves a differential crosslinking copolymerization between the areas of higher and lower intensity. Because of the different reactivity parameters, the rates of incorporation of the monomer structures are different, and a modulation of the chemical composition of the final fully polymerized material is created. The novelty of these materials consists in the use of a formulation containing, among others, a highly multifunctional acrylate monomer with such a high reactivity that on preilluminating the layer with a homogeneously distributed beam, it forms almost instantaneously a sponge-like structure in the channels and cavities of which the various remaining monomers diffuse freely during the subsequent holographic illumination. The presence of this network since the very beginning of the recording process allows the coupling between photochemical conversion and transport to be more efficient in the sense that it simultaneously presents the advantage of a liquid formulation over diffusion taking place and that of solid layers over the shrinkage and the resulting hydrodynamical strain therefrom. Finally, the refractive index modulation arising from the modulation of the microscopic chemical composition is much larger than the one paralleling a modulation of segment density or crosslinking average length.

A series of materials (called PHG-###), formulated according to these general lines and having sensitivities from 450 to 800 nm, were proposed [118–127]. The sponge-forming monomer was typically a multifunctional acrylate monomer (pentaerythritol tri- or tetra-acrylate, dipentaerythritol pentracrylate). It was associated with various low functionality acrylates, a xanthene or polymethine dye sensitizer, an amine cosynergist and, occasionally, specific additives or transfer agents that improved the recording characteristics. Diffraction efficiencies of about 90% were achieved with an almost flat frequency response from 500 to 2500 lines/mm. The application fields such as holographic interferometry (both real-time and time-average) [121, 126], holographic images [123], computer-generated holograms [124], multiple holograms, and recording in chopped mode [119] can be quoted in illustration.

25.5.4.3 Photocrosslinkable Polymers

Photocrosslinkable materials are another important class of polymer recording materials that include two categories of systems: (1) polymers undergoing crosslinking through intercalation of monomer units or by simple coupling of the linear polymer chain and (2) polymers undergoing crosslinking through a complexation process involving a metal ion (Table 25.3).

The Monomer/Polymer System

An attractive example of such systems uses poly N-vinylcarbazole as the base polymer, a carbonyl initiator, and a sensitizer. The mechanism of recording involves,

Table 25.3 Characteristics of the Recording Systems Involving the Crosslinking of a Polymer Structure

Reference	Conditioning	Thickness (μm)	Recording wavelength (nm)	Sensitivity (mJ/cm^2)	Resolution (lines/mm)	Diffraction efficiency (%)
p-Vinylcarbazole	Solid film on glass	2.5–7	488	50–500	800–2500	80
PMMA	Solid film on glass	100–200	488	7000	2000	~100
DCPVA	Solid film on glass	30–60	488	500	3000	~70
DCPAA	Solid film on glass	60	488	200	3000	~65
FePVA	Solid film on glass	60	488	>15 000	3000	80

first, a holographic exposure that induces photocrosslinking to the polymer base in proportion to the illumination received by every part of the recording layer. A modulation of density is thus created, and a hologram is recorded. This record is then pretreated with a good solvent of the initiator–sensitizer system, to remove these components. The next step consists in swelling the record in a good solvent of the polymer base; the final treatment involves the deswelling, by dipping in a bad solvent of the polymer base. This swelling–deswelling treatment causes the cross-linked regions to reconfigurate, shrink, and finally partially crystallize. Consequently, the attainable modulation of refractive index exceeds by far the values reported in other systems (up to 0.32). Several practical applications were developed with this class of materials, the main feature of which is an almost infinite durability under extreme environmental conditions (at 70°C and 95% RH) [128].

Another system that held the attention of several groups used the photoinitiated coupling of linear PMMA. A radical initiator capable of abstracting hydrogen atoms creates macroradical sites on the polymer backbone, that decay by geminate recombination. The pattern of interference fringes is thus transferred as a pattern of crosslinking density. Typically, such systems exhibit a poor energy sensitivity (several J/cm^2) but their angular discrimination capability is excellent over a spatial frequency range extending from 100 to 2000 lines/mm [129].

The Metal Ion/Polymer Systems

All the systems categorized under this title were developed on the model of dichromated gelatine (DCG). Although the base component of this type of material that has been used as a recording medium ever since 1968 [130] is a biopolymer, it is generally listed and described under a specific heading in the reviews dealing with holographic recording. The recording mechanism prevailing in DCG implies the crosslinking of the gelatin backbone by Cr(III) ions photogenerated through the reduction of dichromate [Cr(VI)] centers. As the result of a chemical post-treatment under carefully controlled conditions, a large refractive index modulation can be achieved (0.08), thus making it possible to record both transmission and reflection volume phase holograms with near-ideal diffraction efficiencies. DCG is, besides, endowed with specially desirable properties, such as uniform spatial frequency response over a broad range of frequencies (from 100 to about 6000 lines/mm) and reprocessing capacity [131–134].

In spite of these outstanding advantages, DCG suffers from some typical drawbacks that may temper the enthusiasm of the holographers. Among others, the fluctuation of many characteristic parameters from batch to batch, the need for a fine-tuning of the prehardening treatment, the complex procedure of development and fixing that requries a specially accomplished know-how, and the sensitivity of the record to environmental factors are not of lesser importance. In addition to these, DCG needs to be sensitized by incorporation of a suitable blue dye to allow extension of its sensitivity to the red. A fair amount of research has been carried out with regard to the optimization and characterization of these sensitized systems as well as to their use in various applications such as holographic elements, head-up display, laser scanning systems, fiber-optic couplers, optical interconnects, and dichoric filters [135–140]. Since the basic mechanism of the phototransformation of gelatin is still a controversial question and owing to its typical shortcomings, several DCG-mimetic systems likely to bring about significant improvements were studied.

Polyvinylalcohol/Cr(III) Systems (DCPVA)

One of the most popular systems uses polyvinylalcohol (PVA) as the polymer base and Cr(III) as the crosslinking agent [141]. It has received continuous attention from many research groups who studied various aspects of its holographic recording performance: exposure energy, beam ratio, polarization, MTF, angular selectivity, acidity of the medium, effect of electron donors, effect of plasticizing substances, and molecular weight of the polymer base. Even though this material exhibits a high real-time diffraction efficiency ($\sim 65\%$ for an exposure of about $1\,J/cm^2$) several chemical developing methods allow the final efficiency to be improved (up to 70% for an exposure level of $200\,mJ/cm^2$) [142, 143]. In a continuing effort to penetrate the secrets of the recording mechanism of the PVA/Cr(III) systems, many physicochemical studies were carried out. They suggest, convergingly, the intermediacy of Cr(V) in the photoreduction of Cr(VI) to Cr(III). Such systems were also reported to lend themselves to sensitization in the red by incorporation of a blue dye (e.g., methylene blue) [144–154].

Polyacrylic Acid/Cr(III) (DCPAA)

Polyacrylic acid (PAA) was also used as a possible substitute for gelatin to record volume holograms. Although much less work has been devoted to the optimization of this material, a great similarity with DCPVA was found. Efficiencies exceeding 65% at an exposure of $200\,J/cm^2$ were obtained [152, 154].

Polyvinylalcohol/Fe(II) (FePVA)

As a continuation of the ongoing search for DCG-mimetic systems, the feasability of substituting Fe(II) for Cr(III) in PVA-based materials was examined. After optimization of the initial Fe(III) content, the electron donor structure, and the recording conditions, a diffraction efficiency of about 80% at an exposure of $25\,J/cm^2$ was achieved [147].

25.5.5 Doped Polymer Materials

During the recording step, all of the materials described in this review are the seat of transformations involving the creation of chemical bonds between monomer or polymer structures. In essence, the recording process is thus a nonreversible process. The possibility of obtaining erasable and reusable polymer materials capable of recording a large number of write/read/erase (WRE) cycles was examined by several research groups. Several approaches were proposed that employ matrixes of homopolymers or copolymers (linear or weakly crosslinked) and dye dopants. The selection of the dye dopant is essentially dependent on its photochromic character and on its photostability. Materials doped with azo dyes or spiropyran derivatives were thus reported to be suitable for real-time reversible amplitude or phase hologram recording. Some of the corresponding systems are capable of more than 10^4 cycles without noticeable fatigue, with thermal self-erasing times of a few seconds. Efforts were also made at developing similar materials where the photochromic active molecule is chemically intercalated in the polymer structure or bound to a side group. Since linearly polarized light induces optical anisotropy in the sensitive film due to reorientation and alignment of the absorbing molecules, these doped materials were mainly used for polarization holographic recording [155–158].

A similar approach was used by several authors who formulated recording materials in doping PVA with bacteriohodopsin (Table 25.4). This structure is a protein transducer involved in the mechanism of mammalian vision. Upon absorption of light energy, this purple ground state sensitizer passes through several intermediate states to a blue light absorbing photoproduct that under standard biological conditions thermally reverts to the ground state material within a decay time of about 10 ms. This back-process can be drastically slowed down by addition of chemical agents. This product was used as a dopant in holographic recording media that can be read out at wavelengths between 620 and 700 nm in some interesting applications such as dynamic filtering with a spatial light modulator or optical pattern recognition (159–162].

25.5.6 Spectral Hole Burning

Hole burning is a phenomenon that takes places when molecules undergoing a photoreaction generating photoproducts absorbing at different spectral positions are excited by a sharply monochromatic light. With such assumptions, a "hole" is left in the inhomogeneously broadened absorption band. If this operation is repeated at different wavelengths within the absorption band, every hole can be associated

Table 25.4 Holographic Characteristics of the Dye-Doped Recording Systems

Reference	Conditioning	Thickness (μm)	Recording wavelength (nm)	Sensitivity (mJ/cm^2)	Resolution (lines/mm)	Diffraction efficiency (%)
Methyl orange/ PVA	Solid film on glass	15–30	442–488	> 300	500–4000	35
Bacteriorhodopsin	Membrane	150–500	360–600	—	> 5000	11

with a bit of information and a high storage capacity becomes available. The feasability of this concept was demonstrated using an oxazine dye in a polyvinylbutyral matrix at 4.1 K [163, 164].

25.6 CONCLUSION

The present review takes stock of all the different approaches reported in the specialized literature to formulate polymer or polymerizable classes of materials for holographic recording. The time to achieve an ideal holographic recording material is, no doubt, still a long way off but, in some degree, several existing formulations, nevertheless, meet some of the required properties needed for application.

Even though polymer materials will never excel silver halides in terms of energy sensitivity owing to intrinsic limitations, their high storage capability, rapid access, excellent signal-to-noise ratio and, occasionally, self-processing character, open up new vistas.

Whatever directions in which attempts at perfecting these materials are carried out by multidisciplinary research groups, it must be kept in mind that the key question in designing new polymerizable systems for holographic recording is not concerned with the tailoring of more or less exotic initiator or monomer structures likely to undergo faster curing. It is of paramount importance to realize that the main issue to be dealt with is definitely to gain a fuller insight into the coupling between photochemically induced monomer conversion, shrinkage, and mass transfer. In this respect, the ongoing activity devoted to their improvement from a thorough investigation of the recording mechanism is a cheerful omen.

ACKNOWLEDGMENTS

We would like to thank Eugenio R. Mendez (CICESE, Ensenada, c.p. 22860, Mexico) for writing the section devoted to photoresist.

S. Calixto would like to acknowledge Z. Malacara and M. Scholl for fruitful discussions. Thanks are given to Raymundo Mendoza for the drawings.

REFERENCES

1. Mees, C. E. K. and T. H. James, eds, *The Theory of the Photographic Process*, 3rd Edn, Macmillan, New York, 1966.
2. Kosar, J., *Light Sensitive Systems: Chemistry and Applications of Non-Silver Halide Photographic Processes,*" John Wiley, New York, 1965.
3. Smith, H. M., "Holographic Recording Materials," in *Topics in Applied Physics*, Vol. 20, Springer-Verlag, New York, 1977.
4. Bazhenov, V., S. Yu Marat, V. B. Taranenko, and M. V. Vasnetsov, "Biopolymers for Real Time Optical Processing," in *Optical Processing and Computing*, H. H. Arsenault, T. Zoplik, and Macukow, eds, Academic Press, San Deigo, 1989, pp. 103–144.
5. Eder, J. M., *History of Photography*, Dover, New York, 1978, Chapters 1, 2 and 64.
6. Carrol, B. H., G. C. Higgins, and T. H. James, *Introduction to Photographic Theory. The Silver Halide Process*, John Wiley and Sons, New York, 1980.
7. Todd, H. N., *Photographic Sensitometry a Self Teaching Text*, John Wiley, New York, 1976.

8. *Scientific Imaging with Kodak Films and Plates*, Publication P-315, Eastman Kodak, Rochester, New York, 1987.
9. ANSI PH 2.16 – 1984, "Photographic Density: Terms, Symbols and Notation."
10. ANSI PH 2.19 – 1986, "Photographic Density Part 2: Geometric Conditions for Transmission Density Measurements."
11. ISO 5/1 – 1984, "Photographic Density: Terms, Symbols and Notation."
12. ISO 5/2 – 1984, "Photographic Density Part 2: Geometric Conditions for Transmission Density Measurements."
13. ISO 5/3 – 1984, "Photographic Density, Part 3: Spectral Conditions for Transmission Density Measurements."
14. Swing, R. E., "Microdensitometry," *SPIE Press*, **MS 111** (1995).
15. Niederpruem, C. J., C. N. Nelson, and J. A. C. Yule, "Contrast Index," *Phot. Sci. Eng.*, **10**, 35–41 (1966).
16. ISO 6 – 1975, "Determination of Speed Monochrome, Continuous-Tone Photographic Negative Materials for Still Photography."
17. ANSI PH 2.5 – 1979, "Determination of Speed Monochrome, Continuous-Tone Photographic Negative Materials for Still Photography."
18. ANSI PH 3.49 – 1971, R1987.
19. Thomas, W., *SPSE Handbook of Photographic Science and Engineering*, Wiley, New York, 1973.
20. ANSI PH 2.40 – 1985, R1991.
21. Altman, J. H., "Photographic Films," in *Handbook of Optics*, Michel Bass, ed., McGraw-Hill, New York, 1994.
22. Dainty, J. C. and R. Shaw, *Image Science*, Academic Press, London, 1974.
23. ISO PH 6328 – 1982, "Method for Determining the Resolving Power of Photographic Materials."
24. ANSI Ph 2.33 – 1983, "Method for Determining the Resolving Power of Photographic Materials."
25. ANSI PH 2.39 – 1984, "Method of Measuring the Modulation Transfer Function of Continuous-Tone Photographic Films."
26. Crane, E. M., "Acutance and Graunlance," *Proc. Soc. Photo-Opt. Instrum. Eng..*, **310**, 125–130 (1981).
27. Bachman, P. L., "Silver Halide Photography," in *Handbook of Optical Holography*, Caulfield, J., ed., Academic Press, London, 1979, pp. 89–125.
28. Walls, H. J. and G. G. Attridge, *Basic Photoscience, How Photography Works*, Focal Press, London, 1977.
29. Grant Haist, *Modern Photographic Processing*, Wiley Interscience, New York, 1979, Chapter 6, pp. 284, 324.
30. *1994 Kodak Professional Catalog*, Publication L-9, Eastman Kodak, Rochester, New York, 1994.
31. Bjelkhagen, H. I., *Silver-Halide Recording Materials for Holography and Their Processing*, Springer-Verlag, New York, 1993.
32. Gibson, H. Lou, "Photographic Film," in *The Infrared and Electrooptical Systems Handbook*, Acetta, J. S. and D. L. Shumaker, eds, Spie Optical Engineering Press, Washington, 1992, pp. 519–539.
33. *Kodak Color Films*, Publication E-77, Eastman Kodak, Rochester, New York, 1977.
34. Collier, R. J., C. Burckhardt, and L. H. Lin, *Optical Holography*, Academic Press, New York, 1971, Chapter 10, p. 271.
35. Agfa-Gevaert N.V., B-2510, Mortsel, Belgique, 1983. Holographic Materials. Technical Bulletin.
36. Friesem, A. A., A. Kozma, and G. F. Adams, "Recording Parameters of Spatially Modulated Coherent Wavefronts," *Appl. Opt.*, **6**, 851–856 (1967).

37. Lin, L. H., "Method of Characterizing Hologram – Recording Materials," *J. Opt. Soc. Am.*, **61**, 203–208 (1971).
38. Nassenstein, H., H. J. Metz, H. E. Rieck, and D. Schultze, "Physical Properties of Holographic Materials," *Phot. Sci. Eng.*, **13**, 194–199 (1969).
39. Horne, D. F., "Microcircuit Production Technology," Adam Hilger, Bristol, 1986, pp. 13–19.
40. Hutley, M. C., *Diffraction Gratings*, Academic Press, London, 1982, Chapter 4.
41. Popov, E. K., L. V. Tsonev, and M. L. Sabeva, "Technological Problems in Holographic Recording of Plane Gratings," *Opt. Eng.*, **31**, 2168–2173 (1992).
42. Mello, B. A., I. F. da Costa, C. R. A. Lima, and L. Cescato, "Developed Profile of Holographically Exposed Photoresist Gratings," *Appl. Opt.*, **34**, 597–603 (1995).
43. Bartolini, R. A., "Characteristics of Relief Phase Holograms Recorded in Photoresists," *Appl. Opt.*, **13**, 129–139 (1974).
44. Bartolini, R. A., "Photoresists," in *Holographic Recording Materials*, Smith, H. M., ed., Springer-Verlag, New York, 1977, pp. 209–227.
45. Haidner, H., P. Kipfer, J. T. Sheridan, *et al.*, "Polarizing Reflection Grating Beamsplitter for the 10.6-μm Wavelength," *Opt. Eng.*, **32**, 1860–1865 (1993).
46. Habraken, S., O. Michaux, Y. Renotte, and Y. Lion, "Polarizing Holographic Beam Splitter on a Photoresist," *Opt. Lett.*, **20**, 2348–2350 (1995).
47. O'Donnell, K. A. and E. R. Méndez, "Experimental Study of Scattering from Characterized Random Surfaces," *J. Opt. Soc. Am.*, **A4**, 1194–1205 (1987).
48. Méndez, E. R., M. A. Ponce, V. Ruiz-Cortés, and Zu-Han Gu, "Photofabrication of One-Dimensional Rough Surfaces for Light Scattering Experiments," *Appl. Opt.*, **30**, 4103–4112 (1991).
49. Gabor, D., *Nature*, **161**, 777 (1948).
50. Maiman, T. H., *Nature*, **187**, 493 (1960).
51. Kastler, A., *J. Phys. Rad.*, **11**, 255 (1950).
52. Leith, E. N. and J. Upatnieks, *J. Opt. Soc. Am.*, **52**, 1123 (1962).
53. Denisyuk, Y. N., *Soviet Phys-Doklady*, **7**, 543 (1963).
54. Allen, N. S., ed., *Photopolymerization and Photoimaging Science and Technology*, Elsevier, New York, 1989.
55. Fouassier, J. P., *Makromol. Chem., Makromol. Symp.*, **18**, 157 (1988).
56. Decker, C., *Coating Technology*, **59**(751), 97 (1987).
57. Crivello, J. V. and J. W. H. Lam, *Macromolecules*, **10**, 1307 (1977).
58. Lougnot, D. J., in *Techniques d'utilisation des photons*, DOPEE ed., p. 245–334 Paris (1992).
59. Decker, C., in Fouassier, J. P. and J. F. Rabek (eds) *Radiation Curing in Polymer Science and Technology*, Elsevier Applied Science, 1973.
60. Hariharan, P., *Optical Holography: Principle, Technology and Applications*, Cambridge University Press, Cambridge, 1984, pp. 88–115.
61. Francon, M., *Holographie*, Masson Ed., Paris, 1969.
62. Kogelnik, H., *Bell. Syst. Tech. J.*, **48**, 2909 (1969).
63. Margerum, J. D., *Polymer Preprints for the 160th ACS Meeting*, 1970, p. 634.
64. Close, D. H., A. D. Jacobson, J. D. Margerum, R. G. Brault, and F. J. McClung, *Appl. Phys. Lett.*, **14**, 159 (1969).
65. Jenney, J. A., *J. Opt. Soc. Am.*, **60**, 1155 (1970).
66. Jenney, J. A., *Appl. Opt.*, **11**, 1371 (1972).
67. Jenney, J. A., *J. Opt. Soc. Am.*, **61**, 116 (1971).
68. Jenney, J. A., "Recent Developments in Photopolymer Holography," *Proc. SPIE*, **25**, 105 (1971).
69. Booth, B. L., *Appl. Opt.*, **11**, 2994 (1972).
70. Haugh, E. F., US Patent 3 658 526 (1972) assigned to E. I. Dupont de Nemours and Co.

71. Baum, M. D. and C. P. Henry, US Patent 3 652 275 (1972) assigned to E. I. Dupont de Nemours and Co.
72. Colburn, W. S. and K. A. Haines, *Appl. Opt.*, **10**, 1636 (1971).
73. Wopschall, R. H. and T. R. Pampalone, *Appl. Opt.*, **11**, 2096 (1972).
74. Van Renesse, R. L., *Opt. Laser Tech.*, **4**, 24 (1972).
75. Sukegawa, K., S. Sugawara, and K. Murase, *Electron. Commun. Jap.*, **58**, 132 (1975).
76. Sukegawa, K., S. Sugawara, and K. Murase, *Rev. Electr. Commn. Labs.*, **25**, 580 (1977).
77. Sukegawara, S. and K. Murase, *Appl. Opt.*, **14**, 378 (1975).
78. Sadlej, N. and B. Smolinksa, *Ont. Laser Techn.* **7**, 175 (1975)
79. Tomlinson, W. J., I. P. Kaminow, E. A. Chandross, R. L. Fork, and W. T. Silfvast, *Appl. Phys. Lett.*, **16**, 486 (1970).
80. Chandross, E. A., J. Tomlinson, and G. D. Aumiller, *Appl. Opt.*, **17**, 566 (1978).
81. Colburn, W. S., R. G. Zech, and L. M. Ralston, *Holographic Optical Elements*, Tech. Report AFAL, TYR-72-409 (1973).
82. Verber, C. M., R. E. Schwerzel, P. J. Perry, and R. A. Craig, *Holographic Recording Materials Development*, N.T.I.S. Rep. N76-23544 (1976).
83. Collier, R. J., C. B. Burckhardt, and L. H. Lin, *Optical Holography*, Academic Press, New York, 1971, pp. 265–336.
84. Peredereeva, S. I., V. M. Kozenkov, and P. P. Kisilitsa, *Photopolymers Holography*, Moscow, 1978, p. 51.
85. Smith, H. M., *Holographic Recording Materials*, Springer-Verlag, Berlin, 1977.
86. Gladden, W. and R. D. Leighty, "Recording Media," in *Handbook of Optical Holography*, H. J. Caulfield, ed., Academic Press, New York, 1979, pp. 277–298.
87. Solymar, L. and D. J. Cooke, *Volume Holography and Volume Gratings*, Academic Press, New York, 1981, pp. 254–304.
88. Delzenne, G. A., "Organic Photochemical Imaging Systems," in *Advances in Photochemistry*, Pitts, J. N., Jr., G. S. Hammond, and K Gollnick, eds, Wiley-Interscience, New York, 1980, Vol. 11, pp. 1–103.
89. Tomlinson, W. J. and E. A. Chandross, "Organic Photochemical Refractive-Index Image Recording Systems," in *Advances in Photochemistry*, Pitts, J. N., Jr., G. S. Hammond, and K. Gollnick, eds, Wiley-Interscience, New York, 1980, Vol. 12, pp. 201–281.
90. Monroe, B. M. and W. K. Smothers, "Photopolymers for Holography and Wave-Guide Applications," in *Polymers for Lightwave and Integrated Optics: Technology and Applications*, Hornak, L. A., ed., Marcel Dekker, New York, 1992, pp. 145–169.
91. Monroe, B. M., "Photopolymers, Radiation Curable Imaging Systems," in *Radiation Curing: Science and Technology*, Pappas, S. P., ed., Plenum Press, New York, 1992, pp. 399–434.
92. Lougnot, D. J., "Photopolymer and Holography," in *Radiation Curing Polymer Science and Technology, Vol. 3, Polymerization Mechanism*, Fousassier, J. P. and J. F. Rabek, eds, Chapman and Hall, Andover, 1993, pp. 65–100.
93. Monroe, B. M., W. K. Smothers, D. E. Keys, *et al.*, *J. Imag. Sci.*, **35**, 19 (1991).
94. Monroe, B. M., Eur. Patent 0 324 480 (1989) assigned to E. I. Dupont de Nemours and Co.
95. Keys, D. E., Eur. Patent 0 324 482 (1989) assigned to E. I. Dupont de Nemours and Co.
96. Monroe, B. M., *J. Imag. Sci.*, **35**, 25 (1991)
97. "Heat is on with New Dupont Photopolymers," *Holographic International*, Winter, 26 (1989).
98. Monroe, B. M., US Patent 4,917,977 (1990), assigned to E. I. Dupont de Nemours and Co.
99. Fielding, H. L. and R. T. Ingwall, US Patent 4,588,664 (1986), assigned to Polaroid Corp.

100. Ingwall, R. T. and M. Troll, *Opt. Eng.*, **28**, 86 (1989).
101. Ingwall, R. T. and M. Troll, "Holographic Optics: Design and Applications," *Proc. SPIE*, **883**, 94 (1998).
102. Ingwall, R. T., M. Troll, and W. T. Vetterling, "Practical Holography II," *Proc. SPIE*, **747**, 67 (1987).
103. Fielding, H. L. and R. T. Ingwall, US patent 4 588 664 (1986) assigned to Polaroid Corp.
104. Fielding, H. L. and R. T. Ingwall, US patent 4 535 041 (1985) assigned to Polaroid Corp.
105. Ingwall, R. T. and M. A. Troll, US patent 4,970,129 (1990), assigned to Polaroid Corp.
106. Withney, D. H. and R. T. Ingwall, in "Photopolymer Device Physics, Chemistry, and Applications," *Proc. SPIE*, **1213**(1), 8–26 (1990).
107. Ingwall, R. T. and H. L. Fielding, *Opt. Eng.*, **24**, 808 (1985).
108. Zhang, C., M. Yu, Y. Yang, and S. Feng, *J. Photopolymer Sci. Tech.*, **4**, 139 (1991).
109. Luckemeyer, T. and H. Franke, *Appl. Phys. Lett.*, **B46**, 181 (1988).
110. Mikaelian, A. L. and V. A. Barachevsky, "Photopolymer Device Physics, Chemistry and Applications 11," *Proc. SPIE*, **1559**, 246 (1991).
111. Boiko, Y. and E. A. Tikhonov, *Sov. J. Quantum Electron.*, **11**, 492 (1981).
112. Boiko, Y., V. M. Granchak, I. I. Dilung, V. S. Solojev, I. N. Sisakian, and V. A. Sojfer, *Proc. SPIE*, **1238**, 253 (1990).
113. Boiko, Y., V. M. Granchak, I. I. Dilung, and V. Y. Mironchenko, *Proc. SPIE*, **69**, 109 (1990).
114. Gyulnazarov, E. S., V. V. Obukhovskii, and T. N. Smirnova, *Opt. Spectrosc.*, **69**, 109 (1990).
115. Gyulnazarov, E. S., V. V. Obukhovskii, and T. N. Smirnova, *Opt. Spectrosc.*, **67**, 99 (1990).
116. Calixto, S., *Appl. Opt.*, **26**, 3904 (1987).
117. Boiko, Y., V. S. Solojev, S. Calixto, and D. J. Lougnot, *Appl. Opt.*, **33**, 797 (1994).
118. Lougnot, D. J. and C. Turck, *Pure Appl. Opt.*, **1**, 251 (1992).
119. Lougnot, D. J. and C. Turck, *Pure Appl. Opt.*, **1**, 269 (1992).
120. Carre, C. and D. J. Lougnot, *J. Optics*, **21**, 147 (1990).
121. Carre, C., D. J. Lougnot, Y. Renotte, P. Leclere, and Y. Lion, *J. Optics*, **23**, 73 (1992).
122. Lougnot, D. J., *Proc. OPTO'92*, 99 (1992).
123. Carre, C. and D. J. Lougnot, *Proc. OPTO'91*, 317 (1991).
124. Carre, C., C. Maissiat, and P. Ambs, *Proc. OPTO'92*, 165 (1992).
125. Carre, C. and D. J. Lougnot, *Proc. OPTO'90*, 541 (1990).
126. Noiret-Roumier, N., D. J. Lougnot, and I. Petitbon, *Proc. OPTO'92*, 104 (1992).
127. Baniasz, I., D. J. Loungot, and C. Turck, "Holographic Recording Material," *Proc. SPIE*, **2405** (1995).
128. Yamagishi, Y., T. Ishizuka, T. Yagashita, K. Ikegami, and H. Okuyama, "Progress in Holographic Applications," *Proc. SPIE*, **600**, 14 (1985).
129. Matsumoto, K., T. Kuwayama, M. Matsumoto, and N. Taniguchi, "Progress in Holographic Applications," *Proc. SPIE*, **600**, 9 (1985).
130. Shankoff, T. A., *Appl. Opt.*, **7**, 2101 (1968).
131. Chang, B. J. and C. D. Leonard, *Appl. Opt.*, **18**, 2407 (1979).
132. Meyerhofer, D., *RCA Review*, **33**, 270 (1976).
133. Chang, B. J., *Opt. Commun.*, **17**, 270 (1976).
134. Changkakoti, R. and S. V. Pappu, *Appl. Opt.*, **28**, 340 (1989).
135. Pappu, S. V. and R. Changkakoti, "Photopolymer Device Physics, Chemistry and Applications," *Proc. SPIE*, **1223**, 39 (1990).
136. Cappola, and N. and R. A. Lessard, "Microelectronic Interconnects and Packages: Optical and Electrical Technologies," *Proc. SPIE*, **1389**, 612 (1990).

137. Cappola, N., R. A. Lessard, C. Carre, and D. J. Lougnot, *Appl. Phys.*, **B52**, 326 (1991).
138. Calixto, S. and R. A. Lessrad, *Appl. Opt.*, **23**, 1989 (1984).
139. Horner, J. L. and J. E. Ludman, "Recent Advances in Holography," *Proc. SPIE*, **215**, 46 (1980).
140. Wang, M. R., G. J. Sonek, R. T. Chen, and T. Jannson, *Appl. Opt.*, **31**, 236 (1992).
141. Zipping, F., Z. Juging, and H. Dahsiung, *Optica Acta Sinica*, **4**, 1101 (1984).
142. Lelievre, S. and J. J. A. Couture, *Appl. Opt.*, **29**, 4384 (1990).
143. Lelievre, S., "Holographie de Polarisation au Moyen de Films de PVA Bichromate," MO dissertation, Université Laval, Québec, Canada, 1989.
144. Manivannan, G., R. Changkakoti, R. A. Lessard, G. Mailhot, and M. Bolte, "Nonconducting Photopolymers and Applications," *Proc. SPIE*, **1774**, 24 (1992).
145. Manivannan, G., R. Changkakoti, R. A. Lessard, G. Mailhot, and M. Bolte, *J. Phys. Chem.*, **71**, 97 (1993).
146. Trepanier, F., G. Manivannan, R. Changkakoti, and R. A. Lessard, *Can. J. Phys.*, **71**, 423 (1993).
147. Changkakoti, R., G. Manivannan, A. Singh, and R. A. Lessard, *Opt. Eng.*, **32**, 2240 (1993).
148. Couture, J. J. A. and R. A. Lessard, *Can. J. Phys.*, **64**, 553 (1986).
149. Solano, C., R. A. Lessard, and P. C. Roberge, *Appl. Opt.*, **24**, 1189 (1985).
150. Lessard, R. A., R. Changkakoti, and G. Manivannan, "Photopolymer Device Physics, Chemistry and Applications 11," *Proc. SPIE*, **1559**, 438 (1991).
151. Lessard, R. A., R. Changkakoti, and G. Manivannan, *Optical Memory and Neutral Networks*, **1**, 75 (1992).
152. Manivannan, G., R. Changkakoti, and R. A. Lessard, *Opt. Eng.*, **32**, 665 (1993).
153. Lessard, R. A., C. Malouin, R. Changkakoti, and G. Manivannan, *Opt. Eng.*, **32**, 665 (1993).
154. Manivannan, G., R. Changkakoti, and R. A. Lessard, *Polym. Adv. Technologies*, **4**, 569 (1993).
155. Kakichashvili, Sh. D., *Opt. Spektrosk.*, **33**, 324 (1972).
156. Todorov, T., N. Tomova, and L. Niklova, *Appl. Opt.*, **23**, 4309 (1984).
157. Todorov, N. Tomova, and L. Nikolova, *Appl. Opt.*, **23**, 4588 (1984).
158. Couture, J. J. A. and R. A. Lessard, *Appl. Opt.*, **27**, 3368 (1988).
159. Gross, R. B., K. Can Izgi, and R. R. Birge, "Image Storage and Retrieval Systems," *Proc. SPIE*, **1662**, 1 (1992).
160. Birge, R. R., C. F. Zhang, and A. L. Lawrence, *Proceedings of the Fine Particle Society*, Santa Clara, CA (1988).
161. Brauchle, and N. Hampp, *Makromol. Chem., Macromol. Symp.*, **50**, 97 (1991).
162. Hampp, N., in *Photochemistry and Polymeric Materials*, Kelly, J. J., C. B. McArdle, and M. J. de F. Maudner, *Roy. Soc. Chem.*, Cambridge, 1993.
163. Renn, A. and U. P. Wild, *Appl. Opt.*, **26**, 4040 (1987).
164. Wild, U. P., A. Renn, C. De Caro, and S. Bernet, *Appl. Opt.*, **29**, 4329 (1987).

26

Optical Fabrication

DAVID ANDERSON
Rayleigh Optical Corporation, Tucson Arizona

JIM BURGE
The University of Arizona, Tucson, Arizona

26.1 INTRODUCTION

The goal of most optical engineering is to develop hardware that uses optics, and invariably optical components such as lenses, prisms, mirrors, and windows. The purpose of this chapter is to summarize the principles and technologies used to manufacture these components to help the optical engineer understand the relationships between fabrication issues and specifications. For detailed information on how to actually make the optics, we provide references to other books and articles that give a more complete treatment.

26.1.1 Background

The field of optical fabrication covers the manufacture of optical elements, typically from glass, but also from other materials. Glass is used for nearly all optical elements because it is highly stable and transparent for light in the visible range of wavelengths. Glass optics can be economically manufactured to high quality in large quantities. Glass also can be processed to give a near-perfect surface, which transmits light with minimal wavefront degradation or scattering.

Additional materials besides glass are also frequently used for optics. Plastic optics have become increasingly common for small (< 25 mm) lenses and for irregularly shaped optics with reduced accuracy requirements. Metal mirrors are used for numerous applications, especially those with stringent dynamic requirements. Optics made from crystals are used for special purpose lenses as well as prisms.

The optical engineer who is specifying optics needs to understand how the size and quantity affect the manufacturing process and thus the final parts. Special tooling is required for large and difficult parts, which drives the cost up. However, special tooling can also lead to an efficient process, reducing the per-item cost for parts made in large quantities. Like any industrial process, optical fabrication has significant economy of scale, meaning that items can be mass-produced much more efficiently than they can be made one at a time. The key item here is the tradeoff between the improved efficiency and cost of the tooling. (We define tooling here to be special equipment used for manufacturing an item. Tooling is not used up in the process so it can be used repeatedly.) If only a few elements are needed, then it does not make sense to spend more on tooling than it would cost to make the few parts by a less efficient method.

The most difficult aspect for many optical components comes from the tight tolerances specified for optics. These tolerances need to be assigned by the system engineer to balance what is good enough for the design with what can be made in the shop. The tolerances relate to uncertainties in the final parts, which affect the system performance analysis, so the engineer wants to make sure the tolerances are tight enough but not unnecessarily so. For a particular project, the fabrication process is usually selected to achieve the specified tolerances and, nearly always, parts with tighter requirements are more expensive and take longer.

The ability to fabricate optics to extreme accuracy is usually limited by the ability to measure the part to sufficient accuracy. Also, much optical testing is done in the shop as part of the fabrication process, so the fields of optical fabrication and optical testing are coupled. For information an optical testing, even as it relates to optical fabrication, we refer the reader to the other chapters in this book, and to the comprehensive reference on this topic, *Optical Shop Testing*, edited by D. Malacara [32].

26.1.2 Overview of this Chapter

The field of optical fabrication is much too broad to be covered completely, even in summary, in this one chapter. Instead, our goal is to assist the optical engineer in understanding fabrication issues by providing:

1. a description of the common procedures that are used for making optics
2. a list of references for more detailed study
3. insight into the relationships between quantity, quality, tolerances, material properties, and cost for fabrication of optical elements.

It is a general goal of this chapter to help the optical design engineer understand enough of the fabrication issues to make good design decisions. More important, we hope to educate the optical system engineer about the general manufacturing issues, so he knows to discuss particular issues of the project with the optician. It is only through the communication between the designer and the fabricator that optimal specifications can be developed and implemented.

This chapter is divided into four sections, including this introduction, giving the basics of traditional fabrication, followed by special, and more modern methods, and concluding with a discussion on important fabrication issues that should be considered by the system designer. The purpose of this chapter is *not* to enable the engineer to make good designs that he can throw over a wall to the fabricator, but to provide information to help the engineer initiate a dialog with the fabricator so they can jointly work out an optimal set of specifications.

In Section 26.2, we give an overview of the traditional methods of optical fabrication. The steps for manufacturing common optical elements are outlined and some of the key issues are described. Modern shop practices build on the rich heritage for fabricating optical elements from glass. Modern shops use numerous process improvements that take advantage of new machines and materials to give better performance and lower cost, but the same basic principles are used that have been developed over generations.

Section 26.3 discusses fabrication of aspheric surfaces and introduces some advanced fabrication methods that have been developed in the past few decades, which are now readily available for production parts. Here we describe common methods for making optical components that do not follow the more traditional approach given in the Section 26.2. This section on advanced fabrication methods discusses molded optics of glass and plastic, single-point diamond turning, computer-controlled surfacing, and replication.

Section 26.4 summarizes some of the relationships between fabrication methods and cost. It is impossible to establish any absolute quantitative rules about how specifications, tolerances, materials, size, and quantity affect cost. We do offer some rules of thumb, which serve as helpful starting points for getting at the real relationships. Here again, this serves as a starting point for discussions between the fabricator and the designer.

26.1.3 General References

Despite the variety and the economic importance of optical manufacturing, there is very little published about this field. Most of the workers in an optics shop were trained on the job as an apprentice under a more skilled master optician. The basic operations required for making most optics have changed little in the past 100 years. However, improved materials and machine tools have allowed these steps to be performed more economically, relying less on the optician's craft.

There are a few excellent references available on the topic of optical fabrication. Hank Karow's book *Fabrication Methods for Precision Optics* [27] is the most modern and complete, and provides an excellent reference for most common techniques and equipment. Outstanding review articles are given as chapters in *Applied Optics and Optical Engineering* by Parks [40] and Scott [50].

Some classic books in this field that have good descriptions of the basics for hands-on-work are *Amateur Telescope Making, Volumes 1–3*, by Ingalls [24], *Prism and Lens Making* by Twyman [56], and *How To Make A Telescope* by Texereau [55]. A large number of interesting solutions to tough fabrication problems are given in *The Optics Cooke Book*, edited by S. Fontane [13].

Also, there are some other excellent references, now out of print, which you may find in the library. A classic German reference by Zschommler *Precision Optical Glassworking* has been translated to English [62]. This book gives complete step-by-step instructions for manufacturing some common optics. *Optical Production Technology* by Horne [22] includes aspects of setting up a production shop with a good overview of optical manufacturing technologies in the production shop. *Generation of Optical Surfaces* by Kumanin [30] gives an excellent reference on the machining and grinding of glass.

26.2 TRADITIONAL METHODS OF OPTICAL FABRICATION

26.2.1 Introduction

Current optical fabrication methods are a curious blend of the very old, dating back several centuries when the use of pitch for polishing was introduced, and the modern, particularly through the application of computer control, laser based interferometry, and advanced materials. There is still a considerable "art" component to these methods in most optical shops, especially in the custom optics area requiring a high level of expertise that takes years to develop. However, with the recent wide application of computer controls to fabrication machines and the development of more deterministic shaping methods and processes, a revolution is well underway in their application to both custom optics and production optics in terms of efficiency and value.

The optician's expertise must now include computer literacy, a requirement now shared by many industries. Research has led to a greater understanding of the ground and polished surface and in the ways of producing them. Diamond-turning and grinding technology in combination with computer-controlled machines has had a large impact in both glass and metal fabrication methods. Pitch polishing is no longer tha only way to finish a high-quality optical surface. A great deal of work and progress continues to be made in the production of aspheric optics, utilizing advances in all areas of fabrication. The direct milling of glass and other brittle materials is now accomplished not only with diamonds but also with streams of ions. We must note that much of the progress has resulted from various new or improved testing methods, particularly computer-controlled interferometry and pro-filometry. The advanced measurement techniques, along with developments in fabrication described here, have made the production of optical surfaces possible for optics that cover virtually the entire electromagnetic spectrum.

The explosion of new materials and processes available to the engineer and the fabricator has fragmented the industry to a large extent. Expertise is no longer to be found in a single "optics house" for all optics needs. Neither can a single chapter begin to review all the existing methods and materials. The scope of this section will, therefore be limited to a discussion of some of the more common methods used in the fabrication of glass optics. In the next section, we summarize some manufacturing methods for other materials. We present here only a summary of optical fabrication methods and issues, and provide references for more detail.

Generally, there are a few steps common for making most optical elements, although each step will be done differently depending on the optic and the quantity:

- *Rough shaping*: The initial blank is manufactured, typically to within a few millimeters of final dimensions.
- *Support*: The optic must be held for the subsequent operations. Much of the difficulty in fabrication comes from the requirements of the support.
- *Generating*: The blank is machined, typically with diamond tools, to within 0.1–1.0 mm of finished dimensions.
- *Fining*: The optical surfaces are ground to eliminate the layer of damaged glass from generating and to bring the surface within 1–5 μm from the finished shape.

- *Polishing*: The optical surfaces are polished, providing a specular surface, accurate to within 0.1 μm. Through repeated cycles of polishing, guided by accurate measurements, surfaces can be attained with 0.005 μm accuracy.
- *Centering and Edging*: The optic is aligned on a spindle and the outer edge is cut.
- *Cleaning*: The finished elements are cleaned and prepared for coating.
- *Bonding*: Frequently, lenses and prisms are cemented to form doublets (two lenses) or triplets (three lenses).

Subsequent coating and mounting are usually handled by a different group of people and are not generally considered part of optical fabrication.

Surface size, quantity, form, and finish are probably the four most important factors in determining the methods to be used in the fabrication of common optics. In this section we summarize some of the methods used in fabricating optics in three basic size ranges: from 5 to 50 mm in diameter; from 50 to 500 mm; and > 500 mm. These choices are somewhat arbitrary, but methods and available machinery favor these basic sizes. We will also look at how the vast variety of "off-the-shelf" optics are produced economically as well as the very great influence quality has on both the choice of methods and the cost of fabrication.

Most fabrication methods deal with the production of spherical surfaces. Since a sphere has no optical axis but only a radius (or, equivalently, a center of curvature) that defines its shape, any section of that sphere looks like any other section (of the same size), as shown in Fig. 26.1. This fact has important consequences in how these surfaces are produced. The basic idea is that when two surfaces of nearly equal size are rubbed together two spherical surfaces result, having opposite curvatures. Note that a flat surface is simply a surface having infinite radius. Aspheric surfaces, on the other hand, lack this symmetry, as described in the next section, making these surfaces much more difficult to fabricate. Off-axis aspherics have no rotational

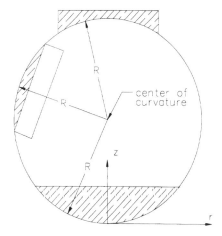

Figure 26.1 A spherical surface is defined only by its radius of curvature R and its size. The surface profile $z(r)$ is defined as $z(r) = R^2 - \sqrt{R^2 - r^2}$. These surfaces do not have a unique optical axis.

symmetry and are perhaps the most difficult to fabricate. Note also that there is no such thing as an "off-axis sphere," just one that is tilted.

26.2.2   Initial Fabrication of the Blank

Optical materials can be procured in many forms. The initial piece of material that has roughly the correct shape is called the blank. In subsequent processing steps, material is removed from the blank to yield the finished optic. The choice of material is obviously dictated by the final application, but the initial form of the blank depends on the fabrication method.

Glass is purchased in several forms – rolled plate, blocks, strips, pressings, gobs, slabs, and rods. The choice of the bulk glass is made according to the fabrication plan and the material specifications. In general, glass for mass-produced optics is supplied in the near net shape to minimize the cost of additional processing. Glass blanks for production lenses and prisms are produced in large quantities as pressings oversized and irregular by about 1 mm. Precision pressings are available at higher cost, requiring as little as 0.1 mm glass to be removed to shape the part. These are shown in Fig. 26(a).

Glass for very high-performance systems must be carefully selected to get the highest quality. Glass with tight requirements on internal quality is provided in blocks, shown in Fig. 26.2(b). These blocks are then polished on two or more surfaces and are inspected and graded for inclusions, striae, birefringence, and refractive index variation. The blanks for the optics are then shaped from the glass blocks by a combination of sawing, cutting, and generating.

26.2.3   Support Methods for Fabrication

Most optical fabrication processes begin with the extremely important consideration of the method of holding on to the part during subsequent fabrication steps. The support method must be chosen with numerous factors to consider, including part size, thickness, shape, expansion coefficient, direction, and magnitude of applied forces. The general notion is to hold on to the part with the least amount of deformation to the part so that when the part is finished and unmounted (or "deblocked") a minimum amount of "springing" or change in shape results. However, the part must be held with enough rigidity to resist the forces of the various surfacing methods. Many times the support will be changed as the part progresses to reflect changing requirements on the precision of the surface. Here, we discuss several of the more common methods of blocking both small and large parts. More detailed analysis and discussion of these and other methods can be found in the references provided.

Most modern fabrication begins with fixed diamond abrasive on high-speed spindles as discussed in Karow [27] and Piscotty *et al.* [42]. The lateral forces can be large, so the part must be held quite firmly to a rigid plate or fixture. This plate, usually called the blocking body, or simple "block," can be made of various materials, but is usually aluminum, steel, cast iron, or glass, with rigidity being the most important factor. The two principal methods for holding the part to the block are to use adhesives or mechanical attachments at the edge.

The ideal adhesive for blocking glass to the block would provide a rigid bond with little stress, and should allow the part to be easily removed. Most adhesives

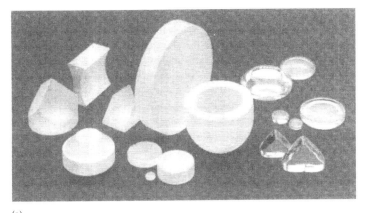

(a)

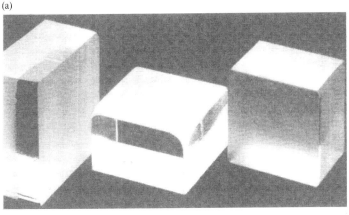

(b)

Figure 26.2   Optical glass is commonly procured in (a) pressings or (b) blocks. (a) Pressings, hot molded an annealed: may have rough or fire-polished surfaces. (b) Block glass, with two opposite faces polished for test purposes. Other common forms are slabs (six worked surfaces), rods, strips and rolled sheet (unworked surfaces, cut to length), and gobs (roughly cylindrical). (*Source*: Schott Glass Technologies, Inc.)

cannot achieve all three requirements equally well, so choices must be made by the optician, depending on which is most critical for any particular process. For the generation processes, using high-speed diamond tools, rigidity and ease of removal are usually the dominant criteria, with higher stress being allowed. The effects of this stress are then removed in the subsequent processes of grinding and polishing, where a less stressful blocking method is employed.

   Blocking of plano and spherical parts up to around 100 mm in diameter is commonly done with a variety of waxes, both natural and synthetic. These are heated to a liquid before applying to the block, or applied, then heated by the block itself. The glass parts are then warmed and placed on the waxed block. For heat-sensitive materials the wax can be dissolved in solvent before applying to the block. The great advantage of waxes is that they hold the glass quite firmly and are also easily removable by dissolving them in common liquid or vapor solvents. Most waxes, however, due to their shrinkage, impart large stresses, requiring parts to be deblocked after generating and subsequently reblocked with a less stressful substance for grinding and polishing.

Pitch remains the blocking material of choice when the parts cannot be highly stressed. Pitch is an outstanding material, and is used in the optics shop both for blocking, and for facing polishing tools. Brown gives an excellent reference on properties of pitch [8]. Pitch is a viscoelastic material that will flow when stress is applied, even at room temperatures, so parts blocked with pitch will become stress-relieved if left long enough for the pitch to flow.

Cements such as epoxies and RTVs bond very well, but are extremely difficult to deblock and remove. There also some UV-curable cements that can provide low-stress blocking and can be removed with hot water. More information about the cements is available from the manufacturers.

An old and interesting blocking method that is still used when the surface to be blocked needs to be held in close reference to the block to maintain a particular wedge or angle as in the production or assembly of windows or prisms is the optical contact method. The block is usually made from the same material as the part, and the mating surfaces must be polished and cleaned. When the two surfaces that have are brought close together (with a little finger pressure to force out the air) they will pull each other together in a very close bond due to molecular forces. This blocking method can be used with very thin or thick parts but is quite difficult to apply, particularly to large surfaces dues to the required cleanliness.

In production optics, where many parts with the same radius of curvature need to be produced, a number of the parts are blocked together, as shown in Fig. 26.3(a). Many times the block is very carefully machined so that each part can be loaded into it, sometimes automatically under computer control, to a very precise position relative to the block's center. This type of block is called a spot block and has gained widespread use in production shops. These spots can be machined directly into the

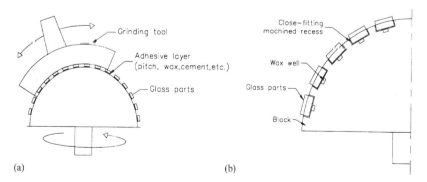

(a)   (b)

Figure 26.3 Multiple parts may be made on the same block (a) or a spot black (b) by adhering them to a common spherical block using some type of adhesive. A more accurate and repeatable method is to use a spot block where premachined holes are provided for the lens blanks. The usual method for working the block is to have the block rotating while a matching spherical tool is stroked across it. This can also be inverted.

block, as shown in Fig. 26.3(b), or separate lens seats can be machined that are screwed onto the block. The spot blocks are costly to make but they can be used efficiently for making numerous runs of the same lens.

Block size limitations are based both on machine size limitations and on the radius of curvature. Most generators and grinding/polishing machines cannot easily handle anything beyond a hemisphere, which limits the number of parts to a block. Plano parts have no theoretical limitation based on their radius but only on the capacity of the machines in the shop. Literally hundreds of small plano parts can be fabricated on a single block.

Aspherics cannot be fabricated on blocks because, unlike spherical surfaces, the aspheric surface has a particular optical axis. This implies that only a part that is centered to the machine spindle can be turned into an asphere. This is one reason aspherics are more expensive than spherical surfaces. Note, however, that off-axis aspherics can be made as a block! This is, in fact, how most off-axis aspheres are made: by making a block, called the parent, that is large enough to encompass the required off-axis section pieces. The parent is then aspherized in a symmetric way (as discussed below), after which the completed parts are removed from the correct position on the block. Usually the parent is manufactured into a single piece of glass and the off-axis sections are cut from the parent after aspherizing, but if many are required a spot-type blocking methods can be utilized.

The blocking techniques mentioned are used for production optics or where a relatively large number of parts are to be made. Even if only one part is required it is usually wise to block many together so that spares are available. It generally doesn't pay to make just one spherical part of it is small, say less than 100 mm. Indeed, the designers should always try to use off-the-shelf optics for optics in this size range.

Most optics larger than this are generally supported mechanically without the use of adhesives of any sort. Mechanical supports for larger optics have the same requirements as their adhesive counterparts in that they must hold the part firmly while introducing little stress. As with smaller optics, large optics can be supported differently for different fabrication processes where the conflicting requirements of high rigidity and low stress are considered relatively more or less important.

Mechanical supports during diamond generating must be quite rigid, as the forces placed on the part by the high-speed diamond tools are large. The generating support can, however, allow larger distortions, which will be corrected later in grinding and polishing. Most generating machines have turntables that incorporate either magnetic or vacuum systems to hold moderately sized (up to about 500 mm) parts firmly on the table or support ring. A magnetic system, commonly found on Blanchard-type machines, uses steel plates that are placed around the periphery of the part. The electromagnetic turntable is switched on, firmly holding the steel plates and thus the part in position. In vacuum systems, the part is held on a shallow cup having an "O" ring seal. A vacuum is pulled on the cup and the part is held in place by friction against the turntable.

For larger optics the part may rest on a three or more multipoint support system that is adjustable in tilt, and held laterally by three adjustable points at the edge of the part. These support systems can introduce rather large figure errors that need to be eliminated in subsequent grinding and polishing. Some machine turntables are turned to be extremely flat, even diamond turned in some cases, to reduce the amount of induced deformation.

During grinding and polishing, large parts are again commonly supported axially using pitch or other viscoelastic materials (such as or Silly Putty) depending on the stiffness of the part. This type of support can flow to eliminate any induced stresses in the part. There are also several methods of achieving a well-defined set of axial forces where the part is supported at a number of discrete points. Hindle type (or "whiffle-tree") supports or hydrostatic supports use mechanics or hydraulics to provide a unique, well-defined, set of support forces. [61]. The number and arrangement of the support points required can be accurately predicted using finite-element analysis so that the deformation between support points can be reduced to an acceptable amount. To resist the lateral forces of grinding and polishing the part is generally held either with tape for smaller parts or with three mechanical blocks or tangent rods for larger.

26.2.4 Diamond Machining and Generating

The initial fabrication step following the blocking is generally machining the part using a variety of diamond-impregnated tools to rapidly bring the part to its near-final shape, thickness, and curvature with the optical surface at least smooth enough for fine grinding or even direct polishing. These tools have exposed diamond particles that basically chip away at the glass on scales of tens of microns (micrometers, μm). Additional information on specific aspects of generating is provided in the references [42, 36, 53, 22.]

Most tools have a steel body, onto which is bonded a layer of material impregnated with diamond particles of a particular size distribution. The size is usually specified as a mesh number, which is approximately equal to 12 mm divided by the average diamond size (Fig. 26.4). A 600 mesh wheel has 20 μm diamonds. The specifications for the absolute sizes of the diamonds and their distribution are not standard and should be obtained from the vendor.

There are two basic configurations for diamond tooling, as shown in Fig. 26.5: a peripheral tool, where the diamond is bonded to the outer circumference of the tool; and a cup tool, where the diamond is bonded to the bottom of the tool in a ring. Peripheral wheels are generally used for shaping operations on the edge of the part such as edging, sawing, and beveling, while cup wheels are generally used for working on the surface of the part, such as cutting holes and for generating curvature.

A small optic (< 20 cm) or block of lenses is usually generated using a cup wheel wheel where the axis of rotation of the wheel is tilted with respect to the part, so that it passes through the desired center of curvature (Fig. 26.6). If the axis of rotation does pass through the center of curvature of the part it will cut a perfectly spherical shape into the part or block. Since all the parts on a block of lenses share the same center of curvature, they will all be cut to the same radius and thickness if properly blocked. This fact is key to the production of large quantities of smaller optics blocked together.

Plano optics are generated using a Blanchard-type geometry. Here a cup wheel is used with the axis of rotation aligned to be perpendicular to the linear axis of a tool bed. The parts are translated under the spinning diamond wheel and are ground flat to high precision. Multiple operations of this type must be performed for the different faces of prisms, and the relative orientation of the different cuts determines

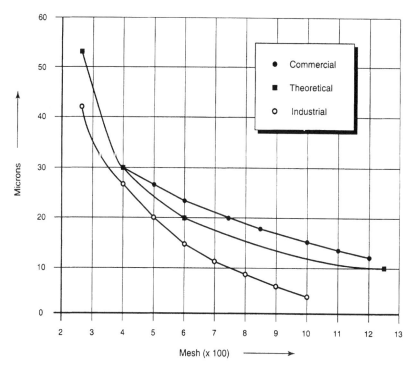

Figure 26.4 Correlation between mesh sizes and micron sizes. (From Ref. 27.)

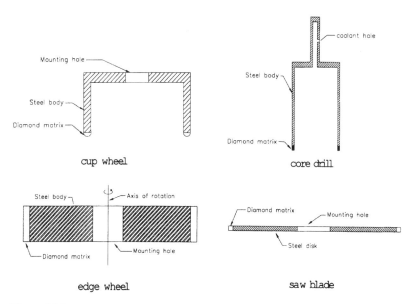

Figure 26.5 Diamond generating wheels. There are two basic types of diamond tooling used for cutting and generating depending on where the diamond is placed: on the face or on the edge. The cup wheel and core drill are the most common face wheels used in cutting radii and drilling holes. The peripheral wheels having diamond on their edges are used to edge and to saw cut.

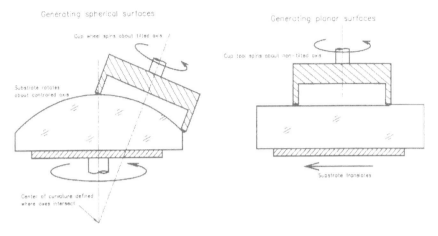

Figure 26.6 Generating with cups tools. Spherical surfaces are cut by tilting the axis of the cutting wheel so it intersects the axis of rotation of the part. This will cut a spherical surface with center of curvature at this intersection. Plane parts are milled by translating the optic in a direction perpendicular to the cup wheel.

the accuracy of the prism (Fig. 26.7). Precision blocks having accurately machined surfaces can be used for this purpose.

Diamond tools are also quite versatile and are used for many different operations. For example, a simple perpheral wheel can be used to cut the curvature into a part. The tool can be moved slowly across a spinning part. The shape of the cut is determined by the tool motion, which can be run on a numerically controlled (NC) machine, or it can be driven to follow a template attached to the machine.

A large emphasis has been placed on progress in both diamond tooling and in the machines that use them because accurately cut fine surfaces reduces the time

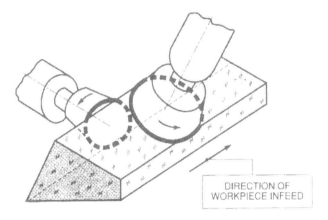

Figure 26.7 Generating a prism profile with two mill heads. (Courtesy Karow [27], Fig. 5.38.)

spent in the traditional loose abrasive grinding step that follows. Machining of the part on high-speed machines is very rapid, with removal rates up to several cubic inches per minute, two orders of magnitude faster than even very coarse loose abrasive grinding.

Unfortunately, generating tends to create significant damage to the glass under the surface, which must be removed in subsequent grinding and polishing operations. One current area of interest is investigating how to perform diamond generating to produce finer surfaces and more accurate shapes. Any type of abrasive action on glass occurs due to small fractures that form when an abrasive particle is pushed against it with enough force. When enough fractures intersect, small pieces of glass pop out, leaving small pits. Underneath the pits are larger fractures that continue some distance depending on the materials. This is called subsurface damage and it must be removed in subsequent processing steps. The structure of glass as it is typically abraded is shown in Fig. 26.8. Generally, the smaller the diamond and the softer the matrix material the less damage will be imparted to the surface. However, the finer the diamond the lower the removal rate and a softer matrix allows greater tool wear, which can reduce accuracy.

Most diamond surfacing methods use at least two if not three different diamond wheels to rapidly produce a fine surface. Using a computer-controlled machine, flat, spherical, and even aspherical parts can be rapidly surfaced. If the find diamond tool and machine have sufficient accuracy the tool can be brought to bear on the surface with a low enough force that the glass does not fracture but material is removed by plastic flow with no subsurface damage introduced and the surfaces produced are specular. This process is called micromachining or ductile regime machining. [6,16–19] Currently, surfaces capable of being polished directly can be routinely produced on production machines bypassing all loose-abrasive grinding. Most of these machines produce optics less than 100 mm in diameter, however. With larger parts, fine diamond machining is only performed on a few specialized machines limited primarily by both the stiffness in the machine and of the mount. Microgrinding is just beginning to gain more widespread use in industry and the combination of numerically controlled machines and diamond tooling will undoubtedly have a large impact on fabrication methods of the future.

Cross section of ground surface

Figure 26.8 An abraded optical surface consists of two components: the surface damage layer and a subsurface damaged (cracked) layer. For loose-abrasive ground surfaces the surface damage layer is about the size of the grit size and the maximum subsurface damage is about twice that. Diamond-generated surfaces typically have less surface damage but a large subsurface component.

Another application of fixed diamond abrasive is in pel grinding of generated surfaces. [52]. This operation generates the surface using a large tool with bound diamond in cylindrical pellet form. The tool is made to match the shape to be generated, and is then driven at high speeds. The fine diamonds generate the surface directly, leaving little subsurface damage, allowing polishing without any subsequent fining operations.

The pellets are bonded to a curved tool having the proper radius of curvature inverse to that of the part, as shown in Fig. 26.9. The tool is rapidly rotated while the part is stroked over the tool in the same fashion as loose-abrasive grinding only with higher speed and with higher pressure, which very quickly grinds the surface smooth enough for polishing. This method is very efficient for the high-speed production of thousands of the same part but costly to set up, as a new tool is required for each radius. Hence, more traditional loose abrasive grinding, described below, is employed for low-volume production to bring a generated surface to the polishing stage.

26.2.5   Fining

As described above, the typical diamond machined surface has a rather smooth surface, below which is a layer of material riddled with fractures. These fractures, if left in the final polished part, are quite visible under bright illumination and cause the surface to scatter light. Since these fractures can extend quite deeply, up to 100 microns or so depending on the process used to generate, sufficient material must be removed to reduce them to only a few microns, which can be taken out in polishing [46, 47]. As we have seen, the use of fine diamonds on high-quality machines can

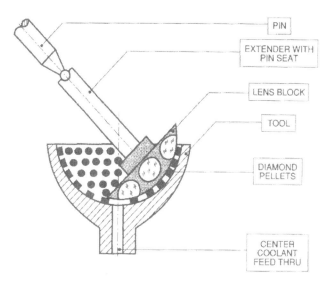

Figure 26.9 Typical configuration for using a pel grinder for working convex surfaces. (Courtesy Karow [27], Fig. 5.62).

accomplish this quickly. However, the high cost of these machines and tools limits their use.

The more traditional and less costly method of bringing a surface to the polishing stage is through loose abrasive grinding. In this method, the part and the tool are rubbed together while an abrasive powder, usually in an aqueous solution, is maintained between them. The abrasive particles cause tiny fractures in the glass, which results in material removal as the fractures intersect. This abrasive action itself causes subsurface damage but the sizes of the particles are chosen to reduce the amount of damage in a series of steps, generally reducing the damage by a factor of 2 with each grade. A rule of thumb for loose abrasive grinding would be that the maximum surface pits left after grinding will be on the order of the size of the grain, while the maximum subsurface fractures extend to about twice that. A typical sequence might be to diamond generate, then remove 100 microns with a 40 micron abrasive, then remove 50 microns with a 25 micron abrasive, followed with the removal of 25 microns with a 9 micron abrasive. This surface can be polished, but up 20 microns of materials needs to be polished off to eliminate the remaining damage.

It would be possible to remove all of the generating subsurface damage by polishing directly, skipping the step of loose abrasive grinding. It would, however, take an unacceptably long time to remove 100 microns of material by polishing. It only makes sense to polish a surface with a few microns of surface damage remaining.

Other factors that contribute to the amount of damage produced include the hardness of the abrasive, the material being ground, and the tool, and to some extent the shape of the abrasive grains. Harder abrasive grains or tools will remove material more rapidly material at the price of imparting more damage. The more plate-like grains found in modern aluminum oxides appear to produce less damage than their blockier counterparts like garnet.

Tool materials for loose abrasive grinding range from cast iron, which is quite hard, to brass, glass, and aluminum on the softer side. The harder tools will grind faster and retain their form somewhat longer than the softer tools but at the cost of more subsurface damage. Most production tools are made from cast iron because they keep their shape so well.

Loose abrasive grinding is almost always used for fining large optical surfaces, and build-up or layered tools are the rule. Tools for larger optics are usually made from some soft, workable material such as aluminum, wood, or plaster. The curvature of the tool is either machined or cast into the tool that is then faced with ceramic or glass tiles that are bonded to the tool's surface. When the ceramic layer grinds down it is simply replaced with another layer or a fresh layer is bonded to the first.

The point at which grinding stops and polishing begins depends on a number of factors. While glass surfaces can be ground to a very fine finish, to perhaps 1 micron grit size, which minimizes subsurface damage and speeds polishing, other factors generally limit the final grit size to around 5 microns. For very small particle size, the intimacy between the part surface and the tool can cause the tool to seize on the part, which makes the two virtually inseparable without major forces being applied that will end up damaging the part. The larger or more costly the part, the more this risk becomes unacceptable. Also, the risk of scratching the surface increases with small grit, especially when using very hard tools such as cast iron or steel. On many surfaces it is a good compromise to perform the final grinding with a softer tool material such as brass, aluminum, or glass, and use a slightly larger grit size. The softer tool material will result in less damage and reduce the risks.

Compared with generating with bound abrasive wheels, loose abrasive grinding is performed at much lower speeds. Very small parts are ground at a few hundred rpm whereas large parts are ground at only a few rpm. This lower speed largely accounts for the tremendous removal rate difference between the two methods. At much higher speeds, the loose abrasive slurry mixture would be simply flung off the tool or part.

26.2.6 Polishing

The aim of polishing an optical surface is to bring its surface figure or form into compliance with a specification depending on its intended use, at the same time as its finish or microroughness is also reduced to an acceptable value. Polishing is a seemingly magical process, which uses a combination of mechanical motion and chemistry to produce surfaces smooth to molecular levels [21].

Most high-quality optical surfaces are polished with a tool, or lap, similar to the grinding tool except that it is faced with viscoelastic pitch or polyurethane. This tool is stroked over the surface with an aqueous slurry of polishing compound. In a region a few microns thick that includes the glass surface, the lap surface, and the slurry in-between, a complicated interaction occurs involving chemical and mechanical effects that produces a polished surface. Polishing is partially a chemical process, so different substances must be used to polish different materials: no one substance is ideal for all materials. Some polishing compounds for common optical materials are cerium oxide, zirconium oxide, alumina, and colloidal silica. These are available in proprietary mixtures from several suppliers.

Pitch laps are frequently used for high-quality surfaces. Pitch is a generic term describing a group of substances made from the distillation of tar derived from wood or petroleum. It is very soft compared with glass so it will not scratch, and has a low melting temperature of about $50-100^\circ$C. Its viscosity, usually in the range of 10^8-10^{11} poise, allows the pitch to slowly flow near room temperature so that it takes the shape of the part being polished and remains in very close contact.

Pitch laps usually give the best performance, but they require considerable maintenance. Production parts are usually polished using laps faced with polishing pads made of polyurethane. These synthetic pads works well with particular polishing compounds that have been optimized for use with the pad. Laps faced with these polishing pads are extremely stable and they can be run at higher pressures than pitch laps, so they polish more quickly. However, unlike pitch, these laps do not naturally flow to conform to the shape of the optic, so the pads must be applied to a precision machined surface. Again, this special tooling is highly efficient, but will only work for a particular radius of curvature.

Metals, plastics, and crystals can be polished the same was as glass, but using different polishing pads and compounds. Metals are polished best with cloth polishers and polishing compound with very fine chrome oxide or diamond [10]. The quality of the finish depends on the hardness, porosity, and inclusions of the metal substrate. Plastic optics such as acrylics are polished with aluminum or tin oxide with soft synthetic polishing pads. Most crystals are polished using synthetic pads with a compound of colloidal silica, find diamond, or alumina (aluminum oxide).

The macrotopography or surface figure, and microtopography or surface finish, are generally the two most difficult specifications to meet in the fabrication process and are typically the biggest cost drivers. The figure of the surface is commonly specified as an average (rms) or absolute value (peak–valley) height difference between the actual surface height and the ideal theoretical surface. This difference is usually specified in units of waves, or fraction of a wave, at the wavelength at which it is used or tested. Typical figure tolerances are in the 0.05–0.2 waves rms at the widely used measurement wavelength of the He–Ne laser at 632.8 nm.

Control of the figure comes from the geometry of the polisher – how the lap is stroked and the table is rotated. High-quality parts are time consuming to make because they require the optician to measure the part, usually with interferometry, then to adjust the fabrication process to correct the errors in the surface. The cost of the optic will depend on the efficiency of the optician to converge on the final specification, and the ultimate accuracy will depend on both the residual errors that the optician measures and any errors in the optical test.

Surface figure specifications are usually made as peak-to-valley (P–V) or rms departure of the surface from ideal. Peak-to-valley specifications are becoming less popular (particularly to the fabricators) since they relate only to very local regions of a surface. This specification makes sense for optical surfaces measured by inspection with a test plate. In this case, the optician evaluates the large-scale distortions of the interferogram and gives a limit to the irregularity that he sees, defining this as the P–V distortion in the surface.

For the case of computerized phase-measuring methods, however, the P–V error is strictly the difference between the maximum point and the minimum point in the data. The high resolution of these instruments will provide surface maps with 30,000 points, so any two points are not statistically significant. In fact, the minimum and maximum of the data will usually be driven only by measurement noise. It is not uncommon to relate a P–V specifications to an equivalent rms value by applying a simple rule of thumb – the allowable rms figure can be estimated by dividing the P–V specification by a factor of 5. Nonetheless, P–V specifications in the $\frac{1}{2}$ to 1/20 wave are still common.

Clearly, higher-quality surfaces require more time to make and are more expensive to produce. Flats and spheres can be produced by conventional methods down to 0.01 waves rms or better, perhaps to 0.002 waves rms using special methods and depending on the size and surface shape. Aspheres are considerably more difficult to figure and will be discussed in the next section.

The surface finish is the local roughness of a surface compared with a perfectly smooth surface and is usually specified as an average (rms) surface height irregularity over spatial scales of a few tens of microns. Unlike the figure, the surface finish comes from the process itself – the type of lap, polishing compound, pressures, and speeds. These processes are derived before starting the production parts, so the optician does not typically adjust the polishing based on measured results, as he does for the figure.

Interferometric techniques are now generally used to measure both the figure and finish of optical surfaces to very high precision. Using computer-controlled phase-measuring interferometric methods, surface figures can be measured to a few nanometers and surface finish to a few tenths of a nanometer. The ability to precisely measure these quantities has resulted in improved polishing processes with correspondingly better surfaces. However, the understanding of the polishing process, particularly of glass, remains mired in its great complexity.

High-quality optical surfaces will generally be finished less than 20 angstroms (2 nm) rms, down to a few angstroms rms. For most optics, the standard pitch polish, giving about 10 angstrom rms roughness is more than adequate. Special applications require superpolished surfaces, with roughness below 2 angstroms rms. Special effort is required to produce such surfaces, and very few fabricators have developed this capability.

Producing a high-quality optical figure is perhaps the most cost-sensitive aspect of fabrication. Various methods have been developed to create high-quality surfaces on different types of surface shapes. Here, we describe some of the methods used to produce flat surfaces, spherical surfaces, and aspheric surfaces, looking at both tried and true methods still widely used, as well as some more modern methods being developed.

Polishing of Spherical Surfaces

Spherical surfaces are probably the simplest of all to fabricate because of their symmetry. The grinding process tends to produce spherical surfaces. The fact that the tool and the workpiece are not full spheres, but are segments, allows variations in wear across the surfaces that can be used to change their radius of curvature. This basic method has been used for literally thousands of years to produce spherical surfaces, and spheres are by far the most commonly used optical surface shape. Most optical system designs utilize optics with all spherical surfaces due to the relative ease of manufacture over aspherical ones, although from a design perspective an aspherical surface would simplify the optical design.

Conventional methods for grinding and polishing spheres use one surface, usually the optic, to rotate on a turntable and the other usually the lap, to move over it, always remaining in close contact. Overarm machines, shown in Fig. 26.10, stroke the tool over the part using an arm that attaches to the tool through a ball joint. Also, the roles can be reversed and the block with the optics can be attached to the arm and driven over a rotating tool. By adjusting the length of the stroke and the relative speeds of the rotation vs. the stroke as well as the length of the stroke, the radius of the two surfaces can be made to move longer or shorter in radius or stay the same. With operator skill, the figure can be brought spherical to very high accuracy with very low surface roughness.

As mentioned, various blocking methods can be utilized to increase production volume, such as the use of spot blocks or other multiple element blocking, as shown in Fig. 26.11. Simply running the machine faster using highly controlled slurry feeds dramatically increases production rates. Finally, using high pressure between the tool and block in conjunction with materials that work well at high pressures such as diamond pellet tools for grinding and polyurethane pads for polishing, very high production volume can be achieved. Fabrication of spherical surfaces where the process parameters have been fine tuned to produce predictable results have resulted in very high yield rates and efficient production. Most catalog items fall into this category of production optics and make them the designer's first choice when designing an optical system.

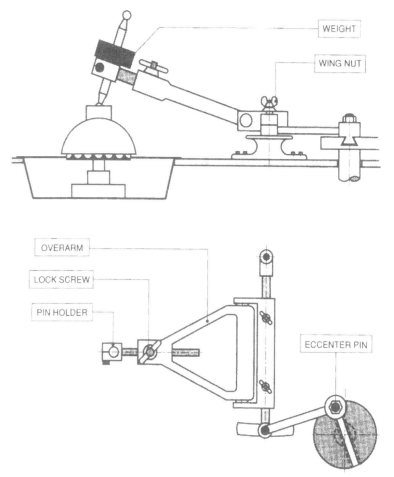

Figure 26.10 Overarm polishing machine. Production shops use machines with numerous spindles running simultaneously. (Courtesy Karow [27], Fig. 5.105).

When custom optics are desired the story changes dramatically. Individual parts must be blocked individually and tools and test plates must be fabricated for each surface. Many optics houses keep a large range of both tools and test plates used in prior work to minimize the initial expenses. If catalog optics cannot be used in a design it is always cost-effective to choose radii for the spherical surfaces that are in the test plate inventory of a manufacturer. These lists are usually supplied as a data base.

Large spherical surfaces (> 100 mm) are produced in the same way as small ones. The tooling and machines become proportionally larger but the basic method of rubbing two spheres together is the same. However, controlling the shape becomes increasingly difficult as the part diameter becomes larger. It is also increasingly difficult to handle the very large tools. Generally, it is necessary to use a large

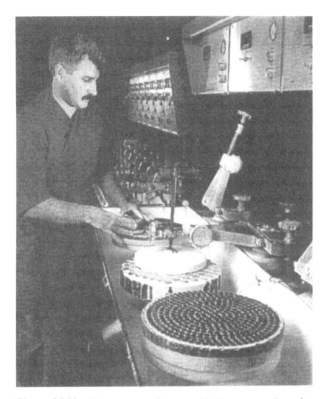

Figure 26.11 Numerous small parts with the same radius of curvature can be blocked together and processed simultaneously. (Courtesy Newport Corp.)

tool (large meaning 60–100% of the part diameter) initially after the part has been generated with a diamond tool to smooth out large surface errors. Following large tool work, smaller diameter tools are usually used to figure the surface to high accuracy. The use of smaller tools, however, can have some effect on the surface slope errors since a small tool is working locally and usually leaves behind local wear patterns. This becomes increasingly important as the tools get smaller. With skill and experience an optician can keep these errors very small by not dwelling too long at any one location on the surface.

Polishing of Flats
The production of a flat surface used to be difficult due to the fact that the tolerance on the radius of the surface is usually the same as the tolerance on the irregularity, i.e., power in the surface is an error to be polished out. This changed with the development of what is usually referred to as a continuous polishing (CP) machine (Fig. 26.12). A continuous polisher is just that: it is a large lap (at least three times the size of the part) in the shape of an annulus, that turns continuously, onto which smaller parts are placed. The parts are carried in holders (usually called septums) that are fixed in place on the annulus and are driven so that they cause the part to

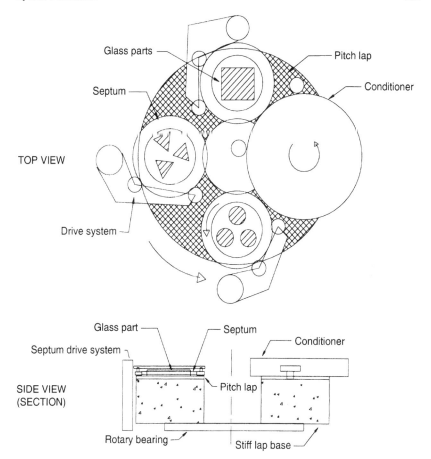

Glass parts

Pitch lap

Septum

Conditioner

TOP VIEW

Drive system

Glass part

Septum

Septum drive system

Conditioner

SIDE VIEW
(SECTION)

Pitch lap

Rotary bearing

Stiff lap base

Figure 26.12 The continuous polishing (CP) machine can polish both flats and long radius spheres to very high surface figure quality and surface finish. As long as the parts do not pass over the edge of the lap and are rotated synchronously with the lap, they will experience uniform wear. The conditioner is a large disk that keeps the lap flat and also rotates synchronously with the lap.

turn in synchronous motion with the lap. It can be shown that if the part is in synchronous rotation with the lap and always remains in full contact with the lap, then the wear experienced by the part's surface will be uniform across the surface. If the wear is uniform and the lap is flat, then parts that are not initially flat very rapidly become so.

The lap of the continuous polisher is kept flat by the use of a large flat called a conditioner or bruiser, having a diameter as big as the radius of the lap. The conditioner rides continuously on the lap and is caused to rotate at a more or less synchronous rate. By adjusting the conditioner's raidal position the lap can be brought to a very flat condition and remain there for long periods. Slight adjustments in its position are made as parts are found to be slightly convex or concave.

Since these machines run continuously 24 hours a day, their throughput can be very large. Careful attention must be paid to environmental control and slurry control to maintain consistent results. Because the contact between lap and part is exceptionally good on these machines, they routinely can produce excellent surfaces with very little roll at the edge.

The fact that there is uniform wear is not dependent on the part's shape. This means that plano parts with highly unusual shapes can be fabricated to high quality right to their edges or corners. The only other variable that needs to be controlled to produce uniform wear is the pressure. Some parts with large thickness variations and low stiffness need to have additional weights added so that the pressure is nearly uniform across the part. If the figure is seen to be astigmatic, weights can be distributed on the back of the part to counteract any regions of decreased wear.

Continuous grinding is also performed on these types of machines where brass or other metal or ceramic surfaces are used in place of the pitch. Sometimes, polyurethan or other types of pads are substituted for, or are set on top of the pitch to polish different types of materials. Pad polishers do not require as much maintenance as a pitch lap and can produce excellent surfaces with the proper materials and conditions.

This technique has been extended to parts having two polished parallel faces such as semiconductor wafers and various types of optical windows, by fabricating both faces at the same time on what are called twin polishers. In this case there is a lap on top and bottom with the parts riding in septums in between. These machines can rapidly grind and polish windows to very high flatness, low wedge, and critical thickness.

Spherical parts can also be fabricated on a continuous polisher by cutting a radius into the lap and maintaining the radius with a spherical conditioner. In this way parts all have exactly the same radius can be manufactured very economically. This works well with a part whose radius is long compared with its diameter, i.e., parts with large focal ratios. If the focal ratio becomes too small, the uniform wear condition is invalid due to an uncompensated angular velocity term in the wear equation. This term would cause a small amount of spherical aberration to be maintained in the part which must be removed through pressure variation or some other means.

Continuous polishing machines have been built up to almost 4 m in diameter capable of producing flats 1 m or somewhat larger in diameter. To produce flats larger than this a more conventional polishing machine is used such as a Draper type, overarm type, or swing-arm type [28]. In this case, the situation is reversed from a CP where the smaller part rides on top of the much larger lap. Here, the mirror is placed on a suitable support on the turntable of the polishing machine and ground and polished with laps that a smaller than the part; this is a more conventional process, but one where it is difficult to achieve the smoothness and surface figure quality that the CP provides.

26.2.7 Centering and Edging

After polishing both sides of lenses, the edges are cut to provide an outer cylinder and protective bevels. The lens are aligned on a rotary axis so both optical surfaces spin true, meaning that the centers of curvature of the spherical surfaces lie on the axis of rotation. This line, through the centers of curvature, defines the true optical axis of the lens. When the lens is rotated about the optical axis, the edge is cut with a peripherial diamond wheel. This insures that the newly cut edge cylinder, which now defines the mechanical axis of the part, is nominally aligned to the optical axis.

There are two common centering methods shown in Fig. 26.13 – one optical and the other mechanical. The lens can be mounted onto a spindle that allows light to go through the center. As the lens is rotated, any misalignment in the lens will show up as wobble for an image projected through the lens. The lens is centered by sliding the optic in its mount and watching the wobble. When the wobble is no longer discernable, the part is centered and can be waxed into place for edging.

Also, the centering can be automated using two coaxial cups that squeeze the lens. Here, the lens will naturally slide to the position where both cups are making full ring contact on the surface and will thus be aligned (at least as good as the alignment of the two cups). This method of bell chucking is self-centering so it is naturally adapted to automatic machines. It is important that the edges of the chucks are rounded true, polished, and clean so they will not scratch the glass surfaces.

When the optical element is centered and rotated about its optical axis, the outer edge is cut to the final diameter with a diamond wheel. This operation can be guided by hand, with micrometer measurements of the part, and it can also be performed automatically using numerically controlled machines.

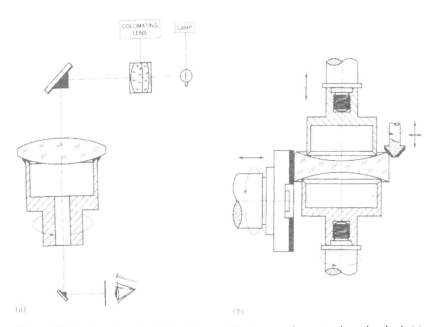

Figure 26.13 Centering the edging of lenses. The lens can be centered on the chuck (a) optically by moving the element to null wobble of the image, or (b) automatically using a bell chuck. Once centered on the spindle, the edge and bevels are cut with diamond wheels. (From Ref. 27, Figs. 5.126 and 5.128.)

When cutting the edge, a protective bevel should always be added to protect the corners from breakage. A sharp, nonbeveled edge is easily damaged and the chips may extend well into the clear aperture of the part. A good rule of thumb for small optics is that bevels should be nominally 45° with face width of 1% of the part diameter.

Large optics, which are made one at a time, are frequently manufactured differently. The blanks are edged first, then the optical surfaces are ground and polished, maintaining the alignment of the optical axis with the mechanical axis. Also, optics with loose tolerance for wedge can be edged first, then processed as described above.

26.2.8 Cleaning

The finished parts must be cleaned thoroughly to remove any residue of pitch, wax, and polishing compounds. The optics are typically cleaned in solvent baths with methyl alcohol or acetone. Optics can be cleaned one at a time by carefully wiping with solvent-soaked tissues, or they can be cleaned in batches in large vapor decreasing units followed up with ultrasonic bath in solvent. Parts that were not edged after polishing tend to have stained bevels and edges from the polishing process. This can be difficult to clean and this residual compound can contaminate the coating chambers.

26.2.9 Cementing

Lenses and prisms are commonly bonded to make doublets or complex prisms. The bonded interface works extremely well optically as long as the cement layer is thin and nearly matches the refractive index of the glasses. The bonded surface allows two glasses to be used to compensate chromatic effects, and this interface introduces negligible reflection or scattering.

Most cementing of optics is performed using a synthetic resin, typically cured with UV light. The procedure for cementing lenses is first to clean all dust from the surfaces. Then the cement is mixed and outgassed, and a small amount is dispensed into the concave surface. The mating convex surface is then gently brought into press the cement out. Any air bubbles are forced to the edge and a jig is used to align the edges or the optical surfaces to center the lenses with respect to each other. Excess cement is cleaned from the edge using a suitable solvent. When the lens is aligned, the cement is cured by illuminating with UV light, such as from a mercury lamp.

26.2.10 Current Trends in Optical Fabrication

Through the use of various types of motors, sensors, switching devices, and computers, automation has begun to have a major impact on the productivity of fabrication equipment. Numerically controlled machining has made the fabrication of tooling and the shaping of parts much more rapid and less costly. Generating has become more automated with the application of position encoders and radius measuring hardware and software integrated with the machine. Grinding/polishing machines are slowly having most of their subsystems automated, although the basic process has remained as described above. For most precision optics made today the optician's skill in the operation of the polishing machine still has a large impact on the results. However, automation is currently making large strides in making the fabrication process much less skill dependent and more deterministic, the often used buzz word of modern optical fabricators.

Machines are currently being developed that use a somewhat different approach to fabricating custom optics that may become so efficient that they eventually will outperform current production methods for spherical optics [44]. These machines such as the OPTICAM (Optics Automation and Management) being developed at the Center for Optics Manufacturing at the University of Rochester, (on the World Wide Web at www.opticam.rochester.edu) apply advanced NC machining technology to the fabrication of small optics. The idea is to have a single, high-precision machine very rapidly generate, grind, polish, and shape a single lens at a time. Metrology for each stage is integrated to the machine and corrections are applied automatically. Very stiff, high-precision spindles allow for precise generation of curves and the production of surfaces with low subsurface damage using fixed diamond abrasive wheels, operating down to very shallow cut depths that allow the rapid production of polished surfaces. Ring tool polishers are utilized to bring the surfaces to final figure and finish. Although the machines are currently expensive compared with conventional labor-intensive methods, the future of production optics clearly lies in this direction. The development of these machines has driven a wide range of deeper investigations into the grinding and polishing of glass that will inevitably lead to further developments in the automation of optics production.

26.3 FABRICATION OF ASPHERES AND NONTRADITIONAL MANUFACTURING METHODS

In the previous section, we give the basic steps for making spherical and plano optics by following the conventional processes, although frequently these steps are made with advanced machinery. In this section we describe the fabrication of aspheric surfaces and introduce a variety of methods that are in practice for making non-classical optics.

Aspheric optical surfaces, literally any surfaces that are not spherical, are much more difficult to produce than the spheres and flats above. Since these nonspherical surfaces lack the symmetry of the spheres, the method of rubbing one surface against another simply does not converge to the desired shape. Aspheric surfaces can be polished, but with difficulty, and one at a time. The difficulty of making aspherics greatly limits their use, which is unfortunate since a single aspheric surface can often replace a number of spherical surfaces in an optical system.

26.3.1 Definitions for Common Aspheric Surfaces

Many aspheric surfaces can be approximated as conic sections of revolution, although some are manufactured as off-axis pieces from the ideal parent. Conic sections are generally easier to test than a general asphere (there are geometrical null tests for conics). The general shape for a conic aspheric surface is

$$z(r) = \frac{r^2}{R + \sqrt{R^2 - (K + 1)r^2}} \tag{26.1}$$

$z(r) =$ surface height

$r =$ radial position $(r^2 = x^2 + y^2)$

$R =$ radius of curvature

$K =$ conic constant $(K = -e^2$ where e is eccentricity).

The types of conic surfaces, determined by the conic constant, are as follows, and are shown in Fig. 26.14:

$K < -1$	Hyperboloid
$K = -1$	Paraboloid
$-1 < K < 0$	Prolate ellipsoid (rotated about its major axis)
$K = 0$	Sphere
$K > 0$	Oblate ellipsoid (rotated about its minor axis).

Equation (26.1) gives the sag, which is equivalent to deviation of the optical surface from a plane. But, in optical fabrication, we are concerned with the deviation of this surface from spherical. Using a Taylor expansion, the aspheric departure is

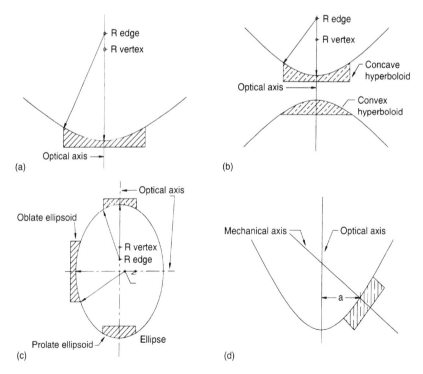

Figure 26.14 Common aspherical surfaces, defined as comic sections of revolution: (a) paraboloidal surface $(K = -1)$; (b) hyperboloidal surface $(K < -1)$; (c) ellipsoidal surfaces $(K.0,$ or $0 < K < -1)$; (d) off-axis paraboloid.

given in Eq. (26.2) (sometimes optics are given additional aspheric polynomial coefficients, which would add the coefficients on $r^2, r^4, \ldots$):

$$S(r) = \frac{Kr^4}{8R^3} + \frac{1.3[(K+1)^2 - 1]r^6}{2^3 3! R^5} + \frac{1 \cdot 3 \cdot 5[(K+1)^3 - 1]r^8}{2^4 4! R^7} + \ldots . \qquad (26.2)$$

Although many aspherics are specified as conic surfaces and polynomial aspherics, there are some other common aspheric surfaces:

- Toroids – These surfaces are part of a torus, having a different radius of curvature for two orthogonal directions on the optical surface. They are used for astigmatism correction in eyeglasses, and are used at grazing incidence for focusing high-energy radiation. Toroids are made with special generators, and polished with a variation of the process for making spherical optics.
- Axicons – These surfaces are basically cones, generated by a tilted line rotated about an optical axis. Axicons, which are used in unstable resonator laser cavities and for special alignment tooling, are nearly always made by molding or single-point diamond tooling.

26.3.2 Conventional Methods for Fabricating Aspherics

There is tremendous experience behind the traditional fabrication methods that were presented in Section 26.2. These methods can be applied for making aspheric surfaces, with a few adjustments. Since the methods work best for spheres, we define the difficulty of an asphere by its aspheric departure, or the difference between the aspheric shape and a close fitting sphere.

The Difficulty with Aspherics
Spherical surfaces are used for most optics because they are easy to describe, easy to manufacture, and easy to test. The spherical surface can be specified by a single parameter – its radius of curvature R. The spherical surface is the easiest to make because of its symmetry. The lap and the part tend to wear on the high spots, and since both are in constant motion about several axes, they will both tend to be spherical. Any other shape would present a misfit, which would tend to be worn down. Testing of spherical surfaces also takes advantage of the symmetry. (Testing is described elsewhere in this handbook.)

When figuring optical surfaces by lapping, the optician uses two different effects to control the surface – natural smoothing and directed figuring. Small-scale features, much smaller than the lap, tend to be removed by natural smoothing. This is the same process as using a sanding block to get a smooth texture in wood. As long as the block is rigid, any bumps in the wood will see large forces and will be quickly removed. This effect, for polishing and sanding, is diminished for features larger than the tool, or for the case where the tool is not rigid and easily conforms to the surface. Using good shop practices and large, rigid tools, optics can be finished spherical to about 0.2 μm of ideal, using only natural smoothing. The symmetry of the spherical surface insures that the tool will fit the surface well everywhere.

Features on optical surfaces larger than the polishing tools can be shaped using directed figuring. This is simply controlling the process based on surface measurements to target the high areas on the optic and hit them directly. In its simplest form, an optician will use directed figuring by making a small tool and running it on the high regions of the optic, as determined by an optical test. In polishing, any combination of speed, dwell times, and pressure variation may be used, but the premise is the same.

Aspherizing

The traditional steps for making an aspheric surface are to first generate and grind to a spherical surface using the methods described in the previous section. Then the surface is "aspherized" by grinding or polishing with a specially designed tool, stroke, or machine [41]. For small departures, say a few tens of microns, this can be polished in; otherwise it is generally ground into the surface and the entire surface is polished with small or flexible tools.

There are a variety of methods for aspherizing. Full-size compliant tools can be used which have petals defined in the working area to give the desired removal as the part rotated underneath. Full-size metal tools with the inverse aspheric curve are used for "plunge grinding" of small parts. Most commonly, smaller laps are used and the dwell is adjusted based on the aspheric curve to be ground in. The aspherizing process is usually monitored with mechanical measurements such as spherometry or profilometry.

Polishing

Once the part has been aspherized, it is polished and figured using a combination of large, semi-compliant tools and small tools. The optical test is critical for this process, as the optician will work the part based only on the measurement. Unlike making spheres, there is no tendency for the process to give the correct shape. The optician iteratively measures the surface and works the surface until it meets the specification.

Mild aspheres have surface slopes that are only a few microns over the diameter of the part. In that instance, large tools can still be used to produce the asphere and very smooth aspheric surfaces can be made. When slopes become larger, say tens of microns over the part diameter, a single large tool cannot be used and small tools become the rule. For fast aspherics, where local slopes can become greater than several microns per millimeter, very small tools or other methods must be employed. The usual result from using a tool that is too large to fit the local surface is that the tool wears in a restricted region and produces ripples or zones in the surface. These zones can become very sharp and are often very difficult to get rid of once acquired. Zones can be prevented or removed by using a properly sized tool or by making the tool flexible enough to bend into the global shape of the surface but still retain some local stiffness. Much experience and hard-gained knowledge has traditionally been required to produce high-quality aspheres; however, more deterministic methods are currently being developed and are discussed below.

Tools for Working Aspherics

The difficulty in polishing aspheric surfaces is due to the fact that a large rigid tool cannot fit everywhere on the surface. If the tool fits one place, it will not fit at a different position or orientation, and will lose the ability for natural smoothing. Opticians deal with this in two ways, both at the expense of large-scale natural smoothing – they can make the tool smaller, until the misfit is no longer important, or they can make the tool compliant, so it will always fit. In fact, most opticians will use a combination of both types of tools for any single asphere.

For analysis of the tool misfit, we treat the case shown in Fig. 26.15, with a circular lap, diameter $2a$, a distance b from the optical axis of the parent asphere. The misfit of the lap can be represented in several modes, which take the same form as optical aberrations. Power corresponds to a radius of curvature mismatch. Astigmatism gives the curvature difference in the two principal directions. Coma has a cubic form and spherical aberration (SA) has a quartic dependence on position.

We give the lap misfit for a few common conditions (Table 26.1):

1. Lap fits a spherical surface with radius of curvature R
2. Lap is revolving
3. Lap is rotated a small amount $\Delta\theta$
4. Lap is translated a small amount Δb

Note that the spherical aberration term has no effect for the real cases (2, 3, 4). This is because the spherical aberration of the asphere is constant on the lap for any position. It is only the change of suface aberrations that affects polishing. Also, most of the terms for the aberrations in Table 26.1 can be neglected for two common cases: for a large tool with a small stroke near the axis, the coma term dominates; and for a small tool, off axis by an amount much larger than the tool size, the astigmatism and power terms dominate. The astigmatism and power are coupled, so the P–V misfit for the case of the stroking tool will be equal to the sum of the power and astigmatism terms.

The relationships in Table 26.1 giving lap misfit are used to design the equipment for grinding and polishing. In grinding, the shape errors should be less than the size of the grit in the grinding compound, and in polishing, the lap should fit to a few microns. (The better the lap fit, the better the finish.) The laps designed for aspheric

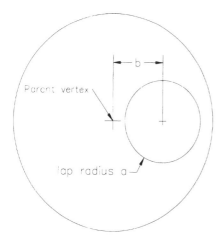

Figure 26.15 The lap misfit is calculated for a polishing surface with diameter $2a$, offset from the vertex of the parent asphere by an amount b.

Table 26.1 P–V Lap Misfit for the Cases Described

	Power	Astig	Coma	SA
1. Spherical lap	$\dfrac{Ka^4}{8R^3} + \dfrac{Ka^2b^2}{2R^3}$	$\dfrac{Ka^2b^2}{2R^3}$	$\dfrac{Ka^3b}{3R^3}$	$\dfrac{Ka^4}{32R^3}$
2. Revolving lap	0	$\dfrac{Ka^2b^2}{R^3}$	$\dfrac{2Ka^3b}{3R^3}$	0
3. Small rotation $\Delta\theta$	0	$\dfrac{Ka^2b^2}{R^3}\Delta\theta$	$\dfrac{Ka^3b}{3R^3}\Delta\theta$	0
4. Small translation Δb	$\dfrac{Ka^2b}{R^3}\Delta b$	$\dfrac{Ka^2b}{R^3}\Delta b$	$\dfrac{Ka^3}{3R^3}\Delta b$	0

surfaces use a combination of small size, small stroke, and compliance to maintain the intimate contact required.

26.3.3 Computer-Controlled Polishing

Most aspheric surfaces are still produced using small tools by highly skilled opticians using conventional machinery. There are, however, a number of methods being developed that integrate computer technology with radically different polishing methods that can rapidly produce aspheric surfaces. The first of these is the computer-controlled polishing (CCP) method [4,25,26]. This is essentially a traditional small tool method where the tool is driven in an orbital motion producing, on average, a known wear profile. This wear profile is applied to the measured errors in a surface to produce a tool path that essentially rubs longer on the high areas and less on the low areas, but in a precise relative way that can rapidly improve the figure. Sophisticated optimal machine proprietary computer algorithms are used to determine the motions from the surface measurement and removal function.

Another method that radically departs from traditional polishing methods is the ion figuring method [1,34]. Here, the already polished surface is bombarded by ions from an ion gun to remove material in a very deterministic way. The removal function of the ion gun is well established prior to use. Just like the CCP process, a tool path is developed from the measured surface errors to produce a dwell time function for the surface. The surface figure can be rapidly improved due to the relatively high removal rate of the ion gun verses polishing. The process is highly deterministic, so many parts can be finished with a single run in the ion mill. Ion figuring is only used to remove about a micron from the surface, because it can degrade surface finish.

A figuring process that utilizes the etching of glass is the PACE or plasma assisted chemical etching method [7]. This is also a small tool, dwell time dependent figuring method with an unusual removal mechanism. Here a small confined plasma, which is reactive with the glass substrate, is moved over the surface and material is removed proportional to dwell time. By choosing a suitable tool path as in CCP figuring, the surface can be figured without introducing high spatial frequency errors into the surface. The tool size can be adjusted to produce the most appropriate removal profiles for the particular surface error. As with ion figuring, this method also demonstrates very high removal rates and excellent figure convergence.

Another deterministic polishing method has been developed and is used for small aspheres that uses a lap made with a magnetorheological substance [18]. This tool gives a well-defined removal profile, which can be modulated with a magnetic field. The parts are rotated under the lap and the magnetic field is adjusted under computer control according to the measured surface.

The finishing of optics with such computer-controlled methods has been limited to large companies or research groups. These techniques provide excellent results when everything is worked out correctly. However, it takes many hours to polish a large optic with a small tool and, if something goes wrong in this process, the polisher can drive a small low region into the part. If this happens, the entire surface must be driven down to meet this low spot, so one must have confidence in the process to use small tool figuring on production parts. Also, these methods rely on good, computer-acquired data which is mapped carefully to the surface. If the polishing run is shifted slightly, relative to ideal, the polisher can drive low spots right next to the high spot it was intending to hit. Even with these difficulties, the large optics companies have developed excellent processes and equipment for computer-controlled polishing.

Large-tool polishing is also possible for aspheric surfaces if the tool itself is controlled by computer. Several groups have developed large active polishing tools that polish aspheric surfaces under computer control by changing the lap shape or force. The stressed lap polisher [35,57] uses a large, rigid polishing tool that is actively bent under computer control to take the shape of the aspheric surface. This retains the advantage of large tools to provide passive smoothing, even on steep aspheres. [11]. A different concept has been demonstrated that uses a membrane lap with the polishing force dynamically controlled by computer [29]. This allows the use of a large tool, although there is little gain by passive smoothing.

One last semi-conventional method for making aspheres is the bend and polish technique [12,31]. Here the substrate itself is carefully distorted by applying external forces or moments. The distortion is controlled and the part is polished spherical in its distorted state. The optic should then relax into the desired aspheric shape.

26.3.4 New Methods of Asphere Fabrication

In this section we study the new methods that have been developed for the fabrication of aspherical optical surfaces.

Molding

Many small aspheric lenses, such as camera lenses, are made by the direct molding of glass or plastic into an aspheric mold. The molds have the exact opposite shape of the finished asphere and are made from materials that can withstand the required elevated temperature. These optics are readily mass-produced by the million with astonishingly good quality. [2]

Small lenses are directly molded in glass using a method call precision glass molding (PGM) [43]. The lenses are directly formed into the final shape by being pressed into a die at high temperature. This method economically produces small (< 25 mm) spherical and aspherical optics in a variety of glasses with diffraction limited performance and with excellent surface finish. These lenses are used in high-volume goods such as optical disk drives and pocket cameras. Larger condenser lenses, which have reduced requirements, for projectors are also made this way.

High-quality plastics optics are mass-produced by the process of injection molding [20]. Liquid plastic is forced into a heated mold cavity at high pressures. The plastic solidifies to take the inverse shape of the mold. By carefully controlling the pressure and temperature profiles, high-quality lenses up to 50 mm can be produced. The tooling to produce these lenses is quite expensive, but it enables a low-cost process that produces lenses by the thousand. Plastic optics find use in the same type of applications as the molded glass lenses. Advantages to plastic optics are the reduced weight and the ability to have complex mounting features integrated into the optic.

Replicating

In addition to molding, optical surfaces are created by replicating against a master. Compression molding of plastics is used to make large, flat optics such as Fresnel lenses [39]. A thermoplastic blank is pressed between two platens and heated. Parts as large as 1.5 m have been made using this method.

Optical surfaces, especially gratings, are often replicated into epoxy. Typically, the epoxy is cast between two glass surfaces – the master and the final substrate. A release is applied to the master surface, so the result is a replicated inverse of this master, held fast to the final substrate. Diffraction–limited accuracy can be obtained for parts made using a carefully controlled process.

Metal optics are electroformed against precision mandrels to make good, smooth optics. [14] Electroforming is simply electroplating on to a surface with a suitable release. After completion, the thin metal "electroplate" can be removed and used as a reflective optic. Reflectors for high-power light sources are made by electroforming a thin reflective layer of nickel or rhodium on to a convex mandrel. Then, several millimeters of copper is electroformed on top to give the part structural rigidity. These optics are quite smooth, but can have larger figure errors.

Single-Point Diamond Turning

In recent years, high-performance machines have been produced that use sharp diamond tools to turn optical surfaces directly to finished tolerances. These machines use accurate motions and rigid mounts to cut the optical surface with a single diamond point, just as one would machine the part on a lathe. This has the obvious advantage that aspheric surfaces can be cut directly into the surface, without the need for special laps or metrology. In fact, some optical surfaces, such as axicons, would be nearly impossible with conventional processes. Single-point diamond turning (SPDT) is, in fact not new, but only in recent years has it become economical for production parts. Some references on the subject are in Arnold *et al.*, [3], Gerchman [15], Rhorer and Evans [45], and Sanger [48].

A variety of materials have been fabricated using SPDT. The best results are for ductile metals like aluminum, copper, nickel, and gold. Crystalline materials used for infrared applications such as ZnSe, ZnS, and germanium are also diamond turned with excellent results. Diamond turning does not work well for glass materials because they are brittle, although under carefully controlled conditions, glass can be cut in a ductile mode [6].

The surface structure obtained from diamond turning is different from conventional processes. Polished optics have no systematic structure in them, and they can be made perfect to a few angstroms. Diamond-turned surfaces always have residual grooves from the diamond tool. These can be made quite small (10 nm) by making a final light cut with fine pitch. The surface scattering from these grooves limits the application of most diamond-turned optics to infrared applications, which are not sensitive to such surface effects. In some cases it is possible to post-polish the diamond-turned part to smooth out these grooves [5].

There are two common configurations for diamond-turning machines – as a precise lathe with the part spinning and the diamond bit carefully controlled and as a fly cutter with the part fixed and the diamond bit moving on a rotating arm. The lathe-type machines produce both axisymmetric surfaces and off-axis optics (by mounting the optic off the axis of rotation). The fly cutter geometry is used to produce flats, especially for crystals that are difficult to polish and for multifaceted prisms where the relative angle from one facet to the next can be controlled.

26.4 FABRICATION ISSUES FOR THE OPTICAL DESIGNER

When the optical designer is first developing the system concept, he should ask the question: "How is this going to be made?" It makes no sense to design with components that cannot be manufactured accurately enough to meet the technical specifications or economically enough to meet the cost goals. This section discusses some of the fabrication issues that face optical designers. Different manufacturing methods, and commonly different shops, are used depending on whether the order is for thousands of optics, or only a few. Tolerances on the components can drive the fabrication methods, so these must be carefully thought out. Size also plays an important role in deciding the fabrication methods and the reasonable specifications. The choice of material for the optics can also limit the choices for fabrication methods.

26.4.1 Fabrication issues Related to Quantity

As part of the overall system design and optimization, the optical engineer must decide how the components will be fabricated. The previous sections describe common fabrication methods employed by many shops. Quantity is the most important parameter for defining which techniques make sense and which shop to use. Some optics will necessarily be expensive because they are one-off items that require special attention. At the other end of the spectrum, considerable savings can be made for large production runs by taking advantage of the technologies that reduce the reliance on highly skilled labor.

Since a large portion of the cost for fabricating optics is in the setup and tooling, one should always start by finding what optics are already in production and attempting to use them, rather than setting up a new line with optics only slightly different. The optical designers frequently have flexibility to push the designs around to work with existing lenses. The lens catalogs for the largest suppliers are included in the optical design software libraries for this reason.

Plastic optics, replicated by injection molding provide the lowest cost option for large quantities of small lenses. Also, small molded glass optics are produced at high volume for low cost per part. So the designer for a system in mass production should think first about plastic, then molded aspheres, then, if necessary, conventional glass optics. The rest of this discussion is limited to the care of custom-manufactured glass optics.

Ordering the Glass

The first step in fabrication is to order the glass. For most cases, it is far better to specify the optics, along with glass requirements, to the fabricator and have them order the glass, rather than to purchase and supply the glass yourself. The fabricators are used to dealing with the glass companies and they will know best what form the material should come in. The fabricator will know how much glass to buy to cover samples for setup, tooling, process development, etc., as well as the inevitable losses due to parts outside of tolerances. By letting the fabricator purchase the glass, you also reduce the number of interfaces for the project. The fabricators will take seriously their responsibility for the performance of the optic, including the glass. If you supply the glass yourself, the tendency for the fabricator is to treat the optic as a set of surfaces being made on a substrate, which they have no control over.

The selection of the material is driven by the quantities to be made and by the required quality. For large numbers of lenses or prisms, the blanks can be supplied as pressings, which are molded to about 1 mm over the final dimensions. These can be blocked and processed by going straight to grinding, skipping the rough shaping step. For large orders, the glass company can supply these blanks with tighter control on the variability of the refractive index from one part to another.

Glass supplied in pressings will usually be of high quality, but it is impossible to inspect, due to the rough surface. The glass for low-volume, high-performance optics is specified in blocks so the internal quality can be assessed. The glass is melted, annealed, cut into blocks, polished for inspection, then graded according to the measured quality. This gives the customer data showing each piece of glass, and premium prices are paid for the highest-quality glass. It is interesting that when you buy standard quality glass that has not been graded, you know only that the process results in good material, and you may have excellent substrates. However, if you buy standard glass that has been inspected, you know exactly what you have and you can be assured that you do *not* have excellent quality material. It has been identified and sold at a higher price.

The glass will be supplied with a melt sheet, which gives the pedigree of the material. The refractive index will be measured for samples from each melt and interferograms will be provided for glass with high-quality refractive index homogeneity. If there is any variability for different melts, it is important to develop a good system to track lenses through fabrication, so it is known which finished parts came from which melt. They all look the same.

Support for Fabrication

Traditionally, optics are blocked to the support with pitch or wax. This can be a labor-intensive process and requires skill and experience to be done correctly. However, this type of support does not require special tooling, so it makes sense for low-volume production.

One of the largest cost savings for volume production parts comes from the use of spot blocks, described in Section 26.2.3. These allow the blanks to be inserted directly into machined holders on the block, which does not need the highly skilled expertise of an optician. The spot blocks are expensive to make, but they can be used repeatedly for parts with the same radii. The cost per part obviously decreases with the number of parts per block and the number of times the block is used. The breakeven point for the spot blocks depends on the particular shop practices, but it is fair to say that spot blocks are economical for the case where numerous (rather than a few) blocks of the same element will be made.

Rough Shaping and Generating

As described above, volume optics can be supplied as pressings with sufficient accuracy that no roughshaping is required. For odd-shaped optics that require initial shaping, cost savings can be made using automated NC machines.

The generating of optics in low volume uses careful alignment of the diamond cup wheel with the part. The optician controls the radius by measuring the optic (or block of optics) with a spherometer, then adjusting the tilt of the wheel to give the desired curve. The thickness of the parts must be monitored separately. This type of generating works well, but it requires a highly skilled operator.

Optics made in large quantities can skip the generating directly, and be worked with pel grinders, which are rigid grinding tools, faced with diamond pellets. These high-speed diamond tools work the surface quickly from the rough shape, to a fine grind, ready for polish. The radii of the parts are controlled simply by maintaining the radius of the tool. Here again, the tooling is expensive, but can be used for multiple blocks of the same radius.

Polishing

Pitch tools are used for high-precision optics and for runs of small quantities. The pitch tools require skilled labor to make and maintain, but they do not require any expensive or difficult machining. Optics made in large numbers are frequently polished using pads of special fabric or textured polyurethane. These tools can be expensive because they must be carefully manufactured so the radius of the tool matches the ground surface to be polished. Once this is achieved, however, the tool can be used repeatedly for multiple blocks of lenses. Also, the polishing speeds and pressures can be increased for these synthetic laps to speed up polishing.

26.4.2 Relationships Between Tolerances and Fabrication Issues

The importance of effective tolerancing for optical components cannot be overstated. It is in the specification of the tolerances that the optical engineer must know something about the fabrication. A tendency for optical designers is to come up with a lens design that performs well according to simulations, then to expect the real optics to behave as the ideal ones do in the computer. The tolerances for the system are often assigned as an afterthought to the design and they tend to be tight. A better way to design is to anticipate the fabrication limitations in the design of the lens. This way the designer balances sensitivity to expected errors as part of the optical system design.

The optimal value for the tolerances can only be found by communication between the fabricator and the designer. The designer always wants tighter tolerances because they will give improved performance. These come at a cost, however, because the fabricator must work harder to meet these tolerances. So how good is "good enough"? The designer cannot decide this on his own because it depends on the incremental fabrication costs. The fabricator cannot define this, because he does not have sufficient information to know how the manufacturing errors affect the system performance.

To perform system tolerance analysis, the engineer will assume some tolerances and perturb the simulation of the optical system to determine the effect of each parameter (such as radius of curvature or lens thickness) being at the edge of tolerance. The overall performance is estimated by combining the effect of all of the terms as a root sum square. Here is where the fun begins. Usually, the optical designer finds one of two things from this exercise – either the system has excellent performance, in which case the assumed tolerances are too tight, or the performance is not acceptable, in which case the assumed tolerances must be tightened. Now the designer should go to the fabricator and discuss which tolerances to adjust to give acceptable performance without driving up manufacturing costs.

Because the effects of the separate tolerances are uncorrelated and added as RSS, only the few largest terms contribute to the total. If the designer looks carefully at the individual terms, tolerances that do not affect the performance can be made looser than would be otherwise. Also, only a few critical parameters will need to be controlled to high accuracy.

The key to good tolerancing is to know the relationships between tolerances and cost. Unfortunately, this information is hard to get and it can vary significantly from one shop to the other and over time. This relationship depends on two things – how much extra work is required to achieve the tolerance and whether special equipment is required.

A simple example is the angle for a prism. using standard shop practices, and paying no particular attention, the angle will be good to about 5 arc min. If the optician takes special care using common tools, the angle can be controlled to 1 arc min. The added expense here is only the additional time required by the optician to measure the angle and adjust the process. To get to accuracy of 10 arc secs, the optician will need more sophisticated measuring equipment and it will take more iterations of the measure – adjust cycle. Now if the angle must be made to 1 arc sec, only an experienced optician with very good metrology can spend a lot of time to get there. Optics with requirement of 0.1 arc sec will require a research effort to come up with a way of both making this part and validating it, so the cost may be extremely high and the delivery time quite long.

In some cases the cost curves do not change with tolerance until the capacity of a machine is exceeded. A good example here is machining with NC machines. A good machine will give 10 μm accuracy over small distances, independent of the tolerance assigned by the designer. There would be no cost savings for assigning a looser tolerance. There would, however, be a sharp cost increase if a 9 μm tolerance is assigned and the machine is certified to 10 μm. This would drive the fabricator to another method, which may cost several times more. It is important to feel these things out with your fabricator.

We provide some rules of thumb for tolerancing optics (Tables 26.2–26.4). Like any rules of thumb, these serve as useful guidelines, but the particular circumstances may be well outside these assumptions. Many of the numbers come from some excellent articles on this subject [37,38,51,58,59,60].

We define several classes of tolerances:

- *base* – this is what the manufacturing process gives, without any special effort
- *precision* – most shops can do this, at a cost increase of roughly 25% for that operation
- *high precision* – at the limit for most shops, cost could increase 100% for that operation.

26.4.3 Size Effects for Fabrication

The effects of size on optical fabrication are quite interesting. There are numerous methods and plenty of shops that make production lenses to 50 mm. Optics in the range of 50–500 mm are not uncommon, but they require special tooling and they are usually made as one-off parts [with the exception of flats processed on a continuous polisher (CP)]. Optics greater than 500 mm nearly always mirrors, are in a class by themselves and there are only a few places with equipment and expertise to handle these.

The advantage of using optics smaller than 50 mm is that there are so many of them! There are large numbers of companies set up to make these optics to high quality at good prices. The parts are small enough that many optics can be processed economically on a common block. The infrastructure is in place for grinding, polishing, edging, cleaning, and coating optics of this size. In fact much of the processing can be totally automated.

Things get more difficult for larger optics. The market has not supported the development of efficient tools and processes for mass-producing optics in the 50–500 mm range. In fact, each new part in this range will need a special polishing support and set of polishing tools. These parts need to be processed one at a time, so they require significantly more labor than the small parts. The size of these parts is

Table 26.2 Rules of Thumb for Optical Element Tolerances

Parameter	Base	Precision	High precision
Lens diameter	100 μm	12 μm	6 μm
Lens thickness	200 μm	50 μm	10 μm
Radius of curvature (tolerance on sag)	20 μm	1.3 μm	0.5 μm
Wedge (light deviation)	6 arc min	1 arc min	15 arc sec
Surface irregularity	5 fringes	1 fringe	0.25 fringe
Surface finish	50 Å rms	20 Å rms	5 Å rms
Scratch/dig	160/100	60/40	20/10
Dimension tolerances for complex elements	200 μm	50 μm	10 μm
Angular tolerances for complex elements	6 arc min	1 arc min	15 arc sec
Bevels (0.2 to 0.5 mm typical)	±0.2 mm	±0.1 mm	±0.02 mm

Table 26.3 Rules of Thumb for Optical Element Mounting Tolerances

Parameter	Base	Precision	High precision
Spacing (manual machined bores or spacers)	200 μm	25 μm	6 μm
Spacing (NC machined bores or spacers)	50 μm	12 μm	2.5 μm
Concentricity (if part must be removed from chuck between cuts)	200 μm	100 μm	25 μm
Concentricity (cuts made without de-chucking part)	200 μm	25 μm	5 μm

such that they can usually be simply supported, either on a few defining points or on a compliant pad.

In addition, the metrology for these larger optics can drive the cost up. Small optics are easily measured with test plates. The larger optics may need to use auxiliary optics for testing. The testing is not just for qualification, but it is an integral part of the fabrication sequence. The optician works these optics according to the results from the optical test.

Large optics (> 50 cm) are almost always mirrors, and have other unique difficulties due to their size, and often due to the surface requirements. (For the same optical performance, a mirror surface must be four times better than a refractive surface.) For large optics, each processing and handling operation requires custom tooling. The support for large optics becomes difficult and extremely sensitive. Often, separate supports must be used for holding the optics during polishing than can be used for testing. The polishing forces from large laps can be substantial and must be resisted by the support. The self-weight deflection of large mirrors alone will quickly dominate the shape if it is not accommodated in the support.

Table 26.4 Optical Material Tolerances

Parameter	Base	Precision	High precision
Refractive index departure from nominal	±0.001 (Standard)	±0.0005 (Grade 3)	±0.0002 (Grade 1)
Refractive index measurement	$\pm 3 \times 10^{-5}$ (Standard)	$\pm 1 \times 10^{-5}$ (Precision)	$\pm 0.5 \times 10^{-5}$ (Extra Precision)
Dispersion departure from nominal	±0.8% (Standard)	±0.5% (Grade 3)	±0.2% (Grade 1)
Refractive index homogeneity	$\pm 1 \times 10^{-4}$ (Standard)	$\pm 5 \times 10^{-6}$ (H2)	$\pm 1 \times 10^{-6}$ (H4)
Stress birefringence (depends strongly on glass)	20 nm/cm	10 nm/cm	4 nm/cm
Bubbles/inclusions ~ (> 50 μm) (Area of bubbles per 100 cm^3)	0.5 mm^2 (class B3)	0.1 mm^2 (class B1)	0.029 mm^2 (class B0)
Striae (based on shadow graph test)	Normal quality (has fine striae)	Precision quality (no detectable striae)	Precision quality (no detectable striae)

Reference: Schott glass catalog.

The shear size of large mirrors presents a challenge. Opticians may need to climb out on to the optical surface to clean and inspect a large mirror. Every handling operation must be carefully thought out and all of the tooling must be tested before it can be safely used. Unlike picking up small optics, large optics can be extremely heavy, so the forces are large, and the parts are extremely valuable, so all efforts to make sure every operation is 100% safe are justified.

It is much more difficult to estimate the costs for large optics than for small ones because of the difficulties with large optics and the fact that each one is special. Large optics are only processed in a limited number of shops, so the costs will often depend on the current workload in the shop as much as it will on the technical difficulties. The best advice here is to plan ahead and design for optics that are identical to others already in production. Much of the cost for large optics is in the equipment, so considerable savings can be made by using existing tooling. A good example is the lightweight mirrors made at the University of arizona. Fig. 26.16 shows a primary mirror blank that is 8.4 m across, which will be used as one of the twin telescopes in the Large Binocular Telescope. A large fraction of the cost of this mirror is due to the engineering and fabrication of all the equipment to process and handle this glass. Much of this equipment is specifically designed for this mirror and could not be used for an optic with a different shape.

26.4.4 Fabrication Issues Relating to Material Properties

The choice of material clearly influences the method of fabrication and the selection of the appropriate shop. There is not a wide variation for fabricating most of the optical glasses. Some glasses stain, and require specific polishing compounds, and others are relatively hard and require more time for processing. But these are not large issues. The choice of glass will mostly affect cost directly, by the price of the glass itself.

The big differences come from more exotic materials, such as crystals and special metals. Some of these materials are extremely useful in optical systems, but their material properties make them difficult and expensive to fabricate [35,54]. The most important material properties for the fabricator are:

- CTE coefficient of thermal expansion, which will drive blocking and thermal requirements
- Thermal conductivity (or diffusivity) which will define the thermal time constant and potential for thermal shock
- hardness or softness, which will define polishing methods
- solubility, which can limit the polishing and cleaning solutions
- ductility, which will define whether the material can be diamond turned.

The best advice for difficult materials is to find a shop that specializes in processing that type of optic. Again, it is important to talk to the potential fabricator early in the design phase because some materials will impose hard size or quality constraints that need to be folded in from the start. Also, you may be pleasantly surprised to find that there are better alternatives to the material or process you originally assumed.

Figure 26.16 An 8.4-m diameter, $f/1.1$ primary mirror blank for the Large Binocular Telescope. This optic, the largest in the world, requires considerable engineering and tooling to support each operation in the shop. This image shows the backside of the honeycomb mirror as it is supported vertically in the shop. (Photo by Lori Stiles, University of Arizona).

There are steep cost curves for fabricating difficult materials that depend largely on equipment and the state of the market. Like large optics, these markets are not large enough to have a wide selection of vendors competing for your business. The expertise for fabricating optics from less common materials tends to be with small companies that have developed particular specialities.

A different issue is the choice of substrate material for reflective optics. The light does not care what substrate the mirror is made of, because it reflects off a coating on the surface and never goes through the mirror. So the mirror substrate can be chosen according to the operating environment. Frequently, mirrors are made from low-expansion glass because this takes an excellent polish and it minimizes the sensitivity to thermal effects. Mirror substrates can be procured as lightweighted structures to reduce the self-weight deformation.

26.5 CONCLUSION

This chapter has given a summary of the most common fabrication methods in use today. Most optics are made by modern variants on classical methods, but the highest performance optics rely on more advanced techniques. Clearly, there are numerous fabrication methods for speciality optics that lie outside the scope of what has been presented here.

We present this information to the optical engineer to give some understanding of limitations and alternatives in the shop. An engineer who knows the basic issues can work directly with the fabricator to design cost-effective systems. Clearly, the system cannot be optimized for either performance or cost if the fabricator is not involved in the decisions. Remember, without the fabricator, the optical engineer would have nothing but a pile of computer printouts and some sand!

REFERENCES

1. Allen, L. N. and R. E. Keim, "Ion Figuring System for Large Optic fabrication," *Proc. SPIE*, **1168**, 33–50 (1989).
2. Aquilina, T. "Characterization of Molded Glass and Plastic Aspheric Lenses," *Proc. SPIE*, **896**, 167–170 (1988).
3. Arnold, J. b., R. E. Sladky, P. J. Steger, N. D. Woodall, and T. Saito "Machining Nonconventional-Shaped Optics", *Opt. Eng.*, **16**, 347–354 (1977).
4. Bajuk, D. J. "Computer Controlled Generation of Rotationally Symmetric aspheric Surfaces," *Opt. Eng.*, **15**, 401–406 (1976).
5. Bender, J. W., S. R. Tuenge and J. R. Bartley, "Computer-Controlled Belt Polishing of Diamond-Turned Annular Mirrors," *Proc. SPIE*, **966**, 29–38 (1988).
6. Bifano, T. G., T. A. Dow, R. O. Scattergood, "Ductile-Regime Grinding of Brittle Materials: Experimental Results and the Development of a Model," *Proc. SPIE*, **966**, 108–115 (1988).
7. Bollinger, D. *et al.*, "Rapid, Non-Contact Optical Figuring of Aspheric Surfaces with Plasma Assisted Chemical Etching", *Proc. SPIE*, **1333**, 44–57 (1990).
8. Brown, N. J. "Optical Polishing Pitch," Preprint UCRL-80301 (Lawrence Livermore National Laboratory, 1977).
9. Brown, N. J., "Computationally Directed Axisymmetric Aspheric Figuring," *Opt Eng.*, **17**, 602–620 (1978).
10. Brown, N. J., P. C. Baker, and R. T. Maney, "Optical Polishing of Metals," *Proc. SPIE*, **306**, 42–57 (1981).
11. Burge, J. H. "Simulation and Optimization for a Computer-Controlled Large-Tool Polisher," *OSA Trends in Optics and Photonics, Vol. 24, Fabrication and Testing of Aspheres*, Taylor, J. S., M. Piscotty, and A. Lindquist (eds), Optical Society of America, Washington, DC, 1999.
12. Everhart, E., "Making Corrector Plates by Schmidt's Vacuum Method," *Appl. Opt.*, **5**, 713–715 (1966) and *Errata in Appl. Opt.* **5**, 1360 (1966).
13. Fontane, S. D., *Optics Cooke Book*, Optical Society of America, 1991.
14. George, R. W. and L. L. Michaud, "Optical Fabrication by Precision electroform," *Proc. SPIE*, **818**, 325–332 (1987).
15. Gerchman, M., "Specifications and Manufacturing Considerations of Diamond Machined Optical Components," *Proc. SPIE*, **607**, 2–13 (1986).
16. Golini, D. and S. D. Jacobs, "The Physics of Loose Abrasive Microgrinding," *Appl. Opt.*, **30**, 2761–2777 (1991).

17. Golini, D. and S. D. Jacobs, "Transition between Brittle and Ductile Mode in Loose Abrasive Grinding," *Proc. SPIE*, **1333**, 80–93 (1990).

18. Golini, D., S. D. Jacobs, and W. Kordonsky, "Fabrication of Glass Aspheres Using Deterministic Microgrinding and Magnetorheological Finishing," *Proc. SPIE*, **2536**, 208–211 (1995).

19. Golini, D. and W. Czaijkowski, "Center for Optics Manufacturing Deterministic Microgrinding," *Proc. SPIE*, **1752**, 146–152 (1992).

20. Hoff, A. M., "Basic Considerations for Injection Molding of Plastic Optics," *Proc. SPIE*, **2600**, 2–5 (1995).

21. Holland, L., *The Properties of Glass Surfaces*, John Wiley and Sons, New York, 1964.

22. Horne, D. F., *Optical Production Technology*, Adam Hilger, Bristol, 1983.

23. Horne, D. F., "Loose Abrasives, Impregnated Diamonds and Electro-Plated Diamonds for Glass Surfacing," *Proc. SPIE*, **109**, 12–18 (1977).

24. Ingalls, A. G., *Amateur Telescope Making, Volumes 1–3*, Willmann-Bell, Richmond, 1996.

25. Jones, R. A., "Grinding and Polishing with Small Tools Under Computer Control," *Opt. Eng.*, **18**, 390–393 (1979).

26. Jones, R. A. and W. J. Rupp, "Rapid Optical Fabrication with CCOS," *Proc. SPIE*, **1333**, 34–43 (1990).

27. Karow, H. H., *Fabrication Methods for Precision Optics*, Wiley, New York, 1993.

28. Khaladji, J. E., "Quantitative Analysis of Polishing Efficiency Obtained on Standard Machines," *Proc. SPIE*, **109**, 31–38 (1977).

29. Korhonen, T. and T. Lappalainen, "Computer-Controlled Figuring and Testing," *Proc. SPIE*, **1236**, 685–691 (1990).

30. Kumanin, K. G., *Generation of Optical Surfaces*, The Focal Library, london, 1962.

31. Lubliner, J. and J. Nelson, "Stressed Mirror Polishing," *Appl. Opt.,*, **19**, 2332–2340 (1980).

32. Malacara, D., *Optical Shop Testing*, 2nd edn, Wiley, New York, 1992.

33. Martin, H. M. D. S. Anderson, J. R. P. Angle, R. H. Nagel, S. C. West, and R. S. Young, "Progress in the Stressed-Lap Polishing of a 1.8-m f/l Mirror", *Proc. SPIE*, **1236**, 682–690 (1990).

34. Meinel, A. B., S. Bushkin, and D. A. Loomis, "Controlled Figuring of Optical Surfaces by Energetic Ionic Beams," *Appl. Opt.*, **4**, 1674 (1965).

35. Musikant, S., *Optical Materials*, Marcel Dekker, New York, 1985.

36. Ohmori, H., "Ultraprecision Grinding of Optical Materials and Components Applying ELID (Electrolytic In-Process Dressing)," *Proc. SPIE*, **2576**, 26–45 (1995).

37. Parks, R. E., "Optical Component Specifications," *Proc. SPIE*, **237**, 462–463 (1980).

38. Parks, R. E., "Optical Specifications and Tolerances for Large Optics," *Proc. SPIE*, **406**, 98–105 (1983).

39. Parks, R. E., "Overview of Optical Manufacturing Methods," *Proc. SPIE*, **306**, 2–12 (1981).

40. Parks, R., "Traditional Optical Fabrication Methods," in *Applied Optics and Optical Engineering, Vol. X*, Shannon S. R. and J. C. Wyant, ed., Academic Press, San Diego, 1987.

41. Paxton, K. B., *et. al.*, "Uniform Polishing of Convex Aspheres with an Elastic Lap," *Appl. Opt.*, **14**, 2274–2279 (1975).

42. Piscotty, M. A., J. S. Taylor, and K. L. Blaedel, "Performance Evaluation of Bound Diamond Ring Tools," *Proc. SPIE*, **2536**, 231–247 (1995).

43. Pollicove, H. M., "Survey of Present Lens Molding Techniques," *Proc. SPIE*, **896**, 158–159 (1988).

44. Pollicove, H. M. and D. T. Moore, "Automation for Optics Manufacturing", *Proc. SPIE*, **1333**, 2–6 (1990).

45. Rhorer, R. L. and C. J. Evans, "Fabrication of Optics by Diamond Turning," in *Handbook of Optics*, OSA, 1995.
46. Rupp, W., "Loose Abrasive Grinding of Optical Surfaces," *Appl. Opt.*, **11**, 2797–2810 (1972).
47. Rupp, W., "Surface Structure of fine Ground Surface," *Opt. Eng.*, **15**, 392–396 (1976).
48. Sanger, G. M., "The Precision Machining of Optics," in *Applied Optics and Optical Engineering*, Vol. 10, Academic Press, New York, 1987.
49. Sanger, G. M., ed., "Contemporary Methods of Optical Manufacturing and Testing," *Proc. SPIE* **433** 2–18 (1983)
50. Scott, R. M., "Optical Manufacturing," in *Applied Optics and Optical Engineering*, Vol. III. Lingslake, R., ed., Academic Press, New York, 1965.
51. Smith, W. J., "Fundamentals of Establishing an Optical Tolerance Budget," *Proc. SPIE*, **531**, 196–204 (1985).
52. Spira, m. W., "Precision Grinding with Pellets and High-Speed Polishing by Means of Synthetic Material," *Proc. SPIE*, **109**, 27–30 (1977).
53. Stowers, I. F., *et al.* "Review of Precision Surface Generation Processes and Their Potential Application to the Fabrication of Large Optical Components," *Proc. SPIE*, **966**, 62–73 (1988).
54. Sumner, R. "Working Optical Materials," in *The Infrared Handbook*, Wolfe, W. L. and G. J. Zissis, eds, Office of Naval Research, Arlington, VA, 1978.
55. Texereau, J., *How to Make a Telescope*, Willmann-Bell, Richmond, 1984.
56. Twyman, F., *Prism and Lens Making*, 2nd edn, Adam Hilger, Bristol, 1988.
57. West, S. C., H. M. Martin, R. H. Nagel, R. S. Young, W. B. Davison, and T. J. Trebisky, S. T. DeRigne, and B. B. Hille, "Practical Design and Performance of the Stressed Lap Polishing Tool", *Appl. Opt.* **33**, 8094 (1994).
58. Willey, R. R., "The impact of tight tolerances and other factors on the cost of optical components," *Proc. SPIE* **518** 106–111 (1984).
59. Willey, R. R. and R. E. Parks, "Optical Fundamentals," in *Handbook of Optomechanical Engineering*, Ahmad, A., ed., CRC Press, Boca Raton, Florida, 1997.
60. Willey, R. R., R. George, J. Odell, and W. Nelson, "Minimized Cost Through Optimized Tolerance Distribution in Optical Assemblies," *Proc. SPIE*, **389**, 12–17 (1983).
61. Yoder, P. R., *Opto-mechanical Systems Design*, 2nd edn. Marcel Dekker, New York, 1993.
62. Zschommler, W., *Precision Optical Glassworking*, MacMillan, London, 1984; also published as *Proc. SPIE*, **472** (1984).

Index

Printed and bound by CPI Group (UK) Ltd, Croydon, CR0 4YY

23/10/2024

01778237-0020